The third edition of this book has been completely revised. It is intended for first- and second-year undergraduates in chemistry, and for undergraduates in other science and engineering subjects which require an understanding of chemistry.

The author gives more attention to the solid and liquid states than is found in most other books on physical chemistry, and introduces topics such as computer simulation and quasicrystals. Each chapter concludes with a set of problems designed to lead the reader to familiarity with the subject and its application in new situations. Computer programs designed to assist the reader are downloadable from the Worldwide Web (www.cup.cam.ac.uk). Detailed solutions to the problems are available on the same internet, while brief answers are contained within the book itself. Stereoviews are presented for three-dimensional structures and instructions for viewing them are provided. The book assumes only pre-degree mathematics, and special mathematical and other topics are contained in appendices.

This modern text on physical chemistry will be of interest to undergraduate students in chemistry and also to students in other areas of science and engineering requiring a familiarity with the subject.

Introduction to physical chemistry

3rd Edition

To the memory of
Dr W H Lee

Introduction to physical chemistry

3rd Edition

MARK LADD DSc (Lond) FRSC FInstP

Department of Chemistry, University of Surrey

CAMBRIDGE
UNIVERSITY PRESS

PUBLISHED BY THE PRESS SYNDICATE OF THE UNIVERSITY OF CAMBRIDGE
The Pitt Building, Trumpington Street, Cambridge CB2 1RP, United Kingdom

CAMBRIDGE UNIVERSITY PRESS
The Edinburgh Building, Cambridge, CB2 2RU, United Kingdom
40 West 20th Street, New York, NY 10011-4211, USA
10 Stamford Road, Oakleigh, Melbourne 3166, Australia

First published 1998

Printed in the United Kingdom at the University Press, Cambridge

Typeset in 10/12pt Times [SE]

A catalogue record for this book is available from the British Library

Library of Congress Cataloguing in Publication data
Ladd, M F C (Marcus Frederick Charles)
Introduction to physical chemistry/Mark Ladd.–3rd ed.
p. cm.
Includes index.
ISBN 0 521 48000 0 (hardbound). ISBN 0 521 57881 7 (pbk)
1. Chemistry, Physical and theoretical. I. Title.
QD453.2.L33 1997
541–dc21 96-37799 CIP

ISBN 0 521 48000 0 hardback
ISBN 0 521 57881 7 paperback

CONTENTS

PREFACE

This book is a complete revision of an earlier work of the same title by the author and his late colleague, Dr W H Lee. It is intended to meet the requirements of students in their first and second years of a degree course in chemistry, or in those sciences for which chemistry forms a significant part, and so to prepare the ground for more advanced final-year studies in this subject.

The mathematical arguments that have been employed in the book should lie within the scope of any chemistry degree student, who will have studied mathematics to A-level, or its equivalent, at least. Physical chemistry is not a nonmathematical subject: any attempt to make it appear so will probably detract from the elegance that attends those approaches that have served this subject well. Some less familiar mathematical topics are discussed in appendices.

Each chapter has been provided with a set of problems designed to enhance the reader's appreciation of the subject matter and its application to new situations. A suggested scheme for solving problems has been given in Appendix 1. The ready availability of computers means that much more extensive data sets can be handled than would be reasonable with hand calculators.

To this end, a number of computer programs has been written, as outlined in Appendix 1. They have been made available on the Worldwide Web internet and may be accessed at web site www.cup.cam.ac.uk; the set of programs, with notes, may be obtained also from the author. The reader is encouraged to make full use of these facilities. The set of programs includes the derivation of point groups and practical point-group recognition, which have been used successfully by the author in teaching these subjects over many years, and the calculation of Madelung constants, in addition to other, more numerical procedures.

The detailed solutions to the problems are available on the same web site, while brief answers are provided within the book itself.

The SI system of units is used throughout the book. However, there are several instances, for example, wavenumber (cm^{-1}) or ionization energy (eV), for which current practice demands that these alternative units should, at least, appear. Competency in more than one system of units will enhance the reader's appreciation of the subject and its literature.

In the sections that discuss three-dimensional topics, such as Bravais lattices and crystal structures, many illustrations have been provided as stereoscopic views and directions for viewing them are given in Appendix 2. Sections on liquid structure, quasicrystals and Wigner–Seitz cells have been introduced, in order to broaden the scope of the treatment of the subject of physical chemistry.

The author is most appreciative of those publishers and authors who have given their permission to reproduce those illustrations that carry the appropriate acknowledgements. The author is very greatly indebted to colleagues who have assisted in the preparation of this book: to Professor S F A Kettle, Professorial Fellow at the University of East Anglia, for reading the work in manuscript and for making a number of helpful suggestions; to Professor J R Jones, Head of the Department of Chemistry at the University of Surrey, for a careful reading of the book in proof, thereby eliminating some unforced errors; and finally to the publishers for their assistance and cooperation in bringing the work to a state of completion. Any infelicities that might remain are the sole responsibility of the author.

Mark Ladd

Note added at proof

The reader is cautioned to distinguish carefully between the italic 'vee' and the Greek *nu*. An example of their occurrence together is in equation (A3.5), p. 458, where the first of these characters is a 'vee'.

PHYSICAL CONSTANTS AND OTHER NUMERICAL DATA

These data have been selected, or derived, from the compilation of E R Cohen and B N Taylor, *J. Phys. Chem. Ref. Data* (1988) **17**, 1795–1803; the values are reported in SI units. The figures in parentheses after each value represent the standard deviation to be applied to its last two digits; the values of c and ϵ_0 are *defined.* Although the data are presented here with their full precision, we shall rarely need to employ more than about the first four or five significant figures.

Speed of light in a vacuum	c	2.99792458	$\times 10^{8}$	m s^{-1}
Permittivity of a vacuum	ϵ_0	8.854187817	$\times 10^{-12}$	F m^{-1}
Permeability of a vacuum	μ_0	4π	$\times 10^{-7}$	H m^{-1}
Planck constant	h	6.6260755(40)	$\times 10^{-34}$	J Hz^{-1}
Elementary charge	e	1.60217733(49)	$\times 10^{-19}$	C
Avogadro constant	L	6.0221367(36)	$\times 10^{23}$	mol^{-1}
Atomic mass unit	u	1.6605402(10)	$\times 10^{-27}$	kg
Bohr magneton	μ_B	9.2740154(31)	$\times 10^{-24}$	J T
Rydberg constant	R_∞	1.0973731534(13)	$\times 10^{7}$	m^{-1}
Rydberg constant for hydrogen	R_H	1.0967758772(13)	$\times 10^{7}$	m^{-1}
Bohr radius	a_0	5.29177249(24)	$\times 10^{-11}$	m
Boltzmann constant	k_B	1.380658(12)	$\times 10^{-23}$	J K^{-1}
Molar gas constant	$\mathscr{R}$	8.314510(70)	J K^{-1} mol^{-1}	
		0.0820577(7)	dm^3 atm K^{-1} mol^{-1}	
Molar volume of ideal gas at 273.15 K and 101 325 Pa	V_m	22.41410(19)	$\times 10^{-3}$	m^3 mol^{-1}
Compton wavelength (electron)	λ_c	2.42631058(22)	$\times 10^{-12}$	m
Rest mass of electron	m_e	9.1093897(54)	$\times 10^{-31}$	kg
Rest mass of proton	m_p	1.6726231(10)	$\times 10^{-27}$	kg
Rest mass of neutron	m_n	1.6749286(10)	$\times 10^{-27}$	kg
Reduced mass of proton and electron pair	μ	9.1044313(54)	$\times 10^{-31}$	kg
Faraday	$\mathscr{F}$	9.6485309(29)	$\times 10^{3}$	C mol^{-1}
Ice-point temperature	T_{ice}	273.1500(01)		K

(Farad F=C V^{-1}; Tesla T=10^4 G (gauss)=J C^{-1} m^{-2} s; Henry H=J C^{-2} s^2)

Some of the more important additional units are listed below.

Length

1 Å (ångström unit)=10^{-10} m=10 nm

Energy
1 eV (electronvolt)$\equiv$1.60217733(49)$\times 10^{-19}$ J
1 cal (calorie)=4.184 J$\equiv$96.485309(29) kJ mol^{-1}
1 cm$^{-1}\equiv$1.9864673(4)$\times 10^{-23}$ J$\equiv$1.1962568(5)$\times 10^{-2}$ kJ mol^{-1}
Pressure
1 atm (atmosphere)=101 325 Pa (N m^{-2})=760 Torr$\equiv$760 mmHg
Dipole moment
1 D (debye)=3.33564$\times 10^{-30}$ C m

PREFIXES TO UNITS

The following prefixes to units are in common use.

femto	pico	nano	micro	milli	centi	deci	kilo	mega	giga
f	p	n	μ	m	c	d	k	M	G
10^{-15}	10^{-12}	10^{-9}	10^{-6}	10^{-3}	10^{-2}	10^{-1}	10^{3}	10^{6}	10^{9}

CHAPTER 1

Structure, energy, mechanism

1.1 Introduction

Physical chemistry is concerned with the structures of chemical compounds, the mechanisms by which these compounds react and the energy changes that accompany the reactions between the chemical species. Studies in these fundamental aspects of the subject are based largely upon experimental measurements, but theoretical and computer-simulation techniques provide powerful additional methods of investigation. Notwithstanding the subject has, for convenience here, been subdivided, the sections are inevitably linked, so that the order in which they are treated is simply a matter of preference.

All chapters are provided with sets of problems, for which detailed solutions are available on the Internet (see Preface), that have been designed to enhance the reader's appreciation of the subject matter. The reader is encouraged to attempt these problems, and the material described in Appendix 1 should be of assistance in this aspect of the study.

1.2 Structure

The term structure embraces a wide range of properties, among which we may include the stereochemistry of a molecule, the lengths of bonds between its atoms, the angles between pairs of bonds, the vibrations of atoms and groups of atoms, the distribution of electron density, the arrangement of molecules in the condensed state and the contact, or nonbonded, distances between species.

In the water molecule H_2O, for example, each hydrogen atom is linked to the oxygen atom by a bond that is mainly covalent, with O−H bond lengths of 0.096 nm and an H−O−H bond angle of 104.4°; the distance between the two hydrogen atoms, the intramolecular proton separation, is 0.152 nm. These distances depend to a small extent on the method used to measure them. For example, X-ray diffraction provides distances between the electron density maxima of atoms, whereas neutron diffraction methods measure the distances between nuclei; generally, there will be a small but significant difference between the two results for one and the same bond length (see also Section 2.7.1).

Hydrogen and oxygen atoms have, formally, one and eight electrons respectively, but there is a shift of electronic charge when the atoms are combined, as in the water molecule. Theoretical calculations have shown that the charges on the hydrogen and oxygen atoms in a water molecule are approximately $0.16e$ and $-0.32e$ respectively, where e is the charge on an electron; the water molecule is said to be polar.

In the liquid state, the polarity of the water molecules leads to association between them. Relatively strong hydrogen bonds are set up: a hydrogen atom acts as an electrostatic link between two oxygen atoms, one in the same molecule as the hydrogen atom and the other in

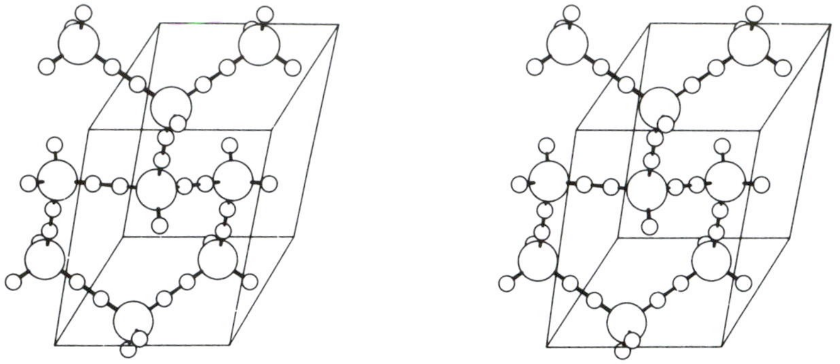

Figure 1.1 Stereoview of the unit cell and environs of the hydrogen-bonded structure of ice at 90 K. The circles represent, in decreasing order of size, oxygen and statistical half-hydrogen atoms. A tetrahedral disposition of bonds exists around any oxygen atom but, in any one molecule, only two of its four tetrahedral directions carry hydrogen atoms.

an adjacent molecule. These hydrogen bonds are continually breaking and reforming, and there exist constantly changing regions of water molecules in localized, approximately tetrahedral environments.

In ice at 90 K, there is nearly total hydrogen-bond formation between the water molecules, leading to a four-coordinate structure for ice. Figure 1.1 is a stereoview of the ice structure at 90 K; a discussion on stereoviewing is given in Appendix 2. In Figure 1.1, each small circle represents a statistically distributed half-hydrogen atom. At 313 K, the average number of hydrogen bonds is reduced to approximately one half of that at 90 K. The structure of ice is relatively open, or loosely packed, and a volume contraction occurs on melting. The non-bonded distance between adjacent oxygen atoms in ice is approximately 0.276 nm.

Hydrogen bonds are known to exist between hydrogen and several other atomic species, but are strongest in combination with fluorine, oxygen and nitrogen. The hydrogen bonds in water are responsible for many of the properties of this substance that appear to be anomalous when compared with related hydrides. The important chemical and biological functions of water depend upon the existence of its hydrogen bonds. Table 1.1 lists the boiling-point for a series of hydrides of periodic groups 15, 16 and 17. The values for the compounds in the first row are higher than would have been expected because, normally, the boiling-point temperature increases with an increase in the molar mass.

The existence of hydrogen-bonding in ammonia has been challenged on the grounds of spectroscopic and crystallographic evidence[1]. It is clear that, in the combination of ammonia with hydrogen fluoride, the lone pair of electrons on nitrogen bond to hydrogen; hence, a changing role for ammonia in combination with itself has been considered to be possible. From this premise, it would follow that liquid ammonia should behave differently from water and liquid hydrogen fluoride, but it is clear that more evidence is needed to resolve this question fully.

Many compounds do not exist as discrete molecules. Sodium chloride, for example, exists as ions in the solid state, linked together by Coulombic forces, but with no two ions preferentially associated with each other. The formula NaCl expresses the molar proportions of sodium to chlorine in the solid. If sodium chloride is heated, it melts at 1074 K to form a clear liquid containing sodium and chloride ions. The liquid boils at 1686 K, and the vapour consists of mainly covalent molecules of NaCl and Na_2Cl_2.

[1] D D Nelson *et al. Science* **238**, 1670 (1987).

Table 1.1 Boiling-points/K for some hydrides

H_3N	240	H_2O	373	HF	254
H_3P	186	H_2S	213	HCl	188
H_3As	218	H_2Se	232	HBr	206
H_3Sb	256	H_2Te	271	HI	238

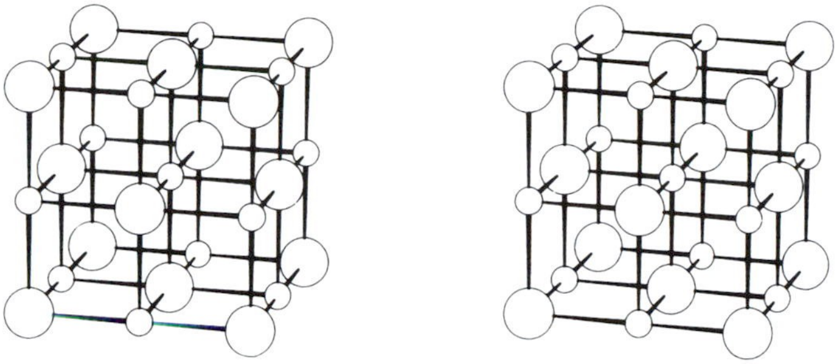

Figure 1.2 Stereoview of the face-centred (F) cubic unit cell and environs of the sodium chloride structure. The circles represent, in decreasing order of size, Cl^- and Na^+ ions. Each ion is coordinated by six ions of the opposite type, forming the apices of a regular octahedron about the given ion. There are four Na^+Cl^- pairs per unit cell.

The structure of crystalline sodium chloride is illustrated by Figure 1.2. Pairs of sodium and chloride ions are associated with each point of a face-centred cubic unit cell (see Figure 3.13c later) of side 0.564 nm. These pairs $Na^+ \ldots Cl^-$ are in identical vector orientations throughout the structure. The closest distances of ions are 0.282 nm for $Na^+ \ldots Cl^-$ and 0.399 nm for $Na^+ \ldots Na^+$ and $Cl^- \ldots Cl^-$. The ideal macroscopic crystal of sodium chloride may be considered to be formed by packing unit cells together such that each face is common to, and shared by, two adjacent unit cells.

Calcium sulfate dihydrate (gypsum) $CaSO_4 \cdot 2H_2O$ has a quite different structure, Figure 1.3. The doubly charged sulfate anions are essentially covalently bonded entities, with S–O bond lengths of 0.16 nm, O–S–O angles of 109.5° and a tetrahedral geometry; they are linked to calcium cations by Coulombic forces. In addition, hydrogen-bonding occurs between the water molecules and the oxygen atoms of the sulfate ions; in fact, the hydrogen bonds maintain coherence of the structure in one direction.

1.3 Energy

Reactions take place with differing degrees of completeness. On the one hand, the reaction between hydrogen and oxygen to form liquid water is, for all practical purposes, complete and is accompanied by an enthalpy (heat content) change of approximately -286 kJ mol^{-1}; the negative sign implies that heat is *liberated* in the *exothermic* combination reaction. The energy of the hydrogen bonds in water accounts for 6 to 7% of this total enthalpy change. On the other hand, the dissolution of silver iodide in water at 298 K is negligible; its solubility at 298 K is 1.02×10^{-8} mol dm^{-3}.

Between these extreme examples, all other stages of completion may be encountered with reacting systems. As an example, we consider the synthesis of ammonia by the Haber process,

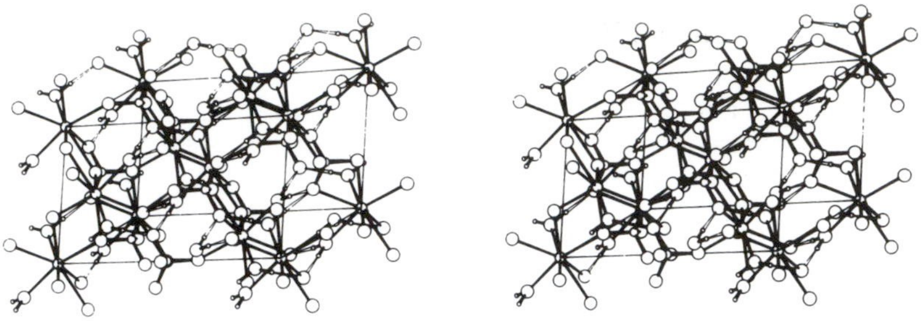

Figure 1.3 Stereoview of the unit cell and environs of the structure of calcium sulfate dihydrate. The circles represent, in decreasing order of size, O, Ca, S and H. The hydrogen bonds, shown in double lines (a hand-lens may help), are responsible for cohesion of the structure along the right-left direction in the illustration.

in which nitrogen and hydrogen at approximately 450°C and 200 atm are combined over a catalyst of α-iron, in the presence of small amounts of oxides of iron, silicon and magnesium; the traces of oxides promote the reaction by increasing the active surface area of the catalyst.

The reaction is exothermic for the production of ammonia:

$$\tfrac{1}{2}N_2(g)+\tfrac{3}{2}H_2(g)\rightleftharpoons NH_3(g) \qquad (1.1)$$

The $\rightleftharpoons$ sign indicates that an equilibrium is attained between the components of the reaction at any given temperature and pressure. Tables 1.2 and 1.3 list some experimental results for this equilibrium. The equilibrium constant K_p, which is a measure of the completeness of a reaction, is given here by

$$K_p=\frac{p_{NH_3}}{p_{N_2}^{1/2}\,p_{H_2}^{3/2}} \qquad (1.2)$$

where p represents partial pressure; evidently, K_p depends only on the temperature and decreases with an increase in this variable for the formation of ammonia by (1.1).

The percentage of ammonia at equilibrium increases with an increase in pressure, but decreases with an increase in temperature. These conclusions are in accord with the Le Chatelier–Braun principle, which states that *a system at equilibrium, when subjected to a perturbation, responds so as to tend to annul the effect of the perturbation.*

The production of ammonia, the forwards reaction in (1.1), is accompanied by a decrease in the *free energy* ΔG of the system, given by

$$\Delta G=\Delta H-T\Delta S \qquad (1.3)$$

A negative value of ΔG for (1.1) is taken to indicate that the formation of ammonia is energetically feasible. The terms on the right-hand side of (1.3) represent respectively the change in enthalpy ΔH and the change in entropy ΔS, at the temperature T; ΔG is a sort of compromise parameter between ΔH and ΔS, which often act in opposition to each other.

At the temperatures considered in Tables 1.2 and 1.3 ΔG is actually positive. It would appear also that, at 298 K, the percentage of ammonia would be appreciably greater than that at, say, 700 K. This view is supported by the equilibrium equation[2]

$$\Delta G_m^{-0-}=-\mathcal{R}T\ln K_p \qquad (1.4)$$

where $\mathcal{R}$ is the gas constant.

[2] The superscript −0− refers to *standard* conditions (see Section 4.5); 'ln' stands for *logarithmus naturalis* (logarithm to the base e).

Table 1.2 Nitrogen–hydrogen–ammonia equilibrium: percentage of ammonia at equilibrium as a function of temperature and pressure

p/atm	10	50	100
T/K			
623	10.4	25.1	37.1
673	3.85	15.1	24.9
723	2.04	9.17	16.4
773	1.20	5.58	10.4

Table 1.3 Nitrogen–hydrogen–ammonia equilibrium: equilibrium constant K_p as a function of temperature[3]

p/atm	10	50	100
T/K			
623	0.0266	0.0278	0.0288
673	0.0129	0.0130	0.0137
723	0.00659	0.00690	0.00725
773	0.00381	0.00387	0.00402

Using Table 1.3, the molar free energy change ΔG_{m} at 723 K is 30.2 kJ mol^{-1}; at 298 K, $\Delta G_{\mathrm{m}}^{-0-}$ is -16.5 kJ mol^{-1}, and the corresponding equilibrium constant is 780. However, the rate of reaction at 298 K is extremely slow and optimum working conditions have to be chosen. We see how considerations of kinetics are now introduced. Notwithstanding the energetics at 298 K are favourable, the reactants must acquire a certain amount of energy, the activation energy, before reaction will take place at a measurable rate. Although a reaction at 298 K may be thermodynamically feasible, it need not take place immediately the reactants are mixed. We shall consider this reaction again in the chapter on thermodynamics.

1.4 Mechanism

The mechanisms of chemical reactions include considerations of rate, molecularity and reaction pathway, and are usually derived from a study of the kinetics of reactions. For example, the formation of methoxyethane from iodomethane and sodium ethanoate in alcoholic solution occurs as a result of collisions between reactant molecules:

$$CH_3I + CH_3CH_2ONa \xrightarrow{EtOH} CH_3OCH_2CH_3 + NaI \qquad (1.5)$$

The rate of this reaction depends first upon how often the reactants meet each other, which events are proportional to their concentrations, and whether or no they collide with suffi-

[3] The small drift in K_p with pressure occurs because the partial pressure has been used in (1.2) rather than the fugacity (see Section 4.9.3).

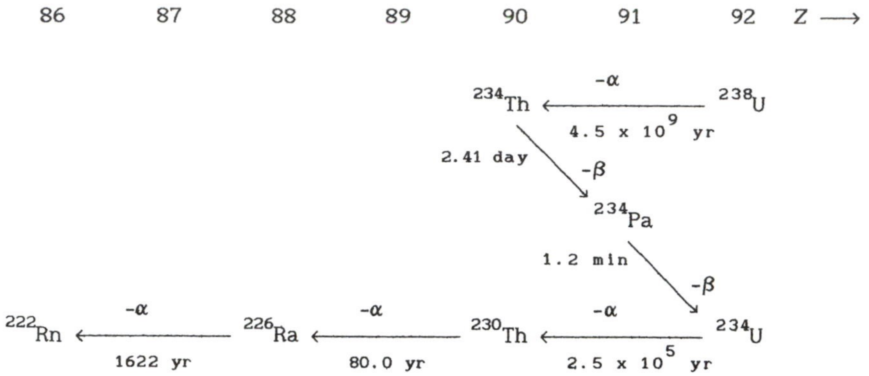

Figure 1.4 Radioactive decay in the ^{238}U series between atomic numbers Z 86 and 92: $-\alpha$ and $-\beta$ indicate emission of the corresponding radiations; the half-lives of the species are also given.

cient energy to drive the process forwards. Reaction (1.5) is bimolecular, that is, there are two molecular species involved in the process. It is also of second-order kinetics, because the rate is proportional to the concentrations of both reactants raised to the power of unity.

The decay of the nuclei of elements of atomic number greater than 83 (bismuth) is spontaneous and energy is emitted as α or β radiation, Figure 1.4. The rate of radioactive decay depends only on the concentration of the radioactive species present; it is a first-order reaction.

The apparently straightforward reaction between hydrogen and chlorine

$$\tfrac{1}{2}H_2(g)+\tfrac{1}{2}Cl_2(g)\rightarrow HCl(g) \tag{1.6}$$

is very slow unless activated by radiation of sufficient energy:

$$\tfrac{1}{2}Cl_2(g)+h\nu\rightarrow Cl^{\bullet}(g) \tag{1.7}$$

Here, $Cl^{\bullet}$ is a chlorine free radical, that is, a species having an unpaired electron; $h\nu$ represents a photon, a packet of radiant energy, where h is the Planck constant and ν is the frequency of the radiation, in this case in the ultraviolet (UV) region of the energy spectrum (Figure 1.5). A chain mechanism is set up, in which more free radicals are generated. If uncontrolled, the reaction will proceed at an explosive rate:

$$Cl^{\bullet}(g)+H_2(g)\rightarrow HCl(g)+H^{\bullet}(g) \tag{1.8}$$

$$H^{\bullet}(g)+Cl_2(g)\rightarrow HCl(g)+Cl^{\bullet}(g) \tag{1.9}$$

Reactions (1.5) and (1.6) differ in the way in which they acquire the energy necessary to initiate reaction: (1.5) by warming in alcoholic solution, (1.6) by UV irradiation.

The bond dissociation enthalpies (or energies, if at 0 K) of hydrogen and chlorine molecules are approximately 436 kJ mol^{-1} and 243 kJ mol^{-1} respectively. It follows that (1.7) will be initiated if the system is supplied with energy somewhat greater than 243 kJ mol^{-1}, or 4.04×10^{-19} J (2.5 eV) per molecule. This amount of energy corresponds to radiation of wavelength 492 nm; hence, bright visible light or UV radiation brings about a ready reaction.

Hydrogen and oxygen may be caused to combine at an explosive rate. Again, free radicals are involved and the process is complex. Branching reactions occur, in which the number of free radicals is increased greatly in the course of the reaction:

$$O_2(g)+H^{\bullet}(g)\rightarrow HO^{\bullet}(g)+O^{\bullet}(g) \tag{1.10}$$

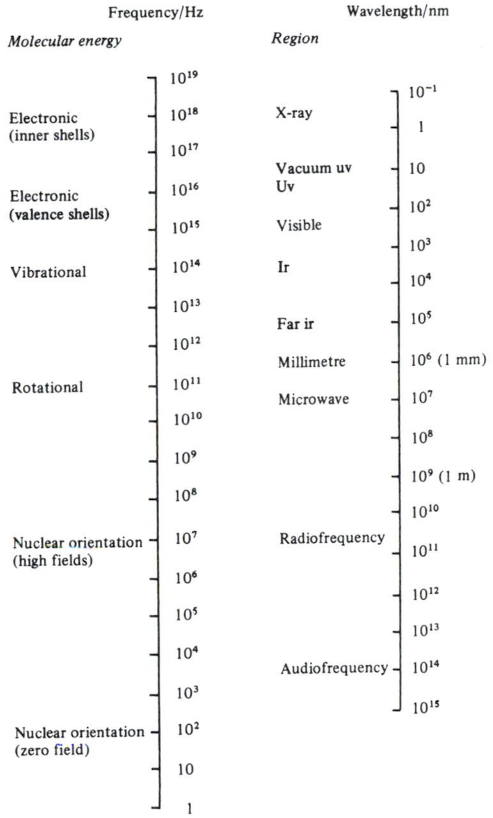

Figure 1.5 Electromagnetic spectrum, with frequencies and wavelengths indicated for molecular energies throughout the spectrum.

Branching leads to a large increase in the rate of reaction, so that an explosion occurs when a mixture of hydrogen and oxygen is subjected to a high-energy radiation, such as an electric spark. Other reactions exist that terminate the chain process, such as

$$HO^{\bullet}(g)+H^{\bullet}(g)\rightarrow H_2O(g) \qquad (1.11)$$

We discuss the kinetics of chemical reactions in more detail in Chapter 10.

Since kinetic studies involve the mechanisms of reactions, which may themselves be dependent upon structure, and because spontaneity of reaction depends upon energy changes, it is evident that the fundamental topics of structure, energy and mechanism are closely linked in the reactions of chemical species. In the ensuing chapters, we enlarge on these themes as we pursue our study of physical chemistry.

Problems 1

1.1 In the molecule of ClO_2, the Cl—O bond distance is 149 pm and the distance between the two oxygen atoms is 254 pm. Calculate the O—Cl—O bond angle.

1.2 Assuming that no hydrogen-bonding existed in water, what would be its approximate boiling-point (refer to Table 1.1)?

1.3 In the sulfate ion, the excess charge associated nominally with each oxygen atom is $-0.95e$. What would be the corresponding nominal charge on the sulfur atom?
1.4 What is the second shortest sodium–chlorine distance in the sodium chloride structure; the cube side a equals 0.564 nm? Calculate the density of solid sodium chloride.
1.5 The energy needed to initiate a given chemical reaction is approximately 400 kJ mol^{-1}. What would be the wavelength of a suitable radiation source and in what region of the electromagnetic spectrum would the radiation occur?
1.6 The volume of gaseous helium, at 273 K and 1 atm, liberated by the radioactive disintegration of 1 g of radium during a period of 1 year is 0.043 cm^3. In the same time period, the number of α particles emitted in the disintegration of the same amount of radium is 116×10^{16}. Assuming that each α particle yields one atom of helium, calculate a value for the Avogadro constant N_A.

Atoms, molecules and their structures

2.1 Introduction

Until the turn of the present century, it was believed that the behaviour of atoms was described by the classical mechanics of Newton, as was the case with macroscopic bodies. However, several experimental results seemed to be at variance with this assumption and in 1926 the new science of quantum mechanics was developed to explain the behaviour of microscopic particles. We shall consider in this chapter some aspects of quantum mechanics and see how classical mechanics is, in fact, a special case of this more general theory.

2.2 Classical mechanics

The behaviour of a classical particle can be described by two basic equations. In the first of them, the total energy E of a particle of mass m and speed v, at a position x and time t with respect to a given origin, is equal to the sum of its kinetic energy $\frac{1}{2}mv^2$ and potential energy $V(x)$:

$$E=\tfrac{1}{2}mv^2+V(x) \tag{2.1}$$

where v and x are functions of time t. We may also write

$$E=p^2/(2m)+V(x) \tag{2.2}$$

where p is the linear momentum. If we consider a constant potential energy, independent of x, we have

$$m\frac{\mathrm{d}x}{\mathrm{d}t}=[2m(E-V)]^{1/2} \tag{2.3}$$

whence by integration

$$x_t=x_0+[2(E-V)/m]^{1/2}t \tag{2.4}$$

Using (2.2), we obtain

$$x_t=x_0+p_0t/m \tag{2.5}$$

and, since the total energy E is constant,

$$p_t=p_0 \tag{2.6}$$

Hence, for a given value of E, the values at any time both of x and of p, which constitute the trajectory of the particle, may be determined. The second basic equation depends on Newton's 2nd law (force=mass×acceleration=rate of change of momentum):

$$F = m\frac{d^2x}{dt^2} = \frac{dp}{dt} \tag{2.7}$$

from which it may be shown that, for a given constant force, a particle may acquire any value for the energy E.

2.3 Conflict with experiment

The experimental results to which we referred at the start of this chapter conflicted both with the demand for continuous energy ranges and with the simultaneous knowledge of precise position and momentum that followed from the application of classical mechanics to atoms. We shall consider some of these results next, to see how they led towards the development of an improved atomic theory.

2.3.1 Black-body radiation

An ideal *black body* is able to absorb or emit radiation of all frequencies (or energies) and is approximated well by a container with a pin-hole in one wall. If the container is heated, its temperature is indicated by its colour, because the radiation emitted from the hole is at the same temperature as the container, having been absorbed by and re-emitted from the interior of the walls many times; the emitted radiation is in thermal equilibrium with the walls of the black body. As the temperature of the radiator is increased, the frequency of the emitted radiation increases from the infrared end of the spectrum, through the visible range and into the ultraviolet region.

Figure 2.1 is a plot of an energy distribution $E(\nu)$ as a function of the frequency ν. The value ν_{max} at which $E(\nu)$ is a maximum moves to higher frequencies as the temperature T is increased. Similar curves are obtained in terms of wavelength λ, but with λ_{max} moving to lower wavelengths as the temperature is increased.

Experiments on black-body radiation by Stefan (1879) showed that the energy density $\mathscr{E}$, the total energy density per unit volume emitted over all wavelengths, followed the equation

$$\mathscr{E} = aT^4 \tag{2.8}$$

where the constant a was independent of the nature of the material of the body. In another set of experiments, Wien (1894) found

$$T\lambda_{max} = b \tag{2.9}$$

where the experimentally determined value of b was 2.9×10^{-3} m K.

Rayleigh considered that black-body radiation was emitted by classical molecular oscillators, one for each frequency ν; the intensity of the radiation was taken to be proportional to the amplitude of oscillation. He determined the number density $N(\nu)\,d\nu$ of oscillators with frequencies lying between ν and $\nu + d\nu$ inside a cubical enclosure of side c/λ, where c is the speed of light in a vacuum. The result, as amended later by Jeans, was

$$N(\nu)\,d\nu = (8\pi\nu^2/c^3)\,d\nu \tag{2.10}$$

We show in Appendix 3 that the mean energy of a classical oscillator at a temperature T is k_BT, where k_B is the Boltzmann constant. Hence, the required energy density, as given by the Rayleigh–Jeans equation, is

$$E(\nu)\,d\nu = N(\nu)k_BT\,d\nu = (8\pi\nu^2 k_BT/c^3)\,d\nu \tag{2.11}$$

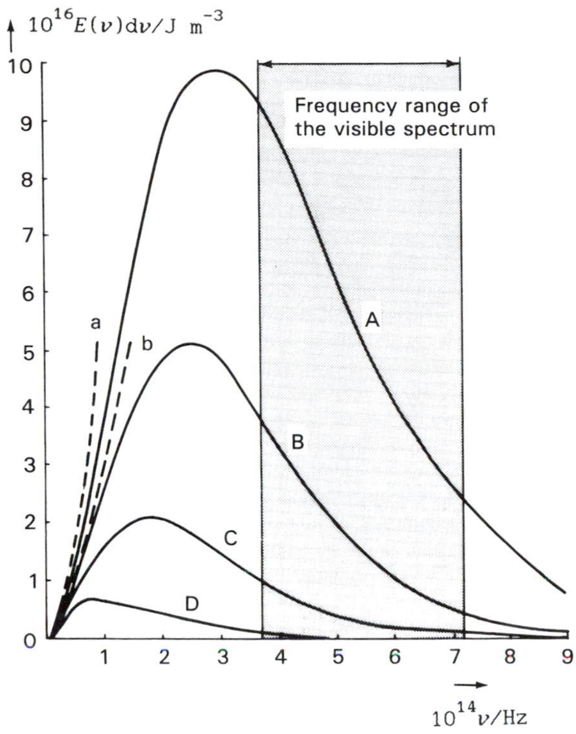

Figure 2.1 Black-body radiation. Energy density versus frequency plots for: Rayleigh–Jeans equation at (a) 5000 K and (b) 4000 K; the curves continue indefinitely and, if followed, would lead to the 'ultraviolet catastrophe'. Planck equation at (A) 5000 K, (B) 4000 K, (C) 3000 K and (D) 2000 K; the contributions from the very high frequencies are very significantly reduced by the exponential term in the Planck equation.

In terms of wavelength λ, remembering that $d\nu=(-c/\lambda^2)\,d\lambda$ and neglecting the negative sign because we are interested in the absolute value of the energy density, we obtain

$$E(\lambda)\,d\lambda=(8\pi k_B T/\lambda^4)\,d\lambda \tag{2.12}$$

Now, (2.11), or (2.12), conflicts with Wien's observation (2.9), and Figure 2.1 shows that (2.11) would lead to an accumulation of radiation in the high-energy region, the so-called 'ultraviolet catastrophe', since $E(\nu)$ would increase without limit as the frequency was increased.

Planck (1900) made a fundamental revision to the Rayleigh–Jeans equation. He was able to account for the experimental results on radiation by postulating that an oscillator of frequency ν could radiate an energy that was an integral multiple of $h\nu$, the single quantum of energy:

$$E=h\nu \tag{2.13}$$

where h is the Planck constant. Radiation was considered to consist of particles called *photons*, each of energy $h\nu$. Consider a sodium-vapour lamp of power 100 W emitting radiation with wavelength λ of 589.3 nm. Since 1 W is 1 J s^{-1}, the energy output of the lamp, per second, is 100 J. The frequency of the radiation (c/λ) is 5.087×10^{14} Hz, where c is the speed of light *in vacuo*; hence, the number of photons emitted in 1 s is 100 J s^{-1}/(5.087×10^{14} Hz×6.6261×10^{-34} J Hz^{-1}), or 2.97×10^{20}.

Classical theory permitted electromagnetic oscillators of all frequencies, even very high values, to be activated in the black body by absorbing radiation from the material of its walls. The quantum restriction allows oscillators to be activated only if they can acquire energy $nh\nu$ (n=1, 2, 3, ...). This limitation reduces the number of high-frequency oscillators, because they cannot readily acquire sufficiently large energy quanta from the black body.

The Planck equation (2.14) reconciled the equations of Rayleigh–Jeans, Wien and Stefan, all of which may be seen as special cases of the Planck equation:

$$E(\nu)\,d\nu=(8\pi h\nu^3/c^3)\{\exp[h\nu/(k_BT)]-1\}^{-1}\,d\nu \tag{2.14}$$

or in terms of wavelength

$$E(\lambda)\,d\lambda=(8\pi hc/\lambda^5)\{\exp[hc/(\lambda k_BT)]-1\}^{-1}\,d\lambda \tag{2.15}$$

and the presence of the exponential term in (2.14), or (2.15), means that only small contributions from the high-energy oscillations are feasible.

We can compare the Rayleigh–Jeans equation (2.12) and the Planck equation (2.15) by calculating the energy density of radiation in the wavelength range 590 nm to 600 nm in a radiator of volume 10 cm^3 at a temperature of 500 °C.

The mean wavelength may be taken as 595 nm and the wavelength range as 10 nm. Hence, from (2.12)

$$E(\lambda)\,d\lambda=(2.1402\times10^6\ \mathrm{J\ m^{-4}})\,(10\times10^{-9}\ \mathrm{m})$$
$$=2.14\times10^{-2}\ \mathrm{J\ m^{-3}}$$

whereas from (2.15)

$$E(\lambda)\,d\lambda=\{6.6947\times10^7\ \mathrm{J\ m^{-4}}/[\exp(31.275)-1]\}(10\times10^{-9}\ \mathrm{m})$$
$$=1.75\times10^{-14}\ \mathrm{J\ m^{-3}}$$

whereupon the limitation of the exponential term under the given conditions is very evident.

It is easy to show that (2.14) may be written as

$$E(\nu)\,d\nu=\frac{8\pi h\nu^3k_BT}{c^3h\nu}\left[1+\left(\frac{h\nu}{k_BT}\right)/2!+\left(\frac{h\nu}{k_BT}\right)^2/3!+\dots\right]^{-1}d\nu \tag{2.16}$$

Classical conditions are attained for $h\nu/(k_BT)\ll1$, that is, at low frequencies or high temperatures. Typically, $\nu=10^{11}$ Hz and T=3000 K, whence $h\nu/(k_BT)=0.0016$, whereupon (2.16) becomes

$$E(\nu)\,d\nu\simeq(8\pi h\nu^2k_BT/c^3)\,d\nu \tag{2.17}$$

which compares with (2.11). At high frequencies or low temperatures $h\nu/(k_BT)\gg1$; typically, $\nu=5\times10^{14}$ Hz and T=500 K, whence $h\nu/(k_BT)=48$. Then, (2.14) may be written as

$$E(\nu)\,d\nu\approx(8\pi h\nu^3/c^3)\exp[-h\nu/(k_BT)]\,d\nu \tag{2.18}$$

Differentiating (2.15) with respect to λ and setting the derivative to zero, we obtain

$$hc/(5\lambda_{max}k_BT)=1-\exp[hc/(\lambda_{max}k_BT)] \tag{2.19}$$

from which, by successive approximations,

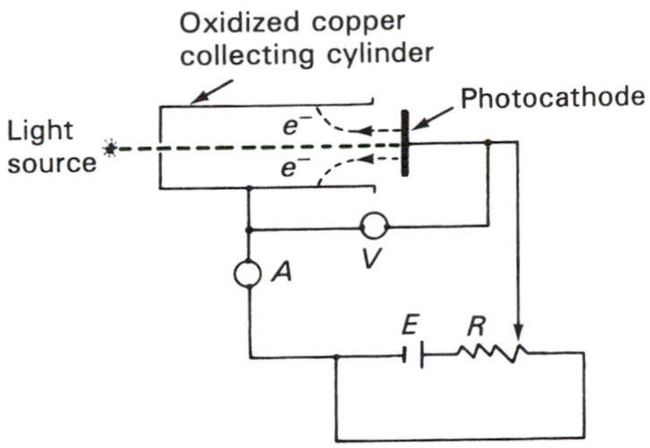

Figure 2.2 Schematic diagram for the production of a photoelectric effect: A ammeter, V voltmeter, R rheostat, E external EMF source; the photocathode and collecting cylinder are *in vacuo*. A potential difference exists between the photocathode and the cylinder, such that the cylinder is at a positive potential. Electrons e^- are expelled from the photocathode and travel to the collecting cylinder, thus completing an electrical circuit through V.

$$\lambda_{max}T=0.20140hc/k_B=2.898\times10^{-3}\text{ m K} \tag{2.20}$$

in very good agreement with experiment. Thus, we see that Wien had observed quantum behaviour, whereas Rayleigh and Jeans addressed the black body under classical conditions.

2.3.2 Photoelectric effect

Figure 2.2 indicates an apparatus for obtaining a photoelectric effect. A monochromatic light source incident upon a metal cathode in a vacuum results in the ejection of electrons from the cathode, provided that the external EMF E is at least a certain minimum value V_0, dependent upon the nature of the cathode. The electrons are attracted to the anode, thus leading to current flow.

Experimentally, several conditions were found to exist, of which the following two are of most significance:

(a) the mean kinetic energy of the electrons emitted was proportional to the frequency of the incident light;
(b) no electrons were emitted unless the frequency of the incident light exceeded a certain minimum value, whereupon even a low light intensity was sufficient to cause emission.

If the light source comprises photons of energy $h\nu$ then, by conservation of energy, the kinetic energy of the electrons emitted should follow the law

$$\tfrac{1}{2}mv^2=h\nu-\phi_M \tag{2.21}$$

When the incident radiant energy $h\nu$ transferred to an electron exceeds ϕ_M, an electron is ejected with kinetic energy equal to the excess of $h\nu$ over ϕ_M. The term ϕ_M is the *work function* of the metal, which represents the energy required to dislodge an electron from the metal of the cathode. Some values of ϕ_M for a few metals are listed below in eV (1 eV$=1.6022\times10^{-19}$ J).

	Li	Na	K	Mg	Cu	Ag
ϕ_M/eV	2.42	2.3	2.25	3.7	4.8	4.3

In this experiment, the light beam appears to be particulate in nature, with photons having a momentum p given by

$$p=mc \tag{2.22}$$

where m is the mass associated with a light photon. Using the Einstein equation that relates mass and energy

$$E=mc^2 \tag{2.23}$$

together with (2.13), we obtain

$$p=h\nu/c \tag{2.24}$$

which relates the momentum p directly to the energy of irradiation.

2.3.3. Compton effect

When light interacts with electrons in a material, it undergoes a wavelength shift that depends on the angle of scattering and is independent of the wavelength of the incident light. If we assume that a photon is a particle of momentum $h\nu/c$, then collision between a photon and an electron of mass m_e, with conservation both of energy and of momentum, leads to an increase in wavelength $\delta\lambda$:

$$\delta\lambda=[h/(m_e c)](1-\cos\theta_C) \tag{2.25}$$

where θ_C is the angle of Compton scattering; $h/(m_e c)$ is the Compton wavelength for electrons λ_C and its calculated value agrees well with experiment.

2.3.4 Diffraction of electrons

Experiments by Davisson and Germer in 1925 showed that electrons could be diffracted from metallic nickel, similarly to the way in which light is diffracted from a ruled grating. This initial experiment was, to some extent, fortuitous because an increase in temperature during the course of the experiment caused the powdered specimen to recrystallize and so diffract the electrons strongly (see also Section 3.6.11ff).

The experiment was repeated successfully with many other substances and led to the view that electrons could exhibit wave properties, with a wavelength given by the de Broglie equation (2.29), as well as the particulate properties discussed above.

2.3.5 Atomic spectra of hydrogen

The spectral emission from hydrogen atoms excited by an electric discharge includes light at four frequencies in the visible region, Figure 2.3. Balmer showed (1895) that this spectrum could be represented by an equation of the type

$$\bar{\nu}=R_H(1/m^2-1/n^2) \tag{2.26}$$

where $\bar{\nu}$ is the wavenumber ($1/\lambda$) of the spectral line, R_H is the Rydberg constant for hydrogen, m is equal to 2 and n is another integer, greater than 2. The expression of a frequency (or wavenumber) by the difference between two terms is known as the Ritz combination principle.

It was a feature of the Bohr atomic theory that it could explain the spectra of atomic

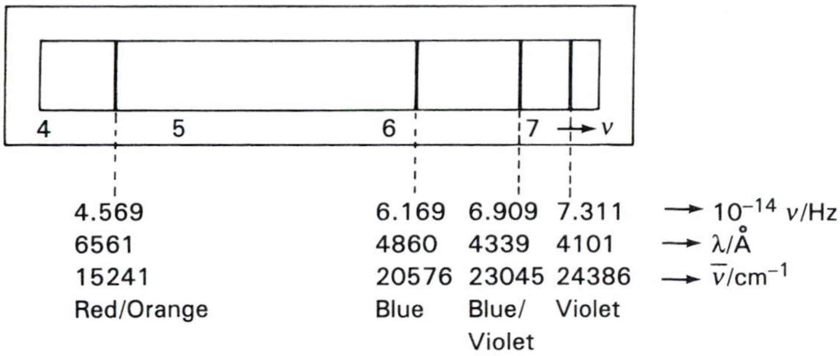

Figure 2.3 Balmer line spectrum for atomic hydrogen; other series exist in the IR and UV regions, and are fitted by (2.26) with different values of the integers m and n ($n>m$).

hydrogen in terms of transitions between two energy levels, assuming the Bohr restriction that an electron emitted radiation only when moving from one energy level E_2 to a level of lower energy E_1; the frequency of the radiation was then given by

$$\nu=(E_2-E_1)/h \tag{2.27}$$

This equation represented a fundamental break with the classical theory, which required that a charge moving along a circular path, thus under acceleration, emitted energy continuously. The theory predicted the energy levels for hydrogen correctly:

$$E_n=\mu e^4/(32\pi^2\hbar^2\epsilon_0^{\,2}n^2)\ (n=1,2,3\ldots) \tag{2.28}$$

where μ is the reduced mass (see Appendix 4) of the system of one proton and one electron, $\hbar$ ('cross-h') is $h/(2\pi)$ and ϵ_0 is the permittivity of a vacuum. Evaluating $[\mu e^4/(8h^2\epsilon_0^{\,2})]/(hc)$ gives 1.0967759×10^7 m^{-1} for R_H, in excellent agreement with the experimental value of 1.096776×10^7m^{-1} for this constant.

Bohr's theory of the atom was based on classical mechanics, with electrons orbiting the nucleus rather like planets around the sun. Quantum conditions were imposed in order to fit the experimental results and to explain how electrons did not follow a spiral path into the nucleus, as implied in Rutherford's earlier theory. It was, however, a patchwork theory and, more seriously, it could not explain the spectra of ground state (see Section 3.6.5.1) atomic species other than hydrogen, nor even that spectrum in the presence of a magnetic field (see Section 3.6.2.3).

The principal objections to applying classical mechanics to an electron in an atom are that it defines precisely both the position and momentum of the electron, and follows its trajectory (see Section 2.2). The uncertainty principle, which arises from quantum mechanics (see Section 2.5.2), shows that these parameters cannot all be determined with precision simultaneously: that they can be for planets, with satisfactory results, depends on the differing sizes of planets and electrons in relation to the methods used for their observation. If, then, we are not to obtain the desired success by hypotheses based on the particulate nature of the electron, it seems not unreasonable to consider instead its wave properties.

2.4 Wave–particle duality

The experiments that we have just discussed indicate that electrons exhibit both a particle nature, as in the photoelectric effect and the Compton effect, and a wave nature, as in the diffraction experiments. One is led to the conclusions that, on the atomic scale, particles can exhibit the features of waves and that waves can show the characteristics of particles, according to the types of experiments performed. These conclusions were reconciled by de Broglie (1924), who suggested that any particle travelling with a linear momentum p had an associated wave character, where

$$p=h/\lambda \tag{2.29}$$

a relationship implicit in (2.24). We turn our attention next to quantum mechanics.

2.5 Quantum mechanics of particles

The Schrödinger equation, proposed in 1926, leads to a wavefunction ψ for any system of particles. For a particle of mass m that is free to move in one dimension, the Schrödinger equation may be written as

$$[-\hbar^2/(2m)]\,\mathrm{d}^2\psi(x)/\mathrm{d}x^2+V(x)\psi(x)=E\psi(x) \tag{2.30}$$

where $\psi(x)$ is the one-dimensional wavefunction, $V(x)$ the potential energy of the particle and E its total energy. A convenient notation for (2.30) is

$$\mathcal{H}_1\psi=E\psi \tag{2.31}$$

where $\mathcal{H}_1$ is a one-dimensional Hamiltonian operator given by

$$\mathcal{H}_1=[-\hbar^2/(2m)]\,\mathrm{d}^2/\mathrm{d}x^2+V \tag{2.32}$$

V and ψ are then understood to mean $V(x)$ and $\psi(x)$ respectively. Although (2.32) cannot be derived, its form may be justified by a straightforward argument. If a particle of mass m is subjected to a constant potential V, (2.31) may be written *in extenso* as

$$\frac{-\hbar^2}{2m}\frac{\mathrm{d}^2\psi}{\mathrm{d}x^2}=(E-V)\psi \tag{2.33}$$

a simple solution of which is (see Appendix 10)

$$\psi=\exp(\mathrm{i}kx)=\cos(kx)+\mathrm{i}\sin(kx) \tag{2.34}$$

where

$$k=[2m(E-V)/\hbar^2]^{1/2} \tag{2.35}$$

If the potential energy is zero,

$$k=(2mE/\hbar^2)^{1/2} \tag{2.36}$$

and from (2.2) it follows then that

$$p=k\hbar \tag{2.37}$$

Since $\cos(kx)$ or $\sin(kx)$ represents a wave of wavelength λ equal to $2\pi/k$, we have

$$\lambda=h/p \tag{2.38}$$

which is de Broglie's equation (2.29). If V is constant but nonzero, then at a fixed total energy E we have, from (2.35)

$$\lambda=h/[2m(E-V)]^{1/2} \tag{2.39}$$

and as the kinetic energy $(E-V)$ decreases the wavelength increases, becoming infinite when the particle is at rest.

If $V(x)$ changes linearly with x, the particle is subject to a force that is proportional to the potential gradient $-\mathrm{d}V(x)/\mathrm{d}x$. Since $V(x)$ decreases with increase in x, the kinetic energy increases, λ decreases and momentum increases. The situation is similar to that of a particle under acceleration from an applied constant force, its motion being determined by Newton's second law. Newtonian mechanics, we have already stated, is a special case of quantum mechanics.

We may note here that the wave equation (2.30) may be obtained from the classical equation (2.2) if we make the substitution

$$p=(\mathrm{i}\hbar)\mathrm{d}\,/\mathrm{d}x \qquad (2.40)$$

where the term on the right-hand side of (2.40) is the linear momentum operator. Many solutions to (2.30) can be formulated. For a free particle, $\psi=A\exp(\mathrm{i}kx)$ is a solution, where A is a constant, as double differentiation can confirm. Hence, k and E can take on any value. However, we need an interpretation of ψ that will both be physically meaningful and eliminate certain values of E, as demanded by quantization.

2.5.1 Born's interpretation of the wave equation

In the corpuscular theory of light, the intensity of the light at any instant is determined by the number of photons present, whereas in wave theory it is governed by the square of the amplitude of the wavefunction. In Born's interpretation of the wavefunction for particles, $\psi\psi^*\,\mathrm{d}x$, where ψ^* is the complex conjugate of ψ, represents the probability of finding the particle between the limits x and $x+\mathrm{d}x$; if the wavefunction is real, $\psi=\psi^*$. In three dimensions, the linear element $\mathrm{d}x$ is replaced by the volume element $\mathrm{d}\tau$, or $\mathrm{d}x\,\mathrm{d}y\,\mathrm{d}z$, in Cartesian space.

In order to investigate the meaning of the Born interpretation, let the particle be an electron in an atom of hydrogen, described by a wavefunction $\psi(r)$ given by $\psi(r)=(1/\pi a_0^{\,3})^{1/2}\exp(-r/a_0)$, where r is the distance of the electron from the nucleus and a_0 is the *Bohr radius* for hydrogen, approximately 52.9 pm. We will estimate the probability of locating the electron within a 1 pm volume centred (a) at the nucleus, (b) 25 pm from the nucleus and (c) 100 pm from the nucleus.

(a) $|\psi|^2\,\mathrm{d}\tau=1/(52.9^3\pi\ \mathrm{pm}^3)\exp(-2\times0/52.9)\times1\ \mathrm{pm}^3=2.2\times10^{-6}$
(b) $|\psi|^2\,\mathrm{d}\tau$ evaluates to 8.4×10^{-7}
(c) $|\psi|^2\,\mathrm{d}\tau$ evaluates to 4.9×10^{-8}

We see that, as the distance from the nucleus increases, the probability of finding the electron falls off rapidly (see also Section 2.7.1).

2.5.2 Uncertainty principle

If a particle were to occupy an exact position, then its wavefunction would have a large amplitude at that position and zero value elsewhere. Such a wavefunction can be built up by a superposition of waves having different amplitudes and wavelengths, but of correct relative phases. Figure 2.4a shows the result of summing a Fourier series with three and eight waves. In each case, the most probable position for the particle is $x=0.35$, but the probability, proportional to the square of the amplitude (the wavefunction here is real), is clearly much greater with the larger number of waves.

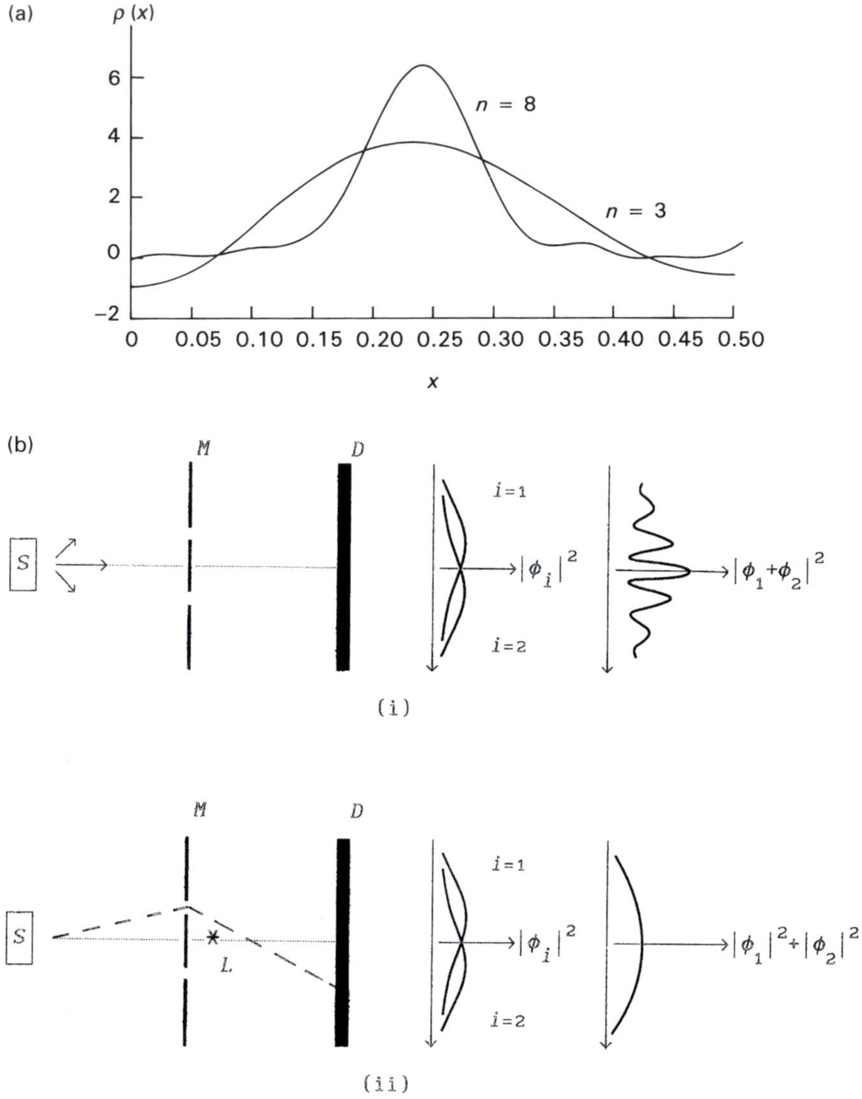

Figure 2.4 (a) Fourier summation $\rho(x)$ of waves over a fractional repeat period 0 to $\frac{1}{2}$ along x for a particle: (a) three waves, (b) eight waves. As the number of waves increases, the location of the particle becomes more precise, but simultaneously the momentum (h/λ) becomes less well defined. (b) Two-slit 'thought' experiment. (i) The atoms are unperturbed, and they behave as coherent waves: $|\phi_i|^2$ ($i=1, 2$) are the intensities of the separate waves; $|\phi_1+\phi_2|^2$ is the intensity of the interference pattern resulting from addition of the wave amplitudes ϕ_i from each slit. (ii) The use of the light source L to monitor the path of the atoms destroys the interference pattern; the atoms now appear to behave as particles and the resultant curve is the sum of the separate distributions $|\phi_i|^2$.

Each individual wave in the summation has a definite momentum, given by the de Broglie equation (2.29) but, in superposition, the sum of the waves presents an indefinite wavelength and, hence, an indefinite momentum. The more the waves that are included in the summation, the better the precision of the position and the more indeterminate the momentum. The theoretical, ideal localization of a position requires an infinite number of waves, whereupon the momentum of the resultant is totally indeterminate. The result of the superposition of waves is the formation of a *wave packet*: the size of the packet may be taken as the width of the base of the peak on the x axis, and this length is, in fact, the uncertainty in position, δx.

Position and momentum are said to be *complementary*, that is, only one of them may be specified with precision at a given time. This result is one statement of Heisenberg's uncertainty principle, which may be formulated as

$$\delta x\, \delta p_x \geq \hbar/2 \tag{2.41}$$

where δx and δp_x are the uncertainties in position and momentum respectively, along the x direction. This result is true generally, but its effect is negligible in dealing with macroscopic objects.

An interesting 'thought' experiment that has a bearing on both the uncertainty principle and wave–particle duality is illustrated by Figure 2.4b. Atoms from a source S impinge on a mask M containing two narrow, closely spaced slits and give rise to an interference pattern at the detector screen D, Figure 2.4b(i). The pattern would not arise if each atom passed through only one of the slits. However, because the atoms hit the screen at clearly defined points, they appear to behave as particles. In order to determine which slit an atom passes through, the slits can be illuminated, as shown in Figure 2.4b(ii), so that an atom reflects light as it passes through the slit.

Figure 2.4b(ii) indicates that the light will change the trajectory of the atom unpredictably because of the uncertainty principle. The effect would not be reduced by using a less intense illuminating light source; only the chance of the atom–light interaction would be decreased. Each photon that strikes an atom would have the same effect as before, according to (2.13). The atoms that are hit do not contribute to the interference pattern; the atoms that are not impinged upon by the light continue to produce the interference pattern, but the slit through which they travelled cannot be determined. Light of lower frequency may be employed, so as to perturb the interference pattern by a negligible amount. However, a low frequency implies a long wavelength, and the short distance between the slits would then not be resolved.

It must be concluded that, when atoms (or elementary particles) are showing interference effects, they are behaving like waves, not like particles. This result has received a recent experimental verification.[1]

2.5.3 Normalization and quantization

The Born interpretation of the wavefunction implies that

$$N^2 \int_{-\infty}^{\infty} \psi\psi^* \,\mathrm{d}\tau = 1 \tag{2.42}$$

since the probability of finding the electron under consideration somewhere in space must be unity. A wavefunction that satisfies this criterion is called *normalized* and N is its normalization constant.

As an example of the application of (2.42), consider a radial wavefunction $\exp(-r/a_0)$. Since this wavefunction is real, we shall use $|\psi|^2$ in (2.42) and, with reference to Appendix 5 for spherical polar coordinates, we have

[1] M S Chapman *et al.* (1995), *Phys. Rev. Lett.* **75**, 3783.

$$4\pi N^2\int_0^\infty r^2\exp(-2r/a_0)\,\mathrm{d}r=1$$

Following Appendix 6, we may write this expression in the form

$$\pi N^2 a_0^3/2\int_0^\infty t^2\exp(-t)\,\mathrm{d}t=1$$

where $t=-2r/a_0$. Since the integral is $\Gamma(3)$, N becomes $[1/(\pi a_0^3)]^{1/2}$, as we have used earlier, in Section 2.5.1.

A wavefunction must also satisfy the conditions of being finite everywhere, single-valued and continuous both in itself and in its slope. Otherwise physically unacceptable solutions would arise; for example, a wavefunction that was infinite would imply a zero value for N, from (2.42). The restrictions on the wavefunctions ensure solutions only for certain values of the energy E, as required by normalization. We shall see that discrete values for energies follow directly from the wave equation, whereas Bohr had to introduce them empirically.

The important solutions of (2.30) are called *stationary states*: the energy E is invariant with time (a *conservative* system) and, hence, $\psi\psi^*$ is also time-independent. The stationary states are our main interest; their wavefunctions are called *eigenfunctions* and the corresponding energies are *eigenvalues*.

We show that a function $f(x)$ is an eigenfunction of an operator Ω by first operating on $f(x)$ with Ω and then checking whether or no the result can be expressed as $cf(x)$, where c is a constant, the eigenvalue. For example, $\sin(ax)$ is an eigenfunction of the operator $\mathrm{d}^2/\mathrm{d}x^2$, with an eigenvalue of $-a^2$, but not of the operator $\mathrm{d}/\mathrm{d}x$. Most quantum mechanical equations that will be met are eigenfunctions of Hamiltonian operators.

2.5.4 Particle in a one-dimensional box: quantization of translational energy

Let a particle of mass m be constrained to linear motion in a one-dimensional box (something like a single bead on a bead-frame) of length a, and let the box be terminated by a potential barrier (walls) of infinite height. Then, the potential energy V is zero for $0\leq x\leq a$, but infinite for $0>x>a$. The wave equation for this system is given by (2.30) with $V=0$, with the general solution

$$\psi=A\exp(\mathrm{i}kx)+B\exp(-\mathrm{i}kx) \tag{2.43}$$

where A and B are constants, and k is given by (2.36); this solution may be confirmed by its double differentiation. Using de Moivre's theorem, (2.43) can be expanded to

$$\psi=C\cos(kx)+D\sin(kx) \tag{2.44}$$

where C and D are constants. The particle is confined to the box, which means that ψ must be zero at $x=0$, so that $C=0$, and at $x=a$, which means that $ka=n\pi$ $(n=1, 2, 3, \ldots)$. Hence, generally, under these boundary conditions,

$$\psi_n=D\sin(n\pi x/a) \tag{2.45}$$

and using (2.36), or (2.35) with $V=0$, we find for the energies of the particle

$$E_n=n^2\pi^2\hbar^2/(2ma^2)=n^2h^2/(8ma^2) \tag{2.46}$$

The energy is now quantized and determined by the single integral quantum number n. It follows that the lighter the particle or the closer together the walls become, the greater the separation of successive energy levels.

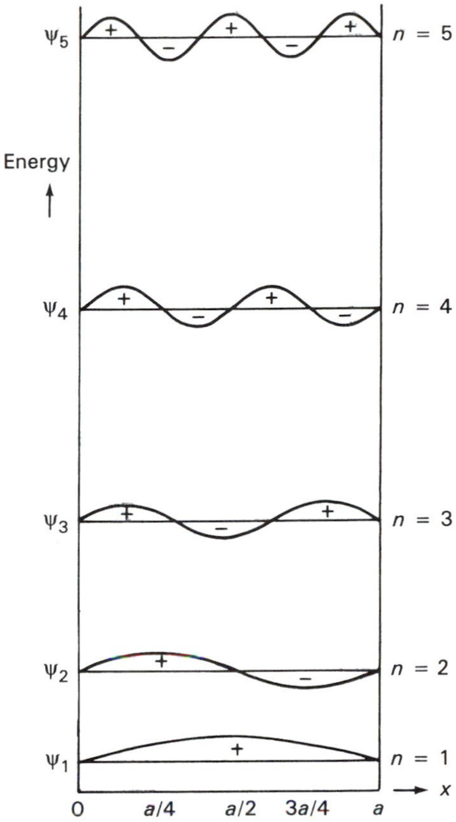

Figure 2.5 Wavefunctions ψ_n for the first five solutions of the particle in a one-dimensional box; the number of nodes at the nth level is $n-1$ and the energy level is proportional to n^2, the curvature of ψ_n increasing with increase in n.

The lowest state ψ_1 has an energy E_1, the *zero-point energy*, of magnitude $h^2/(8ma^2)$. This energy is solely kinetic. Even in the lowest energy state the particle is in motion: this property is entirely wave-mechanical and obeys the uncertainty principle: for if the minimum energy were zero, h would also be zero and the energy would not be quantized.

The probability of finding the particle lying in the interval x to $x+\mathrm{d}x$, between 0 and a, is unity and, from (2.42), we can readily show that

$$D=(2/a)^{1/2} \tag{2.47}$$

Figure 2.5 illustrates the shapes of the wavefunctions for $n=1$ to 5; they resemble the fundamental (ψ_1) and first four overtone vibrations (ψ_2 to ψ_5) of a stretched string. As n increases, the curvature, or second derivative, of the wavefunction increases, which implies an accompanying increase in kinetic energy.

The separation of any two neighbouring energy levels is given by

$$\Delta E=E_{n+1}-E_n=(2n+1)h^2/(8ma^2) \tag{2.48}$$

As a increases ΔE becomes smaller; in the limit as $a\rightarrow\infty$, $\Delta E\rightarrow 0$. Thus, a completely free particle has unquantized energy, for which reason atoms and molecules involved in laboratory experiments behave with unquantized translational energy.

The probability of finding the particle in an interval between x and $x+\mathrm{d}x$ is

$$(2/a)\int_{x}^{x+\mathrm{d}x} \sin^2(n\pi x/a)\,\mathrm{d}x \tag{2.49}$$

Thus, it varies with x, the more so as n increases. At high n, the probability is approximately constant for all values of x and, in the limit as $n\rightarrow\infty$, the probability is the same for all values of x. In other words, the particle exhibits classical behaviour at very high quantum numbers, which is one statement of the *correspondence principle*. Simple plots of $|\psi|^2$ from (2.45) verify these results: at very high values of n, results from classical mechanics tend to quantum values.

2.5.4.1 Tunnelling

If the confining walls are not infinitely high, then the potential energy of a particle at $x=0$ or $x=a$ is finite, and the kinetic energy term $(E-V)$ for the particle is positive there. In this situation, the wavefunction penetrates the walls to a small extent, so that there is a finite probability that a particle may be found outside the box. This effect is known as *tunnelling*; it is another quantum mechanical effect, since classical physics would predict that the particle remains within the box in all circumstances. The probability of tunnelling is related to $\exp(-\alpha m)$, where α is a constant and m is the mass of the particle. Thus, only particles of low mass, such as electrons or protons, are able to execute significant tunnelling.

2.5.5 Boxes of higher dimensions

We can extend the above discussion to a two-dimensional box, for which the appropriate wave equation (for $V=0$) may be written as

$$[-\hbar^2/(2m)][\partial^2\psi(x,y)/\partial x^2+\partial^2\psi(x,y)/\partial y^2]=E\psi(x,y) \tag{2.50}$$

This differential equation is separable, that is, the eigenfunctions can be written as products in x and y (see Appendix 11). This means that, by analogy with (2.45), and including boundary conditions as before, we write

$$\psi_{n_x,n_y}=2/(ab)^{1/2}\sin(n_x\pi x/a)\sin(n_y\pi y/b) \tag{2.51}$$

where n_x and n_y are integral quantum numbers, and the total energy E_{n_x,n_y} is given by

$$E_{n_x,n_y}=E_{n_x}+E_{n_y} \tag{2.52}$$

From (2.50) and (2.51),

$$E_{n_x,n_y}=[h^2/(8m)](n_x^2/a^2+n_y^2/b^2) \tag{2.53}$$

The lowest energy state is given by $n_x=n_y=1$. The next highest states are $\psi_{1,2}$ and $\psi_{2,1}$ (Figure 2.6), with energies

$$E_{1,2}=[h^2/(8m)](1/a^2+4/b^2) \tag{2.54}$$

$$E_{2,1}=[h^2/(8m)](4/a^2+1/b^2) \tag{2.55}$$

For a square two-dimensional box, $a=b$ and $E_{1,2}=E_{2,1}$; the energies are then said to be *degenerate*. In a three-dimensional box, the energy levels are characterized by three integer quantum numbers; it is not difficult to show that, for a rectangular box,

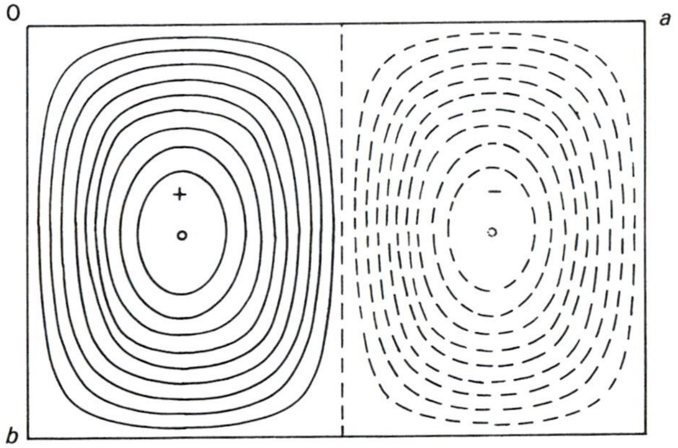

Figure 2.6 Particle in a two-dimensional box: the wavefunction $\psi_{2,1}$, with $a/b=1.5$. The contours of ψ rise to a maximum (full lines) and descend to a minimum (dashed lines); the vertical dashed line represents $\psi_{2,1}=0$.

$$E_{n_x,n_y,n_z}=[h^2/(8m)](n_x^2/a^2+n_y^2/b^2+n_z^2/c^2) \tag{2.56}$$

For a cube, $a=b=c$, and multiply degenerate energy states now exist. For example, for quantum numbers n_x, n_y and n_z equal to 1, 2 and 2 respectively, three degenerate states exist with energies given by

$$E_{1,2,2}=9h^2/(8ma^2) \tag{2.57}$$

2.6 Vibrational and rotational motion

In this section, we shall introduce some general aspects of the quantum mechanics of vibration and rotation that will be important in our subsequent study of molecules.

2.6.1 *Vibrational motion*

A particle of mass m undergoing simple harmonic motion about the position $x=0$, such as at the end of a stretched spring, experiences a restoring force $-kx$, proportional to the displacement x; k is a force constant and $\frac{1}{2}kx^2$ is the potential energy V of the particle (since force$=-\mathrm{d}V/\mathrm{d}x$). The Schrödinger equation for this motion takes the form

$$[-\hbar^2/(2m)]\,\mathrm{d}^2\psi/\mathrm{d}x^2+\tfrac{1}{2}kx^2\psi=E\psi \tag{2.58}$$

The energies are again quantized, because a boundary condition requires that ψ tends to zero as x becomes large, since the potential energy rises rapidly as x increases. The acceptable solutions of (2.58) give for the energies of the harmonic oscillator

$$E_\upsilon=\left(\upsilon+\tfrac{1}{2}\right)\hbar\omega \tag{2.59}$$

where υ can take the values 0, 1, 2, …, and $\omega=(k/m)^{1/2}$. The separation of successive vibrational energy levels is constant for all values of υ, provided that the motion remains harmonic:

$$\Delta E=E_{\upsilon+1}-E_\upsilon=\hbar\omega \tag{2.60}$$

The zero-point energy of vibration ($\upsilon=0$) corresponds to $\hbar\omega/2$. Classical mechanics allows the spring to be completely at rest (an energy of zero), but quantum mechanics requires that it has a residual vibrational energy of $\frac{1}{2}\hbar\omega$, albeit infinitesimally small. The energy width $\hbar\omega$ becomes important for microscopic particles. For example, the mechanical spring may have, typically, a vibrational frequency of 1 Hz, which corresponds to an energy $\hbar\omega$ of 6.6×10^{-34} J ($\omega=2\pi\nu$). A vibrating molecule of hydrogen chloride has a frequency of approximately 10^{14} Hz, which corresponds to an energy width $\hbar\omega$ of 6.6×10^{20} J. The energy width for the spring is negligibly small and energy is transferred in an apparently continuous manner. In the case of the hydrogen chloride molecule, however, the width corresponds to an energy change of 40 kJ mol^{-1}, which is an experimentally significant quantity. For the hydrogen chloride vibrator, the zero-point energy is approximately 3.3×10^{-20} J, or 20 kJ mol^{-1}.

2.6.2 Rotational motion

A particle of mass m moving along a circular path of radius r has an angular momentum J equal to pr, where p is the tangential linear momentum at any point on the circle. The kinetic energy, $p^2/(2m)$, may be replaced by $J^2/(2I)$, where I is the moment of inertia mr^2 of the system; the energy (kinetic) now becomes

$$E=J^2/(2I) \tag{2.61}$$

Since only certain values for the wavelength λ of the motion are permitted E is quantized, and a number n of wavelengths must fit exactly the circumference of the circular path. This closure leads to a cyclic boundary condition (Figure 2.7), such that λ has only the values $2\pi r/n$ ($n=0, 1, 2, \ldots$), which implies that $p=h/\lambda=n\hbar/r$. Hence, $J=n\hbar$, and the permitted energies are given by

$$E=n^2\hbar^2/(2I) \tag{2.62}$$

Furthermore, the momentum p may be directed in one of two ways, corresponding to a clockwise or an anticlockwise rotation of the particle. By convention, angular momentum is indicated by a vector along the z axis, perpendicular to the plane of rotation, and the angular momentum is quantized in units of $\hbar$:

$$J_z=m_l\hbar \ (m_l=0, \pm1, \pm2, \ldots) \tag{2.63}$$

Positive values of m_l correspond to anticlockwise rotation as viewed in the direction of $-z$, towards the plane of rotation (Figure 2.7). The energies of the particle are better given by

$$E_m=m_l^2\hbar^2/(2I) \tag{2.64}$$

from which it is evident that the rotational energy is, not surprisingly, independent of the direction of rotation. These results may be confirmed by solving the (time-independent) Schrödinger equation (2.50), sensibly in polar coordinates (here, θ is constant):

$$[-\hbar^2/(2m)](1/r^2)\,\partial^2\psi/\mathrm{d}\phi^2=E\psi \tag{2.65}$$

for which the solutions are the travelling waves

$$\psi_{m_l}=N\exp(im_l\phi) \ (m_l=0, \pm1, \pm2, \ldots) \tag{2.66}$$

It is left to the reader to derive (remember that ψ_{m_l} is complex) the normalization constant N in (2.66) and to show that the quantized energies derived from (2.65) and (2.66) are as given in (2.64). Since ψ must fit the circumference of the circle in order to satisfy the boundary condi-

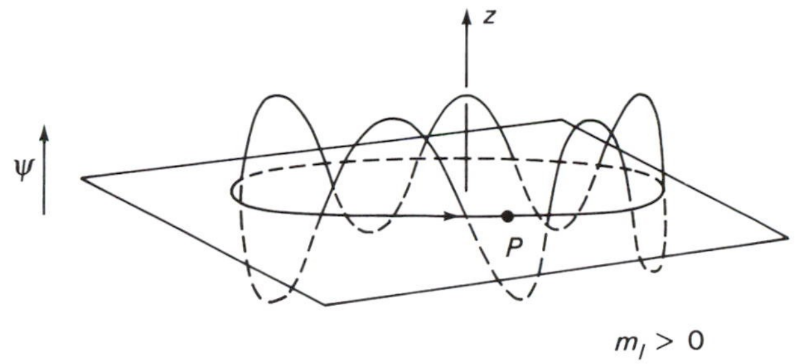

Figure 2.7 Particle P on a ring: the z axis corresponds to the direction of ψ for the wavefunction and to the direction of the J_z angular momentum vector. For the direction of rotation indicated, m_l is conventionally positive.

tion, ψ must be the same at ϕ and at $\phi+2\pi$, which situation obtains with integral values for m_l.

In three dimensions, the wavefunction must match along the equator of a sphere and along any path across the poles. The kinetic energy and angular momentum are determined by rotation about the three Cartesian axes x, y and z, and a further quantum number is needed to govern the new boundary condition. Solution of the Schrödinger equation for this situation provides the following results.

Quantization of rotational energy:

$$E_l = l(l+1)\hbar^2/2I \ (l=0, 1, 2, \ldots) \tag{2.67}$$

Quantization of total angular momentum:

$$J = [l(l+1)]^{1/2}\hbar \ (l=0, 1, 2, \ldots) \tag{2.68}$$

Quantization of angular momentum along z:

$$J_z = m_l\hbar \ (m_l=0, \pm 1, \pm 2, \ldots) \tag{2.69}$$

The angular wavefunctions are the products of two functions, one depending upon the spherical polar coordinate θ and the other depending upon ϕ, and are referred to as *spherical harmonics*, $Y_{l,m_l}(\theta,\phi)$; they are listed in Table 2.1 for l up to 2.

2.6.3 Space and spin quantization

Quantization of J_z means that the vector representing angular momentum (Figure 2.7) is restricted in its spatial position, by the integral nature of m_l. The momentum vector is normal to the plane of the path of the rotating particle, from which it follows that this plane can have only certain orientations in space, a *space quantization*.

This property was confirmed by the Stern–Gerlach (1921, 1922) experiments, in which a beam of silver atoms was directed through an inhomogeneous magnetic field. The field set up across the beam interacts with the field set up by the rotating charged particles (electrons) in the silver atoms and, because of the imhomogeneity of the field, the direction followed by an atom depends upon orientation. According to classical theory, a continuous band of atoms should emerge from the confines of the applied magnetic field, whereas two discrete, narrow bands were observed, in line with the wave mechanical prediction of the orientation quantization of angular momentum.

However, with $2l+1=2$, two bands could arise only if the quantum number l were equal

Table 2.1 $Y_{l,m_l}(\theta,\phi)$ functions (l=0–2)

l	m_l	$Y_{l,m_l}(\theta,\phi)$ (normalized)
0	0	$(1/4\pi)^{1/2}$
1	0	$(3/4\pi)^{1/2}\cos(\theta)$
1	+1	$-(3/8\pi)^{1/2}\sin(\theta)\exp(\mathrm{i}\phi)$
1	−1	$(3/8\pi)^{1/2}\sin(\theta)\exp(-\mathrm{i}\phi)$
2	0	$(5/16\pi)^{1/2}[3\cos^2(\theta)-1]$
2	+1	$-(15/8\pi)^{1/2}\cos(\theta)\sin(\theta)\exp(\mathrm{i}\phi)$
2	−1	$(15/8\pi)^{1/2}\cos(\theta)\sin(\theta)\exp(-\mathrm{i}\phi)$
2	+2	$(15/32\pi)^{1/2}\sin^2(\theta)\exp(\mathrm{i}2\phi)$
2	−2	$(15/32\pi)^{1/2}\sin^2(\theta)\exp(-\mathrm{i}2\phi)$

to $\frac{1}{2}$. This dilemma was resolved by the suggestion that the Stern–Gerlach experiments revealed not an orbital angular momentum of the electron around the nucleus, but a spin angular momentum of the electron effectively spinning on an axis through it.

For an electron, the spin angular momentum quantum number s corresponds to a spin angular momentum magnitude of $[s(s+1)]^{1/2}\hbar$, where s has the single value of $\frac{1}{2}$. The component of the spin angular momentum along the z axis is quantized in units of $m_s\hbar$, where $m_s=s, s-1, \ldots, -s$. An electron with $m_s=+\frac{1}{2}$ is often called an α electron, and that with $m_s=-\frac{1}{2}$ a β electron. Two electrons with their spins antiparallel (one $+\frac{1}{2}$ and the other $-\frac{1}{2}$) are said to be paired, and their total spin angular momentum is zero.

Spin angular momentum is a fundamental and fixed property of the electron, arising from a relativistic treatment of the Schrödinger equation. The later experiments of Uhlenbeck and Goudschmidt (1925) associated the electron spin with its magnetic moment; in fact, the electron behaves like a tiny bar magnet. In the experiment with silver, each atom has one unpaired electron and, hence, a spin angular momentum that can assume only two spatial orientations, those observed earlier by Stern and Gerlach.

2.7 Structure of the hydrogen atom

The wavefunction for the hydrogen atom may be written as

$$\mathcal{H}\psi=E\psi \tag{2.70}$$

where the three-dimensional Hamiltonian operator $\mathcal{H}$ is given by

$$[-\hbar^2/(2\mu)]\nabla^2-V(r) \tag{2.71}$$

μ is the reduced mass of an electron–proton pair of separation r, $V(r)$ is the potential energy of the electron in the field of the proton and ∇^2 is the Laplacian operator:

$$\nabla^2=\partial^2/\partial x^2+\partial^2/\partial y^2+\partial^2/\partial z^2 \tag{2.72}$$

The potential energy of an electron of charge $-e$ in the field of a proton of charge $+e$ distant r from it is the Coulomb energy $-e^2/(4\pi\epsilon_0 r)$. When expressed in polar coordinates (Appendix 5), the wave equation takes the form

Table 2.2 Normalized hydrogenic radial functions $R_{n,l}(r)$ to $n=3$

Type	n	l	$R_{n,l}(r)$
1s	1	0	$(Z/a_0)^{3/2}2\exp(-\rho/2)$
2s	2	0	$(Z/a_0)^{3/2}1/2\sqrt{2}\,(2-\rho)\exp(-\rho/2)$
2p	2	1	$(Z/a_0)^{3/2}1/2\sqrt{6}\,\rho\exp(-\rho/2)$
3s	3	0	$(Z/a_0)^{3/2}1/9\sqrt{3}\,(6-6\rho+\rho^2)\exp(-\rho/2)$
3p	3	1	$(Z/a_0)^{3/2}1/9\sqrt{6}\,(4\rho-\rho^2)\exp(-\rho/2)$
3d	3	2	$(Z/a_0)^{3/2}1/9\sqrt{30}\,\rho^2\exp(-\rho/2)$

$$\frac{-\hbar/(2\mu)}{r^2}\left[\frac{\partial}{\partial r}\left(r^2\frac{\partial\psi}{\partial r}\right)+\frac{1}{\sin^2(\theta)}\frac{\partial^2\psi}{\partial\phi^2}+\frac{1}{\sin(\theta)}\frac{\partial}{\partial\theta}\left(\sin(\theta)\frac{\partial\psi}{\partial\theta}\right)\right]$$

$$-[e^2/(4\pi\epsilon_0 r)]\psi=E\psi \qquad (2.73)$$

This equation is separable (see Appendix 11) into three terms, depending respectively on r, θ and ϕ. The wavefunction $\psi(r,\theta,\phi)$ may then be given the form $R(r)\Theta(\theta)\Phi(\phi)$. The $\Theta\Phi$ products are the spherical harmonics listed in Table 2.1 whereas the $R(r)$ functions refer to the radial distribution in the wavefunction.

The exact solution of (2.73) is complex and is not considered here; the reader is referred to a quantum mechanical text[2] for the necessary detail. We note here that the solution of (2.73), within the Born interpretation (2.42), leads to an expression for the wavefunction involving the three quantum numbers n, l and m_l, so that the wavefunction can be denoted ψ_{n,l,m_l}.

Normally, energies are governed by all the quantum numbers, but in the particular case of the hydrogen atom n only is involved and we obtain for the energies of the hydrogen atom

$$E_n=-\mu e^4/(32\pi^2\hbar^2\epsilon_0 n^2)\ (n=1,2,3,\ldots) \qquad (2.74)$$

in complete agreement with the values given by (2.28). The negative sign indicates bound states of the electron, for which energy is conventionally a negative quantity rising to zero as r tends to infinity.

One should not be tempted to regard (2.74) as supporting evidence for the Bohr theory: the hydrogen atom is a special case because of the simple form of the potential energy function, and the integer parameter n does not play exactly the same role in the two treatments. From (2.74) the lowest permitted energy for the hydrogen atom ($n=1$) is -2.17867×10^{-18} J, or 1312.0 kJ mol^{-1}, which corresponds to the first ionization energy of atomic hydrogen.

The normalized radial wavefunctions for hydrogen-like (one-electron) species (H, He^+, Li^{2+} and so on) are listed in Table 2.2 for n up to three; $\rho=2Zr/(na_0)$ and Z is the atomic number of the species; the Bohr radius a_0 is given by $4\pi\hbar^2\epsilon_0/(\mu e^2)$.

2.7.1 Atomic orbitals

An atomic orbital is the wavefunction of an electron in a bound state of an atom. One-electron wavefunctions form a basis for describing the structures of atoms and molecules generally.

[2] See, for example, L Pauling and E B Wilson *Introduction to Quantum Mechanics* (McGraw-Hill, 1935).

The products of the $R(r)$ and $Y(\theta,\phi)$ functions give the total wavefunctions, or atomic orbitals, for the hydrogen atom. For example, with $n=2$, $l=1$ and $m=\pm1$, we have

$$\psi_{2,1,\pm1}=\mp1/(8\sqrt{\pi})(Z/a_0)^{3/2}\rho\exp(-\rho/2)\sin(\theta)\exp(\pm i\phi) \quad (2.75)$$

Since it is preferable to work with a real function, the spherical harmonics in (2.75) may be written as

$$Y_{1,1}=\Phi^+=-3/(8\pi)^{1/2}\sin(\theta)[\cos(\phi)+i\sin(\phi)] \quad (2.76)$$

$$Y_{1,-1}=\Phi^-=3/(8\pi)^{1/2}\sin(\theta)[\cos(\phi)-i\sin(\phi)] \quad (2.77)$$

Subtraction of (2.76) from (2.77) gives

$$(\Phi^- - \Phi^+)=2N'[3/(8\pi)]^{1/2}\sin(\theta)\cos(\phi) \quad (2.78)$$

where N' is an additional normalizing constant for the combined function ($2p_x$ in Table 2.3). Using (2.42), we have

$$N'^2[3/(2\pi)]\int_0^{\pi}\sin^3(\theta)\,d\theta\int_0^{2\pi}\cos^2(\phi)\,d\phi=1 \quad (2.79)$$

from which $N'=2^{-1/2}$. A similar function ($2p_y$ in Table 2.3) can be constructed from the combination $(\Phi^+ + \Phi^-)/i$; the three p orbitals (and five d orbitals) are degenerate. The complete hydrogenic wavefunctions (up to $n=3$) are listed in Table 2.3; the p and d functions are strongly directional.

The Born equation (2.42) shows that the probability of finding an electron within a small volume of space $d\tau$ is given by $\psi\psi^*\,d\tau$. We can illustrate this probability by defining a boundary surface within which, say, 95% of the electron density is distributed as a sort of electron cloud. This cloud has different shapes for different energy states of the electron, and we can discover the shape of any given cloud by plotting the wavefunction in a suitable form. In order to represent ψ_{n,l,m_l}, it would be necessary to plot in four dimensions; more conveniently, we can separate the radial function $R_{n,l}$ for constant θ and ϕ, and the angular function Y_{l,m_l} for constant r.

The atomic orbitals $\psi_{n,0,0}$ are spherically symmetrical (s orbitals); they have no angular dependence, and decrease in amplitude exponentially with the distance r of the electron from the nucleus. Orbitals for which l is greater than zero have an important angular dependence (p and d orbitals).[3]

The probability of finding the electron of a hydrogen atom between distances r and $r+dr$ from the nucleus is given by the electron probability density $\psi\psi^*$, or $|\psi|^2$, in the volume enclosed by a spherical shell of radii r and $r+dr$, that is, the volume $4\pi r^2 dr$. For the $\psi_{1,0,0}$ orbital this probability is, from (2.42) and Table 2.3, $(4/a_0^3)\exp(-2r/a_0)\,r^2\,dr$. It may be noted that the volume element $4\pi r^2\,dr$ evolves from (2.42), in spherical coordinates, as $r^2\,dr\int_0^{\pi}\sin(\theta)\,d\theta\int_0^{2\pi}d\phi$.

Figure 2.8 shows the radial function R, the probability density R^2 and the radial distribution $4\pi r^2R^2$, each plotted as a function of r for $R_{1,0}$ (designated 1s), and also for $R_{2,0}$ (designated 2s) and $R_{2,1}$ (designated 2p).

The exponential form of $R_{1,0}$ indicates that the most probable place for the electron should be at the nucleus ($r=0$). That it is not actually there depends upon a balance between the kinetic energy of the radial motion of the electron and its potential energy with respect to

[3] There are also f, g and h orbitals, corresponding to $l=3$, 4 and 5, respectively.

Table 2.3 Normalized one-electron wavefunctions ψ_{n,l,m_l}

Function	$R_{n,l}(r)\ Y_{l,m}(\theta,\phi)$	Orbital type
$\psi_{1,0,0}$	$(\pi)^{-1/2}\left(\frac{Z}{a_0}\right)^{3/2}\exp(-\rho/2)$	1s
$\psi_{2,0,0}$	$(32\pi)^{-1/2}\left(\frac{Z}{a_0}\right)^{3/2}(2-\rho)\exp(-\rho/2)$	2s
$\psi_{2,1,0}$	$(32\pi)^{-1/2}\left(\frac{Z}{a_0}\right)^{3/2}\rho\exp(-\rho/2)\cos(\theta)$	$2p_z$
$\psi_{2,1,\pm 1}$	$(32\pi)^{-1/2}\left(\frac{Z}{a_0}\right)^{3/2}\rho\exp(-\rho/2)\sin(\theta)\cos(\phi)$	$2p_x$
$\psi_{2,1,\pm 1}$	$(32\pi)^{-1/2}\left(\frac{Z}{a_0}\right)^{3/2}\rho\exp(-\rho/2)\sin(\theta)\sin(\phi)$	$2p_y$
$\psi_{3,0,0}$	$(972\pi)^{-1/2}\left(\frac{Z}{a_0}\right)^{3/2}(6-6\rho+\rho^2)\exp(-\rho/2)$	3s
$\psi_{3,1,0}$	$(648\pi)^{-1/2}\left(\frac{Z}{a_0}\right)^{3/2}(4\rho-\rho^2)\exp(-\rho/2)\cos(\theta)$	$3p_z$
$\psi_{3,1,\pm 1}$	$(648\pi)^{-1/2}\left(\frac{Z}{a_0}\right)^{3/2}(4\rho-\rho^2)\exp(-\rho/2)\sin(\theta)\cos(\phi)$	$3p_x$
$\psi_{3,1,\pm 1}$	$(648\pi)^{-1/2}\left(\frac{Z}{a_0}\right)^{3/2}(4\rho-\rho^2)\exp(-\rho/2)\sin(\theta)\sin(\phi)$	$3p_y$
$\psi_{3,2,0}$	$(7776\pi)^{-1/2}\left(\frac{Z}{a_0}\right)^{3/2}\rho^2\exp(-\rho/2)\,[3\cos^2(\theta)-1]$	$3d_{z^2}$
$\psi_{3,2,\pm 1}$	$(648\pi)^{-1/2}\left(\frac{Z}{a_0}\right)^{3/2}\rho^2\exp(-\rho/2)\sin(\theta)\cos(\theta)\cos(\phi)$	$3d_{xz}$
$\psi_{3,2,\pm 1}$	$(648\pi)^{-1/2}\left(\frac{Z}{a_0}\right)^{3/2}\rho^2\exp(-\rho/2)\sin(\theta)\cos(\theta)\sin(\phi)$	$3d_{yz}$
$\psi_{3,3,\pm 2}$	$(2592\pi)^{-1/2}\left(\frac{Z}{a_0}\right)^{3/2}\rho^2\exp(-\rho/2)\sin^2(\theta)\cos(2\phi)$	$3d_{x^2-y^2}$
$\psi_{3,2,\pm 2}$	$(2592\pi)^{-1/2}\left(\frac{Z}{a_0}\right)^{3/2}\rho^2\exp(-\rho/2)\sin(\theta)\sin(2\phi)$	$3d_{xy}$

Note:
m_l values may not be assigned to the real functions of x, y, xz, yz, xy and x^2-y^2 because, as quoted, they are linear combinations of $+m_l$ and $-m_l$, formulated as in (2.76) and (2.77), for example.

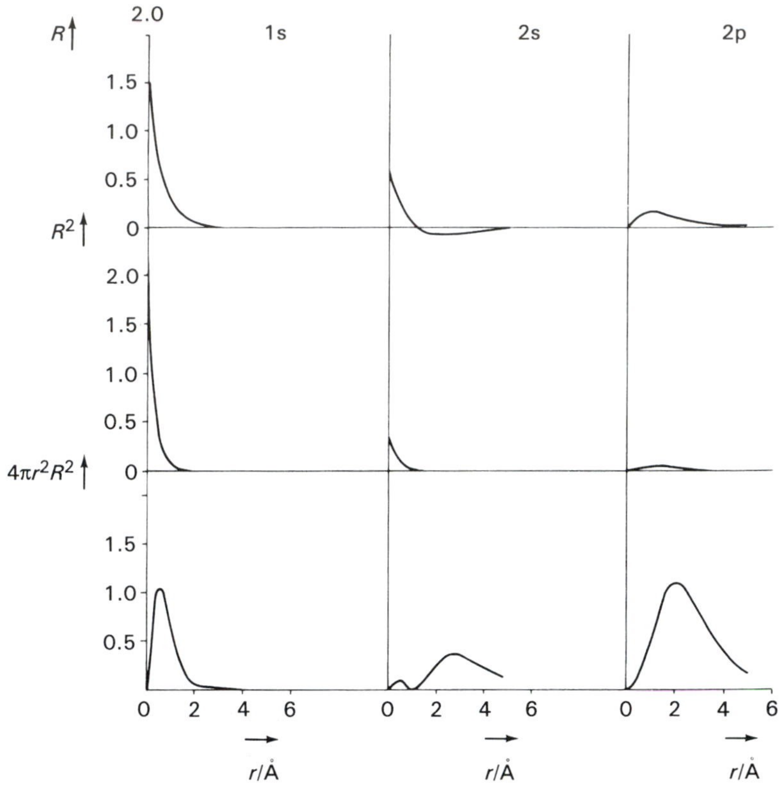

Figure 2.8 Radial part R of the normalized one-electron wavefunctions plotted as a function of distance r (1 Å$=10^{-10}$ m). The maximum of the function $4\pi r^2 R^2$ occurs at $r=a_0$, or 0.5292 Å, for the 1s orbital; the corresponding values for 2s and 2p may be confirmed by differentiation of the appropriate wavefunctions in Table 2.2.

the nucleus.[4] The position of the maximum in the 1s radial distribution function shown by Figure 2.8 may be confirmed by differentiating $4\pi r^2 R^2_{1,0}$ with respect to r and setting the derivative to zero, whence the maximum is seen to occur at r equal to a_0, the Bohr radius.

The next function $R_{2,0}$ behaves in a similar manner (Figure 2.8), except that the 'size' of the function is increased and the radial function has more than one maximum; the numbers of nodes, or zeros in R, is equal to $n-l-1$.

When we consider $R_{2,1}$, we have to consider also its accompanying angular function. In energy states having finite angular momenta ($l>0$), the z component is given a precise orientation (Section 2.6.2). From the uncertainty principle, the angular position of the electron around the z axis, that is, in the x–y plane, is indefinite. This situation may be illustrated by Figure 2.9. The slant length of a cone is a measure of angular momentum, given by $[l(l+1)]^{1/2}\hbar$, and its direction shows a possible but undetermined orientation for the angular momentum vector components in the x–y plane.

A state with $m_l=0$ (and $l>0$) has zero angular momentum along the z axis. The corre-

[4] In a K-electron capture reaction, a K electron enters the nucleus: $^{54}_{25}Mn + ^{0}_{-1}e \rightarrow ^{54}_{24}Cr + \gamma$. The vacancy in the K shell is filled by an electron transition from a higher level, and is accompanied by the emission of characteristic (γ) radiation (of Cr in this example).

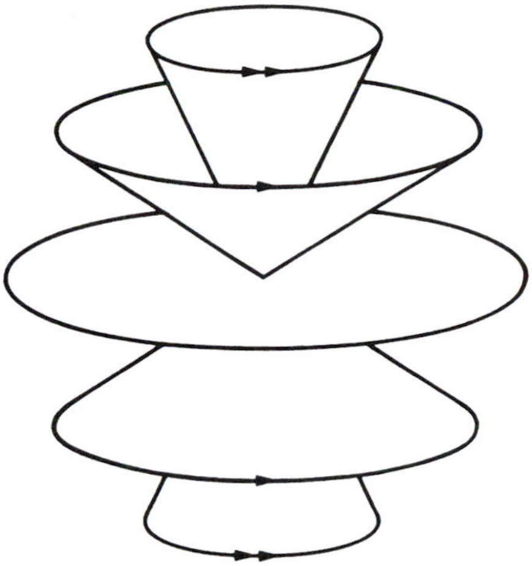

Figure 2.9 Vector model for electron orbital angular momentum: the uncertainty principle does not permit specification of the position of the angular momentum vector components in the x–y plane; for a given m_l, it could be the projection of any generator of the corresponding cone, including the circle (limiting cone of semi-angle 90°).

sponding wavefunction is designated p_z; a nodal surface exists in the x–y plane and through the nucleus, over which ψ (and $|\psi|^2$)=0 (Figure 2.10); the sign of the function is indicated by the ± signs in the lobes.[5] This situation does not arise immediately for the other degenerate states, with $l=1$, $m_l \geq +1$, but we have shown in (2.76)–(2.79) how to form other, real solutions that are linear combinations of those first given. If two or more functions are eigenfunctions of a linear operator,[6] then linear combinations of the functions are also eigenfunctions of the operator, provided that they are degenerate, or nearly so; the appropriate normalizing constants must be applied.

A simplification of the presentation of atomic orbitals may be achieved by drawings such as those shown in Figure 2.11, the three-dimensional surfaces of which are assumed to enclose a large fraction, say 0.95, of the electron density cloud $|\psi|^2$; in other words, there is a 95% probability of finding the electron within the surface described. Regions of the lobes have been shaded (light +, dark −) to indicate the changes in the signs of the corresponding wavefunctions, ψ.

Another descriptive illustration of an orbital is a density contour diagram, shown in Figure 2.12 for the $2p_z$ orbital of carbon. We shall not normally employ this type of representation, but we may note that it is similar to that used in X-ray electron density maps.

However, X-ray methods do not resolve the electron density for separate orbitals, but rather an averaged electron density for each of the species in a structure. For this reason, the X-ray electron density contour maximum of an atom (Figure 2.13) does not usually coincide with the position of its nucleus, a significant factor in a discussion of the meaning of bond lengths between atoms (see also section 1.2). For example, the C−H bond length in sucrose was found to be 0.109 nm by neutron diffraction, but only 0.098 nm by X-rays; O−H bonds have been found to be approximately 0.098 nm and 0.080 nm, respectively, by the two

[5] In Schrödinger quantum mechanics, an electron can cross a nodal plane only if the corresponding orbital is in a bonding situation. In Dirac relativistic quantum mechanics, there are no nodal planes.

[6] If $\Omega(\phi_1+\phi_2)=\Omega\phi_1+\Omega\phi_2$, where ϕ_1 and ϕ_2 are any two functions, then the operator Ω is a linear operator: compare the effects of the operators d/dx and sine on the function $(\alpha x^2+\beta x)$.

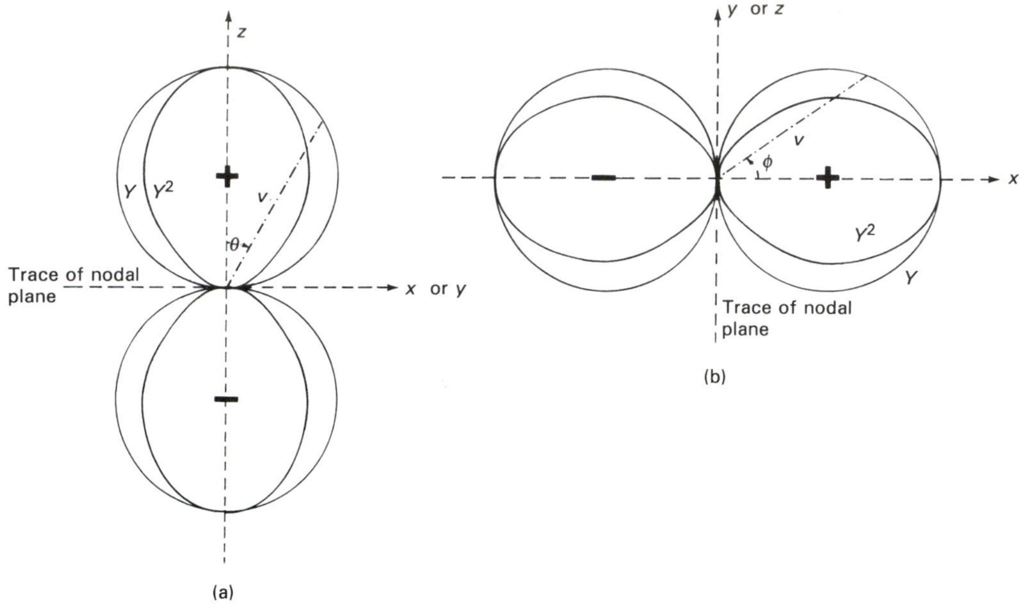

Figure 2.10 p Atomic orbitals: (a) p_z; the length v of any radial line is $|\cos(\theta)|$ and the locus of this line of constant θ about the z axis defines a (circular) section of the orbital about its axis. The x (or y) and z coordinates of Y are $v\cos(\theta)$ and $\pm v\sin(\theta)$; θ=0–180°. For Y^2 the magnitude of v is $\cos^2(\theta)$. The nodal plane x–y passes through the nucleus (z=0). (b) p_x Orbital; the section illustrated corresponds to a curve of $\cos(\phi)$ with $\theta=\pi/2$; the nodal plane is y–z. The section ϕ=0 is the curve of $|\sin(\theta)|$. The p_y orbital has its lobes directed along the y axis with x–z as the nodal plane, but is otherwise similar to p_x. The curves designated Y^2 are the squares of the corresponding Y functions, and are always positive in sign (see also Figure 2.11).

methods. Useful plots of ψ and $|\psi|^2$ functions that show variations in density may be obtained readily by the computer programs (RADL and PLOT) described in Appendix 1.

2.7.2 *Orbital terminology*

The atomic orbitals for a given value of the principal quantum number n constitute a *shell*, whereupon the orbitals of varying l within a shell form a *sub-shell*. Thus, we obtain the traditional terminology, listed here for n up to 4:

n	1	2	3	4
Shell	K	L	M	N
l	0	1	2	3
Sub-shell	s	p	d	f

Thus, for n=2, the L shell, the atomic orbitals in sub-shells are 2s (one orbital) and 2p (three degenerate orbitals). In general, a shell of number n contains a total of n^2 orbitals. The alphabetical notation derives from spectroscopic usage. In particular, s, p, d and f were used to describe spectral transitions (*sharp*, *principal*, *fundamental* and *diffuse*) involving the sub-shells with l=0 to 3.

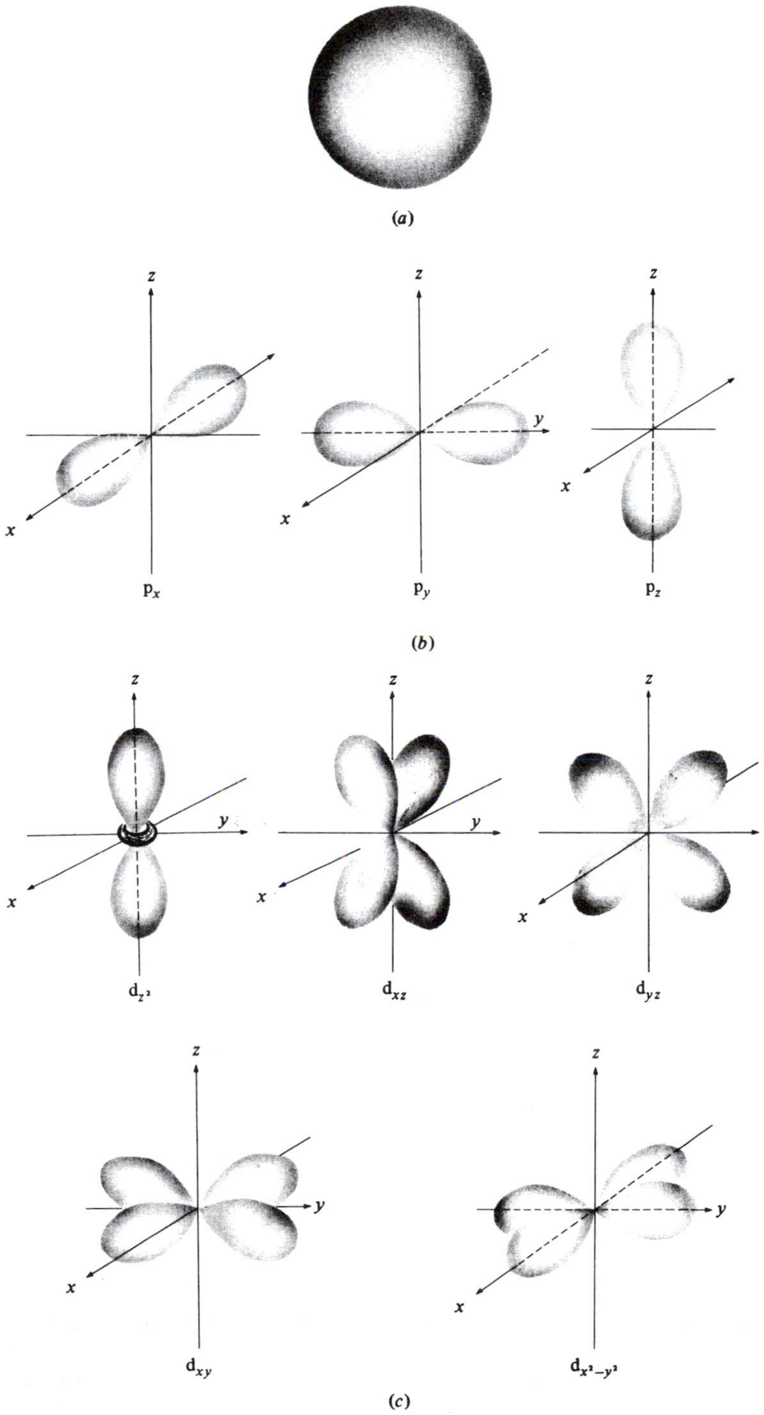

Figure 2.11 Atomic orbitals illustrated by surfaces of constant $|\psi|^2$ such that 95% of the electron density lies within them: (a) s; (b) p; (c) d. The notation for p orbitals is straightforward. With the d orbitals the lobes of d_{z^2} lie along the z axis, and those of $d_{x^2-y^2}$ are along x and y. The lobes of d_{xz}, d_{yz} and d_{xy} lie in the corresponding planes. The shading indicates how the corresponding ψ functions change sign: light +, dark −.

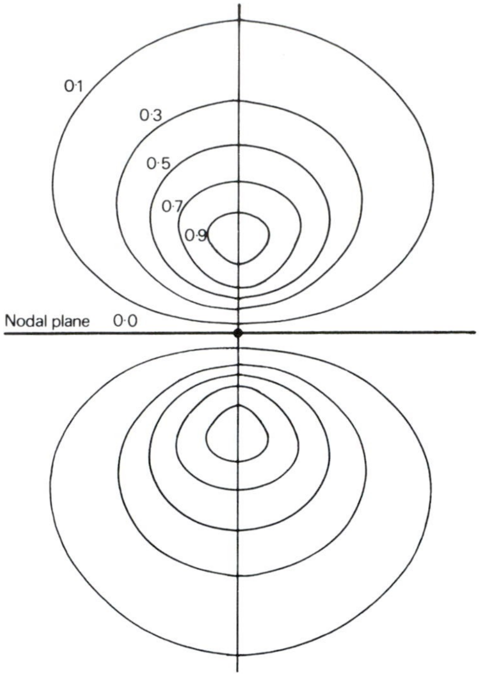

Figure 2.12 Electron density contours for a $2p_z$ atomic orbital of carbon as fractions of the value of $|\psi_{max}|^2$; the nodal plane is x–y, for which the density is zero. The 0.1 contour surface encloses approximately 66% of the electron density; 90% is enclosed by the 0.03 contour (not shown).

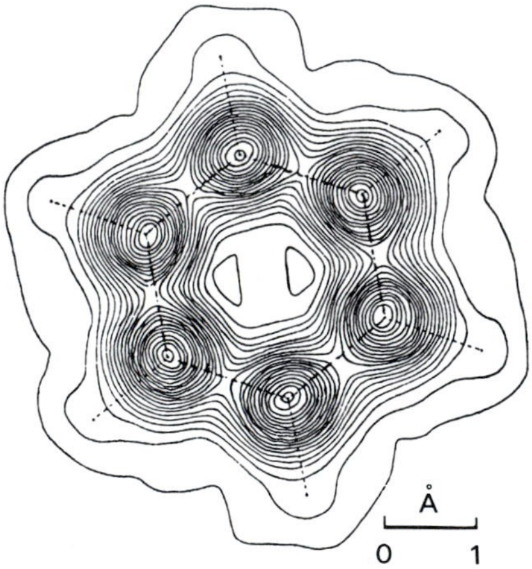

Figure 2.13 Electron density contour map of a molecule of benzene, obtained from X-ray diffraction studies on crystalline benzene. The contour intervals are 0.25 e Å^{-3}; there is a tendency for the hydrogen atoms to be just resolved. (After E G Cox, D W J Cruickshank and J A S Smith, *Proc. Roy. Soc.* **A247**, 1 (1958), and reproduced by permission of the Royal Society, London.)

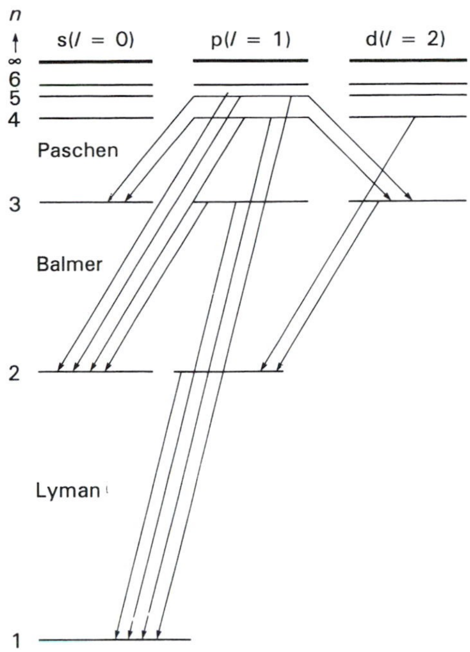

Figure 2.14 Grotrian diagram showing some of the permitted spectral transitions in atomic hydrogen; changes in n are unrestricted, but those in l and m_l must be ± 1 and ± 1 or 0, respectively. The transitions indicated would not all be of the same intensity.

2.7.3 Selection rules for atoms

We saw in Section 2.3.5 that the spectra of atomic hydrogen could be explained by transitions of electrons between energy levels. However, not all possible transitions are allowed. If a photon is expelled from an atom by an electron transition, conservation of angular momentum requires that the electron angular momentum must change to compensate for that carried away by the photon. A photon has an intrinsic spin angular momentum of unity; thus, a p electron (l=1) can fall to an s orbital (l=0), of lower energy, with emission of radiation; changes in n are not so restricted because n governs energy rather than angular momentum. Hence, we obtain the selection rules for atoms

$$\Delta n = 1, 2, 3, \ldots \tag{2.80}$$

$$\Delta l = \pm 1 \tag{2.81}$$

$$\Delta m_l = 0, \pm 1 \tag{2.82}$$

Possible transitions that conform to these selection rules may be depicted on a Grotrian diagram, of which Figure 2.14 is a simple example.

2.8 Atoms with more than one electron

Helium, the next atom after hydrogen in the periodic table, has two electrons and the Schrödinger equation cannot be solved exactly for this or other more complex species. The electronic Hamiltonian for helium contains the term $e^2/(4\pi\epsilon_0 r_{12})$, representing electronic interaction, which makes separation of the variables r_1 and r_2 mathematically intractable. It becomes necessary to make an approximation in order to achieve a solution.

Each electron in the helium atom may be represented by its one-electron wavefunction; then, if the wavefunction for helium is written as $\psi(1,2)$, the approximation becomes

$$\psi(1,2)=N\psi(1)\psi(2) \tag{2.83}$$

where N is a normalizing constant for the combined function. The individual atomic orbitals $\psi(1)$ and $\psi(2)$ are hydrogenic, but the nuclear charge in each is modified to take account of both electrons. Atoms with two or more electrons are subject to the *Pauli exclusion principle*, which arises from the quantum mechanical Pauli principle (Section 2.11.8). Electrons are *fermions*, governed by Fermi–Dirac statistics (see Section 6.5.4 and Appendix 15). Interference between two wave packets is destructive, so that no two electrons can occupy one and the same quantum state. A general statement of the exclusion principle is that *an atomic orbital can accommodate a maximum of two electrons with opposed, or paired, spins.*

The ground (lowest energy) state of hydrogen is written as (1s), or $(1s)^1$. Thus, helium becomes $(1s)^2$; the K shell now contains two electrons and is full or *closed.* In lithium we anticipate a start with $(1s)^2$: the K shell is closed and the third electron occupies the next highest (energy) orbital $(2s)^1$.

Each electron is fully determined by the quantum numbers n, l, m_l and m_s, the latter being $\pm\frac{1}{2}$, as we have seen. Thus, the configuration $(1s)^2$ in helium implies a full 1s orbital, with paired (α, β) spins. Electrons with paired spins have zero resultant spin angular momentum.

2.8.1 *Screening*

The order of energies of atomic orbitals is, normally,

$$1s<2s<2p<3s<3p<3d<4s\ldots \tag{2.84}$$

However, electron–electron repulsion can modify this order in situations in which the energy levels of orbitals are close, as for the 3d and 4s levels of the first transition series of elements, for example. Figure 2.15 shows how the relative energies E of atomic orbitals vary with atomic number Z.

Electron–electron repulsion reduces Z to an *effective atomic number* Z_{eff} which expresses the net nuclear attraction:

$$Z_{\text{eff}}=Z-\sigma \tag{2.85}$$

where σ is a quantum mechanical screening constant that may be calculated approximately by Slater's rules (Appendix 7).

2.9 *Aufbau* principle

The atomic orbitals are linked closely with the periodic table of the elements (Table 2.4). If we continue the process that we have carried out with hydrogen, helium and lithium, we find that the electron configurations for beryllium and boron are straightforward, being written as $(1s)^2$ $(2s)^2$ and $(1s)^2$ $(2s)^2$ $(2p)^1$, respectively. With carbon, however, the two p electrons could occupy either a single orbital with paired spins or two orbitals with unpaired spins.

The alternatives are governed by Hund's multiplicity rule, namely, that *in the ground state, degenerate or near-degenerate orbitals tend to be occupied singly*; hence, the configuration

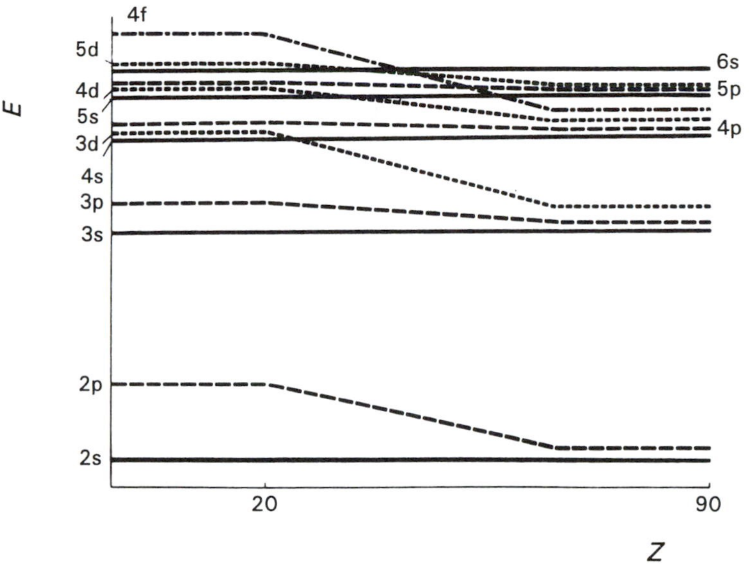

Figure 2.15 Variation of the energy E with atomic number Z for atomic orbitals up to 6s. For a given value of n, the energies of all orbitals except s vary with Z. In particular, the nd orbitals lie between the $(n+1)$s and $(n+1)$p for $n \geq 3$ and $Z<20$, and cross the $(n+1)$s to lower energies for $Z>20$.

with the greater number of unpaired spins is preferred. Hund's rule depends upon the wave mechanical property of *spin correlation*, which describes the tendency of electrons with unpaired spins to repel each other less than do those with paired spins. Spin correlation causes a slight decrease in atomic volume, thereby enhancing the attraction between the electron and the nucleus.[7] Thus, the carbon atom electron configuration may be written as $(1s)^2(2s)^2(2p_x)^1(2p_y)^1$, or just $(1s)^2(2s^2)(2p)^2$ with the understanding that the p electrons occupy separate orbitals.

The order of building up of electronic configurations, subject to Hund's multiplicity rule and to the order of energies shown by Figure 2.15, is known as the *Aufbau*[8] principle. Figure 2.16 illustrates the result of applying this procedure to the elements hydrogen to argon; an unpaired electron is indicated by ↑ and paired electrons by ↑↓. A shorthand notation may be used for the written configurations: for example, sodium may be given either as $(1s)^2(2s)^2(2p)^6(3s)^1$ or as $(Ne)(3s)^1$, where (Ne) implies an inner configuration equivalent to that of neon.

The periodic table is not completed exactly in the manner suggested by Figure 2.16. Between calcium, $(Ar)(4s)^2$, and zinc, $(Ar)(3d)^{10}(4s)^2$, for example, there exists the first transition series of the elements. Here, the 3d orbitals are filled progressively from scandium to zinc, because the energy of the 3d orbital lies below that of 4p for atomic numbers greater than about 20 (Figure 2.15). Even this process is not as simple as it seems; on the basis of experimental evidence, Cu is written as $(Ar)(3d)^{10}(4s)^1$, rather than the expected $(Ar)(3d)^9(4s)^2$.

We recall that the stability of the noble-gas configuration was the pillar of Lewis' (1916)

[7] For a discussion of spin correlation see, for example, F L Pilar, *Elementary Quantum Chemistry*, 2nd Edition (McGraw-Hill, 1990).

[8] German: *Aufbau* means 'construction', in the sense of building up.

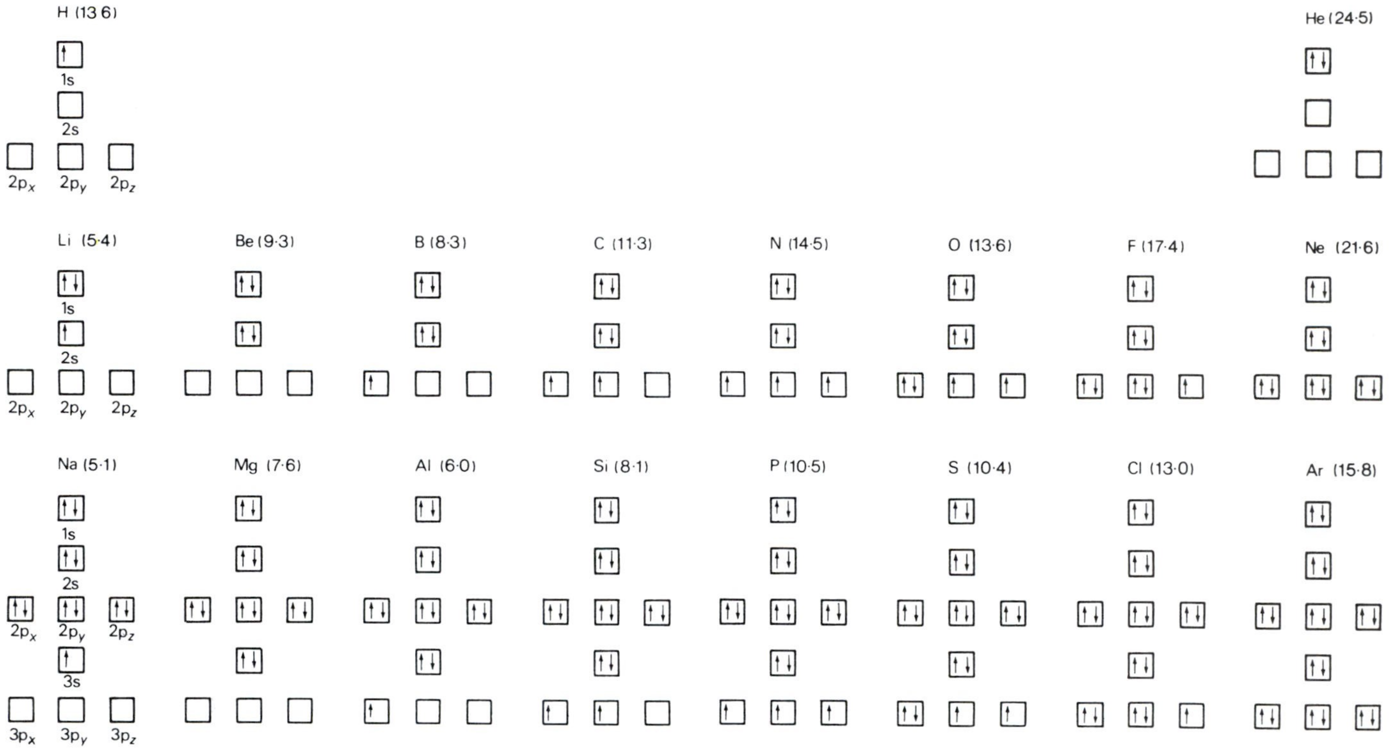

Figure 2.16 The elements hydrogen to argon, with the atomic orbitals occupied according to the *Aufbau* and Pauli exclusion principles. The sign ↑↓ indicates paired electrons in a fully occupied orbital, whereas ↑ indicates occupancy by a single unpaired electron. The numbers in parentheses after the element symbol are the first ionization energies in eV.

electron-pair bond hypothesis. From Figure 2.16, we see that this configuration is obtained with sets of orbitals fully occupied up to and including those governed by the principal quantum number n, that is, with the outermost configurations $(ns)^2(np)^6$. Table 2.4 shows a completed periodic table, giving atomic numbers, relative atomic masses and electronic configurations.

2.10 Ionization energy

The ionization energy[9] is the energy required just to remove an electron from an atom in its ground state at $T=0$. The first ionization energy I_1 refers to the energy change for the process

$$M(\mathrm{g}) \rightarrow M^+(\mathrm{g}) + \mathrm{e}^- \tag{2.86}$$

where M is a chemical species and e^- is an electron; subsequent ionization energies may be defined in a similar manner.

At any other temperature, the process of (2.86) is characterized by an ionization enthalpy, since the gaseous products will contain an additional energy term. From thermodynamic arguments, we have for reaction (2.86):

$$\Delta H_\mathrm{i}^{\ominus} = I + \Delta n \mathcal{R} T \tag{2.87}$$

where $^{\ominus}$ refers to the thermodynamic standard state at a temperature T and Δn is the number of moles of products *minus* the number of moles of reactants. Ionization energies may be obtained from spectral measurements, and we shall consider how this may be done for hydrogen. Some of the transitions in the Lyman spectral series for atomic hydrogen (see Figure 2.14) are listed below:

n	2	3	4	5	6
$\bar{\nu}$/cm^{-1}	82258	97491	102822	105290	106631

The Lyman series corresponds to (2.26) with $m=1$ and $n \geq 2$, and it is clear that a graph of $\bar{\nu}$ against $1/n^2$ should be a straight line of slope $-R_\mathrm{H}$ and intercept $\bar{\nu}_\infty$, the *series limit*. The plot is shown in Figure 2.17, and by least squares (Appendix 8) we obtain $R_\mathrm{H} = 109676.9$ cm^{-1} and $\bar{\nu}_\infty = 109677.2$ cm^{-1}. The latter quantity corresponds to the ionization energy, which we obtain by multiplication by hc, giving 2.1787×10^{-18} J, or 1312.0 kJ mol^{-1}, in good agreement with that calculated from (2.74). It should be noted that, in spectroscopy, it is quite common to refer to measurements on spectral lines in wavenumbers (cm^{-1}) as their 'frequencies', or even as their 'energies'.

The energy of an atomic orbital is associated with the ionization energy of the atom; values of first ionization energies have been given in Figure 2.16. They show an upward trend from lithium to neon and from sodium to argon. We can calculate the screening constant for the outer electron in each element across a period, using Slater's rules:

I_1/eV	Li	Be	B	C	N	O	F	Ne
Z	3	4	5	6	7	8	9	10
σ	1.70	2.05	2.40	2.75	3.10	3.45	3.80	4.15
Z_eff	1.3	1.95	2.60	3.25	3.90	4.55	5.20	5.85

[9] Sometimes referred to as ionization 'potential' (see also Section 4.5).

Table 2.4 Periodic table of elements

This periodic table numbering follows the recommendatons of the IUPAC. The elements in groups 1 and 2, with helium, form the s block, those in groups 13–18 the p block, the lanthanides and actinides the f block and the remaining, transition elements the d block. Groups 1–7 were formerly IA–VII A, 8–10 were group VIII, 11–17 were I B–VIII B and 18 was group 0. Each box contains the chemical

1	2	3	4	5	6	7	8	9
1 1.0079 **H** $(1s)^1$								
3 6.941(2) **Li** $(2s)^1$	4 9.0122 **Be** $(2s)^2$							
11 22.980 **Na** $(3s)^1$	12 24.305 **Mg** $(3s)^2$							
19 39.098 **K** $(4s)^1$	20 40.078(4) **Ca** $(4s)^2$	21 44.956 **Sc** $(3d)^1(4s)^2$	22 47.88(3) **Ti** $(3d)^2(4s)^2$	23 50.942 **V** $(3d)^3(4s)^2$	24 51.996 **Cr** $(3d)^5(4s)^1$	25 54.938 **Mn** $(3d)^5(4s)^2$	26 55.847(3) **Fe** $(3d)^6(4s)^2$	27 58.933 **Co** $(3d)^7(4s)^2$
37 85.468 **Rb** $(5s)^1$	38 87.62 **Sr** $(5s)^2$	39 88.906 **Y** $(4d)^1(5s)^2$	40 91.224 **Zr** $(4d)^2(5s)^2$	41 92.906 **Nb** $(4d)^4(5s)^1$	42 95.94 **Mo** $(4d)^5(5s)^1$	43 98.906 **^{99}Tc** $(4d)^5(5s)^2$	44 101.07(2) **Ru** $(4d)^7(5s)^1$	45 102.91 **Rh** $(4d)^8(5s)^1$
55 132.91 **Cs** $(6s)^1$	56 137.33 **Ba** $(6s)^2$	71 174.97 **Lu** $(4f)^{14}(5d)^1(6s)^2$	72 178.49(2) **Hf** $(5d)^2(6s)^2$	73 180.95 **Ta** $(5d)^4(6s)^2$	74 183.85(3) **W** $(5d)^4(6s)^2$	75 186.21 **Re** $(5d)^4(6s)^2$	76 190.2 **Os** $(5d)^6(6s)^2$	77 192.22(3) **Ir** $(5d)^7(6s)^2$
87 223.02 **^{223}Fr** $(7s)^1$	88 226.03 **^{226}Ra** $(7s)^2$	103 262.11 **^{262}Lr** $(5f)^{14}(6d)^1(7s)^2$	104 (260) **Ku** $(5f)^{14}(6d)^2(7s)^2$	105 (261) **Ha** $(5f)^{14}(6d)^3(7s)^2$				

57 138.91 **La** $(4f)^0(5d)^1(6s)^2$	58 140.12 **Ce** $(4f)^1(5d)^1(6s)^2$	59 140.91 **Pr** $(4f)^3(5d)^0(6s)^2$	60 144.24(3) **Nd** $(4f)^4(5d)^0(6s)^2$	61 146.92 **^{147}Pm** $(4f)^5(5d)^0(6s)^2$	62 150.36(3) **Sm** $(4f)^6(5d)^0(6s)^2$	63 151.96 **Eu** $(4f)^7(5d)^0(6s)^2$
89 227.03 **^{227}Ac** $(5f)^0(6d)^1(7s)^2$	90 232.04 **Th** $(5f)^0(6d)^2(7s)^2$	91 231.04 **Pa** $(5f)^2(6d)^1(7s)^2$	92 238.029 **U** $(5f)^3(6d)^1(7s)^2$	93 237.05 **^{237}Np** $(5f)^4(6d)^1(7s)^2$	94 239.05 **^{239}Pu** $(5f)^6(6d)^0(7s)^2$	95 241.06 **^{241}Am** $(5f)^7(6d)^0(7s)^2$

[a] *Nomenclature of Inorganic Chemistry* (1989), Butterworth.
[b] *Pure & Applied Chemistry* Vol 63, pp. 987–988 (1991).

symbol of the element, its atomic number, relative atomic mass and outermost electronic configuration. The elements are arranged by group number and period (principal quantum number of outermost electron/s). The atomic masses are those recommended by the IUPAC 1989[b]; they are relative values, being scaled to $^{12}C=12$. The precision is ± 1 in the last digit quoted, unless indicated otherwise.

10	**11**	**12**	**13**	**14**	**15**	**16**	**17**	**18**
								2 4.0026 **He** $(1s)^2$
			5 10.811(5) **B** $(2s)^2(2p)^1$	6 12.011 **C** $(2s)^2(2p)^2$	7 14.007 **N** $(2s)^2(2p)^3$	8 15.999 **O** $(2s)^2(2p)^4$	9 18.998 **F** $(2s)^2(2p)^5$	10 20.180 **Ne** $(2s)^2(2p)^6$
			13 26.982 **Al** $(3s)^2(3p)^1$	14 28.086 **Si** $(3s)^2(3p)^2$	15 30.974 **P** $(3s)^2(3p)^3$	16 32.066(6) **S** $(3s)^2(3p)^4$	17 35.453 **Cl** $(3s)^2(3p)^5$	18 39.948 **Ar** $(3s)^2(3p)^6$
28 58.963 **Ni** $(3d)^8(4s)^2$	29 63.546(3) **Cu** $(3d)^{10}(4s)^1$	30 65.39(2) **Zn** $(3d)^{10}(4s)^2$	31 69.723 **Ga** $(4s)^2(4p)^1$	32 72.61(2) **Ge** $(4s)^2(4p)^2$	33 74.922 **As** $(4s)^2(4p)^3$	34 78.96(3) **Se** $(4s)^2(4p)^4$	35 79.904 **Br** $(4s)^2(4p)^5$	36 83.80 **Kr** $(4s)^2(4p)^6$
46 106.42 **Pd** $(4d)^{10}(5s)^0$	47 107.87 **Ag** $(4d)^{10}(5s)^1$	48 112.41 **Cd** $(4d)^{10}(5s)^2$	49 114.82 **In** $(5s)^2(5p)^1$	50 118.71 **Sn** $(5s)^2(5p)^2$	51 121.76 **Sb** $(5s)^2(5p)^3$	52 127.60(3) **Te** $(5s)^2(5p)^4$	53 126.90 **I** $(5s)^2(5p)^5$	54 131.29(2) **Xe** $(5s)^2(5p)^6$
78 195.08(3) **Pt** $(5d)^9(6s)^1$	79 196.97 **Au** $(5d)^{10}(6s)^1$	80 200.59(2) **Hg** $(5d)^{10}(6s)^2$	81 204.38 **Tl** $(6s)^2(6p)^1$	82 207.2 **Pb** $(6s)^2(6p)^2$	83 208.98 **Bi** $(6s)^2(6p)^3$	84 209.98 **^{210}Po** $(6s)^2(6p)^4$	85 209.99 **^{210}At** $(6s)^2(6p)^5$	86 222.02 **^{222}Rn** $(6s)^2(6p)^6$

64 157.25(3) **Gd** $(4f)^7(5d)^1(6s)^2$	65 158.93 **Tb** $(4f)^9(5d)^0(6s)^2$	66 162.50(3) **Dy** $(4f)^{10}(5d)^0(6s)^2$	67 164.93 **Ho** $(4f)^{11}(5d)^0(6s)^2$	68 167.26(3) **Er** $(4f)^{12}(5d)^0(6s)^2$	69 168.93 **Tm** $(4f)^{13}(5d)^0(6s)^2$	70 173.04(3) **Yb** $(4f)^{14}(5d)^0(6s)^2$	**Lanthanides**
96 244.06 **^{244}Cm** $(5f)^7(6d)^1(7s)^2$	97 249.08 **^{249}Bk** $(5f)^9(6d)^0(7s)^2$	98 252.08 **^{252}Cf** $(5f)^{10}(6d)^0(7s)^2$	99 252.08 **^{252}Es** $(5f)^{11}(6d)^0(7s)^2$	100 257.10 **^{257}Fm** $(5f)^{12}(6d)^0(7s)^2$	101 259.10 **^{258}Md** $(5f)^{13}(6d)^0(7s)^2$	102 259.10 **^{259}No** $(5f)^{14}(6d)^0(7s)^2$	**Actinides**

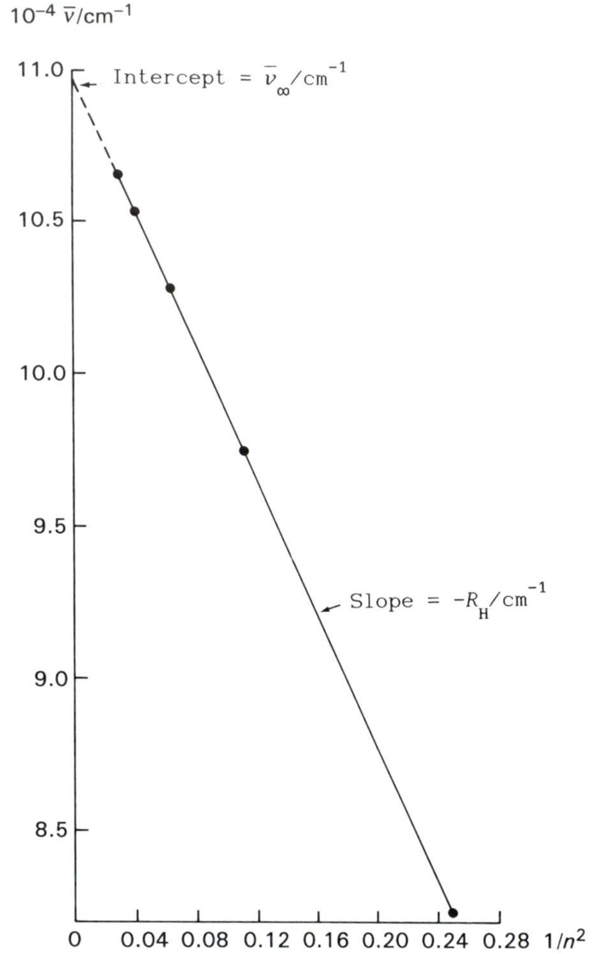

Figure 2.17 Plot of wavenumber $\bar{\nu}$ as a function of $1/n^2$ for the Lyman spectral series ($m=1$, $n \geq 2$) of atomic hydrogen. The least-squares line has a slope of $-R_H/cm^{-1}$ and an intercept of $\bar{\nu}_{\infty}/cm^{-1}$.

The ionization energies I_1 for lithium to neon are plotted in Figure 2.18 as a function of atomic number Z. The general upwards trend reflects the increased attraction of the nucleus for an outermost electron, as Z_{eff} increases from $0.43Z$ in lithium to $0.59Z$ in neon.

The decrease at boron occurs as an electron enters a higher energy 2p orbital; the drop at oxygen arises as a 2p atomic orbital becomes fully occupied by two electrons with paired spins, thus increasing the electron–electron repulsion. Note that the higher (*more positive*) the electronic energy of a species, the less is the work (*smaller value of* I_1) required to ionize it.

Table 2.5 completes the data in Figure 2.16 on the first ionization energies for the elements, other than those of the transition and lanthanide series. The reader may wish to construct plots similar to that of Figure 2.18, so as to observe the periodicity in I_1, and to consider any other variations in the light of the electron configurations of the elements.

Table 2.5 First ionization energies/eV for lithium to radon, excluding the transition series

K	Ca	Ga	Ge	As	Se	Br	Kr
4.34	6.11	6.00	7.90	9.82	9.75	11.82	14.00
Rb	Sr	In	Sn	Sb	Te	I	Xe
4.18	5.70	5.79	7.35	8.64	9.01	10.45	12.13
Cs	Ba	Tl	Pb	Bi	Po	At	Rn
3.89	5.21	6.11	7.42	7.29	8.4	9.6	10.7

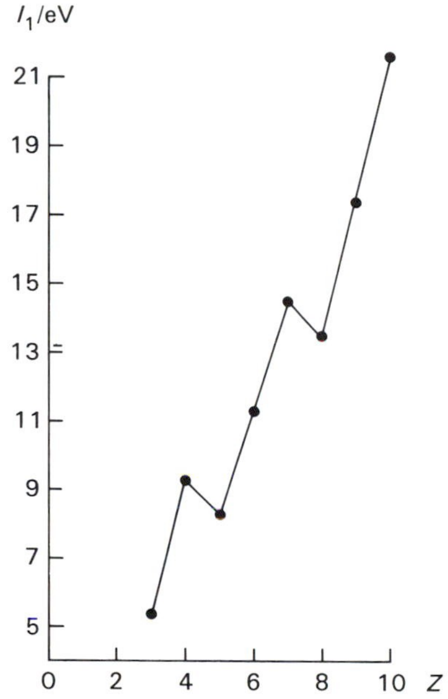

Figure 2.18 Plot of the first ionization energy as a function of atomic number for the elements Li to Ne. The fall at boron occurs with the first occupancy of a 2p orbital, whereas that at oxygen arises as a 2p orbital becomes fully occupied by two spin-paired electrons.

2.11 Structures of molecules

In a molecule, each electron comes under the influence of the potential fields of the nucleus and the other electrons. In the simple example of the hydrogen molecule, six such interactions occur, as indicated in Figure 2.19. In the wave equation (2.70), $\mathcal{H}$ is now given by

$$\mathcal{H}=-[\hbar^2/(2M)](\nabla_A^2+\nabla_B^2)-[\hbar^2/(2\mu)](\nabla_1^2+\nabla_2^2)-\\ \{[e^2/(4\pi\epsilon_0)][1/r_{A1}+1/r_{A2}+1/r_{B1}+1/r_{B2}-1/r_{12}-1/r_e]\} \quad (2.88)$$

where M is the mass of the nucleus. The first term on the right-hand side of (2.88) describes the kinetic energy of the nuclei A and B, and the second term that of the electrons 1 and 2; the third term lists the interparticle Coulombic attractions and repulsions. The presence of terms such as $1/r_{12}$ precludes an exact solution of the Schrödinger equation in species with more than one electron, as indicated in Section 2.8.

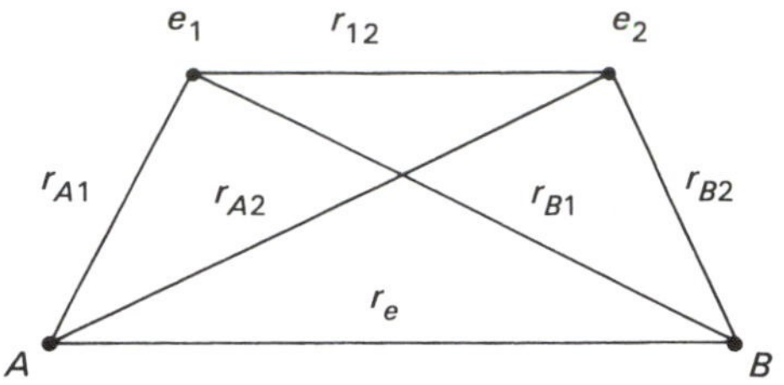

Figure 2.19 Schematic illustration of the six particle interactions between two nuclei A and B and two electrons e_1 and e_2 in the hydrogen molecule; r_e corresponds to the minimum energy configuration.

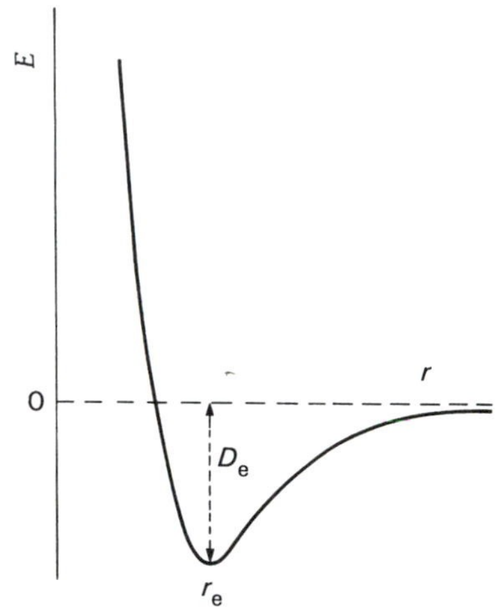

Figure 2.20 Variation of energy E with internuclear distance r in a diatomic molecule: the minimum theoretical energy (D_e), at $r=r_e$, differs from the experimental value (D_0) by the value of the zero-point energy.

Since M is approximately 1840 times greater than μ, we may neglect the term in $1/M$ in comparison with the other two terms. The electrons move so rapidly that the nuclei are effectively stationary relative to them. This treatment is known as the *Born–Oppenheimer approximation* and will be assumed to apply hereinafter. Effectively, it separates the kinetic energy of the nuclei from the electronic energy and permits calculation of the electronic energy for fixed values of the internuclear distance r. The total energy is then given by the sum of the eigenenergy E from (2.70) and the repulsion energy of the two nuclei, at each selected value of r. It may be noted that, in the absence of a Born–Oppenheimer approximation, the nuclear positions become probabilities, cloud-like, just like electrons.

2.11.1 Variation method

Since the Schrödinger wave equation cannot be solved exactly for species with two or more electrons, approximate methods must be invoked in order to obtain the configuration of minimum energy. The energy of a diatomic molecule varies with the internuclear distance in the manner shown by Figure 2.20. The minimum in the curve corresponds to the equilibrium internuclear distance r_e and the energy minimum for the molecule. The minimum corresponds to the theoretical bond dissociation energy (D_e) which differs from the experimental dissociation energy (D_0) by the magnitude of the zero-point energy of vibration (see also Section 3.6.5.1).

The variation method, an adaptation of a procedure due to Rayleigh (1880), leads to an estimate of the equilibrium parameters E and r_e. In applying this method, we first multiply both sides of (2.70) by ψ, or by ψ^* if ψ is complex (we shall use ψ here), and integrate over the space of the variables (see Appendix 5):

$$E=\int\psi\mathcal{H}\psi\,\mathrm{d}\tau/\int\psi^2\,\mathrm{d}\tau \tag{2.89}$$

It should be remembered that the Hamiltonian $\mathcal{H}$ is an operator acting on ψ and the order of the operations cannot be changed.

A trial function ψ_t is adopted for ψ, so that (2.89) becomes

$$E=\int\psi_t\mathcal{H}\psi_t\,\mathrm{d}\tau/\int\psi_t^2\,\mathrm{d}\tau \tag{2.90}$$

The approximation leads to an expectation value for E that will normally be greater (more positive) than the true value and ψ_t is subjected to modifications[10] so as to obtain the best estimate of E.

We will apply the variation procedure to the ground state of the hydrogen atom, the result for which we know already. We shall postulate a trial function

$$\psi_t=\exp(-\zeta r) \tag{2.91}$$

where ζ is a constant. Since this function is independent of θ and ϕ, we need consider only the radial part of the Hamiltonian in (2.73), which, with (2.91), may be written as

$$\begin{aligned}&[-\hbar^2/(2\mu)][\partial^2\psi_t/\partial r^2+(2/r)\,\partial\psi_t/\partial r]-[e^2/(4\pi\epsilon_0 r)]\psi_t\\&=\{[-\hbar^2/(2\mu)](\zeta^2-2\zeta/r)-[e^2/(4\pi\epsilon_0 r)]\}\exp(-\zeta r)\end{aligned} \tag{2.92}$$

whence, from (2.90),

$$E=\frac{\displaystyle\int_0^\infty \exp(-\zeta r)\,[-\hbar^2/(2\mu)](\zeta^2-2\zeta/r)-[e^2/(4\pi\epsilon_0 r)]\,\exp(-\zeta r)\,r^2\,\mathrm{d}r}{\displaystyle\int_0^\infty \exp(-2\zeta r)\,r^2\,\mathrm{d}r} \tag{2.93}$$

Following the methods in Appendix 6, it is straightforward to show that E reduces to

$$E=\hbar^2\zeta^2/(2\mu)-e^2\zeta/(4\pi\epsilon_0) \tag{2.94}$$

We now differentiate E with respect to ζ and set the derivative to zero, leading to the value for ζ:

$$\zeta=\mu e^2/(4\pi\epsilon_0\hbar^2) \tag{2.95}$$

which turns out to be the reciprocal of the Bohr radius, a_0. Hence, from (2.94) and (2.95), we obtain

$$E=-\mu e^4/(32\pi^2\hbar^2\epsilon_0^2) \tag{2.96}$$

which is equivalent to (2.74) with $n=1$. This excellent result has been obtained because we made a particularly fortunate (or cunning) choice for the trial wavefunction and it serves to give confidence in the variation method.

[10] Strictly, only when $\mathcal{H}$ is the exact Hamiltonian operator.

2.11.2. Linear combination of atomic orbitals

A procedure that is preferable to guessing an appropriate trial wavefunction combines one-electron atomic wavefunctions by the *linear combination of atomic orbitals* (LCAO) technique. Let a molecular wavefunction Ψ be postulated as the sum of a *basis set* of atomic orbitals ψ_n in the proportions c_n:

$$\Psi = c_1\psi_1 + c_2\psi_2 + \ldots + c_n\psi_n \tag{2.97}$$

where c_1 to c_n are variable parameters that are chosen so as to minimize the energy E, that is, $\partial E/\partial c_i = 0$, $(i=1, 2, 3, \ldots, n)$.

The LCAO equation may be justified by the following argument. For a given eigenfunction ψ_i, we may write

$$\mathcal{H}\psi_i = \epsilon\psi_i \tag{2.98}$$

where ϵ is the eigenvalue of the function ψ_i. Let a linear combination for Ψ be set up, such that

$$\Psi = \Sigma_i c_i \psi_i \tag{2.99}$$

Then

$$\mathcal{H}\Psi = \Sigma_i c_i \mathcal{H}\psi_i = \epsilon\,\Sigma_i c_i \psi_i = \epsilon\psi \tag{2.100}$$

which shows that ϵ is also the eigenvalue of the linear combination Ψ.

If we substitute (2.97) into (2.90), restricting n to 2 for conciseness, we obtain, since[11] $\int \psi_1 \mathcal{H}\psi_2\,\mathrm{d}\tau = \int \psi_2 \mathcal{H}\psi_1\,\mathrm{d}\tau$ in this example,

$$E = \frac{c_1^2\int \psi_1\mathcal{H}\psi_1\,\mathrm{d}\tau + 2c_1c_2\int \psi_1\mathcal{H}\psi_2\,\mathrm{d}\tau + c_2^2\int \psi_2\mathcal{H}\psi_2\,\mathrm{d}\tau}{c_1^2\int \psi_1^2\,\mathrm{d}\tau + 2c_1c_2\int \psi_1\psi_2\,\mathrm{d}\tau + c_2^2\int \psi_2^2\,\mathrm{d}\tau} \tag{2.101}$$

If we write $\int \psi_i \mathcal{H}\psi_j\,\mathrm{d}\tau = H_{ij}$, and $\int \psi_i\psi_j\,\mathrm{d}\tau = S_{ij}$, then

$$E = \frac{c_1^2 H_{11} + 2c_1c_2H_{12} + c_2^2H_{22}}{c_1^2 S_{11} + 2c_1c_2S_{12} + c_2^2S_{22}} \tag{2.102}$$

If ψ_i and ψ_j are separately normalized, then $\int \psi_i\psi_j\,\mathrm{d}\tau \leqslant 1$, the equality sign holding when $i=j$.

It is a straightforward matter to form $\partial E/\partial c_1$ and $\partial E/\partial c_2$, and equate them to zero, whence

$$c_1(H_{11} - ES_{11}) + c_2(H_{12} - ES_{12}) = 0 \tag{2.103}$$

$$c_1(H_{12} - ES_{12}) + c_2(H_{22} - ES_{22}) = 0 \tag{2.104}$$

These equations are known as the *secular equations* for the system, and from the theory of homogeneous linear equations it follows that a solution for which not all of the c_n are zero must satisfy the determinantal equation

$$\begin{vmatrix} H_{11} - ES_{11} & H_{12} - ES_{12} \\ H_{12} - ES_{12} & H_{22} - ES_{12} \end{vmatrix} = 0 \tag{2.105}$$

[11] By symmetry, at least, for homonuclear diatomic molecules.

As a general condition, a set of secular equations of the form of (2.103) has nontrivial values for c_n provided that the corresponding secular determinant has a zero value. The secular determinant is solved for E, which is then substituted back into the secular equations to obtain the ratios of the c_n coefficients. The normalizing condition

$$\left(\sum_n (c_n/c_1)^2\right)^{1/2} = N \tag{2.106}$$

is then used in obtaining the individual c_n values. The procedures for solving a secular determinant and obtaining the c_n values are explored in Appendix 9, with respect to buta-1,3-diene.

2.11.3 Overlap integral

The overlap integral S_{ij} measures the lack in orthogonality (see Section 2.11.5) of a pair of orbitals, or the extent to which one is projected on to the other. If we consider the $2p_x$ and $2p_y$ functions on one and the same atom, we have from the foregoing the overlap integral given by

$$S_{x,y} = \int \psi_{2p_x} \psi_{2p_y} \, d\tau \tag{2.107}$$

In polar coordinates we have from Table 2.3

$$S_{x,y} = \frac{(Z/a_0)^5}{32\pi} \int_0^{\infty} r^4 \exp(-Zr/a_0)\, dr \int_0^{\pi} \sin^3(\theta)\, d\theta \int_0^{2\pi} \cos(\phi)\sin(\phi)\, d\phi \tag{2.108}$$

Without more ado, it is evident that the integral over ϕ is zero; hence, the functions are orthogonal; there is zero overlap between them.

In a bonding situation, typical values of S_{ij} range between 0.2 and 0.3 and, qualitatively, the overlap integral may be thought of as the extent of overlap of one orbital with another. Figure 2.21 illustrates differing extents of overlap between two 1s orbitals and between 1s and 2p orbitals; S_{ij} is larger and leads to stronger bonding for the greater degree of overlap. The net overlap between 1s and 2p is zero in the orientation shown here because of the opposite signs of the wavefunction in the two lobes of the p orbital.

The calculation of overlap integrals is complex,[12] and Appendix 12 illustrates the calculation of the simplest of them:

$$S_{1s,1s} = \exp(-r/a_0)(1 + r/a_0 + r^2/3a_0^2)$$

Other aspects of the overlap integral will emerge from Sections 3.4.6.5 and 3.6.7.

2.11.4 Coulomb integral

In studying diatomic molecules, it is commonplace to let $H_{11} = \alpha_1$, $H_{22} = \alpha_2$ and $H_{12} = \beta$; the α terms are called Coulomb integrals and β is the resonance integral. A Coulomb integral measures the energy of an electron when it occupies its own orbital; in a homonuclear diatomic molecule, $\alpha_1 = \alpha_2$. The resonance integral is negative in a bonding situation, but vanishes to zero when the two orbitals do not overlap.

[12] See, for example, H Eyring, J Walter and G E Kimball *Quantum Chemistry* (Wiley, 1963).

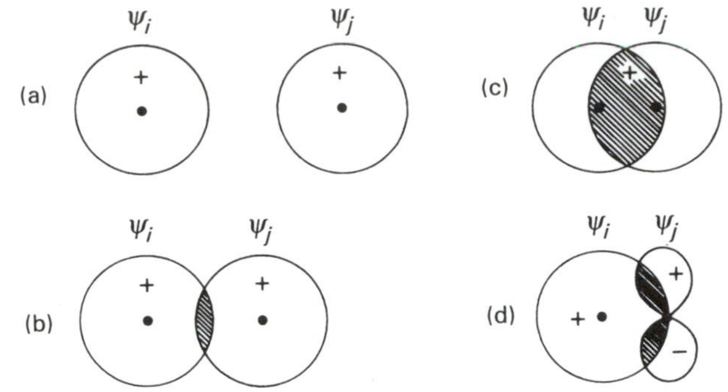

Figure 2.21 Bond formation by overlap of atomic orbitals ψ_i and ψ_j: (a) two separated 1s orbitals; $S=0$; (b) two 1s orbitals with a small degree of overlap; (c) two 1s orbitals overlapping strongly; (d) overlap of 1s and 2p in this orientation has zero value for S, because of the cancelling effect of the ± regions of the p orbital.

2.11.5 Orthogonality

Atomic orbitals are designed so as to be orthogonal, a property that enables them to be discussed individually; otherwise one would have a component on another. Mathematically, orthogonality is defined by

$$\int_{\substack{\text{all}\\\text{space}}} \psi_i \psi_j \, d\tau = \delta_{ij} \tag{2.109}$$

where δ_{ij} is the Kronecker delta; for normalized orbitals δ_{ij} is zero for $i \neq j$ and unity otherwise. Two different orbitals on one and the same atom are always orthogonal. The overlap integral for s and p orbitals depends on their relative orientation. Figures 2.21b and 2.21c show different degrees of overlap between two 1s orbitals; in Figure 2.21d there is zero resultant overlap between 1s and 2p. If the orientation is changed such that the axis of the p orbital is the internuclear axis, the overlap integral is finite and given *mutatis mutandis* by (2.108).

2.11.6 Methods for molecules

The models that have been developed for applying quantum mechanics to molecules include the valence-bond and the molecular-orbital approximations. The molecular-orbital method is similar to that developed for atoms, with molecular orbitals that encompass the whole molecule replacing the atomic orbitals of the component atoms. It lends itself readily to computational methods, which is one reason for the intense development that it has received.

The valence-bond and molecular-orbital methods both seek to determine a structure by finding its minimum energy configuration. They both involve the concept of an accumulation of overlap density in the internuclear region as the factor determining bonding. The valence-bond technique leans strongly towards Lewis' electron-pair bonding concept. However, the number of valence-bond canonical forms (electron-paired structures) that can be drawn up increases rapidly with increasing numbers of atoms: five for benzene (a to e; f is a conventional representation, but not a canonical form), but over one thousand for naphthalene. Neither method treats spin correlation, the property of electrons to repel one another, in a totally satisfactory manner and the reader is referred to discussions of more detailed calculations.[13]

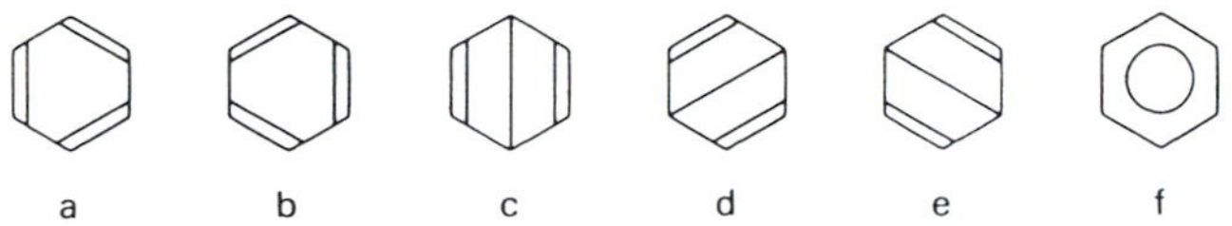

Benzene structures

2.11.7 Hydrogen-molecule ion

We confine our attention to the molecular-orbital model and introduce it with reference first to the one-electron hydrogen-molecule ion H_2^+, which plays a role for molecules similar to that of atomic hydrogen for atoms. Under the Born–Oppenheimer approximation, a single electron moves in the field of two stationary nuclei, and the Schrödinger equation may be written as

$$[-\hbar^2/(2\mu)]\nabla^2\Psi-[e^2/(4\pi\epsilon_0)](-1/r_1-1/r_2)\Psi=E\Psi \tag{2.110}$$

where r_1 and r_2 represent the distances of the electron from the two nuclei. This equation can be solved exactly, but here the LCAO molecular-orbital technique will be used.

The lowest energy configuration of a hydrogen atom is 1s and we assume that the ground state molecular orbital for H_2^+ is similar. Let ψ_1 and ψ_2 be normalized 1s hydrogen-like orbitals for the electron in the neighbourhoods of nuclei 1 and 2 respectively. Following (2.97), we have

$$\Psi=c_1\psi_1+c_2\psi_2 \tag{2.111}$$

In this species the two functions are equivalent by symmetry and, since probabilities are proportional to Ψ^2, it follows that $c_1^2=c_2^2$. The two solutions may then be written as

$$\Psi_\pm=c_\pm(\psi_1\pm\psi_2) \tag{2.112}$$

Applying the variation method, and because of the identity of ψ_1 and ψ_2, we write $H_{11}=H_{22}=\alpha$, $H_{12}=\beta$, $S_{11}=S_{22}=1$ and $S_{12}=S$, leading to the secular equations

$$c_1(\alpha-E)+c_2(\beta-ES)=0 \tag{2.113}$$

$$c_1(\beta-ES)+c_2(\alpha-E)=0 \tag{2.114}$$

with the secular determinant

$$\begin{vmatrix}\alpha-E & \beta-ES\\ \beta-ES & \alpha-E\end{vmatrix}=0 \tag{2.115}$$

Expanding this determinant and solving the ensuing quadratic equation gives the two values for the energy E

$$E_+=\frac{\alpha+\beta}{1+S} \tag{2.116}$$

$$E_-=\frac{\alpha-\beta}{1-S} \tag{2.117}$$

[13] See, for example, J N Murrell, S F A Kettle and J M Tedder *The Chemical Bond* (Wiley, 1985), and C A Coulson, revised by R McWeeney *Valence* (O U P, 1979).

Substituting these values of E in (2.113) and (2.114) shows that $c_1 = \pm c_2$, as anticipated, and $c_\pm$ is obtained by normalizing the molecular wavefunction Ψ according to (2.42). Thus

$$c_\pm^2 \int (\psi_1 \pm \psi_2)^2 \, d\tau = c_\pm^2 \int (\psi_1^2 + \psi_2^2 \pm 2\psi_1\psi_2) \, d\tau = 1 = c_\pm^2(1+1\pm 2S) \tag{2.118}$$

whence $c_\pm$, the normalizing constant, becomes $1/(2 \pm 2S)^{1/2}$ for the two wavefunctions (2.112). The value for S in H_2^+ at its internuclear separation of 0.11 nm is 0.56, as can be verified from (2.107). This value is atypically large, as it is also in H_2.

2.11.7.1 Bonding and antibonding orbitals

The two values $c_+ = 0.566$ and $c_- = 1.066$ lead to two wavefunctions for H_2^+:

$$\Psi_+ = 0.566\,(\psi_1 + \psi_2) \tag{2.119}$$

$$\Psi_- = 1.066\,(\psi_1 - \psi_2) \tag{2.120}$$

with corresponding energies given by (2.116) and (2.117), respectively. Since α and β are both negative, (2.119) corresponds to the lower (more negative) energy E_+ and constitutes a *bonding* molecular orbital; (2.120) corresponding to the energy E_- is an *antibonding* molecular orbital. Both of these orbitals have cylindrical symmetry about the internuclear axis and are called *σ orbitals* and, more fully, 1sσ orbitals, since they have been formed from 1s atomic orbitals.

The probability density is proportional to the square of the wavefunction. For the 1sσ bonding orbital, we have

$$\Psi^2 = 0.566^2(\psi_1^2 + \psi_2^2 + 2\psi_1\psi_2) \tag{2.121}$$

which may be expanded to the proportionality

$$\Psi^2 \propto \{\exp(-2r_1/a_0) + \exp(-2r_2/a_0) + 2\exp[-(r_1+r_2)/a_0]\} \tag{2.122}$$

Figure 2.22a shows the sum of the first two terms on the right-hand side of (2.122). It is the superposition of ψ_1^2 and ψ_2^2, separated by a distance of 0.11 nm, the equilibrium internuclear distance r_e in H_2^+, where $r_1 = |r + r_e/2|$ and $r_2 = |r - r_e/2|$. Figure 2.22b shows the overlap density $2\psi_1\psi_2$, which represents a build-up of the density in the internuclear region over that shown in Figure 2.22a; because of the constructive interference of the overlap of the two 1s atomic orbitals of positive amplitude ψ, it remains constant over the distance r_e. Figure 2.22c is the sum of all three terms in (2.122), and shows clearly the enhancement of density in the internuclear region compared with the sum of the separate densities in Figure 2.22a.

Detailed calculations on this system give $r_e = 0.13$ nm (experimental 0.11 nm) and $E =$ 1.8 eV (experimental 2.6 eV), which are acceptable for a first approximation wavefunction. Figure 2.23 shows the variation of energy E with internuclear distance r for the two wavefunctions, (2.119) bonding and (2.120) antibonding; the curve for the exact solution is given for comparison.

Repeating the calculations for the antibonding 1sσ* orbital leads to the need to change the sign in front of the third term on the right-hand side of (2.122). The subtraction of this contribution constitutes a destructive interference and leads to a reduction in the probability density compared with that of the sum of the separate densities. The result is shown in Figure 2.22d, from which it is clear that the density falls to zero between the two nuclei. The source of cohesion between the nuclei has been lost and comparison with Figure 2.22a shows that the 1sσ* molecular orbital is less stable than the system of two separate nuclei and one electron.

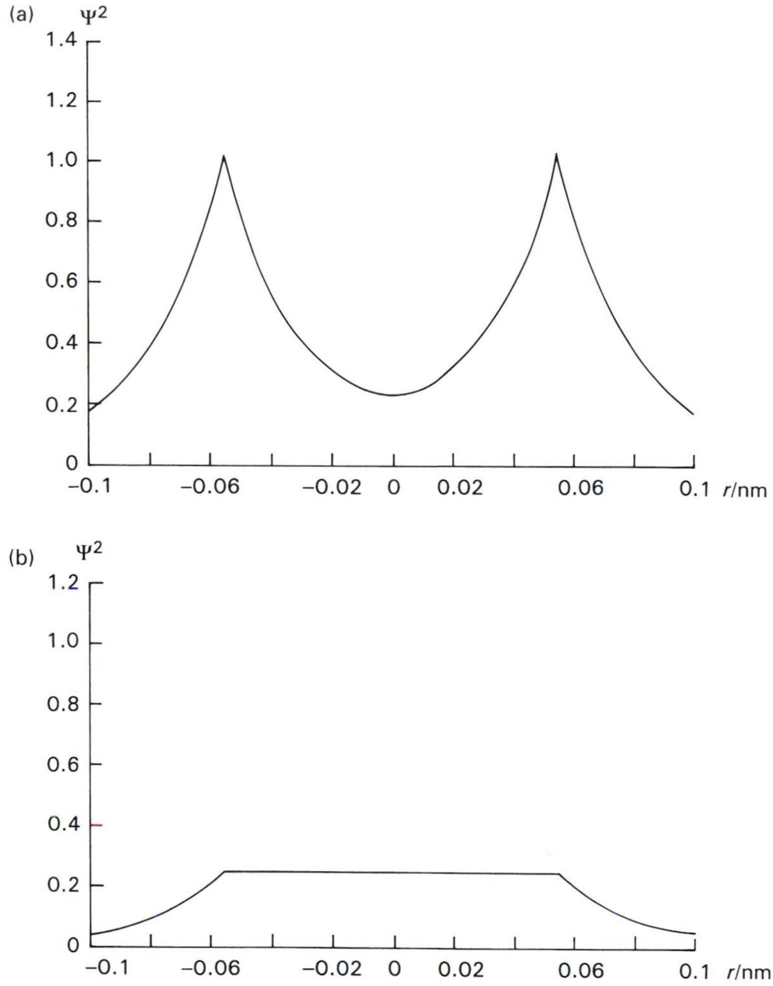

Figure 2.22 Hydrogen-molecule ion probability density function ψ^2: (a) superposition of ψ_1^2 and ψ_2^2; (b) the function $2\psi_1\psi_2$, a constant for $(r_1+r_2)\leq r_e$; (c) the superposition (a)+(b), leading to a bonding orbital, with enhancement of the density in the internuclear region (±0.055 nm) compared with (a); (d) the superposition (a)−(b), leading to an antibonding orbital, with a significant decrease in density in the internuclear region compared with (a).

2.11.8 Pauli principle

The Pauli principle states that *a total wavefunction is antisymmetric*. Thus, when two (identical) electrons change places, the total wavefunction changes sign. Symbolically, this interchange means that

$$\psi(1,2) = -\psi(2,1) \tag{2.123}$$

The application to electrons is reasonable: electrons are indistinguishable, so their interchange should not change the calculated properties of the system, such as the electron density or the energy.

The functions $\Psi_\pm$ in (2.112) are symmetrical with respect to interchange of electrons, so that their antisymmetry must reside in the spin. Furthermore, the + sign leads to a single

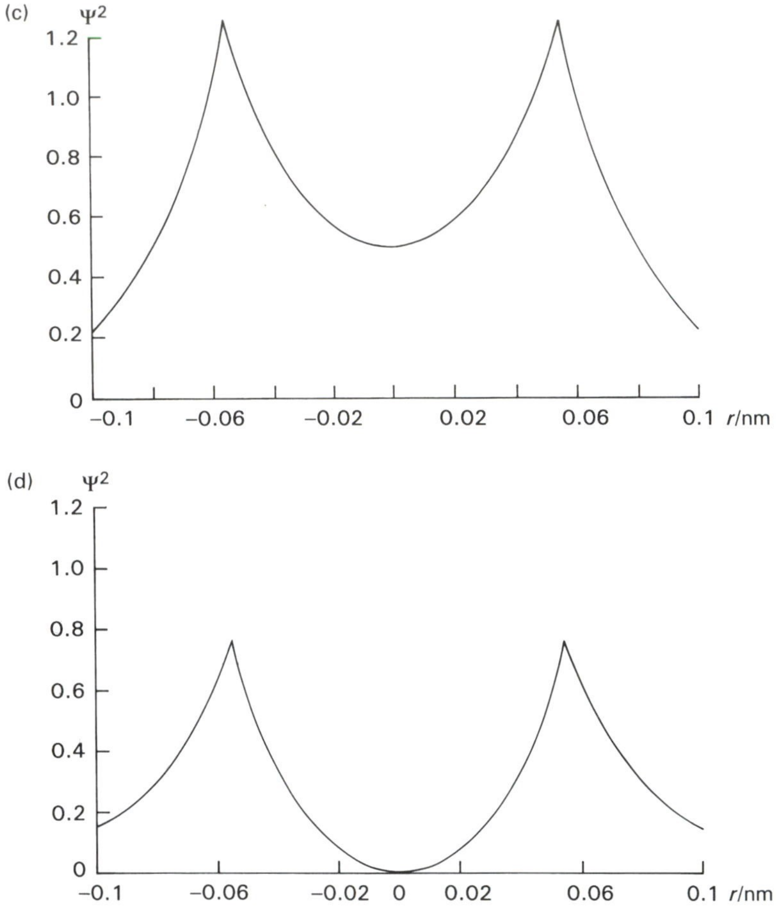

Figure 2.22 (*cont.*)

energy state (singlet), whereas the − sign leads to a triplet energy state (triplet), as explained with the aid of Table 2.6.

Let electron 1 have spin α and electron 2 spin β; the spin wavefunction will be designated $\alpha(1)\beta(2)$. We must also consider $\alpha(2)\beta(1)$ and, since the two are degenerate, the linear combinations $\alpha(1)\beta(2) \pm \alpha(2)\beta(1)$ are permissible functions. The total wavefunctions are summarized in Table 2.6. Interchanging electrons 1 and 2 throughout the total functions shows that those involving spin terms (a) are multiplied by −1 (antisymmetric), whereas those involving spin terms (b) are unchanged (symmetric). Heisenberg showed (1926) that only antisymmetric wavefunctions for electrons explained the spectra of the so-called *ortho*-helium (triplet) and *para*helium (singlet) states for this element. Thus, we reject spin terms (b) and the bonding wavefunction becomes

$$\Psi_{+} = \psi(1) + \psi(2) + [\alpha(1)\beta(2) - \alpha(2)\beta(1)] \tag{2.124}$$

The spin term corresponds to a total spin of zero, because the spins are paired in each spin wavefunction.

The Pauli exclusion principle that we have already discussed (see Section 2.8) arises from the general Pauli principle. If certain electrons of an atom occupy different orbitals, the Pauli

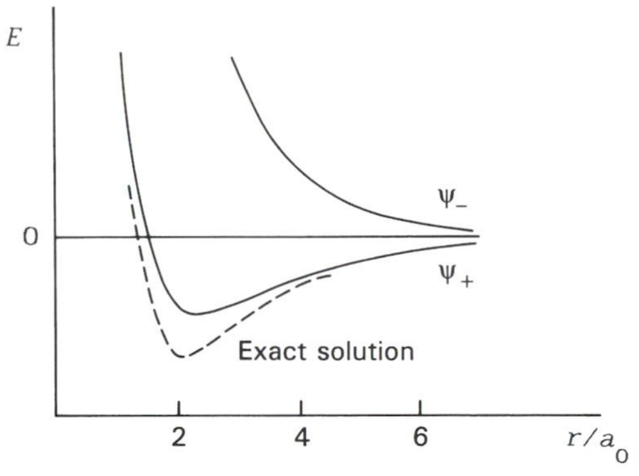

Figure 2.23 Hydrogen-molecule ion: variation of energy E with inter-nuclear distance r (in units of r/a_0) for the bonding ψ_+ and antibonding ψ_- orbitals; the exact solution is shown in dashed lines, and is lower in energy than that given by the approximation for Ψ_+. The energy for the antibonding orbital does not fall below zero.

Table 2.6 Total wavefunction for H_2

Space terms	Spin terms (a)	Spin terms (b)
$\psi_1(1)+\psi_2(2)$	$\alpha(1)\beta(2)-\alpha(2)\beta(1)$	$\alpha(1)\alpha(2)$ $\beta(1)\beta(2)$ $\alpha(1)\beta(2)+\alpha(2)\beta(1)$
$\psi_1(1)-\psi_2(2)$	$\alpha(1)\alpha(2)$ $\beta(1)\beta(2)$ $\alpha(1)\beta(2)+\alpha(2)\beta(1)$	$\alpha(1)\beta(2)-\alpha(2)\beta(1)$

exclusion principle becomes inoperative for them, but the Pauli principle of antisymmetry still applies to all electrons.

2.11.9 Homonuclear diatomic molecules

The discussion on the hydrogen-molecule ion can be carried over to diatomic molecules. In the hydrogen molecule, a molecular orbital can be formed from two 1s atomic orbitals, as with the hydrogen-molecule ion, and we obtain $1s\sigma$ and $1s\sigma^*$ molecular orbitals. We stress that a molecular orbital, like an atomic orbital, is a mathematical function; the overlapping of atomic orbitals to form molecular orbitals is a pictorially convenient way of expressing the mathematical functions.

The energies of the two $1s\sigma$ molecular orbitals for hydrogen are given approximately by (2.116) and (2.117) with multiplying factors of 2, since there are now two electrons present. Other factors such as electron–electron repulsion need to be taken into account in detailed calculations. We shall not introduce that degree of sophistication here, but record that at the internuclear distance of 0.074 nm in H_2 (smaller than that in H_2^+) an early calculation of the energy was 3.6 eV (experimental 4.7467 eV) and that the most detailed calculations have given an energy in exact agreement with the observed value.[14]

[14] W Kolos and L Wolniewicz, *J. Chem. Phys.* **49**, 404 (1968).

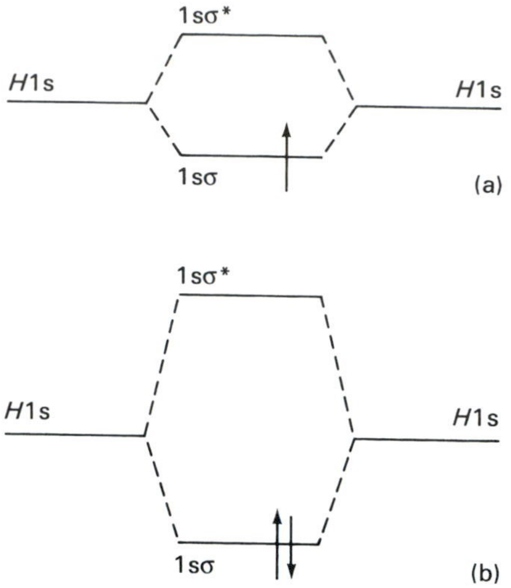

Figure 2.24 Molecular-orbital energy-level diagrams: (a) H_2^+, with a single electron in the 1sσ molecular orbital; (b) H_2, with the 1sσ molecular orbital fully occupied by two electrons with paired spins.

The similarity between H_2^+ and H_2 can be brought out by a molecular-orbital energy-level diagram, Figure 2.24. In H_2^+, the 1sσ orbital is occupied by a single electron, whereas in H_2 the 1sσ orbital contains two electrons, following the *Aufbau* principle, with paired spins according to the Pauli exclusion principle.

It is now possible to see that He_2 cannot form a stable molecule. The ground state configuration would be $(1s\sigma)^2\ (1s\sigma^*)^2$: since, as we have seen (Figures 2.22a and 2.22d), an antibonding orbital more than negates a bonding orbital, He_2 would be less stable than two separate helium atoms. How would this discussion apply to Li_2?

2.11.10 Symmetry of orbitals

Diatomic molecules, and their orbitals, exhibit symmetry, a property by which they may be usefully classified:

(a) orbitals that are symmetrical about the internuclear axis are called σ molecular orbitals, Figures 2.25a–d;
(b) orbitals that have a nodal plane containing the internuclear axis are called π molecular orbitals, Figures 2.25e and 2.25f;
(c) orbitals may be described as even g (German, *gerade*), or as odd u (German *ungerade*), with respect to inversion across the centre of the internuclear axis, Figures 2.25a–f. The subscripts g and u are called the *parity* of the orbital.

We note that, in Figure 2.25, σ_g and π_u molecular orbitals are bonding whereas σ_u and π_g are antibonding, so that σ_u and σ^* represent like types of orbital, and similarly with π_g and π^*. The u and g descriptors cannot be applied to heteronuclear diatomics because these molecules do not possess a centre of symmetry.

For species with n=2, the outer, or valence, electrons are 2s and 2p; the 1s electrons form

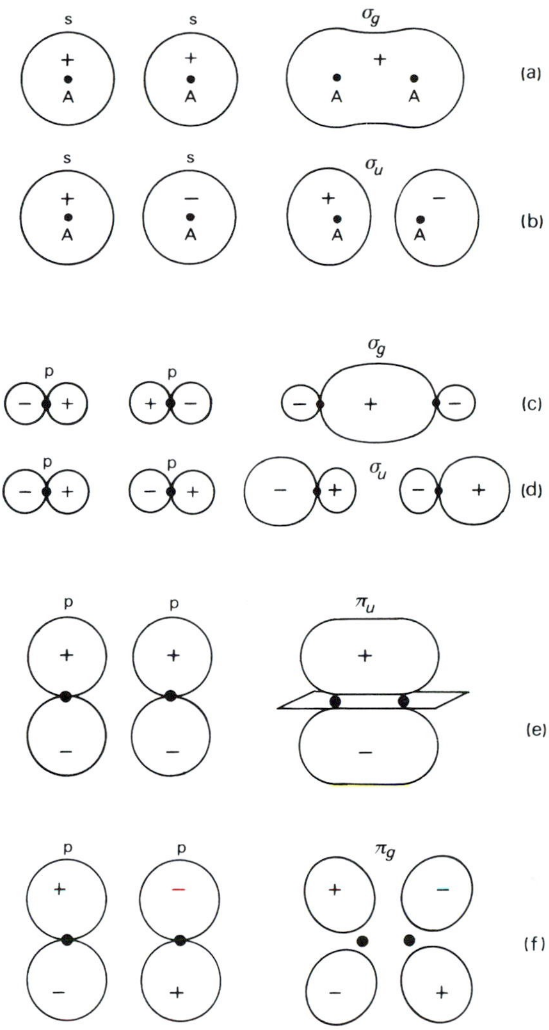

Figure 2.25 Schematic illustrations of LCAO molecular orbitals for homonuclear diatomic molecules: sσ and pσ molecular orbitals are g (bonding) and u (antibonding); pπ molecular orbitals are u (bonding) and g (antibonding).

a *core* which interacts only slightly with the valence electrons and may be neglected in a first approximation. We can envisage 2s−2s overlap to give sσ molecular orbitals of the type just discussed (Figures 2.25a and 2.25b); $2p_z-2p_z$ overlap (z conventionally along the internuclear axis) to give pσ (Figures 2.25c and 2.25d); and $2p_x$ and/or $2p_y$, perpendicular to the internuclear axis, overlap to give pπ (Figures 2.25e and 2.25f). Another type is 2s−2p: $2s-2p_z$ overlap contributes to the pσ molecular orbital, with significant overlap given by $\int\psi(2s)\psi(2p)\,d\tau$, but the $2s-2p_x$ and $2s-2p_y$ overlaps are zero (see Figure 2.21d).

Molecular-orbital energy-level diagrams for homonuclear diatomic molecules up to and including nitrogen, and for oxygen and subsequent species are shown by Figures 2.26 and 2.27: the 2pσ and 2pπ energies are reversed in order after nitrogen; for molecules in later

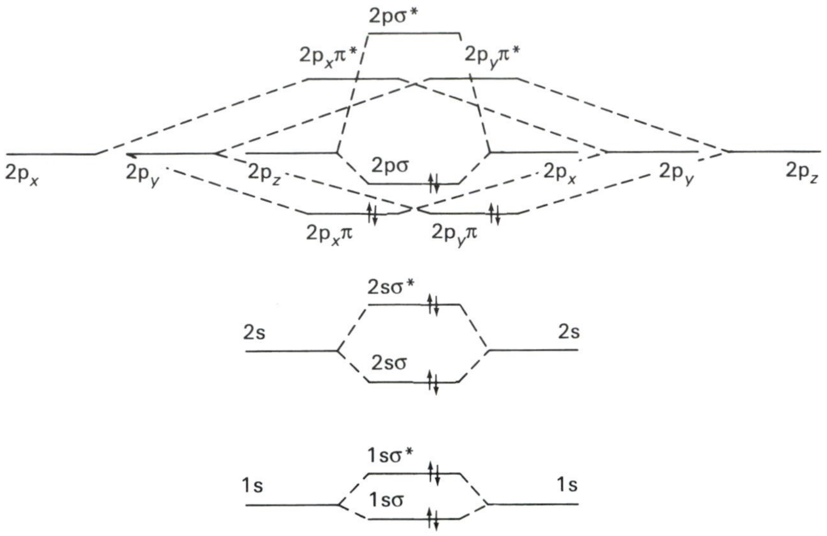

Figure 2.26 Molecular-orbital energy-level diagram for N_2: the configuration is $(1s\sigma)^2$ $(1s\sigma^*)^2$ $(2s\sigma)^2$ $(2s\sigma^*)^2$ $(2p\pi)^4$ $(2p\sigma)^2$. The arrows represent electrons, and in N_2 they are all paired; the net bonding parameter is 3.

horizontal periods the pσ is always below the pπ in energy. The change-over depends upon the separation of the 2s and 2p energies, which increases with increasing atomic number. Hence, we write the molecular-orbital configurations for N_2 and O_2 as follow:

$$N_2\ (1s\sigma)^2(1s\sigma^*)^2(2s\sigma)^2(2s\sigma^*)^2(2p\pi^4)(2p\sigma)^2$$

$$O_2\ (1s\sigma)^2(1s\sigma^*)^2(2s\sigma)^2(2s\sigma^*)^2(2p\pi)^4(2p\sigma)^2(2p_x\pi^*)^1(2p_y\pi^*)^1$$

The separation here of the pπ^* molecular orbitals in oxygen, rather than writing $(2p\pi^*)^2$, highlights the single occupancy of these orbitals; it is this feature that is responsible for the paramagnetic property of molecular oxygen.

We may define a *net bonding parameter* κ, or bond order, by

$$\kappa=(n_e-n_e^*)/2 \tag{2.125}$$

where n_e and n_e^* are respectively the numbers of electrons in the bonding and antibonding molecular orbitals of a species. Thus, the net bonding parameters are 2 and 3 for O_2 and N_2 respectively, corresponding to the classical formulations O=O and N≡N for these molecules.

2.11.11 Heteronuclear diatomic molecules

In a diatomic molecule containing different chemical species, the electron distribution is polarized: there is an accumulation of negative charge on one species, the more electronegative, and a corresponding depletion on the other; the bond is then said to be polar (see also Section 1.2). In hydrogen fluoride the fluorine atom is the more electronegative and carries a partial negative charge of approximately $-0.4e$ (see also Table 2.7 later), with a balancing charge of $+0.4e$ on the hydrogen atom; the molecule possesses a *dipole moment*.

The dipole moment is an important physical parameter of a molecule. If a diatomic molecule is represented by numerical point charges $\pm q$ separated by a distance r, its dipole moment μ is defined by

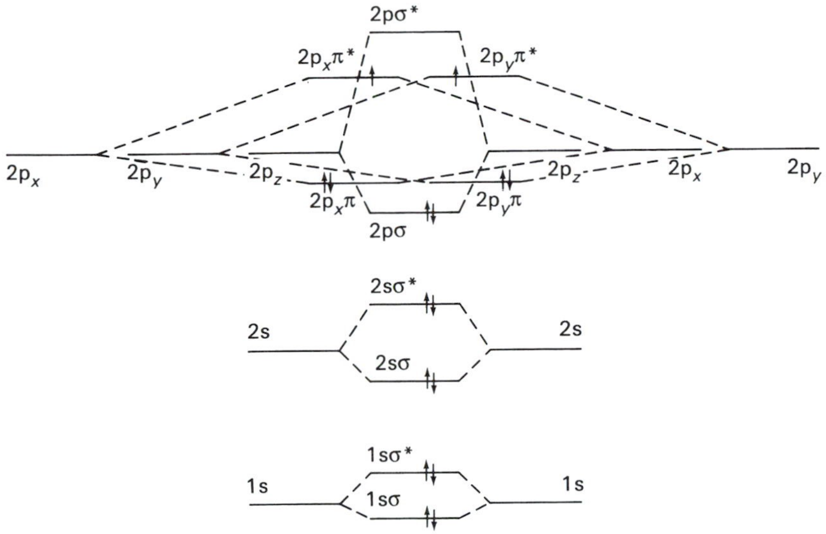

Figure 2.27 Molecular-orbital energy-level diagram for O_2: the configuration is $(1s\sigma)^2$ $(1s\sigma^*)^2$ $(2s\sigma)^2$ $(2s\sigma^*)^2$ $(2p\sigma)^2$ $(2p\pi)^4$ $(2p\pi^*)^2$; the positions of $2p\sigma$ and $2p\pi$ are reversed relative to N_2. The arrows represent electrons, and in O_2 they are paired except for the $2p\pi^*$, for which the single occupancy leads to paramagnetism; the net bonding parameter is 2.

$$\mu=|q|er \tag{2.126}$$

where e is the charge on an electron. The experimental value for the dipole moment[15] of the hydrogen fluoride molecule is 1.82 D and the bond length is 0.0927 nm; thus, q is 0.41.

The electron configuration of atomic fluorine is $(1s)^2(2s)^2(2p)^5$, and we can expect bond formation between the H(1s) and F($2p_z$) atomic orbitals (Figure 2.28). The LCAO molecular orbitals may be written as

$$\Psi_{\pm}=c_H\psi(\mathrm{H,1s})\pm c_F\psi(\mathrm{F,2p}) \tag{2.127}$$

Because of the lack of symmetry between H and F, $c_H\neq\pm c_F$. The values for the coefficients have been determined by the variation method; $c_H^2/c_F^2=0.11/0.89$ in the bonding orbital and in the antibonding orbital the reciprocal of this ratio applies.

In studying heteronuclear diatomic molecules we continue to neglect the core electrons, for reasons already discussed. Bonding overlap arises only for atomic orbitals of the same symmetry and of similar energy. Ionization energies can provide a useful guide to the orbitals that are likely to be involved in bonding:

	F(2s)	F(2p)	H(1s)
I/eV	40.2	18.6	13.6

Hence, the most probable (normalized) wavefunctions for HF will be chosen as

$$\Psi_{+}=0.331\psi(\mathrm{H,1s})+0.944\psi(\mathrm{F,2p}) \tag{2.128}$$

$$\Psi_{-}=0.944\psi(\mathrm{H,1s})-0.331\psi(\mathrm{F,2p}) \tag{2.129}$$

since $c_H^2+c_F^2=1$.

[15] The commonly used unit for dipole moment is the debye, D; 1 D=3.3356×10^{-30} C m.

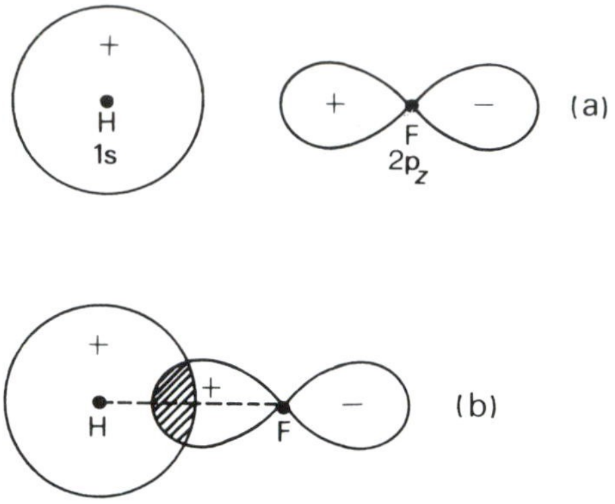

Figure 2.28 Bond-forming atomic orbitals for HF: (a) H(1s) and F(2p$_z$), of similar energy, oriented along the z axis for maximum overlap; (b) overlap leading to bonding, principally through a 2pσ molecular orbital.

The secular determinant for HF is, following (2.115) and assuming that overlap is relatively negligible,

$$\begin{vmatrix} \alpha_H - E & \beta \\ \beta & \alpha_F - E \end{vmatrix} = 0 \tag{2.130}$$

This equation reduces to

$$E = \frac{(\alpha_H + \alpha_F) \pm [(\alpha_H - \alpha_F)^2 + 4\beta^2]^{1/2}}{2} \tag{2.131}$$

When the difference in atomic orbital energies is large, E tends to the values α_H and α_F, the energies of the individual orbitals, which leads to a nonbonding situation. If the energy difference is small, E tends to the values $(\alpha_H + \alpha_F) \pm 2\beta$, corresponding to bonding (+) and antibonding (–) molecular orbitals.

The electronic configuration of HF may be written as $[(1s\sigma)^2(2s\sigma)^2(2p\pi)^4](3p\sigma)^2$. The configuration in the brackets represents the core, and the net bonding parameter may be reckoned in terms of the 3pσ molecular orbital alone; thus, $\kappa = 1$.

2.11.12 Electronegativity

In the hydrogen halides the partial charge on the halogen decreases from fluorine to iodine, with consequent reductions in the value for the dipole moments of the species. The partial charge implies a degree of ionic character λ in the bond, where

$$\lambda^2 = q/(1+q) \tag{2.132}$$

We summarize the results for the hydrogen halides in Table 2.7.

The electronegativity χ is a measure of the power of an atom to attract electrons in compound formation: the greater the value of χ, the greater is the attraction of electrons. Pauling's scale of electronegativities (1932) was drawn up by relating the difference in electronegativity of two species $\Delta\chi_{AB}$, given by $|\chi_A - \chi_B|$, for the compound AB to the excess of the bond energy of AB over the geometric mean of the bond energies for A_2 and B_2 in the diatomic species A_2 and B_2.

Table 2.7 Dipole moment μ and partial charge q in the hydrogen halides

	HF	HCl	HBr	HI
$10^{30}\mu$/C m	6.07	3.44	2.77	1.50
$10^{10}d$/m	0.927	1.27	1.41	1.61
q	0.41	0.17	0.12	0.058
q_χ	0.40	0.20	0.12	0.089
λ	0.83	0.45	0.37	0.25

Table 2.8 Pauling electronegativities

H						
2.20						
Li	Be	B	C	N	O	F
0.98	1.57	2.04	2.55	3.04	3.44	3.98
Na	Mg	Al	Si	P	S	Cl
0.93	1.31	1.61	1.90	2.19	2.58	3.16
K	Ca	Ga	Ge	As	Se	Br
0.82	1.00	1.81	2.01	2.18	2.55	2.96
Rb	Sr	In	Sn	Sb	Te	I
0.82	0.95	1.78	1.96	2.05	2.10	2.66
Cs	Ba	Tl	Pb	Bi		
0.79	0.89	2.33	2.02			

For hydrogen and fluorine, we have $E(\text{HF})=565$ kJ mol^{-1}, $E(\text{H}_2)=436$ kJ mol^{-1} and $E(\text{F}_2)=155$ kJ mol^{-1}. Hence, $\Delta\chi_{\text{HF}}=(0.102\times\text{mol}^{1/2}\ \text{kJ}^{-1/2})\{(565\times\text{kJ mol}^{-1})-[(436\ \text{kJ mol}^{-1})\times(155\times\text{kJ mol}^{-1})]^{1/2}\}^{1/2}=1.78$, and this is the difference between the Pauling electronegativities of hydrogen and fluorine. (Pauling's original proportionality factor was 0.208 mol$^{-1/2}$ kcal$^{1/2}$, because he used energies in kcal mol^{-1}; $(0.208\times\text{mol}^{-1/2}\ \text{kcal}^{1/2})/(4.184\times\text{J cal}^{-1})^{1/2}=0.102\ \text{mol}^{1/2}\ \text{kJ}^{-1/2}$.)

The more recent Allred–Rochow scale sets χ equal to $(0.00359\ Z_{\text{eff}}/r^2)+0.744$, where r is the covalent radius in nanometres, and Z_{eff} is calculated for an outermost electron; the results are similar to those of Pauling in Table 2.8.

The relationship between electronegativity and partial charge q has been studied, and the best equation is that given by Hannay and Smith; it is most reliable for q less than about 0.5:

$$q=0.16\,\Delta\chi+0.035\,\Delta\chi^2 \tag{2.133}$$

The results of applying this equation to the hydrogen halides are the q_χ values in Table 2.7; the agreement with q from dipole moment data is acceptable.

2.11.13 Hybridization

The lithium hydride molecule provides a slightly more complex situation. We note first the relevant ionization energies:

	Li(1s)	Li(2s)	Li(2p)	H(1s)
I/eV	75.6	5.4	3.7	13.6

The values for Li(2s) and Li(2p) are commensurate, and a bonding wavefunction may be written as

$$\Psi = c_{\mathrm{Li(s)}}\psi(\mathrm{Li,2s}) + c_{\mathrm{Li(p)}}\psi(\mathrm{Li,2p}) + c_{\mathrm{H}}\psi(\mathrm{H,1s}) \tag{2.134}$$

Application of the variation method leads to the result

$$\Psi = 0.323\psi(\mathrm{Li,2s}) + 0.231\psi(\mathrm{Li,2p}) + 0.685\psi(\mathrm{H,1s}) \tag{2.135}$$

and the configuration for LiH is $[(1s\sigma)^2](2s\sigma)^2$, with an infusion of Li 2p character into the $2s\sigma$ molecular orbital; $[(1s\sigma)^2]$ is the core of Li $(1s)^2$, with negligible interaction because the energy of the Li 1s electron state is well below that of the H 1s level. The net bonding parameter κ for the Li–H bond is 1.

In order to adjust this result to the picture of overlap between two atomic orbitals, we introduce the concept of *hybridization*, or quantum mechanical ‘mixing’, of the 2s and 2p atomic orbitals on lithium. Thus, we may define a hybrid atomic orbital for lithium by this mixing of 2s and 2p:

$$\psi(\mathrm{Li,sp}) = \psi(\mathrm{Li,2s}) + \lambda\psi(\mathrm{Li,2p}) \tag{2.136}$$

where λ is (0.231/0.323), or 0.715; this hybrid then overlaps the $\psi(\mathrm{H,1s})$ to give a hybrid bonding molecular orbital:

$$\Psi = 0.323\psi(\mathrm{Li,sp}) + 0.685\psi(\mathrm{H,1s}) \tag{2.137}$$

where the sp hybrid has 66% s character and 34% p character. We should note that hybridization is, again, something that we do to attempt a description of that which exists, rather than a statement of a real situation.

The ratio $c_{\mathrm{X}}^2/(c_{\mathrm{X}}^2 + c_{\mathrm{H}}^2)$, where X is another element that forms a diatomic molecule HX, determines the average extent to which electron density is transferred from hydrogen to the species X in the molecule. Hybridization modifies the shapes of the participating atomic orbitals and, hence, those of the resulting molecular orbitals. Figure 2.29 shows electron density contour maps for the heteronuclear hydrides from lithium to fluorine, which may be taken to correspond to the probability density $|\Psi|^2$ for the hybrid molecular orbital. It is clear that the molecular orbital for each species becomes ‘fatter’ around the nonhydrogen species as the series progresses from lithium to fluorine; electronegativity increases in the same direction, too.

2.11.14 Polyatomic molecules

We consider next some simple polyatomic molecules, including water, ammonia and hydrocarbons, the latter leading to an introductory discussion on Hückel molecular-orbital theory. Polyatomic molecules are built up in the same way as the diatomic molecules but more orbitals are involved and a greater range of shapes is possible.

2.11.14.1 Water

Experimental studies have shown that the H–O bond length is 0.096 nm, and that the H–O–H bond angle is 104.4°. The electronic configuration of oxygen, $(1s)^2(2s^2)(2p_z^2)(2p_x)^1(2p_y)^1$, implies divalency, with bonding between O(2p) and H(1s) atomic

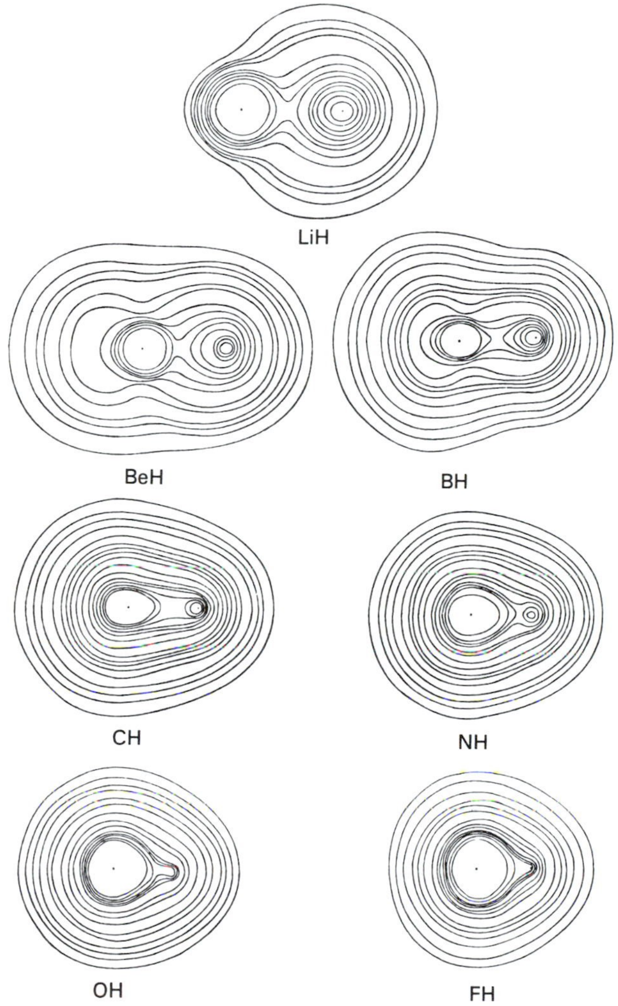

Figure 2.29 Electron density contour diagrams for the diatomic molecules LiH to FH, on a common scale; the innermost contours have been omitted for clarity. The H species is on the right-hand side in each drawing. Along the series from Li to F, the increasing electronegativity draws progressively more of the hydrogen electron density towards these species. (After R F W Bader, I Keaveney and P E Cade, *J. Chem. Phys.* **47**, 3381 (1967) and reproduced by permission of the American Chemical Society.)

orbitals. The directional character of p orbitals should lead to a bond angle of 90°. In a simplistic way we could then say that the hydrogen atoms repel each other, which causes an opening out of the bond angle to the observed value.

Consider instead two similar p orbitals, say p_u and p_v, with their axes making an angle θ_{uv} with each other, and let another similar orbital p_w, in the plane of p_u and p_v, be directed orthogonally to p_u (Figure 2.30). Then we have

$$p_v = p_u \cos(\theta_{uv}) + p_w \sin(\theta_{uv}) \tag{2.138}$$

A similar orbital p'_v at an angle $-\theta_{uv}$ to p_u is given by

$$p'_v = p_u \cos(\theta_{uv}) - p_w \sin(\theta_{uv}) \tag{2.139}$$

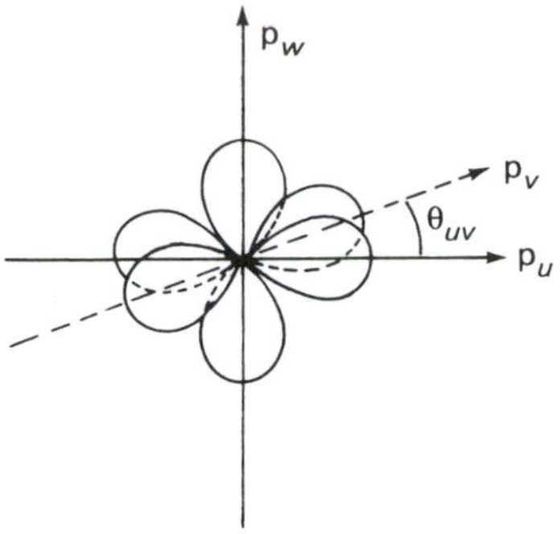

Figure 2.30 Formation of hybrid orbitals: p_u and p_w are two orthogonal p orbitals, with another orbital p_v at an angle θ_{uv} to p_u; a similar p'_v orbital can be defined at an angle $-\theta_{uv}$ to p_u. The vector directions of p_v and p'_v may be resolved along p_u and p_w in the usual way.

We postulate two hybrid orbitals h and h', using s and p atomic orbitals:

$$h = c_s s + c_p p_v \tag{2.140}$$

$$h' = c_s s + c_p p'_v \tag{2.141}$$

It is straightforward to show that normalization of h and h' needs $\int h^2 d\tau = \int h'^2 d\tau = c_s^2 + c_p^2 = 1$, and that the orthogonality criterion gives $\int hh'\, d\tau = c_s^2 + c_p^2 \int p_v p'_v\, d\tau = c_s^2 + c_p^2 \cos(2\theta_{uv}) = 0$. Hence, we find for the fractional characters c_s and c_p that

$$c_s^2 = \cos(\Theta)/[\cos(\Theta) - 1] \tag{2.142}$$

$$c_p^2 = \frac{1}{1 - \cos(\Theta)} \tag{2.143}$$

where $\Theta = 2\theta_{uv}$, and is the angle between the hybrids h and h', in the direction of p_v and p'_v.

A hybrid molecular wavefunction may be written as

$$\Psi = c_s \psi(s) + c_p \psi(p) \tag{2.144}$$

For common values of Θ we have the following character values:

Θ/deg	90	109.47	120	180
c_p^2	0	3/4	2/3	1/2
c_s^2	1	1/4	1/3	1/2
Type	p	sp^3	sp^2	sp

A hybrid atomic orbital for an oxygen atom in the water molecule may be written as

$$\psi_h = c_s \psi(s) + c_p \psi(p) \tag{2.145}$$

From the known H−O−H bond angle, $c_s^2 = 0.20$ and $c_p^2 = 0.80$, whence $c_s = 0.45$ and $c_p = 0.89$. The p character of the hybrid is 80% and the s character 20%; the hybrid type may be written as $s^{0.20}p^{0.80}$, or sp^4. The valence electrons no longer belong to any one atom; they are spread around, or *delocalized,* over the molecular orbitals. We have used part of the filled 2s orbital of oxygen in making each sp hybrid orbital. Energy has to be expended in unpairing electrons in the fully occupied 2s orbital, but it is more than compensated by the increased overlap of O(hybrid) with H(1s) compared to that of O(2p) and H(1s).

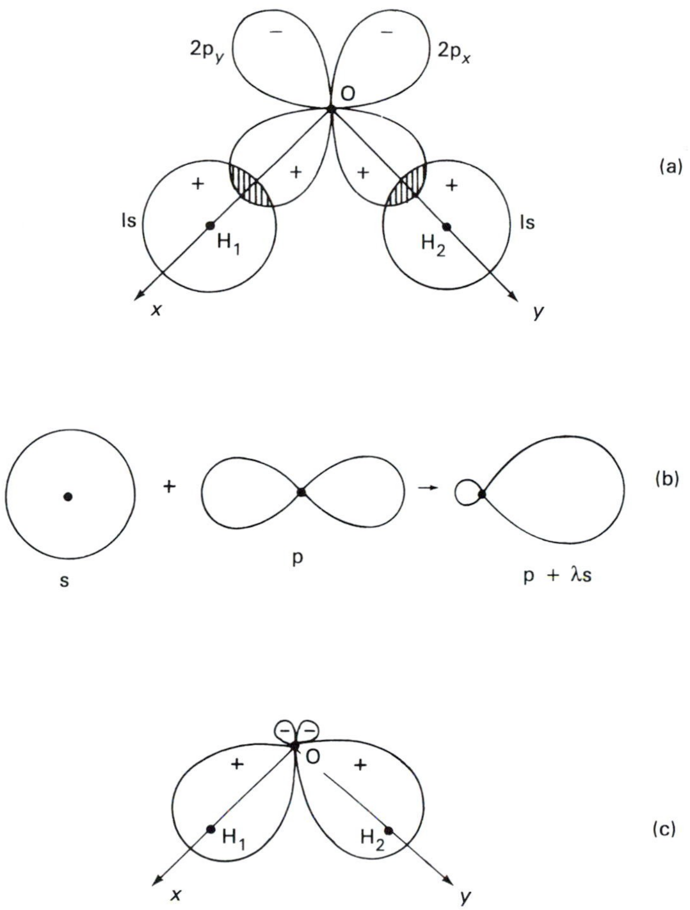

Figure 2.31 Water molecule: (a) bonding by overlap of O(2p) and H(1s) would lead to a bond angle of 90°; (b) mixing of s and p to form the hybrid p+λs, where $\lambda = c_s/c_p$; (c) bonding between O(sp) and H(1s) leads to strong overlap and a bond angle of 104.4° in the equilibrium configuration. Orthogonality of the hybrids is supported experimentally by the fact that in CH_3OH the O–H bonding electrons are not significantly changed, thus providing a basis for characteristic bond lengths.

The remaining 2s, $2p_x$ and $2p_y$ electron density, together with the $2p_z$ occupy two other hybrid orbitals. They are equivalent to two *lone pairs* of electrons which, with the bonding hybrids, are directed almost tetrahedrally. This configuration means that the lone pairs make a major contribution to the polar nature of the water molecule. The dipole moment of 1.8 D for the water molecule compared with 0.2 D for F_2O supports this view; the larger polarity of the F–O bond, directed away from the lone pairs, reduces their contributions to the dipole moment. Figure 2.31 illustrates some aspects of the discussion of the water molecule. Another way of looking at the water molecule structure is to assume the use of all 2s and 2p electrons of oxygen to form initially four equivalent sp^3 hybrid orbitals at 109.47° to one another. Two of them overlap with H 1s orbitals and the remaining nonbonding hybrids, each containing two lone pair electrons, repel each other so as to modify the angles from the tetrahedral value to those observed. In this way, the water molecule is treated as a modified tetrahedron, similar to the Bernal model for the water molecule in liquid water.

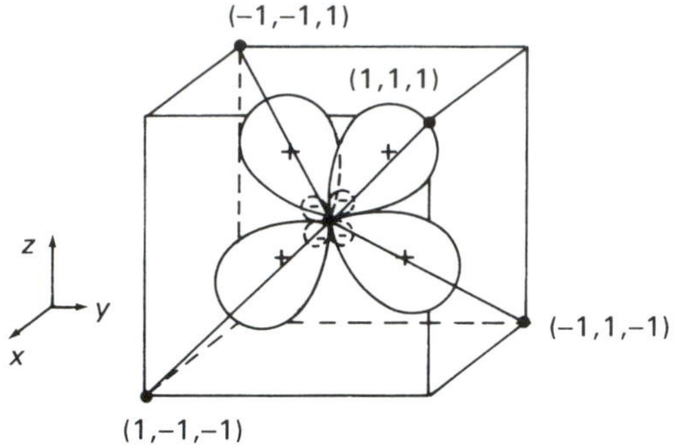

Figure 2.32 Methane sp^3 hybrid orbitals are directed to four corners of a cube, along the C_3 symmetry axes of the cube (and methane). Carbon is at the centre (the point 0,0,0), and by joining all pairs of hydrogen atoms at the apices (six lines) a regular tetrahedron may be outlined.

In his essentially qualitative approach, Bernal treated the water molecule as a regular tetrahedron with two charges of $+\frac{1}{2}$ and two of $-\frac{1}{2}$ at the four apices. The positive apices corresponded to the bonded hydrogen atoms and the negative apices to the nonbonding electrons (see also Section 3.6.7). This model can be useful in considering the role of the water molecule in ice, hydrates and other structures containing water molecules. We shall consider the water molecule when discussing group theory, particularly in Section 3.4.6.3.

2.11.14.2 Methane

The electron configuration for carbon is $(1s)^2(2s)^2(2p)^2$. Four bonds with hydrogen would need all the 2s and 2p electrons of carbon. In order that they may be equivalent, hybridization is invoked: since all four electrons are involved, plus the three 2p orbitals as well as the 2s, the s : p character will be in the ratio of 1 : 3; the orbitals are called sp^3 hybrid atomic orbitals. Following the previous arguments, the hybrid wavefunction may be written as

$$\psi(sp^3) = c_s\psi(s) + c_p\psi(p) \tag{2.146}$$

In order to achieve tetrahedral directions between the sp^3 hybrids, $c_p^2=0.75$ and $c_s^2=0.25$. Hybrid orbitals are often referred to as $s^{c_s^2}\,p^{c_p^2}$, which then becomes sp^3 in methane.

The energy needed to promote carbon to the valence state has been calculated to be approximately 4.0 eV, whereas the average energy (Section 4.5.1) of a C−H bond is 4.3 eV (-415 kJ mol^{-1}); hence, bond formation through hybrid orbitals is energetically feasible. The relationship of the sp^3 hybrid directions to a cube is shown in Figure 2.32. The hybrids lie along the body diagonals of the cube, the directions of the four C_3 symmetry axes; the bisectors of any pair of these directions give rise to three S_4 symmetry axes, and the lines joining the midpoints of opposite edges of the cube form six C_2 axes (see Chapter 3).

2.11.14.3 Delocalized systems

We have referred briefly to delocalization when discussing the water molecule. Here, we consider some simple unsaturated hydrocarbons, conjugated hydrocarbons and benzene as a representative of the aromatic system. We use these compounds also to introduce the important Hückel molecular-orbital (HMO) theory and some features that follow from it.

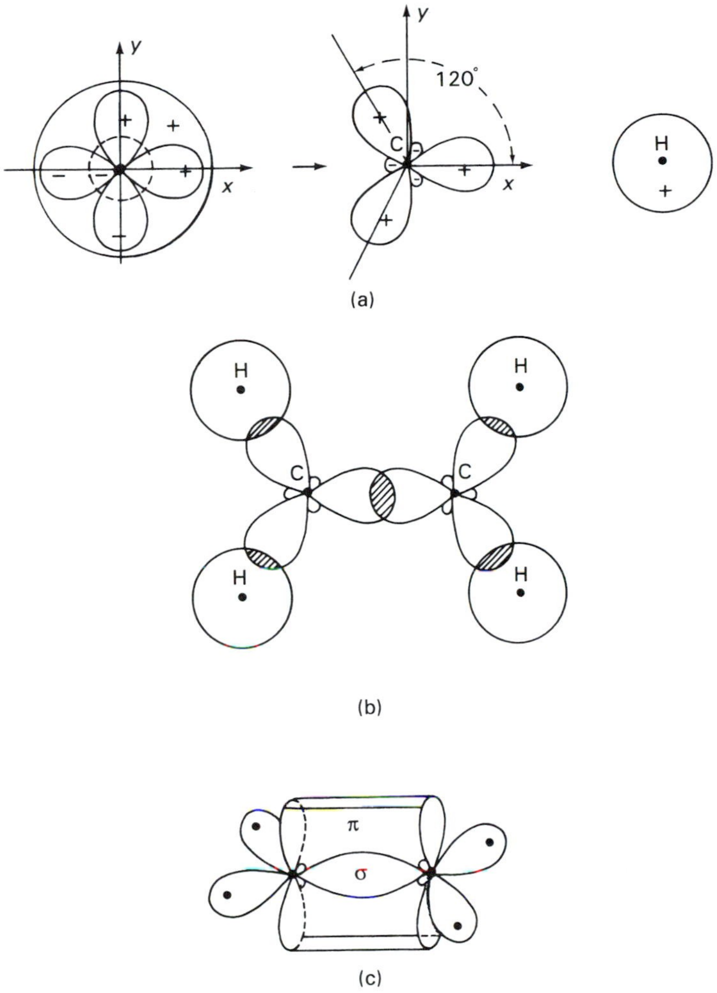

Figure 2.33 Ethene: (a) formation of carbon sp^2 hybrid orbitals, from 2s and 2p atomic orbitals, to bond with H(1s); (b) overlap of C(sp^2) with H(1s) to form C_2H_4; (c) σ bonds exist for C−C and C−H; the $2p_z$ atomic orbitals on the adjacent carbon atoms overlap to form a π molecular orbital.

Arguments similar to those given above show that sp^2 hybrid bonds, such as in ethene, have ideal 120° valence angles. The C−C and C−H bonds are σ type, and the $2p_z$ atomic orbital of carbon lies normal to the molecular plane. Two of these orbitals on adjacent carbon atoms overlap to form a π bond (Figure 2.33; see also Figure 2.25e). In ethyne, the C−C and C−H bonds are again σ, but two π bonds are now formed from the 2p atomic orbitals on adjacent carbon atoms (Figure 2.34). The single, double and triple C−C bonds in ethane, ethene and ethyne form an order of increasing average bond dissociation enthalpy[16] $\Delta\overline{H}_d$:

[16] Often called bond *energy* but, strictly, it is an enthalpy because it relates to the standard state of 298.15 K and 1 atmosphere.

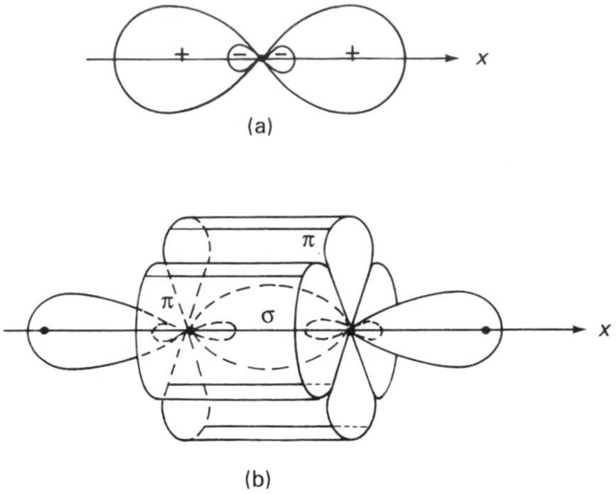

Figure 2.34 Ethyne: (a) linear sp hybrid orbitals for carbon; (b) overlap of C(sp) and H(1s) to form C_2H_2; a dot represents the overlap of H(1s). Two pairs of 2p orbitals overlap to form two π molecular orbitals.

	C_2H_6	C_2H_4	C_2H_2
$\Delta\overline{H}_d$/kJ mol^{-1}	+344	+613	+831
ΔH_f^{-0-}/kJ mol^{-1}	−84.6	+52.3	+227

The π bond is the source of reactivity in organic compounds: for example, the values above show that two single C−C bonds are more stable than one double bond by approximately 75 kJ mol^{-1}, which is an energetic prerequisite for the ready reactions of unsaturated compounds. The thermodynamic stability of these compounds lies in the reverse order, as the enthalpies of formation ΔH_f^{-0-} indicate. A $\sigma^2\pi^2$ configuration in multiple bonds confers a degree of rigidity on a molecule by inhibiting rotation about the σ bond. Thus, rotation can take place about the C−C bond in ethane, but not in ethene or ethyne; in ethene, geometrical isomers (*cis* and *trans*) of the type CA_2B_2 can exist.

2.12 Hückel molecular-orbital theory

Hückel molecular-orbital (HMO) theory treats the π orbitals in a multiple bond system and assumes that the σ orbitals form a rigid molecular core, rather like the nonbonding electrons of fluorine in hydrogen fluoride, for example. The carbon atoms are assumed to be identical, so that all Coulomb integrals α take the same value. Although of most interest in conjugated molecules, it is of value to develop the theory in terms of the simplest molecule with a multiple bond, ethene.

Let the two carbon atoms of ethene be identified by subscripts 1 and 2; then the wavefunction Ψ_π of the bonding π molecular orbital is given by

$$\Psi_\pi = c_1\psi_1 + c_2\psi_2 \tag{2.147}$$

where ψ_1 and ψ_2 represent $2p_z$ atomic orbitals of carbon. Following the variation method developed earlier (Section 2.11.1ff), the secular determinant may be written as

$$\begin{vmatrix} \alpha-E & \beta-ES \\ \beta-ES & \alpha-E \end{vmatrix} = 0 \tag{2.148}$$

In the Hückel approximation, three assumptions are made:

(a) all overlap integrals S can be equated to zero;
(b) all resonance integrals other than those between adjacent carbon atoms can be equated to zero;
(c) all other resonance integrals have the same value β in hydrocarbons.

Then (2.148) is reduced to

$$\begin{vmatrix} \alpha-E & \beta \\ \beta & \alpha-E \end{vmatrix} = 0 \tag{2.149}$$

for which the roots are

$$E_{\pm} = \alpha \pm \beta \tag{2.150}$$

E_+ corresponds to the fully occupied bonding π molecular orbital (2.147), whereas E_- corresponds to an empty π^* antibonding molecular orbital. We recall that both α and β are negative quantities: β is used as a measure of the delocalization energy of the π electrons in the molecule, and the largest value of β corresponds to the molecular orbital of lowest energy. The pair of molecular orbitals in (2.150) form the *frontier molecular orbitals* of ethene. More generally, the frontier molecular orbitals are the highest energy occupied molecular orbital (HOMO) and the lowest energy unoccupied molecular orbital (LUMO). Frontier molecular orbitals are important in studying chemical reactivity in unsaturated compounds.

An important molecule in HMO theory is buta-1,3-diene, C_4H_6; it is an example of a conjugated system, that is, a molecule containing, formally, alternating double and single bonds:

$$CH_2{=}CH{-}CH{=}CH_2$$

The experimentally determined molecular geometry shows that the bond lengths/nm are

$$C \overset{0.135}{\text{———}} C \overset{0.146}{\text{———}} C \overset{0.135}{\text{———}} C$$

from which it is clear that the double bonds are slightly longer than the standard double-bond length of 0.133 nm (ethene), whereas the single bond is shorter than the standard value of 0.154 nm (ethane). The π electrons are said to be partially delocalized over the molecule.

We can apply HMO theory to butadiene. Following the previous procedure, we may write a π molecular orbital wavefunction as

$$\Psi_\pi = c_1\psi_1 + c_2\psi_2 + c_3\psi_3 + c_4\psi_4 \tag{2.151}$$

With the approximations as before, the secular determinant is written as

$$\begin{vmatrix} y & 1 & 0 & 0 \\ 1 & y & 1 & 0 \\ 0 & 1 & y & 1 \\ 0 & 0 & 1 & y \end{vmatrix} = 0 \tag{2.152}$$

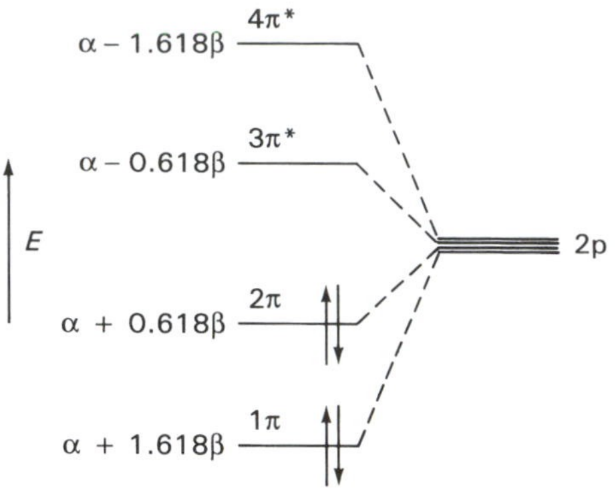

Figure 2.35 Molecular-orbital energy-level diagram for buta-1,3-diene; the four π electrons occupy the two lowest (bonding) molecular orbitals. The delocalization energy against ethene is 0.472β, or 33 kJ mol^{-1}.

where $y=(\alpha-E)/\beta$. Expansion of the determinant (see Appendix 9) leads to the equation

$$y^4-3y^2+1=0 \tag{2.153}$$

whence, through the substitution $p=y^2$, we obtain $y^2=2.618$ and 0.382, hence, the energies of the four π molecular orbitals are

$$E=\alpha\pm1.618\beta,\ \alpha\pm0.618\beta \tag{2.154}$$

Figure 2.35 is a molecular-orbital energy-level diagram for buta-1,3-diene. The four bonding π electrons occupy the 1π and 2π molecular orbitals, and the frontier orbitals are 2π (HOMO, bonding) and $3\pi^*$ (LUMO, antibonding).

2.12.1 Delocalization energy

The total π binding energy E_π is the sum of the energy of each state multiplied by the number of electrons in that state. For ethene, E_π is $2\alpha+2\beta$. In butadiene, we have

$$E_\pi=2(\alpha+1.618\beta)+2(\alpha+0.618\beta)=4\alpha+4.472\beta \tag{2.155}$$

If butadiene contained two ethenic double bonds, the π energy would be expected to be $4\alpha+4\beta$. Conjugation results in butadiene being more stable than a simple diene would be by 0.472β. This extra stability is called *delocalization energy* D_π; since β has been calculated to be approximately -76 kJ mol^{-1}, the delocalization energy in butadiene is about -36 kJ mol^{-1}. The final wavefunctions for butadiene require the coefficients c_1 to c_4 in (2.151), and their derivation is set out in Appendix 9. With the values obtained there, we have

$$\begin{aligned}\Psi_1&=0.3717\psi_1+0.6015\psi_2+0.6015\psi_3+0.3717\psi_4\\ \Psi_2&=0.6015\psi_1+0.3717\psi_2-0.3717\psi_3-0.6015\psi_4\\ \Psi_3&=0.6015\psi_1-0.3717\psi_2-0.3717\psi_3+0.6015\psi_4\\ \Psi_4&=0.3717\psi_1-0.6015\psi_2+0.6015\psi_3-0.3717\psi_4\end{aligned} \tag{2.156}$$

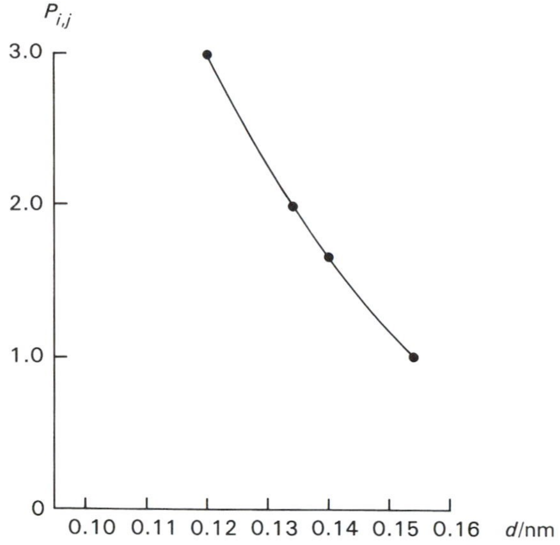

Figure 2.36 Plot of total bond order $P_{i,j}$ against bond length d, for orders 1, 1.667, 2 and 3 in ethane, benzene, ethene and ethyne respectively.

2.12.2 *π-Bond order*

The π electrons in conjugated hydrocarbons are not localized in pairs, so that classical double bonds do not exist as such. Each electron contributes to the bonding and the relative π bonding between pairs of adjacent atoms i,j is related to the coefficients c_n by the π-bond order $p_{i,j}$:

$$p_{i,j}=\Sigma_k \eta_k c_{ik} c_{jk} \qquad (2.157)$$

where η_k is the number of electrons in the kth occupied π molecular orbital, i in c_{ik} refers to atom i in molecular orbital k and j in c_{jk} refers to atom j in molecular orbital k, the sum being taken over all k occupied π molecular orbitals.

Applying this equation to the molecular orbital between atoms 1 and 2 of butadiene, noting that the antibonding orbitals are not occupied in this species, we have with (2.156)

$$p_{1,2}=2\times0.3717\times0.6015 + 2\times0.3717\times0.6015=0.8943$$

and in this way we can build up a picture for π-bond orders in butadiene ($p_{1,2}=p_{3,4}$):

$$\mathrm{CH_2}\overset{0.894}{\text{———}}\mathrm{CH}\overset{0.447}{\text{———}}\mathrm{CH}\overset{0.894}{\text{———}}\mathrm{CH_2}$$

If we assume a bond order of unity for the σ bonds, the total bond order $P_{i,j}$ becomes $1+p_{i,j}$, which is comparable with the net bonding parameter κ:

$$\mathrm{CH_2}\overset{1.894}{\text{———}}\mathrm{CH}\overset{1.447}{\text{———}}\mathrm{CH}\overset{1.894}{\text{———}}\mathrm{CH_2}$$

Bond order correlates well with bond length; Figure 2.36 has been drawn on the basis of the standard bond lengths given below. The π-bond orders calculated for butadiene interpolate at 0.136 nm and 0.144 nm for bond length, and agree well the experimental results.

i,j	Molecule	d/nm	P_{ij}
C−C	C_2H_6	0.154	1.00
C−C	C_6H_6	0.140*	1.667
C=C	C_2H_4	0.134	2.00
C≡C	C_2H_2	0.120	3.00

*Aromatic

2.12.3 Free-valence index

The free-valence index $\mathcal{F}_i$ for an atom i measures the extent to which that atom is bonded to its neighbours; a large value corresponds to weak bonding. This result has a bearing on the reactivity of a species. The free-valence index is defined by

$$\mathcal{F}_i = M_i - \Sigma_k P_{i,k} \tag{2.158}$$

where M_i is the maximum bonding power of the ith atom and the sum is taken over the k occupied orbitals. We can show that M_i=4.732 for carbon, with the known but unstable trimethylenemethane:

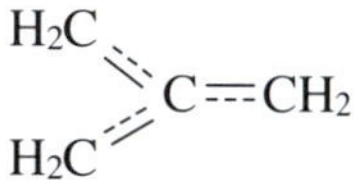

The central carbon atom has three σ bonds and three π bonds to its neighbours. An HMO calculation gives $1/\sqrt{3}$ for each π-bond order. Hence the total C−C bond order is $3+3/\sqrt{3}=4.732$. In butadiene,

$$\mathcal{F}_1 = \mathcal{F}_4 = 4.732 - (2+1+0.894) = 0.838$$

$$\mathcal{F}_2 = \mathcal{F}_3 = 4.732 - (3+0.447+0.894) = 0.391$$

thus completing our picture of the butadiene molecule:

$\mathcal{F}$	0.838		0.391		0.391		0.838
$P_{i,j}$		1.894		1.447		1.894	
	CH_2	——	CH	——	CH	——	CH_2
d_{calc}/nm		0.136		0.144		0.136	
d_{expt}/nm		0.135		0.146		0.135	

$$D_\pi = 0.472\beta$$

The preferential 1,4 addition of bromine to butadiene hinges on the higher free-valence index at these positions compared with that at the 3,4 positions.

$$Br_2 + CH_2{=}CH{-}CH{=}CH_2 \rightarrow CH_2Br{-}CH{=}CH{-}CH_2Br$$

2.12.4 Aromatic systems

Complete delocalization of the π electrons is achieved in benzene, C_6H_6. The six carbon atoms are linked to one another and to six hydrogen atoms by sp^2 σ-bonds that give the

C−C−C and C−C−H bond angles their characteristic value of 120° (Figure 2.37a). The remaining $2p_z$ atomic orbitals are directed normally to the molecular plane and overlap to give six π molecular orbitals, including a 'double streamer' (Figure 2.37b) all around the ring. Figures 2.37c–e show the three lowest energy, occupied molecular orbitals as seen in a direction normal to the molecular plane; Figure 2.37c is another view of Figure 2.37b.

The HMO treatment of benzene leads to the secular determinant

$$\begin{vmatrix} y & 1 & 0 & 0 & 0 & 1 \\ 1 & y & 1 & 0 & 0 & 0 \\ 0 & 1 & y & 1 & 0 & 0 \\ 0 & 0 & 1 & y & 1 & 0 \\ 0 & 0 & 0 & 1 & y & 1 \\ 1 & 0 & 0 & 0 & 1 & y \end{vmatrix} \tag{2.159}$$

where $y=(\alpha-E)/\beta$. Solution of this determinant (Appendix 9) gives the six values

$$E=\alpha\pm2\beta,\ \alpha\pm\beta,\ \alpha\pm\beta \tag{2.160}$$

Figure 2.38 is a π molecular-orbital energy-level diagram for benzene; the $\alpha+\beta$ (HOMO) and $\alpha-\beta$ (LUMO) energy states each have a twofold degeneracy. If we compare benzene energetically with a formal triene, the benzene delocalization energy is

$$D_\pi=2(\alpha+2\beta)+4(\alpha+\beta)-3(2\alpha+2\beta)=2\beta \tag{2.161}$$

which is approximately -152 kJ mol^{-1}. The stable nature of aromatic compounds arises both from the strain-free character of the benzene ring and because the complete delocalization of the π electrons, with all π electrons in bonding molecular orbitals, leads to large values of the delocalization energy.

The π-bond order calculations show that all C−C bonds in benzene are equivalent, with a total π-bond order of 1.667 each. The free-valence index for each atom is 0.399, indicating low reactivity; it may be compared with the values of 0.730 for ethene, 0.894 for butadiene and 1.15 for the unstable trimethylenemethane. We should note that the free-valence index refers to a molecule in isolation, whereas reactivity depends also upon the energy of activation and kinetics for the reaction in question.

2.12.5 Charge distribution

If a carbon atom forms three σ bonds and is also π bonded, it will remain neutral if there is an average of 1 electron in its π orbital. The charge distribution q_i may be taken as the deviation from neutrality for the ith atom and is defined by

$$q_i=1-\Sigma_k\eta_k c_{ik}^2 \tag{2.162}$$

where η_k is the number of electrons in the kth occupied molecular orbital Ψ_k, c_{ik} is the coefficient of ψ_i in Ψ_k and the sum is taken over the k occupied molecular orbitals. It is left as an exercise for the reader to show that, for benzene, the π-bond orders $p_{i,j}$ are $\frac{2}{3}$, with a charge q_i of zero and the free-valence index $\mathcal{F}_i$ of 0.399 for each atom, given the following final π molecular-orbital wavefunctions (see also Problem 3.8):

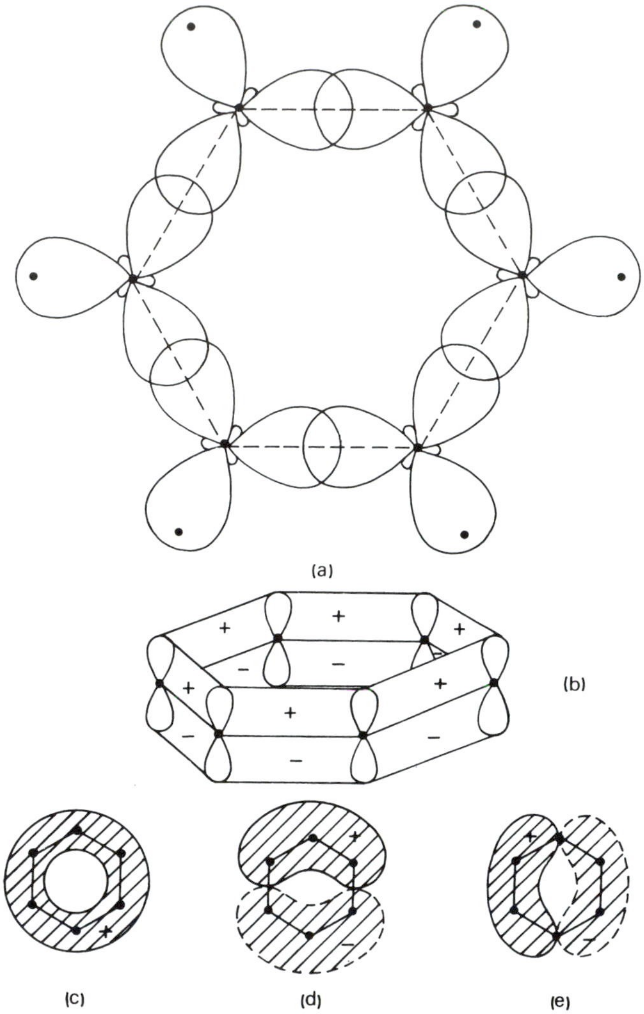

Figure 2.37 Bonding molecular orbitals for benzene: (a) $sp^2\sigma$-bonds for C−C and C−H; the $2p_z$ atomic orbitals are directed normal to the molecular plane, and a dot represents the overlap of H(1s). (b) Double streamer π molecular orbital $(\alpha+2\beta)$ formed by the overlap of the $2p_z$ orbitals; (c), (d), (e) the same double streamer and the two degenerate π molecular orbitals of next higher energy $(\alpha\pm\beta)$ respectively, shown normal to the molecular plane.

$$\begin{aligned}
\Psi_1 &= (1/\sqrt{6})(\psi_1+\psi_2+\psi_3+\psi_4+\psi_5+\psi_6)\\
\Psi_2 &= (1/\sqrt{12})(2\psi_1+\psi_2-\psi_3-2\psi_4-\psi_5+\psi_6)\\
\Psi_3 &= (1/2)(-\psi_2-\psi_3+\psi_5+\psi_6)\\
\Psi_4 &= (1/\sqrt{12})(2\psi_1-\psi_2-\psi_3+2\psi_4-\psi_5-\psi_6)\\
\Psi_5 &= (1/2)(-\psi_2+\psi_3-\psi_5+\psi_6)\\
\Psi_6 &= (1/\sqrt{6})(\psi_1-\psi_2+\psi_3-\psi_4+\psi_5-\psi_6)
\end{aligned} \tag{2.163}$$

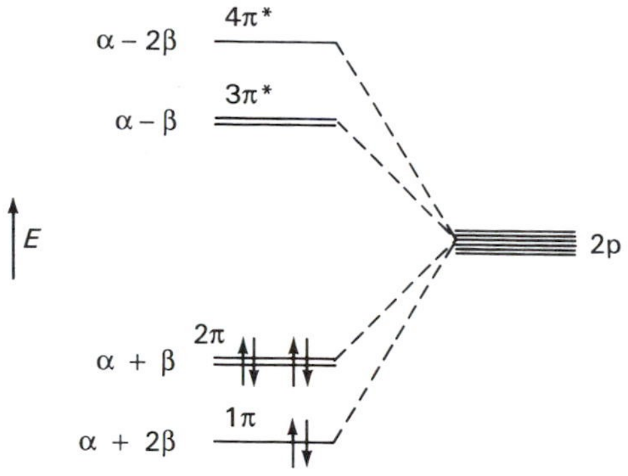

Figure 2.38 Molecular-orbital energy-level diagram for benzene: $E_\pi=6\alpha+8\beta$, and $D_\pi=2\beta$, or $-152\ \text{kJ mol}^{-1}$.

The HMO charge distributions for two other aromatic molecules, aniline and pyridine, are shown below:

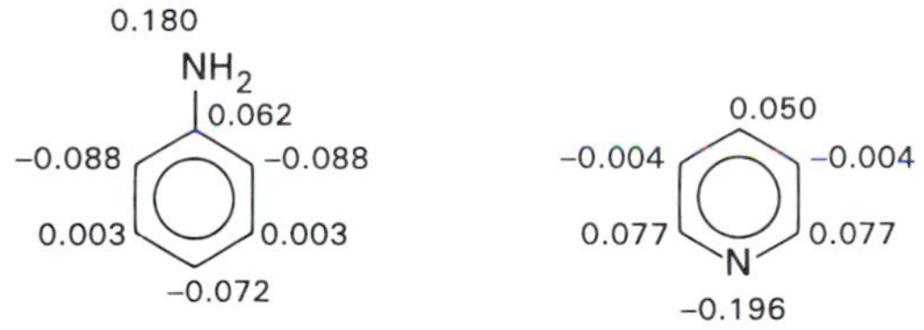

Formulae for aniline and pyridine

Aniline reacts with bromine water to give 2,4,6-tribromoaniline. The active brominating agent in bromine water is the Br^+ species, which seeks preferentially the relatively negative 2, 4 and 6 centres. In a similar way the nitration of pyridine by the nitronium ion, NO_2^+, leads to 3-nitropyridine; a nucleophilic reagent, however, seeks the 2 position in pyridine:

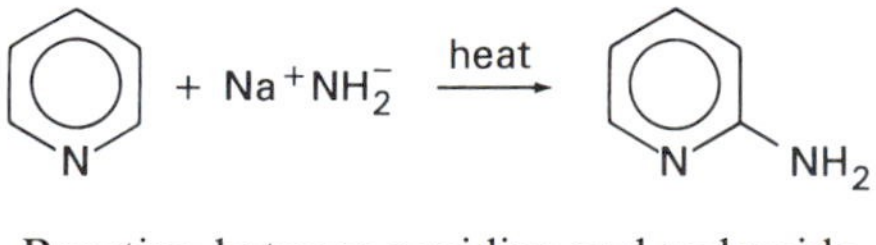

Reaction between pyridine and sodamide

2.13 Valence-shell electron-pair repulsion theory

The valence-shell electron-pair repulsion theory (VSEPR) model attempts to predict the shapes of simple polyatomic molecules in terms of the repulsion between pairs of electrons, particularly lone pairs, on the component atoms. It begins with the Lewis bonding-pair/lone-pair model and assumes that a molecule adopts that shape which minimizes the repulsions between pairs of electrons: in other words, it seeks to place electron pairs as far apart as possible.

In a molecule of the type MX_n each electron pair may be represented by a point on a sphere with the species M at its centre, and the positions of the n X species arranged so as to minimize the repulsive energy. The basic structural arrangements are as follow:

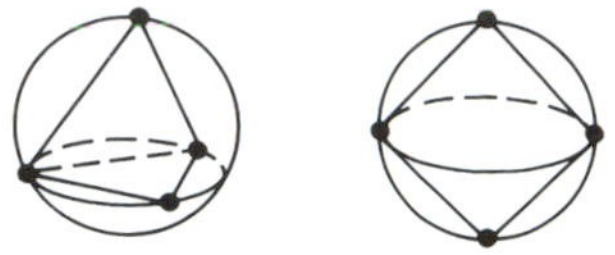

(a) (b)

Figure 2.39 VSEPR point-charge model for MX_4 species: (a) preferred tetrahedral configuration; (b) square-planar configuration. The separation of adjacent points in (a) and (b) is in the ratio of 1.15 : 1.

Number of electron pairs	Configuration
2	Linear
3	Trigonal planar
4	Tetrahedral
5	Trigonal bipyramidal
6	Octahedral

Figure 2.39 shows the tetrahedral arrangement of four electron pairs. An alternative square-planar arrangement is possible, but it would place the electron pairs closer together.

The model may be refined by allowing each electron pair to take up the space of a sphere, in order to imitate a fully occupied atomic orbital. It then has the advantage of showing domains in space that may not be overlapped by other similar domains. The point-charge model adopts the following order for repulsion of electron pairs:

$$\text{lone pair/lone pair} > \text{lone pair/bonding pair} > \text{bonding pair/bonding pair}$$

The domain model then superimposes qualitative size factors on to this energy order. The formula for molecules may be given generally as MX_nL_m, where L_m now refers to m lone pairs on the species M. We note the following conditions:

(a) a bonding domain involves the M and X valence electrons, both of which are attracted to the nuclei; the nonbonding L domain belongs only to M;
(b) classical double and triple bonds involve two and three shared pairs respectively; the sizes of the domains increase in the order single bond<double bond<triple bond;
(c) electronegativities govern the extent to which electrons may be transferred from M to X.

A lone pair domain L is a larger sphere than a bonding domain X and, because it belongs to M, it will tend to be closer to M than are the X species. If we consider NH_3 (MX_3L type) and H_2O (MX_2L_2type), the total of four domains would be expected to lead to a tetrahedral arrangement. However, if the L domain is larger than that for X and more strongly attracted to M, then the X−M−X angle will be less than the tetrahedral value. In fact for NH_3 it is 107°, whereas for H_2O it is 104.4° as we have discussed already.

The molecules of SF_2 and SCl_2 (MX_2L_2 type) resemble the water molecule, so we expect distorted tetrahedral arrangements of M, X and L. Furthermore, because fluorine is more electronegative than chlorine, it will draw relatively more charge from the sulfur atom; this in turn leads to a smaller bond angle in SF_2 (SF_2 98°, SCl_2 102°). The point-charge VSEPR model would postulate that the type MX_2 is formally linear, but that the lone pairs on M would tend to repel the X species and so produce a bent molecule. The result is similar, except that it does not permit comparison of the relative sizes of the X−M−X angle.

Sulfur dioxide has multiple bonds, with the formal structure shown. Thus, it may be classed as an MX_2L species; larger domains are assumed for multiple bonds and lead to a bond angle of more than the 120° found in the trigonal-planar arrangement. In the case of SO_3 the symmetry is restored and the trigonal-planar shape is obtained. In carbon dioxide, multiple bonds give rise to large bonding domains. However, there are no lone pairs to disturb the symmetry of a linear arrangement for this molecule, which corresponds to minimum-energy configuration.

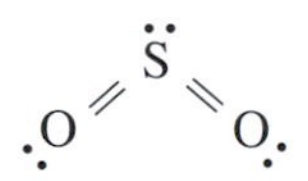

VSEPR is a simple method for predicting the shapes of small molecules containing a central atom. It is particularly applicable to main-group elements in the periodic table, but less satisfactory with transition-metal compounds because the central atom domain does not retain an implicitly spherical shape. The d^0, d^5 and d^{10} configurations, however, do respond to the simple treatment. For example, $TiCl_4$ (d^0) is predicted to be tetrahedral, $[CoF_6]^{2-}$ (d^5) to be octahedral, and $[Ag(NH_3)_2]^+$ (d^{10}) linear. The VSEPR model is discussed in most modern books on inorganic chemistry.[17]

2.14 Ligand-field theory

Ligand-field theory is used to describe the structures of transition-metal compounds in which the bonding to ligand groups takes place through the d orbitals of the central metal atom. The first transition series from scandium to zinc is characterized by the change in configurations $3d^1$ to $3d^{10}$ as the d orbitals are progressively filled. The d orbitals were described in Section 2.7.1, their wavefunctions listed explicitly in Table 2.3 and the shapes of their $|\psi|^2$ density functions given in Figure 2.11.

Ligand-field theory is a molecular-orbital theory, with molecular orbitals built from the atomic orbitals of ligands and the d orbitals of a central atom that has a very symmetrical (often cubic) environment. A typical transition-metal compound, also called transition-metal complex or coordination compound, is hexamminecobalt(III) chloride $[Co(NH_3)_6]Cl_3$. This compound has the octahedral symmetry $\mathbb{O}_h$ ($m3m$), and the environment of the cation is shown in Figure 2.40.

In crystal-field theory, an approximation to ligand-field theory, the electrostatic effects of the ligands are considered in terms of electron repulsions, particularly with respect to lone pairs, which are particularly important on the nitrogen atoms in the example compound. Ligand-field theory encompasses both the electrostatic effects of the crystal field and the molecular orbital properties that we have discussed in this chapter.

It is clear from Figures 2.11 and 2.40 that the d electrons of the metal atom in the d_{z^2} and $d_{x^2-y^2}$ orbitals have less favourable energies for (greater repulsions to) the approach of lone pairs than do the electrons in the other three d orbitals, because they are directed towards the ligand positions. The result is that the d orbitals of the metal atom are split into a group of two, the e_g orbitals, and a group of three, the t_{2g} orbitals, of which the t_{2g} has the lower energy. The significance of this notation will become clearer in the next chapter.

The energy difference between the e_g and t_{2g} levels is called Δ, or $10Dq$, the ligand-field splitting energy parameter. Figure 2.41 illustrates the situation for a d orbital compound in the field of a ligand.

[17] See, for example, R J Gillespie and I Hargittai, *The VSEPR Model of Molecular Geometry* (Prentice-Hall, 1991); D F Shriver, P W Atkins and C H Langford, *Inorganic Chemistry* (OUP, 1990).

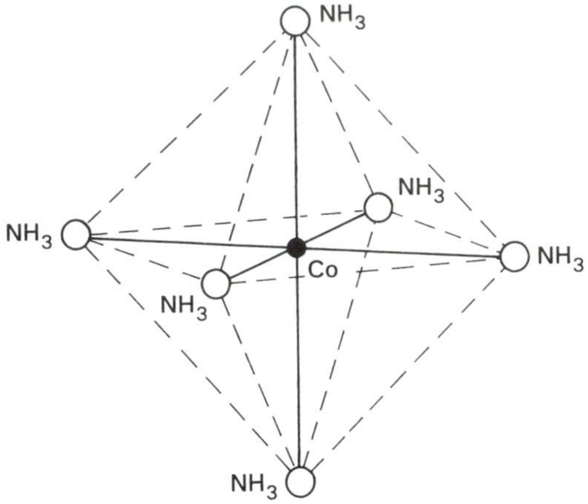

Figure 2.40 Geometry of the hexamminecobalt(III) ion, $[Co(NH_3)_6]^{3+}$; the dashed lines form the edges of a regular octahedron, point-group symmetry $\mathbb{O}_h$ (see Figure 2.11c). The Co−NH_3 ligand directions are the axes of the bond-forming d orbitals.

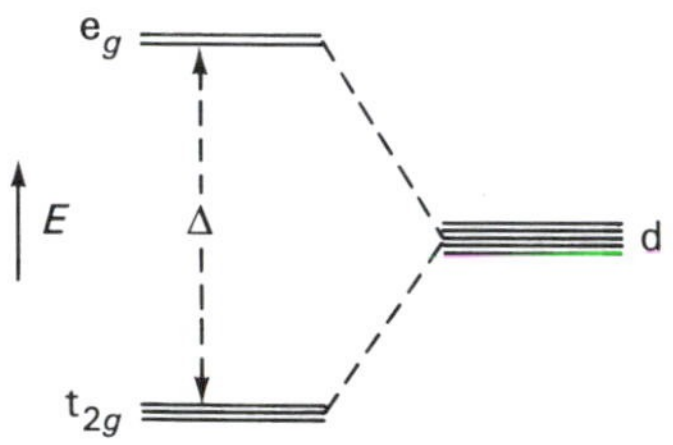

Figure 2.41 Molecular-orbital energy-level diagram for the five degenerate d orbitals of a transition metal that are split by a ligand field. A group of three degenerate levels, the t_{2g}, and a group of two degenerate levels, the e_g, are formed; the energy difference is Δ (also called $10Dq$), the ligand-field splitting energy parameter.

The approach of the six NH_3 ligands to the positively charged cobalt ion is an energetically favourable process. The t_{2g} molecular orbitals are occupied preferentially, and the *Aufbau* principle predicts that they will be occupied by the six d electrons of cobalt, with paired spins. There is, however, an important dependence on the magnitude of Δ.

The electrons from the six lone pairs on the nitrogen atoms and the d electrons of cobalt are fed into the bonding molecular orbitals a_{1g}, t_{1u} and e_g, in that order of increasing energy. The electrons cannot actually be localized according to the crystal-field approximation: the molecular orbitals of octahedral symmetry are spread over the whole complex ion and the electrons are delocalized among them.

Six bonds are formed with the ammonia ligands and the remaining six electrons occupy nonbonding t_{2g} molecular orbitals on the metal. If Δ is large and the strength of the field of the ligands is not high, as with ammonia, the nonbonding t_{2g} molecular orbitals are occupied by six pairs of electrons. It follows that the hexamminecobalt(III) ion will be diamagnetic because there are no unpaired electron spins. The configuration of cobalt is given here as $(t_{2g})^6$. This type of compound is often called a *strong-field* or *low-spin* complex; it is characterized by a *ligand-field stabilization energy* (LFSE) of approximately 2.5 eV, or 241 kJ mol^{-1}.

In a weak-field compound, with ligands such as Br^- or SCN^-, the LFSE may be only about one-fifth of the above value. It is then commensurate with the repulsion energy

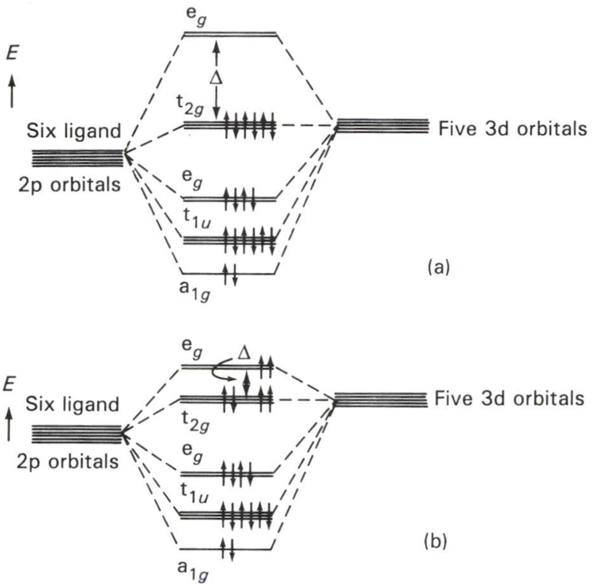

Figure 2.42 Molecular-orbital energy-level diagram for octahedral coordination. (a) Low-spin case, such as $[Co(NH_3)_6]^{3+}$. The bonding orbitals a_{1g}, t_{1u} and e_g are fully occupied by paired electrons, as are the nonbonding t_{2g} molecular orbitals; the species is diamagnetic, with the configuration $(t_{2g})^6$. (b) High-spin case, such as $[Co(SCN)_6]^{3-}$: the smaller splitting energy parameter Δ is close to the repulsion energy of paired electrons, and Hund's rule leads to singly occupied, nonbonding e_g molecular orbitals. The configuration for the species is $(t_{2g})^4\,(e_g)^2$ and it is paramagnetic. The unoccupied nonbonding molecular orbitals are present in both examples, but are of little interest in the ground state.

between paired electrons, so that a configuration with orbitals singly occupied by electrons having parallel spins may be preferred. Then, the six nonbonding electrons would enter the t_{2g} and e_g orbitals with, say, one t_{2g} orbital fully occupied and two t_{2g} and two e_g orbitals each with one unpaired electron, a high-spin configuration leading to paramagnetism. The molecular orbitals that we have described are summarized in Figures 2.42a and 2.42b.

Although placed here, this section may usefully be re-read after the study of group theory in the next chapter.

2.14.1 Magnetic properties

A magnetic moment may arise either from electron spin or from a combination of spin and orbital motion. Table 2.9 indicates the relationship between numbers of unpaired electrons and the total theoretical and observed magnetic moments for a selection of transition-metal species.

Another common type of complex is tetrahedral, as in the $[Zn(NH_3)_4]^{2+}$ ion (point-group symmetry $\mathcal{T}_d$, or $\bar{4}3m$), for example. We shall consider this type briefly by means of a problem; the reader is directed to more detailed texts for further discussion of coordination compounds.[18]

[18] See, for example, D F Shriver, P W Atkins and C H Langford *Inorganic Chemistry* (OUP, 1990); F A Cotton and G Wilkinson *Comprehensive Inorganic Chemistry* (Wiley, 1988), and S F A Kettle *Physical Inorganic Chemistry* (Spektrum, 1996).

Table 2.9 Theoretical and experimental magnetic moments

Species	Number of unpaired e^-	μ_B(spin)	μ_B(spin+ orbital)	μ_B(experiment)
V(IV)	1	1.73	3.00	1.7–1.8
Cu(II)	1	1.73	3.00	1.7–2.2
V(III)	2	2.83	4.47	2.6–2.8
Ni(II)	2	2.83	4.47	2.8–4.0
Cr(III)	3	3.87	5.20	≃3.8
Co(II)	3	3.87	5.20	4.1–5.2
Fe(II)	4	4.90	5.48	5.1–5.5
Co(III)	4	4.90	5.48	≃5.4
Mn(II)	5	5.92	5.92	≃5.9
Fe(III)	5	5.92	5.92	≃5.9

Note:
Values are in Bohr magneton units: 1 $\mu_B = e\hbar/(2m_e) = 9.274\times10^{-24}$ J T^{-1}. The Mn(II) and Fe(III) species have no orbital component of magnetic moment.

The mass magnetic susceptibility[19] χ of a species is given, by analogy with the Langevin–Debye equation[20] for polarization, by

$$\chi = \mathcal{N}\mu_0[\zeta + m^2/(3k_BT)] \tag{2.164}$$

where $\mathcal{N}$ is the number density of the species in m^{-3}, μ_0 is the permeability of a vacuum in H m^{-1} (J C^{-2} s^2 m^{-1}), ζ is the magnetizability and m is the magnetic moment of the species, both in J T^{-1} (C m^2 s^{-1}). In practice, since $\mathcal{N} = N_A/V_m$, where V_m is the molar volume, we multiply both sides of (2.164) by V_m and refer to a derived molar magnetic susceptibility χ_m.

The magnetic susceptibility may be measured with a Gouy balance. In this method, a sample is suspended from a sensitive balance between the poles of a magnet. A paramagnetic sample becomes oriented with the field, whereupon its apparent weight is increased. The balance must be calibrated with a material of known magnetic susceptibility. The magnetic moment can be obtained practically by plotting values of χ_m, from measurements at different temperatures, against $1/T$, according to (2.164). The inverse dependence on temperature is known as the (empirical) Curie law.

In an experiment with a cobalt compound, the graph of χ_m against $1/T$ had a slope of 3.18×10^{-5} K. From (2.164), $m = 4.17\times10^{-23}$ J T^{-1} or, in units of the Bohr magneton, 4.5. From Table 2.9, this value indicates cobalt(II), with three unpaired electrons.

At a low temperature known as the Curie point, a paramagnetic compound may undergo a phase transition to a state in which large domains exhibit parallel spins. This alignment is called *ferromagnetism* and gives rise to strong magnetization. Alternatively, the spins may be aligned in an alternating manner (*antiferromagnetism*) which has zero magnetization; the temperature at which the antiferromagnetic transition takes place is called the Néel temperature.

A compound with no unpaired electrons is diamagnetic; in a Gouy balance experiment a diamagnetic sample moves out of the field and appears to weigh less. All molecules have a diamagnetic contribution to their susceptibilities, but if unpaired electrons are present the diamagnetic component is swamped by the paramagnetism.

[19] Volume magnetic susceptibility=mass magnetic susceptibility×density.

[20] See, for example, M F C Ladd, *Structure and Bonding in Solid State Chemistry* (Ellis Horwood, 1979).

2.15 Apparently abnormal valence

Certain compounds appear to exhibit abnormal valences. For example, silicon in the $[SiF_6]^{2-}$ ion is often stated to have an 'expanded octet' of 12 valence electrons. In the boron hydrides, there appear to be insufficient valence electrons for the number of bonds formed. These topics are explained conveniently by molecular-orbital theory.

Although the electronic configuration of silicon is $(Ar)(3s)^2(3p)^2$, this atom may make use of 3d orbitals in bond formation; the energy of these levels is not too high. However, the involvement of d orbitals is not essential, because there are sufficient atomic orbitals from the Si and the six F atoms to form six occupied, bonding molecular orbitals. The ability to pack the six fluorine atoms around the central silicon atom, without large repulsion effects, is also a significant factor.

Consider next the so-called electron deficient compound B_2H_6. From the atomic orbitals of the eight atoms a total of 14 molecular orbitals can be drawn up, four from each boron atom, excluding the use of $(1s)^2$ core, and one from each hydrogen atom. There can then be seven bonding and seven antibonding molecular orbitals. The 12 valence electrons then fill the six lower bonding molecular orbitals in accordance with the *Aufbau* principle, which results in a bonding situation. In the actual structure, there are two B,H,B molecular orbitals that give rise to two three-centre, two-electron bridging bonds. The theory provides neat explanations for compounds such as these, which were insuperable obstacles for the Lewis electron-pair hypothesis, and also not easily resolvable by the valence-bond model.

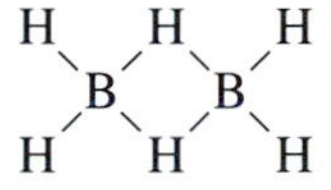

Problems 2

2.1 A particle of mass 0.1 kg, initially at rest, receives an instantaneous force of 1 N which causes it to accelerate along a straight-line path. What are its trajectory (x,p) and kinetic energy after 2 s?

2.2 Calculate the energy and momentum of an electron travelling at one tenth of the speed of light *in vacuo*, if the energy is wholly kinetic.

2.3 At what wavelength is the energy density for a black-body radiator a maximum at 500 K?

2.4 Determine the energy density of light in the wavelength range 780 nm to 800 nm in a black-body cavity at 1500 °C. Is it important to use the Planck equation rather than the Rayleigh–Jeans equation for this calculation?

2.5 Millikan exposed a freshly cut surface of sodium metal, in a vacuum, to monochromatic radiation from a quartz–mercury arc source. The photoelectrons emitted were collected in an oxidized- copper Faraday cylinder. The current obtained for different values of an applied potential difference V was measured with an electrometer.

In different experiments differing wavelengths λ were employed. The deflection θ recorded by the electrometer was proportional to the photoelectric current, and θ increased as V was made more negative at the metal. A field was set up in the space between the dissimilar sodium and oxidized-copper metals, and a contact potential difference existed between them; in Millikan's experiments, it acted from sodium to copper, that is, sodium was positive with respect to the Faraday cylinder. The following results were obtained for three wavelengths:

λ=546.1 nm		λ=365.0 nm		λ=312.6 nm	
V/V	θ/deg	V/V	θ/deg	V/V	θ/deg
−2.257	28	−1.157	67.5	−0.5812	52
−2.205	14	−1.105	36	−0.5288	29
−2.152	7	−1.0525	19	−0.4765	12
−2.100	3	−1.0002	11	−0.4242	5.7
		−0.9478	4	−0.3718	2.5

(a) For each wavelength, plot θ against the independent variable V, and estimate the *minimum* applied voltage V_0 that prevents the fastest-moving photoelectrons from reaching the Faraday cylinder.

(b) An electron moving through a potential difference V acquires an energy of eV eV. This energy is equivalent to the kinetic energy $\frac{1}{2}m_e v^2$, where m_e and v are, respectively, the mass and speed of the electron. Show that an equation $V_0 e = k\nu - \phi$ represents the variation of V_0 with frequency ν. Find the value of the constant k. What does k represent in this context?

2.6 X-rays of wavelength 100 pm, incident upon a material, are scattered with a Compton angle of 45°. What is the wavelength of the scattered radiation?

2.7 In a certain electron diffraction experiment, electrons of wavelength 0.5 nm were found to be desirable. What would be the velocity of such electrons?

2.8 The Balmer series in the spectrum of atomic hydrogen was analysed originally in terms of wavelength λ, through the equation $\lambda = K[n^2/(n^2-4)]$, where K is a constant and n is an integer greater than 4. Show that this equation is equivalent to (2.26) and find K in terms of R_H. What is the energy associated with the spectral line in the Balmer series, nearest to the red end of the spectrum (Figure 2.3)?

2.9 A proton and an electron, taken as point charges, are held at a distance from each other no greater than 10^{-15} m by Coulombic forces. Is this system feasible in the light of the uncertainty principle?

2.10 Calculate the probability that a particle in a one-dimensional box of length 20 nm will lie between (a) 5 nm and 15 nm and (b) 9 nm and 11 nm, for (i) the ground state and (ii) the first harmonic (excited) state of the function.

2.11 What is the smallest value of the kinetic energy for an electron in a cubical box of side 10^{-15} m?

2.12 Calculate the probability of finding a hydrogen 1s electron in the volume bounded by $r=1.10a_0$ to $1.11a_0$, $\theta=0.20\pi$ to 0.21π and $\phi=0.60\pi$ to 0.61π. The $\psi(1s)$ wavefunction may be assumed to be constant over the small volume considered. Use the value of the wavefunction given in Table 2.3 and the integral formula (A12.7) from Appendix 12.

2.13 Write the electron configurations for (a) N, (b) Al, (c) Cl^- and (d) K.

2.14 Calculate the energy of the species He^+ in the ground state.

2.15 Set up as fully as possible, but do not attempt a solution of, the Schrödinger equation for the helium atom.

2.16 By considering the orthogonality criterion determine whether or no, and under what restraints if any, the linear combination of hydrogen-like atomic orbitals (a) 1s+2s and (b) 1s+2p will lead to bonding molecular orbitals.

2.17 The overlap integral for two hydrogen-like 1s atomic orbitals is given by $S_{1s,1s} = \exp(-\rho)(1+\rho+\rho^2/3)$, and for 1s and 2p it is $S_{1s,2p} = \rho\exp(-\rho)(1+\rho+\rho^2/3)$, where ρ is r/a_0, r being the internuclear distance. Plot both $S_{1s,1s}$ and $S_{1s,2p}$ for $r=0$ to $3a_0$. Determine the value of the overlap integral $S_{1s,1s}$ for (a) H_2^+ (r_e=0.11 nm) and (b) H_2 (r_e=0.074 nm). (c) At what value of (r/a_0) is the 1s–2p overlap a maximum? Confirm this result by differentiation of the appropriate overlap function.

2.18 What are the ground state electronic configurations for (a) Be_2 and (b) C_2. Which of these species is likely to be more stable than the corresponding separate atoms?

2.19 Draw molecular-orbital energy-level diagrams for (a) NO and (b) CN, and give the ground state electronic configuration and net bonding parameter for each species. Which of the species is likely to be more stable as either a singly negative or a singly positive ion? Give reasons.

2.20 The dipole moment of the water molecule is 1.8 D. Given that O–H=0.096 nm and H–O–H=104.4°, calculate the partial charges on oxygen and hydrogen.

2.21 Show that sp^2 hybridization of atomic orbitals leads to σ bond angles of 120°.

2.22 Show that an sp^x hybrid from normalized atomic orbitals has a normalized wavefunction given by

$$\psi(sp^x)=(1+x)^{-1/2}[\psi(s)+x^{1/2}\psi(p)].$$

2.23 (a) Use the result from Problem 2.22 to show that an sp^3 hybrid orbital has a normalized wavefunction given by

$$\psi(sp^3)=\tfrac{1}{2}[\psi(s)+\sqrt{3}\psi(p)]$$

(b) Refer to Figure 2.32. For the corner 1,1,1, p may be resolved into components p_x, p_y and p_z. Show that

$$\psi(sp^3)=\tfrac{1}{2}(s+p_x+p_y+p_z)$$

using s to represent the $\psi(s)$ atomic orbital, and so on. Formulate the other three sp^3 wavefunctions in a similar manner and show that they are mutually orthogonal.

2.24 Determine and solve the secular determinant for cyclobutadiene – remember that atoms 1 and 4 are adjacent. Determine the energies E_π of the first four π molecular orbitals and, hence, find the delocalization energy D_π for cyclobutadiene. Which π orbitals constitute the frontier molecular orbitals in this molecule?

2.25 The polymethene dye

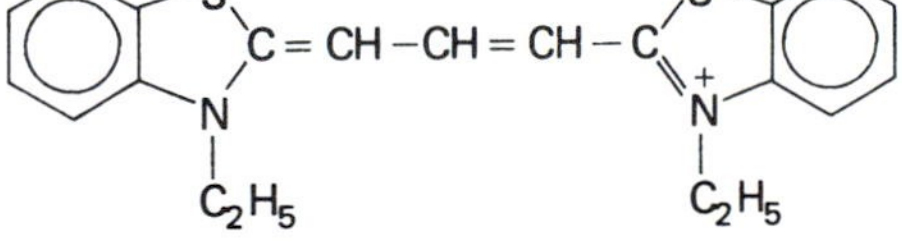

may be treated as the 'box'

$$-\ddot{N}-C=C-C=C-C=\overset{+}{N}-$$

where the mean bond length is 140 pm. The box contains $2N+2$ π-electrons (two from each double bond and two from the neutral nitrogen atom), where N is the number of double bonds; these electrons occupy the first $N+1$ molecular orbitals. The colour of the dye arises from the transition of an electron between the $N+1$ and $N+2$ orbitals. Show that the wavelength of the transition is given by

$$\lambda=3.297\times10^{12}L^2/(2N+3)$$

where L is the length of the box. Calculate λ and state the colour of the dye.
2.26 Write down the molecular orbital configurations for Ne_2 and LiH, and comment upon the results.
2.27 Calculate the percentage of ψ(S,3s) character in hydrogen sulfide, given that the H–S–H bond angle is 92°.
2.28 Use the HMO approximation to obtain the complete wavefunctions, molecular orbital and delocalization energies, bond orders, free-valence parameters and charge distributions for bicyclobutadiene:

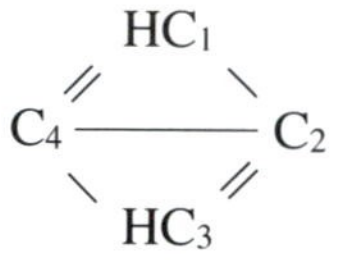

2.29 For the transition-metal ions $(3d)^1$ to $(3d)^9$ write down the d electron configuration and number of unpaired electrons in (a) a weak ligand field and (b) a strong ligand field.
2.30 $[Zn(NH_3)_4]^{2+}$ has regular tetrahedral symmetry (point group $\mathcal{T}_d$, or $\bar{4}3m$). The d orbitals are split into e_g and t_{2g} as before; $\Delta_{tet} \simeq 0.44\Delta_{oct}$ and is less than about 0.8 eV. By considering how a regular tetrahedron is related in symmetry to an octahedron (both can be placed inside a cube so that corresponding symmetry axes coincide), and by comparison with Figure 2.40, show what d orbital splitting is to be expected for tetrahedral complexes. Would the zinc compound be diamagnetic or paramagnetic?
2.31 A coordination compound has a density of 2512 kg m^{-3} and a molar mass of 0.2822 kg mol^{-1}. Its magnetic susceptibility at 298 K was found to be 1.64×10^{-3}. Given that the diamagnetic susceptibility is negligible, determine the probable number of unpaired electrons in the species.
2.32 Use VSEPR theory to predict probable structures for (a) F_2O, (b) NF_3, (c) $SiCl_4$ and (d) SF_4.

Determination of structure

3.1 Introduction

The determination of the structures of chemical species is an important and absorbing aspect of physical chemistry. Many methods have been devised as the subject of structure determination has developed, but among the most rewarding techniques currently in practice are those based either on spectroscopic or on diffraction methods. The extent of each of these fields is considerable, and here we shall introduce only parts of them. It is important in structure determination to acquire a knowledge of symmetry, and the application of symmetry principles in chemistry is far reaching. We give first a brief introduction to the symmetry of three-dimensional, finite bodies and its application in chemistry.

3.2 Symmetry concepts

We can see evidence of symmetry all around us; it is not just a feature of molecules and crystals. The emblems associated with the National Westminster Bank plc, the Isle of Man and Mercedes-Benz cars all reveal threefold symmetry; the latter shows other symmetry as well. The splendid Dobermann in Figure 3.1 illustrates reflection symmetry about a vertical, medial plane through her.

A *symmetry operation* applied to a molecule, or other body, moves it into a state that is indistinguishable from its initial state, thereby *revealing* the symmetry inherent in the body. We may link a symmetry operation with a *symmetry element*, which is a conceptual geometrical entity (point, line or plane) with respect to which a symmetry operation may be said to be performed. A *symmetry operator* is the symbol for a symmetry operation, and may be represented mathematically by a matrix. For example, in three-dimensional space, a threefold rotation operation may be associated with a line, say the z axis, about which the operation may be envisaged; or it may be associated with a matrix A which takes any point $\boldsymbol{x}$ (x,y,z) to the position $\boldsymbol{x}'$ (x', y', z'), where

$$A\boldsymbol{x}=\boldsymbol{x}' \tag{3.1}$$

If x, y and z are Cartesian axes, then A would be given by the matrix

$$\begin{bmatrix} -\frac{1}{2} & -\frac{1}{2}\sqrt{3} & 0 \\ \frac{1}{2}\sqrt{3} & -\frac{1}{2} & 0 \\ 0 & 0 & 1 \end{bmatrix}$$

Thus, the point 0.1, 0.2, 0.3 transforms under A to -0.2232, -0.01340, 0.3.

A molecule may possess one or more symmetry elements, which are known collectively as

Figure 3.1 'Vijentor Seal of Approval at Valmara': an everyday example of σ_V symmetry.

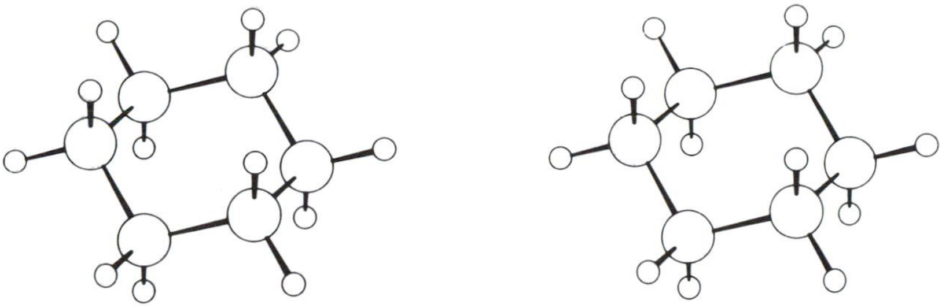

Figure 3.2 Stereoscopic view of a molecule of the *chair* form of cyclohexane, C_6H_{12}. The axis (z) normal to the 'plane' of the molecule is a coincident C_3 and S_6; σ_v planes pass through opposite pairs of carbon atoms; C_2 axes pass through the mid-points of opposite pairs of C−C bonds; the molecule has a centre of symmetry i and its point group is $\mathscr{D}_{3d}$.

a *point group*. The action of the symmetry operations of a point group leaves at least one point unmoved; this point is the origin of the reference axes for the molecule.

3.3 Symmetry elements and symmetry operations

The underlying spatial feature of all symmetry operations on a body is that *their action leaves the body in an orientation that is indistinguishable from its orientation before the operation*. We use now the term 'orientation' rather than the more general word 'state' because the symmetry operations on a finite body produce no translational motion of that body.

In three-dimensional, finite space there are six symmetry operations that we need to consider, although they are not all independent. We shall adopt the conventional use of the Schönflies symmetry notation in this section, although we shall refer to the Hermann–Mauguin notation because it is employed in studying crystal symmetry. One day, perhaps, a single notation will suffice.

3.3.1 Rotation

A rotation operation $\mathbf{C}_n$ (n=1, 2, 3, ..., ∞) moves a body from a given orientation to an indistinguishable orientation for a rotation of $(360/n)^\circ$ about the C_n axis. In Figure 3.2, a C_3 rotation axis is normal to the cyclohexane ring.[1] Two symmetry operations are associated with the C_3 axis: $\mathbf{C}_3(\mathbf{C}_3^+)$ corresponds to an anticlockwise rotation of 120°, whereas $\mathbf{C}_3^2(\mathbf{C}_3^-)$ is an anticlockwise rotation of 240° (a clockwise rotation through 120°). An anticlockwise (positive) rotation acts in the sense $x \rightarrow y$ in the $x-y$ plane as observed along the direction $-z$, where the Cartesian x, y and z form a right-handed set, Figure 3.3. It may be noted that this convention corresponds to that used for the angular momentum vector resolved along the z axial direction (see Section 2.6.2). A molecule may possess more than one rotation axis, and that corresponding to the highest value of n is called the *principal* axis.

In setting reference axes in a molecule, it is usual to choose z as the principal symmetry axis and, if there are more than one of them, z intersects the maximum number of atoms. If the molecule is planar and z lies in that plane, the x axis is normal to that plane, that is, the molecular plane is $y-z$. If the molecule is planar and z is normal to the plane, then y lies in

[1] That is, normal to a plane containing three carbon atoms related by threefold rotational symmetry.

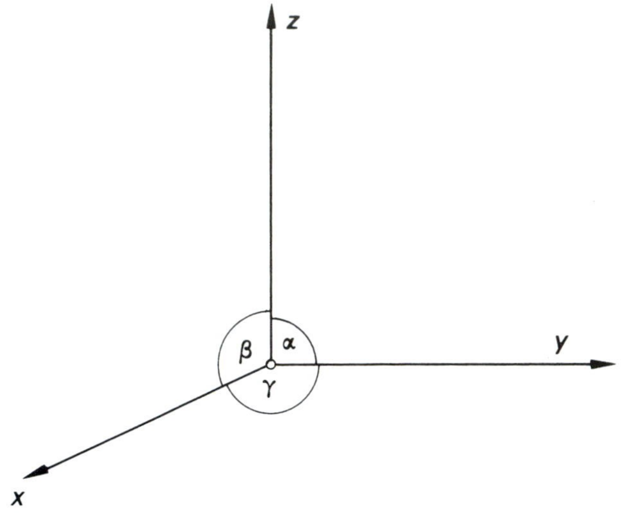

Figure 3.3 Conventional right-handed Cartesian reference frame, with the axes mutually perpendicular. If the thumb and first two fingers of the left hand are held mutually perpendicular, the thumb is x, the second finger y and the first finger z. In crystallography, however, the interaxial angles may not always be 90°, in which case the notation shown here for them is conventional.

the plane and passes through the maximum number of atoms. The x axis is perpendicular to both y and z, as shown in Figure 3.3.

In the Hermann–Mauguin notation, the symbol n alone is used in place of C_n, but with the same meaning.

3.3.2 Reflection

A reflection symmetry plane σ relates the halves of a molecule, across that plane, as an object is related to its mirror image. If the reflection plane contains the principal axis it is called *vertical*, σ_v; if it is normal to the principal axis it is *horizontal*, σ_h. In Figure 3.2, three σ planes exist; they pass through the 1,4, 2,5 and 3,6 pairs of carbon atoms, and they are equivalent under the $\mathbf{C}_3$ rotation. Although they are vertical they are symbolized by σ_d, because they lie between twofold axes that are in a plane normal to the principal axis[2]. Strictly, the symbol σ_d should be used only in this way. However, it is common practice to use σ_d when there is more than one form (q.v.) of vertical reflection planes, the planes in each form being related by the principal axis, as in $\mathscr{C}_{4v}$. The operation $\boldsymbol{\sigma}_h$ is eqivalent to $\mathbf{S}_1$ (see below).

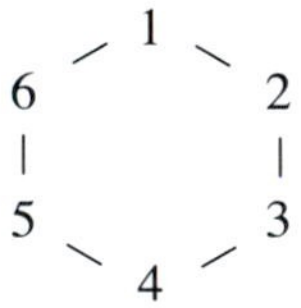

[2] See R S Mulliken *J. Chem. Phys.* **23**, 1997 (1955).

The Hermann–Mauguin notation uses the single symbol *m*: as we shall see in Section 3.5.3, the reason that only one *m* symbol is needed lies in the full description of the symbol in that notation.

3.3.3 Roto-reflection

A roto-reflection operation is a *single* operation comprising rotation by $(360/n)^\circ$ about an axis S_n *plus* reflection across a plane normal to that axis. It is important to note that this plane may or may not be a σ_h plane in the molecule, but it will be equidistant from points related by the $\mathbf{S}_n$ operation, in the direction of the S_n axis. The vertical axis in Figure 3.2 is also an S_6 axis, but the plane normal to it that is involved in the symmetry operation $\mathbf{S}_6$ is *not* a σ_h plane in the molecule. The operations $\mathbf{S}_1$ and $\mathbf{S}_2$ are equivalent to $\boldsymbol{\sigma}_h$ and $\boldsymbol{i}$, respectively.

The roto-reflection symbol is not used in the Hermann–Mauguin notation.

3.3.4 Roto-inversion

This operation is used in the Hermann–Mauguin symmetry notation, and we include it here for completeness. Roto-inversion is a *single* operation comprising rotation about an axis *n* through an angle of $(360/n)^\circ$ *plus* inversion through the origin. Generally S_n (Schönflies)$\not\equiv\bar{n}$ (Hermann–Mauguin), but $S_4\equiv\bar{4}$, for example.[3] In Figure 3.2 the vertical axis is a $\bar{3}$ axis; $\bar{3}$ (Hermann-Mauguin)$\equiv S_6$ (Schönflies).

3.3.5 Inversion

The operation of inversion $\boldsymbol{i}$ through a point corresponds to a centre of symmetry and involves taking each part of the molecule, in a straight line, through the point to an equal distance on the other side. For example, in Figure 3.2 atom 1 inverts through the origin (at the centre of the molecule) to atom 4 and so on. Although the inversion operation $\boldsymbol{i}$ is equivalent to $\mathbf{S}_2$, the former symbol is usual. It is important to note that the configuration of the object is inverted, too: thus, a right hand inverts to a left hand.

The Hermann–Mauguin equivalent symbol is $\bar{1}$.

3.3.6 Identity

The identity operation **E** consists in doing nothing, or a rotation of 360° about any axis (C_1) through the body in question. All molecules, and other finite bodies, possess identity symmetry; some such as CHFClBr (Figure 3.4) possess no other. The **E** operation, although apparently trivial, is fundamental to group theory.

The Hermann–Mauguin notation for identity is 1.

3.3.7 Symmetry and chirality

The operations of $\boldsymbol{\sigma}$, **S** and $\boldsymbol{i}$ relate enantiomorphous (change-of-hand) aspects of the molecule, whereas an operation **C** relates congruent aspects. It follows that a *chiral* molecule can have, at most, only pure rotation symmetry elements. Thus, lactic acid, $CH_3CHOHCO_2H$, is

[3] Strictly, $\bar{4}\equiv S_4^3$, that is, three successive operations of S_4, However, when the operations of the point groups $\bar{4}$ and S_4 are completed, the same set of representative points is obtained. In a similar manner, $\bar{3}\equiv S_6^5$ and $\bar{6}\equiv S_3^5$. The *point groups* $\bar{4}$ and $\mathscr{S}_4$ are identical, as are $\bar{3}$ and $\mathscr{S}_6$, and $\bar{6}$ and $\mathscr{S}_3$.

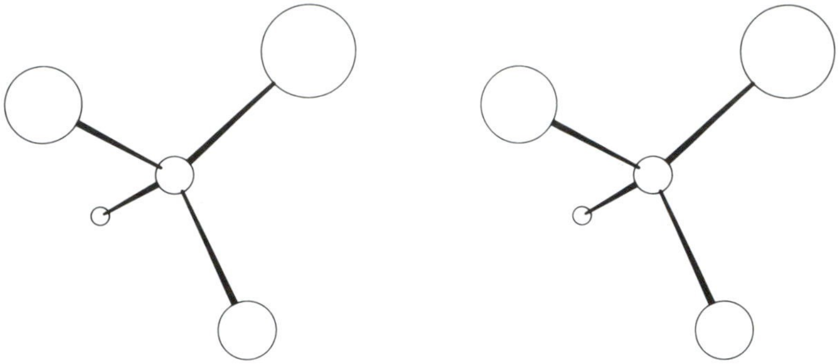

Figure 3.4 Stereoscopic view of a molecule of fluorochlorobromomethane, CHFClBr; only identity symmetry is present, and the point group of the molecule is $\mathscr{C}_1$.

a chiral molecule, but propanoic acid $CH_3CH_2CO_2H$ is not chiral because a symmetry plane σ can be defined through the $-CH_3$ and $-CO_2H$ groups and the central carbon atom. A *polar* molecule may have only symmetry elements C_n or C_{nv} where C_n lies along the dipole axis, or symmetry σ where the symmetry plane is the molecular plane. The NH_3 and HF molecules, point groups $\mathscr{C}_{3v}$ and $\mathscr{C}_{\infty v}$ respectively, are examples of polar species.

3.4 Group theory

Sets of symmetry operations constitute mathematical groups. Any two operations in a set carried out in succession are equivalent to another single operation, also in the set, starting from the same initial state. The combination of two operations is called their *product*, whatever the process of combination. In Figure 3.2, $\mathbf{C}_3$, acting vertically through the centre of the molecule (normal to a plane of atoms 1, 3 and 5), on the carbon atom in a position 1 takes it to position 5; $\boldsymbol{\sigma}_d$ (through atoms 1 and 4) operating on the carbon atom moved to position 5 takes it then to position 3. Now $\boldsymbol{\sigma}''_d$ (through atoms 2 and 5) acting on carbon 1 takes it directly to position 3, so that we may write

$$\boldsymbol{\sigma}_d\mathbf{C}_3 = \boldsymbol{\sigma}''_d \tag{3.2}$$

which means that $\mathbf{C}_3$ *followed by* $\boldsymbol{\sigma}_d$ is equivalent to another reflection operation $\boldsymbol{\sigma}''_d$; the order of the operations is important. Note that, although we were confining our attention to certain atoms for the purpose of the discussion, all atoms move under the operation of any symmetry element present.

3.4.1 Group postulates

A group consists of a set of mathematical objects, such as *a*, *b*, *c* …, called *members* of the group, which satisfy the following conditions:

(a) products such as *aa* and *ab* are also members of the set;
(b) the associative law holds, that is, $(ab)c = a(bc)$;
(c) the set contains the identity member *E*, and $Ea = aE = a$;
(d) each member of the set has an inverse that is also a member of the set, that is, for each *a* there is a member a^{-1} such that $aa^{-1} = E$.

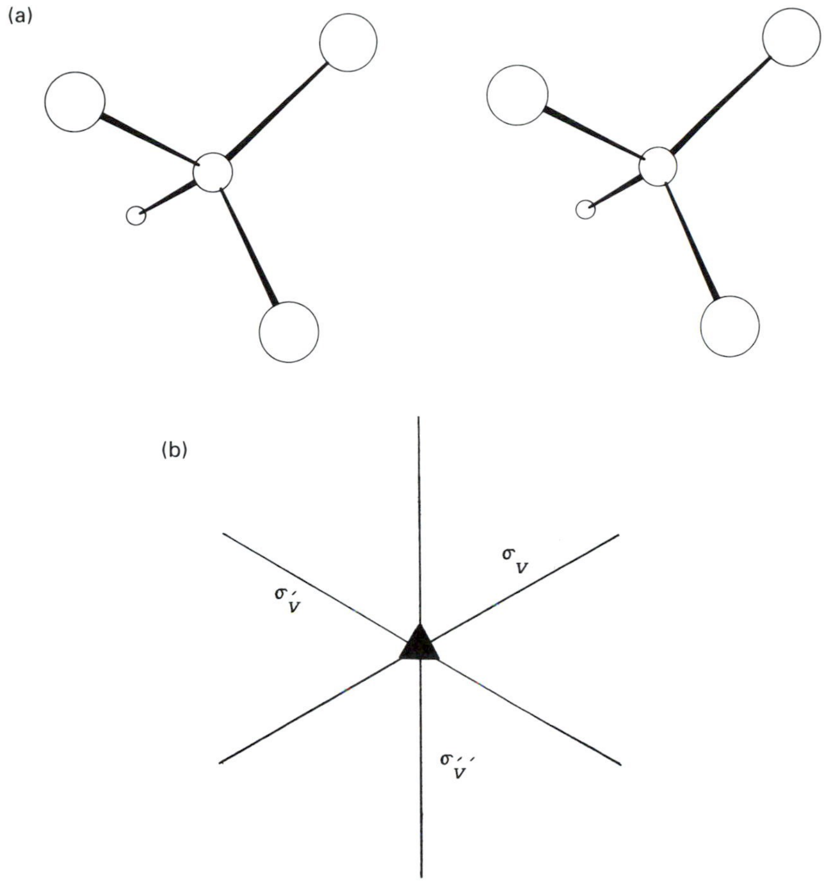

Figure 3.5 Trichloromethane, $CHCl_3$: (a) stereoview of the molecule, point group $\mathscr{C}_{3v}$; (b) the C_3 axis (z), indicated by the equilateral triangle, acts along the C−H bond of the molecule, normal to the plane of the drawing; each σ_v plane contains this axis, and passes through carbon, hydrogen and one chlorine atoms.

A group consisting of a single member a and its powers a^2, a^3, …, a^p is a *cyclic* group of *order p*. Point group $\mathscr{C}_3$ is a cyclic group of order 3 and comprises the operators $\mathbf{C}_3$, $\mathbf{C}_3^2$ and $\mathbf{E}$ ($\mathbf{C}_3^3$); $\mathbf{C}_3^2$ may also be written as $\mathbf{C}_3^{-1}$; the *order* of a group is the total number of symmetry operations in that group. Group members need not commute, but if they do the group is Abelian, that is, $ab=ba$. Appendix 1 refers to programs that have been found useful in understanding the derivation of the crystallographic point groups, and in identifying the point groups of molecules. It provides also a list of point groups in both the Schönflies and the Hermann–Mauguin symmetry notations.

3.4.2 Group multiplication tables

The properties of members of a group are presented conveniently by a group multiplication table; we describe this procedure with reference to the trichloromethane molecule. This substance belongs to point group $\mathscr{C}_{3v}$, Figure 3.5.

The group multiplication table for $\mathscr{C}_{3v}$ is given below. The top row of the table represents a first operator, like $\mathbf{C}_3$ in (3.2), and the extreme left-hand column a second operator, whereupon their combination in that order is given by their intersection in the body of the table.

$\mathscr{C}_{3v}$	$\mathbf{E}$	$\mathbf{C}_3$	$\mathbf{C}_3^2$	$\boldsymbol{\sigma}_v$	$\boldsymbol{\sigma}'_v$	$\boldsymbol{\sigma}''_v$
$\mathbf{E}$	$\mathbf{E}$	$\mathbf{C}_3$	$\mathbf{C}_3^2$	$\boldsymbol{\sigma}_v$	$\boldsymbol{\sigma}'_v$	$\boldsymbol{\sigma}''_v$
$\mathbf{C}_3$	$\mathbf{C}_3$	$\mathbf{C}_3^2$	$\mathbf{E}$	$\boldsymbol{\sigma}'_v$	$\boldsymbol{\sigma}''_v$	$\boldsymbol{\sigma}_v$
$\mathbf{C}_3^2$	$\mathbf{C}_3^2$	$\mathbf{E}$	$\mathbf{C}_3$	$\boldsymbol{\sigma}''_v$	$\boldsymbol{\sigma}_v$	$\boldsymbol{\sigma}'_v$
$\boldsymbol{\sigma}_v$	$\boldsymbol{\sigma}_v$	$\boldsymbol{\sigma}''_v$	$\boldsymbol{\sigma}'_v$	$\mathbf{E}$	$\mathbf{C}_3^2$	$\mathbf{C}_3$
$\boldsymbol{\sigma}'_v$	$\boldsymbol{\sigma}'_v$	$\boldsymbol{\sigma}_v$	$\boldsymbol{\sigma}''_v$	$\mathbf{C}_3$	$\mathbf{E}$	$\mathbf{C}_3^2$
$\boldsymbol{\sigma}''_v$	$\boldsymbol{\sigma}''_v$	$\boldsymbol{\sigma}'_v$	$\boldsymbol{\sigma}_v$	$\mathbf{C}_3^2$	$\mathbf{C}_3$	$\mathbf{E}$

The table shows clearly that $\boldsymbol{\sigma}_v\, \mathbf{C}_3 = \boldsymbol{\sigma}''_v$, whereas $\mathbf{C}_3\, \boldsymbol{\sigma}_v = \boldsymbol{\sigma}'_v$; the three σ_v operations are symmetry-related under $\mathbf{C}_3$ but are, nevertheless, individual members of the group $\mathscr{C}_{3v}$.

3.4.3 Similarity transformations

Two symmetry operations a and b are said to be of the same *class*[4] if there exists an operation r such that

$$r\, a\, r^{-1} = b \tag{3.3}$$

where b is the *similarity transformation* of a by r. The operation $\boldsymbol{\sigma}_v$ is its own inverse; hence, we may perform a similarity transformation on $\mathbf{C}_4$ by finding the operation equivalent to $\boldsymbol{\sigma}_v\, \mathbf{C}_4\, \boldsymbol{\sigma}_v$. Let $\mathbf{C}_4$ be along z and $\boldsymbol{\sigma}_v$ normal to x. Then, we have

$$\underset{\mathbf{C}_4}{\begin{bmatrix} 0 & -1 & 0 \\ 1 & -0 & 0 \\ 0 & -0 & 1 \end{bmatrix}} \underset{\boldsymbol{\sigma}_v}{\begin{bmatrix} -1 & 0 & 0 \\ 0 & 1 & 0 \\ 0 & 0 & 1 \end{bmatrix}} = \underset{\mathbf{C}_4\ \boldsymbol{\sigma}_v}{\begin{bmatrix} -0 & -1 & 0 \\ -1 & -0 & 0 \\ -0 & -0 & 1 \end{bmatrix}} \tag{3.4}$$

$$\underset{\mathbf{C}_v}{\begin{bmatrix} -1 & 0 & 0 \\ -0 & 1 & 0 \\ -0 & 0 & 1 \end{bmatrix}} \underset{\mathbf{C}_4\, \boldsymbol{\sigma}_v}{\begin{bmatrix} 0 & -1 & 0 \\ 1 & -0 & 0 \\ 0 & -0 & 1 \end{bmatrix}} = \underset{\boldsymbol{\sigma}_v \mathbf{C}_4 \boldsymbol{\sigma}_v = \mathbf{C}_4^{-1}}{\begin{bmatrix} -0 & 1 & 0 \\ -1 & 0 & 0 \\ -0 & 0 & 1 \end{bmatrix}} \tag{3.5}$$

Thus, $\mathbf{C}_4$ and $\mathbf{C}_4^{-1}$ are in the same class; $\mathbf{C}_4^{-1} = \mathbf{C}_4^3$ and is equivalent to performing $\mathbf{C}_4$ three times in succession.

In a similar manner, we can show that

$$\mathbf{C}_4 \boldsymbol{\sigma}_v \mathbf{C}_4^{-1} = \boldsymbol{\sigma}'_v \tag{3.6}$$

where $\boldsymbol{\sigma}'_v$ is normal to y. This result is to be expected, since $\boldsymbol{\sigma}_v$ and $\boldsymbol{\sigma}'_v$ are equivalent under $\mathbf{C}_4$. In the case of $\mathbf{C}_4^2$ we can show that

[4] Here, 'symmetry class' must not be confused with 'crystal class'.

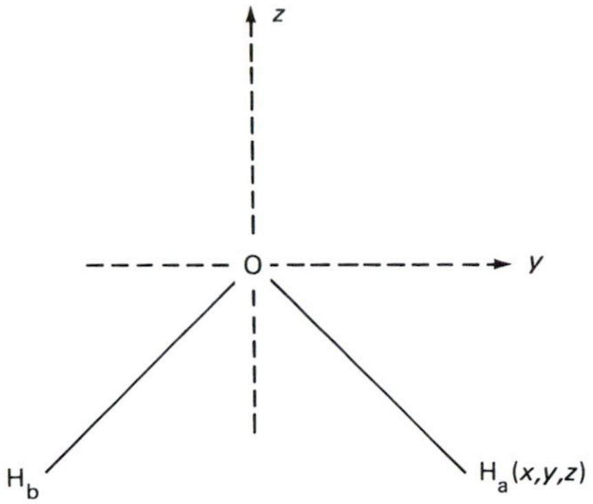

Figure 3.6 Schematic drawing of the water molecule, with the oxygen atom at the origin of a Cartesian reference frame. The C_2 axis is along z and x is normal to the plane of the molecule; y is perpendicular to both x and z; σ_v is the plane x–z and σ'_v the plane of the molecule, y–z. The line of intersection of σ_v and σ'_v is the C_2 axis. The atoms H_a and H_b are interconverted by the operations C_2 and σ_v, but are invariant under σ'_v and E.

$$\mathbf{C}_4\mathbf{C}_4^2\mathbf{C}_4^{-1} = \mathbf{C}_4^2 = (\mathbf{C}_4^2)^{-1} \tag{3.7}$$

so that $\mathbf{C}_4^2$ is its own inverse, which is not surprising since $\mathbf{C}_4^2 \equiv \mathbf{C}_2$; $\mathbf{C}_4^2$ is not of the same class as $\mathbf{C}_4$ and is referred to as $\mathbf{C}_2$.

In point group $\mathscr{C}_{4v}$, there is a second set of vertical mirror planes, at 45° to the first set, σ_v and σ'_v. They are not of the same class as σ_v and σ'_v; we identify them as σ_d and σ'_d, and find that they themselves are in one and the same class (see Section 3.3.2). Thus, the complete set of classes for $\mathscr{C}_{4v}$ is

$$\mathbf{E}\ 2\mathbf{C}_4\ \mathbf{C}_2\ 2\boldsymbol{\sigma}_v\ 2\boldsymbol{\sigma}_d$$

and the order of the group is 8. It is usually unnecessary to work out similarity transformations in order to assign symmetry operations in classes. Operations belong to one and the same class if they are equivalent under a symmetry operation of the group. In the NO_3^- ion, point group $\mathscr{D}_{3h}$, we find the operations $\mathbf{E}$, $\mathbf{C}_3$, $\mathbf{C}_3^{-1}$, $\mathbf{C}_2$, $\mathbf{C}'_2$, $\mathbf{C}''_2$, $\mathbf{S}_3$, $\mathbf{S}'_3$, $\boldsymbol{\sigma}_h$, $\boldsymbol{\sigma}_v$, $\boldsymbol{\sigma}'_v$, $\boldsymbol{\sigma}''_v$, which may be arranged readily in their classes by considering which symmetry operations are equivalent under, say, $\mathbf{C}_4$. The final result is

$$\mathbf{E}_4\ 2\mathbf{C}_3\ 3\mathbf{C}_2\ 2\mathbf{S}_3\ \boldsymbol{\sigma}_h\ 3\boldsymbol{\sigma}_v$$

3.4.4 Representations

The water molecule has point group $\mathscr{C}_{2v}$ and reference axes are set up as in Figure 3.6. The operations of $\mathscr{C}_{2v}$ are given by the following multiplication table, in which σ_v is the plane x–z normal to y and σ'_v is the plane y–z, the plane of the molecule. The group is Abelian, of order 4.

$\mathscr{C}_{2v}$	E	$\mathbf{C}_2$	$\boldsymbol{\sigma}_v$	$\boldsymbol{\sigma}'_v$
E	E	$\mathbf{C}_2$	$\boldsymbol{\sigma}_v$	$\boldsymbol{\sigma}'_v$
$\mathbf{C}_2$	$\mathbf{C}_2$	E	$\boldsymbol{\sigma}'_v$	$\boldsymbol{\sigma}_v$
$\boldsymbol{\sigma}_v$	$\boldsymbol{\sigma}_v$	$\boldsymbol{\sigma}'_v$	E	$\mathbf{C}_2$
$\boldsymbol{\sigma}'_v$	$\boldsymbol{\sigma}'_v$	$\boldsymbol{\sigma}_v$	$\mathbf{C}_2$	E

The $\mathbf{C}_2$ operation acts on the hydrogen atom H_a as follows:

$$x \rightarrow -x \qquad y \rightarrow -y \qquad z \rightarrow z$$

Similar results can be obtained for the other operations and the set may be summarized in the following table:

$\mathscr{C}_{2v}$	$\mathbf{E}$	$\mathbf{C}_2$	$\boldsymbol{\sigma}_v$	$\boldsymbol{\sigma}'_v$
x	1	−1	1	−1
y	1	−1	−1	1
z	1	1	1	1

The effect of the symmetry operations on product functions, such as x^2 or xy, can be obtained in a like manner. In all there are four ways in which functions of the coordinates may transform under $\mathscr{C}_{2v}$. Tabulating them fully, we have

$\mathscr{C}_{2v}$	$\mathbf{E}$	$\mathbf{C}_2$	$\boldsymbol{\sigma}_v$	$\boldsymbol{\sigma}'_v$		
A_1	1	1	1	1	T_z	x^2; y^2; z^2
A_2	1	1	−1	−1	R_z	xy
B_1	1	−1	1	−1	T_x, R_y	zx
B_2	1	−1	−1	1	T_y, R_x	yz

where T refers to translational, R to rotational and x^2, xy, ... to product operations, or displacements. There are no other ways in which functions can transform, since x^3 transforms like x, x^4 like x^2 and so on. The four transformations are the *irreducible representations* of $\mathscr{C}_{2v}$, the table is the *character table* of the group, and the numbers in the table are the *characters (of the irreducible representations)* of $\mathscr{C}_{2v}$. The A and B labels are conventional symbols, and we say, for example, that T_x belongs to the B_1 representation of $\mathscr{C}_{2v}$; T_x is symmetrical under $\mathbf{E}$ and $\boldsymbol{\sigma}_v$ but antisymmetrical under $\mathbf{C}_2$ and $\boldsymbol{\sigma}'_v$, and T_z is symmetrical under all operations of $\mathscr{C}_{2v}$. It should be noted that the form of a character table may depend upon the setting of the symmetry planes with respect to the symmetry axes of the group under consideration. The reader may inspect this feature by interchanging the x and y reference axes for the water molecule and reworking this section to this point.

It follows that the p_x, p_y and p_z atomic orbitals transform under $\mathscr{C}_{2v}$ as B_1, B_2 and A_1 respectively. Although the d orbitals (Figure 2.11) are a little more complex, we can show, for example, that in the same point group d_{z^2} belongs[5] to A_1 and d_{yz} to B_2.

The full analysis for the water molecule leads to the representation

$\mathscr{C}_{2v}$	$\mathbf{E}$	$\mathbf{C}_2$	$\boldsymbol{\sigma}_v$	$\boldsymbol{\sigma}'_v$	
Γ	9	−1	1	3	$=3A_1+A_2+2B_1+3B_2$.

From the character table, we see that the three translational and three rotational displacements of the molecule (see Section 3.6.5.3) are represented by $A_1+A_2+2B_1+2B_2$. Hence, the *remaining*, internal molecular displacements are expressed by the *reducible* representation Γ':

[5] From Table 2.3, $3\cos^2(\theta)-1=3\cos^2(\theta)-[\sin^2(\theta)+\cos^2(\theta)]=2\cos^2(\theta)-\sin^2(\theta)$. Using Appendix 5 (A5.1), it follows that $3\cos^2(\theta)-1=(2z^2-x^2-y^2)/r^2$. In most groups, z^2 and (x^2+y^2) transform alike, so that the compact notation d_{z^2} is used.

	E	C_2	σ_v	σ'_v
Γ'	3	1	1	3
A_1	1	1	1	1
A_1	1	1	1	1
B_2	1	−1	−1	1
$2A_1+B_2$	3	1	1	3

We shall consider the water molecule again in Section 3.4.6.3, in the context of symmetry-adapted orbitals.

In more complex examples of the reduction of representations, a formula can be used. If N is the number of times that a given irreducible representation appears in a reducible representation, then N is given by

$$N=\frac{1}{p}\sum_{R}\chi_r\chi N_s \tag{3.8}$$

where p is the order of the group, χ_r and χ are, respectively, the characters of the reducible and irreducible representations, N_s is the number of symmetry operations in a given symmetry class, and the sum is taken over all symmetry classes R. For the last example, we have

$$N_{A_1}=\tfrac{1}{4}\{[9\times1\times1]+[(-1)\times1\times1]+[1\times1\times1]+[3\times1\times1]\}=3$$

$$N_{A_2}=\tfrac{1}{4}\{[9\times1\times1]+[(-1)\times1\times1]+[1\times(-1)\times1]+[3\times(-1)\times1]\}=1$$

$$N_{B_1}=\tfrac{1}{4}\{[9\times1\times1]+[(-1)\times(-1)\times1]+[1\times1\times1]+[3\times(-1)\times1]\}=2$$

$$N_{B_2}=\tfrac{1}{4}\{[9\times1\times1]+[(-1)\times(-1)\times1]+[1\times(-1)\times1]+[3\times1\times1]\}=3$$

3.4.5 Degenerate representations

Consider the pentafluoroantimonate ion, Figure 3.7, point group $\mathscr{C}_{4v}$; its classes have been derived in Section 3.4.3. We take the z axis along the apical Sb−F bond, with x and y along two Sb−F bonds in the plane normal to z forming a right-handed set (Figure 3.3).

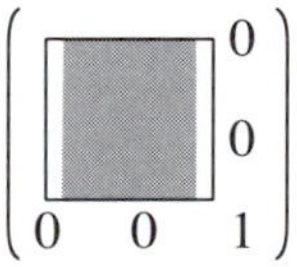

The representations for the p orbitals on antimony are three-dimensional, but of the form shown by the matrix since each of the symmetry operations of $\mathscr{C}_{4v}$ leaves the direction of z invariant, and p_z does not mix with p_x and p_y. Thus, the basis functions can be divided into two parts. The z axis direction belongs to the totally symmetrical A_1 symmetry species (irreducible representation):

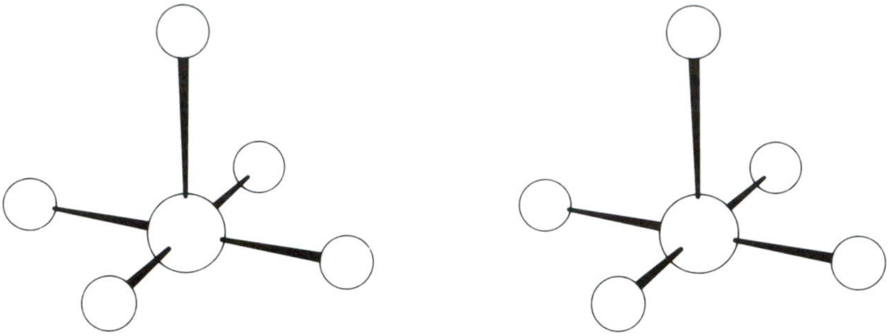

Figure 3.7 Stereoview of the pentafluoroantimonate ion, $[SbF_5]^{2-}$. The vertical direction (z) is a C_4 axis; σ_v planes pass through opposite pairs of fluorine atoms in the x–y plane; σ_d planes lie at 45° to x and y and contain the C_4 axis; the point group is $\mathscr{C}_{4v}$.

$\mathscr{C}_{4v}$	**E**	$2\mathbf{C}_4$	$\mathbf{C}_2$	$2\boldsymbol{\sigma}_v$	$2\boldsymbol{\sigma}_d$
A_1	1	1	1	1	1

The x and y directions are interconverted under $\mathbf{C}_4$ and $\boldsymbol{\sigma}_d$, but not under $\mathbf{C}_2$ and $\boldsymbol{\sigma}_v$. The x, y relationships here are conveniently expressed by two-dimensional matrices, such as

$$\begin{bmatrix} 0 & -1 \\ 1 & -0 \end{bmatrix} \begin{bmatrix} x \\ y \end{bmatrix} = \begin{bmatrix} -y \\ x \end{bmatrix} \qquad (3.9)$$

$$\mathbf{C}_4 \qquad \mathbf{x} \qquad \mathbf{x}'$$

The *trace* of a square matrix is the sum of its principal diagonal terms and, if the matrix represents an element of a group, the trace is also the character for that element; it is zero for the operation $\mathbf{C}_4$. Continuing the building of these matrices leads to the result

$\mathscr{C}_{4v}$	**E**	$2\mathbf{C}_4$	$\mathbf{C}_2$	$2\boldsymbol{\sigma}_v$	$2\boldsymbol{\sigma}_d$
Γ	2	0	−2	0	0

In a nondegenerate representation, the matrices are all 1×1 and each number[6] is its own trace χ. The matrix for the degenerate representation E has a trace greater than unity and is 2 in this example. The complete character table for $\mathscr{C}_{4v}$ is shown below.[7] In general, the degeneracy of a representation is given by the trace of its identity matrix, and degenerate functions, such as T_x,T_y and yz, zx are enclosed by parentheses and transform as pairs:

$\mathscr{C}_{4v}$	**E**	$2\mathbf{C}_4$	$\mathbf{C}_2$	$2\boldsymbol{\sigma}_v$	$2\boldsymbol{\sigma}_d$		
A_1	1	1	1	1	1	T_z	z^2; x^2+y^2
A_2	1	1	1	−1	−1	R_z	
B_1	1	−1	1	1	−1		x^2-y^2
B_2	1	−1	1	−1	1		xy
E	2	0	−2	0	0	(T_x, T_y), (R_x, R_y)	(yz, zx)

[6] In some groups, such as $\mathscr{C}_6$ (see Problem 3.8), a character may be a complex number.

[7] The choice of B_1 and B_2 depends upon the setting of the x and y axes. Here, σ_v is the plane of x–z (and y–z).

The tetrahedral $\mathcal{T}_d$ and octahedral $\mathcal{O}_h$ groups are important in coordination chemistry. The symmetry operations of a regular tetrahedron can be studied with reference to Figure 2.32. It is a straightforward matter to deduce the symmetry classes in $\mathcal{T}_d$:

$$\mathbf{E}\ 8\mathbf{C}_3\ 3\mathbf{C}_2\ 6\mathbf{S}_4\ 6\boldsymbol{\sigma}_d$$

The group has the order 24, and the T representations are threefold degenerate; the character table is given below:

$\mathcal{T}_d$	$\mathbf{E}$	$8\mathbf{C}_3$	$3\mathbf{C}_2$	$6\mathbf{S}_4$	$6\boldsymbol{\sigma}_d$		
A_1	1	1	1	1	1		$x^2+y^2+z^2$
A_2	1	1	1	−1	−1		
E	2	−1	2	0	0		$(2z^2-x^2-y^2, x^2-y^2)$
T_1	3	0	−1	1	−1	(R_x, R_y, R_z)	
T_2	3	0	−1	−1	1	(T_x, T_y, T_z)	(xy, yz, zx)

The octahedral group will be discussed later.

3.4.6 Some applications of group theory

We can apply some of the principles that we have discussed first to some simple chemical species, such as the carbonate ion and methane, and then to the LCAO approximation for molecules.

3.4.6.1 Carbonate ion

The carbonate ion $[CO_3]^{2-}$ belongs to point group $\mathcal{D}_{3h}$, of order 12; it may be shown readily that this species has the following symmetry classes:

$$\mathbf{E}\ 2\mathbf{C}_3\ 3\mathbf{C}_2\ \boldsymbol{\sigma}_h\ 2\mathbf{S}_3\ 3\boldsymbol{\sigma}_v$$

Figure 3.8a illustrates the bond directions in $[CO_3]^{2-}$, with the C_3 axis along z, normal to the plane of the ion. We can use a triplet of vectors $\boldsymbol{j}_1, \boldsymbol{j}_2$ and $\boldsymbol{j}_3$ *along the* C−O σ *bonds* as a geometrical basis on which to generate a reducible representation for $\mathcal{D}_{3h}$; the symmetry elements of this group are indicated in Figure 3.8b, which is drawn normal to the z axis.

Next we set up a matrix for each symmetry class and find its character. Thus, for the identity $\mathbf{E}$ we have

$$\begin{bmatrix} 1 & 0 & 0 \\ 0 & 1 & 0 \\ 0 & 0 & 1 \end{bmatrix} \begin{bmatrix} j_1 \\ j_2 \\ j_3 \end{bmatrix} = \begin{bmatrix} j_1 \\ j_2 \\ j_3 \end{bmatrix} \tag{3.10}$$

with a character χ_E of 3; for $\mathbf{C}_3$ we have

$$\begin{bmatrix} 0 & 1 & 0 \\ 0 & 0 & 1 \\ 1 & 0 & 0 \end{bmatrix} \begin{bmatrix} j_1 \\ j_2 \\ j_3 \end{bmatrix} = \begin{bmatrix} j_2 \\ j_3 \\ j_1 \end{bmatrix} \tag{3.11}$$

with $\chi_{C_3}=0$. Any vertical symmetry plane, passing through the carbon atom and one oxygen atom, leaves one $\boldsymbol{j}$ vector invariant, so that $\chi_{\sigma_v}=1$. Proceeding in this manner, we build up the representation

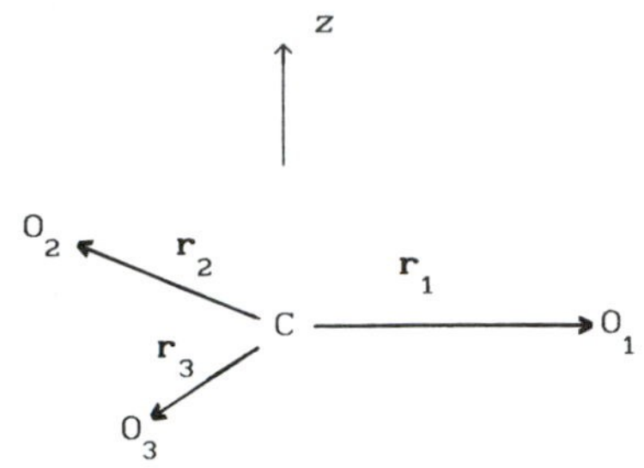

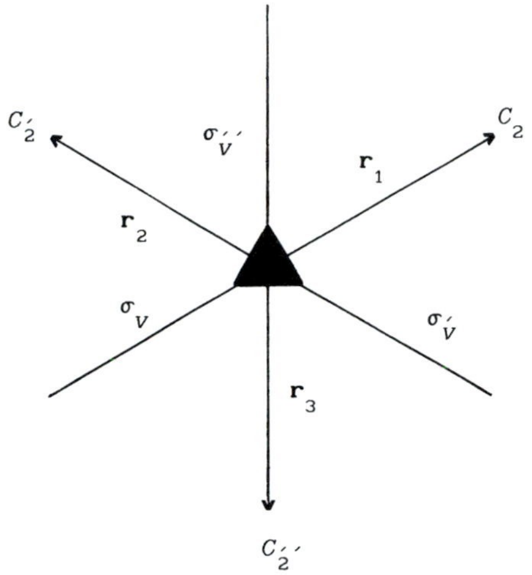

Figure 3.8 Carbonate ion, point group $\mathscr{D}_{3h}$: (a) C−O_j bond vectors $\boldsymbol{r}_j$ (j=1, 2, 3) form a basis for generating a reducible representation for $\mathscr{D}_{3h}$. (b) Symmetry elements of point group $\mathscr{D}_{3h}$ as seen on the x–y plane, normal to the vertical C_3 (z) axis; the σ_v planes are the same as in Figure 3.5b and the C_2 axes lie on them, in the x–y plane; the x–y plane is a σ_h plane in this point group.

$\mathscr{D}_{3h}$	**E**	$2C_3$	$3C_2$	σ_h	$2S_3$	$3\sigma_v$
Γ_σ	3	0	1	3	0	1

The character table for $\mathscr{D}_{3h}$ is given below:

$\mathscr{D}_{3h}$	**E**	$2C_3$	$3C_2$	σ_h	$2S_3$	$3\sigma_v$		
A_1'	1	1	1	1	1	1		z^2; x^2+y^2
A_2'	1	1	−1	1	1	−1	R_z	
E'	2	−1	0	2	−1	0	(T_x, T_y)	(xy, x^2-y^2)
A_1''	1	1	1	−1	−1	−1		
A_2''	1	1	−1	−1	−1	1	T_z	
E''	2	−1	0	−2	1	0	(R_x, R_y)	(yz, zx)

The single and double primed symbols are used for irreducible representations that are, respectively, symmetrical and anti-symmetrical for reflection in the σ_h plane. It is clear that Γ_σ is not an irreducible representation in $\mathcal{D}_{3h}$, and following (3.8) we have

$$N_{A'_1} = \tfrac{1}{12}\,[3+0+3+3+0+3] = 1$$

Similarly, $N_{A'_2}=0$, $N_{E'}=1$ and so on, such that $\Gamma_\sigma = A'_1 + E'$. Hence, the bonding orbitals available are s or d_{z^2} for A'_1; for E' there are either p_x and p_y, or $d_{x^2-y^2}$ and d_{xy} to consider. Possible combinations are (s, p_x, p_y), (s, d_{xy}, $d_{x^2-y^2}$), (d_{z^2}, p_x, p_y) and (d_{z^2}, d_{xy}, $d_{x^2-y^2}$). The more stable (lower energy) configuration is preferred, that is, s for A'_1 and p_x, p_y for E'; only orbitals of the same symmetry may mix (see Section 3.4.6.3).

There is a total of 24 electrons in $[CO_3]^{2-}$, excluding the carbon $(1s)^2$ and oxygen $(1s)^2\,(2s)^2$ cores. Since there are two lone pairs on each oxygen, six valence electrons remain, to be accommodated in π molecular orbitals. The same basis directions as before can be used, but now we have p_z orbitals that are $\pm$ in sign across the σ_h plane. Taking this factor into account, the Γ_π representation becomes

$\mathcal{D}_{3h}$	**E**	**2C$_3$**	**3C$_2$**	$\boldsymbol{\sigma}_h$	**2S$_3$**	**3$\boldsymbol{\sigma}_v$**
Γ_π	3	0	−1	−3	0	1

We see that the signs against the characters for $\boldsymbol{\sigma}_h$ and $\mathbf{C}_2$ have changed with respect to Γ_σ, because these operations act across the x–y plane.

A simple way of obtaining the set of characters forming a representation is to note that the matrix of a given operation on the orbitals gains a contribution to its diagonal (trace) as follows:

1 if the orbital is mapped on to itself;
0 if the orbital is moved in position;
−1 if the orbital is inverted in sign.

Thus, with σ_v, for example, two of the p orbitals are moved in position and one is mapped on to itself; thus, the corresponding character is 1. Each operation in a class behaves similarly, so that only one of them need be considered.

The Γ_π representation reduces to $A''_2 + E''$, and A''_2 interacts with the carbon p_z orbital (A''_2) with E'' remaining nonbonding. Figure 3.9a shows a molecular-orbital energy-level diagram for bonding in the carbonate ion.

3.4.6.2 Methane

We referred to this molecule in Chapter 2 and the character table for its point group $\mathcal{T}_d$ has been given in Section 3.4.5 above. Using the four C−H bonds (Figure 2.32) as a basis for generating a representation, we obtain

$\mathcal{T}_d$	**E**	**8C$_3$**	**3C$_2$**	**6S$_4$**	**6$\boldsymbol{\sigma}_h$**
Γ_σ	4	1	0	0	2

From the character table we deduce that this representation is reducible to $\Gamma_\sigma = A_1 + T_2$, which corresponds to s and either p^3 or d^3 orbitals; chemical arguments indicate the use of the lower energy p^3 orbitals, to form four equivalent sp^3 hybrids.

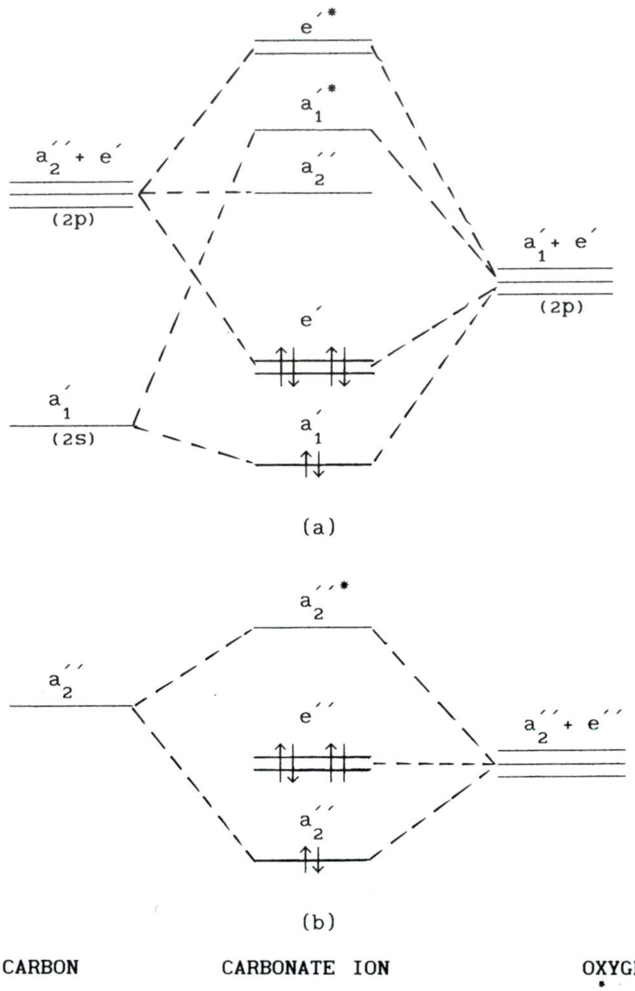

Figure 3.9 Molecular-orbital energy-level diagrams for the carbonate ion: (a) σ-bonding with a'_1 and e' molecular orbitals in the carbonate ion; (b) π-bonding with a''_2 and e'' orbitals.

3.4.6.3 LCAO approximations

In Section 3.4.4, we generated a representation for the water molecule by referring the atoms to a Cartesian coordinate frame. In the present context of the LCAO, we use the O−H bond vectors $\boldsymbol{r}$ to generate the representation

$\mathscr{C}_{2v}$	**E**	$\mathbf{C}_2$	$\boldsymbol{\sigma}_v$	$\boldsymbol{\sigma}'_v$
Γ_r	2	0	0	2

which reduces to $\Gamma_r = A_1 + B_2$. Let ψ_a be a generating atomic orbital on hydrogen atom H_a (Figure 3.6). Then for $\mathscr{C}_{2v}$ we can obtain, by symmetry

$\mathscr{C}_{2v}$	**E**	$\mathbf{C}_2$	$\boldsymbol{\sigma}_v$	$\boldsymbol{\sigma}'_v$
ψ_a	ψ_a	ψ_b	ψ_b	ψ_a
a_1	1	1	1	1
b_2	1	−1	−1	1

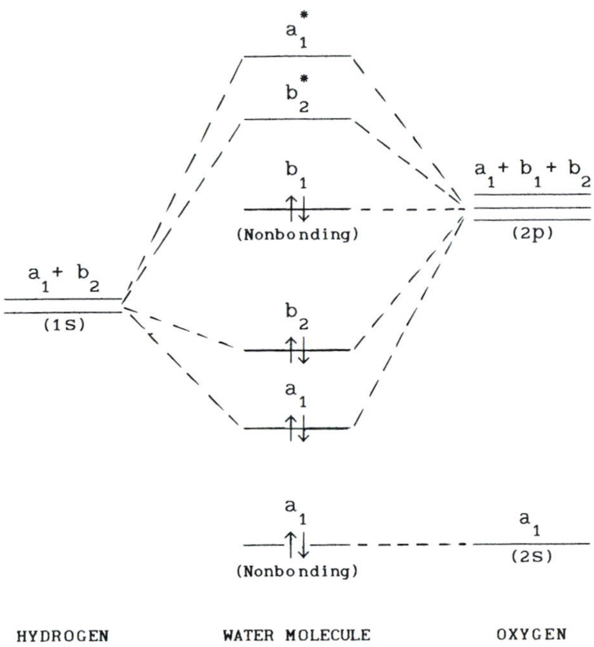

Figure 3.10 Molecular-orbital energy-level diagram for the water molecule: σ bonds are formed with the a_1 and b_2 orbitals; nonbonding lone pairs reside in a_1 of oxygen (core) and in the b_1 orbitals of the water molecule ion.

where the lower case letters refer to one-electron functions (Mulliken notation) corresponding to the irreducible representations. For a_1 we form the sum $2\psi_a+2\psi_b$, but since we need only the relative proportions of ψ_a and ψ_b, we may write simply $\psi_a+\psi_b$; similarly for b_2 we have $\psi_a-\psi_b$. The normalizing factor for each of these combinations is easily seen to be $1/\sqrt{2}$. The resultant functions are symmetry-adapted linear combinations, or symmetry-adapted orbitals. They have the correct symmetry and are appropriate bases for the irreducible representations: the a_1 combination is unchanged for all operations of $\mathscr{C}_{2v}$, whereas the b_2 combination is reversed in sign under $\mathbf{C}_2$ and $\boldsymbol{\sigma}_v$.

If we generate similar linear combinations for the A_2 and B_1 representations, we obtain

$$a_2\text{: } \psi_a+\psi_b-\psi_b-\psi_a=0$$

$$b_1\text{: } \psi_a-\psi_b+\psi_b-\psi_a=0$$

which shows that these functions are not allowed for the water molecule. Symmetry-adapted orbitals must be orthogonal: this condition is fulfilled by a_1 and b_2, because multiplication of corresponding terms in the two orbitals followed by summation results in zero.

Figure 3.10 shows a molecular orbital energy level diagram for the H_2O molecule. The valence electrons fill the bonding and nonbonding (lone pair) molecular orbitals to give a stable molecular configuration.

In using the bond vectors to generate a representation, we neglected the bond angle H−O−H, which is a third internal coordinate. No symmetry operation of $\mathscr{C}_{2v}$ acting on the bond angle causes a change in the O−H vectors. The angle α forms an independent basis for a 1×1 A_1 representation. Thus, we have Γ_r and Γ_α:

$\mathscr{C}_{2v}$	**E**	$\mathbf{C_2}$	$\boldsymbol{\sigma}_v$	$\boldsymbol{\sigma}'_v$
Γ_r	2	0	0	2
Γ_α	1	1	1	1
Γ_{int}	3	1	1	3

which sum to give Γ as found in Section 3.4.4.

3.4.6.4 Projection operators

An alternative, powerful approach to deriving symmetry-adapted linear combinations (SALCs) is by use of the *projection operator* **P.** We refer the reader to a standard text[8] on group theory for the derivation of the projection operator, and give here the required formulation for an SALC with a generating function ψ:

$$\mathbf{P}\psi = \left\{\frac{1}{p}\sum_{R}\chi_R(\Omega)\,\Omega\right\}\psi \tag{3.12}$$

where p is the order of the group and $\chi_R(\Omega)$ is the character of symmetry operation Ω in the representation R. We shall evaluate **P** for the B_2 representation in point group $\mathscr{C}_{2v}$. We may ignore the $1/p$ and any subsequent multiplying factors since they will be absorbed in a final normalization, and we are concerned here with relative values for the SALCs:

$$\begin{aligned}\mathbf{P}_{B_2} &= \{\chi_{B_2}(\mathbf{E})\mathbf{E} - \chi_{B_2}(\mathbf{C}_2)\mathbf{C}_2 - \chi_{B_2}(\boldsymbol{\sigma}_v)\boldsymbol{\sigma}_v + \chi_{B_2}(\boldsymbol{\sigma}'_v)\boldsymbol{\sigma}'_v\} \\ &= \{\mathbf{E} - \mathbf{C}_2 - \boldsymbol{\sigma}_v + \boldsymbol{\sigma}'_v\}\end{aligned} \tag{3.13}$$

and the SALC is given by

$$\mathbf{P}_{B_2}\psi = \{\mathbf{E} - \mathbf{C}_2 - \boldsymbol{\sigma}_v + \boldsymbol{\sigma}'_v\}\psi \tag{3.14}$$

In a similar manner $\mathbf{P}_{A_1}\psi = \{\mathbf{E} + \mathbf{C}_2 + \boldsymbol{\sigma}_v + \boldsymbol{\sigma}'_v\}\psi$, whereas $\mathbf{P}_{B_1} = \mathbf{P}_{A_2} = 0$ and they do not form representations for the water molecule. If we consider the atomic orbital ψ_a on H_a as a generating function, we may derive the following scheme:

$\mathscr{C}_{2v}$	**E**	$\mathbf{C_2}$	$\boldsymbol{\sigma}_v$	$\boldsymbol{\sigma}'_v$
ψ_a	ψ_a	ψ_b	ψ_b	ψ_a
a_1	1	1	1	1
b_2	1	−1	−1	3

Thus, the a_1 combination, $\psi_a+\psi_b$, when normalized is $(1/\sqrt{2})(\psi_a+\psi_b)$ and the b_2 combination $(1/\sqrt{2})(\psi_a-\psi_b)$, which are in agreement with the SALCs established in the previous section.

We consider again the carbonate ion, as an example in which degenerate representations are important in bonding. With wavefunctions based on the three C−O bond vectors, we can derive symmetry-adapted orbitals. We have already obtained the representation

$\mathscr{D}_{3h}$	**E**	$2C_3$	$3C_2$	σ_h	$2S_3$	$3\sigma_v$	
Γ_σ	3	0	1	3	0	1	$=A'_1+E'$

In applying (3.13), we need to consider *all* operations of the group with nonzero characters, which we did implicity in the above example in $\mathscr{C}_{2v}$. Referring again to Figure 3.8, we align

[8] See D M Bishop, *Group Theory and Chemistry* (Clarendon Press, 1973).

the three $\mathbf{C}_2$ and three $\boldsymbol{\sigma}_v$ operators with the $\boldsymbol{r}$ vectors, as shown. Then we can work out the following relationships for a typical vector $\boldsymbol{r}_1$:

$\mathcal{D}_{3h}$	$\mathbf{E}$	$\mathbf{C}_3$	$\mathbf{C}_3^2$	$\mathbf{C}_2$	$\mathbf{C}_2'$	$\mathbf{C}_2''$	$\boldsymbol{\sigma}_h$	$\mathbf{S}_3$	$\mathbf{S}_3^2$	$\boldsymbol{\sigma}_v$	$\boldsymbol{\sigma}_v'$	$\boldsymbol{\sigma}_v''$
$\boldsymbol{r}_1$	$\boldsymbol{r}_1$	$\boldsymbol{r}_2$	$\boldsymbol{r}_3$	$\boldsymbol{r}_1$	$\boldsymbol{r}_3$	$\boldsymbol{r}_2$	$\boldsymbol{r}_1$	$\boldsymbol{r}_2$	$\boldsymbol{r}_3$	$\boldsymbol{r}_1$	$\boldsymbol{r}_3$	$\boldsymbol{r}_2$

If we use this pattern of displacements with the characters for A_1' and apply (3.12) to wavefunctions ψ_i, in place of vectors $\boldsymbol{r}_i$, we obtain $(\psi_1+\psi_2+\psi_3)$, and the normalized result for a_1' is

$$\psi_{\mathrm{a}_1'}=(1/\surd 3)(\psi_1+\psi_2+\psi_3)$$

In a similar manner, we obtain with the E' characters

$$\psi_{\mathrm{e}_1'}=(1/\surd 6)\,(2\psi_1-\psi_2-\psi_3)$$

Another SALC must be generated to give a pair required under E' symmetry in $\mathcal{D}_{3h}$. If we use $\boldsymbol{r}_2$ as a typical vector, we obtain $\psi_{\mathrm{e}}^*=(1/\surd 6)\,(2\psi_2-\psi_3-\psi_1)$, which is not orthogonal to $\psi_{\mathrm{e}_1'}$. A similar result is obtained with $\boldsymbol{r}_3$ in place of $\boldsymbol{r}_2$. If we choose the linear combination $\psi_{\mathrm{e}}^*+\psi_{\mathrm{e}_1'}/2$, then under the nonzero operations of Γ_σ we obtain the following scheme:

$\mathcal{D}_{3h}$	$\mathbf{E}$	$\mathbf{C}_2$	$\mathbf{C}_2'$	$\mathbf{C}_2''$	$\boldsymbol{\sigma}_h$	$\boldsymbol{\sigma}_v$	$\boldsymbol{\sigma}_v'$	$\boldsymbol{\sigma}_v''$
$\boldsymbol{r}_2-\boldsymbol{r}_3$	$\boldsymbol{r}_2-\boldsymbol{r}_3$	$\boldsymbol{r}_3-\boldsymbol{r}_2$	$\boldsymbol{r}_2-\boldsymbol{r}_1$	$\boldsymbol{r}_1-\boldsymbol{r}_3$	$\boldsymbol{r}_2-\boldsymbol{r}_3$	$\boldsymbol{r}_3-\boldsymbol{r}_2$	$\boldsymbol{r}_2-\boldsymbol{r}_1$	$\boldsymbol{r}_1-\boldsymbol{r}_3$

which, when multiplied by the characters of E' and summed, gives $6(\boldsymbol{r}_2-\boldsymbol{r}_3)$. Proceeding as before, leads to the normalized SALC

$$\psi_{\mathrm{e}_1'}=(1/\surd 2)\,(\psi_2-\psi_3)$$

which is orthogonal to both $\psi_{\mathrm{a}_1'}$ and $\psi_{\mathrm{e}_1'}$.

The reader is invited to show that the SALCs for π bonding in the $[CO_3]^{2-}$ ion ($\psi_{\mathrm{a}_2''}$ and $\psi_{\mathrm{e}''}$) are of the same form as $\psi_{\mathrm{a}_1'}$ and $\psi_{\mathrm{e}'}$, respectively. An energy-level diagram for this π bonding has been given in Figure 3.9b.

Finally here, we consider the octahedral hexacyanoferrate(II) ion, $[Fe(CN)_6]^{4-}$. In the first transition series of elements, atomic orbitals are formed from the 3d, 4s and 4p electrons of the metal atom. The point group of the ion is $\mathbb{O}_h$, the octahedral group of order 48: it is centrosymmetric, and the subscript notation g or u is added to the representations, as shown in its character table below; g implies that the representation is symmetric under $\boldsymbol{i}$, whereas u implies antisymmetry under $\boldsymbol{i}$. The operation $\mathbf{C}_2'$ acts along the same line as $\mathbf{C}_4$, but it is in a different symmetry class from $\mathbf{C}_2$.

$\mathbb{O}_h$	$\mathbf{E}$	$8\mathbf{C}_3$	$6\mathbf{C}_2$	$6\mathbf{C}_4$	$3\mathbf{C}_2'$	$\boldsymbol{i}$	$6\mathbf{S}_4$	$8\mathbf{S}_6$	$3\boldsymbol{\sigma}_h$	$6\boldsymbol{\sigma}_d$		
A_{1g}	1	1	1	1	1	1	1	1	1	1		$x^2+y^2+z^2$
A_{2g}	1	1	−1	−1	1	1	−1	1	1	−1		
E_g	2	−1	0	0	2	2	0	−1	2	0		$(2z^2-x^2-y^2,\ x^2-y^2)$
T_{1g}	3	0	−1	1	−1	3	1	0	−1	−1	(R_x, R_y, R_z)	
T_{2g}	3	0	1	−1	−1	3	−1	0	−1	1		(xy, yz, zx)
A_{1u}	1	1	1	1	1	−1	−1	−1	−1	−1		
A_{2u}	1	1	−1	−1	1	−1	1	−1	−1	1		
E_u	2	−1	0	0	2	−2	0	1	−2	0		
T_{1u}	3	0	−1	1	−1	−3	−1	0	1	1	(T_x, T_y, T_z)	
T_{2u}	3	0	1	−1	−1	−3	1	0	1	−1		

In ligand-field theory, which is an expansion of the electrostatic crystal-field theory, we may use the six Fe–CN bond vectors to show, albeit somewhat laboriously, that the σ bond representation Γ_σ is given by

$\mathcal{O}_h$	**E**	$8C_3$	$6C_2$	$6C_4$	$3C_2'$	i	$6S_4$	$8S_6$	$3\sigma_h$	$6\sigma_d$
Γ_σ	6	0	0	2	2	0	0	0	4	2

which reduces to $A_{1g}+T_{1u}+E_g$. The ligand symmetry orbitals are denoted ψ_1 to ψ_6 with respect to the directions $\pm x$, $\pm y$ and $\pm z$, and the symmetry-adapted orbitals are set out below.

Representation	Metal atom orbitals	Normalized symmetry orbitals
A_{1g}	s	a_{1g}: $(1/\sqrt{6})(\psi_1+\psi_2+\psi_3+\psi_4+\psi_5+\psi_6)$
T_{1u}	p_z	t_{1u}: $(1/\sqrt{2})(\psi_1-\psi_6)$
	p_x	$(1/\sqrt{2})(\psi_2-\psi_4)$
	p_y	$(1/\sqrt{2})(\psi_3-\psi_5)$
E_g	d_{z^2}	e_g: $(1/\sqrt{12})(2\psi_1-\psi_2-\psi_3-\psi_4-\psi_5+2\psi_6)$
	$d_{x^2-y^2}$	$(1/2)(\psi_2-\psi_3+\psi_4-\psi_5)$

The molecular-orbital energy-level diagram for this complex is shown in Figure 3.11. The three t_{2g} orbitals have no matching representations among the six orbitals of the cyanide ions and are nonbonding, holding electrons originating from the d_{xy}, d_{yz} and d_{zx} orbitals of the central atom. The 12 other valence electrons occupy the bonding orbitals a_{1g}, t_{1u} and e_g with paired spins.

3.4.6.5 Vanishing integrals

We encountered this subject implicitly with overlap integrals, and more explicitly with orthogonality. We need to determine whether or no an integral

$$\mathcal{I}=\int \psi_1\psi_2\,d\tau \tag{3.15}$$

is finite or zero. If $\mathcal{I}=0$, then ψ_1 and ψ_2 do not overlap to form a bond. Since $\mathcal{I}$ and $d\tau$ are each unchanged under any symmetry operation, it follows that $\mathcal{I}$ will be finite only if $\psi_1\psi_2$ is nonzero, that is, the product must be a basis for an A_1 irreducible representation.

In studying the carbonate ion, we derived the representation

$$\Gamma_\pi \quad 3 \quad 0 \quad -1 \quad -3 \quad 0 \quad 1 \quad = A_2''+E''$$

for the π bonding. To test for a nonzero value of (3.15) here, we examine each of the representations of $\mathcal{D}_{3h}$ against Γ_π in turn as follows:

Γ_π	3	0	−1	−3	0	1
A_1	1	1	1	1	1	1

Multiplying and adding, we obtain $3+0-1-3+0+1=0$. Hence, there can be no overlap between the hybrid π orbitals of oxygen and s orbitals of carbon (which have an A_1 representation). In fact, it turns out that the only nonzero integral is between Γ_π and carbon p_z, as we have shown; E'' remains nonbonding.

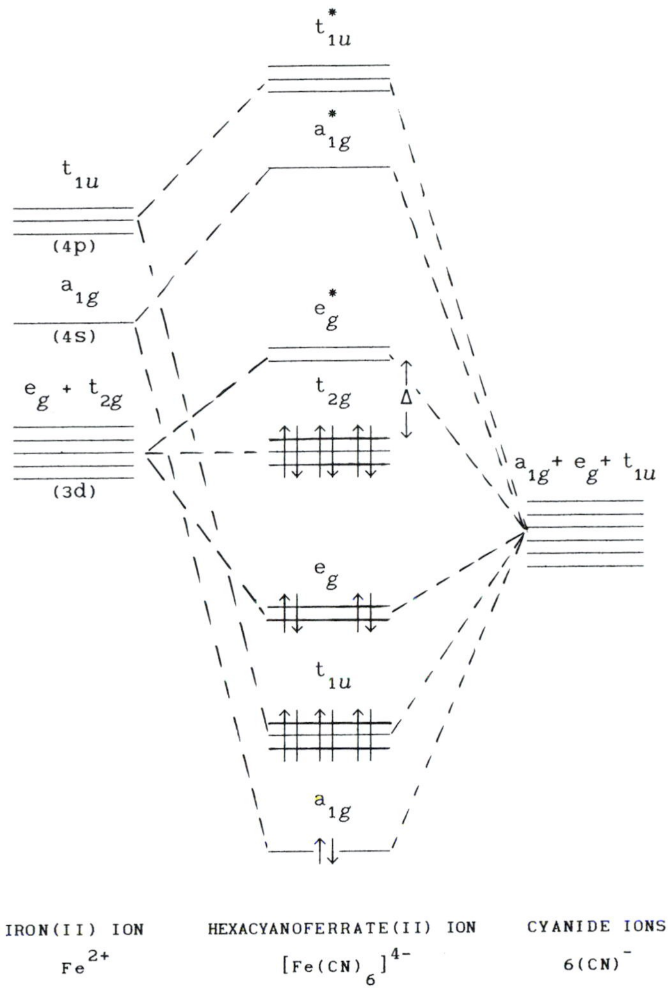

Figure 3.11 Molecular-orbital energy-level diagram for the hexacyanoferrate(II) ion, $[Fe(CN)_6]^{4-}$; sp^3d^2 bonding takes place through the a_{1g}, t_{1u} and e_g molecular orbitals of the complex ion, with nonbonding electrons in t_{2g} that undergo π interaction.

A linear molecule, such as HF, has the infinite point group $\mathscr{C}_{\infty v}$, and the character table for this group is shown below; ϕ is an angle of infinitesimal value, which may be clockwise or anticlockwise:

$\mathscr{C}_{\infty v}$	**E**	$2C_{\infty}^{\phi}$	σ_v		
A_1	1	1	1	T_z	z^2; x^2+y^2
A_2	1	1	−1	R_z	
E_1	2	$2\cos(\phi)$	0	(T_x, T_y); (R_x, R_y)	
E_2	2	$2\cos(2\phi)$	0		(yz, zx)
E_3	2	$2\cos(3\phi)$	0		(x^2-y^2, xy)
. .	. .	. .	. .		

Rotation about the z axis, the H−F bond vector direction, by any angle ϕ modifies x and y by $\cos(\phi)$ and leaves z unchanged. Since the hydrogen 1s atomic orbital is the fully symmetrical a_1, bonding with fluorine is only through the fully symmetrical fluorine a_1 (p_z orbital); no other product of of A_1 and any other representation in $\mathscr{C}_{\infty v}$ contains A_1, whatever the value of ϕ.

We shall have recourse to some of these group-theory procedures in studying other areas of physical chemistry.

3.5 Symmetry of crystals

We conclude our discussion on symmetry with an introduction to the symmetry of crystals and crystal structures. Crystals possess one of the point groups that we have considered already. Because of microscopic defects in real crystals, some individual unit cells (Section 3.5.1ff.) may differ from others in minute detail, but an experimental average result over a very large number of unit cells corresponds to one of the 32 crystallographic point groups.

A cube has symmetry $m3m$ ($\mathbb{O}_h$), and a tetrahedron $\bar{4}3m$ ($\mathscr{T}_d$). Here, we are using the Hermann-Mauguin symmetry notation, which is conventional in crystallography because of its greater value in space-group theory. In Appendix 1, we give a table of comparisons of the Schönflies and Hermann–Mauguin notations for point groups.

In the Hermann–Mauguin notation, a point-group symbol generally defines either one direction or three, in relation to the x, y and z *crystallographic* axes,[9] which arises from the fact that two interacting symmetry elements introduce a third, according to (3.2). Thus, point group $2/m$ ($\mathscr{C}_{2h}$) belongs to the monoclinic system, and the (single) symbol means that there is a twofold axis (2) along y, with a reflection plane (m) normal to it, the plane x–z (Figure 3.12a). Similarly, mmm ($\mathscr{D}_{2h}$) is orthorhombic (Figure 3.12b) and the symbol indicates m planes normal to the x, y and z axes, in that order. In the trigonal system, the combination of the operations m and 3 introduces other m planes: but they are of the same symmetry class and the symbol $3m$ is used; $3mm$ is irrelevant and incorrect. For a full description of the Hermann–Mauguin notation and crystal symmetry, the reader is directed to a standard text on the subject.[10]

3.5.1 Lattices and unit cells

The geometrical basis of a crystal structure is its (Bravais) lattice, which is *a regular arrangement of points in space, of infinite extent, such that each point has the same vector environment as every other point*. Figures 3.13a, b and c show unit cells of a monoclinic P lattice, orthorhombic I lattice and cubic F lattice. Strictly, labels such as these refer to the unit cell that is chosen to represent the lattice, but it is customary to speak of, say, an I lattice, with this understanding. It is important to think of the unit cell of a lattice *not* as a finite body, but as a representative portion of a three-dimensional spatial arrangement, either of lattice points or of units of structure. Thus, the lattice points may be shared among adjacent unit cells, corner points among eight (Figure 3.13a) and face-centring points among two unit cells and so on. We may apply space-group theory in practice because the ratio of the size of an experimental crystal to that of the unit cell is very large, approximately 10^{19}.

[9] Crystallographic axes are chosen in relation to the symmetry elements of a crystal. Thus, in the orthorhombic system, they are a Cartesian set, along the twofold symmetry axes. In the hexagonal system, x and y are at 120° to each other and normal to z.

[10] See, for example, M F C Ladd *Symmetry in Molecules and Crystals* (Ellis Horwood, 1991), and M F C Ladd and R A Palmer *Structure Determination by X-Ray Crystallography* (3rd Edition, Plenum Press, 1994).

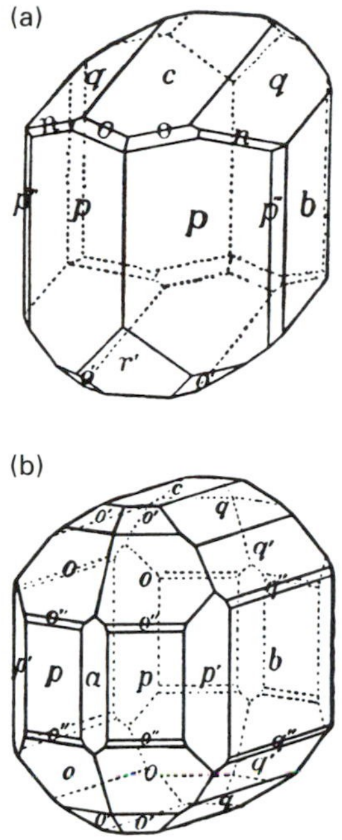

Figure 3.12 Examples of crystals: (a) monoclinic, point group $2/m$; the m plane is parallel to the face lettered b, and the twofold axis intersects this plane normally, at the centre of the crystal. Faces carrying the same letter are related across the m plane. (b) Orthorhombic, point group mmm; the m planes are parallel to the faces lettered a, b and c, and intersect in a point ($\bar{1}$) at the centre of the crystal. Notice that the faces occur in groups of 2, 4 and 8. How do they relate to the point-group symmetry?

There are fourteen Bravais lattices distributed, unequally, among seven crystal systems. Each system has a characteristic, minimum symmetry and particular (conventional) unit cell relationships, as Table 3.1 shows.

We list three important features of this table.

(a) The inversion axis $\bar{n}$ is used in crystallography, rather than the alternating axis S_n, and the symbol $\neq$ should be read as 'not constrained by symmetry to equal', rather than just 'not equal to'.
(b) A two-dimensional unit cell with $a=b$ and $\gamma=120°$ is compatible with sixfold and threefold symmetry. Hence, the hexagonal lattice (P unit cell) may be used for crystals in the trigonal system, with 3 along the line [0, 0, z]. However, the presence of threefold axes in the unit cell, along [$\frac{2}{3}$, $\frac{1}{3}$, z] and [$\frac{1}{3}$, $\frac{2}{3}$, z], introduces a lattice which is trigonal but has a triply primitive unit cell R_{hex}, with unique points at 0,0,0 and $\pm\frac{2}{3},\frac{1}{3},\frac{1}{3}$. This cell may be transformed to the primitive unit cell R (listed in the above table).
(c) The unit cell designations imply the unique lattice points per unit cell listed below:

P (R)	0, 0, 0
C	0, 0, 0; $\frac{1}{2}$, $\frac{1}{2}$, 0
I	0, 0, 0; $\frac{1}{2}$, $\frac{1}{2}$, $\frac{1}{2}$
F	0, 0, 0; 0, $\frac{1}{2}$, $\frac{1}{2}$ $\frac{1}{2}$, 0, $\frac{1}{2}$; $\frac{1}{2}$, $\frac{1}{2}$, 0

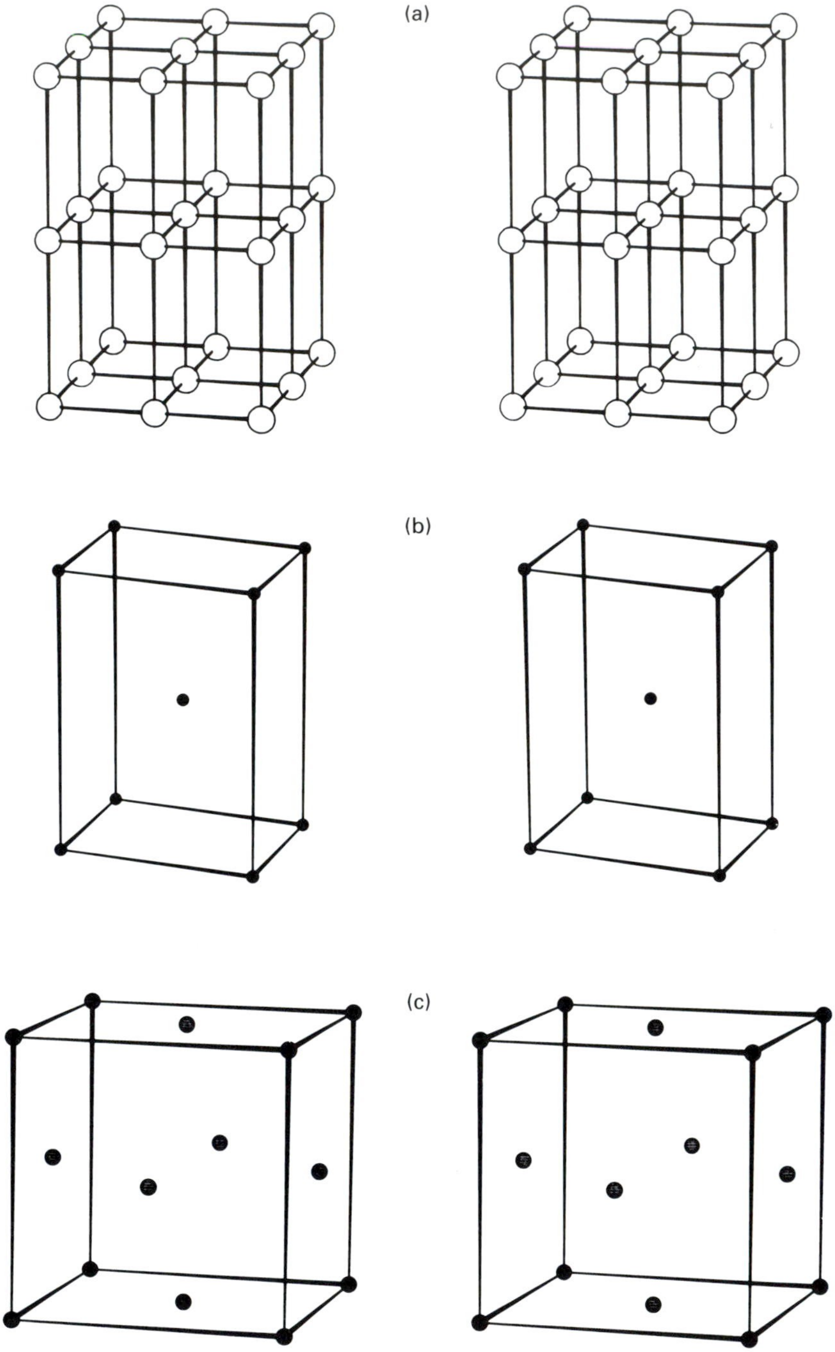

Figure 3.13 Stereoviews of unit cells and environs of lattices: (a) monoclinic *P* unit cell, showing eight adjacent unit cells, with *one* lattice point per unit cell volume; (b) orthorhombic *I* unit cell, with *two* points per unit cell; (c) cubic *F* unit cell, with *four* points per unit cell.

Table 3.1 Crystal systems and Bravais lattices *(The axial relationships refer to the conventional unit cell)*

System	Characteristic symmetry	Unit cell symbols	Axial relationships	Symmetry at points
Triclinic	None	P	$a \neq b \neq c$ $\alpha \neq \beta \neq \gamma \neq 90°, 120°$	$\bar{1}$
Monoclinic	2 or $\bar{2}$ along y	P, C	$a \neq b \neq c$ $\alpha = \gamma = 90°$ $\beta \neq 90°, 120°$	$2/m$
Orthorhombic	Mutually ⊥2 or $\bar{2}$ along x, y and z	$P, C,$ I, F	$a \neq b \neq c$ $\alpha = \beta = \gamma = 90°$	mmm
Tetragonal	4 or $\bar{4}$ along z	P, I	$a = b \neq c$ $\alpha = \beta = \gamma = 90°$	$\frac{4}{m}mm$
Cubic	Four 3 along ⟨111⟩	P, I F	$a = b = c$ $\alpha = \beta = \gamma = 90°$	$m3m$
Hexagonal	6 or $\bar{6}$ along z	P	$a = b \neq c$ $\alpha = \beta = 90°$ $\gamma = 120°$	$\frac{6}{m}mm$
Trigonal	3 along [111]	R	$a = b = c$ $\alpha = \beta = \gamma \neq 90°$ and $< 120°$	$\bar{3}m$

Any lattice point may be taken as the origin O of the lattice, and the distance $\boldsymbol{r}$ of any other point from O is given by

$$\boldsymbol{r} = U \cdot \boldsymbol{a} + V \cdot \boldsymbol{b} + W \cdot \boldsymbol{c} \tag{3.16}$$

where a, b and c are the unit cell sides, and U, V and W are the coordinates of the lattice point. Since the integers U, V and W may take both positive and negative values, every lattice, and its unit cell, is centrosymmetrical: $[UVW]$ defines a *direction* in the lattice, and $<UVW>$ is a set (form) of such directions related by symmetry.

Within a unit cell, the vector $\boldsymbol{d}$ from O to any point x,y,z is given by

$$\boldsymbol{d} = x \cdot \boldsymbol{a} + y \cdot \boldsymbol{b} + z \cdot \boldsymbol{c} \tag{3.17}$$

where $x < a$, $y < b$ and $z < c$. It is conventional to work with fractional coordinates; thus, $x = X/a$ and so on.

3.5.1.1 Translation unit cells

In the fields of solid-state spectroscopy and physics, it is normal practice to work with a primitive unit cell, whether or no the conventional unit cell is primitive. It is always possible to delineate a primitive cell in any latttice. The totality of these unit cells may be referred to as *translation unit cells*,[11] notwithstanding seven of them are the same as the conventional unit cells.

[11] M F C Ladd, 'The Language of Lattices and Cells', *J. Chem. Educ.* **74**, 461 (1997).

The face-centred unit cell (Figure 3.13c), for example, may be transformed to a translation unit cell, rhombohedral in shape, by the matrix

$$\begin{array}{cc} & F \\ P & \begin{bmatrix} -\frac{1}{2} & \frac{1}{2} & \frac{1}{2} \\ \frac{1}{2} & -\frac{1}{2} & \frac{1}{2} \\ \frac{1}{2} & \frac{1}{2} & -\frac{1}{2} \end{bmatrix} \end{array}$$

It is important to note that, if this rhombohedral cell is considered in isolation, the symmetry at each lattice point is $\bar{3}m$ ($\mathcal{D}_{3d}$). Thus, this choice of cell may tend to disguise the fact that it derives from a higher, cubic class, which has symmetry $m3m$ ($\mathcal{O}_h$) at each point; $\bar{3}m$ is a subgroup[12] of $m3m$. The conventional unit cell may be derived from its translation unit cell by a Delaunay reduction,[13] and would show the true cubic symmetry.

3.5.2 Reciprocal lattice

For each Bravais lattice there is a corresponding reciprocal lattice of the same symmetry, which may be derived geometrically by the following construction. Figure 3.14a shows a monoclinic P lattice in projection on the x–z plane. Families of parallel equidistant planes, shown by their traces in the figures, may be described through the Bravais lattice points. From the origin O, lines are constructed normal to the families of planes (hkl).[14] Points are marked off along each of these lines such that the distance of any first point from O is inversely proportional to the corresponding interplanar spacing, that is,

$$d^*(hkl) = K/d(hkl) \tag{3.18}$$

where K is a constant, normally either unity or an X-ray wavelength. Thus, the first point along OP, the normal to the (100) family of planes in real space, is labelled 100 in reciprocal space.

The reciprocal lattice is shown in projection in Figure 3.14b; the particular reciprocal lattice points 100, 010 (the first point in reciprocal space along the normal to the $(0k0)$ family) and 001 define the reciprocal unit cell:[15]

$$a = K\frac{bc\sin(\alpha)}{V} \tag{3.19}$$

$$\cos(\alpha^*) = \frac{\cos(\beta)\cos(\gamma) - \cos(\alpha)}{\sin(\beta)\sin(\gamma)} \tag{3.20}$$

with corresponding equations for other sides and angles; the volume V of the unit cell is given by

$$V = abc[1 - \cos^2(\alpha) - \cos^2(\beta) - \cos^2(\gamma) + 2\cos(\alpha)\cos(\beta)\cos(\gamma)]^{1/2} \tag{3.21}$$

[12] A subgroup is a subset of elements of a group of higher order, the subset also being a group.

[13] See, for example, *International Tables for X-Ray Crystallography*, Volume I, edited by N F M Henry and K Lonsdale (Kynoch Press, 1965).

[14] The Miller indices are integers h, k and l that are the ratios of the fractional intercepts on a, b and c of the first plane in the family, measured from the origin of the unit cell.

[15] More generally $\boldsymbol{a}^* = (\boldsymbol{b}\times\boldsymbol{c})/(\boldsymbol{a}\cdot\boldsymbol{b}\times\boldsymbol{c})$, with $\boldsymbol{b}^*$ and $\boldsymbol{c}^*$ by cyclic permutation, where $\boldsymbol{a}$, $\boldsymbol{b}$ and $\boldsymbol{c}$ are the delineating vectors of the translation unit cell in real space. Here, $\boldsymbol{b}\times\boldsymbol{c} = \boldsymbol{i}bc\sin(\boldsymbol{b}\wedge\boldsymbol{c})$, where $\boldsymbol{i}$ is a unit vector normal to the plane of $\boldsymbol{b}$–$\boldsymbol{c}$ such that $\boldsymbol{b}$, $\boldsymbol{c}$ and $\boldsymbol{i}$ form a right-handed vector triplet; $\boldsymbol{a}\cdot\boldsymbol{b}\times\boldsymbol{c}$ is the volume of the translation unit cell.

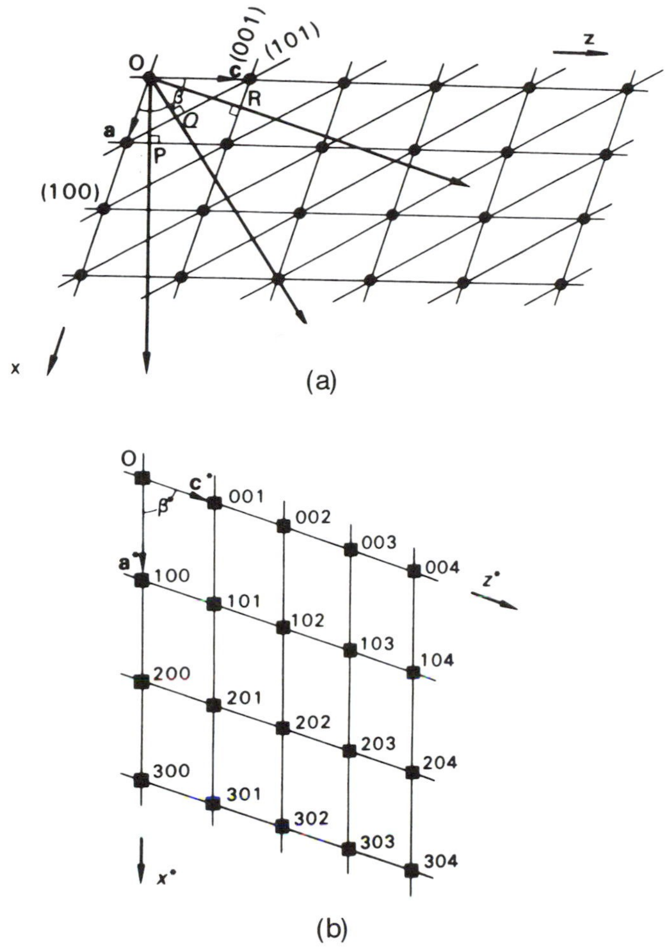

Figure 3.14 Geometrical development of a reciprocal lattice: (a) monoclinic *P* unit cells, as seen in projection along the $-y$ direction in the lattice, showing the (100), (001) and (101) families of parallel, equidistant planes; (b) corresponding monoclinic *P* reciprocal unit cell in the reciprocal lattice; the geometry of planes has been transformed to a geometry of points.

We shall make use of the reciprocal lattice concept in studying X-ray diffraction in Section 3.6.11ff.

3.5.3 Space groups

If a pair of ions $Na^+ \ldots Cl^-$, of fixed separation, be placed in identical orientation at each point of the cubic *F* lattice (Figure 3.13c), the structure of sodium chloride (Figure 1.2) is obtained. Thus, a structure comprises a *pattern motif*, often called the *asymmetric unit* of structure, and a *repeat mechanism*, the space group, operating over three-dimensional space to an indefinite extent. Space groups differ from point groups in possessing *translational* symmetry. Following (3.1), we may write the more general equation for a symmetry operation on a point $\boldsymbol{x}$ (x,y,z)

$$A\,\boldsymbol{x}+\boldsymbol{t}=\boldsymbol{x}' \tag{3.22}$$

where $\boldsymbol{t}$ is a translation vector, identically zero in point groups. If we take the point 0.1, 0.2, 0.3, then the effect of body-centring translation is exemplified by (3.22) *in extenso*:

$$\underset{A}{\begin{bmatrix} 1 & 0 & 0 \\ 0 & 1 & 0 \\ 0 & 0 & 1 \end{bmatrix}} \underset{\boldsymbol{x}}{\begin{bmatrix} 0.1 \\ 0.2 \\ 0.3 \end{bmatrix}} + \underset{\boldsymbol{t}}{\begin{bmatrix} 1/2 \\ 1/2 \\ 1/2 \end{bmatrix}} = \underset{\boldsymbol{x}'}{\begin{bmatrix} 0.6 \\ 0.7 \\ 0.8 \end{bmatrix}} \tag{3.23}$$

In general, translational symmetry includes screw axes and glide planes. A screw axis is of the form n_p (p=2, 3, 4, 6; $p<n$)[16] and is a *single* operation consisting of rotation by 360°/*n* *plus* a fractional translation of (*p*/*n*) times the repeat distance in the direction of the screw axis. The spiral staircase in Figure 3.15a is an example of a 6_1 (vertical) screw axis. If it were of infinite length, and assuming its geometrical perfection, each 6_1 operation would map the staircase on to itself.

Glide plane symmetry is shown in Figure 3.15b. A rectangular unit cell may be outlined by joining the tips of the noses of four nearest 'light' men, all facing to the right. Then the light man within the unit cell is related to the other four by the operation of reflection across a (vertical) glide plane *plus* a translation in a direction contained by the plane. Generally, translations are of the form *a*/2, (*b*+c)/2, (*c*+*a*)/4 and so on. Similar arguments apply to the 'dark' man in Figure 3.15b, too. Operation of the space group symmetry maps the whole pattern on to itself.

If we regard Figure 3.15b as representing a structure in the monoclinic system, as seen along the direction -*c*, then its space-group symbol would be *Pa*: *P* signifies a primitive unit cell (or lattice), and *a* symbolizes an *a* glide plane, with a translational component of *a*/2, as in Figure 3.15b.[17]

3.5.3.1 Space groups $P2_1/c$ and *Imma*

We consider two examples of space groups in a little more detail: first, $P2_1/c$, a monoclinic space group of frequent occurrence, and then the orthorhombic space group *Imma.*

Figure 3.16 is a conventional representation of the symmetry elements and representative points for this space group. The power of the Hermann–Mauguin notation can be seen in this symbol. Evidently, $P2_1/c$ is related to point group 2/*m*: Thus, by the notation, a 2_1 axis (graphic symbol ⇀, translation *b*/2) is parallel to the *y* axis, with a *c* glide plane (graphic symbol ……; translation *c*/2) normal to *y*.[18] The symmetry elements are then positioned so that a centre of symmetry $\bar{1}$ (graphic symbol o) coincides with the origin of the unit cell.

The coordinates of the general equivalent positions (4, set e, symmetry 1) are listed. By considering how each *general equivalent position* may be reached from any other one of them by a *single* symmetry operation, all symmetry elements in the space group are invoked in turn.

16 n_p with $p=n$ gives rise to the pure rotation *n*.

17 *Pa* is equivalent to *Pc* (standard symbol) under the transformation $\boldsymbol{a}'=\boldsymbol{c}$, $\boldsymbol{b}'=\boldsymbol{b}$, $\boldsymbol{c}'=-\boldsymbol{a}$.

18 In general, *n*/*m* implies a symmetry plane *m normal* to a symmetry axis *n*. For a general discussion of the Hermann–Mauguin symmetry notation, see, for example, M F C Ladd, *Symmetry in Molecules and Crystals*, (Horwood 1992), and M F C Ladd and R A Palmer, *Structure Determination by X-Ray Crystallography*, 3rd edition (Plenum, 1994).

(a)

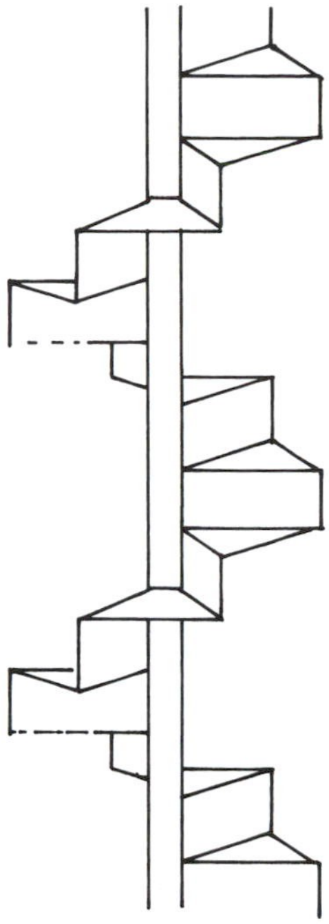

(b)

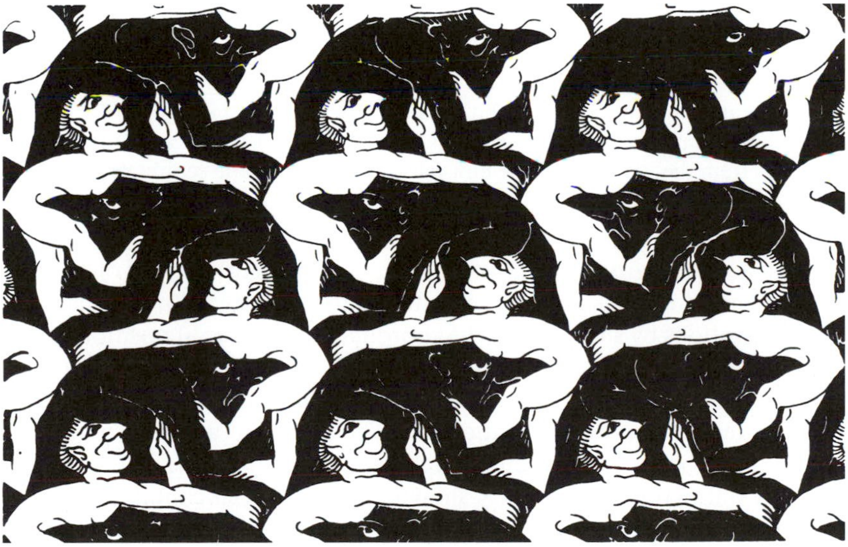

Figure 3.15 Translational symmetry elements: (a) spiral staircase, an illustration of a (vertical) 6_1 (6_5) screw-axis; (b) a (vertical) glide-plane relates the 'light' or the 'dark' men to themselves. (Reproduced from C H MacGillavry, *Symmetry Aspects of M C Escher's Periodic Drawings* by permission of A Oosthoek, Utrecht.)

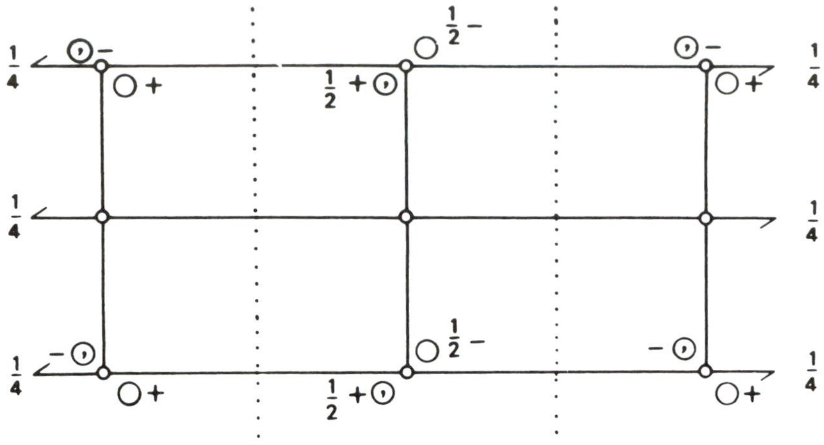

Origin on $\bar{1}$

4	e	1	x, y, z;	$x, \frac{1}{2} - y, \frac{1}{2} + z$
			$\bar{x}, \bar{y}, \bar{z}$;	$\bar{x}, \frac{1}{2} + y, \frac{1}{2} - z$
2	d	$\bar{1}$	$\frac{1}{2}, 0, \frac{1}{2}$;	$\frac{1}{2}, \frac{1}{2}, 0$
2	c	$\bar{1}$	$0, 0, \frac{1}{2}$;	$0, \frac{1}{2}, 0$
2	b	$\bar{1}$	$\frac{1}{2}, 0, 0$;	$\frac{1}{2}, \frac{1}{2}, \frac{1}{2}$
2	a	$\bar{1}$	$0, 0, 0$;	$0, \frac{1}{2}, \frac{1}{2}$

Figure 3.16 General equivalent positions and symmetry elements in space group $P2_1/c$ (the standard drawings for space groups have the *x* axis running top to bottom, *y* running left to right and *z* directed upwards from the *x–y* plane). The unit cell contains four more centres of symmetry and two more 2_1 axes, unique to the unit cell, at 1/2 and 3/4, respectively, along the *z* axis.

Four sets of *special equivalent positions* (2, sets a–d, symmetry $\bar{1}$), sited on *nontranslational* symmetry elements, are also listed. It is easy to see that the special positions form a subset of the general positions by substituting the coordinates of a special position, say, $\bar{1}$ at 0,0,0, into the general set. It should be noted that centres of symmetry and 2_1 screw axes repeat at 1/2 and 3/4 along the *z* axis; they are not immediately obvious from the diagram.

Each unique general equivalent position could be the site of an atom, a molecule or part of a molecule. Biphenyl belongs to this space group. Measurements of the unit cell dimensions (by X-ray diffraction) and density show that the unit cell contains two molecules of biphenyl. From the space-group data, it follows that they lie in special positions on centres of symmetry, say 0,0,0 and $0,\frac{1}{2},\frac{1}{2}$. Thus, the molecule is centrosymmetric and, hence, planar. This result is interesting because the stable configuration for the free molecule has the ring planes at an angle of about 45° to each other. In the crystal, the planar configuration is stabilized by π-electron overlap between adjacent molecules.

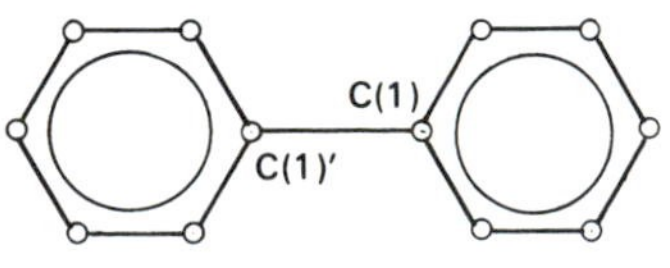

To solve the structure of this compound, we need to determine the positions for only one of the biphenyl rings; the space-group symmetry can then generate a crystal of macroscopic dimensions.

The orthorhombic space group *Imma* is related to point group *mmm*; hence, there are, by the notation, *m* planes normal to the *x* and *y* axes and an *a* glide plane normal to *z*; the unit cell is body centred. The planes are arranged such that a centre of symmetry (2/*m*) is at the origin. This condition requires that $m_{\perp x}$ intersects the *x* axis at *a*/4, $m_{\perp y}$ the *y* axis at 0 and $a_{\perp z}$ the *z* axis at 0, as shown. Since the *full* symbol for point group *mmm* is $\frac{2}{m}\frac{2}{m}\frac{2}{m}$, we can expect 2 and/or 2_1 symmetry axes parallel to the *x*, *y* and *z* axes. Note the other interleaving planes and axes. They arise on account of the *I* centring condition; draw diagrams for *Pmma* and examine its differences from *Imma*.

Figure 3.17a shows the general equivalent positions, and Figure 3.17b the symmetry elements and the coordinates of the general and special positions for *Imma*. The two sets of translations at the head of the list of coordinates are applied to all the coordinates listed below them to take account of the *I* centring.

Potassium dimercury KHg_2 crystallizes with this space group, and there are four molecules in the unit cell. Considerations of the atomic dimensions relative to the unit cell enabled sets (c), (d), (f), (g), (i) to be ruled out, leaving three models for the Hg atom positions:

I 4 Hg in (a)+4 Hg in (e)
II 4 Hg in (b)+4 Hg in (e)
III 8 Hg in (h)

Further analysis showed that model III was correct, with 4 K in positions (b).

These two examples, as well as introducing space-group symmetry, show that a knowledge of the space group of a crystal is a very important aid in the determination of its structure.

3.6 Spectroscopic methods in structure determination

In this section we shall consider some of the spectroscopic techniques used in determining structural and physical data on chemical species. First, however, we consider in outline how spectral data are obtained.

3.6.1 Experimental procedure in spectroscopy

When a species is irradiated it can absorb energy and change to an excited, or higher energy, state. On subsequent reversion to its original state, the excitation energy is degraded into heat, lattice vibrations or radiation emission, depending upon the nature of the species under investigation. When radiation is emitted, it has a wavenumber $\overline{\nu}$ given by

$$\overline{\nu}=(E_2-E_1)/hc \tag{3.24}$$

where E_1 and E_2 are, respectively, the lower and higher energy states involved in the transition.

Absorption and emission spectroscopy are both governed by (3.24), the former being the more commonly used technique; emission methods are usually confined to the ultraviolet and visible regions of the spectrum. The radiation source in an absorption spectrometer may be a heated ceramic filament coated with rare-earth oxides (a Nernst emitter) for the infrared region, a heated tungsten filament for the visible region, or a hydrogen discharge lamp for the ultraviolet region of the spectrum; a laser is used where there is a need for a more intense radiation source of monochromatic or coherent radiation.

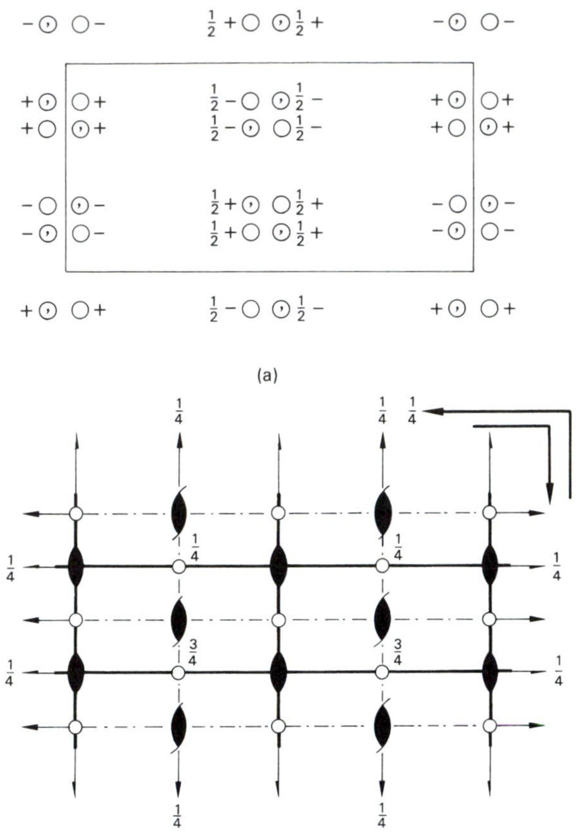

(a)

(b)

Origin on 2/*m*

$(0, 0, 0; \tfrac{1}{2}, \tfrac{1}{2}, \tfrac{1}{2})+$

16	j	1	$x, y, z;$	$x, \bar{y}, \bar{z}$	$x, \tfrac{1}{2}+y, \bar{z};$	$x, \tfrac{1}{2}-y, z$
			$\bar{x}, \bar{y}, \bar{z};$	$\bar{x}, y, z;$	$\bar{x}, \tfrac{1}{2}-y, z;$	$\bar{x}, \tfrac{1}{2}+y, \bar{z}$
8	i	*m*	$x, \tfrac{1}{4}, z;$	$\bar{x}, \tfrac{3}{4}, \bar{z};$	$\bar{x}, \tfrac{1}{4}, z;$	$x, \tfrac{3}{4}, \bar{z}$
8	h	*m*	$0, y, z;$	$0, \bar{y}, \bar{z};$	$0, \tfrac{1}{2}+y, \bar{z};$	$0, \tfrac{1}{2}-y, z$
8	g	2	$\tfrac{1}{4}, y, \tfrac{1}{4};$	$\tfrac{3}{4}, \bar{y}, \tfrac{3}{4};$	$\tfrac{3}{4}, y, \tfrac{1}{4};$	$\tfrac{1}{4}, \bar{y}, \tfrac{3}{4}$
8	f	2	$x, 0, 0;$	$\bar{x}, 0, 0;$	$x, \tfrac{1}{2}, 0;$	$\bar{x}, \tfrac{1}{2}, 0$
4	e	*mm*2	$0, \tfrac{1}{4}, z;$	$0, \tfrac{3}{4}, \bar{z}$		
4	d	2/*m*	$\tfrac{1}{4}, \tfrac{1}{4}, \tfrac{3}{4};$	$\tfrac{3}{4}, \tfrac{1}{4}, \tfrac{3}{4}$		
4	c	2/*m*	$\tfrac{1}{4}, \tfrac{1}{4}, \tfrac{1}{4};$	$\tfrac{3}{4}, \tfrac{1}{4}, \tfrac{1}{4}$		
4	b	2/*m*	$0, 0, \tfrac{1}{2};$	$0, \tfrac{1}{2}, \tfrac{1}{2}$		
4	a	2/*m*	$0, 0, 0;$	$0, \tfrac{1}{2}, 0$		

Figure 3.17 Space group *Imma*: (a) general equivalent positions; (b) symmetry elements. The centring of the unit cell introduces *b* glide and *n* glide planes and 2_1 screw axes. There are further symmetry elements at 1/4, 1/2 and 3/4 along the *z* axis. What are they?

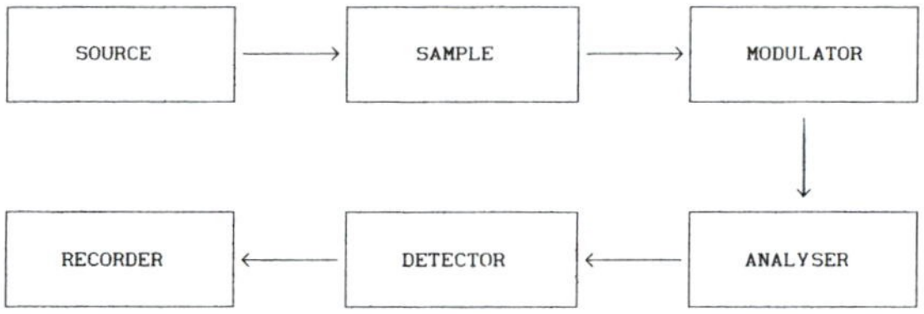

Figure 3.18 Schematic diagram of the apparatus for absorption spectrometry. The 'white' source radiates over a wide range of frequencies; the modulator converts a signal of frequency $\overline{\nu}$ to pulsed character, so that stable AC electronics may be employed in the subsequent stages; the analyser may be a grating or a Michelson-type interferometer; the detected signal is amplified and passed to the recorder, which may be a chart or an on-line computer.

The variation of absorption with frequency is determined, traditionally, by analysing the spectral radiation by means of a prism or a diffraction grating. The prism functions through the variation of refractive index with wavelength, and the diffraction grating by an interference process.

A schematic diagram of a modern absorption spectrometer for use between the infrared and ultraviolet regions of the spectrum is shown in Figure 3.18. The source radiation is reflected on to the sample and thence to an analyser, so that the individual frequencies can be selected in turn and passed to the detector. The detector converts the spectral radiation into an electric signal that is passed to a recording device operating synchronously with the analyser, thus producing either a trace on a chart recorder or a computer record of the spectrum. A modulator is introduced to convert the signal to alternating character. This procedure enables more stable AC electronics to be employed in the recording stages. In the microwave region the source frequency is varied and an analyser is not necessary.

3.6.1.1 Fourier-transform spectroscopy

It is common practice nowadays to use a Fourier-transform procedure in spectroscopy, particular with infrared and nuclear magnetic resonance techniques. In Fourier-transform infrared spectroscopy, a Michelson-type interferometer is used to analyse the spectrum. It functions by producing an interferogram which is the superposition of a series of waves, each of which represents a component in the spectrum in terms of intensity and wavenumber. A Fourier transformation of the interferogram then produces a well-resolved absorption spectrum of the species, with a good signal to noise ratio.

The spectral radiation is detected by converting it into an electric signal by means of a semiconducting device, or by use of photographic procedures, particularly with the ultraviolet and visible spectra. Figures 3.19a and 3.19b show portions of a NMR spectrum of a protein obtained both by normal and by Fourier-transform techniques. The improvement in resolution is clearly evident.

3.6.1.2 Spectral intensities

The intensities of spectral lines are determined by the probabilities of transitions, which are governed by quantum mechanics; by the population of energy levels, which are governed by a Boltzmann distribution; and by the path length of the sample, which is related to the spectral intensity I by the Beer–Lambert equation

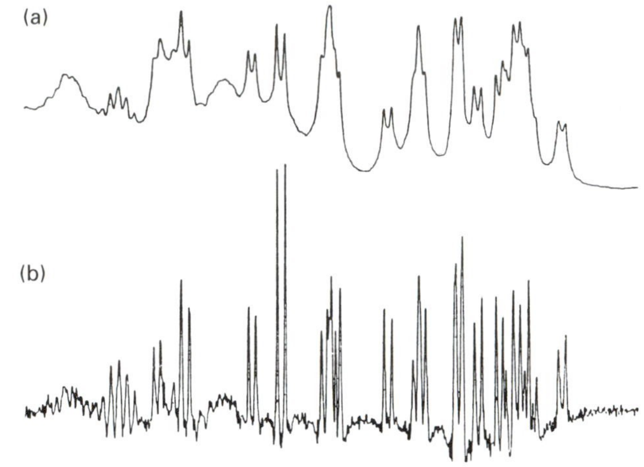

Figure 3.19 NMR spectra of a protein: (a) normal spectrum, (b) Fourier transform spectrum, showing greatly enhanced resolution. (After A de Marco and K Wüthrich, *J. Magn. Res.* **24**, 201, 1976.)

$$I = I_{inc} \exp(-\kappa l c) \tag{3.25}$$

where I_{inc} is the incident intensity, l and c are the path length and concentration of the sample, and κ is a constant.

3.6.2 Spectra of atoms

We introduced atomic spectra in Chapter 2, where we saw that the energy levels for the hydrogen atom given by the Bohr theory were confirmed, albeit fortuitously, by quantum mechanics. The spectra of atoms of higher atomic numbers are more complex, and we consider here the spectrum of sodium in the visible region of the spectrum.

Figure 3.20 is a photograph of a portion of the *P* series absorption spectrum for sodium. The first two members of this series are the closely spaced sodium D yellow lines. Subsequent members lie in the ultraviolet region, converging on a series limit $\overline{\nu}_\infty$, after which an energy continuum is formed. At the series limit an expelled electron has zero kinetic energy at 0 K, so that $\overline{\nu}_\infty$ is related to the ionization energy for that electron.

The wavenumber of a line in this spectrum is no longer dependent in a simple way on the quantum number *n*, but is well represented by the equation

$$\overline{\nu} = \overline{\nu}_\infty - R_\infty/(n-\delta)^2 \tag{3.26}$$

where R_∞ is the Rydberg constant, *n* is the principal quantum number of the higher level involved in the absorption and δ is a constant known as the quantum defect. We shall explore this equation with a recorded absorption spectrum.

3.6.2.1 Ionization energy of atomic sodium

The wavenumbers for eight lines from the spectrum of atomic sodium are listed in Table 3.2. Although both $\overline{\nu}_\infty$ and δ are unknown initially, it follows from (3.26) that a plot of $\overline{\nu}$ against $1/n^2$ will lead to a first estimate of δ. Then, by fitting $\overline{\nu}$ to $1/(n-\delta)^2$ by a least-squares method,

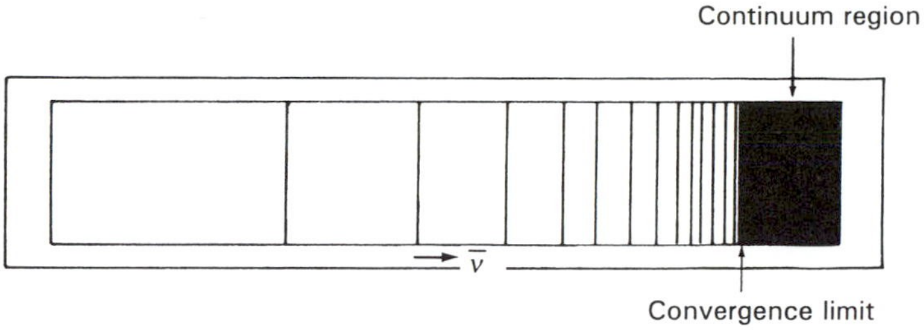

Figure 3.20 Portion of the absorption spectrum of atomic sodium, starting from $n=10$. The convergence limit (series limit) corresponds to $\bar{\nu}_\infty$ and is followed by an energy continuum.

improved estimates of both $\bar{\nu}_\infty$ and δ are obtained. This procedure is repeated until constant values are obtained for both $\bar{\nu}_\infty$ and δ; these stages are set out in Table 3.2.

The series limit is found to be 41449.71 cm^{-1}, and multiplication by hc gives the value of 5.1390 eV, or 495.84 kJ mol^{-1}, for the first ionization energy of atomic sodium. On account of the high precision obtainable with spectroscopic measurements, the precision of the ionization energy is ± 0.001 kJ mol^{-1}. Table 3.3 lists current values for the first ionization energies of the alkali metal atoms.

3.6.2.2 Electron affinities of the halogens

Another property that may be determined from the spectra of atomic species is the electron affinity E(X), relating to the process

$$X^-(g) + h\nu \rightarrow X(g) + e^- \tag{3.27}$$

Alkali-metal halides, such as RbI or CsCl, when heated ultrasonically by shock waves produce a vapour containing I^- or Cl^- species in abundance. Their ultraviolet spectra under these conditions are continua with sharp, low-energy thresholds that correspond to the photodetachment of electrons, in accordance with (3.27). The threshold frequencies for the halogen spectra have been assigned precisely and the results are listed in Table 3.4. The values of E(X) are accorded a negative sign because the attachment of a single electron, the reverse process of (3.27), is normally thermodynamically spontaneous for single-electron affinities.

3.6.2.3 Zeeman effect

The application of a strong magnetic field to a species undergoing spectral transitions removes the degeneracy of atomic energy levels and so provides information about the levels involved in transitions. For example, the p→d red line transition of cadmium at 643.85 nm in the absence of a magnetic field is a single line, but on application of the field three lines are observed, corresponding to $m_l=0, \pm 1$. The line frequencies are ν_0, the unchanged frequency, and $\nu_0 \pm \Delta\nu$; the doublet components are circularly polarized,[19] with $\nu_0 - \Delta\nu$ being left-handed and $\nu + \Delta\nu$ right-handed. This splitting is known as the Zeeman effect. Its magnitude is small; a field of about 1 T (10 kG) produces a line separation of approximately 0.03 nm in a wavelength of approximately 600 nm.

[19] Optical activity is the ability of a substance to rotate the plane of linearly polarized light. Linearly polarized light is made up of two waves that are *circularly polarized* in opposite directions, but each with the same frequency.

Table 3.2 Spectral wavenumbers[a] and the first ionization energy for sodium

n	$\bar{\nu}$/cm^{-1}	$[R_\infty/(n-\delta)^2]$/cm^{-1}	$\bar{\nu}_\infty$/cm^{-1}	δ	
		δ=0.81			
10	40137.2	1299.3		0.867	
11	40383.2	1056.8		0.871	0.87
12	40566.0	876.4		0.875	
13	40705.7	738.5		0.880	
14	40814.5	630.8		0.888	
15	40901.1	545.0		0.896	
16	40971.2	475.6		0.905	
17	41028.7	418.7		0.914	
			41452.7		
		δ=0.87			
		1316.5		0.853	
		1069.4		0.852	
		885.9		0.850	0.85
		745.8		0.848	
		636.5		0.847	
		549.6		0.845	
		479.4		0.842	
		421.8		0.838	
			41448.4		
		δ=0.85			
		1310.7		0.857	
		1065.2		0.858	
		882.7		0.859	
		743.4		0.859	
		643.6		0.861	0.86
		548.1		0.862	
		478.1		0.863	
		421.8		0.863	
			41450.1		
		δ=0.856[b]			
		1312.4		0.856	
		1066.4		0.857	
		883.6		0.856	
		744.1		0.855	
		635.2		0.857	
		548.5		0.857	
		478.5		0.857	
		421.0		0.856	
			41449.71		

Note:
[a] 1 cm^{-1}=100 m^{-1}.
[b] Subjective estimate from iterations 2 and 3.

Table 3.3 First ionization energies of the alkali metal atoms; their precision is ± 0.00001 eV atom^{-1}, or ± 0.001 kJ mol^{-1}

	I/eV atom^{-1}	I/kJ mol^{-1}
Li	5.39163	520.211
Na	5.13899	495.835
K	4.34062	418.804
Rb	4.17712	403.029
Cs	3.89382	375.695

Table 3.4 Electron affinities for the halogen atoms; their precision is ± 0.003 eV atom^{-1}, or ± 0.3 kJ mol^{-1} except where otherwise specified

	E/eV atom^{-1}	E/kJ mol^{-1}
F	-3.399 ± 0.002	328.0 ± 0.2
Cl	-3.613	-348.6
Br	-3.363	-324.5
I	-3.063	-295.5

The anomalous Zeeman effect refers to a splitting of multiplets into several components. The spin of an electron creates a magnetic moment of $g_e \gamma m_s \hbar$, where γ is the gyromagnetic ratio (see Section 3.6.9) and the g_e factor emerges from a relativistic treatment of the effect of an applied magnetic field as approximately 2.00232. Since the spin magnetic moment is larger than expected by the factor g_e, the total magnetic moment is not collinear with the total angular momentum and extra spectral lines appear.

When the applied magnetic field is very large, the spin and orbit moments uncouple and each aligns with the magnetic field. Transitions then occurring relate only to the orbital angular momentum, because optical frequencies do not affect electron spin. The spectrum then exhibits a normal Zeeman effect and this process of reversal is known as the Paschen–Back effect.

3.6.3 Spectra of molecules

Molecules, like atoms, undergo transitions by emission and absorption of electromagnetic radiation. However, unlike atoms, they possess energy of rotation of the whole molecule and energy of vibration among the component parts of the molecule. Not surprisingly, molecular spectra are more complex than those of atoms.

In both atoms and molecules, an interaction between any species and an incident radiation requires the presence of an electromagnetic effect in the species that can be influenced by the radiation. Normally, a spectrally active species has a magnetic or electric moment that can respond to irradiation. Different spectroscopic techniques invoke differing electromagnetic responses.

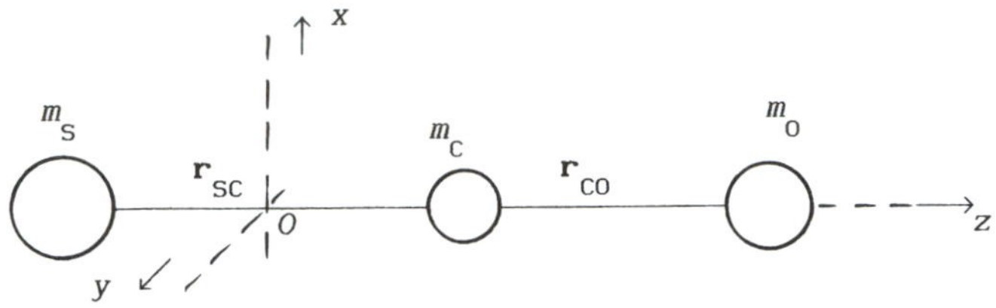

Figure 3.21 Axes for the moments of inertia of a linear molecule, illustrated here for the COS species; the molecular axis is along z.

3.6.4 Rotational spectra

A molecule at temperatures above absolute zero is able to rotate. Although the rotations take place about random axes, they can be resolved into components along three Cartesian axes set in a fixed relation to the molecule. The energy of a linear molecule follows from (2.67), but with l replaced by J, the rotational quantum number for the molecule that represents the result of the coupling of individual angular momenta:

$$E_{\text{rot}} = J(J+1)\hbar^2/2I \qquad (3.28)$$

J is equal to $(J_x^2+J_y^2+J_z^2)^{1/2}$, where J_x, J_y and J_z are the quantum numbers for the resolved rotational motion.

3.6.4.1 Moments of inertia

The moment of inertia of a molecule depends on its geometry. In general, it may be resolved into three components along Cartesian axes: I_z along the molecular axis, I_x perpendicular to the molecular axis and passing through the centre of mass O of the molecule, and I_y mutually perpendicular to and intersecting both I_x and I_z (Figure 3.21). Molecules may then be characterized as follows:

Linear	$I_z=0;\ I_x=I_y$	HCl, COS
Spherical top	$I_z=I_x=I_y$	CH_4, SF_6
Symmetric top	$I_z\neq I_x=I_y$	NH_3, CH_3F
Asymmetric top	$I_z\neq I_x\neq I_y$	H_2O, HCHO

3.6.4.2 Microwave spectra of linear molecules

A molecule that has a dipole moment, such as HBr, changes the direction of its dipole on rotation about any axis other than that of the molecule. In a pure rotational (unaccompanied by vibration) displacement, the centre of mass of the molecule is invariant, and the orientation of the dipole oscillates in the manner of the vectors of an electric field. Since rotational transitions are observed in the microwave region of the spectrum, such species are termed microwave active. Molecules such as H_2 are apolar and, consequently, microwave inactive.

We shall consider here only linear molecules. Equation (3.28) may be recast as

$$\bar{\nu}_{\text{rot}} = J(J+1)B \qquad (3.29)$$

where the rotational constant B is $\hbar/(4\pi cI)$. Selection rules on angular momentum require that $\Delta J=\pm 1$; it follows that a rotational spectrum should consist of equally spaced lines.

Table 3.5 Microwave spectral data for the COS molecule

$J-1$	→	J	ν/MHz	B/MHz
1		2	24325.92	
				12162.90
2		3	36488.82	
				12162.82
3		4	48651.64	
				12162.44
4		5	60814.08	
				12162.37
7		8	97301.19	
				12161.72
9		10	121624.63	
				12161.08
11		12	145946.79	
				12160.35
13		14	170267.49	

Table 3.5 list some lines in the microwave absorption spectrum of carbon oxysulfide, COS.

The variations in B with increasing frequency are significantly greater than the errors in the spectral measurements, and arise from the rigid rotor assumption used in (3.29), which was that bond lengths are invariant under radiation-induced rotation. In fact, a centrifugal distortion arises that must be taken into account in a more exact explanation of the spectra. For the purpose of showing how bond lengths may be determined, we focus attention on the first four lines in Table 3.5, among which the variation in B is small.

From the definition of moment of inertia for a body consisting of masses m_i of perpendicular distances r_i from the centre of mass O

$$I=\Sigma m_i r_i^2 \tag{3.30}$$

and of the centre of mass

$$\Sigma m_i r_i=0 \tag{3.31}$$

it may be shown (using Figure 3.21) for COS, albeit somewhat tediously, that

$$I=\frac{m_S m_C r_{SC}^2+m_O m_C r_{OC}^2+m_S m_O(r_{CS}+r_{CO})^2}{m_S+m_C+m_O} \tag{3.32}$$

Since the atomic masses are known, we have two unknowns that may be resolved through sets of spectra, $CO^{32}S$ and $CO^{34}S$. Thus, values of I are obtained from two equations (3.32), which may then be solved for the bond lengths. Results obtained in this way for the COS molecule are $r_{SC}=0.1559$ nm and $r_{CO}=0.1161$ nm.

3.6.4.3 Stark effect

Spectra in all frequency regions undergo a frequency shift on application of an electric field. This change is known as the Stark effect and its magnitude depends on the dipole moment of the species. Hence, rotational spectra may be used to obtain accurate data on dipole moments, a useful alternative to measurement through the relative permittivity of a bulk sample.

3.6.5 Vibrational spectra

In this section we shall consider mainly diatomic molecules, which behave something like a pair of point masses m_1 and m_2, of reduced mass μ, linked by a mechanical spring. The potential energy curve for a diatomic molecule is shown by Figure 2.20 (Section 2.11.1) and the energy zero corresponds to the equilibrium separation r_e. In the vicinity of r_e the potential energy V varies parabolically, and we have

$$V=\Bbbk(r-r_e)^2/2 \tag{3.33}$$

where $\Bbbk$ is the bond force-constant. Solving the Schrödinger equation of the form of (2.58) leads to the equation

$$E_{vib}=(\upsilon+\tfrac{1}{2})\hbar\omega_0 \tag{3.34}$$

where the vibrational quantum number υ takes the values 0, 1, 2, ..., and $\omega_0=(\Bbbk/\mu)^{1/2}$; in the harmonic approximation, ω_0 $(=2\pi\nu_0)$ is the normal frequency of oscillation. The selection rule in the harmonic approximation is $\Delta\upsilon=\pm1$, so that the vibrational harmonic energy levels would be equally spaced.

3.6.5.1 Anharmonic motion

As r diverges from the value of r_e the anharmonicity of the motion becomes evident (Figure 2.20). A more exact representation of this diagram is given by the Morse equation:

$$V=D_e\{1-\exp[-a(r-r_e)]\}^2 \tag{3.35}$$

where $a=\omega_e(\mu/2D_e)^{1/2}$. The difference between D_e and D_0, the dissociation energy referred to the ground state, is the zero-point energy, given by (3.34) with $\upsilon=0$. Using the Morse potential in the Schrödinger equation leads to the expression

$$E_{vib}=(\upsilon+\tfrac{1}{2})\hbar\omega_e-(\upsilon+\tfrac{1}{2})^2x_e\hbar\omega_e \tag{3.36}$$

where the anharmonicity constant x_e is equal to $\hbar\omega_e/(4D_e)$, and its value often lies between 0.01 and 0.02 for diatomic molecules.

Readers who are by now challenged to solve the type of differential equation to which we have referred may wish to consider Appendix 10.

Writing (3.36) as

$$E_{vib}=\hbar\omega_e[1-x_e(\upsilon+\tfrac{1}{2})](\upsilon+\tfrac{1}{2}) \tag{3.37}$$

it follows that

$$\omega_0=\omega_e[1-x_e(\upsilon+\tfrac{1}{2})] \tag{3.38}$$

Hence, as $\upsilon\to0$, so $\omega_0\to\omega_e$, because $x_e/2\ll1$; thus, ω_e is the oscillator frequency for infinitely small vibrations about the equilibrium position r_e. The energy difference for a transition $\upsilon_1\to\upsilon_2$ is readily shown, from (3.36), to be

$$\Delta E_{vib}=\hbar\omega_e\,\Delta\upsilon[1-x_e(1+\upsilon_1+\upsilon_2)] \tag{3.39}$$

where $\Delta\upsilon=\upsilon_2-\upsilon_1$. For $\upsilon_2=\upsilon_1+1$, $\Delta E_{vib}=\hbar\omega_e[1-2x_e(\upsilon_1+1)]$, and the term $-2x_e(\upsilon_1+1)$

expresses the difference from the harmonic approximation. It follows that the energy separation of successive levels actually decreases as v increases. The selection rule for anharmonic vibration is $\Delta v = \pm 1, \pm 2, \pm 3, \ldots$; thus, larger jumps are possible, but ± 2 is normally the largest found in practice.

We can consider a practical implication of these transitions in terms of the Boltzmann equation, to which we referred in Section 3.6.1.2 (see also Appendix 17). The spacing between adjacent vibrational energy levels is of the order of 2500 cm^{-1}, so that at 298 K we have for levels $v=0$ and $v=1$

$$N_{v=1}/N_{v=0} = \exp\{(-6.626\times10^{-34}\ \text{J Hz}^{-1})\ (2.998\times10^{8}\ \text{m s}^{-1})\times (2.5\times10^{5}\ \text{m}^{-1})/[(298\ \text{K})\ (1.3807\times10^{-23}\ \text{J K}^{-1}]\}$$

so that the population at $v=1$ is approximately 6×10^{-6} times that at $v=0$, whereas that at $v=2$ is much smaller still. Hence, the important transitions are from $v=0$, the ground state.

The first two transitions for $H^{79}Br$ (see Figure 3.22) occur at (ω-values) 2560 cm^{-1} and 5030 cm^{-1}. It follows from (3.39) that

$$\omega_e(1-2x_e) = 2560\ \text{cm}^{-1} \quad (3.40)$$

$$2\omega_e(1-3x_e) = 5030\ \text{cm}^{-1} \quad (3.41)$$

whence $\omega_e = 2650.1$ cm^{-1} and $x_e = 0.0170$.

3.6.5.2 Bond force-constants and dissociation energies

Studies of vibration spectra may be used to obtain values for bond force-constants and dissociation energies of molecules. For example, from the fundamental vibration wavenumber for $H^{79}Br$ and the reduced relative molar mass (0.99525), the force constant k evaluates ($4\pi^2\omega_e^2c^2\mu$) to 412 N m^{-1}. We note that the frequency ω_e corresponds to the infrared region of the spectrum, and that the value of μ shows that the bromine atom in hydrogen bromide is almost stationary relative to the hydrogen atom. Data for the hydrogen halides are compared in Table 3.6; the longer the bond the numerically smaller is the force-constant and the bond energy (D_0).

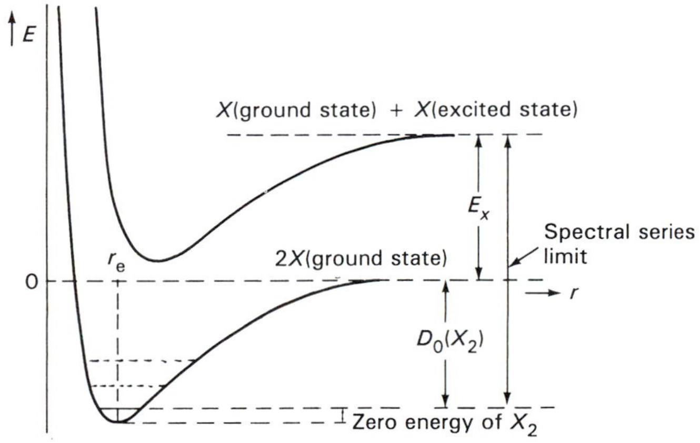

Figure 3.22 Curves of energy E for the vibration of a diatomic molecule in the ground and first excited states, as a function of the internuclear distance r; the dashed lines within the ground state curve refer to the levels $v=1$ and $v=2$, and r_e is the equilibrium internuclear distance in the ground state.

Table 3.6 Bond force-constants, internuclear distances and dissociation enthalpies (energies referred to 298.15 K) for the hydrogen halides, from infrared spectroscopy

	k/N m^{-1}	r_e/nm	D_0/kJ mol^{-1}
HF	966	0.0927	565
HCl	516	0.1274	431
HBr	412	0.1414	366
HI	314	0.1609	299

Table 3.7 Dissociation enthalpies D_0 for the halogen molecules; their precision is ±0.0001 eV molecule^{-1}, or ±0.01 kJ mol^{-1} except where otherwise specified

	D_0/eV molecule^{-1}	D_0 /kJ mol^{-1}
F_2	1.630±0.04	157.3±4.0
Cl_2	2.5180	242.95
Br_2	2.3191	223.75
I_2	2.2150	213.72

When a series of transitions is measurable, an extrapolation technique may be used to determine dissociation energies with high precision. In the example of the chlorine molecule, the convergence limit of its absorption spectrum has been found to be 21189 cm^{-1}, which corresponds to an energy of 2.6270 eV per molecule (Figure 3.22). However, the dissociated atoms are not in the ground state, and an excitation energy E_x amounting to 0.1090 eV per molecule must be deducted, giving a value of 2.5180 eV molecule^{-1}, or 242.95 kJ mol^{-1}. The results for the halogens are listed in Table 3.7.

3.6.5.3 Vibrations of polyatomic molecules

In order to specify the positions of N atoms in a molecule, $3N$ coordinates are needed in general. We may allocate three coordinates to specify the centre of mass of the molecule, that is, its translational motion. Three more coordinates are needed to specify the orientation of a molecule, that is, its rotational motion. It follows that there are $3N-6$ vibrational coordinates, or degrees of freedom. In the case of a linear molecule, rotational motion about the internuclear axis does not affect the position or orientation of the molecule. Hence, there are $3N-5$ vibrational degrees of freedom for linear species. The number of vibrational modes increases rapidly with the size of the molecule.

Infrared activity arises from dipole changes that are brought about by vibration. The CO_2 molecule has four vibrational degrees of freedom, as shown opposite.

The symmetric stretch does not alter the value of the dipole moment and this vibration is inactive in the infrared; the other three modes are all active. The value given above for $\bar{\nu}_1$ in the inactive symmetric stretch is based on bond force-constants, and is approximate.

←O———C———O→ Symmetric stretch $\bar{\nu}_1 \simeq 1330\ \mathrm{cm}^{-1}$

←O———C→←O Antisymmetric stretch $\bar{\nu}_3 \simeq 2349.3\ \mathrm{cm}^{-1}$

↑ ↑
O – C – O Two bends, in and normal to the plane $\bar{\nu}_2 = 667.3\ \mathrm{cm}^{-1}$
↓

Normally, bending modes have smaller wavenumbers (energies) than do stretching modes; it is easier to bend a bond than to stretch it.

We shall not be studying vibration and rotation spectra *per se* further; the reader is directed to more specialized texts on this subject for greater detail on these topics (see Bibliography). However, many vibrational modes give rise to spectra in the infrared region, and may be used in identifying structure, as we shall discuss shortly.

3.6.6 Raman spectroscopy

Molecular vibrations may be studied also by Raman spectroscopy. Molecular motions, vibration or rotation, are Raman active if the polarizability of the species changes during the motion. The positions of Raman lines depend on the excitation frequency, but they are encountered most often in the infrared region of the spectrum. Certain symmetrical molecules may show lines in the Raman spectrum that are absent from the infrared and *vice versa*: in particular, if a molecule has a centre of symmetry, a mutual exclusion rule requires that no given frequency occurs in both spectra, as long as the centrosymmetric portions of the molecule that are giving rise to the particular spectra are vibrationally coupled. A given set of vibrational coordinates will usually give rise to both infrared and Raman active modes. They differ in frequency because of intramolecular coupling, but if this coupling is effectively zero the infrared and Raman lines coincide.

The energy levels of a molecule are investigated by passing a beam of monochromatic light through the sample and collecting the scattered radiation in a direction perpendicular to the incident beam. The monochromatic source may be a mercury discharge tube or, preferably, a laser, so that an intense irradiating beam is provided. Photons from the incident beam collide with molecules of the sample, then to emerge with either a higher or a lower energy. The low-energy scattered photons are known as Stokes radiation and the high-energy photons as anti-Stokes radiation. The energy differences and, hence, the shifts in the lines are small and are best monitored by a photomultiplier detector. Analysis for functional groups may be carried out, and tables given for infrared spectra (Section 3.6.8.1) apply to Raman spectra as well.

3.6.7 Spectral activity and symmetry

Spectral activity may be studied by means of the symmetry rules that we discussed in the early sections of this chapter; we shall indicate a procedure by reference to the specific example of methane.

In Figure 3.23, the methane molecule is pictured in a cube, so that the symmetry elements of the molecule coincide with those of the cube. There are 15 molecular displacements, of which nine are vibrational modes. We need to consider how the displacements are affected by the symmetry classes of methane; its point group is $\mathcal{T}_d$, and the relevant character table has been given in Section 3.4.5.

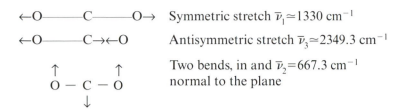

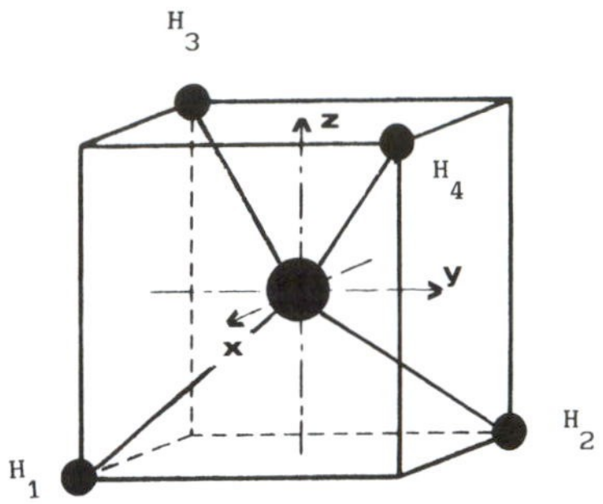

Figure 3.23 Methane molecule placed in a cube such that the symmetry elements of methane and the cube coincide; *x*, *y* and *z* are the directions for the resolved displacements of the atoms from their mean positions. The hydrogen atoms 1, 2, 3 and 4 correspond to the directions $[1\bar{1}\bar{1}]$, $[\bar{1}1\bar{1}]$, $[\bar{1}\bar{1}1]$ and [111], respectively.

Symmetry **E** leaves the molecular displacements invariant. No coordinates change sign, so that the transformation matrix is diagonal with each element equal to unity; thus, the character $\chi_E=15$. Under $\mathbf{C}_3$ all displacement coordinates change and $\chi_{C_3}=0$. The $\mathbf{C}_2$ operation about *z* changes only the sign of the *x* and *y* coordinates of the carbon atom, so that $\chi_{C_2}=-1$. Under $\mathbf{S}_4$ the only resultant change is in the *z* coordinate of carbon and $\chi_{S_4}=-1$. The $\boldsymbol{\sigma}_d$ operation leaves the signs of *z* unchanged for carbon and for the hydrogen atoms in the symmetry plane; of the other coordinates for these atoms, three are unchanged and three changed, so that $\chi_{\sigma_d}=3$. As we saw earlier, it is not necessary to consider the effects of the other symmetry operations on the molecule because the symmetry elements of one and the same class operate in an identical manner. This general procedure has been summarized by the rules given in Section 3.4.6.1.

We thus arrive at a representation $\Gamma=15\ 0\ -1\ -1\ 3$. Reference to the character table shows that Γ is reducible and applying (3.8) shows that $\Gamma=A_1+E+T_1+3T_2$. Furthermore, the translations and rotations correspond to T_2 and T_1 respectively, so that the vibrations are represented by A_1+E+2T_2.

A vibration is infrared active when there is an interaction between the radiation and the dipole moment of the vibrating molecule, that is, the integral $\int \psi_i \boldsymbol{\mu} \psi_f \mathrm{d}\tau$ must be finite, where i and f refer to initial and final states of the species. The integral will be zero unless it contains the fully symmetrical representation A_1, for which the character is +1 for all symmetry operations.

From the product functions in the character table (Section 3.4.5), it follows that Raman activity alone in methane is related to A_1 and E, whereas T_2 is both infrared and Raman active. If we generate representations with first, the C−H bonds (see also Section 3.4.6.2) and secondly, the H−C−H bond angles, we obtain the representations

$\mathcal{T}$	E	$8\mathbf{C}_3$	$3\mathbf{C}_2$	$6\mathbf{S}_4$	$6\boldsymbol{\sigma}_d$
Γ_{CH}	4	1	0	0	2
Γ_{HCH}	6	0	2	0	2

which reduce to

$$\Gamma_{CH}=A_1+T_2$$

$$\Gamma_{HCH}=(A_1)+E+T_2=E+T_2$$

The (A_1) representation in Γ_{HCH} is superfluous: although the C−H bond lengths can change independently, only five of the six H−C−H bond angles are independent.

Raman frequency shifts are very small. being of the order of 1 : 3000 cm^{-1} and 1 : 1400 cm^{-1} for the two T_2 vibrations.

3.6.7.1 *Direct product*

We made an implicit reference to the direct product in Section 3.4.6.5; here it is considered in a little more detail. A dipole moment vector **μ** can be resolved into three components, μ_j ($j=x$, y, z) and it is sufficient for any one to be nonzero for infrared activity to be present. The unperturbed, initial vibrational state ψ_i has spherical symmetry; μ_j has the symmetry of the translation vector T_j along the same axis (see the character table); and ψ_f has the symmetry of the vibrational mode involved. Thus, we form *direct products* between the totally symmetric, translation vectors and the vibrational mode representations. The direct product of two representations is obtained by multiplying together the characters, for each operation in turn, in the two irreducible representations. We continue with methane as an example.

A_1 vibration: $A_1 \times T_2 \times A_1 = T_2$
E vibration: $A_1 \times T_2 \times E = T_2 \times E = T_1 + T_2$
T_2 vibrations: $A_1 \times T_2 \times T_2 = A_1 + E + T_1 + T_2$.

We have used implicitly certain general multiplication properties for irreducible representations in $\mathcal{T}_d$ (and $\mathcal{O}_h$):

$$E \times T_1 = E \times T_2 = T_1 + T_2$$

$$T_1 \times T_1 = T_2 \times T_2 = A_1 + E + T_1 + T_2$$

$$T_1 \times T_2 = A_2 + E + T_1 + T_2$$

We can see how they are obtained. For example, consider $E \times T_2$:

$\mathcal{T}_d$	**E**	**8C$_3$**	**3C$_2$**	**6S$_4$**	**6σ$_d$**
E	2	−1	2	0	0
T_2	3	0	−1	−1	1
$E \times T_2$	6	0	−2	0	0

which reduces to $T_1 + T_2$. We see now that only the T_2 vibrations, containing the fully symmetrical A_1, will be infrared active.

In the case of Raman spectra, which we considered briefly in Section 3.6.6, active modes require a nonzero integral of the form $\int \psi_i \alpha \psi_f \, d\tau$, where α is the polarizability of the molecule. In the general case, the polarizability is a tensor $\boldsymbol{\alpha}_{jk}$ (j, $k = x$, y, z) and a nonzero integral is needed with any one of the six combinations x^2, y^2, z^2, xy, yz, and zx. Consider the direct product $A_1 \times E \times E$. It is clear that $A_1 \times E = E$; hence, we need $E \times E$:

	E	**8C$_3$**	**3C$_2$**	**6S$_4$**	**6σ$_d$**
E	2	−1	2	0	0
E	3	−1	−2	0	0
$E \times E$	6	1	−4	0	0

which reduces to $A_1 + A_2 + E$, and is Raman active because it contains A_1 ($r^2 = x^2 + y^2 + z^2$ in A_1). We summarize the results for methane:

A_1 Raman active
E Raman active
T_2 Infrared and Raman active.

3.6.8 Infrared spectra in structure determination

We discussed in Chapter 2 some reasons for the approximate constancy of bond lengths and angles for given moieties among various compounds. Small changes occur in these parameters according to the other species present in the molecule. These factors give rise to characteristic ranges of wavenumbers and intensities for functional groups in molecules. Assignment of these wavenumbers in an infrared spectrum often provides a rapid method for the identification of a compound.

In practice, an examination may be conducted on a gas, liquid or solid material. A liquid, or solution in CCl_4, may be formed into a thin film between plates of sodium chloride or potassium bromide, which are transparent to infrared radiation in the approximate range 600–4000 cm^{-1}. The sample is then examined in the infrared spectrometer and an infrared spectrum obtained.

Table 3.8 Wavenumber ranges for some functional groups

Group	Mode	Range/cm^{-1}	Intensity
$>CH_2$	C−H (s)	3000–2850	*s*
$-CH_3$	C−H (s)	3000–2850	
$\geq C-H$	C−H (s)	2900–2880	*w*
$>CH_2$	C−H (b)	1470–1430	*m*
$-CH_3$	C−H (b)	1470–1430	
$-CH_3$	CH_3 (sb)	1390–1370	*m*
$>CH_2$	CH_2 (r)	700–750	*w*
$-C\equiv C-H$	C−H (s)	3300	*s*
$RHC=CHR$ (*trans*: R and H on each C, R opposite)	C−H (s)	960–980	*s*
$>C=C<$	C=C (s)	1680–1620	*s*
$-O-H$	O−H (b)	1410–1260	*s*
$\geq C-OH$	C−OH (s)	1150–1050	*s*
$>N-H$	N−H (ss/as)	3500–3300	*m*
$=N-H$	N−H (ss/as)	3500–3300	
$>C=O$	C=O (s)	1780–1680	*vs*
$\geq C-O-C\leq$	C−O (s)	1300–1150	*s*

Key: s, stretch; b, bend; sb, symmetric bend; r, rock; ss, symmetric stretch; as, asymmetric stretch. *vs*, very strong; *s*, strong; *m*, medium; *w*, weak.

From studies over a wide range of compounds, charts have been developed that give the vibrational frequency ranges for most functional groups in different environments. Table 3.8 lists some of the assignments for a selection of the commoner functional groups encountered in organic chemistry.

A more comprehensive list of infrared frequencies is presented in Figure 3.24 (see also Bibliography). Data of this nature are used for speedy qualitative analysis of compounds.

3.6.8.1 Examination of an infrared spectrum

Figure 3.25 shows the infrared spectrum between 700 cm^{-1} and 3500 cm^{-1} of an aliphatic compound of molecular formula $C_6H_{10}O_2$, determined by standard analytical procedures. The formula shows that the number of double bonds[20] is two. A possible model could contain C=C and C=O. The principal features of the spectrum may be identified as follows:

Peak	$\bar{\nu}/cm^{-1}$	Assignment
A	2980–2900 (3 lines)	$CH_3-C\lessgtr$
B	1720	$>C=O$
C	1650	$>C=C<$
D	1440	$CH_3-C\lessgtr$
E	1370	$CH_3-C\lessgtr$
F	1310	O – H
G	1270	O – H
H	1190	$>C=O$
I	1100	$\gtrless C$
J	1040	$\gtrless C$
K	970	$R\backslash C=C\backslash R'$ (*trans*)
L	840 }	$\gtrless C-H$
M	680 }	$\gtrless C-H$

[20] In an aliphatic $C_aH_bO_cN_d$ compound, the number of double bonds is $[(2a+2)-(b+d)]/2$.

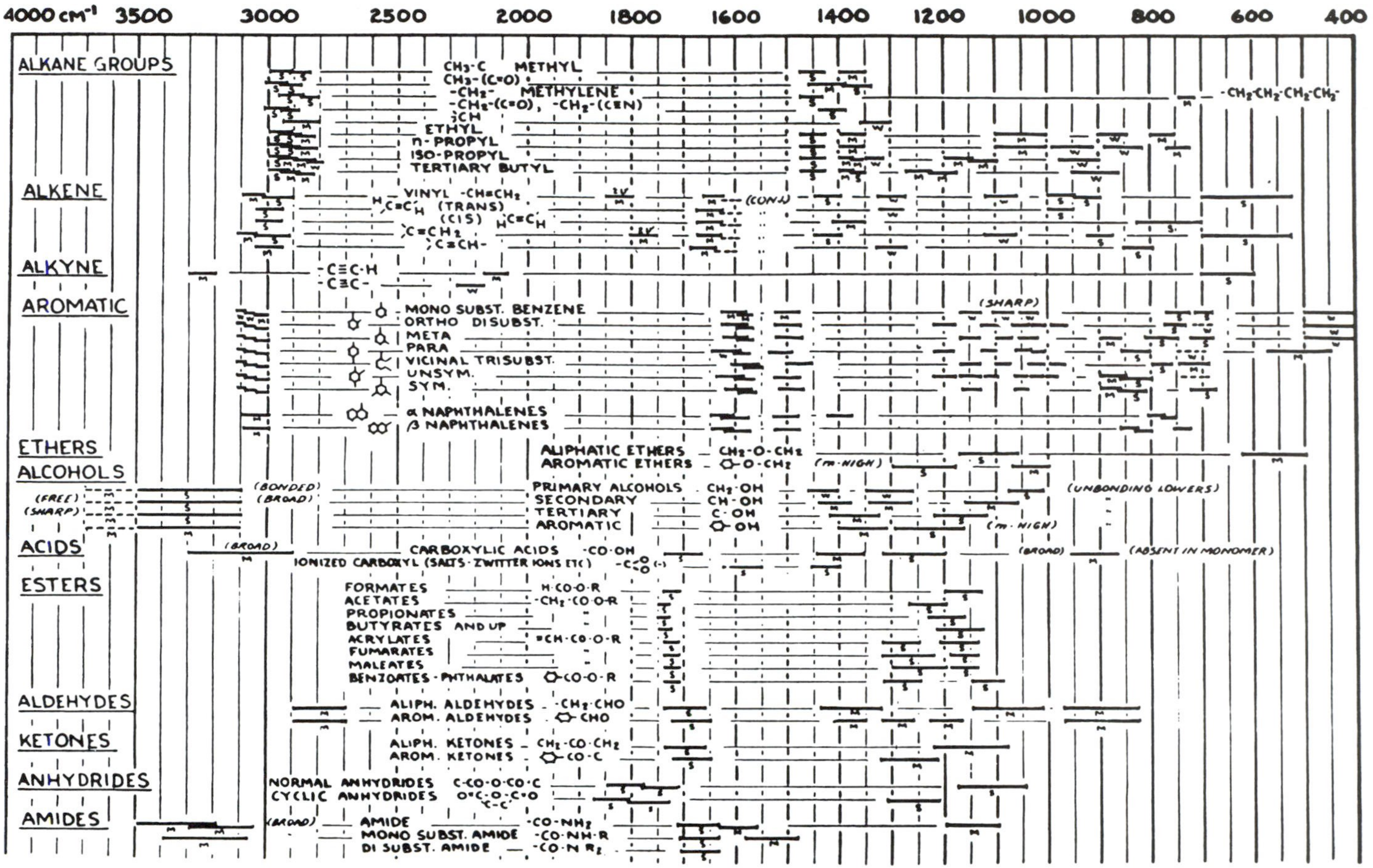

4000 cm⁻¹ 3500 3000 2500 2000 1800 1600 1400 1200 1000 800 600 400
ALKANE GROUPS
CH3-C METHYL
CH3-(C=O)
-CH2- METHYLENE
-CH2-(C=O), -CH2-(C≡N)
-CH2-CH2-CH2-CH2-
ETHYL
n-PROPYL
ISO-PROPYL
TERTIARY BUTYL
ALKENE
VINYL -CH=CH2
(TRANS)
(CIS)
(CONJ)
ALKYNE
-C≡C-H
-C≡C-
AROMATIC
MONO SUBST. BENZENE
ORTHO DISUBST.
META
PARA
VICINAL TRISUBST.
UNSYM.
SYM.
(SHARP)
α NAPHTHALENES
β NAPHTHALENES
ETHERS
ALIPHATIC ETHERS CH2-O-CH2
AROMATIC ETHERS
(m-HIGH)
ALCOHOLS
(FREE)
(SHARP)
(BONDED)
(BROAD)
PRIMARY ALCOHOLS CH2-OH
SECONDARY CH-OH
TERTIARY C-OH
AROMATIC
(UNBONDING LOWERS)
ACIDS
(BROAD)
CARBOXYLIC ACIDS -CO-OH
IONIZED CARBOXYL (SALTS-ZWITTER IONS ETC)
(ABSENT IN MONOMER)
ESTERS
FORMATES H-CO-O-R
ACETATES -CH2-CO-O-R
PROPIONATES
BUTYRATES AND UP
ACRYLATES =CH-CO-O-R
FUMARATES
MALEATES
BENZOATES-PHTHALATES
ALDEHYDES
ALIPH. ALDEHYDES -CH2-CHO
AROM. ALDEHYDES
KETONES
ALIPH. KETONES CH2-CO-CH2
AROM. KETONES
ANHYDRIDES
NORMAL ANHYDRIDES C-CO-O-CO-C
CYCLIC ANHYDRIDES
AMIDES
AMIDE -CO-NH2
MONO SUBST. AMIDE -CO-NH-R
DI SUBST. AMIDE -CO-N R2

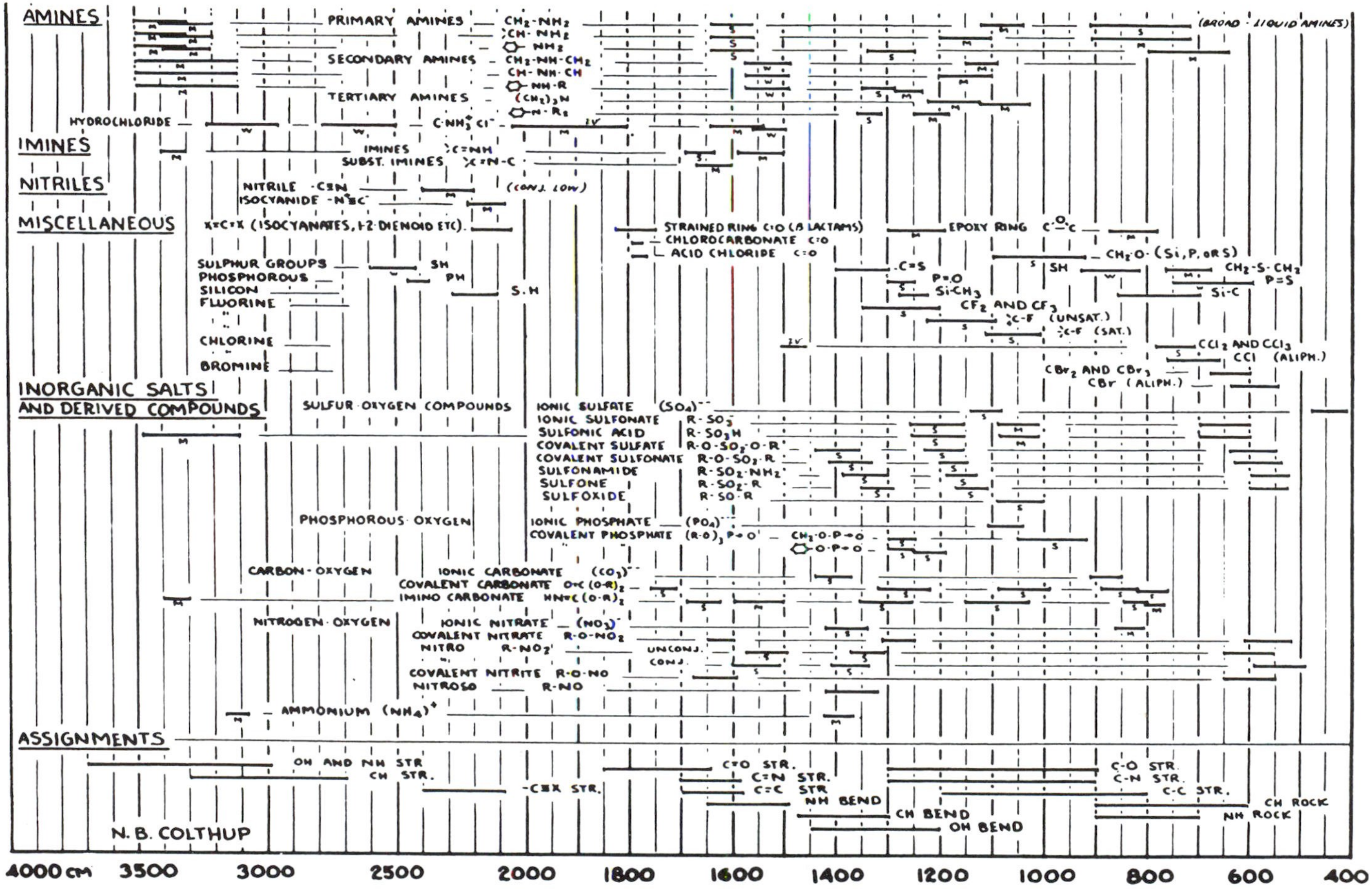

Figure 3.24 Infrared frequencies/cm^{-1} for functional groups and ions. Key: intensity grades are S *strong,* M *medium,* W *weak;* 2ν *overtone vibration;* Str *stretch.* (After N B Colthup, and reproduced by permission of the Optical Society of America.)

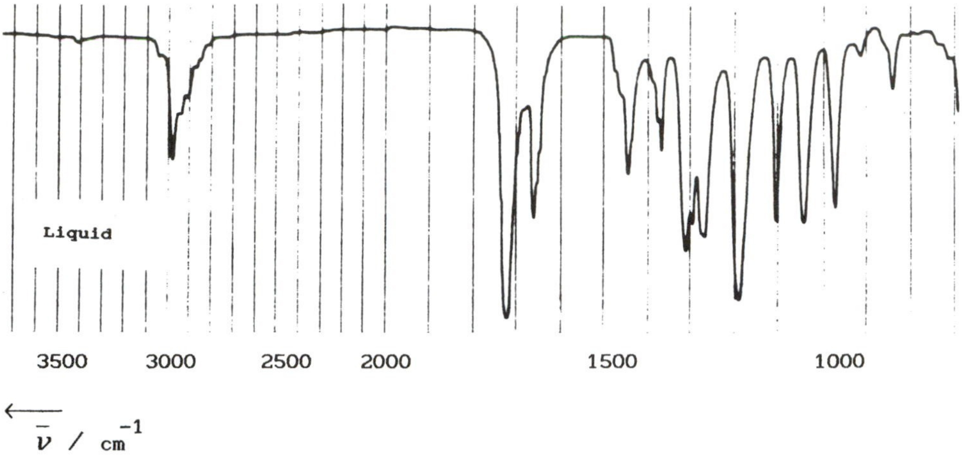

Figure 3.25 Infrared spectrum for a substance of molecular formula $C_6H_{10}O_2$, between wavenumbers 700 cm^{-1} and 3500 cm^{-1}.

If we put together the information from these assignments, the probable formula $CH_3CH{=}CHCOOC_2H_5$ emerges. However, assignments F and G are not included by this formula; their frequencies correspond to the range for O−H bending vibrations, which are of strong intensity. It is unlikely that the $-CO_2H$ group is present, because that moiety is characterized by a group of bands in the 2500–3000 cm^{-1} region. It is possible that some alcoholic contaminant may be present, and further investigations will be needed in order to confirm the findings; we shall return to this compound shortly.

3.6.9 Nuclear magnetic resonance spectroscopy

Nuclear magnetic resonance (NMR) spectra are obtained by studying the behaviour of molecules containing magnetic nuclei, such as ^{1}H, ^{13}C and ^{31}P, in a magnetic field, whereupon the magnetic species come into resonance at particular frequencies. When ^{1}H is the species in question, the technique is often called proton magnetic resonance (PMR).

A species with nonzero spin possesses a magnetic moment m, and its component along the z axis is given by

$$m_z = \gamma \hbar m_I \tag{3.42}$$

where $m_I = I, I-1, \ldots -I$; I is the spin number of the magnetic species, and γ is the gyromagnetic ratio. A given value for m_I corresponds to a particular orientation of the magnetic moment, and the energy is given by

$$E_{m_I} = -m_z B = -\gamma \hbar B m_I \tag{3.43}$$

where B is the magnetic field strength. We shall consider only PMR spectra in this survey, so that the relevant values of m_I are $\pm\frac{1}{2}$.

The energy separation for the two states corresponding to $m_I = \frac{1}{2}$ is

$$\Delta E = \gamma \hbar B \tag{3.44}$$

If the species is subjected to a radiation of frequency ν, a given energy will resonate when

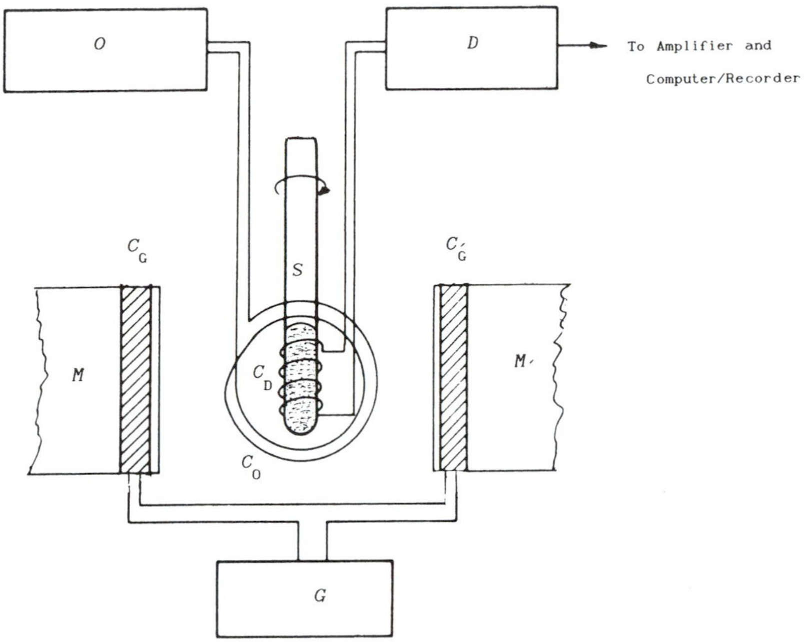

Figure 3.26 Schematic diagram of a NMR spectrometer: O oscillator of frequency 100 MHz; C_O oscillator coil for energizing the enclosed sample S; C_D detector coil for transmitting the resonant signal to the detector D; M and M′ poles of powerful permanent magnet; C_G and C'_G sweep-frequency coils energized by the generator G. The signal from the detector is amplified and then passed to a chart recorder or, in Fourier-transform NMR, to a computer.

$$h\nu = \gamma\hbar B \tag{3.45}$$

Then, the nuclear spins couple, and a strong absorption occurs in the energy transition between the states of $I = \pm\frac{1}{2}$. Calculations show that ^{1}H resonance at 100 MHz occurs in a field of nearly 2.35 T. The allowed transitions for a single magnetic nucleus are given by $\Delta m_I = 1$, and the intensity of the resonance signal is proportional to $c(I+1)m^2\nu/T$, where c is the concentration of the species, and T is the temperature.

A continuous-wave (CW) NMR spectrometer is shown schematically in Figure 3.26. The specimen S is contained in a glass tube, together with a trace of tetramethylsilane (TMS) as a reference standard. A spectrum may be recorded by immersing the sample in 100 MHz radiation from a radio-frequency oscillator coil C_O placed vertically, setting the field strength to 2.35 T, and sweeping the field by means of the generator G feeding the coils C_G and C'_G around the permanent magnet M.

A species in resonance absorbs energy from the oscillator, later to emit it through coil C_D for collection at the detector, followed by amplification and recording. An alternative frequency-sweep method may also be used. It is now normal practice to collect all the emitted radiation simultaneously, and analyse it by Fourier-transform techniques. Sweep coils are not then needed, and the exciting radiation for PMR would be 100 MHz at 2.35 T, but with a frequency range sufficient to cover the ^{1}H resonances that vary slightly according to their position in the molecule. Figure 3.27 shows a Fourier transform PMR spectrum of ethanol.

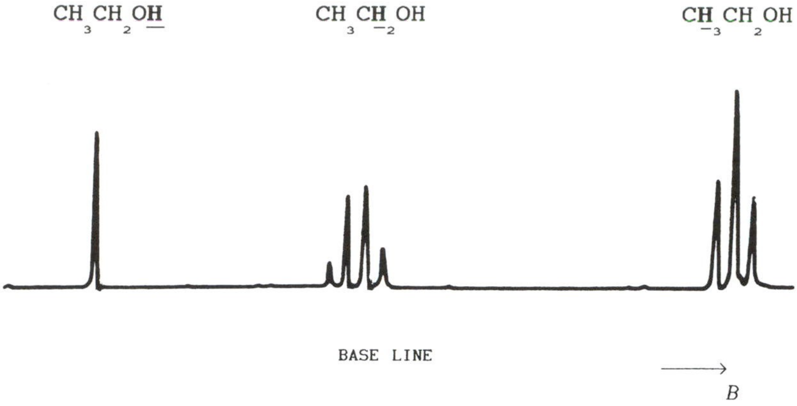

Figure 3.27 PMR spectrum of ethanol: the methyl protons (CH_3) form a group with a chemical shift (q.v.) $\delta=1$, the methylene protons (CH_2) form a group at $\delta=3$ and the hydroxyl proton (OH) is at $\delta=4$; the intensities (areas under the peaks) are in the ratio 3 : 2 : 1, in the same order. The patterns of the groups, which arise from coupling, are in the ratios 1 : 2 : 1 for CH_3 and 1 : 2 : 2 : 1 for CH_2.

3.6.9.1 Shielding and the chemical shift

The frequency ν at which a proton resonates is given by (3.45). The local field B experienced by a proton will differ from the applied field B_0 because of shielding by electrons in the molecule. We define a shielding parameter σ through

$$B=B_0(1-\sigma) \tag{3.46}$$

whence

$$\nu=\gamma B_0(1-\sigma)/(2\pi) \tag{3.47}$$

The shielding parameter σ may be positive or negative; for $\sigma>0$ the nucleus is said to be shielded, and for $\sigma<0$ it is deshielded. In a closed shell atom, σ is given by the Lamb equation:

$$\sigma=\frac{e^2\mu_0}{3m_e}\int_0^\infty r\psi^2(r)\,\mathrm{d}r \tag{3.48}$$

where μ_0 is the permeability of a vacuum. For ^{1}H, we may use the Slater orbital $\psi(r)=(1/\pi a_0^3)^{1/2}\exp(-r/a_0)$. Solving the integral gives $\sigma=10^{-7}e^2/(3m_e a_0)$, which evaluates to 1.78×10^{-5}.

Protons in differing environments may be brought into resonance successively by the field sweep. The positions of the proton resonances are defined with respect to zero for the strong resonance of the TMS reference standard. The *chemical shift* parameter δ is then given by

$$\delta=10^6(\nu_{\text{sample}}-\nu_{\text{TMS}})/\nu_{\text{TMS}} \tag{3.49}$$

The parameter δ is similar to σ, but it is a practical property indicating shielding with respect to the TMS standard. A given group of magnetic nuclei for which $\delta>0$ has a resonant frequency higher than that for the TMS standard; thus, the local field B is stronger for the given group than for TMS. The choice of TMS as a standard ensures that these conditions are normal. Values of δ for a selection of atomic groupings are listed in Table 3.9.

Table 3.9 Chemical shifts δ for ^{1}H NMR spectroscopy of aliphatic C, H, O, Br and Cl species

	δ/ppm		δ/ppm
CH_3R	1.0	CH_3-C-O	1.3
$R-CH_2-R$	1.4	$>CH-R$	1.5
$CH_3-C=C$	1.6	$R-CH_2-C-O$	1.9
$>CH-C-O$	2.0	CH_3-CO-R	2.2
$R-CH_2-CO-OR$	2.2	$R-CH_2-C=C$	2.3
$R-CH_2-CO-R$	2.4	$>CH-O-OR$	2.5
$>CH-CO-R$	2.7	CH_3-OR	3.3
$R-CH_2-OR$	3.4	$R-CH_2-Br$	3.5
$R-CH_2-OH$	3.6	$R-CH_2-Cl$	3.6
$CH_3-O-CO-R$	3.7	$>CH-OR$	3.7
$>CH-OH$	3.9	$>CH-Cl$	4.2
$>CH-Br$	4.3	$>C=CH-$	4.5–6.0
$>CH-O-CO-R$	4.8	$>C=CHCO-$	5.8–6.7
$-OCHO$	8.0–8.2	$R-CHO$	9.4–10.0

3.6.9.2 Coupling

The spectrum of ethanol in Figure 3.27 shows fine structure, that is, the lines of the spectrum are split, so that groups of two or more lines occur. It arises because there are two or more magnetic nuclei present that interact with one another. The splitting is expressed by a spin-spin coupling constant J, which is a property of the two magnetic nuclei under consideration. In the example of ethanol, the CH_3CH_2 group is involved in splitting, and J_{HH} is approximately 7 Hz. For the purpose of analysing this coupling, we symbolize the hydrogen on the methylene group by H′.

Initially, we assume no coupling; then we would have two singlets, at δ values of 1 and 3 (Figure 3.28a) in the intensity ratio of 3:2 for CH_3:CH′. Next, let H couple to H′, whereupon the H_3 resonance line is split into two; the intensity of each line in the doublet is 3/2, because of the conservation of energy, and their width is J_{HH}. Each line in the doublet then couples with the second methylene hydrogen H′ to give a quartet of intensity 3/4 each. However, because of the constancy of J_{HH}, the two 'inner' lines of the quartet coincide to give a resultant intensity of 3/2. Figures 3.28b and 3.28c illustrate the last two stages, leading to a triplet.

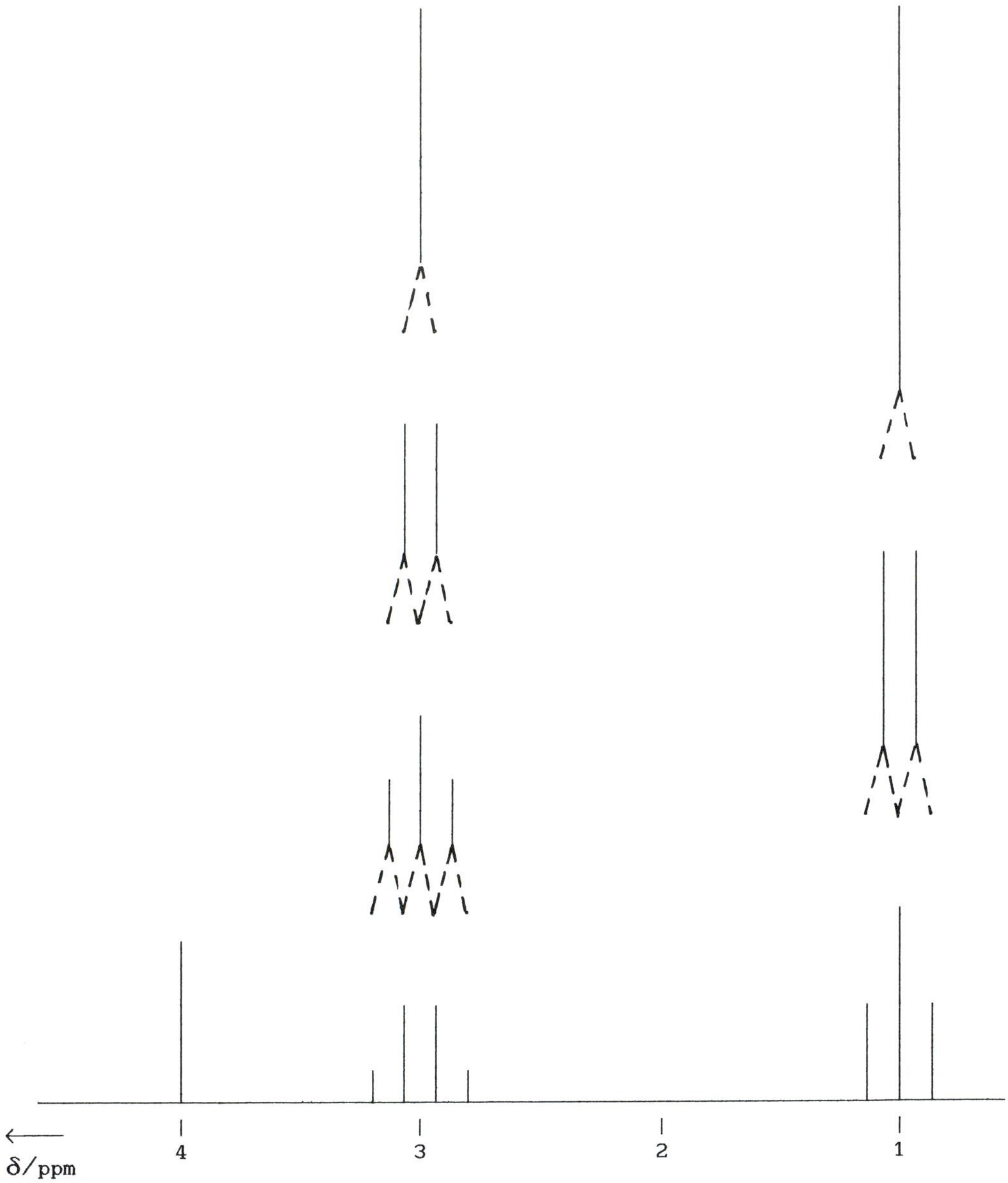

Figure 3.28 Development of the coupling patterns for the PMR spectrum of ethanol in Figure 3.27: (a) no coupling; singlets at $\delta=1$ and 3. (b) Coupling between H (CH_3) and H′ (CH_2) gives a doublet at $\delta=1$ which is split (J_{HH}) by the second H′ to give a triplet; two lines coincide to give the 3 : 2 ratio. A similar argument leads to the quartet (CH_2) at $\delta=3$.

Applying the same reasoning to the splitting of CH′, but with an extra stage because of the three hydrogen atoms on the methyl group, leads to a quartet. The line for −OH has been added for completeness, but it is too distant to couple with either $-CH_2$ or $-CH_3$.

In general, a group of j equivalent nuclei splits an adjacent group into $j+1$ lines, with intensities proportional to the coefficients in a Pascal triangle:

$j=1$					1		1				
$j=2$				1		2		1			
$j=3$			1		3		3		1		
$j=4$		1		4		6		4		1	
$j=5$	1		5		10		10		5		1
…											
…											
$j=N$			Select coefficients from $(1+x)^N$								

3.6.9.3 Examination of a NMR spectrum

Finally in this section, we consider again the compound $C_6H_{10}O_2$, to which we assigned the structure $CH_3CH{=}CHCOOC_2H_5$ on the basis of the infrared spectrum, while noting the possible presence of an alcoholic compound. The PMR spectrum is shown in Figure 3.29. From a study of Table 3.9, the following assignments can be made:

Shift d/ppm	State	Group
1.3	Triple }	$-CH_2CH_3$
4.1	Quartet }	
1.9	Doublet }	$R-CH_2-CO-R$; $\rangle CH-C-O$
5.8	Singlet }	$C=CHCO-$
5.9	Singlet }	$C=CH-$
6.9	Singlet group	$CH_3-CH=CH$
3.9	Singlet	$\rangle CH-OH$

These data confirm the assignment $CH_3CH{=}CHCOOC_2H_2$. The small peak at $\delta=3.9$ again suggests an alcoholic group, but in view of the other assignments it must be treated as a contaminant.

3.6.9.4 Other NMR techniques

The complexity of a NMR spectrum can be reduced if it is displayed with respect to two δ axes. This result is achieved by *two-dimensional* NMR. In effect, spin couplings and chemical shifts are recorded on perpendicular axes, thereby simplifying the appearance and analysis of the spectrum.

In analysing solid samples by NMR, a difficulty arises from a low resolution of the lines. The line width is determined both by the magnetic interaction between nuclear spins, and by the anisotropy of the chemical shift. The effects are proportional to the term $-[3\cos^2(\theta)-1]$ that appears in the dipole-dipole interaction potential (see Section 5.4.4). In solution, this

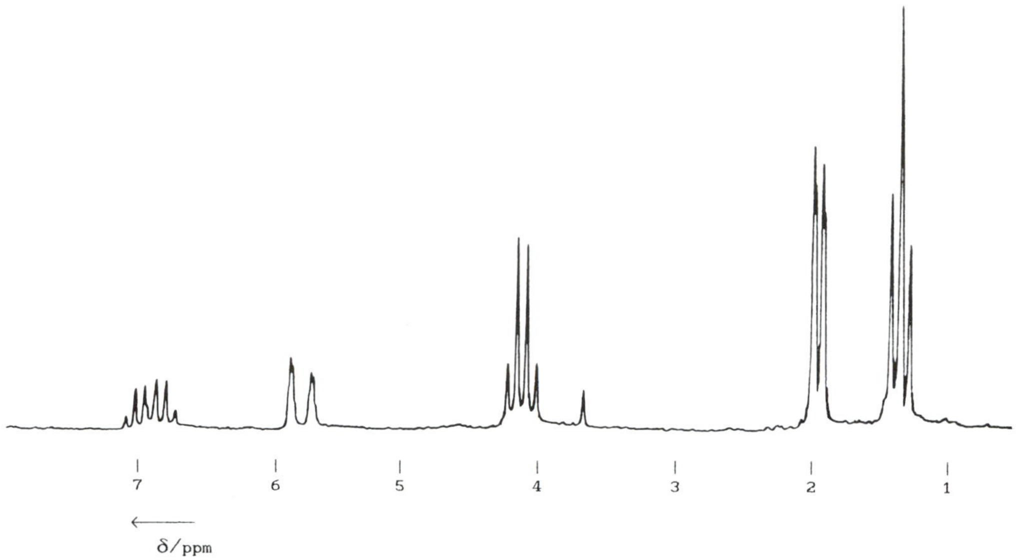

Figure 3.29 PMR spectrum for the compound $C_6H_{10}O_2$; the assignments listed indicate that the peak at $\delta=3.9$ most probably relates to a contaminant.

anisotropy averages to zero, but not in a solid, where resonances depend upon orientation.

One way of reducing the line width is to spin the sample at an angle θ to the field given by $-[3\cos^2(\theta)-1]=0$; thus, $\theta=54.74°$, or one-half of the tetrahedral angle. At this value of θ, the dipole anisotropy averages to zero, and the technique is referred to as *magic-angle spinning*.

Other techniques in NMR spectroscopy include the use of ^{13}C, ^{19}F and ^{31}P nuclei, and deuterated and tritiated compounds. These topics, and more comprehensive lists of chemical shifts and coupling constants, together with electron spin and quadrupole resonance methods are available in the more specialized texts on magnetic resonance spectroscopy listed in the Bibliography.

3.6.10 Mass spectrometry

Although mass spectrometry is not strictly a spectroscopic technique, it is included here because it can be used to obtain useful structural data, or to supplement other physical methods of structure determination.

Mass spectrometry is used to analyse compounds by separating singly charged positive group-ions from a volatilized sample, according to their mass/charge (m/z) ratios.[21] A schematic diagram of the basic components of a mass spectrometer is shown in Figure 3.30. The vapour or a solution of the substance is introduced into the ionization and acceleration chamber *I*. Volatile materials may be ionized by an electron stream from a heated filament, whereas nonvolatile compounds may be ionized by bombardment with a stream of xenon atoms of approximately 10 keV energy. The positive ions are then accelerated by an electric potential of strength V, and pass into the semicircular magnetic analyser.

[21] Generally, we are using q for the magnitude of a charge on a species, but in mass spectrometry it is customary to use z.

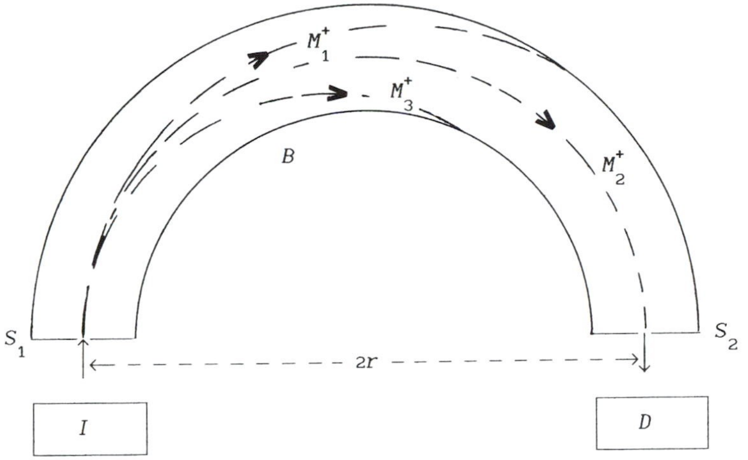

Figure 3.30 Schematic diagram of the basis of a mass spectrometer: the ionizer I produces ionized vapour of the sample and accelerates it into the circular magnetic field, of strength *B* and directed normal to the diagram; for given values of *B* and accelerating voltage *V*, an ion M_2^+ traverses the path of radius *r* and enters the detector D, whereas other ions, such as M_1^+ and M_3^+ are trapped. By sweeping through values of *B* at a constant *V*, the complete spectrogram may be recorded.

In the simplest type of mass analysis, all ions that differ in relative mass by one unit are separated, that is, m/z values of 91 and 92 are separately collectable. This analysis is achieved by a circular magnetic path of radius *r* and field strength *B*. In Figure 3.30 the poles of the magnet are arranged such that the field is directed normal to the plane of the illustration.

A given positive ion of mass *m* accelerated by an electric potential *V* has a kinetic energy given by the equation

$$\tfrac{1}{2}mv^2 = Vze \tag{3.50}$$

where *v* is its speed at any instant, *e* is the charge on an electron and *z* is the number of such charges. The magnetic force on a charge *ze* moving with a linear speed *v* is $Bvze \sin\theta$, where θ is the angle between the directions of the motion and the magnetic field. Since the force is normal to *v* and *B*, the charge moves in a circular path with an acceleration of v^2/r, and we have

$$Bvze = mv^2r \tag{3.51}$$

and it follows that

$$m/z = B^2r^2e/(2V) \tag{3.52}$$

By varying the value of the field *B*, ions of all possible m/z ratios can be caused to pass into the detector *D*, thus building up a complete spectrogram. In a higher-resolution, double-beam mass spectrometer, the ion beam passes through an electrostatic analyser before entering the magnetic analyser. Relative masses may be obtained to a precision of 1 in 10^6.

3.6.10.1 Isotopic variation

Isotopes have a variation of one or more relative mass units, and many occur naturally in given relative proportions. For example, any carbon compound containing ^{12}C

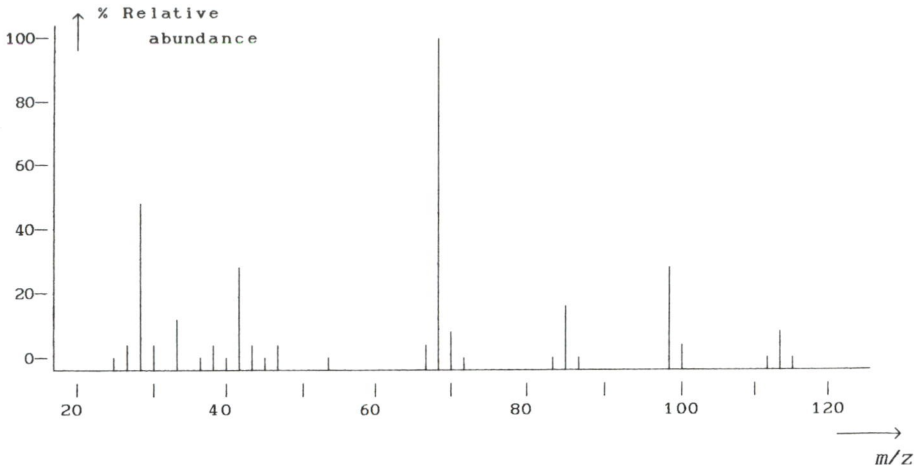

Figure 3.31 Mass spectrogram for the compound $C_6H_{10}O_2$; the six important peaks, in order from right to left, occur at *m/z* values 114, 99, 86, 69, 44 and 28; the first of them corresponds to the molecule-ion.

would contain ^{13}C also, to an extent of 1.1%. In the case of natural chlorine, isotopes ^{35}Cl and ^{37}Cl occur in the proportions of 3 to 1 respectively. It is possible for more than one isotope of a given atom to be present in one and the same ion. These differences may be resolved by the experimental measurement of the relative abundances of the differing ion peaks.

3.6.10.2 Examination of a mass spectrogram

Figure 3.31 illustrates the mass spectrogram of the compound $C_6H_{10}O_2$ which has already been used in the example infrared and NMR analyses. Through them we determined the formula $CH_3CH{=}CHCOOC_2H_5$ which has a relative mass of 114 (the sum of the mass numbers of the principal isotopes). Examination of the spectrogram shows a peak at 114 which is the molecule-ion. Other peaks of high relative abundance are indicated, and their probable interpretation is given in Table 3.10.

Without concerning ourselves about favoured breakdown paths or energetics of ionization, it is clear that the evidence here supports the previous conclusions. The strong peak at $m/z=28$ probably corresponds to $(C{-}O{-})^+$, but this fragment is too small to assist in confirming the molecular formula.

The combination of NMR, IR and mass techniques is very powerful, and can be used with success on compounds of considerable complexity. However, they do not generally provide quantitative data on molecular geometry. We have obtained some information of this nature on simple molecules by microwave spectroscopy, but with more complex compounds diffraction methods are required. We shall complete this section with a study of single-crystal X-ray analysis.

3.6.11 X-ray crystallographic analysis

X-ray crystallography is the most powerful tool for the determination of the detailed molecular geometry of chemical species. A single crystal of a chemical substance behaves towards

Table 3.10 Analysis of the mass spectrogram of $C_6H_{10}O_2$

m/z	Fragment
114	$(CH_3CH=CHCO_2C_2H_5)^+$
99	$(-CH=CHCO_2C_2H_5)^+$
86	$(>CHCO_2C_2H_5)^+$
69	$(CH_3CH=CHC-O-)^+$
41	$(>CH-C-O-)^+$

X-rays of wavelength approximately 0.1 nm something like a three-dimensional diffraction grating. By a combined diffraction and interference process, X-rays interact constructively with a crystal to give a series of discrete, spot spectra that represent the crystal structure in terms of both the nature of the species present and their relative spatial positions. A careful analysis of these spectra permits a re-creation of the image of the crystal structure. We shall examine in detail an example of this procedure, wherein we shall draw upon studies of crystal symmetry from the early part of this chapter, but first we consider some basic aspects of X-ray diffraction from crystals.

3.6.11.1 Bragg equation

In Section 3.5ff, we discussed the geometry of crystals, and Figure 3.14a was used to illustrate the concept of families of parallel equidistant planes, designated by their Miller indices (*hkl*). An infinite number of planes can be drawn in a crystal lattice but, as we shall see, a limited number of them is accessible by experiment.

Consider any family of planes of interplanar spacing d. Figure 3.32 shows two adjacent planes of the family, with X-rays incident on them at the Bragg angle[22] θ. Two typical rays, normal to an incident wavefront, have a path difference δ after diffraction given by QA_2+A_2R, and it may be shown readily that

$$\delta=A_1A_2\cos(\phi_1)+A_1A_2\cos(\phi_2)=2A_1A_2\cos\left(\frac{\phi_1+\phi_2}{2}\right)\cos\left(\frac{\phi_1-\phi_2}{2}\right) \tag{3.53}$$

The interplanar spacing d is the length A_1P and simple algebraic manipulation shows that

$$\delta=2d\sin(\theta) \tag{3.54}$$

The rays will reinforce each other under the condition

$$2d\sin(\theta)=n\lambda \tag{3.55}$$

where n is an integer and λ is the wavelength of the X-radiation. It is customary to incorporate the order n of diffraction as d/n into $d(hkl)$, the spacing of the particular family of (hkl)

[22] It may be noted that the Bragg angle is the complement of the angle of incidence in geometrical optics.

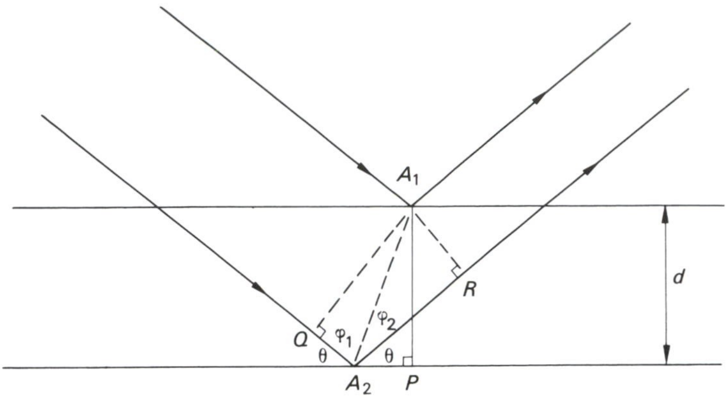

Figure 3.32 Geometry of Bragg X-ray reflection. The two typical rays are in phase at the normals A_1Q and A_1R to the incident and diffracted wavefronts, respectively. The path difference generated from the adjacent planes is (QA_2+A_2R) and is equal to $2d\sin\theta$, where $d=A_1P$.

planes[23] corresponding to the Bragg angle $\theta(hkl)$, so that the modern formulation of the Bragg equation is

$$2d(hkl)\sin[\theta(hkl)]=\lambda \quad (3.56)$$

If (3.56) holds for any two adjacent planes, it will hold also for a family of planes in a crystal. Hence, a crystal will diffract X-rays strongly when (3.56) is satisfied. Since the angle at which the diffracted rays leave the crystal planes is also θ, the term Bragg *reflection*, or just reflection, is commonly used in this context.

3.6.11.2 Ewald's reciprocal space construction

The geometrical interpretation of X-ray diffraction is facilitated by a device due to Ewald. We discussed the reciprocal lattice concept in Section 3.5.2 and we use it here in the following way. In Figure 3.33 a sphere, known as the Ewald sphere, or the sphere of reflection, centred on the crystal C is drawn (in our imagination) with a radius of 1 reciprocal space unit[24] (RU), with the X-ray beam along the diameter AQ. The Bragg construction is superimposed, which shows that a reflected beam CP cuts the Ewald sphere at the point P. The points A, P and Q lie on a circular section of the sphere that passes through the centre C. Hence, it follows that

$$AQ=2 \quad \text{by construction} \quad (3.57)$$

$$A\text{–}P\text{–}Q=90° \quad \text{angle in a semicircle} \quad (3.58)$$

$$QP=2\sin[\theta(hkl)] \quad \text{triangle } APQ \quad (3.59)$$

From the Bragg equation (3.56), $2\sin[\theta(hkl)]=\lambda/d$ and from the definition of reciprocal lattice (Section 3.5.2) P is the reciprocal lattice point hkl, so that QP is $d^*(hkl)$; thus, from (3.18) with $K=\lambda$, we obtain

$$2\sin[\theta(hkl)]=d^*(hkl) \quad (3.60)$$

[23] h, k and l need not be prime with respect to one another.

[24] Reciprocal space is dimensionless if K in (3.18) is equal to λ.

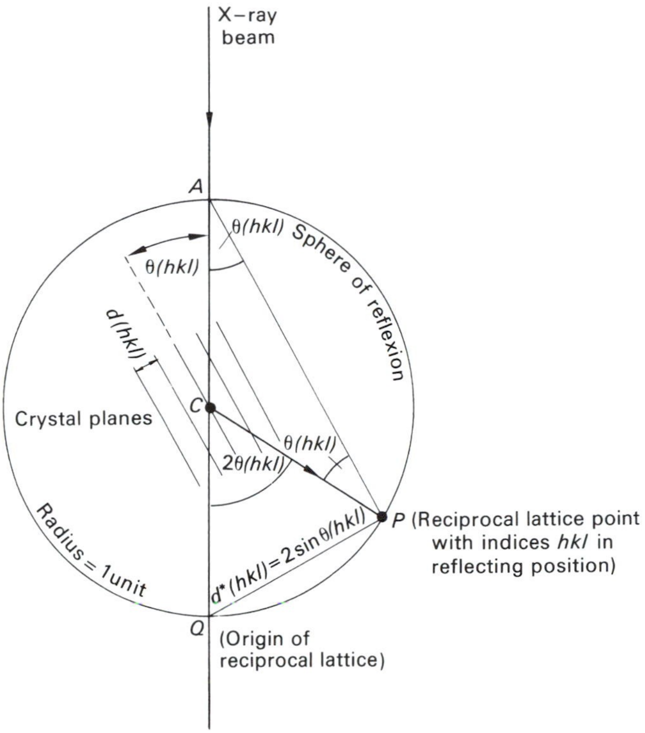

Figure 3.33 Ewald's construction, showing that an X-ray reflection may be considered to arise when a reciprocal lattice point P passes through the Ewald sphere (sphere of reflection) of radius 1. AP is parallel to the traces of the (hkl) planes and the reciprocal lattice vector QP forms a right angle at P. The Bragg equation follows directly from this construction.

We now have a mechanism for predicting the occurrence of possible X-ray reflections. The origin of the reciprocal lattice is at Q, and we may imagine a conceptual crystal at Q, identical to that at C and moving synchronously with it. Then Bragg reflection for the family of (hkl) planes occurs as the corresponding reciprocal lattice point hkl passes through the sphere of reflection, and the direction of the reflection is the vector from the crystal C to that point on the sphere.

3.6.11.3 Recording an X-ray diffraction pattern

It should be clear from the Ewald construction that any photographic film other than a spherical sheet with the crystal at its centre will produce a distorted representation of the reciprocal lattice. The condition for a true representation on a flat film is that the film should be tangential to the Ewald sphere at the moment of each reflection. This situation is achieved by means of an X-ray precession camera, the basic principles of which are shown in Figure 3.34. A crystal axis t precesses about the X-ray beam and the film follows this precession motion, remaining always normal to the axis t.

Figure 3.35 is an example of an X-ray diffraction photograph (idealized) of a single crystal taken by precession photography, using Cu Kα radiation, λ=0.15418 nm. It provides a true picture of the $0kl$ level of the reciprocal lattice and each point is considered to have a weight in proportion to the intensity I of the reflection.

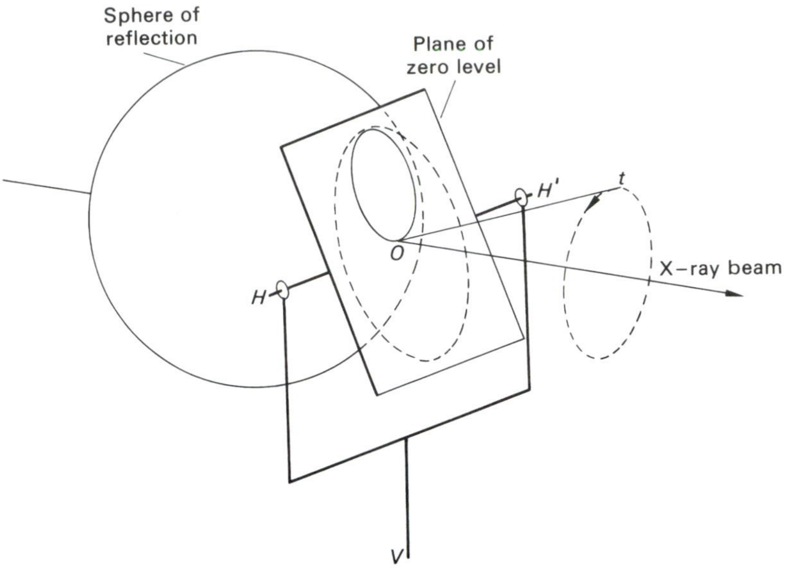

Figure 3.34 Precession geometry: reciprocal lattice zero level (h=0 in this example) with its normal t precessing around the X-ray beam. The film is tangential to the sphere of reflection at the point of emergence of a reflected beam, and an undistorted photograph of the reciprocal lattice is obtained.

The photograph reveals that certain classes of reflections are systematically absent throughout the reciprocal lattice, the significance of which we shall discuss presently. From the camera calibration, the reciprocal unit-cell dimensions are found to be $b^*=0.195$ RU, $c^*=0.197$ RU and $\alpha^*=90°$. Thus, it follows that the unit-cell dimensions in real space are $b=\lambda/b^*=0.791$ nm, $c^*=0.783$ nm and $\alpha=90°$.

Exhaustive experiments with precession photography soon provide the complete unit-cell dimensions and the portion of the weighted reciprocal lattice that is accessible in the given experiment. The Ewald sphere itself sweeps out a sphere of radius 2 RU, the limiting sphere. This fact may be appreciated from Figure 3.33 by considering the crystal stationary and the Ewald sphere rotating about the axis through Q, in the opposite sense to the crystal rotation and taking the X-ray beam with it.

The total number of reflections available, assuming no systematic absences, is given by $V_L/V^*=V_L V/\lambda^3$, where V_L is the volume of the limiting sphere and V is the volume of the unit cell, given by (3.21). Thus, a monoclinic crystal of dimensions a=0.721 nm, b=1.153 nm, c=1.922 nm and β=94.07° examined with Cu Kα radiation could provide a total of $33.51\times(1.594\ \text{nm}^3)/(3.665\times10^{-3}\ \text{nm}^3)=14574$ reflections. However, it is not possible in practice to observe reflections of θ-value up to 90°; the maximum practicable value is approximately 70°, whereupon the number of reflections is multiplied by $\sin^3(\theta_{max})$, reducing it to 12093. In practice, the number of *unique* reflections observable will be further reduced by symmetry, systematic absences and thermal vibrations. For example, if the monoclinic crystal belongs to point group $2/m$, only one quarter of reciprocal space is unique, thus reducing the number of reflections nominally to 3023.

Current practice generally employs a single-crystal diffractometer to collect X-ray data; an example of an instrument is shown in Figure 3.36. From a knowledge of the unit-cell dimensions and the orientation of the crystal, the computer-controlled diffractometer rotates the families of planes into the reflecting position in turn, and their h, k, l and I values

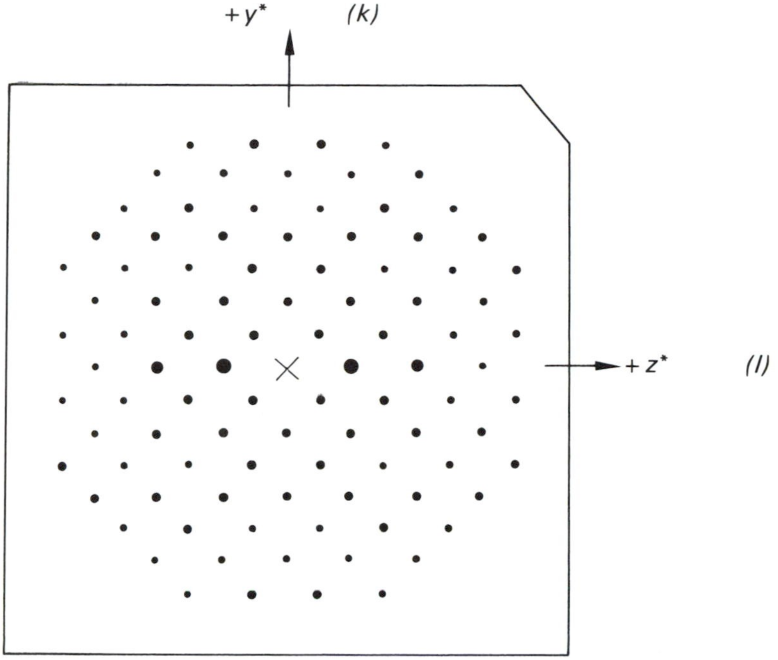

Figure 3.35 Precession zero-level photograph of an orthorhombic crystal precessing about *a*, taken with the arrangement in Figure 3.34. The zero level is the net 0*kl*, permitting b^*, c^* and α^* to be measured. In the particular experimental arrangement, b^* and c^* are magnified by a crystal-to-film distance of 60.00 mm, and the diagram has been scaled by a factor of 0.267; Cu Kα radiation (λ=0.15418 nm) was used; b^*=0.195 RU, c^*=0.197 RU and α^*=90°, so that b=0.791 nm, c=0.783 nm and α=90°. The pattern shows the limiting condition 0*kl*: $k+l=2n$, which corresponds to an *n*-glide plane, of translation $(b+c)/2$, normal to *a*.

are recorded. This procedure is very much faster than photographic techniques and leads to data of high precision.

Whatever the method of collection, we obtain a set of X-ray data, each datum being characterized in position by the indices *hkl*, or by sin (θ), and in strength by an intensity *I*(*hkl*).

3.6.11.4 X-ray scattering by an atom and a unit cell

The crystal structure factor ***F***(*hkl*) represents the reflection scattered by a unit cell, in terms of both amplitude and phase, for the family of planes (*hkl*). Each atom scatters X-rays by its electrons in a manner that depends on its atomic number, sin (θ) and the temperature. The atomic scattering factor f_θ is given by

$$f_\theta = \int_0^\infty \psi^2 \frac{\sin(\boldsymbol{r}\cdot\boldsymbol{S})\, r^2}{\boldsymbol{r}\cdot\boldsymbol{S}}\, \mathrm{d}r \tag{3.61}$$

where $|\boldsymbol{S}| = 4\pi\lambda^{-1}\sin(\theta)$. The f_θ values are calculated for atoms at 0 K (at rest), and modified for other temperatures; f_θ is tabulated as a function of $\lambda^{-1}\sin(\theta)$, and values for all atomic species are available. The simplest temperature factor is of the form $\exp[-B\lambda^{-2}\sin^2(\theta)]$, where *B* is related to the displacement of an atom in a direction normal to the reflecting

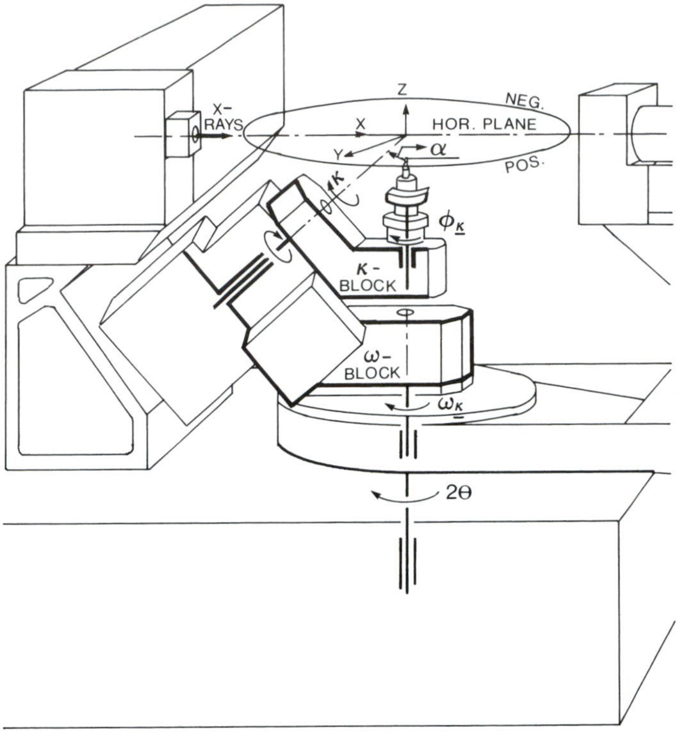

Figure 3.36 Modern four-circle single-crystal CAD4 (kappa) X-ray diffractometer, showing the *X*, *Y*, *Z* instrumental coordinate system and the rotation directions of the circles. Families of planes are brought successively into the reflection position and the reflections monitored by a scintillation counter (shown at the upper right of the diagram). (Reproduced by permission of Enraf-Nonius, Delft.)

plane. Figure 3.37 illustrates the scattering factor curve for an atom at rest and at a finite temperature; at $\sin(\theta)=0, f_\theta=Z$, the atomic number of the species.

In a unit cell, the *N* atoms have coordinates $x_j, y_j, z_j\ (j=1-N)$. The scattering from the *j*th atom may be regarded as a wave of amplitude dependent upon $f_{\theta,j}$, so that the combined scattering is the sum of *N* such waves of varying amplitudes and phases. The resultant may be expressed most simply by an Argand diagram, shown in Figure 3.38 for two waves. Each contribution is of the form $f_{\theta,j}\exp(\mathrm{i}\phi_j)$, so that the resultant is given by

$$\boldsymbol{F}=\sum_{j=1}^{j=2} f_{\theta,j}\exp(\mathrm{i}\phi_j) \tag{3.62}$$

where $\exp(\mathrm{i}\phi_j)$ is the phase of the *j*th atom relative to the positive direction of the real $\mathcal{R}$ axis. It is convenient to think of $\exp(\mathrm{i}\phi_j)$ and an operator that rotates the vector $f_{\theta,j}$ anticlockwise in the complex plane (of the Argand diagram) by an angle of ϕ_j, measured from the positive real axis.

We may generalize this result for a unit cell of *N* atoms. If we write $\boldsymbol{F}=A+\mathrm{i}B$, then the conjugate $\boldsymbol{F}^*=A-\mathrm{i}B$, whence

$$|F|^2=\boldsymbol{F}\cdot\boldsymbol{F}^*=A^2+B^2 \tag{3.63}$$

where

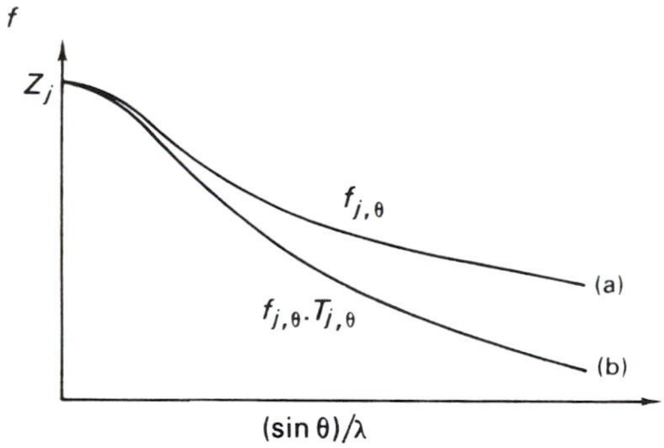

Figure 3.37 Atomic scattering factor f_θ: (a) atom at rest, (b) atom corrected for thermal vibration at a temperature T. The intercept at $\sin(\theta)=0$ corresponds to the atomic number Z of the species.

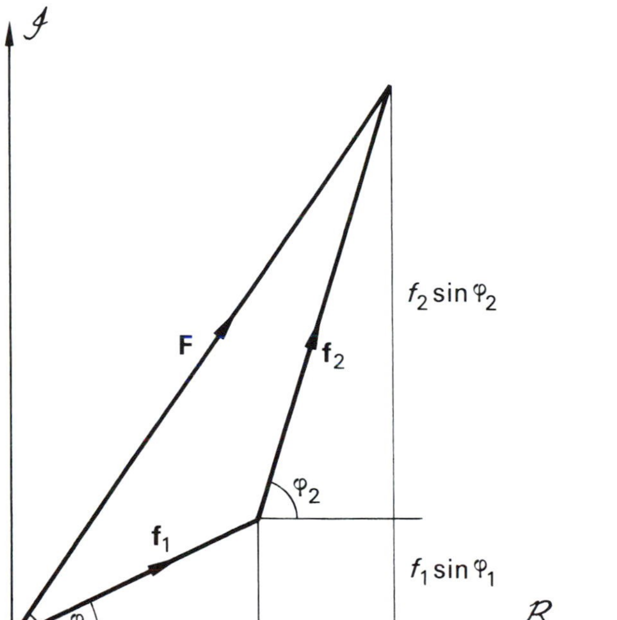

Figure 3.38 Combination of two waves $f_1 \exp(i\phi_1)$ and $f_2 \exp(i\phi_2)$, represented by vectors on an Argand diagram. The resultant is formed from the algebraic sum of the real $\mathcal{R}$ and imaginary $\mathcal{I}$ components. The terms $\exp(i\phi)$ are phases with respect to the positive direction of the $\mathcal{R}$ axis.

$$A=\sum_j f_{\theta,j} \cos(\phi_j) \tag{3.64}$$

$$B=\sum_j f_{\theta,j} \sin(\phi_j) \tag{3.65}$$

The phase angle ϕ of the resultant $\boldsymbol{F}$ is given by

$$\tan(\phi)=B/A \tag{3.66}$$

The path difference δ_j for the jth atom of coordinates X_j, Y_j, Z_j in the reflection *hkl*, relative to the origin, is given by[25]

$$\delta_j = \lambda(hX_j/a + kY_j/b + lZ_j/c) \tag{3.67}$$

It is normal practice to convert this to a phase difference by multiplying by $2\pi/\lambda$ and to introduce fractional coordinates. Thus,

$$\phi_j = 2\pi(hx_j + ky_j + lz_j) \tag{3.68}$$

and the final expression for the structure factor for N atoms in the unit cell is[26]

$$\boldsymbol{F}(hkl) = \sum_j f_{\theta,j} \exp[\mathrm{i}2\pi(hx_j + ky_j + lz_j)] \tag{3.69}$$

We identify the conjugate of $\boldsymbol{F}(hkl)$ as $\boldsymbol{F}(\bar{h}\bar{k}\bar{l})$ and, since the intensity $I(hkl)$, corrected for instrumental and physical factors, is proportional to the square of the amplitude $|F(hkl)|^2$, we have from (3.63)

$$|\mathrm{F}(hkl)|^2 = |\boldsymbol{F}(\bar{h}\bar{k}\bar{l})|^2 = A(hkl)^2 + B(hkl)^2 \tag{3.70}$$

This is a mathematical statement of Friedel's law, which emphasizes the centrosymmetrical nature of every *diffraction pattern*. For the corresponding phases, we have from (3.66)

$$\tan[\phi(\bar{h}\bar{k}\bar{l})] = -\tan[\phi(hkl)] \tag{3.71}$$

These relationships are indicated on the Argand diagram in Figure 3.39 and normally Friedel's law holds within experimental error. Exceptions arise when the resonance absorption of X-rays by an atom in the structure is strong, a feature usually called anomalous scattering.

The equations that we have developed reveal the major problem in X-ray crystal analysis, often called the 'phase problem'. The values of the phase angles ϕ for each *hkl* reflection are lost because only the intensity is measured and, as we see from (3.70), no phase information is directly available. The formation of an image from scattered radiation requires that the scattered beams are recombined in both intensity and phase. A microscope carries out this process during focusing, but diffracted X-rays cannot be focused; they must be recombined by computation, once the phases have been determined. Before beginning this task, we consider some important properties of the structure factor equation.

3.6.11.5 Features of the structure factor equation

Equation (3.69) may be broken down according to (3.64) and (3.65):

$$A(hkl) = \sum_j f_{\theta,j} \cos[2\pi(hx_j + ky_j + lz_j)] \tag{3.72}$$

$$B(hkl) = \sum_j f_{\theta,j} \sin[2\pi(hx_j + ky_j + lz_j)] \tag{3.73}$$

If a crystal is centrosymmetric and the centre of symmetry is at the origin (0,0,0), then the N atoms are grouped into $N/2$ pairs with coordinates $\pm(x_j, y_j, z_j)$. In this situation, it is a straightforward matter to show that

[25] See, for example, M F C Ladd and R A Palmer *Structure Determination by X-ray Crystallography*, 3rd edition (Plenum Press 1994).

[26] It is normal practice to introduce temperature factors when calculating structure factors in a practical situation.

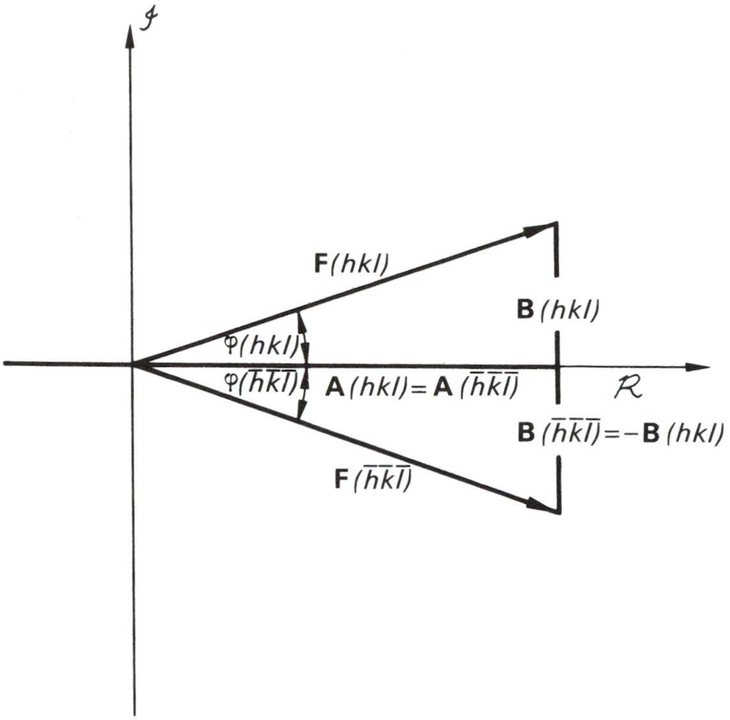

Figure 3.39 Relationship between $\boldsymbol{F}(hkl)$ and its conjugate $\boldsymbol{F}(\bar{h}\bar{k}\bar{l})$, illustrating Friedel's law: $|\boldsymbol{F}(hkl)|=|\boldsymbol{F}(\bar{h}\bar{k}\bar{l})|$.

$$F(hkl)=A(hkl)=2\sum_{j=1}^{N/2} f_{\theta,j}\cos\left[2\pi(hx_j+ky_j+lz_j)\right] \tag{3.74}$$

where the $N/2$ sets of coordinates are those not related by the centre of symmetry; $B(hkl)$ is identically zero. It follows from (3.71) that the only two values for the phase angle are 0 and π, corresponding to $F(hkl)$ positive and negative respectively. Thus, we may speak of the sign s of a centrosymmetric reflection, where $s=\pm 1$. The phase problem is very much simpler for centrosymmetric crystals.

If a unit cell is body-centred, then the N atoms exist as pairs x_j, y_j, z_j and $\frac{1}{2}+x_j$, $\frac{1}{2}+y_j$, $\frac{1}{2}+z_j$. Feeding these data into (3.69) one obtains

$$F(hkl)=\sum_{j=1}^{N/2} f_{\theta,j}\{\exp\left[\mathrm{i}2\pi(hx_j+ky_j+lz_j)\right]+\exp\left[\mathrm{i}2\pi(hx_j+ky_j+lz_j+\tfrac{1}{2}(h+k+l))\right]\} \tag{3.75}$$

The term in braces may be expressed as

$$\exp\left[\mathrm{i}2\pi(hx_j+ky_j+lz_j)\right]\{1+\exp\left[\mathrm{i}\pi(h+k+l)\right]\} \tag{3.76}$$

Since h, k and l are integers, $\{1+\exp[\mathrm{i}\pi(h+k+l)]\}$ will equal 0 or 2 according to whether the sum $(h+k+l)$ is odd or even. In other words, reflections arise from a body-centred unit cell provided that the *limiting condition*[27]

$$hkl:\quad h+k+l=2n \tag{3.77}$$

[27] The term *systematic absences* is also used; then the hkl reflections are *absent* when $h+k+l=2n+1$.

holds, where n is an integer. Conditions such as this exist for all centred unit cells and the pattern of reflection data will reveal the presence of such centring.

We shall consider one more example, namely, a monoclinic centrosymmetric unit cell containing a c glide plane normal to the y axis and a 2_1 screw axis parallel to y; we may apply (3.74). The atoms are related by the c glide plane in pairs as x_j, y_j, z_j and $x_j, \bar{y}_j, \frac{1}{2}+z_j$. Inserting these data into (3.65) and rearranging gives

$$F(hkl)=2\sum_{j=1}^{N/2} f_{\theta,j}\,\{\cos[2\pi(hx_j+ky_j+lz_j)]+\cos[2\pi(hx_j-ky_j+lz_j+l/2)]\} \qquad (3.78)$$

This equation may be expanded to give

$$F(hkl)=4\sum_{j=1}^{N/2} f_{\theta,j}\,\{\cos[2\pi(hx_j+lz_j+l/4)]\cos[2\pi(ky_j-l/4)]\} \qquad (3.79)$$

By choosing the special class of reflections $h0l$ and applying arguments as before, the limiting condition for the c glide plane is

$$h0l:\quad l=2n \qquad (3.80)$$

By a similar analysis, we can show that a 2_1 axis parallel to y, which relates positions x_j, y_j, z_j to $\bar{x}_j, \frac{1}{2}+y_j, \bar{z}_j$, leads to the limiting condition

$$0k0:\quad k=2n \qquad (3.81)$$

If we use the set of general equivalent positions for $P2_1/c$ (Figure 3.16) in (3.74), a single equation emerges that can be used to obtain both limiting conditions for this space group:

$$F(hkl)=4\sum_{j} f_{\theta,j}\cos\{2\pi[hx_j+lz_j+(k+l)/4]\}\cos\{2\pi[ky_j-(k+l)/4]\} \qquad (3.82)$$

where the sum extends over the unique $N/4$ atoms in the unit cell. The reader is invited to derive this equation, and to separate it for the conditions $(k+l)$ odd and even, thence to obtain the limiting conditions for this space group.

Similar results may be obtained for all translational symmetry elements in all space groups, using either (3.69), or (3.72) and (3.73).

3.6.11.6 Representation of electron density by a Fourier series

A set of coordinates for all the atoms in a molecule permits a calculation of its scattering for X-rays. The scattering function is known as the Fourier transform of the molecule, and is given as

$$G(\mathbf{S})=\int_V \rho(\boldsymbol{r})\exp[\mathrm{i}2\pi(\boldsymbol{r}\cdot\boldsymbol{S})]\,\mathrm{d}V \qquad (3.83)$$

where the integral extends over the volume of the molecule; $\boldsymbol{S}$ is a vector of magnitude $2\lambda^{-1}\sin(\theta)$, and $\rho(\boldsymbol{r})$ is the electron density at the vector distance $\boldsymbol{r}$ from the origin of the transform. The function is continuous throughout the space of V, but we cannot observe the transform of a single molecule. In practice, the molecule exists in a unit cell in a crystal, that is, in real space, whereas the transform exists in reciprocal space. Two other functions of these spaces that we have considered already are the lattice of planes in real space and the corresponding reciprocal lattice of points. The X-ray diffraction pattern may be thought of as arising from a superposition of the reciprocal lattice on to the transform of the molecule, so that the transform is recorded, or sampled, only at the reciprocal lattice points, where the Bragg equation (3.56) is satisfied.

In the unit cell of a crystal, we have a number of discrete atoms with coordinates x_j, y_j, z_j, each with a scattering function $f_{\theta,j}$, so that **r** is given by

$$\boldsymbol{r}=x_j\boldsymbol{a}+y_j\boldsymbol{b}+z_j\boldsymbol{c} \tag{3.84}$$

From the Ewald construction, $\boldsymbol{S}$ may be identified with $\boldsymbol{d}^*$, the vector from the origin of the reciprocal lattice to the point hkl, whence

$$\mathbf{S}=h\boldsymbol{a}^*+k\boldsymbol{b}^*+l\boldsymbol{c}^* \tag{3.85}$$

The transform equation (3.83) may now be formulated for the unit cell as

$$G(\boldsymbol{S})=\sum f_{\theta,j}\exp\left[\mathrm{i}2\pi(hx_j+ky_j+lz_j)\right] \tag{3.86}$$

which is comparable with the structure factor equation (3.69). The essential difference is that (3.86) is valid for all values of h, k and l, whereas (3.69) applies only for the *integral* values of h, k and l that correspond to reciprocal lattice points.

The Fourier transform of (3.83) is defined as

$$\rho(\boldsymbol{r})=\int_V G(\boldsymbol{S})\exp\left[-\mathrm{i}2\pi(\boldsymbol{r}\cdot\boldsymbol{S})\right]\mathrm{d}v \tag{3.87}$$

where v is a volume in reciprocal space. Utilizing (3.86), with h, k and l restricted to integral values, and with (3.84) and (3.85), we obtain

$$\rho(x,y,z)=\frac{1}{V}\sum_h\sum_k\sum_l \boldsymbol{F}(hkl)\exp\left[\mathrm{i}2\pi(hx+ky+lz)\right] \tag{3.88}$$

where V is the unit-cell volume and the summations extend over reciprocal space, as in (3.87), but are restricted to the integral values of h, k and l.

Following Figure 3.38, we can write

$$F(hkl)=|F(hkl)|\exp\left[\mathrm{i}\phi(hkl)\right] \tag{3.89}$$

where $|F(hkl)|$ is obtained from the experimental intensity data. By expanding (3.88), and using (3.89), Friedel's law (3.70) and de Moivre's theorem, we obtain

$$\rho(x,y,z)=\frac{2}{V}\sum_{h'}\sum_k\sum_l |F(hkl)|\cos\left[2\pi(hx+ky+lz)\right]\cos\left[\phi(hkl)\right]+ |F(hkl)|\sin\left[2\pi(hx+ky+lz)\right]\sin\left[\phi(hkl)\right] \tag{3.90}$$

where the sum over h' indicates that, with Friedel's law, h (or k, or l)≥ 0. Combining the terms leads to

$$\rho(x,y,z)=\frac{2}{V}\sum_{h'}\sum_k\sum_l |F(hkl)|\cos\left[2\pi(hx+ky+lz)-\phi(hkl)\right] \tag{3.91}$$

This equation highlights the phase problem in X-ray analysis: the correct summation for $\rho(x,y,z)$ requires the correct relative phases $\phi(hkl)$ for each experimental value $F(hkl)$ but, as we have seen, they are not obtained by direct experiment. Thus, we need to consider how they may be obtained, before we can obtain an electron density image from (3.91).

One further simplification will first be introduced. From (3.74) we noted that the phase angle for a centrosymmetric crystal was either 0 or π. In this situation, (3.91) becomes

$$\rho(x,y,z)=\frac{2}{V}\sum_{h'}\sum_{k}\sum_{l}\pm|F(hkl)|\cos[2\pi(hx+ky+lz)] \quad (3.92)$$

and our problem becomes that of finding the *sign* of $|F(hkl)|$, a much less difficult task than the general case, for which $0\leq\phi\leq2\pi$. General computer programs exist for all the principal computations in every space group.

3.6.11.7 Patterson function

The Patterson function for a crystal is defined by

$$P(u,v,w)=\frac{2}{V}\sum_{h'}\sum_{k}\sum_{l}|F(hkl)|^2\cos[2\pi(hu+kv+lw)] \quad (3.93)$$

The coefficients $|F(hkl)|^2$ are obtained directly from the experimental intensity data and no phase information is required in summing the series (3.93). However, the function gives information about interatomic *vectors* in the unit cell, rather than about interatomic positions. A Patterson function always shows centrosymmetry, because both the vectors *ij* and *ji* for a pair of atoms *i*, *j* are included. There is also a large peak at the origin, signifying that every atom is at zero vector distance from itself.

The nonorigin peaks are those of importance, and the information that they provide is frequently sufficient to obtain a first approximation to the phases when the molecule contains one or more heavy atoms, such as bromine in an organic compound; the solution of a crystal structure by this means is usually called the *heavy-atom* method.

Consider a crystal with the space group $P2_1/c$, which we have already studied in some detail. The general equivalent positions in the unit cell are

$$\pm(x,\ y,\ z;\ x,\ \tfrac{1}{2}-y,\ \tfrac{1}{2}+z)$$

It follows that vectors between atoms in these positions will have Patterson coordinates as follows:

u	v	w	Weight
$2x$	$2y$	$2z$	1
$2x$	$\frac{1}{2}$	$\frac{1}{2}-2z$	2
0	$\frac{1}{2}-2y$	$\frac{1}{2}$	2

A weight of two implies that two interatomic vectors have the same Patterson coordinates, in this case formed from the pairs of positions $x,\ \frac{1}{2}-y,\ \frac{1}{2}+z$ plus $x,\ y,\ z$ and $-x,\ \frac{1}{2}+y,\ \frac{1}{2}-z$ plus $-x,\ -y,\ -z$; it is always possible to add or subtract from any coordinate.

Since scattering is dependent on electrons, peaks in a Patterson function will be proportional to Z_iZ_j. If atoms *i* and *j* are bromine atoms, it is evident that the Br–Br vector density would be much greater than that of other vector densities present, such as Br–C or O–O.

With the approximate coordinates of the bromine atom, structure factors are calculated from (3.82), the *calculated* phases are attached to the *experimental* $|F|$ values and an electron-density synthesis performed from (3.92). The resultant contour map should then show the positions of more atoms of the molecule, which can then be added to the next calculation of structure factors. The process is then repeated in order to determine the remainder of the structure; this technique is called successive Fourier synthesis.

We can see how the heavy-atom method can give a good trial structure from the following simple analysis. In a centrosymmetric crystal having, say, two molecules per unit cell with one heavy atom in each molecule, the structure factor may be written from (3.74) as

$$F(hkl)=2f_{\theta,\mathrm{H}}\cos[2\pi(hx_{\mathrm{H}}+ky_{\mathrm{H}}+lz_{\mathrm{H}})]+2\sum_{j} f_{\theta,j}\cos[2\pi(hx_j+ky_j+lz_j)] \tag{3.94}$$

where H refers to the heavy atom, and the sum over j to the remainder of the atoms in the unit cell. The first term on the right-hand side of (3.94) will have a magnitude M and a + or − sign. The second term will have another magnitude N and a + or − sign. On average, about 50% of the data will have $M>N$, so the sign given by the heavy atom will be correct. For the other half, although $M<N$, again on average, 50% of N will have the same sign as that of the heavy-atom contribution, so that in all we might expect about 75% of the $|F|$ data to be given a correct sign. Other factors, such as the atomic number of the heavy atom relative to the rest of the structure, and the extent of the thermal vibrations of the atoms, will modify the conclusions, but they may be expected to lead to a satisfactory initial step.

3.6.11.8 Solution of a crystal structure by the heavy-atom method

We shall consider the solution of the crystal structure of 2-bromobenzo[*b*]indeno[1,2-*e*]pyran, (BBIP) $C_{16}H_9BrO$, prepared by heating an ethanolic solution of equimolar amounts of 3-bromo-6-hydroxybenzaldehyde (I) and 2-oxoindane in the presence of piperidine acetate. Condensation took place with the elimination of water and the product was recrystallized from toluene.

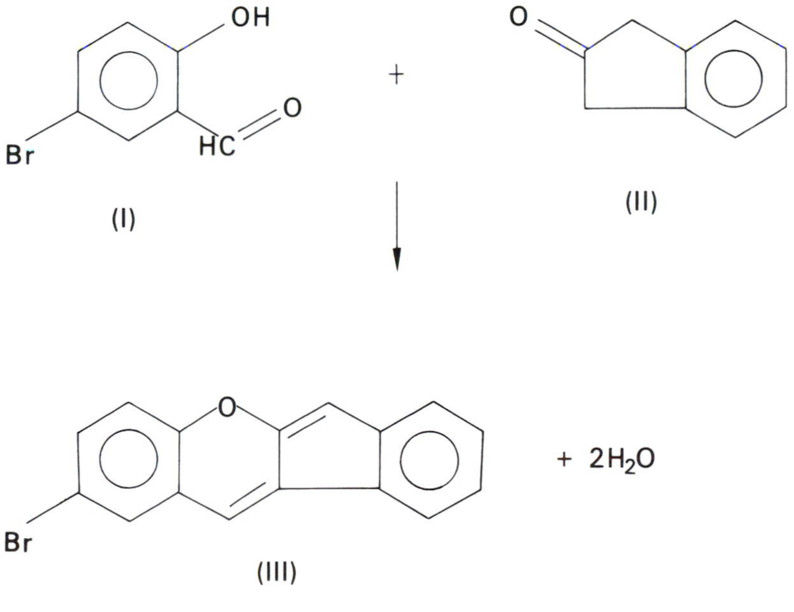

The crystals are monoclinic, and X-ray measurements of the unit-cell dimensions gave $a=0.7508$ nm, $b=0.5959$ nm, $c=2.6172$ nm and $\beta=92.55°$. A total of 1700 data were collected, using Cu Kα radiation ($\lambda=0.154178$ nm), to a value of 60° in θ. From Section 3.6.11.3, the total number of unique data expected was 1700. The density D was measured as 1680 kg m^{-3}. The number Z of molecules per unit cell is given by

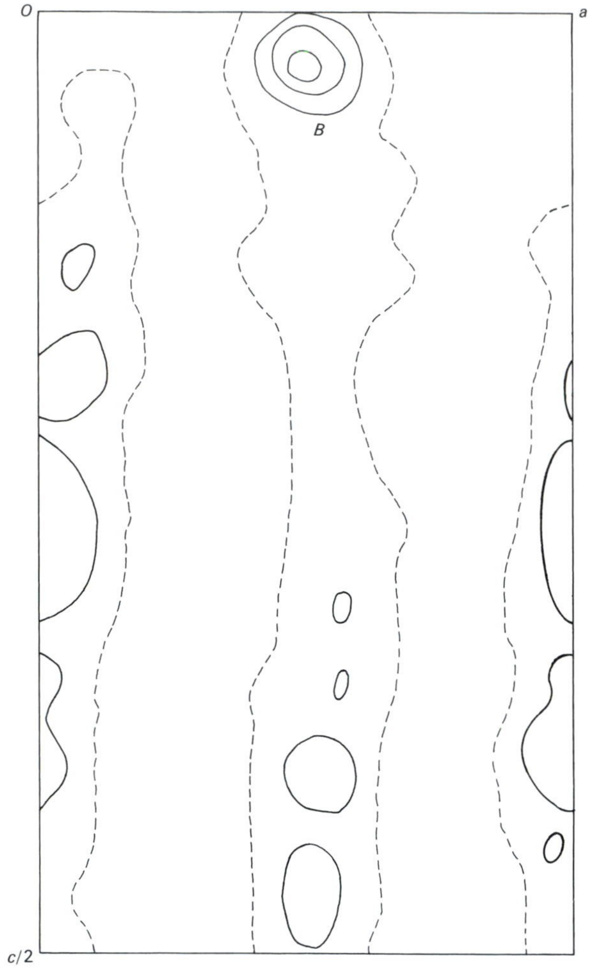

Figure 3.40 A section of the three-dimensional Patterson function $P(u,v,w)$ of BBIP close to v=0.2, the region of the $2x$, $2y$, $2z$ Br−Br vector. The x and z coordinates for the Br atom are 0.245 and 0.015 respectively; y_{Br} was found from the three-dimensional function, along the line through B, to be 0.185.

$$D = ZM_r u / V \tag{3.95}$$

where M_r is the relative molar mass, u the atomic mass unit and V the volume of the unit cell; Z evaluated to 3.98, that is, 4 to the nearest integer. The following limiting conditions existed among the reflection data:

hkl: None

$h0l$: $l=2n$

$0k0$: $k=2n$

Thus, the space group is established as $P2_1/c$. It should come as no surprise that a crystal with this space group has been chosen for this example.

The Patterson function (3.93) was calculated and Figure 3.40 shows the section closest to

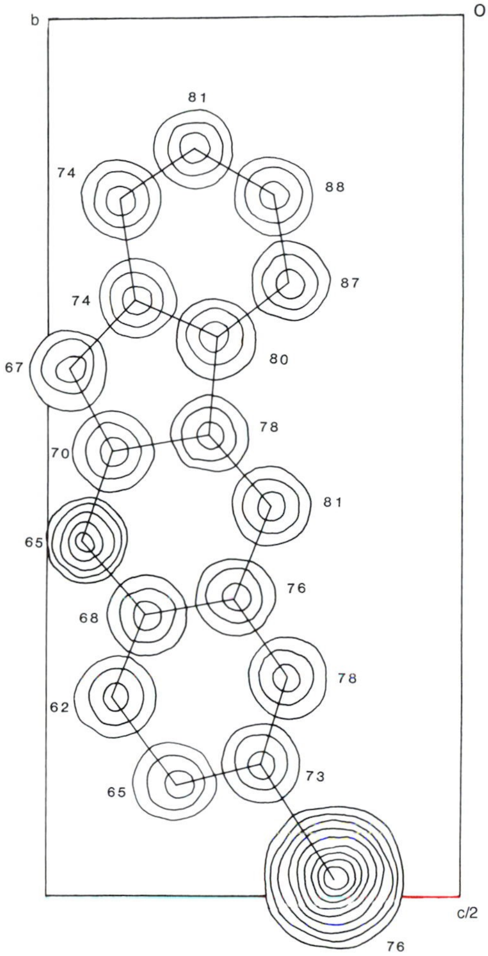

Figure 3.41 Composite three-dimensional electron density map, with the molecule of BBIP (excluding hydrogen atoms) outlined. The view is along a and the numbers represent 100 times the x coordinates. For convenience, a symmetry-related position was chosen for the bromine atom. What is the symmetry operation? (2_1 along the line $\frac{1}{2}, y, \frac{1}{4}$.)

the Patterson v coordinate for bromine. The peak marked B is $2x$, $\simeq 2y$, $2z$ for the bromine atom. Examination of the line through B, parallel to y, gave the total result $x_{Br}=0.245$, $y_{Br}=-0.185$, $z_{Br}=0.015$. These data were used in the structure factor equation to obtain signs for the experimental $|F|$ data, and an electron-density map was computed. Successive Fourier refinement led to the complete structure.

Final adjustments to the coordinates and thermal parameters of the atoms, and to the scale factor for the $|F|$ data were carried out by least-squares techniques. In this structure, the hydrogen atoms were fitted from known geometry and included in the later stages of refinement. Figure 3.41 shows the final electron-density map of the molecule, from which the structure is clearly revealed. For convenience, a symmetry-related ($\overline{x}$, $\frac{1}{2}+y$, $\frac{1}{2}-z$) bromine atom (and molecule) has been drawn.

Figure 3.42 shows the molecular formula deduced from the electron-density map. From the numerical results, bond lengths and angles were calculated by standard methods, and the

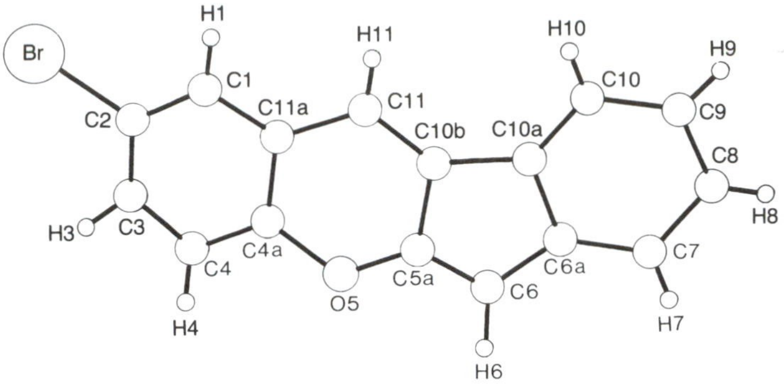

Figure 3.42 Structural formula for the molecule $C_{16}H_9BrO$ (BBIP).

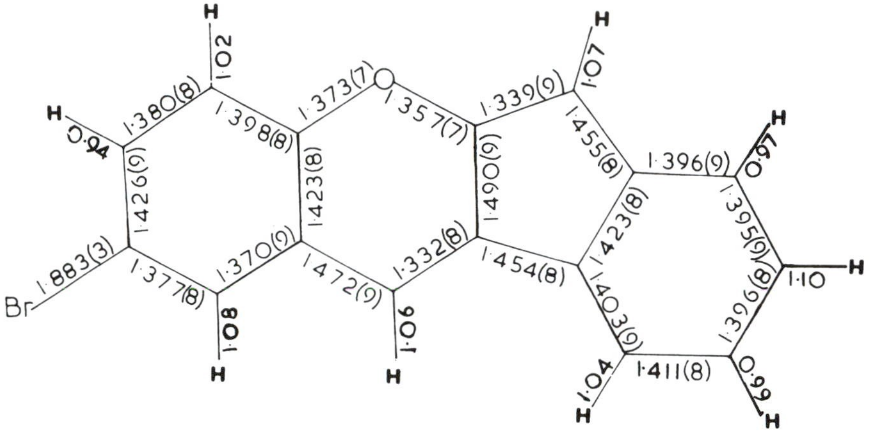

Figure 3.43 Bond lengths in BBIP, with estimated standard deviations in parentheses; the estimated standard deviations for the hydrogen atoms are approximately ten times larger. Hydrogen atoms are located much less precisely because of their relatively small scattering power for X-rays.

packing of molecules in the unit cell determined. Figures 3.43 to 3.45 show the results of these calculations.

The correctness of the structure determination is judged normally by three general criteria:

(a) The R (reliability) factor defined by

$$R=\Sigma||F_o|-|F_c||/\Sigma|F_o|| \tag{3.96}$$

should be small, normally less than about 5% for medium to large structures, indicating good agreement between the experimental data $|F_o|$ and those calculated from the model structure $|F_c|$. It is clear that its value will depend on both the quality of the experimental data, and the correctness of the structure model. The structure discussed above was refined to $R \simeq 6.0\%$.

(b) There should be no significantly negative regions of electron density, nor positive regions that are not indicative of atomic positions. Small positive and negative fluctuations are

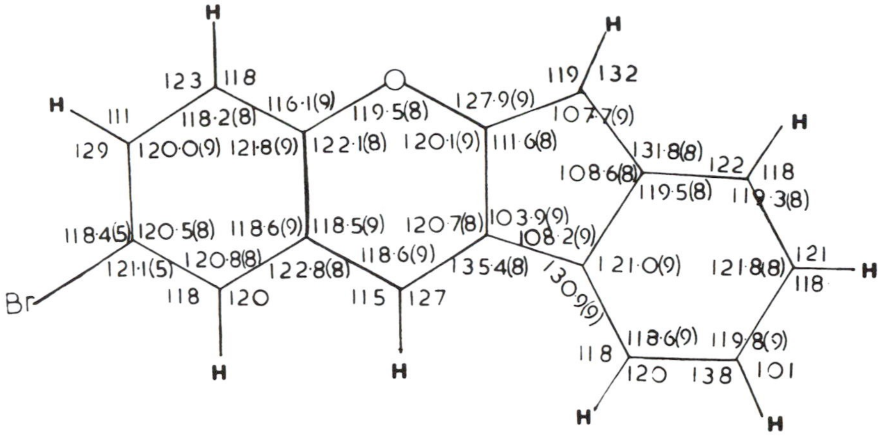

Figure 3.44 Bond angles in BBIP, with estimated standard deviation in parentheses; the estimated standard deviations for hydrogen are approximately ten times larger. Hydrogen atoms may be located precisely, when necessary, by neutron diffraction, for which scattering by hydrogen is relatively large.

Figure 3.45 Stereoview of the packing of the molecules of BBIP in the unit cell and its environs, as seen along *a*. The typical head-to-tail packing of a long molecule is clearly shown.

acceptable because the finite number of terms in the Fourier series (3.91) in any practical situation leads to series termination errors of that nature. Figure 3.41 was significantly free from spurious maxima or minima in the electron density.

(c) The structure should be consistent with the currently accepted chemical knowledge; for example, a C−C single bond that is found to be 0.18 nm or longer would need strong justification if it were not to be regarded as a flaw in the analysis. Figures 3.43 and 3.44 give bond lengths and angles, with standard deviations, that are acceptable for an organic aromatic compound. The standard deviations are not quoted for the bond lengths and angles involving hydrogen; typically they are an order of magnitude greater than those for the heavier atoms. Figure 3.45 shows the molecules with the typical head-to-tail packing common to long molecules. There was no evidence for hydrogen bonding involving the oxygen atom.[28]

For further details about this technique and of related procedures, particularly *direct methods of phase determination*, the reader is directed to the Bibliography.

[28] See M F C Ladd and R A Palmer, *Structure Determination by X-ray Crystallography*, 3rd edition (Plenum Press, 1994, pp. 448–69), and M F C Ladd and D C Povey, *J. Cryst. Mol. Struct.*, 2, 243 (1972).

Problems 3

3.1 List the point-group symbols of and the symmetry elements, other than identity, in molecules of (a) benzene, (b) 1,4-difluorobenzene and (c) 1,2-difluorobenzene.
3.2 Which pairs among the operators $\mathbf{C}_2$, $\mathbf{C}_3$, $\boldsymbol{\sigma}$ and $\mathbf{E}$ commute?
3.3 Draw up group multiplication tables for point groups (a) $\mathscr{C}_{2h}$ and (b) $\mathscr{D}_3$. What is the order of each group? Is either of them Abelian?
3.4 Is $\Gamma = 4 \;\; 1 \; -2$ an irreducible representation in point group $\mathscr{C}_{3v}$? If not, reduce it and express it in terms of irreducible representations of the group. A partial character table is given below:

$\mathscr{C}_{3v}$	$\mathbf{E}$	$2\mathbf{C}_3$	$3\boldsymbol{\sigma}_v$
A_1	1	1	1
A_2	1	1	−1
E	2	−1	0

3.5 By means of similarity transformations, establish the symmetry classes for point group $\mathscr{D}_6$.
3.6 To which irreducible representations would the function $x+y+z^2$ be assigned in point groups (a) $\mathscr{C}_{2v}$, (b) $\mathscr{D}_{3h}$, and (c) $\mathscr{C}_{4v}$?
3.7 Determine the irreducible representations of the most probable hybrid orbitals for the σ bonds in the $[NiF_4]^{2-}$ ion; its point group is $\mathscr{D}_{4h}$. Which symmetries are available for the F σ bonds in $[NiF_4]^{2-}$? Construct a molecular-orbital energy-level diagram for the ion and write its ground state configuration. The character table for $\mathscr{D}_{4h}$ follows.

$\mathscr{D}_{4h}$	$\mathbf{E}$	$2\mathbf{C}_4$	$\mathbf{C}_2$	$2\mathbf{C}_2'$	$2\mathbf{C}_2''$	i	$2\mathbf{S}_4$	$\boldsymbol{\sigma}_h$	$2\boldsymbol{\sigma}_v$	$2\boldsymbol{\sigma}_d$		
A_{1g}	1	1	1	1	1	1	1	1	1	1		z^2; x^2+y^2
A_{2g}	1	1	1	−1	−1	1	1	1	−1	−1	R_z	
B_{1g}	1	−1	1	1	−1	1	−1	1	1	−1		x^2-y^2
B_{2g}	1	−1	1	−1	1	1	−1	1	−1	1		xy
E_{2g}	2	0	−2	0	0	2	0	−2	0	0	(R_x, R_y)	(yz, zx)
A_{1u}	1	1	1	1	1	−1	−1	−1	−1	−1		
A_{2u}	1	1	1	−1	−1	−1	−1	−1	1	1	T_z	
B_{1u}	1	−1	1	1	−1	−1	1	−1	−1	1		
B_{2u}	1	−1	1	−1	1	−1	1	−1	1	−1		
E_u	2	0	−2	0	0	−2	0	2	0	0	(T_x, T_y)	

3.8 (a) By taking a π orbital on each carbon atom of benzene, determine the π molecular orbitals listed in equation (2.163); sp^2 σ bonds on each carbon atom may be assumed. We note the general rule that, *in a cyclic* $(CH)_n$ *molecule containing* C_n *symmetry, there are n* π *molecular orbitals, one for each irreducible representation of* C_n. The following character tables are provided.

$\mathscr{D}_{6h}$	E	$2C_6$	$2C_3$	C_2	$3C_2'$	$3C_2''$	i	$2S_3$	$2S_6$	σ_h	$3\sigma_d$	$3\sigma_v$		
A_{1g}	1	1	1	1	1	1	1	1	1	1	1	1		z^2; x^2+y^2
A_{2g}	1	1	1	1	−1	−1	1	1	1	1	−1	−1	R_z	
B_{1g}	1	−1	1	−1	1	−1	1	−1	1	−1	1	−1		
B_{2g}	1	−1	1	−1	−1	1	1	−1	1	−1	−1	1		
E_{1g}	2	1	−1	−2	0	0	2	1	−1	−2	0	0	(R_x, R_y)	(xz, yz)
E_{2g}	2	−1	−1	2	0	0	2	−1	−1	2	0	0		(x^2-y^2, xy)
A_{1u}	1	1	1	1	1	1	−1	−1	−1	−1	−1	−1		
A_{2u}	1	1	1	1	−1	−1	−1	−1	−1	−1	1	1	T_z	
B_{1u}	1	−1	1	−1	1	−1	−1	1	−1	1	−1	1		
B_{2u}	1	−1	1	−1	−1	1	−1	1	−1	1	1	−1		
E_{1u}	2	1	−1	−2	0	0	−2	−1	1	2	0	0	(T_x, T_y)	
E_{2u}	2	−1	−1	2	0	0	−2	1	1	−2	0	0		

$\mathscr{D}_6$	E	C_6	C_3	C_2	C_3^2	C_6^5		$\epsilon = \exp(2\pi i/6)$
A	1	1	1	1	1	1	T_z, R_z	z^2; x^2+y^2
B	1	−1	1	−1	1	−1		
E_1	$\left\{ \begin{matrix} 1 \\ 1 \end{matrix} \right.$	$\begin{matrix} \epsilon \\ \epsilon^* \end{matrix}$	$\begin{matrix} -\epsilon^* \\ -\epsilon \end{matrix}$	$\begin{matrix} -1 \\ -1 \end{matrix}$	$\begin{matrix} -\epsilon \\ -\epsilon^* \end{matrix}$	$\left. \begin{matrix} \epsilon^* \\ \epsilon \end{matrix} \right\}$	(x, y) (R_x, R_y)	
E_2	$\left\{ \begin{matrix} 1 \\ 1 \end{matrix} \right.$	$\begin{matrix} -\epsilon^* \\ -\epsilon \end{matrix}$	$\begin{matrix} -\epsilon \\ -\epsilon^* \end{matrix}$	$\begin{matrix} 1 \\ 1 \end{matrix}$	$\begin{matrix} -\epsilon^* \\ -\epsilon \end{matrix}$	$\left. \begin{matrix} \epsilon \\ -\epsilon^* \end{matrix} \right\}$		(x^2-y^2, xy)

The SALCs can be determined from the sixfold rotation alone (see Bibliography). (b) Use the Hückel molecular-orbital method to determine the wavefunctions for benzene, and compare the results with those from part (a). (c) Draw up a π molecular-orbital system for this molecule. What is the π delocalization energy?

3.9 Examine a model of a cube and list the symmetry elements present, with their orientations related to the faces, edges and corners of the cube. What is the point-group symbol for the cube in the Hermann–Mauguin and Schönflies notations?

3.10 Consider a tetragonal P unit cell and centre its A faces (the y–z planes). Does the new unit cell still represent a Bravais lattice and, if so, which one?

3.11 A twofold screw axis parallel to y may be represented by

$$\begin{bmatrix} -1 & 0 & 0 \\ 0 & 1 & 0 \\ 0 & 0 & -1 \end{bmatrix} \begin{bmatrix} 0 \\ 1/2 \\ 0 \end{bmatrix}$$

and a c glide plane normal to y by

$$\begin{bmatrix} 1 & 0 & 0 \\ 0 & -1 & 0 \\ 0 & 0 & 1 \end{bmatrix} \begin{bmatrix} 0 \\ 0 \\ 1/2 \end{bmatrix}$$

What is the result of the combination $c2_1$? Is it different from 2_1c?

3.12 Draw a diagram to illustrate the general equivalent positions and symmetry elements in the centrosymmetric space group $P2/c$, with the origin at a centre of symmetry. List the coordinates of the general and special equivalent positions in their correct sets. Determine the condition(s) that limit X-ray reflections for a crystal of this space group if the atoms occupy general equivalent positions in the unit cell; $F(hkl) \propto \cos[2\pi(hx+lz+l/4)]\cos[2\pi(ky-l/4)]$.

3.13 The space group for sodium chloride is $Fm3m$, with a unit-cell dimension a of 0.564 nm; the density is 2164 kg m^{-3}. From these data, and by reference to the space-group data for $Fm3m$ (see Bibliography), give the coordinates for the sodium and chloride ions in the unit cell.

3.14 A spectral emission line has a wavenumber of 500 cm^{-1}. What is the energy change for the associated transition, and in which region of the spectrum would the line be observed?

3.15 Use the data on the water molecule given in Chapter 1 to determine the moment of inertia of the molecule about its twofold symmetry axis (see Figure 3.6). How would the water molecule be classified in terms of its moments of inertia? (^{1}H=1.0078; ^{16}O=15.995.)

3.16 Use the following absorption spectral data to determine the bond lengths r_{CS} and r_{CO} in COS.

	J	1→2	2→3	3→4	4→5
$CO^{32}S$	ν/MHz	24325.92	36488.82	48651.64	60814.08
$CO^{34}S$	ν/MHz	23732.93		47462.40	

^{12}C=12.000; ^{16}O=15.995; ^{32}S=31.972; ^{34}S=33.968. Simultaneous equations in r^2_{CO} and r^2_{CS} arise. One method of solving them is to let $r_{CS}=pr_{CO}$, divide the resulting equations and solve for p.

3.17 The HCl molecule may be described by the Morse equation, with D_e=5.331 eV per molecule, ω_e=2990.6 cm^{-1} and x_e=0.0171. Obtain a value for the dissociation energy for $H^{35}Cl$ (^{1}H=1.0078; ^{35}Cl=34.969).

3.18 How many vibration modes are there for (a) ethyne, (b) benzene and (c) anthracene ($C_{14}H_{10}$)?

3.19 From data in Table 3.6, predict the frequencies (cm^{-1}) of the first two most probable vibrational transitions for hydrogen fluoride, given that the anharmonicity constant is 0.0218 for this molecule (^{1}H=1.0078).

3.20 The following results have been obtained for vibrational energy levels in hydrogen chloride:

υ	1	2	3	4
ϵ_{vib}/cm^{-1}	4367.50	7149.04	9826.48	12399.8

Determine D_e for this species.

3.21 How many vibrational degrees of freedom are there for the water molecule? Show them on sketches of the molecule. Use the direct product method to show which vibration modes are infrared active and Raman active, and comment on them.

3.22 Why is the infrared absorption of 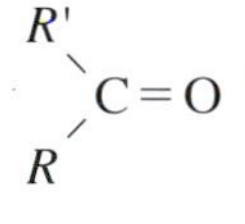 very strong yet that of 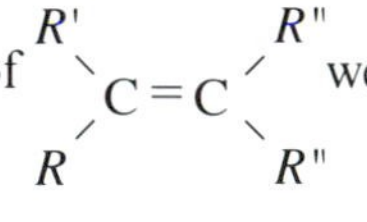 weak?

3.23 Figure P3.1 is an infrared spectrum for a compound C_4H_8O in the range 700–3000 cm^{-1}. Identify the peaks and suggest a possible structural formula for the compound.

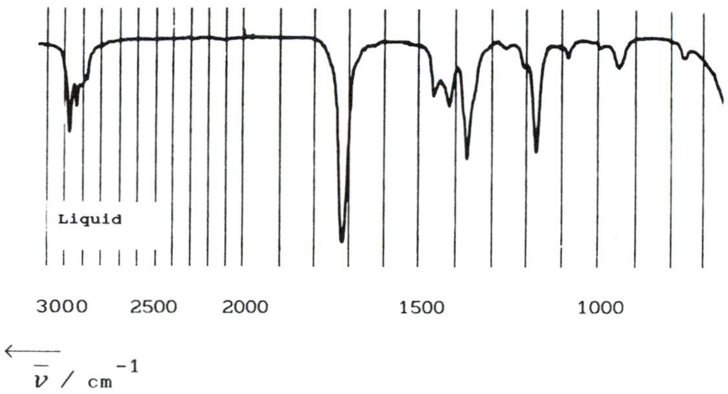

Figure P3.1

3.24 At what frequency, relative to TMS, would a group of protons of chemical shift 3.6 ppm resonate in a NMR spectrometer operating at 100 MHz?
3.25 What principal features would be expected for the PMR spectrum of $(CH_3)_2CHBr$?
3.26 Figure P3.2 shows the principal chemical shifts in the NMR spectrum of the compound C_4H_8O, examined first in Problem 3.23. Assign the chemical shifts and suggest a possible structural formula for the compound.

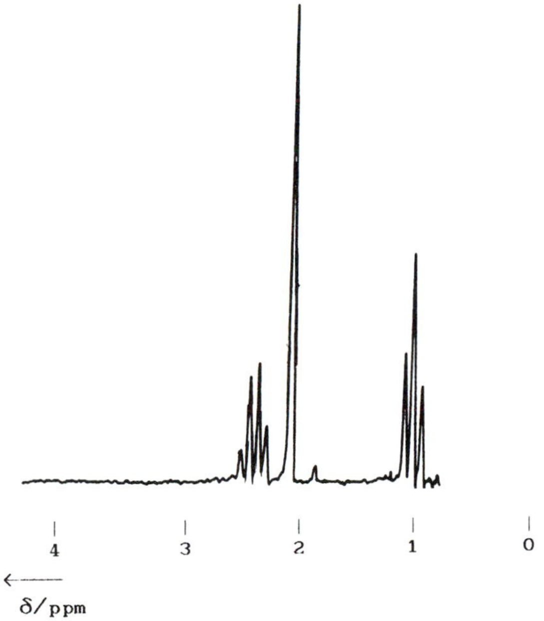

Figure P3.2

3.27 Figure P3.3 is a mass spectrogram for the compound C_4H_8O; the four highest relative abundances occur, in order from right to left, at *m/z* values of 72, 57, 43 and 29. Assign these values to chemical fragments and consider whether or no they reinforce the earlier findings for this compound (Problems 3.23 and 3.26).

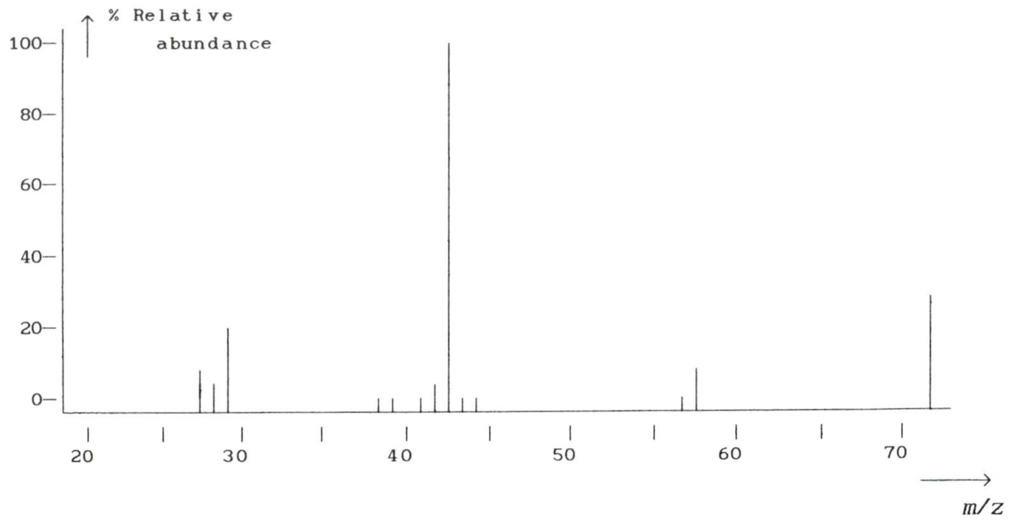

Figure P3.3

3.28 (a) What are the Miller indices for the faces of a cube and of a tetrahedron? (b) In a tetragonal unit cell, a=0.123 nm and c=0.456 nm. Calculate the length from the origin to the reciprocal lattice point 123 ($|\boldsymbol{r}_{123}|$), the *direction* [123], and the angle θ between the vectors $\boldsymbol{r}_{123}$ and $\boldsymbol{r}_{210}$.

3.29 Deduce the limiting conditions for (a) a *C*-face-centred unit cell and (b) a *c* glide plane normal to the x axis and passing through the point x=0 in a centrosymmetric orthorhombic unit cell.

3.30 Euphenyl iodoacetate $C_{32}H_{53}O_2I$ is monoclinic, with the unit-cell dimensions a=0.726 nm, b=1.155 nm, c=1.922 nm and β=94.07°. Determine the reciprocal unit-cell dimensions for X-ray diffraction with Cu Kα radiation (λ=0.15418 nm).

3.31 A crystal with space group $P2_1$ has two molecules per unit cell in the general equivalent positions x, y, z and $\bar{x}, \frac{1}{2}+y, \bar{z}$. Let there be *two heavy atoms per molecule*, with the following fractional coordinates (and symmetry-related values):

x_1	y_1	z_1	x_2	y_2	z_2
0.20	0.15	0.10	0.25	0.35	−0.05

Consider the projection on to the x–z plane and, for convenience, assume that β=90.0°, within the limits of experimental error. (A similar result would be obtained with any other value of β.) By drawing, or otherwise, construct the Patterson vectors for the heavy atoms, around the origin of the unit cell. Comment on the vectors in terms of their geometry and relative weights.

3.32 The crystal in Problem 3.30 belongs to space group $P2_1$, for which the coordinates of general equivalent positions are x, y, z, and $\bar{x}, \frac{1}{2}+\bar{y}, z$; the density is 1230 kg m^{-3}.
(a) Calculate the number of molecules per unit cell, and express the Patterson coordinates of the I−I vectors in terms of x, y and z. (b) On the Patterson map, Figure P3.4, identify the I−I vector and determine the fractional coordinates of the iodine atoms in the unit cell.

(c) Calculate the shortest I–I nonbonded distance in the structure.
(d) Why is the level $v=\frac{1}{2}$ in the Patterson function $P(u,v,w)$ of particular interest?

Figure P3.4

3.33 Hafnium disilicide $HfSi_2$ is orthorhombic, with $a=0.3677$, $b=1.455$, $c=0.3649$ nm and $Z=4$. The space group is *Cmcm*, and the Hf and Si atoms occupy three sets of the special positions

$$\pm(0, y, \tfrac{1}{4}; \tfrac{1}{2}, \tfrac{1}{2}+y, \tfrac{1}{4})$$

The experimental $|F(0k0)|$ data are as follow:

$0k0$	020	040	060	080	0 10 0	0 12 0	0 14 0	0 16 0
$\lvert F(0k0)\rvert$	7	14	18	13	12	<1	10	<1

(a) Form $|F(0k0)|^2$, divide by 10 and round to the nearest integer. Then, calculate the one-dimensional Patterson projection

$$P(v)=A\sum_k |F(0k0)|^2 \cos(2\pi kv)$$

where A is a constant with the dimensions of $(\text{length})^{-1}$ and may conveniently be taken as 0.5 in this example. Plot the function, locate a Hf–Hf vector and, hence, determine y_{Hf}. Interpret all the peaks as far as you are able.

(b) Assuming that the signs of $F(0k0)$ will be governed by the hafnium atoms, show that these signs will be those of $\cos(2\pi k y_{Hf})$, and determine them.

(c) Use the signs with their appropriate $|F|$ values to calculate the one-dimensional electron density function

$$\rho(y)=B\sum_k \pm|F(0k0)|\cos(2\pi ky)$$

where B is a constant of dimensions (length)$^{-1}$ and may be conveniently given the value 1 in this example. Refine the value of y_{Hf} from $P(v)$ and obtain the y coordinates for the two silicon atoms. Can the Patterson projection calculated in (a) be now further interpreted?
(d) Draw a plan of the structure on the x–y plane. Note that the atomic radii of the Hf and Si species are approximately 0.11 nm and 0.13 nm respectively, and determine the shortest, chemically sensible Hf–Si distances.

Notes on computation

Access to a computer enables the calculations to be carried out readily. The intervals of v (or y) can be 60ths of the cell side b, and it is necessary to compute the function only from $v=0$ to $v=15/60$. Reflection symmetry at $v=\frac{1}{4}$ and $v=\frac{1}{2}$ means that a calculated function can be repeated to a full unit-cell dimension, if needed. If a computer is not available, the following tables[29] will enable the summations to be carried out easily. In the case of the electron density, the signs will need to be reversed along any horizontal row for which the corresponding $F(0k0)$ is found to be negative ($\bar{x}$ is written for $-x$).

Note that the term corresponding to $F(000)$ is not included with the data, so that fluctuations below zero will be found for both $P(v)$ and $\rho(y)$. They may be disregarded in this problem; they affect the vertical scale but not the pattern. A constant value of 20 could be added to all values of $\rho(y)$ and 40 to all of $P(v)$, if desired.

		v in 60ths															
$\lvert F\rvert^2$	k	0	1	2	3	4	5	6	7	8	9	10	11	12	13	14	15
5	2	5	5	5	4	3	2	2	1	$\bar{1}$	$\bar{2}$	$\bar{2}$	$\bar{3}$	$\bar{4}$	$\bar{5}$	$\bar{5}$	$\bar{5}$
20	4	20	18	3	6	$\bar{2}$	$\overline{10}$	$\overline{16}$	$\overline{20}$	$\overline{20}$	$\overline{16}$	$\overline{10}$	$\bar{2}$	6	13	18	20
32	6	32	26	$\overline{10}$	$\overline{10}$	$\overline{26}$	$\overline{32}$	$\overline{26}$	$\overline{10}$	10	26	32	26	10	$\overline{10}$	$\overline{26}$	$\overline{32}$
17	8	17	11	$\bar{2}$	$\overline{14}$	$\overline{17}$	8	5	16	16	5	$\bar{8}$	$\overline{17}$	$\overline{14}$	$\bar{2}$	11	17
14	10	14	7	$\bar{7}$	$\overline{14}$	$\bar{7}$	7	14	7	$\bar{7}$	$\overline{14}$	$\bar{7}$	7	14	7	$\bar{7}$	$\overline{14}$
10	14	10	1	$\overline{10}$	$\bar{3}$	9	5	$\bar{8}$	$\bar{7}$	7	8	$\bar{5}$	$\bar{8}$	3	10	$\bar{1}$	$\overline{10}$
Σ_k	=	98	68	9	…												

		v in 60ths															
$\lvert F\rvert$	k	0	1	2	3	4	5	6	7	8	9	10	11	12	13	14	15
7	2	7	7	6	6	5	3	2	1	$\bar{1}$	$\bar{2}$	$\bar{3}$	$\bar{5}$	$\bar{6}$	$\bar{6}$	$\bar{7}$	$\bar{7}$
14	4	14	13	9	4	$\bar{1}$	$\bar{7}$	$\overline{11}$	$\overline{14}$	$\overline{14}$	$\overline{11}$	$\bar{7}$	$\bar{1}$	4	9	13	14
18	6	18	15	6	$\bar{6}$	$\overline{15}$	$\overline{18}$	$\overline{15}$	$\bar{6}$	6	15	18	15	6	$\bar{6}$	$\overline{15}$	$\overline{18}$
13	8	13	9	$\bar{1}$	$\overline{11}$	$\overline{13}$	$\bar{6}$	4	12	12	4	$\bar{6}$	$\overline{13}$	$\overline{11}$	$\bar{1}$	9	13
12	10	12	6	$\bar{6}$	$\overline{12}$	$\bar{6}$	6	12	6	$\bar{6}$	$\overline{12}$	$\bar{6}$	6	12	6	$\bar{6}$	$\overline{12}$
10	14	10	1	$\overline{10}$	$\bar{3}$	9	5	$\bar{8}$	$\bar{7}$	7	8	$\bar{5}$	$\bar{9}$	3	10	$\bar{1}$	$\overline{10}$
$\Sigma_k =$																	

[29] H Lipson and C A Beevers, *Proc. Phys. Soc.* **48**, 772 (1936).

CHAPTER 4

Energy and energetics

4.1 Introduction

The study of the energetics of chemical reactions introduces the topics of thermodynamics and thermochemistry. Thermodynamics is a study of matter in bulk that is concerned with energy changes in systems, chemical or mechanical, whereas thermochemistry is the study of heat changes in chemical reactions and is, thus, a branch of thermodynamics. Thermodynamics allows us to predict whether or not a reaction, say A→B, will occur spontaneously, that is, with a decrease in the energy of the system. It does not indicate the speed of the reaction, or even whether it will take place at all without the assistance of heat, irradiation or a catalyst; for this information we shall study reaction kinetics in a later chapter.

4.1.1 General laws of thermodynamics

Thermodynamics rests upon three fundamental principles, the laws of thermodynamics. The laws apply, in the most general sense, to the universe as a whole, and may be given in the following form.

The *energy* of the universe is constant, whatever processes take place in it.
The *entropy* of the universe increases, whatever processes take place in it, and must, therefore, tend towards a maximum value.
The absolute scale of temperature has a minimum value, *absolute zero*, and the entropy of any perfect, crystalline substance is zero at that temperature.

The development of the subject is based on these laws and we shall investigate how they are used in understanding reactions in chemistry.

4.1.2 Systems, states, properties and processes

Normally, we are concerned to work with a small portion of the universe, a *system*, and we need to specify the contents of the system and the nature of its bounding surfaces. For example, we may enclose pure, dry oxygen gas in a sealed vessel in an initial *state* 1, governed by the *properties* pressure p_1, volume V_1 and temperature T_1. We may carry out a *process* to change the gas from state 1 to another, final state 2 by warming the vessel. Then, a new set of properties p_2, V_2 and T_2 would exist. The change in any property X between two states 1 and 2 of the system is given by its value in the final state *minus* its value in the initial state:

$$\Delta X = X_2 - X_1 \tag{4.1}$$

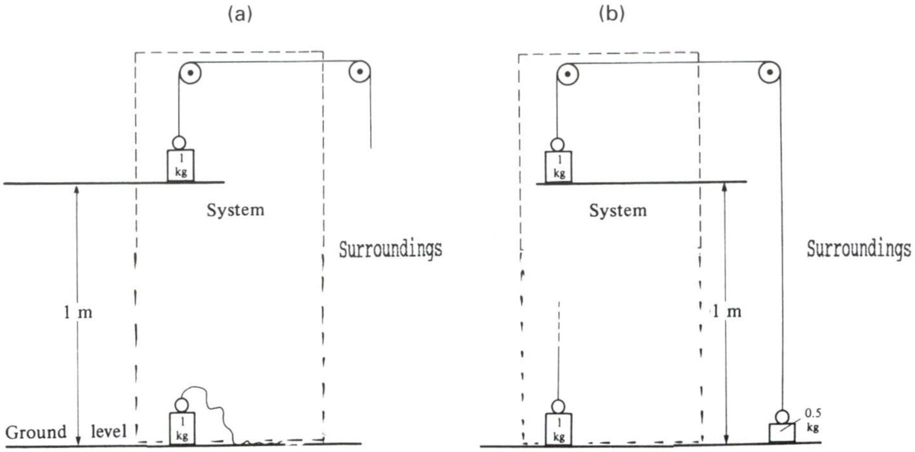

Figure 4.1 Mechanical system and its surroundings; the system is the portion contained by the dashed box, and everything external to the system constitutes the surroundings: (a) no work is done on the surroundings as the 1 kg weight falls freely through 1 m, giving up its potential energy as heat, on contact with the ground, that can be exchanged with the surroundings; (b) half of the potential energy is converted to work in raising the 0.5 kg weight through 1 m, the remainder of the energy being again dissipated as heat which can be exchanged with the surroundings.

This type of system, in which matter in the system, the oxygen gas, is not exchanged with its surroundings, is a *closed* system. If, further, the closed system has no interaction, thermal or mechanical, with the surroundings, it is an *isolated* system. When matter is exchanged between the system and the surroundings, the system is *open*.

4.1.3 Energy, work and heat

Energy is that property of a system that may be utilized to do work. *Work* is done by a system if it raises a weight in the surroundings against gravity, either directly by an arrangement of pulleys, or indirectly by the expansion of gas against an external pressure or by an electric current operating a winding motor. If a weight, of mass m, in the surroundings is raised by a distance z against gravity, work is done *by* the system *on* the surroundings. The amount of work is the product of the force applied to the weight (mg), where g is the gravitational acceleration, and the distance by which it is raised.

In Figure 4.1, the system consists of a brass weight of mass 1 kg attached to a string of negligible mass which passes over frictionless pulleys into the surroundings. In (a) the weight is allowed to fall freely through 1 m to ground level. The change in potential energy of the weight is −9.81 J. No work is done on the surroundings and the energy of the system is dissipated through the walls of the system as heat; the temperature of the surroundings will be raised, albeit only minutely. In (b) the falling weight is used to raise a brass weight of mass 0.5 kg in the surroundings by 1 m. The change in potential energy of the system is still −9.81 J, but it is divided into −4.905 J of work done by the system and −4.905 J dissipated again as heat. Thus, work and heat are both forms of energy.

In this experiment, heat was transferred from the system to the surroundings, a *diathermic* process. When no heat leaves the system during a process, that process is termed *adiabatic;* a process that is very close to being adiabatic may be achieved by conducting it in a

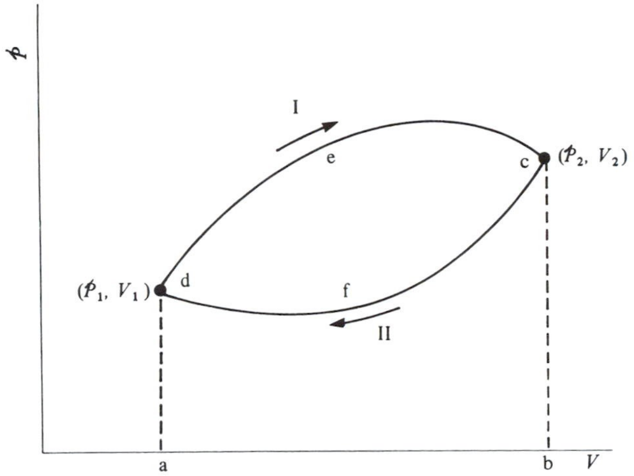

Figure 4.2 A gas in state 1 (p_1, V_1) undergoing a cyclic process to state 2 (p_2, V_2): the outwards path I is *d e c*, and the return path II is *c f d*; ΔU is the same in magnitude for both paths, and over the complete cycle ΔU is zero; q and w are both different for the two paths because, as we shall see, they depend upon the areas under the curves between *a* and *b*.

closed Dewar vessel. In the experiment of Figure 4.1, heat was lost from the system, an *exothermic* process, to the surroundings. When heat is taken in by the system, as in the dissolution of sodium nitrate in water, the process is termed *endothermic*.

Although thermodynamics is, as we have indicated, concerned only with matter in bulk, it is interesting to consider briefly the changes in energy at a molecular level. All molecules possess energy on account of their thermal motion. When work is done on or by a system, the molecular motions in the system are effectively cooperative, as with the rise and fall of the weights. When the weight falls to ground level and heat is dissipated, the cooperative motions of falling are converted into random molecular motions in the surroundings. Some of this heat may, through contact, re-enter the weight, but in that case the molecular motions in the weight would be random: the weight itself does not move spontaneously in any given direction, not even by a fraction of a micrometre. The random *Brownian motion* of microscopically small particles in a suspensory medium originates from the thermal motion of the particle molecules, at any finite temperature (Appendix 3), and from their collisions with one another and with the molecules of the suspensory medium.

4.2 Conservation of energy and the first law of thermodynamics

The principle of the conservation of energy is justified *a posteriori* by the results based on it. The existence of contrary processes such as perpetual motion has never been demonstrated and conservation of energy is accepted universally as a fundamental axiom. It leads to the first law of thermodynamics, which may be stated as *the total energy of a system and its surroundings is constant,* or as *the total energy of an isolated system is constant*. Each of these statements is equivalent to that given in Section 4.1.1, because the universe may be considered to be either any system and its surroundings, or a single isolated system.

The total energy of a system is usually known as its *internal energy U*. We cannot measure the internal energy of a system, but we can measure its change ΔU for any process. Consider a gas in an initial state defined by variables p_1 and V_1, and let it be changed to a final state defined by p_2 and V_2 by following path I (Figure 4.2), and then returned to the initial state by path II. The first law of thermodynamics requires that the internal energy of the gas in state 1 shall be the same after the cycle as it was initially, otherwise energy would not be conserved. The change in internal energy for the process $1 \rightarrow 2$ is given, from (4.1), by

$$\Delta U = U_2 - U_1 \tag{4.2}$$

The magnitude of ΔU will be the same for the process $2 \rightarrow 1$, but its sign will change. A property of a system, such as the internal energy, that depends for its value only on the initial and final states of the system is called a *state property*, or state function.

Consider again the experiment illustrated by Figure 4.1. In (b) the potential energy of the weight was converted into work and heat, and the first law of themodynamics may be formulated as

$$\Delta U = q + w \tag{4.3}$$

By convention, q denotes the heat supplied *to* the system, so that, if the system evolves heat (exothermic) in the process represented by (4.3), q is considered to be negative. Similarly, w denotes work done *on* the system, and, if the system does work on the surroundings in the process, w is negative. Thus, in Figure 4.1b, the potential energy of the brass weight decreased, heat was lost *by* the system to the surroundings and work was done *by* the system *on* the surroundings. Thus, ΔU, q and w were all negative, as we have seen. The change $\Delta U_{1 \rightarrow 2}$ depends only on the values of the variables in the initial and final states, but this situation is not true for q and w; U is a state property, whereas both q and w depend upon the path of the process.

4.2.1 Extensive and intensive properties

The internal energy is an *extensive property* of the system, that is, it depends upon the mass of the system: the first ionization energy of sodium is 495.835 kJ mol^{-1} (Table 3.3); for two moles the corresponding energy is 991.670 kJ. An *intensive property*, such as temperature or pressure, but not volume, is independent of the mass of the system. An intensive property may be defined as a limit of the ratio of two extensive properties. Thus, density D may be formulated as

$$D = \lim_{\delta V \rightarrow 0} \frac{\delta m}{\delta V} \tag{4.4}$$

where δm is the mass of the system in the volume element δV.

4.2.2 Expansion of an ideal gas

An ideal gas has no intermolecular forces between its molecules and, consequently, its internal energy is independent of its volume, depending only upon the amount of substance present and the temperature. Experimental observations on the behaviour of real gases have established (see Section 5.2) that they obey closely the *equation of state*

$$pV = n\mathcal{R}T \tag{4.5}$$

to which the ideal gas conforms exactly. In (4.5), n refers to the number of moles in the volume V and $\mathcal{R}$ is the universal gas constant; the other symbols have the meanings as before. The ideal gas is a useful model on which to discuss and formulate thermodynamic properties; subsequently the departure of real gases from the ideal can be considered.

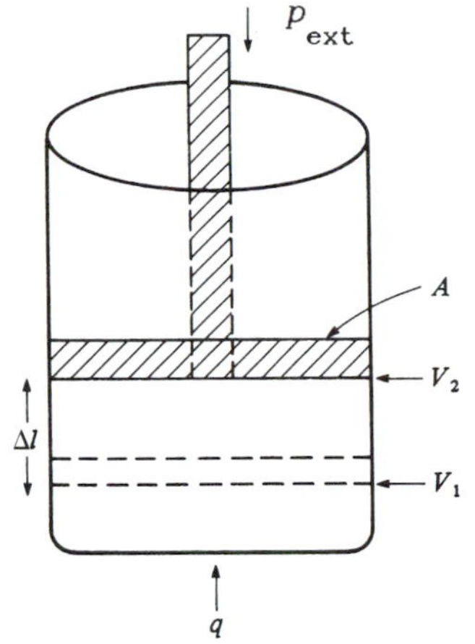

Figure 4.3 Cylinder with a perfectly fitted frictionless piston of negligible mass and area A. Initially enclosing a volume V_1, the piston travels outwards by an amount Δl and the gas expands to a volume V_2 against the external pressure p_{ext} by the application of heat q; the work w done by the system against the external pressure is $-p_{ext}\,\Delta V$, where $\Delta V=(V_2-V_1)=A\,\Delta l$.

Infinitesimal mechanical work $đw$ is done when a force F moves a body an infinitesimal distance dz against an opposing force:

$$đw=F\,dz \tag{4.6}$$

We use $đw$ rather than dw, because work, like heat, is not a state property. In the case of Figure 4.1b, the work done is

$$w=-\int_{z_0}^{z_1} mg\,dz \tag{4.7}$$

where $z_0=0$ (ground level) and $z_1=1$ m, the negative sign implying that work is done by the system on the surroundings.

Consider next an ideal gas confined to a cylinder by a perfect, frictionless piston of negligible mass and of cross-sectional area A, under an external pressure p_{ext} (Figure 4.3). Let the gas be expanded from volume V_1 to V_2, at the same pressure p_{ext}, by applying heat to the cylinder. For an infinitesimal distance of movement dl of the piston the work done $đw$ by the system is $-p_{ext}A\,dl$, where $A\,dl=dV$; hence, for the expansion from V_1 to V_2, we have

$$w=-p_{ext}\int_{V_1}^{V_2} dV=-p_{ext}\,\Delta V=-p_{ext}(V_2-V_1) \tag{4.8}$$

and the negative value of the work corresponds to a weight being raised against gravity, as in Figure 4.1b. The result in (4.8) could be expressed as the area under a curve of pressure against volume, and would be the area of a rectangle of sides p_{ext} and (V_2-V_1). We may now ask whether $p_{ext}\,\Delta V$ represents the maximum amount of work available in the expansion process.

4.2.3 Reversibility

The apparatus in Figure 4.4 is a refinement of the mechanical experiment in Figure 4.1. The spiral spring S carries a pan P to which is attached a needle pointer N all rigidly assembled

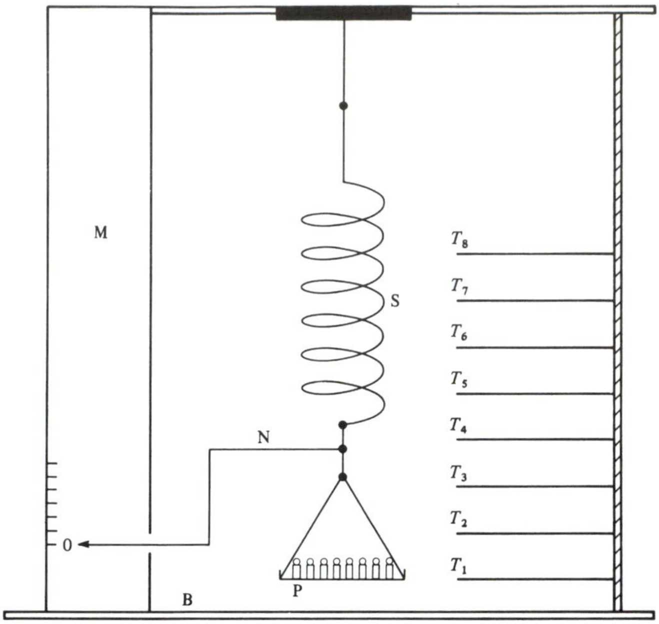

Figure 4.4 Mechanical analogue of a reversible system consisting of the spring S and its attachments, the pan P and the needle pointer N. The remainder of the apparatus is rigidly mounted on the baseboard B and comprises the surroundings. Work is done by the system when weights are raised to varying heights to the trays T_1 to T_8, and is measured by the metre scale M, assuming the applicability of Hooke's law.

on a baseboard B. When the pan is loaded with eight 0.1 kg weights the needle pointer is at zero. We shall assume that the spring and pan have negligible mass and that the extension of the spring follows Hooke's law, with each decrease of 0.1 kg on the pan causing the spring to contract by 0.1 m in vertical length. Eight trays T_1-T_8 are arranged at heights that correspond to the positions of the tray after successive removal of the 0.1 kg weights.

At first the system is fully loaded with 0.8 kg, and we consider how the energy stored in the spring can be caused to do work on the surroundings in raising the weights above the zero of energy corresponding to the level of tray T_1, where N is at zero. If the total of 0.8 kg is moved to T_1, the spring contracts, no work is done by the spring and its stored energy is dissipated as heat. We return to the starting configuration and let 0.4 kg be moved to T_1. The spring carries the remaining 0.4 kg upwards by 0.4 m and the remaining 0.4 kg are transferred to T_5. The work done by the system is that of raising 0.4 kg by 0.4 m, or −1.570 J. We carry out the total procedure again, but in 0.2 kg steps and then in 0.1 kg steps, with the results listed in Table 4.1.

It is evident that the work done increases numerically as the mass transferred per tray decreases. Figure 4.5 shows the plot of total work w against mass Δm transferred. Extrapolation to $\Delta m=0$ leads to a maximum work of approximately −3.14 J, which would be achieved for successive infinitesimally small mass transfers, leading to infinitesimally small displacements of the spring from equilibrium. The experiment could be carried out in the reverse order, whereupon similar but positive values of w would be obtained for work done on the system.

A process that is carried out by a succession of stages, each of which maintains an equi-

Table 4.1 Results from the mechanical analogue of a reversible system

Mass Δm transferred/kg	Tray	Scale movement/m	Work done/J	Total work w/J
0.8	1	0.0	0	
0.4	1	0.0	0	
0.4	5	0.4	−1.570	−1.570
0.2	1	0.0	0	
0.2	3	0.2	−0.392	
0.2	5	0.4	−0.785	
0.2	7	0.6	−1.177	−2.354
0.1	1	0.0	0	
0.1	2	0.1	−0.0981	
0.1	3	0.2	−0.196	
0.1	4	0.3	−0.294	
0.1	5	0.4	−0.392	
0.1	6	0.5	−0.491	
0.1	7	0.6	−0.589	
0.1	8	0.7	−0.687	−2.747

librium condition, constitutes a thermodynamically *reversible* process. It is not, of course, possible to conduct the process in a truly reversible manner. We could approach reversibility closely by using, say 80000 10 mg weights, with 80000 trays: it is left as an exercise for the reader to confirm the truth of this proposition. We may prove the extrapolated result by the following simple argument. The transfer of a small mass δm leads to a work term of $-\delta m\, g\, \delta z$, where δz is the movement of the spring and the negative sign indicates that the movement is upwards. We may write δz as αm, where α is the spring constant, 1 m kg^{-1} in our experiment. Thus, $\delta w = -\alpha g m\, \delta m$ and the total work is given by (g=9.81 m s^{-2})

$$w = -\int_0^{0.8} g\,\mathrm{d}m = -\tfrac{1}{2}\, g m^2 \Big|_0^{0.8} = -3.139 \text{ J}$$

Returning to the ideal gas in the cylinder (Figure 4.3), we may consider next the temperature held constant and the pressure reduced from a value p_1 to a lower value p' by cooling the gas. Because there is no change in volume along the path *de*, no work is involved; since q is negative, ΔU is also negative, from (4.3). Thus, the true position of e does not lie in the p–V plane of the isotherm in Figure 4.6. A succession of small, near-isothermal, near-equilibrium steps like *def*, allowing the gas to return to a state in the p–V plane after each step, would ultimately lead from the state p_1, V_1 to the state p_2, V_2.

To formulate this process, we represent the work $đw$ done by the system in an infinitesimal expansion dV by $đw = -p\,\mathrm{d}V$, where p is the gas pressure, equivalent to p_{ext}, during the expansion. From (4.5), we have $đw = -n\mathcal{R}T\,\mathrm{d}V/V$. Hence, the total work is given by

$$w = -n\mathcal{R}T \int_{V}^{V_2} \mathrm{d}V/V = -n\mathcal{R}T \ln(V_2/V_1) \tag{4.9}$$

Thus, this maximum value of w is the area *abcd* in the p–V plane of the isotherm (Figure 4.6). An irreversible expansion, as discussed with (4.8), against a constant external pressure of the same value (p_2) as that reached in the reversible expansion would lead to less work,

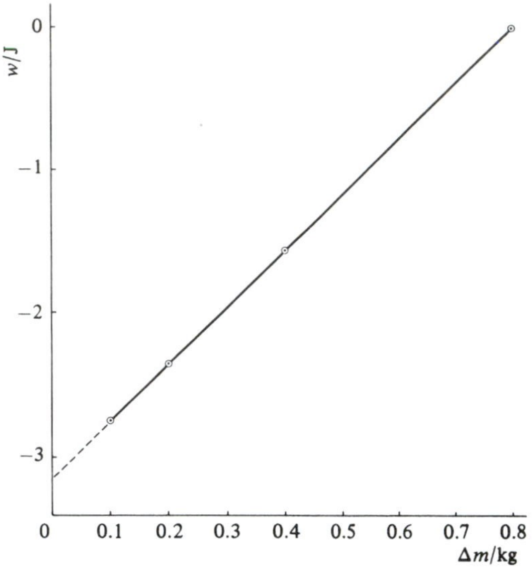

Figure 4.5 Variation of total work w done by the system (Figure 4.4) with mass Δm transferred. The straight line is in accordance with Hooke's law, and the extrapolation to $\Delta m=0$ corresponds to the maximum, reversible work of approximately -3.14 J obtainable from the system for transfer of infinitesimally small masses dm.

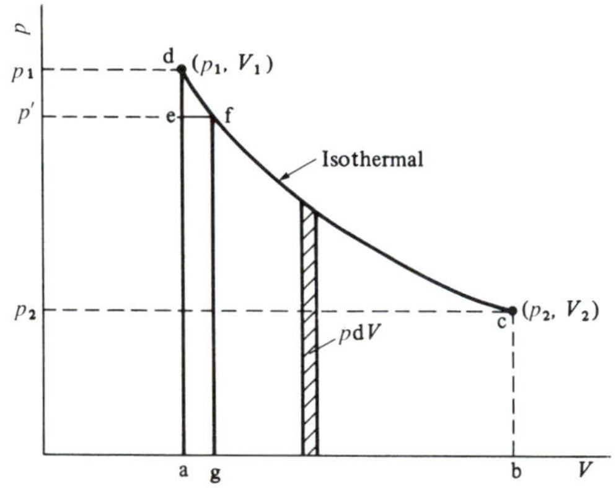

Figure 4.6 Isothermal expansion of an ideal gas from a state p_1, V_1 to a state p_2, V_2: the maximum work (4.9) is represented by the area *abcd*; the irreversible work under the external pressure p_2 is the area of the rectangle from the origin to *b* and of height *bc*.

represented by the area of the rectangle of length *b* on the V axis and height *bc*. For a part of the complete reversible expansion, say *df*, the maximum work is the area *agfd* and the practicable amount is *agef*. Any point, such as *e*, not lying in the p–V plane must be at either a higher or a lower temperature than that for the isotherm. However, if the ratio $(p_1-p')/p_1$ is small, the area *agfd* approximates closely to a rectangle. A succession of such rectangles of width δV then approximates to the area *abcd*.

We may compare the results of the reversible and irreversible expansion of 1 mol of ideal gas at 300 K, with the following data: $p_1=2$ atm, $p_2=1$ atm, $V_1=12.3$ dm^3 and $V_2=24.6$ dm^3. Thus, the irreversible work is -12.3 dm^3 atm and the reversible work is -17.1 dm^3 atm.

The concept of reversibility that we have discussed here both for a purely mechanical system and for the expansion of an ideal gas will find considerable application in subsequent studies.

4.3 State properties

In Figure 4.2, the values of w for the paths I and II are different, being the areas *adecb* and *adfcb*, respectively. Since $\Delta U_{\mathrm{I}} = -\Delta U_{\mathrm{II}}$, there must also be two values for the heat q. As we indicated in Section 4.2, q and w depend on the path and the infinitesimal quantities $đq$ and $đw$ cannot be integrated to give unique results. Thus, we have

$$\oint đw_{\mathrm{I}} \neq \oint đw_{\mathrm{II}} \qquad \oint đq_{\mathrm{I}} \neq \oint đq_{\mathrm{II}} \tag{4.10}$$

where $\oint$ indicates an integral around a cyclic path; but

$$\oint đw_{\mathrm{I}} + \oint đq_{\mathrm{I}} = -\oint đw_{\mathrm{II}} - \oint đq_{\mathrm{II}} \tag{4.11}$$

because

$$\oint \mathrm{d}U_{\mathrm{I}} = -\oint \mathrm{d}U_{\mathrm{II}} \tag{4.12}$$

Thus, the differential $\mathrm{d}U$ is *exact* (see Appendix 13), whereas $đq$ and $đw$ are *inexact*, and (4.3) may be formulated for infinitesimal changes as

$$\mathrm{d}U = đq + đw \tag{4.13}$$

The work $đw$ may represent the work of expansion $đw_{\mathrm{exp}}$ together with any additional work $đw_{\mathrm{a}}$, such as electrical work, but for the moment we shall not divide the work term in this way.

The internal energy U of a real gas in a closed system depends on any two of the variables p, V and T, because they are interconnected by (4.5). Thus, the exact differential $\mathrm{d}U$ may be given by standard calculus as

$$\mathrm{d}U = (\partial U/\partial V)_T \mathrm{d}V + (\partial U/\partial T)_v \mathrm{d}T \tag{4.14}$$

The partial differential $(\partial U/\partial V)_T$ is the rate of change of U with V at constant T. Since the order of second differentiation is immaterial,

$$\left(\frac{\partial}{\partial T}(\partial U/\partial V)_T\right)_v = \left(\frac{\partial}{\partial V}(\partial U/\partial T)_V\right)_T \tag{4.15}$$

To test for the exactness of a differential we may proceed in the following manner. For 1 mol of an ideal gas, $V = \mathcal{R}T/p$, from (4.5). The differential $\mathrm{d}V$ is given by

$$\mathrm{d}V = (\partial V/\partial T)_p \mathrm{d}T + (\partial V/\partial p)_T \mathrm{d}p = (\mathcal{R}/p)\,\mathrm{d}T - (\mathcal{R}T/p^2)\,\mathrm{d}p \tag{4.16}$$

Following (4.15)

$$\left(\frac{\partial}{\partial p}(\mathcal{R}/p)\right)_T = \left(\frac{\partial}{\partial T}(-\mathcal{R}T/p^2)\right)_p = -\mathcal{R}/p^2 \tag{4.17}$$

from which it follows that $\mathrm{d}V$ is exact.

4.4 Expansion under specified conditions

The expansion of a gas may be carried out under various conditions, one of which we have discussed in Section 4.2.3; we turn our attention to three other important cases.

4.4.1 Constant volume processes

The work $đw$ done by a system in expansion is $-p\,dV$. At constant volume $đw=0$ and, if no other work, such as mechanical or electrical work, is done in the process, (4.3) shows that $\Delta U=q_v$ or, for an infinitesimal change,

$$dU=đq_v \tag{4.18}$$

where $đq_v$ is the heat absorbed at constant volume. The heat capacity at constant volume C_v is defined by

$$C_v=đq_v/dT=(\partial U/\partial T)_v \tag{4.19}$$

4.4.2 Constant pressure processes

Constant pressure processes are very common in chemistry. If the only work done by the system is that of expansion against a constant external pressure, we substitute $đw=-p_{ext}\,dV$ into (4.13) and, since p_{ext} is the same as the gas pressure, we may write for an infinitesimal expansion

$$dU=đq_p-p\,dV \tag{4.20}$$

where $đq_p$ is the heat absorbed at constant pressure. For a finite change between two states

$$q_p=\Delta U+p\,\Delta V=(U_2+pV_2)-(U_1+pV_1) \tag{4.21}$$

Since U, p and V are state properties, $(U+pV)$ is also a state property, the *enthalpy* H:

$$H=U+pV \tag{4.22}$$

or, for a measurable change between two states,

$$\Delta H=\Delta U+p\,\Delta V \tag{4.23}$$

where ΔH refers to a process at constant pressure and ΔU to the corresponding process at constant volume. The heat capacity at constant pressure is given by

$$C_p=đq_p/dT=(\partial H/\partial T)_p \tag{4.24}$$

We may write (4.14) as

$$dU=\pi_T\,dV+C_v\,dT \tag{4.25}$$

where π_T may be thought of as the internal pressure of a gas. For an ideal gas π_T is zero; dividing throughout (4.25) by dT and imposing constant pressure conditions, we obtain

$$(\partial U/dT)_p=C_v=(\partial U/\partial T)_v \tag{4.26}$$

It now follows from (4.5) and (4.22) that, for an ideal gas,

$$C_p-C_v=(\partial(pV)/\partial T)_p=n\mathcal{R} \tag{4.27}$$

We shall consider the general case at a later stage.

4.4.3 Adiabatic expansion

In any closed system, an adiabatic expansion between two states involves no loss or gain of heat, that is, q is zero, so that from (4.13) $\mathrm{d}U = đw$. For an ideal gas, (4.19) leads to

$$w = \int_{T_1}^{T_2} C_v \, \mathrm{d}T = C_v \Delta T \tag{4.28}$$

assuming that the variation of C_v with temperature is negligible.

If a sudden (irreversible) expansion of a gas takes place against an external pressure p_{ext} then the work done is $-p_{ext}\Delta V$ and is the same as that given by (4.28); hence,

$$-p_{ext}\Delta V = C_v \Delta T \tag{4.29}$$

Since ΔV is positive ΔT is negative and the temperature of the gas falls.

If the expansion process is reversible (p_{ext} is now equivalent to the gas pressure p),

$$đw = -p\,\mathrm{d}V = \mathrm{d}U = C_v\,\mathrm{d}T \tag{4.30}$$

Using (4.5) we can write $-p\,\mathrm{d}V = -n\mathcal{R}\,\mathrm{d}V/V$ and

$$C_v\,\mathrm{d}T/T = -n\mathcal{R}\,\mathrm{d}V/V \tag{4.31}$$

and integrating between the initial (1) and final (2) states gives

$$C_v \ln(T_2/T_1) = -n\mathcal{R}\ln(V_2/V_1) \tag{4.32}$$

Rearranging (4.32) leads to

$$V_1T_1 = V_2T_2^X \tag{4.33}$$

where $x = C_v/(n\mathcal{R})$. The measurable work of expansion is

$$w = C_v\Delta T = C_vT_1[(V_1/V_2)^{1/X} - 1] \tag{4.34}$$

Using the equation of state (4.5)

$$p_1V_1/p_2V_2 = T_1/T_2 = (V_2/V_1)^{1/X} \tag{4.35}$$

or

$$p_1V_1^{\gamma} = p_2V_2^{\gamma} \tag{4.36}$$

where $\gamma = 1 + 1/x = 1 + n\mathcal{R}/C_v = C_p/C_v$. A monatomic gas has three degrees of freedom, and from Appendix 3 each degree of freedom contributes $\frac{1}{2}\mathcal{R}$T per mole to the internal energy. Hence, for an ideal gas, $C_v = \frac{3}{2}\mathcal{R}$, so that $C_p = \frac{5}{2}\mathcal{R}$ and the heat capacity ratio γ is 5/3. The speed of sound c_s in a gas is related to γ by the equation $c_s = (\mathcal{R}T\gamma/M)^{1/2}$, where M is the molar mass of the gas. The speed of sound in neon has been measured as 455 m s^{-1} at 25 °C; thus, $\gamma = 1.69$, which is very close to the ideal value of 5/3. Measurement of the speed of sound in a gas is a useful experimental method for obtaining the value of C_p/C_v.

Before leaving this section we focus attention on q. In the adiabatic process it is zero: but q is not a state property; so is an adiabatic expansion unique, or is another thermodynamic property involved? From (4.13) and (4.30)

$$đq = \mathrm{d}U - đw = C_v\,\mathrm{d}T + p\,\mathrm{d}V \tag{4.37}$$

Table 4.2 Expansion processes with a gas, between initial (1) and final (2) states; data marked * refer to the ideal gas

	Expansion against constant pressure p_{ext}		
Process	ΔU	q	w
Isothermal	0*	$p_{ext}\,\Delta V$*	$-p_{ext}\,\Delta V$
Adiabatic	$-p_{ext}\,\Delta V$	0	$-p_{ext}\,\Delta V$
	Reversible expansion/compression		
Process	ΔU	q	w
Isothermal	0*	$n\mathcal{R}T\ln\left(\frac{V_2}{V_1}\right)^*$	$-n\mathcal{R}T\ln\left(\frac{V_2}{V_1}\right)^*$
Adiabatic	$C_v\Delta T$*	0	$C_v\Delta T$*

which is simply a statement that $đq=0$ for the adiabatic process. For 1 mol of ideal gas

$$đq=C_v\,\mathrm{d}T+(\mathcal{R}T/V)\,\mathrm{d}V \tag{4.38}$$

Now $đq$ is inexact, as may be confirmed by differentiation of the right-hand side of (4.38). However, on dividing by T we obtain

$$đq/T=C_v\,\mathrm{d}T/T+(\mathcal{R}/V)\,\mathrm{d}V \tag{4.39}$$

The right-hand side is now an exact differential; hence, $\oint đq/T$ is a state property, the entropy change for the process. We shall introduce entropy in considering the second law of thermodynamics. The expansion processes that we have considered in Sections 4.2.2, 4.2.3 and 4.4 are summarized in Table 4.2.

4.5 Thermochemistry

In this section we shall study enthalpy changes in chemical reactions, the measurement of such changes, and the variation of enthalpy with temperature. Consider the reaction[1]

$$\mathrm{Na(s)}+\tfrac{1}{2}\,\mathrm{Cl_2(g)}\rightarrow\mathrm{NaCl(s)} \tag{4.40}$$

The enthalpy change ΔH for this reaction is defined in the usual way for a state property:

$$\Delta H=H(\mathrm{NaCl,s})-[H(\mathrm{Na,s})+\tfrac{1}{2}H(\mathrm{Cl_2,g})] \tag{4.41}$$

Although we can measure ΔH for the reaction, we cannot measure the enthalpy of an individual substance. A standard state is introduced in order to provide a reference level for enthalpy calculations: *the enthalpy of any element in its normal state at 298.15 K and 1 atm is defined to be zero*[2]. Hence, ΔH in (4.41) is the standard enthalpy of formation of sodium chloride, ΔH_f^{-0-} (NaCl,s); the superscript $-0-$ is used to indicate the standard value of a property.

[1] s, solid; 1, liquid; g, gas.

[2] Recent recommendations for standard states use 1 bar in place of 1 atm. In view of the small difference between them in this context, and given that many current texts and compilations of thermodynamic data use the atm, this form is retained here. The exact relations between the bar and the atmosphere are: 1 bar$=10^5$ Pa; 1 atm$=1.01325\times10^5$ Pa (1 Pa$=1$ N m^{-2}).

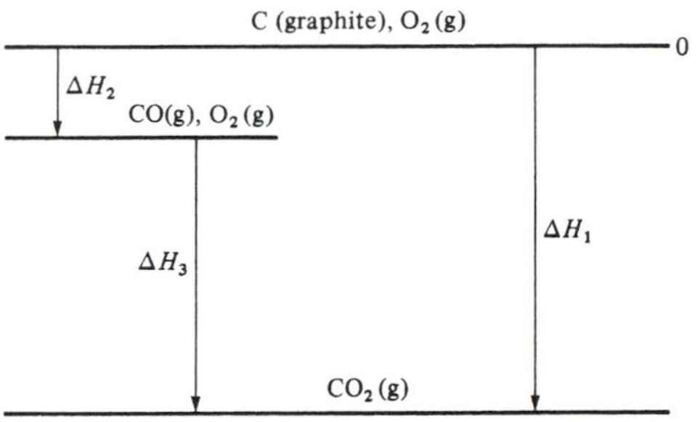

Figure 4.7 Thermochemical cycle for evaluating the standard enthalpy of formation of carbon monoxide CO; the zero level is the thermochemical standard state reference level. From Hess's law, $\Delta H_2+\Delta H_3-\Delta H_1=0$, so that $\Delta H_f(\text{CO,g})=\Delta H_2=\Delta H_1-\Delta H_3$.

The oxidation of carbon (graphite) may be represented by three equations:

$$\text{C(graphite)}+\text{O}_2\text{(g)}\rightarrow\text{CO}_2\text{(g)} \qquad \Delta H_f^{-0-}(\text{CO}_2,\text{g})=-393.5 \text{ kJ mol}^{-1} \tag{4.42}$$

$$\text{C(graphite)}+\tfrac{1}{2}\text{O}_2\text{(g)}\rightarrow\text{CO(g)} \qquad \Delta H_f^{-0-}(\text{CO,g})=-110.5 \text{ kJ mol}^{-1} \tag{4.43}$$

$$\text{CO(g)}+\tfrac{1}{2}\text{O}_2\text{(g)}\rightarrow\text{CO}_2\text{(g)} \qquad \Delta H_f^{-0-}=-283.0 \text{ kJ mol}^{-1} \tag{4.44}$$

The combination of (4.43) and (4.44) is equivalent to (4.42), because enthalpy is a state property. The general case may be stated in terms of Hess's law: *the algebraic sum of the enthalpy changes for the separate stages of a given process, taken in a cyclic order, is equal to the enthalpy change for the overall single-stage process.* Mathematically,

$$\Sigma_{\text{cycle}}\Delta H=0 \tag{4.45}$$

Hess's law is illustrated in Figure 4.7 for the reactions (4.42)–(4.44). The use of such descriptive cycles is preferable to treating the reactions like algebraic equations, and the following strategy is recommended.

(a) Draw up an enthalpy diagram and let an arrow indicate the direction of a reaction for which the corresponding ΔH, including its sign, is defined.
(b) Sum all such ΔH values around the cycle in a given sense, say anticlockwise, but changing the sign of ΔH for any term with an arrow that indicates a movement contrary to the direction of the cyclic summation; equate the sum to zero.
(c) It is permissible to have *elements* present in any amount on any level of the cycle, since their individual contributions to the enthalpy change are zero, from the definition of standard state.

Thus, from Figure 4.7 we obtain,

$$\Delta H_2+\Delta H_3-\Delta H_1=0 \tag{4.46}$$

Figure 4.8 applies to the formation of ethanol in the standard state. The reader is invited to set up the appropriate four chemical equations, one of which involves the complete combustion of ethanol, and evaluate $\Delta H_f^{-0-}(\text{C}_2\text{H}_5\text{OH,l})$.

One stage of the cycle involves the formation of liquid water. If the product had been water vapour, ΔH_f^{-0-} would have been -241.8 kJ mol^{-1}. The difference of 44.0 kJ mol^{-1} represents the standard molar enthalpy of evaporation ΔH_v^{-0-} for liquid water, and highlights the importance of quoting the physical state of each substance in any thermochemical process.

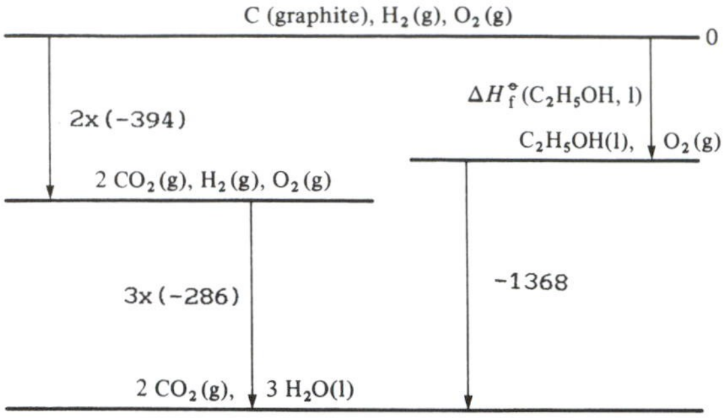

Figure 4.8 Thermochemical cycle for evaluating the standard enthalpy of formation of liquid ethanol C_2H_5OH. From the experimental measurements on enthalpies of formation of CO_2(g) and H_2O(l), and the enthalpy of combustion of ethanol, C_2H_5OH(l), application of Hess's law leads to: $2(-394)+3(-286)-(-1368)-\Delta H_f^{-0-}(C_2H_5OH,l)=0$, so that $\Delta H_f^{-0-}(C_2H_5OH,l)=-278$ kJ mol^{-1}.

The dissolution of a substance, rubidium hydroxide for example, involves a separation of the species Rb^+ and OH^- in the solid followed by their subsequent interaction with (hydration by) water; heat is evolved. If the solution is cooled and more water added, there is a further evolution of heat. Hence, it is necessary to refer the process of dissolution to a standard state and that chosen is the state of infinite dilution (aq). It may be defined as that state of dilution for which a further addition of water produces no measurable heat change.

There are several types of reactions involving enthalpy changes that are named according to the process carried out. Some of them have been discussed already, and Table 4.3 summarizes them.

4.5.1 Bond enthalpies

The enthalpy change for the dissociation of a diatomic molecule is the bond dissociation enthalpy to which we have already referred in Section 3.6.5.2. For polyatomic molecules, an average bond enthalpy (also called bond 'energy') may be defined. The dissociation enthalpy for a particular bond depends on the structure of the molecule or part-molecule. Thus, for H_2O, two values are obtained for the O−H bond enthalpy, namely 499 kJ mol^{-1} and 428 kJ mol^{-1}, from which the average O−H bond enthalpy ΔH_b(O−H) is 464 kJ mol^{-1}.

We can obtain an average C−H bond enthalpy by considering the following sequence of reactions:

$$CH_4(g)\rightarrow C(g)+4H(g) \qquad \Delta H=4\Delta H_b(C-H)$$
$$C(s)\rightarrow C(g) \qquad \Delta H=\Delta H_a(C,s)=717 \text{ kJ mol}^{-1}$$
$$2H_2(g)\rightarrow 4H(g) \qquad \Delta H=2\Delta H_b^{-0-}(H_2,g)=872 \text{ kJ mol}^{-1}$$
$$CH_4(g)\rightarrow C(s)+2H_2(g) \qquad \Delta H=-\Delta H_f^{-0-}(CH_4,g)=75 \text{ kJ mol}^{-1}$$

By constructing a cycle, ΔH_b(C−H)=416 kJ mol^{-1}. In a similar way, we can use data on ethane C_2H_6 to find ΔH_b(C−C)=348 kJ mol^{-1}. These data may be used to calculate heats of formation of compounds, gaseous butane C_4H_{10}, for example. Applying the above method gives −156 kJ mol^{-1}, whereas the experimental value is −126.2 kJ mol^{-1}, a result that demonstrates the approximate nature of average bond enthalpies. Table 4.4 lists average bond enthalpy values; for diatomic molecules they are the values of D_0 (see Section 3.6.5.2) corrected to 298.15 K.

Table 4.3 Named enthalpic reactions

	Initial state	Final state and parameter
Atomization	Element or compound 298.15 K and 1 atm $H_2O(l) \rightarrow 2H(g) + O(g)$	Separated atoms at 298.15 K and 1 atm $\Delta H_a^{\ominus}$ (H_2O,l)
	For an element that vaporizes to a monatomic gas atomization is the same as sublimation: $Li(s) \rightarrow Li(g)$	$\Delta H_a^{\ominus}$ $(Li,s) = \Delta H_s^{\ominus}$ (Li,s)
Combustion	Element or compound at 298.15 K and 1 atm $CH_4(g) + 2O_2(g) \rightarrow CO_2(g) + 2H_2O(l)$	Products of complete combustion in O_2, referred to 298.15 K and 1 atm $\Delta H_c^{\ominus}$ (CH_4,g)
Dissolution	Solid and solvent at 298.15 K $NaCl(s) \rightarrow Na^+(aq) + Cl^-(aq)$	Solvated solid or ions at 298.15 K and infinite dilution $\Delta H_d^{\ominus}$ $(NaCl,s)$
Electron addition	Gaseous atom or compound at 298.15 K and 1 atm $F(g) + e^- \rightarrow F^-(g)$	Singly negative species at 298.15 K and 1 atm $\Delta H_{ea}^{\ominus}$ (F,g)
(At 0 K, this quantity is the *electron affinity*)		
Fusion	Solid element or compound at 298.15 K and 1 atm $H_2O(s) \rightarrow H_2O(l)$	Liquid at 298.15 K and 1 atm $\Delta H_{fu}^{\ominus}$ (H_2O,s)
Formation (Solid)	Elements in their normal state at 298.15 K and 1 atm $Hg(l) + \frac{1}{2} O_2(g) \rightarrow HgO(s)$	Compound at 298.15 K and 1 atm $\Delta H_f^{\ominus}$ (HgO,s)
(Hydrated ions)	Elements in their normal state at 298.15 K and 1 atm $Na(s) + aq \rightarrow Na^+(aq)$	Ions in solution at 298.15 K and infinite dilution $\Delta H_f^{\ominus}$ (Na^+,aq)
Ionization	Gaseous element or compound at 298.15 K and 1 atm $Rb(g) \rightarrow Rb^+(g) + e^-$	Singly positive species at 298.15 K and 1 atm $\Delta H_i^{\ominus}$ (Rb^+,g)
(At 0 K, this quantity is the *ionization energy*)		

Table 4.4 Average standard bond enthalpies $\Delta H_b^{\ominus}/\mathrm{kJ\ mol^{-1}}$

H–H	H–F	H–Cl	H–Br	H–I	H–O
436	565	431	366	299	464
C–H	C–C	C=C	C≡C	C–O	C–N
416	348	615	812	351	292

4.5.2 Variation of enthalpy change with temperature

Integration of (4.24) for a temperature change from T_1 to T_2 gives

$$H_2 = H_1 + \int_{T_1}^{T_2} C_p \, \mathrm{d}T \tag{4.47}$$

In a reaction, (4.47) applies to the reactants and to the products, and the reaction enthalpies at temperatures T_1 and T_2 are given by Kirchoff's equation

$$\Delta H_2 = \Delta H_1 + \int_{T_1}^{T_2} \Delta C_p \, \mathrm{d}T \tag{4.48}$$

where ΔC_p is the sum of C_p for the products *minus* the sum of C_p for the reactants. For a reaction of the general form

$$a\mathrm{A} + b\mathrm{B} \rightarrow c\mathrm{C} + d\mathrm{D} \tag{4.49}$$

ΔC_p takes the form

$$\Delta \mathrm{C}_p = cC_p\,(\mathrm{C}) + dC_p\,(\mathrm{D}) - aC_p(\mathrm{A}) - bC_p(\mathrm{B}) \tag{4.50}$$

In general, C_p itself varies with temperature and it may be represented closely by the function

$$C_p = a + bT + c/T^2 \tag{4.51}$$

The values of the constants a, b and c for several species are listed in Table 4.5.

We shall illustrate the use of these equations with reference to the Haber process; we introduced this reaction in Chapter 1:

$$\tfrac{1}{2}\,\mathrm{N_2(g)} + \tfrac{3}{2}\,\mathrm{H_2(g)} \rightarrow \mathrm{NH_3(g)} \tag{4.52}$$

for which $\Delta H_\mathrm{f}^{-0-}(\mathrm{NH_3,g}) = -46.1$ kcal mol^{-1}; we shall calculate the corresponding value at 598 K. From (4.50) and (4.51),

$$\Delta C_p = [a(\mathrm{NH_3}) - \tfrac{1}{2}\,a(\mathrm{N_2}) - \tfrac{3}{2}\,a(\mathrm{H_2})] + 10^{-3}T[b(\mathrm{NH_3}) - \tfrac{1}{2}\,b(\mathrm{N_2})$$
$$-\tfrac{3}{2}\,b(\mathrm{H_2})] + 10^5/T^2[c(\mathrm{NH_3}) - \tfrac{1}{2}\,c(\mathrm{N_2}) - \tfrac{1}{2}\,c(\mathrm{H_2})]$$

Inserting the data from Table 4.5 gives

$$\Delta C_p = -25.46\ \mathrm{J\,K^{-1}\,mol^{-1}} + (18.325\times10^{-3}\ \mathrm{J\,K^{-2}\,mol^{-1}})T\,\mathrm{K}$$
$$-(2.05\times10^5\ \mathrm{J\,K\,mol^{-1}})/(T^2\mathrm{K}^2)$$

Using (4.48) and integrating between 298 K and 598 K leads to ΔH_{598}=-48.5 kcal mol^{-1}.

4.5.3 Calorimetry

One of the processes included in Figure 4.8 is the complete combustion of ethanol in pure oxygen:

$$\mathrm{C_2H_5OH(l)} + 3\mathrm{O_2(g)} \rightarrow 2\mathrm{CO_2(g)} + 3\mathrm{H_2O(l)} \tag{4.53}$$

The enthalpy change for this and similar reactions may be obtained by means of an adiabatic (bomb) calorimeter, Figure 4.9. In this apparatus, which is isolated thermally from its surroundings, the combustion reaction is carried out both adiabatically and at constant

Table 4.5 Constants of the C_p equation (4.51) for several species

	a/J K^{-1} mol^{-1}	$10^3 b$/J K^{-2} mol^{-1}	10^{-5} c/J K mol^{-1}
Inert gases	20.78	0	0
$H_2(g)$	27.28	3.26	0.50
$O_2(g)$	29.96	4.18	−1.67
$N_2(g)$	28.58	3.77	−0.50
$Cl_2(g)$	37.03	0.67	−2.85
$CO(g)$	28.41	4.10	−0.46
$CO_2(g)$	44.22	8.79	−8.62
$NH_3(g)$	29.75	25.10	−1.55
$H_2O(g)$	30.54	10.29	0
$H_2O(l)$	75.48	0	0

volume, though not reversibly, and the energy change measured is ΔU_c (q_v) for the combustion process. In (4.23), we can replace $p\Delta V$ by $\Delta n\mathcal{R}T$, where Δn refers to the number of moles of gaseous products *minus* the number of moles of gaseous reactants, and so obtain the enthalpy of combustion.

In practice, the apparatus is calibrated by a thermochemical standard substance, usually benzoic acid, so that we may write

$$\frac{\Delta U_c(\text{sample})}{\Delta U_c(\text{standard})} = \frac{\Delta T(\text{sample})}{\Delta T(\text{standard})} \tag{4.54}$$

A precision of approximately 0.01% is attainable and, in a typical experiment with ethanol, ΔU_c was found to be −1365.2 kcal mol^{-1}; since Δn is −1, ΔH_c at 298.15 K is −1367.7 kcal mol^{-1}.

4.6 Second law of thermodynamics

The first law of thermodynamics (Section 4.2) shows that energy is conserved in any process, but it does not indicate the direction of flow of energy, the direction of spontaneous change. In the mechanical system of Figure 4.4, the capacity for work lies in the energy stored in the extended spring. The total transfer of this energy into useful work required the transfer of mass to be carried out in a reversible manner, that is, in a state of continual equilibrium, when the maximum work possible would be done by the system. Thus, $đw$ would be more negative than that for the corresponding process carried out in any irreversible manner:

$$đw_{rev} < đw_{irr} \qquad đq_{rev} > đq_{irr} \tag{4.55}$$

Whether the mechanical process is carried out reversibly or irreversibly, the spontaneous change is that which decreases the potential energy of the system.

When a perfect gas expands into a vacuum, the external pressure is zero and no work is done by the gas, that is $đw=0$. If a metal block at a temperature T_1 is placed in contact with another block at a temperature T_2 ($T_2 > T_1$; Figure 4.10), heat flows from the hotter block to the cooler block. In each case an equilibrium is attained, with respect to pressure in the case of the gas and with respect to temperature in the case of the metal blocks, through a spontaneous change.

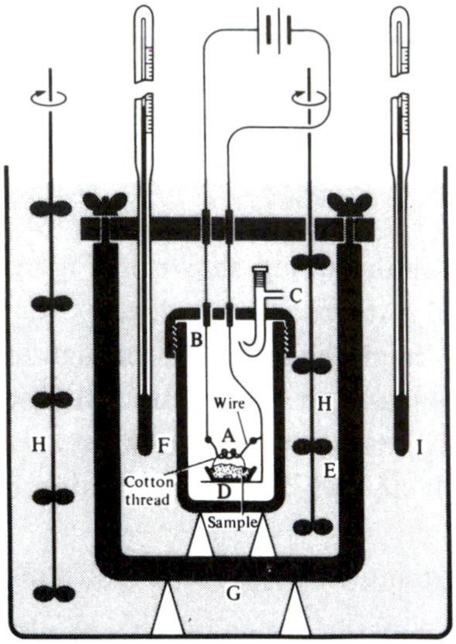

Figure 4.9 Bomb calorimeter: a weighed sample is placed in the crucible D of a steel bomb B which is next filled with oxygen to a pressure of approximately 20 atm through the valve C; it is then immersed in water and allowed to come to temperature equilibrium with the rest of the apparatus. An electric current is passed through the platinum spiral A which ignites a thread of cotton resting in the sample, and so starts the combustion reaction. Heat is evolved, and the associated temperature rise in the water filling the air-jacketed calorimeter E is measured with the Beckmann thermometer F. The calorimeter is contained within a large vessel of water G, all agitated by the stirrers H. The temperature change in G is monitored by another Beckmann thermometer I. A cooling correction must be applied for the loss of heat from E to G. In the *adiabatic* bomb calorimeter, the thermometer I is replaced by an electric heater that is controlled by the temperature difference between E and G, as measured by matched thermistors. Thus, the outer vessel G is kept at the same temperature as the calorimeter E, so eliminating the need for a cooling correction.

The second law of thermodynamics may be stated thus: *if a process that, under suitable conditions, can be made to do work is carried out reversibly, that work is a maximum* – there is a total conversion of energy; but in any natural process the maximum work is not obtained. Other statements of this law will be encountered; that given in Section 4.1.1 will evolve in the next section. The comedy duo Flanders and Swann had a song that contained the couplet

'Heat won't flow from a colder to a hotter
You can try it if you like, but you'd far better notter'

which is probably the only instance of the second law of thermodynamics being set to music.

The mechanical and chemical systems that we have discussed show how the capacity to do work is lost during a spontaneous change. Clearly, the lost capacity is related to $đq$, from (4.13), and to T, since the larger the temperature difference the greater the lost capacity for a given amount of heat. We saw in Section 4.3 that $đq$ would not be a useful measure of the energy $\mathrm{d}U$ unavailable for work, because it is dependent upon the path of the process. However, we have shown in the context of gaseous expansion (Section 4.4.3) that $đq/T$ is an exact differential, and we define the corresponding state function $\oint đq/T$ as the *entropy change* for the expansion process. The entropy change for a system changing reversibly between two states is given by

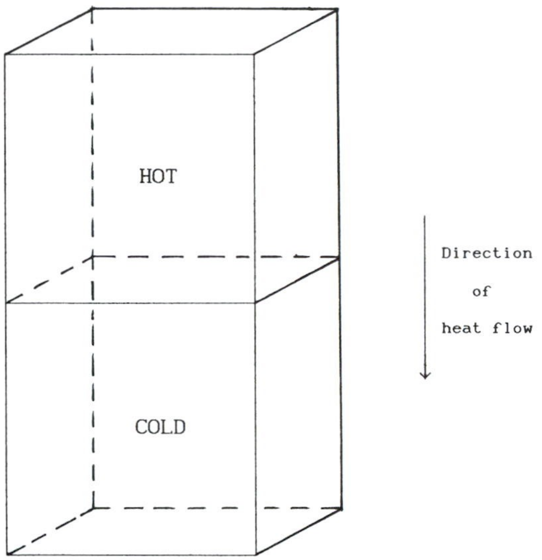

Figure 4.10 Metal blocks, one hot and the other cold, in contact; heat flows across the interface from the hot block to the cold, never spontaneously in the reverse direction.

$$\Delta S=\int_1^2 dS=S_2-S_1=\int_1^2 đq_{rev}/T \tag{4.56}$$

At a constant temperature, it follows from (4.56) that $đq_{rev}=T\,dS$, so that with (4.13)

$$dU=T\,dS+đw_{rev} \tag{4.57}$$

If the work done by the system is that of expansion against a constant external pressure p,

$$dU=T\,dS-p\,dV \tag{4.58}$$

In other words, the change in internal energy in a natural spontaneous process (system and surroundings) is divided between useful work, such as $p\,dV$, and unavailable energy $T\,dS$. Since the entropy change in (4.56) is defined by the initial (1) and final (2) states of the system, it is independent of the path of the process, but is given by (4.56) *only* for a reversible path.

For a spontaneous change, $đq<đq_{rev}$; hence, we obtain the Clausius inequality for a system

$$dS\geq đq/T \tag{4.59}$$

the equality sign providing a general definition of equilibrium. In an isolated system, S is constant, so that $dS=0$. For a spontaneous change,

$$dS\geq 0 \tag{4.60}$$

the equality sign applying again to equilibrium conditions. This equation is a formulation of the definition of the second law in Section 4.1.1, since natural, spontaneous processes are a feature of the universe. Spontaneous, irreversible changes are always accompanied by an overall increase in entropy. Criteria for change in a system must be based on a *non-conservative* property, such as entropy: enthalpy is a conservative property, and a criterion for spontaneity in a system based on enthalpy could be contradicted by observations in the surroundings (see Section 4.1.3), because a decrease in enthalpy of a system leads to a gain in enthalpy by the surroundings.

4.6.1 Heat engines

Consider the flow of heat between the two metal blocks in Figure 4.10 to be interrupted at the interface by a link to a mechanical device, such as a piston and cylinder that can be activated by heat, as in a steam engine. Each cycle of the engine takes heat from the 'source' at a temperature T_h, uses some of it to generate work through the piston and transfers the remainder to the 'sink' at a temperature T_c. The maximum work would be given under reversible conditions, that is, when

$$\left.\begin{aligned}\Delta S_h+\Delta S_c=0\\ |q_h|/T_h=|q_c|/T_c\end{aligned}\right\} \tag{4.61}$$

so that

$$|q_h|/q_c|=T_h/T_c \tag{4.62}$$

Conservation of energy requires that $|q_h|-|q_c|=|w|$; hence,

$$|w|/|q_h|=1-|q_c|/|q_h|=1-T_c/T_h=\frac{T_h-T_c}{T_h}=\eta \tag{4.63}$$

where η is the thermodynamic efficiency of the engine, and is independent of the working substance. A typical generating engine operates with superheated steam as the working substance at approximately 800 K with a sink at about 400 K. Hence, its efficiency is approximately 50%. The other half of the energy of the steam is transferred to the surroundings. Thus, $\Delta S>0$ and the process is spontaneous.

4.6.1.1 Carnot cycle

If the working substance is a perfect gas, the engine cycle may be envisaged with the aid of Figure 4.11 and Table 4.2. Consider the four reversible stages of the cycle in order:

		w	q
AB	Isothermal expansion	$-n\mathcal{R}T_h\ln(V_B/V_A)$	$n\mathcal{R}T_h\ln(V_B/V_A)$
BC	Adiabatic expansion	$C_V(T_c-T_h)$	0
CD	Isothermal compression	$-n\mathcal{R}T_c\ln(V_D/V_C)$	$-n\mathcal{R}T_h\ln(V_D/V_C)$
DA	Adiabatic compression	$C_V(T_h-T_c)$	0

The total work is the sum around the cycle:

$$w=-n\mathcal{R}T_h\ln(V_B/V_A)-n\mathcal{R}T_c\ln(V_D/V_C) \tag{4.64}$$

From (4.33), it is easily shown that $V_B/V_A=V_C/V_D$; hence,

$$|w|=n\mathcal{R}(T_h-T_c)\ln(V_B/V_A) \tag{4.65}$$

The heat delivered from the source at temperature T_h is

$$|q_h|=n\mathcal{R}T_h\ln(V_B/V_A) \tag{4.66}$$

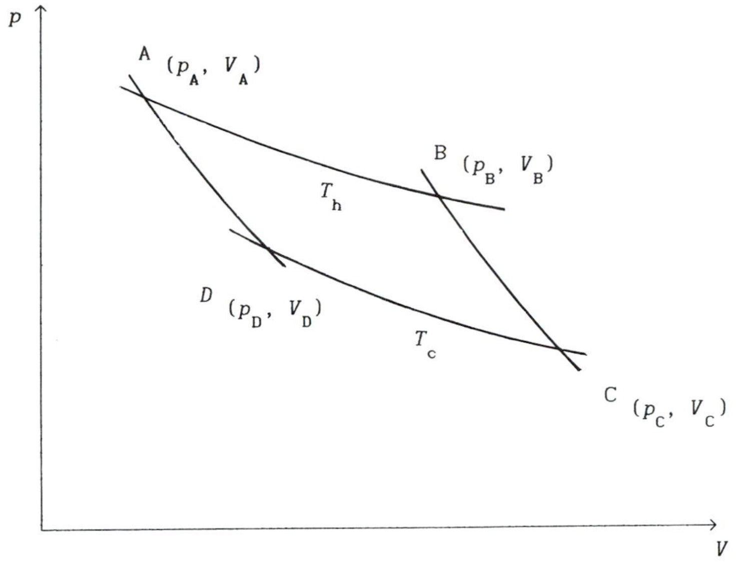

Figure 4.11 Carnot cycle for a perfect gas: *AB*, isothermal expansion at 'source' temperature T_h; *BC*, adiabatic expansion to the 'sink' temperature T_c; *CD*, isothermal compression at T_c; *DA*, adiabatic compression to T_h. The amounts of work done w and heat exchanged q are listed in Table 4.2.

whence the Carnot efficiency η is $(T_h - T_c)/T_h$, as in (4.63).

4.6.1.2 Thermodynamic temperature scale

A heat engine working reversibly may be used to define an absolute thermodynamic scale of temperature. From (4.63), the absolute zero of temperature corresponds to $\eta=1$. The kelvin (K) is defined in terms of the triple point of water, that is, the temperature at which ice, water and water vapour are in equilibrium (see also Section 7.2); this temperature is defined as 273.16 K exactly. Thus, a heat engine with a source at the triple point and a sink at a lower temperature T_c defines this temperature as

$$T_c = 273.16|q_c|/|q_h| \tag{4.67}$$

The *zero of the Celsius scale* is defined as the triple point of water *minus* 0.01 K, that is, 273.15 K.

4.6.2 Entropy at the molecular level

At a molecular level, entropy is related to the degree of order in a system and we may illustrate this concept in the following manner. Imagine N_A molecules of a perfect gas, where N_A is the Avogadro constant, retained by a partition to one-half of the volume V_A of its containing vessel (state A) and then allowed to expand reversibly to occupy the total volume V_B (state B), all at a temperature T. The probability that any given molecule could occupy the first (A) half of the vessel is 1/2; the probability that two molecules would occupy the same half-volume V_A is $(1/2)^2$, since the two events are uncorrelated. The probability W_A that all N_A molecules occupy the volume V_A is then $(1/2)^{N_A}$, that is, the ratio $W_A/W_B=(1/2)^{N_A}$, where W_A is a lower probability than W_B.

Table 4.6 Numbers W of arrangements of five molecules among five energy states

4ϵ	3ϵ	2ϵ	e	0		W	
1	1	1	1	1	5!/1!		120
2		1		2			
2			2	1			
1	2			2			
1		2	2		5!/(2!2!)	30 each	
	2	2		1		i.e.	180
	2	1	2				
1		3		1			
1	1		3		5!/3!	20 each	
	3		1	1		i.e.	80
	1	3	1				
		5			5!/5!		1

The ratio of the two volumes could take any value, so that generally

$$W_A/W_B=(V_A/V_B)^{N_A} \tag{4.68}$$

From Section 4.6.1.1, the heat change q_{rev} in a reversible expansion of a perfect gas is $n\mathcal{R}T\ln(V_B/V_A)$; hence,

$$\Delta S=q_{rev}/T=\mathcal{R}\ln(V_B/V_A) \tag{4.69}$$

In this analysis $n=1$, since we have N_A molecules, and it follows that

$$\Delta S=S_B-S_A=(\mathcal{R}/N_A)[\ln(W_B)-\ln(W_B)] \tag{4.70}$$

Thus, we identify the entropy of a system through the Boltzmann equation

$$S=k_B\ln(W) \tag{4.71}$$

where W is the probability, or number of microstates, of the system and k_B is the Boltzmann constant.

We may look upon W as the number of ways that the inside of a system may be constructed while the outside remains the same. Consider distributing five identical but distinguishable molecules (state P) among a set of energy levels, 0, ϵ, 2ϵ, 3ϵ and 4ϵ, such that the total energy is 10ϵ. The number of arrangements is shown in Table 4.6.

Of the 381 arrangements, one distribution is four times more probable than the next most probable distribution. As the number of molecules increases, the most probable distribution becomes overwhelmingly so, and the properties of this distribution may be taken to be those of the actual system.

Next suppose that the system undergoes a reaction, under conservation of energy, that leads to another state Q in which the energy levels are restricted to the values of 0, 2ϵ and 4ϵ. Only 51 arrangements are now possible (Table 4.7) and the most probable of them is one-quarter as probable as the most probable distribution in state P.

Thus, on probability grounds, the equilibrium

$$P\rightleftharpoons Q \tag{4.72}$$

will lie preponderantly on the side of P. In an actual equilibrium situation, some of the mole-

Table 4.7 Numbers W of arrangements of five molecules among three energy states

4ϵ	2ϵ	0	W	
2	1	2	5!/(2!2!)	30
1	3	3	5!/3!	20
	5		5!/5!	1

cules will be in state Q, but the large majority will be in state P. The equilibrium is, of course, dynamic and the result given here represents an average over a period of time that is long compared with the time of interchange of any molecule between states P and Q.

For (4.72), using (4.71), S is $k_B \ln(4)$, or 1.91×10^{-23} J K^{-1} (11.5 J K^{-1} mol^{-1}). In a purely mechanical system, when a ball is thrown from a distance to land in one of two boxes, one being four times the size of the other, the probability of the ball landing in the larger box is four times the probability of it landing in the smaller box.

The entropy for the most probable system of the molecules is, by the above calculation, 1.91×10^{-23} J K^{-1}. For a mechanical system, the energy is sufficient to determine the position of equilibrium at any normal temperature. The difference with the system of gas molecules is that a workable sample of gas will contain a very large number, of the order of 10^{20}, of molecules.

4.6.3 Entropy calculations

The enthalpy change ΔH_v accompanying the vaporization of benzene at its boiling-point (T_b, 80.1 °C) is 30.8 kJ mol^{-1}. In this process, heat energy is transferred reversibly and isothermally between the system (liquid benzene at T_b) and the surroundings (the laboratory fume cupboard). Since $q_{rev}=\Delta H_v$, $\Delta S_v=\Delta H_v/T_b=87.2$ J K^{-1} mol^{-1}.

Most nonpolar liquids have a value for ΔS_v very close to 85 J K^{-1} mol^{-1} (Trouton's rule). The constancy arises because the increase in disorder when one mole of any liquid vaporizes is due to the same number of molecules being formed in the vapours, liquids being of comparable order. In the case of water, however, the presence of hydrogen-bonding in the liquid leads to a greater degree of disorder upon vaporization, and ΔS_v is 109.1 J K^{-1} mol^{-1}. In the reactions

$$H_2O(s) \rightleftharpoons H_2O(l) \rightleftharpoons H_2O(g) \tag{4.73}$$

the entropy changes are given by $S(H_2O,l)-S(H_2O,s)$, and $S(H_2O,g)-S(H_2O,l)$. The entropy values are, in the order of (4.73), 48.8 J K^{-1} mol^{-1}, 69.9 J K^{-1} mol^{-1} and 188.8 J K^{-1} mol^{-1}, so that $\Delta S_{s\to l}=22.1$ J K^{-1} mol^{-1} and $\Delta S_{l\to g}=118.9$ J K^{-1} mol^{-1}. The degree of disorder in a system increases in the order solid$<$liquid$<$gas.

4.6.4 Measurement of entropy and the third law of thermodynamics

The entropy of a perfect, infinite crystal at 0 K is defined as zero, as expressed by the third law of thermodynamics. The specification 'perfect' implies freedom from any effect that would inhibit the ideal order of a lattice-based structure, such as crystal defects or hydrogen-bonding; the requirement of an 'infinite' crystal implies freedom from any surface effect that would distinguish an atom at the surface from an atom in the bulk of the crystal.

Table 4.8 Entropy S^{-0-} of butane at 298.15 K

Range of T	Procedure	S^{-0-}/J K^{-1} mol^{-1}
0–10	Debye: (1.86/3) J K^{-1} mol^{-1}	0.62
10–107.6	Numerical integration	60.8
107.6	Transition: solid I→solid II ΔH_t=2067 J mol^{-1}	19.2
107.6–134.9	Numerical integration	18.9
134.9	Transition: solid II→liquid ΔH_{fu}=4660 J mol^{-1}	34.5
134.9–272.7	Numerical integration	84.5
272.7	Transition: Liquid→gas ΔH_v=22.39 kJ mol^{-1}	82.1
272.7–293 and extrapolation to 298.15 K	Numerical integration	9.17
0–298.15	Standard entropy of butane	309.8

A molecular basis for the third law may be obtained from (4.71). At 0 K all atoms and molecules are in the lowest energy state. There is only one way in which all atoms can simultaneously be in the lowest energy state and hence, from (4.71) S is zero. We shall assume that, unless otherwise specified, all crystals at 0 K meet the ideal requirement of zero entropy.

From (4.58), we obtain under constant pressure conditions

$$\mathrm{d}U+p\,\mathrm{d}V=\mathrm{d}H=T\,\mathrm{d}S \tag{4.74}$$

However, from (4.24), we have that $\mathrm{d}H=C_p\,\mathrm{d}T$; hence,

$$\mathrm{d}S=C_p/T\,\mathrm{d}T \tag{4.75}$$

$$\mathrm{S}=\int (C_p/T)\,\mathrm{d}T \tag{4.76}$$

which indicates the method of obtaining absolute entropies. When C_p is known through an equation such as (4.51), the integral in (4.76) may be computed readily. Generally, heat capacity may be known through several such equations, each for a specific range of temperature, or by a sequence of individual measurements from which the necessary integration may be performed by a numerical procedure (see Appendix 14). In the low-temperature region, below 10 K, heat capacity measurement is very difficult and a procedure due to Debye is then generally employed. From the Debye theory of heat capacity, it may be shown both that the curve of C_p/T against T tends to zero as T approaches zero and that, at very low temperatures, C_p is proportional to T^3. Thus, if we let $C_p=aT^3$, we have

$$S(0\text{–}T)=\int_0^T aT^2\,\mathrm{d}T=aT^3/3=C_p/3 \tag{4.77}$$

where C_p is the value at the very low temperature T. We will illustrate the calculation of the entropy of butane at 298.15 K from measurements between 10 K and 293 K. The stages involved are set out in Table 4.8.

Table 4.9 Standard entropies, determined calorimetrically, S^{-0-}_{cal} and by statistical mechanics, S^{-0-}_{sm}

	N_2	Cl_2	CO_2	CO	N_2O	H_2O
S^{-0-}_{cal}/J K^{-1} mol^{-1}	192.1	223.1	213.9	193.5	215.3	185.5
S^{-0-}_{sm}/J K^{-1} mol^{-1}	191.6	223.1	213.7	197.7	219.8	188.8

4.6.4.1 Residual entropy

Entropy may be determined calorimetrically, through processes of the type that we have discussed above, and by statistical mechanics, using data from vibrational spectroscopy (see Bibliography). Generally, the results are in good agreement, but in some cases discrepancies arise (Table 4.9).

We consider the case of ice, for which the calorimetric entropy is less than the statistical value by a significant 3.3 J K^{-1} mol^{-1}. The structure of one form of ice was shown in Figure 1.1. In 1 mol of ice there are N_A molecules. There are six ways in which a tetrahedral structural unit may be oriented in space. Each adjacent molecule also has two unoccupied and two occupied tetrahedral directions. The probability that one given direction is available for a hydrogen atom is 2/4, or 1/2, whereas that for both hydrogen atoms being located in the desired orientation is $(1/2)^2$, since the two events are not correlated. Thus, the joint probability of a particular orientation of the given water molecule is $6(1/2)^2$, or 3/2, and for the total of N_A molecules it is $(3/2)^{N_A}$. From (4.71), the entropy associated with this probability is $N_A k_B \ln(3/2)$, or $\mathcal{R} \ln(3/2)$, which is 3.37 J K^{-1} mol^{-1}, in very close agreement with the experimental value for the *residual entropy* of ice.

4.7 Helmholtz and Gibbs free energies

The Helmholtz and Gibbs free energy functions enable equilibrium conditions to be expressed through state properties, rather than by the use of the w and q terms that depend upon the reaction path. For a closed system at constant temperature, in equilibrium with its surroundings, the Clausius inequality (4.59) takes the form

$$\mathrm{d}S = -đq/T \tag{4.78}$$

Using (4.13)

$$\mathrm{d}U = T\mathrm{d}S + đw \tag{4.79}$$

If the work done by the system is that of expansion against a constant external pressure p, then we obtain

$$\mathrm{d}U = T\mathrm{d}S - p\mathrm{d}V \tag{4.80}$$

which equation gives the criterion for spontaneity in a closed system changing reversibly at constant temperature and doing only work of expansion. The amount of work obtainable under these conditions is less than the internal energy change by the amount $T\mathrm{d}S$.

At constant volume (and temperature), (4.80) becomes

$$\mathrm{d}U - T\mathrm{d}S = 0 \tag{4.81}$$

or

$$\mathrm{d}(U - TS) = 0 \tag{4.82}$$

Since (4.82) is an exact differential, it must represent another state property. The Helmholtz free energy A is defined by

$$A = U - TS \tag{4.83}$$

whence, at constant temperature,

$$\mathrm{d}A = \mathrm{d}U - T\mathrm{d}S \tag{4.84}$$

Since A is a function of entropy it, too, is a nonconservative property and thus forms a criterion of spontaneity. For a measurable change, we write

$$\Delta A = \Delta U - T\Delta S \tag{4.85}$$

and a spontaneous change is accompanied by a decrease in the Helmholtz free energy of the system. From (4.79) and (4.85), it is evident that, if $\mathrm{d}A$ is negative (decreases), $\text{đ}w$ is also negative, so that spontaneous changes may be used to perform work.

The Helmholtz free energy $\mathrm{d}A$ measures the maximum work $\text{đ}w_{max}$ that a system can perform. Consider the combustion of ethanol in the bomb calorimeter (Section 4.5.3): $\Delta U_c^{\ominus} = -1365.2$ kJ mol^{-1}. The entropy change for the reaction is given by

$$\Delta S_c^{\ominus} = 2S^{\ominus}(CO_2,g) + 3S^{\ominus}(H_2O,l) - S^{\ominus}(C_2H_5OH,l) - 3S^{\ominus}(O_2,g)$$
$$= 2 \times 213.7\ \mathrm{J\ K^{-1}\ mol^{-1}} + 3 \times 69.9\ \mathrm{J\ K^{-1}\ mol^{-1}}$$
$$-160.6\ \mathrm{J\ K^{-1}\ mol^{-1}} - 3 \times 205.1\ \mathrm{J\ K^{-1}\ mol^{-1}} = -138.8\ \mathrm{J\ K^{-1}\ mol^{-1}}.$$

Hence,

$$\Delta A_c^{\ominus} = [-1365.2\ \mathrm{kJ\ mol^{-1}} - 298.15\ \mathrm{K}\ (-0.1388\ \mathrm{J\ K^{-1}\ mol^{-1}})] = -1323.8\ \mathrm{kJ\ mol^{-1}}.$$

Thus, the complete oxidation of ethanol can, under standard conditions, produce 1364.2 kJ mol^{-1} of energy as heat at constant volume and 1322.8 kJ mol^{-1} of energy as work.

Most chemical reactions are carried out at constant pressure and, at constant temperature and pressure, (4.80) leads to

$$\mathrm{d}(U + pV) = \mathrm{d}(TS) \tag{4.86}$$

or

$$\mathrm{d}(H - TS) = 0 \tag{4.87}$$

which represents another state property. We define the Gibbs free energy G by

$$G = H - TS \tag{4.88}$$

We may note that, in general,

$$\mathrm{d}G = \mathrm{d}(H - TS) = \mathrm{d}U + p\mathrm{d}V + V\mathrm{d}p - T\mathrm{d}S - S\mathrm{d}T \tag{4.89}$$

but at constant temperature and pressure

$$\mathrm{d}G = \mathrm{d}H - T\mathrm{d}S = \mathrm{d}U + p\mathrm{d}V - T\mathrm{d}S \tag{4.90}$$

which, like (4.80) and (4.84), is another fundamental equation.

The condition $dG=0$, at constant pressure (and temperature), is a condition of equilibrium, so that a reaction under these conditions will be spontaneous if dG, or ΔG for a measurable change, is negative. Again, it follows that the work term will be negative, that is, the system can be harnessed to do work. (A similar argument holds for dA, at constant volume.)

If we equate dG to a work term $đw$, then part of this work is that of expansion $(-p\,dV)$, and the remainder $(dU-T\,dS)$ is additional work $đw_a$ (other than that of expansion) obtainable from a system at constant pressure and temperature. For a measurable change, we write

$$\Delta G=\Delta H-T\Delta S \tag{4.91}$$

In the combustion of ethanol that we considered above, the standard free energy change $\Delta G_c^{\ominus}$ is, from (4.91), -1326.3 kJ mol^{-1}. The difference from $\Delta A_c^{\ominus}$, 2.5 kJ mol^{-1}, is the pV work at constant pressure: $\Delta G_c^{\ominus}$ is, in this example, more negative than $\Delta A_c^{\ominus}$ because the work is that of compression; Δn in (4.53) is -1.

The Gibbs free energy G of a substance, like its enthalpy H, cannot be measured by experiment. We define the standard Gibbs free energy of elements in their normal state at 298.15 K and 1 atm to be zero. In the reaction

$$\mathrm{C(graphite)+O_2(g)\rightarrow CO_2(g)} \tag{4.92}$$

$\Delta G^{\ominus}$ is -394.5 kJ mol^{-1}; this value is the standard Gibbs free energy of formation of carbon dioxide, because

$$\Delta G^{\ominus}=\Delta G_f^{\ominus}\,(\mathrm{CO_2,g})-[\Delta G_f^{\ominus}\,(\mathrm{C,graphite})+\Delta G_f^{\ominus}\,(\mathrm{O_2,g})] \tag{4.93}$$

The small difference between $\Delta G_f^{\ominus}$ and $\Delta H_f^{\ominus}$ (-393.5 kJ mol^{-1}) for this reaction arises because there is one mole of gas on each side of (4.92), and $\Delta S^{\ominus}$ is consequently small.

4.8 Maxwell's equations

We summarize first the *fundamental equations* that we have derived so far. By combining (4.13) and the Clausius inequality (4.59) for a closed, reversible system doing only pV work, we obtained (4.80), a combination of the first two laws of thermodynamics; we re-state it here for convenience:

$$dU=T\,dS-p\,dV \tag{4.94}$$

Extracting dH from (4.89) and combining with (4.94) leads to

$$dH=T\,dS+V\,dP \tag{4.95}$$

Since dA is, generally, $dU-T\,dS-S\,dT$, we have, with (4.94)

$$dA=-S\,dT-p\,dV \tag{4.96}$$

Substituting (4.94) into (4.89) leads to

$$dG=-S\,dT+V\,dp \tag{4.97}$$

These four basic equations contain state properties and may be used to derive Maxwell's equations. Each of them is of the form of (A13.2) in Appendix 13; we apply (A13.4) and (A13.5) as follows:

From (4.94),

$$(\partial T/\partial V)_S=-(\partial p/\partial S)_v \tag{4.98}$$

Similarly, from (4.95)–(4.97):

$$(\partial T/\partial p)_S = (\partial V/\partial S)_p \tag{4.99}$$

$$(\partial S/\partial V)_T = (\partial p/\partial T)_v \tag{4.100}$$

$$(\partial S/\partial p)_T = -(\partial V/\partial T)_p \tag{4.101}$$

Thus, we obtain Maxwell's equations, which are useful starting points in deriving many thermodynamic relationships.

4.8.1 Equations of state

From (4.94), at constant temperature and pressure, we see that U is a function of V and S; hence, from the definition of exact differential,

$$\mathrm{d}U = (\partial U/\partial V)_S \mathrm{d}V + (\partial U/\partial S)_v \mathrm{d}S \tag{4.102}$$

so that

$$(\partial U/\partial V)_T = (\partial U/\partial V)_S + (\partial U/\partial S)_v (\partial S/\partial V)_T \tag{4.103}$$

By comparing coefficients between (4.94) and (4.102), we obtain

$$(\partial U/\partial V)_S = -p \qquad (\partial U/\partial S)_v = T \tag{4.104}$$

Also, from (4.14) and (4.25), we define π_T as $(\partial U/\partial V)_T$ whence, from (4.100) and (4.103), we have

$$\pi_T = (\partial U/\partial V)_T = -p + T(\partial p/\partial T)_v \tag{4.105}$$

which is one form of a general thermodynamic equation of state. For an ideal gas (4.5) applies and for 1 mol $(\partial p/\partial T)_v = p/T$, whence π_T is zero, that is, the internal energy of an ideal gas is independent of its volume, as stated in Section 4.2.2.

From (4.94), at constant temperature, we derive

$$(\partial U/\partial p)_T = T(\partial S/\partial p)_T - p(\partial V/\partial p)_T \tag{4.106}$$

and with (4.101), we obtain

$$(\partial U/\partial p)_T = -T(\partial V/\partial T)_p - p(\partial V/\partial p)_T \tag{4.107}$$

which is a second thermodynamic equation of state. This equation may be written for a single substance as

$$(\partial U/\partial p)_T = -TV\alpha + pV\kappa \tag{4.108}$$

where α and κ are, respectively, the (volume) isobaric expansivity $(1/V)(\partial V/\partial T)_p$ and (volume) isothermal compressibility $-(1/V)(\partial V/\partial p)_T$ of the substance.

4.9 Gibbs free energy and chemical potential

In Problem 4.19, we use (4.97) to derive the dependence of the Gibbs free energy on temperature, embodied in the Gibbs–Helmholtz equation:

$$\left(\frac{\partial(\Delta G/T)}{\partial T}\right)_p = -\Delta H/T^2 \tag{4.109}$$

or

$$\left(\frac{\partial(\Delta G/T)}{\partial(1/T)}\right)_p = \Delta H \tag{4.110}$$

Similarly, from (4.97) at constant temperature,

$$(\partial G/\partial p)_T = V \tag{4.111}$$

whence

$$G_2 - G_1 = \int_{p_1}^{p_2} V\,\mathrm{d}p \tag{4.112}$$

Normally, liquids and solids have low compressibilities, and their measurable $V\Delta p$ terms are small compared with those of gases. For example, water at 20 °C compressed from 1 atm to 3 atm undergoes a molar free energy change of (0.01802 kg mol^{-1}/998.2 kg m^{-3})(2 atm)(101325 N m^{-2} atm^{-1}), or 3.7 J mol^{-1}. In the gaseous state, assuming ideality of the vapour, $\Delta G_\mathrm{m} = \int_1^3 (\mathcal{R}T/p)\,\mathrm{d}p = 2.7$ kJ mol^{-1}.

4.9.1 Gibbs free energy and the ideal gas

From (4.111), we write the molar free energy as

$$G_\mathrm{m} = \mathcal{R}T \ln(p) + \mathcal{I} \tag{4.113}$$

The standard state for the ideal gas is defined at a pressure (p^{-0-}) of 1 atm (or 101325 Pa), whereupon $\mathcal{I} = G_\mathrm{m}^{-0-}$ and

$$G_\mathrm{m} = G_\mathrm{m}^{-0-} + \mathcal{R}T \ln(p') \tag{4.114}$$

p' is equal to (p/p^{-0-}), and the logarithmic term is correctly dimensionless; thus,

$$G_\mathrm{m} = G_\mathrm{m}^{-0-} + \mathcal{R}T \ln(p/p^{-0-}) \tag{4.115}$$

4.9.2 Chemical potential

The chemical potential defines the change in Gibbs free energy of a system when a substance is added to it; thus, the chemical potential relates to *open* systems. For a single substance, the chemical potential μ is defined by

$$\mu = (\partial G/\partial n)_{T,p} \tag{4.116}$$

and since, in this case, $G = nG_\mathrm{m}$, it follows that

$$\mu = G_\mathrm{m} \tag{4.117}$$

that is, the chemical potential of a single substance is identical to its Gibbs molar free energy.

Equation (4.97) applies to closed systems. If a system contains n_1 mole of substance 1, n_2 mole of substance 2 and so on, (4.97) may be expanded to give

$$\mathrm{d}G = -S\,\mathrm{d}T + V\,\mathrm{d}p + (\partial G/\partial n_1)_{T,p,n_{(j\neq 1)}} + (\partial G/\partial n_2)_{T,p,n_{(j\neq 2)}} + \ldots \tag{4.118}$$

and the chemical potential of the ith component is

$$\mu_i = (\partial G/\partial n_i)_{T,p,n_{j(j\neq i)}} \tag{4.119}$$

so that (4.118) may be written as the fundamental equation

$$\mathrm{d}G = -S\,\mathrm{d}T + V\,\mathrm{d}P + \sum_i \mu_i\,\mathrm{d}n_i \tag{4.120}$$

and μ_i represents the change in the Gibbs free energy of the system when 1 mole of substance i is added to such a quantity of system that the overall composition is not altered significantly.

The chemical potential is an intensive property and supplies the driving force that impels a system to equilibrium.

4.9.3 Chemical potential and real gases

From (4.115) and (4.117), we have generally

$$\mu = \mu^{-0-} + \mathcal{R}T\ln(p/p^{-0-}) \tag{4.121}$$

Since intermolecular forces exist in real gases (see Sections 5.4 and 5.5), the pressure of the gas will depend on both the translational kinetic energy of the molecules (which is the only factor in the case of an ideal gas) and the intermolecular forces. The effective pressure of a real gas is its *fugacity* f, defined by

$$f = \gamma p \tag{4.122}$$

where γ is the (dimensionless) fugacity coefficient. As p tends to small values, all gases tend towards ideal behaviour, that is, f tends to p. The standard state for a real gas is the hypothetical state of a perfect gas at 1 atm, and with $p=1$ and $f=1$. If we use f in place of p in (4.121), we obtain

$$\mu = \mu^{-0-} + \mathcal{R}T\ln(p/p^{-0-}) + \mathcal{R}T\ln(\gamma) \tag{4.123}$$

Comparison with (4.121), or (4.115), shows that the effects of the intermolecular forces are contained in the term $\mathcal{R}T\ln(\gamma)$ although, in the nature of thermodynamics, the forces themselves are not addressed.

Figure 4.12 is a plot of γ against $\ln(p/p^{-0-})$ for nitrogen. Up to approximately 7 atm, nitrogen, like many other gases, behaves almost ideally; from approximately 7 atm to 150 atm molecular interactions cause the pressure to be less than that for a perfect gas; above 250 atm repulsion forces operate in increasing strength, so that the thermodynamic pressure f becomes much greater than the actual pressure p of the ideal gas under the same conditions.

We can calculate the fugacity on the basis of the equations already developed. From (4.97) and (4.123), we have

$$(\partial\mu/\partial p)_T = V_\mathrm{m} = \mathcal{R}T\{\partial[\ln(f/p^{-0-})]/\partial p\}_T \tag{4.124}$$

Thus

$$\partial[\ln(f/p^{-0-})] = [V_\mathrm{m}/(\mathcal{R}T)]\,\partial p \tag{4.125}$$

For an ideal gas

$$\partial[\ln(p/p^{-0-})] = [V_\mathrm{m}/(\mathcal{R}T)]\,\partial p = \partial p/p \tag{4.126}$$

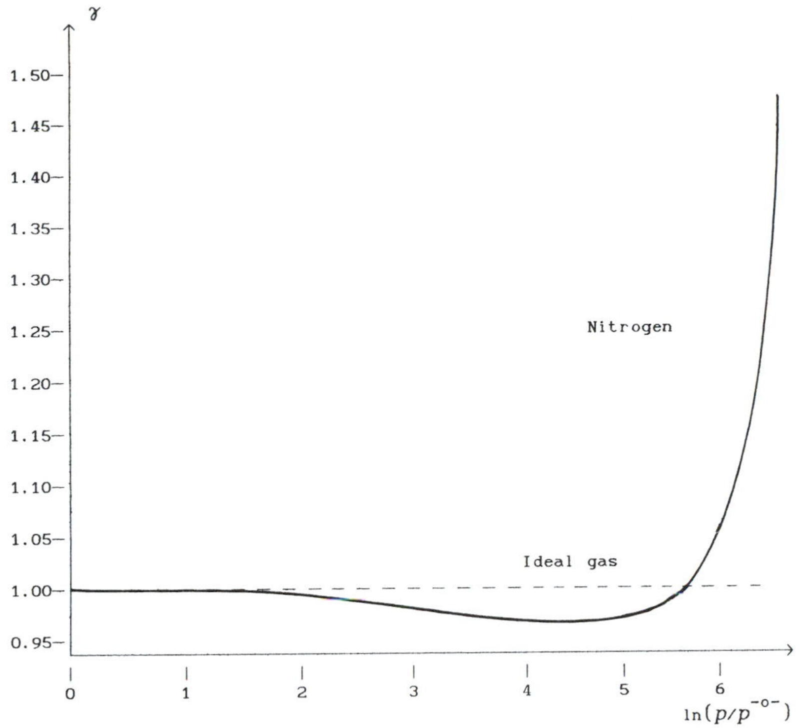

Figure 4.12 Plot of the fugacity coefficient γ as a function of $\ln(p/p^{-0-})$, for p up to approximately 800 atm; for an ideal gas, $\gamma=1$ at all values of p.

whence

$$\ln(f/p)=\int_0 [V_m/(\mathcal{R}T)-(1/p)]\,dp=\int_0 \frac{\mathcal{Z}_m-1}{p}\,dp \tag{4.127}$$

where $\mathcal{Z}_m=pV_m/(\mathcal{R}T)$ is the molar compression factor for the gas (see Sections 5.3.5 and 5.3.7). If the variation of $\mathcal{Z}_m$ with p is known, the fugacity can be determined by numerical integration of (4.127).

4.9.4 Thermodynamics of mixing

Consider n_A mole of a gas A and n_B mole of a gas B, and assume ideality. At a constant temperature T and pressure p, the total Gibbs free energy is given by

$$G=n_A\mu_A+n_B\mu_B \tag{4.128}$$

which, from (4.121), may be written as

$$G=n_A[\mu_A^{-0-}+\mathcal{R}T\ln(p/p^{-0-})]+n_B[\mu_B^{-0-}+\mathcal{R}T\ln(p/p^{-0-})] \tag{4.129}$$

Let the two gases be mixed; each gas develops its own partial pressure, whereupon the Gibbs free energy becomes

$$G'=n_A[\mu_A^{-0-}+\mathcal{R}T\ln(p_A/p^{-0-})]+n_B[\mu_B^{-0-}+\mathcal{R}T\ln(p_B/p^{-0-})] \tag{4.130}$$

The free energy change on mixing is then

$$\Delta G_{mix} = G' - G = n_A \mathcal{R}T \ln(p_A/p) + n_B \mathcal{R}T \ln(p_B/p) \tag{4.131}$$

By combining Dalton's law (5.28) and the gas equation (4.5), we have $p_A = x_A p$ and since $x_A = n_A/n$,

$$\Delta G_{mix} = n\mathcal{R}T[x_A \ln(x_A) + x_B \ln(x_B)] \tag{4.132}$$

In any mixture $x_i < 1$, ΔG_{mix} is negative, and from (4.120), $(\partial G/\partial T)_{p,n} = -S$; hence,

$$\Delta S_{mix} = -n\mathcal{R}[x_A \ln(x_A) + x_B \ln(x_B)] \tag{4.133}$$

It follows now from (4.91) that $\Delta H_{mix} = 0$.

In this chapter, we have developed a number of fundamental equations based upon the three laws of thermodynamics. In subsequent chapters we shall explore them further, as we consider their application to chemical systems, and carry out such further derivations as become necessary. A familiarity with the work of this chapter can be facilitated by working through the problems that follow.

Problems 4

4.1 A system of internal energy -13.3 kJ mol^{-1} changes, at constant pressure, to a state in which its internal energy is -47.8 kJ mol^{-1}. Calculate ΔU and q for the process, and state whether it is endothermic or exothermic.

4.2 Determine the maximum work that can be done by 10 mol of ideal gas expanding against a constant pressure, if the temperature changes by 17 K.

4.3 Six mole of nitrogen gas are compressed isothermally and reversibly from 1.00 m^3 to 0.15 m^3, at 298 K. Calculate the work done on the gas if it obeys the ideal gas equation.

4.4 Calculate the difference between ΔH and ΔU for the evaporation of 1 mol of water at its boiling-point of 373.15 K. The vapour may be assumed to behave ideally, and the volume of liquid may be neglected in comparison with the volume of the gas (which is easy to prove).

4.5 Air is compressed adiabatically from 1 atm to 20 atm, the initial temperature being 298 K. The compression is carried out (a) in one stage, and (b) in two stages, 1 atm to 10 atm and then 10 atm to 20 atm. Show which process needs the least energy for the total compression (C_v (air)=20.9 J K^{-1} mol^{-1}).

4.6 Give and explain a mathematical formulation of Hess's law. Construct a thermochemical cycle to show how the law may be used to obtain $\Delta H_f^{\ominus}$ (CuCl,s), given that $\Delta H_f^{\ominus}$ ($CuCl_2$,s)$=-205$ kJ mol^{-1} and that $\Delta H^{\ominus} = -71$ kJ mol^{-1} for the reaction

$$CuCl(s) + \tfrac{1}{2}Cl_2(g) \rightarrow CuCl_2(s)$$

4.7 In the combustion of benzoic acid $C_6H_5CO_2H$ in a bomb calorimeter, it was found that $\Delta U_c^{\ominus} = -3227.2$ kJ mol^{-1}. Construct an equation for the combustion reaction and calculate the corresponding enthalpy change. Calculate also $\Delta A_c^{\ominus}$ for the reaction, using the following data:

	$\Delta H_f^{\ominus}$/kJ mol^{-1}	$S^{\ominus}$/J K^{-1} mol^{-1}
CO_2(g)	-393.5	213.7
H_2O(l)	-285.8	69.9
O_2(g)	0	205.1
$C_6H_5CO_2H$(s)		167.7

4.8 Obtain a general relation for $C_p - C_v$, and show that for an ideal gas it is equivalent to (4.27).
4.9 The standard enthalpy of formation of carbon dioxide is -393.5 kJ mol^{-1}. Evaluate the enthalpy of formation at 1000 °C, using data on heat capacity from Table 4.5.
4.10 Use the data below to form a thermochemical cycle for determining the enthalpy of formation of rubidium sulfate, and find its value.

	$\Delta H^{\ominus}$/kJ mol^{-1}
$H^+(aq)+OH^-(aq) \rightarrow H_2O(l)$	-55.8
$RbOH(s)+aq \rightarrow Rb^+(aq)+OH^-(aq)$	-62.8
$Rb_2SO_4(s)+aq \rightarrow 2Rb^+(aq)+SO_4^{2-}(aq)$	$+24.3$
$Rb(s)+\frac{1}{2}O_2(g)+\frac{1}{2}H_2(g) \rightarrow RbOH(s)$	-413.8
$H_2(g)+S(s)+2O_2(g) \rightarrow 2H^+(aq)+SO_4^{2-}(aq)$	-909.3
$H_2(g)+\frac{1}{2}O_2(g) \rightarrow H_2O(l)$	-285.8

4.11 Calculate the entropy change for the vaporization of 1 mol of chloromethane at its boiling-point of 294 K; the associated enthalpy change is 21.7 kJ mol^{-1}.
4.12 Calculate $\Delta G^{\ominus}$ for the reaction $N_2O_4(g) \rightarrow 2NO_2(g)$ at 298 K, given the following data:

	$\Delta H_f^{\ominus}$/kJ mol^{-1}	$S^{\ominus}$/J K^{-1} mol^{-1}
$NO_2(g)$	33.2	240.1
$N_2O_4(g)$	9.16	304.3

4.13 Use bond enthalpy data from Table 4.4 and other data from Section 4.5.1 to determine values for the enthalpies of formation for butane and 1-butene. Hence, obtain a value for the enthalpy of hydrogenation of a C=C double bond.
4.14 Use data from Table 4.5 to determine the molar entropy change on heating ammonia from 273 K to 373 K.
4.15 In what circumstances do (a) $đq$ and (b) $đw$ become exact differentials?
4.16 Obtain the relationship $đw = -(n\mathcal{R}T/p)\,dp$ for an ideal gas, and use it to show that $đw$ is an inexact differential.
4.17 Assuming no loss of heat to the surroundings, calculate the increase in temperature of water falling from the top to the bottom of the Niagara Falls (50 m). Use $C_p = 80$ J K^{-1} mol^{-1} and $g = 9.81$ m s^{-2}.
4.18 Two copper cubes of mass 0.5 kg each, at temperatures 500 K and 300 K, are placed in contact and allowed to attain thermal equilibrium. Calculate the total change in entropy of the cubes that occurs, taking the molar heat capacity of copper at constant volume as the classical (Dulong and Petit) value.
4.19 Derive an expression to represent the variation of ΔG with temperature. The standard enthalpy change for the reaction (4.52) is -46.1 kJ mol^{-1}. Calculate $\Delta G^{\ominus}$ and ΔG at 500 K. Comment on the spontaneity of the reaction. ($S^{\ominus}$/J K^{-1}mol^{-1}: H_2, 130.7; N_2, 191.5; NH_3, 192.5.)
4.20 Use the appropriate Maxwell equation to show that the entropy of an ideal gas is a function of its volume.
4.21 Calculate the change in chemical potential of an ideal gas that is compressed isothermally at 25 °C from 2 atm to 20 atm.

4.22 A vessel of volume 10 dm^3 is divided exactly into two by a partition. On one side is hydrogen at 1 atm and 20 °C, and on the other side, nitrogen at the same temperature and pressure. The partition is removed and the gases mix isothermally. Calculate the entropy, enthalpy and Gibbs free energy of mixing for the given conditions.

CHAPTER 5

States of matter: gases and liquids

5.1 Introduction

Matter exists in three states, gaseous, liquid and solid. It is convenient to discuss the properties of matter under these main headings, although certains types of substances, such as vitreous materials and liquid crystals, do not fall clearly into one or other of them.

The gaseous state is the most straightforward to study. The simplicity arises from the fact that the molecules of a gas are, normally, independent of one another, with negligible forces of interaction. Consequently, many of their properties are independent of their chemical nature and may be described by general gas laws; we have made use of this fact in the previous chapter and here we describe these laws in more detail.

Gases are characterized by large volume changes with change of temperature or pressure and by their ability to flow into and fill the space available to them; they are miscible with one another in all proportions. A liquid, like a gas, has no form, but it has a definite volume and takes up the shape of its containing vessel. It has a boundary surface that restricts its extent and which is responsible for many of its properties. The cohesive forces between molecules in a liquid are stronger than those between molecules in a gas, but weaker than those characteristic of solids. Under normal conditions, gases behave as an assemblage of freely moving molecules or atoms, whereas solids form rigid structures. The intermediate nature of the cohesive forces between molecules in a liquid and the lack of long-range structure make the study of a liquid more complex than that of either a gas or a solid.

The so-called colloidal state is not a state of matter but rather a system composed of very small solid or liquid particles, 10 nm to 100 nm in size, known as a disperse phase, in a suspensory phase, the dispersion medium, which itself may be in either the liquid or the gaseous state.

The state of a substance that is obtained at a given temperature and pressure is determined by the result of competition between the intermolecular forces of its component species and the thermal energy that they contain.

5.2 Gases

The relationship between pressure, volume and temperature can be expressed for most gases in terms of the gas laws and their combination. Several well-known laws resulted from experimental studies on gases. In particular, the laws of Boyle and Charles may be combined to form the general gas law.

5.2.1 Combination of Boyle's and Charles' laws

Let 1 mole of gas at p_1, $V_{m,1}$, T_1 be changed isothermally to p_2, V_m, (T_1); it may be imagined to be enclosed within a cylinder by a piston, Figure 5.1. Then, by Boyle's law

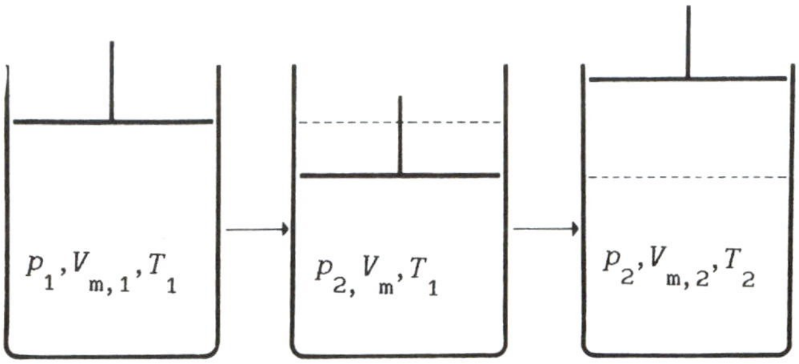

Figure 5.1 Change in conditions of a gas retained in a cylinder by a perfect, frictionless piston: p_1, $V_{m,1}$, $T_1 \rightarrow p_2$, V_m, T_1 and $p_1 V_{m,1} = p_2 V_m$; then p_2, V_m, $T_1 \rightarrow p_2$, $V_{m,2}$, T_2 and $V_m/T_1 = V_{m,1}/T_2$. Overall, we obtain $p_1 V_{m,1}/T_1 = p_2 V_{m,2}/T_2$.

$$p_1 V_{m,1} = p_2 V_m \tag{5.1}$$

Now, let it be changed isopiestically from p_2, V_m, T_1 to (p_2), $V_{m,2}$, T_2. By Charles' law

$$V_m/T_1 = V_{m,2}/T_2 \tag{5.2}$$

Eliminating V_m between (5.1) and (5.2),

$$p_1 V_{m,1}/T_1 = p_2 V_{m,2}/T_2 \tag{5.3}$$

or pV_m/T is a constant, the gas constant $\mathcal{R}$. Since V_m is the volume of 1 mol of gas,

$$pV = n\mathcal{R}T \tag{5.4}$$

where n is the number of moles in the volume V and $\mathcal{R}$ is the universal gas constant. We have met this equation of state for an ideal gas already, in Section 4.2.2. At *standard* temperature and pressure, that is, 0 °C and 1 atm, $V_m = 22.414\ \mathrm{dm^3\ mol^{-1}}$.

5.2.2 Constant-volume gas thermometer

From (5.3), at constant volume

$$p_1/T_1 = p_2/T_2 \tag{5.5}$$

which equation leads to a temperature scale. A constant-volume gas thermometer, Figure 5.2, compares the pressure p of the volume of gas in the thermometer bulb at a temperature T with the pressure p_0 at the triple point of water 273.16 K, at which temperature ice, water and water vapour are in equilibrium. Hence,

$$T = 273.16 p/p_0 \tag{5.6}$$

In practice, readings are taken with successively smaller amounts of gas in the bulb and the results extrapolated to $p = 0$, so that effective ideal behaviour can be ensured. Since (5.5) leads to

$$\Delta V/V_1 = \Delta T/T_1 \tag{5.7}$$

it follows that gases that are behaving ideally all expand or contract equally for a given change in temperature.

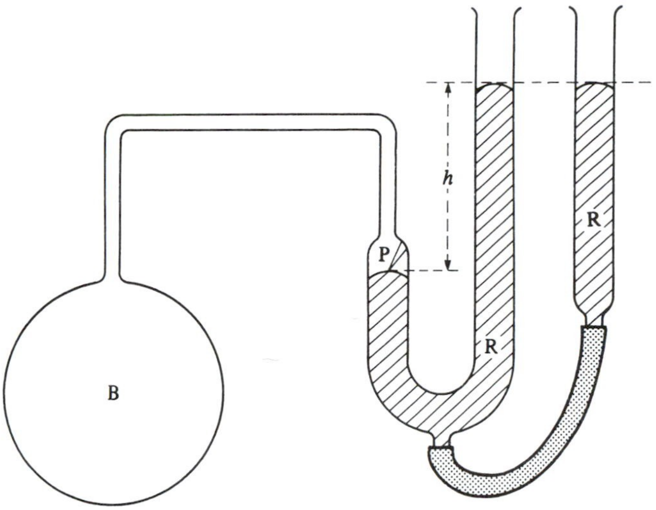

Figure 5.2 Constant-volume gas thermometer: the volume of gas in the bulb B is kept constant by adjusting the mercury levels with the reservoir R such that the pointer P just rests on the surface of the mercury column. The pressure in B is measured in terms of the height h.

5.3 Ideality and the kinetic theory of gases

The ideal gas obeys the kinetic theory of gases. We will develop next some aspects of this theory and show how the experimentally deduced gas laws follow from it. The basis of the kinetic theory of gases may be summarized by the following postulates:

(a) a gas consists of a large number of molecules (or atoms) in continuous random thermal motion, that is, motion related to the translational (kinetic) energy of the molecules at a given temperature and independent of their orientation – an isotropy of motion;
(b) the gas molecules interact only by elastic collisions, that is, those in which the total kinetic energy of the colliding species is conserved;
(c) the gas molecules are of negligible size, that is, their diameters are vanishingly small in relation to the distances travelled by the molecules between collisions.

From these assumptions, we may derive a number of important properties of gases.

5.3.1 Pressure exerted by a gas

Consider a single molecule of a gas of mass m in a rectangular box of sides a, b and c, and let its velocity $\boldsymbol{v}$ be resolved into components v_x, v_y and v_z parallel to a, b and c respectively. Let the molecule travel in the $+x$ direction and undergo an elastic collision with the wall bc. The momentum of the molecule in the x direction is changed from mv_x to $-mv_x$, so that the total momentum transferred to the wall is $2mv_x$, because the momentum of the wall–molecule system is conserved. The molecule will rebound and strike the opposite wall bc' and so on. Collisions with the wall bc will occur every $2a/v_x$ second. Hence, the rate of change of momentum is $2mv_x/(2a/v_x)$ or mv_x^2/a.

From Newton's second law, the rate of change of momentum is the force on the wall bc and, since pressure is force per unit area, the pressure on the wall bc due to the molecule is given by

$$p_x = \frac{mv_x^2/a}{bc} = \frac{mv_x^2}{V} \tag{5.8}$$

where V is the volume of the box. Similar expressions can be generated for p_y and p_z, and they will not, in general, be equal. If we now increase the number of molecules in the box to N then, because it is a postulate that they do not interact, their contribution to the pressure will be additive. Hence, for the wall bc, the total pressure p will be given by the sum over all N molecules:

$$p = \frac{m}{V}\sum_{i=1}^{N} v_{x_i^2} \tag{5.9}$$

and similarly for the other walls of the box. The mean square velocity $\overline{v_x^2}$ is given by

$$\overline{v_x^2} = \frac{1}{N}\sum_{i=1}^{N} v_{x_i^2} \tag{5.10}$$

Hence

$$p = Nm\overline{v_x^2}/V \tag{5.11}$$

When N is large, the impacts with the wall become a steady isotropic pressure. Since no direction is preferred, it follows that $\overline{v_x^2} = \overline{v_y^2} = \overline{v_z^2}$, and the mean square speed $\overline{v^2}$ is then

$$\overline{v^2} = \overline{v_x^2} + \overline{v_y^2} + \overline{v_z^2} \tag{5.12}$$

so that (5.11) becomes

$$pV = \tfrac{1}{3}Nm\overline{v^2} \tag{5.13}$$

This equation assumes implicitly that there is no attraction between the molecules and the wall of their containing vessel, which may be justified by the following argument.

Let there be $\mathcal{N}_x$ molecules of a perfect gas per unit volume at a distance x from a wall and let it be assumed that the molecules are attracted by the wall. If U_x is the attractive energy at a distance x from the wall, the force of attraction by the wall F_x on the molecule is given by

$$F_x = -\mathrm{d}U_x/\mathrm{d}x \tag{5.14}$$

and that on the wall is, by Newton's laws, $\mathrm{d}U_x/\mathrm{d}x$. Then the attractive pressure p_a on the wall is given by

$$p_\mathrm{a} = \int_0^\infty \mathcal{N}_x(\mathrm{d}U_x/\mathrm{d}x)\,\mathrm{d}x \tag{5.15}$$

where the limit ∞ indicates a position in the gas removed from any effect of the wall. If we assume that the molecule density $\mathcal{N}_x$ follows the Boltzmann distribution equation (see Section 3.6.5.1), we may write

$$\mathcal{N}_x = \mathcal{N}_0 \exp[-U_x/(k_\mathrm{B} T)] \tag{5.16}$$

We have from (5.16)

$$\mathrm{d}\mathcal{N}_x/\mathrm{d}x = \mathcal{N}_0 \exp[-U_x/(k_\mathrm{B}T)][-1/(k_\mathrm{B}T)]\,\mathrm{d}U_x/\mathrm{d}x = [-\mathcal{N}_x/(k_\mathrm{B}T)]\,\mathrm{d}U_x/\mathrm{d}x \tag{5.17}$$

whence, with (5.15)

$$p_\mathrm{a} = -k_\mathrm{B}T\int_0^\infty (\mathrm{d}\mathcal{N}_x/\mathrm{d}x)\,\mathrm{d}x = -k_\mathrm{B}T(\mathcal{N}_\infty - \mathcal{N}_0) \tag{5.18}$$

In (5.18), $\mathcal{N}_0$ is the number density at the wall, which relates to the pressure at the wall p_0, given by $\mathcal{N}_0 k_B T$, whereas $\mathcal{N}_\infty$ refers to molecules in the bulk, which may be identified with $\mathcal{N}_w$, the number density at the wall in the absence of attraction, with a corresponding pressure p_w given by $\mathcal{N}_w k_B T$. The actual pressure p due to collision at the wall is the difference between the ideal value p_0 and that of attraction p_a, that is,

$$p=p_0-p_a=\mathcal{N}_0 k_B T+k_B T(\mathcal{N}_\infty-\mathcal{N}_0)=\mathcal{N}_\infty k_B T=p_w \qquad (5.19)$$

Thus, the actual pressure, taking attraction into account, is to be identified with the pressure assuming no attraction. This result applies also to imperfect gases.

5.3.2 Equipartition

The classical distribution of energy in a system of molecules is considered in Appendix 3. Since the average translational kinetic energy $\bar{\epsilon}$ is equal to $\frac{1}{2}mv^2$, it follows from (5.4) and (5.13) that

$$pV=\tfrac{2}{3}N\bar{\epsilon}=n\mathcal{R}T \qquad (5.20)$$

Since $N_A k_B=\mathcal{R}$ and $nN_A=N$, it follows that

$$\epsilon=\tfrac{3}{2}k_B T \qquad (5.21)$$

or $\frac{3}{2}\mathcal{R}T$ per mole. From (5.12), it is evident that each degree of translational freedom of a molecule has an average energy $\frac{1}{2}k_B T$ associated with it.

5.3.3 Gas laws from the kinetic theory

From the foregoing equations we can write the equalities

$$pV=\tfrac{1}{3}\overline{Nmv^2}=\tfrac{1}{3}N_A\overline{nmv^2} \qquad (5.22)$$

For a gas at constant temperature $\overline{v^2}$ is constant and, for a fixed amount n, (5.22) shows that $(pV)_{T,n}$ is constant, which is Boyle's law: *at constant temperature, the volume of a fixed amount of gas varies inversely as its pressure*; the relationship $p\propto 1/V$ is hyperbolic and the corresponding curve is an *isotherm.*

For a gas at constant pressure and amount, and since kinetic energy is proportional to temperature, $(V/T)_{p,n}$ is constant, which is Charles' law: *at constant pressure, the volume of a fixed amount of gas varies directly as its temperature* (see also Section 5.2.2).

For a gas at constant temperature and pressure, $(V/n)_{T,p}$ is constant, which is Avogadro's law: *equal volumes of gases at the same temperature and pressure contain the same number of molecules.*

For a gas at constant pressure, $(\overline{v^2})^{1/2}$ is proportional to $[V/(Nm)]^{1/2}$, or the rate of effusion $R\propto 1/D^{1/2}$, which is Graham's law: *at constant pressure, the rate of effusion (escape from a small hole) of a gas varies inversely as the square root of its density.*

Finally, consider a gas at constant temperature containing N_i mole of gases i $(i=1,2,3,\ldots)$. At equilibrium, we have for the energies

$$\bar{\epsilon}_1=\bar{\epsilon}_2=\bar{\epsilon}_3=\ldots=\bar{\epsilon} \qquad (5.23)$$

and

$$N=\sum_i N_i \qquad (5.24)$$

Each gaseous component follows (5.20), so that

$$p_i V = \tfrac{2}{3} N_i \bar{\epsilon} \tag{5.25}$$

Hence

$$V \sum_i p_i = \tfrac{2}{3} \bar{\epsilon} \sum_i N_i \tag{5.26}$$

and from (5.20) and (5.24)

$$pV = \tfrac{2}{3} N \bar{\epsilon} \tag{5.27}$$

Hence,

$$p = \sum_i p_i \tag{5.28}$$

which expresses Dalton's law of partial pressures: *if two or more gases that do not interact are mixed at a given temperature, then each gas exerts the pressure that it would exert if it alone occupied the whole of the containing volume.*

We have shown how the kinetic theory of gases can explain the experimental gas laws. We discuss next the Maxwell–Boltzmann distributions, so that we may determine mean values of molecular properties and their deviations.

5.3.4 Maxwell–Boltzmann distribution of velocities

The calculation of the pressure of a gas (Section 5.3.1) implied a steady-state distribution of molecular velocities in a gas under equilibrium conditions. The steady-state description refers to average properties, individual variations being a natural consequence of intermolecular collisions.

Let the probability that a molecule has a velocity **v** with components in the range v_x to $v_x + dv_x$, v_y to $v_y + dv_y$ and v_z to $v_z + dv_z$ be $\Phi(v_x, v_y, v_z)\,dv_x\,dv_y\,dv_z$. Since the three components of velocity lie along Cartesian axes they are mutually independent, and we can write

$$\Phi(v_x, v_y, v_z)\,dv_x\,dv_y\,dv_z = \phi(v_x)\phi(v_y)\phi(v_z)\,dv_x\,dv_y\,dv_z \tag{5.29}$$

where $\phi(v_i)$ ($i = x, y, z$) is the individual probability that the ith velocity component lies in the range v_i to $v_i + dv_i$. The kinetic theory postulate of isotropy means that the velocity distribution is independent of orientation, so that $\Phi(v_x, v_y, v_z)$ depends only on the speed v and can be replaced by $\Phi(v)$, where

$$v^2 = v_x^2 + v_y^2 + v_z^2 \tag{5.30}$$

whence

$$\Phi(\boldsymbol{v})\,d\boldsymbol{v} = \Phi(v_x, v_y, v_z)\,dv_x\,dv_y\,dv_z = \phi(v_x)\phi(v_y)\phi(v_z)\,dv_x\,dv_y\,dv_z \tag{5.31}$$

For the one-dimensional distribution along x, partial differentiation of $\Phi(v)$ with respect to v_x leads to

$$\frac{\partial\Phi(v)}{\partial v_x} = \frac{\partial\Phi(v)}{\partial v}\frac{\partial v}{\partial v_x} = \frac{\partial\phi(v_x)}{\partial x}\phi(v_y)\phi(v_z) \tag{5.32}$$

From (5.30), $\partial v/\partial v_x = v_x/v$ and (5.32) becomes

$$\Phi(1/v)\frac{\partial\Phi(v)}{\partial v} = \phi(v_y)\phi(v_z)(1/v_x)\frac{\partial\phi(v_x)}{\partial v_x} \tag{5.33}$$

On dividing both sides of (5.33) by $\Phi(v)$ and using (5.31),

$$\frac{1}{v\Phi(v)}\frac{\partial\Phi(v)}{\partial v}=\frac{1}{v_x\phi(v_x)}\frac{\partial\phi(v_x)}{\partial v_x} \tag{5.34}$$

Since the left-hand side of (5.34) is a function of v alone, whereas the right-hand side is a function of just v_x, each side must separately be equal to a constant, say $-\alpha$. Hence,

$$\frac{\mathrm{d}\phi(v_x)}{\mathrm{d}v_x}[v_x\phi(v_x)]^{-1}=-\alpha \tag{5.35}$$

whence

$$\mathrm{d}\ln[\phi(v_x)]=-\alpha v_x\,\mathrm{d}v_x \tag{5.36}$$

Integrating (5.36) gives

$$\phi(v_x)=\phi_0\exp(-\alpha v_x^2/2) \tag{5.37}$$

where $\ln(\phi_0)$ is the integration constant. The separation constant α in (5.35) was made negative so that $\phi(v_x)$ should not increase indefinitely. In normalizing $\phi(v_x)$, as follows, the limits $\pm\infty$ are used for convenience of integration: the probability of $|v_x|$ being in excess of the speed of light is vanishingly small.

$$\int_{\text{all }v_x}\phi(v_x)\,\mathrm{d}v_x=\phi_0\int_{-\infty}^{\infty}\exp(-\alpha v_x/2)\,\mathrm{d}v_x=1 \tag{5.38}$$

Since v_x^2 is symmetrical about $v_x=0$, we can write (5.38) as

$$2\phi_0\int_0^{\infty}\exp(-\alpha v_x/2)\,\mathrm{d}v_x=1 \tag{5.39}$$

Integrals of this type are solved easily by the use of the gamma function (Appendix 6), and we write

$$(2/\alpha)^{1/2}\phi(0)\int_0^{\infty}t^{1/2}\exp(-t)\,\mathrm{d}t=1 \tag{5.40}$$

where $t=\alpha v_x^2$. The value of the integral is $\Gamma(1/2)$, or $\pi^{1/2}$. Hence, $\phi_0=[\alpha/(2\pi)]^{1/2}$, and we may write the distribution as

$$\phi(v_x)\,\mathrm{d}v_x=(\alpha/2\pi)^{1/2}\exp(-\alpha v_x^2/2)\,\mathrm{d}v_x \tag{5.41}$$

where $\phi(v_x)\,\mathrm{d}v_x$ is the probability that a velocity component along the x axis lies between the values x and $x+\mathrm{d}v_x$.

From the foregoing, $\bar{\epsilon}=\frac{1}{2}m\overline{v^2}=\frac{3}{2}k_BT$, so that

$$\overline{v_x^2}=\tfrac{1}{3}v^2=k_BT/m \tag{5.42}$$

We can now determine α from the distribution function, as follows.

The average value of a parameter X with a distribution function $\phi(X)$ is given generally by

$$\overline{X}=\int X\phi(X)\,\mathrm{d}X/\int\phi(X)\,\mathrm{d}X \tag{5.43}$$

When the distribution function itself is normalized to unity between the limits of integration, the average value X is given by the numerator of (5.43). Hence, we may write

$$\overline{v_x^2} = \int_{-\infty}^{\infty} v_x^2 \phi(v_x)\,\mathrm{d}v_x \tag{5.44}$$

or

$$\overline{v_x^2} = (2\alpha/\pi)^{1/2} \int_0^{\infty} v_x^2 \exp(-\alpha v_x^2/2)\,\mathrm{d}v_x \tag{5.45}$$

Solving the integral as before leads to $[2/(\alpha\pi^{1/2})]\Gamma(3/2)$, or $1/\alpha$, so that from (5.42) it follows that $\alpha = m/(k_B T)$, and (5.41) becomes

$$\phi(v_x)\,\mathrm{d}v_x = [m/(2\pi k_B T)]^{1/2} \exp[-mv_x^2/(2k_B T)]\,\mathrm{d}v_x \tag{5.46}$$

As an example of the use of (5.46), consider metallic sodium heated at 700 K in an oven with a small hole in one wall. We will calculate the average velocity of emergent sodium atoms. The emergent stream is unidirectional; hence, from (5.43) and (5.46)

$$\overline{v_x} = \frac{\int_0^{\infty} v_x \exp[-mv_x^2/(2k_B T)]\,\mathrm{d}v_x}{\int_0^{\infty} \exp[-mv_x^2/(2k_B T)]\,\mathrm{d}v_x}$$

Solving the integrals as before, we obtain

$$\overline{v_x} = (k_B T/m)/[\pi k_B T/(2m)]^{1/2} = [2k_B T/(\pi m)]^{1/2}$$

which, for the given example, is $[(2 \times 1.3807 \times 10^{-23}\ \mathrm{J\,K^{-1}} \times 700\ \mathrm{K})/(\pi \times 22.99 \times 1.6605 \times 10^{-27}\ \mathrm{kg})]^{1/2}$, or $401\ \mathrm{m\,s^{-1}}$.

Results similar to (5.46) can be obtained for $\phi(v_y)$ and $\phi(v_z)$; hence, from (5.31), the probability term $\Phi(\boldsymbol{v})\,\mathrm{d}\boldsymbol{v}$ is given by

$$\Phi(\boldsymbol{v})\,\mathrm{d}\boldsymbol{v} = [m/(2\pi k_B T)]^{3/2} \exp[-mv^2/(2k_B T)]\,\mathrm{d}v_x\,\mathrm{d}v_y\,\mathrm{d}v_z \tag{5.47}$$

The normalization constant for the three-dimensional distribution may be confirmed from

$$\int_{-\infty}^{\infty}\int_{-\infty}^{\infty}\int_{-\infty}^{\infty} \Phi(v_x, v_y, v_z)\,\mathrm{d}v_x\,\mathrm{d}v_y\,\mathrm{d}v_z = 1 \tag{5.48}$$

The exponential term in (5.47) is a form of the Boltzmann equation (see Section 3.6.5.1), with an average molecular energy equal to $mv^2/2$. The distribution of velocities along a single dimension is illustrated in Figure 5.3. It is a normalized Gaussian distribution, with a mean velocity $\overline{v_x}$ of zero and a standard deviation of $(k_B T/m)^{1/2}$.

At the higher temperature, the distribution becomes broader, and more molecules attain higher speeds, and energies, which is an important factor in reaction kinetics. The derivations of (5.46) and (5.47) neglected molecular interactions, consistent with the isotropy postulate of the kinetic theory. More rigorous calculations by statistical mechanics lead to similar results and the conclusions that we have drawn remain valid.

5.3.5 Maxwell–Boltzmann distribution of speeds

In order to obtain a distribution of *speeds*, we can express velocity in terms of magnitude and direction conveniently by a transformation from the Cartesian x, y and z coordinates to spherical polar coordinates v, θ and Ω, where

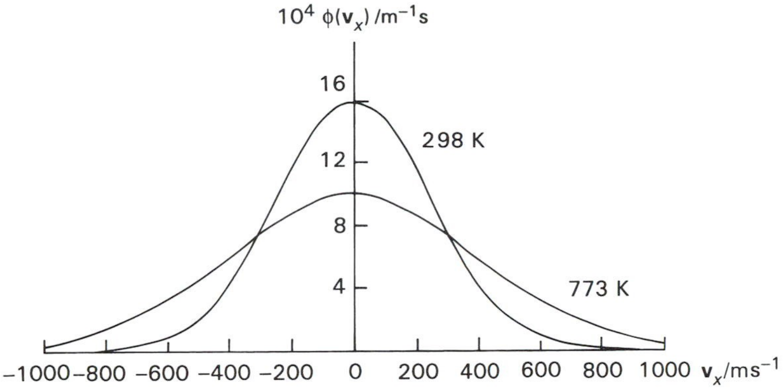

Figure 5.3 Maxwell–Boltzmann distribution of velocities (5.46) for argon at 298 K and 773 K; each distribution shown is a normalized Gaussian, with a mean of zero and a standard deviation of $(k_B T/m)^{1/2}$.

$$\begin{aligned} v_x &= v \sin(\theta) \cos(\Omega) \\ v_y &= v \sin(\theta) \sin(\Omega) \\ v_z &= v \cos(\theta) \end{aligned} \tag{5.49}$$

following Appendix 5, with the distance coordinate r in polar (linear) space replaced by the velocity coordinate v in polar (velocity) space, and the ϕ coordinate replaced by Ω to avoid a confusing notation in this context.

The probability that a molecule has a speed between v and $v+\mathrm{d}v$ is $\Phi(v, \theta, \Omega)\,\mathrm{d}\tau$, where $\mathrm{d}\tau$ is an element of volume in velocity space given by $v^2 \sin(\theta)\,\mathrm{d}v\,\mathrm{d}\theta\,\mathrm{d}\Omega$. The value of $\Phi(v, \theta, \Omega)$ is independent of orientation, so we can integrate over θ and Ω in order to obtain the probability for any region of space between the limits v and $v+\mathrm{d}v$. Hence, from (5.46), (5.30) and Appendix 5, we have

$$\Phi(v)\,\mathrm{d}v = \int_0^{\pi} \sin(\theta)\,\mathrm{d}\theta \int_0^{2\pi} \mathrm{d}\Omega \quad [m/(2\pi k_B T)]^{3/2} v^2 \exp[-mv^2/(2k_B T)]\,\mathrm{d}v \tag{5.50}$$

which may be evaluated to

$$\Phi(v)\,\mathrm{d}v = 4\pi[m/(2\pi k_B T)]^{3/2} v^2 \exp[-mv^2/(2k_B T)]\,\mathrm{d}v \tag{5.51}$$

where $\Phi(v)\,\mathrm{d}v$ represents the probability that a molecular speed lies between the values v and $v+\mathrm{d}v$. This equation may be obtained from (5.47) by noting that the probability $\Phi(v)\,\mathrm{d}v$ of a speed lying between v and $v+\mathrm{d}v$ is $4\pi v^2\,\Phi(v)\,\mathrm{d}v$, where $4\pi v^2\,\mathrm{d}v$ is the volume of a spherical shell defined by radii v and $v+\mathrm{d}v$.

In Figure 5.4, $\Phi(v)$ is plotted as a function of v at 298 K and 773 K, and may be compared with Figure 5.3. Whereas $\Phi(v)$ is proportional to both v^2 and $\exp[-mv^2/(2k_B T)]$, $\phi(v_x)$ is proportional only to $\exp[-mv_x^2/(2k_B T)]$. Although velocity components around zero are the most probable, there are fewer combinations of them so that $\Phi(v)$ is actually small around $v=0$. The most probable speed is that at which $\Phi(v)$ is a maximum. Thus, v_{max} may be found readily by differentiating (5.51) with respect to v and setting the derivative to zero, whence

$$v_{max} = (2k_B T/m)^{1/2} \tag{5.52}$$

The root mean square speed $(\overline{v^2})^{1/2}$ derived from (5.30) and (5.42) is $(3k_B T/m)^{1/2}$; this result can be derived directly from (5.51) and (5.43). These two measures of speed, together with

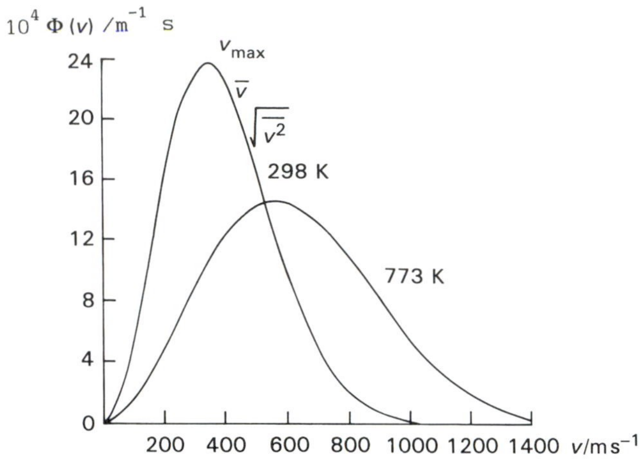

Figure 5.4 Maxwell–Boltzmann distribution of speeds (5.47) for argon at 298 K and 773 K; the position of v_{max}, $\bar{v}$ and $(\overline{v^2})^{1/2}$ are shown for the 298 K curve. An increase in temperature results in a larger proportion of higher speeds (and energies), which is of importance in a study of reaction mechanisms.

the average $\bar{v}$, are indicated on Figure 5.4 for the 298 K curve; it should be noted that $\bar{v}$ is not the same as $(\overline{v^2})^{1/2}$.

As an application of (5.51), we can extend the previous example to determine the mean speed of sodium atoms in the oven at 700 K. The distribution $\Phi(v)$ is normalized to unity between the limits of 0 and ∞. Hence, from (5.43) and (5.51)

$$\bar{v}=4\pi[m/(2\pi k_B T)]^{3/2}\int_0^\infty v^3 \exp[-mv^2/(2k_B T)]\,dv$$

Proceeding as before, the integral is $\frac{1}{2}(2k_B T/m)^2$, so that the mean speed is $[8k_B T/(\pi m)]^{1/2}$ which evaluates to 803 m s^{-1}.

5.3.6 Deviations from ideality: real gases

The kinetic theory applies, strictly, to the ideal gas. All real gases deviate from the model of the ideal gas, to an extent dependent upon the external conditions. At pressures up to approximately 10 atm gases behave approximately ideally; exceptions are the easily liquefiable gases, such as carbon dioxide.

Figure 5.5 shows the graph of $pV_m/(\mathcal{R}T)$ against p for nitrogen at four temperatures; $pV_m/(\mathcal{R}T)$ may be termed the molar compression factor $\mathcal{Z}_m$:

$$\mathcal{Z}_m=pV_m/(\mathcal{R}T) \tag{5.53}$$

For the ideal gas $\mathcal{Z}_m$ is unity at all pressures, but for real gases it is clearly greater at low temperatures and high pressures, conditions that bring the gas molecules closer together. In fact, the equation of state for the ideal gas may be regarded as a limiting case of the general *virial equation of state*

$$pV_m=\mathcal{R}T[1+B(T)/V_m+C(T)/V_m^2+\ldots] \tag{5.54}$$

and we can identify $\mathcal{Z}_m$ with the bracketed part of (5.54). Up to certain pressures, $C(T)$ and higher terms are negligible; the second virial coefficients for argon and nitrogen have the following values:

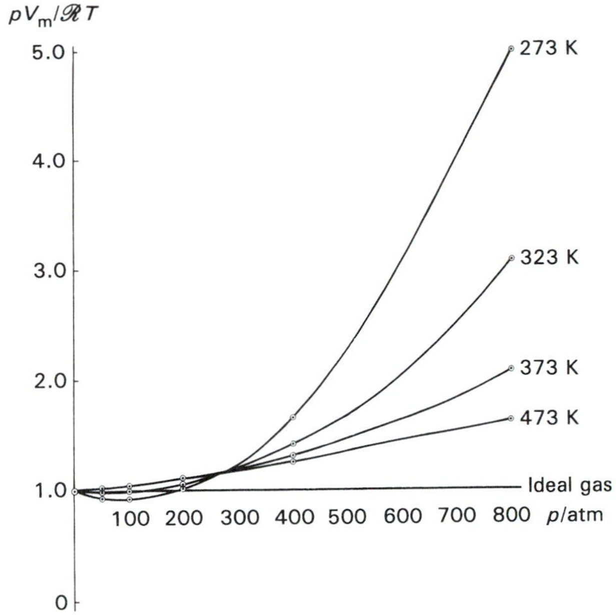

Figure 5.5 Variation in $pV_m/(\mathcal{R}T)$, the molar compression factor $\mathcal{Z}_m$, with pressure p for N_2 at 273 K, 323 K, 373 K and 473 K; the ideal gas line is horizontal, at $pV_m/(\mathcal{R}T)=1$.

	$10^6\ B(T)/m^3\ mol^{-1}$				
T/K	273	323[a]	373	412[b]	873
Ar	−21.7		−4.2	0.0	11.9
N_2	−10.5	0.0	6.2		−21.7

[a]$T_B(N_2)=327.2$ K; [b]$T_B(Ar)=411.5$ K.

From (5.54), at not too high pressures, by neglecting the third and higher virial coefficients and putting $V_m=\mathcal{R}T/p$, we obtain

$$(\partial\mathcal{Z}_m/\partial p)_T=B(T)/(\mathcal{R}T) \qquad (5.55)$$

As the temperature is increased, the dip in the curve of Figure 5.5 becomes smaller and moves to lower values of p. At a particular temperature, the slope is zero up to a relatively high pressure. This temperature is called the Boyle temperature T_B, because Boyle's law is valid over that pressure range. Now, $(\partial\mathcal{Z}_m/\partial p)_T$ must be zero at the Boyle temperature. Hence, from (5.54), $B(T)$ must be negative below T_B and positive above it, as is shown by the data above. It follows that to explain the curves of Figure 5.5 above about 100 atm pressure, the third virial coefficient, at least, would be needed.

5.3.7 Van der Waals' equation of state

Several equations of state have been proposed that endeavour to represent the behaviour of gases under widely varying conditions of temperature and pressure. Real gases experience forces of attraction and repulsion. The pressure of a gas depends upon both the rate of collisions with the walls and the intensity of such collisions (see Section 5.3.1); both of these

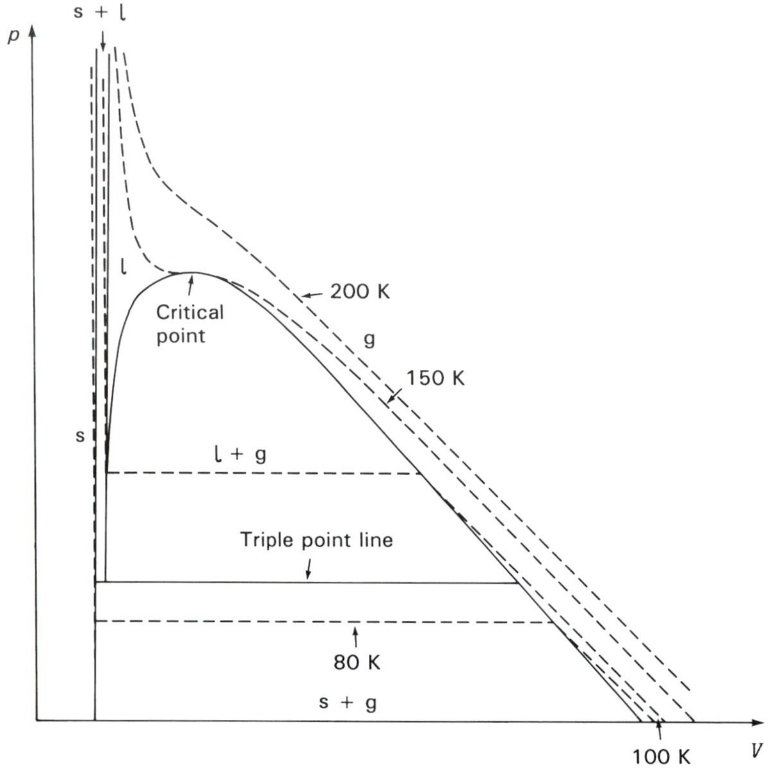

Figure 5.6 p–V–T phase diagram for argon, in projection on to the p–V plane; the isotherms are shown for 80 K, 100 K, 150 K and 200 K; T_c=150.7 K.

effects are decreased by attraction between the molecules. Thus, the pressure has to be corrected by a factor that is proportional to the square of the density n/V (see also the discussion on fugacity in Section 4.9.3).

The molecules of a gas do not have zero volume, so the space available to them has to be reduced by a factor that is proportional to the amount of gas present n. Thus, we obtain the van der Waals' equation of state for a real gas:

$$(p+an^2/V^2)(V-nb)=n\mathcal{R}T \qquad (5.56)$$

where a and b are constants for a given gas. This equation can be written as a power series in V. Thus, for one mole of gas

$$V_m^3-(\mathcal{R}T/p+b)V_m^2+(a/p)V_m-(ab/p)=0 \qquad (5.57)$$

At a temperature known as the critical temperature T_c, (5.57) has only one real root, the critical molar volume $V_{m,c}$; the corresponding pressure is the critical pressure p_c. At a temperature greater than its critical temperature a gas cannot be liquefied, no matter how great the applied pressure. The increase of pressure will increase the density of the gas, but will not cause liquefaction. The liquid–gas interface disappears at the critical temperature and the demarcation between these two states of matter is then indistinct. Figure 5.6 shows the p–V–T diagram for argon, in projection on the p–V plane; isotherms for the system at 80, 100, 150 and 200 K are shown. For argon, T_c=150.7 K, p_c=48 atm and the triple-point temperature is 83 K.

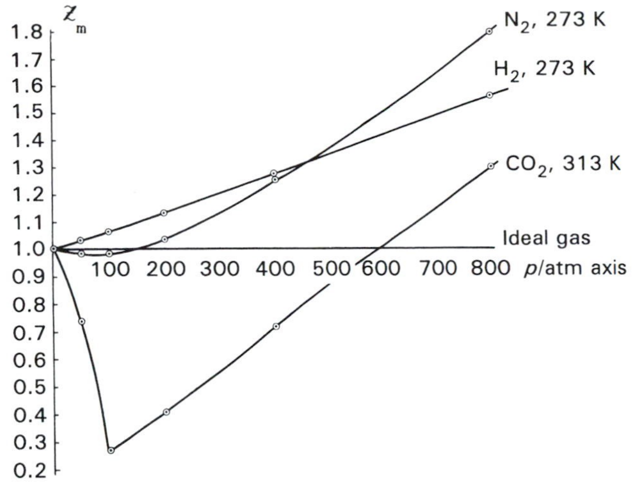

Figure 5.7 Variation in molar compression factor $\mathscr{Z}_m$ with pressure p for N_2 and H_2 at 273 K and CO_2 at 313 K; the ideal gas line is horizontal, at $\mathscr{Z}_m=1$.

5.3.8 Comparing gases

The critical constants of a gas are characteristic of it and they lead to a method whereby gases can be compared. The p, V and T variables are first converted to *reduced* values:

$$p_r=p/p \qquad V_r=V_m/V_{m,c} \qquad T_r=T/T_c \tag{5.58}$$

Figure 5.7 shows a plot of $\mathscr{Z}_m$ against p, for three common gases. If instead of plotting against p we plot against p_r, a single curve suffices for a wide range of gases (see Problem 5.13). The principle of corresponding states expresses the observation that *real gases at the same reduced molar volume and reduced temperature exert (approximately) the same reduced pressure.* This principle is followed most closely for gases composed of spherical, nonpolar molecules.

Equation (5.57) can be expressed in the form

$$(V_m-\alpha_1)(V_m-\alpha_2)(V_m-\alpha_3)=0 \tag{5.59}$$

where α_1, α_2 and α_3 are parameters that have the dimensions of molar volume. At the critical point, $\alpha_1=\alpha_2=\alpha_3=V_{m,c}$ and

$$(V_m-V_{m,c})^3=0 \tag{5.60}$$

or

$$V_m^3-3V_{m,c}V_m^2+3V_{m,c}^2V_m-V_{m,c}^3=0 \tag{5.61}$$

Setting p, V and T in (5.57) to p_c, $V_{m,c}$ and T_c respectively, and then comparing coefficients with (5.61), we obtain

$$V_{m,c}=(\mathscr{R}T_c/p_c+b)/3=(a/3p_c)^{1/2}=(ab/p_c)^{1/3} \tag{5.62}$$

whence

$$p_c=a/(27b^2) \qquad V_{m,c}=3b \qquad T_c=8a/(27\mathscr{R}b) \tag{5.63}$$

Table 5.1 Constants of the van der Waals equation and critical constants for argon, nitrogen and carbon dioxide

	Ar	N_2	CO
$a/\text{dm}^6\ \text{atm mol}^{-2}$	1.355	1.39	3.59
$10^2 b/\text{dm}^3\ \text{mol}^{-1}$	3.22	3.91	4.27
p_c/atm	48.0	33.5	72.8
$V_{m,c}/\text{cm}^3\ \text{mol}^{-1}$	75.2	90.1	94.0
T_c/K	150.7	126.2	304.2
T_B/K	411.5	327.2	714.8
$\mathcal{Z}_{m,c}$	0.292	0.291	0.274

It follows that $\mathcal{Z}_{m,c}=p_c V_{m,c}/(\mathcal{R}T_c)=0.375$. Some values of the constants of the van der Waals equation are listed in Table 5.1.

The van der Waals equation of state is a good model for real gases. It may fail under conditions of very high pressure or very low temperature; other equations of state have been formulated, of which the virial equation (5.54) is the most general. Although the value of $\mathcal{Z}_{m,c}$ in Table 5.1 is different from the predicted value of 0.375, its near-constancy is an indication of good self-consistency of the van der Waals model for the three very different gases considered.

5.3.9 Intermolecular attraction

The departure from ideal behaviour of real gases under certain conditions indicates a degree of failure in the basic postulates of the kinetic theory. The finite volume occupied by molecules is significant in circumstances where the molecules are brought close together. Molecules also exert small forces upon one another, but whereas attractive forces are effective over a relatively long range, repulsive forces become important only at intermolecular distances near to or less than equilibrium intermolecular distances. The effects of molecular interactions are revealed in the compression factor $\mathcal{Z}$ (Figure 5.7) and we shall consider them briefly.

5.3.10 Molecular volume effect

The finite volume of the molecules in a gas reduces the effective molar volume V_m to (V_m-b), and the effect of b is greater the closer the molecules are brought together. The constant b in (5.56) is related to the volume occupied by the gas molecules; it is an *excluded* volume, as the following analysis shows.

Let the molecules of a gas be spherical and of diameter d. If they are regarded as hard spheres, then two molecules cannot approach more closely than twice $d/2$, or twice the sum of their van der Waals' radii (see Section 6.2.2.5). Figure 5.8 shows that the volume of space in which two molecules cannot move freely is a sphere of radius d.

Thus, the volume excluded for a pair of molecules is $4\pi d^3/3$, and that for a single molecule is $4\pi d^3/6$. The actual volume of a single molecule is $4\pi(d/2)^3/3$, or $\pi d^3/6$. Hence, the excluded volume b for a single molecule is four times its own volume. It should be noted that values of b (and a) are normally quoted in molar terms, as in Table 5.1.

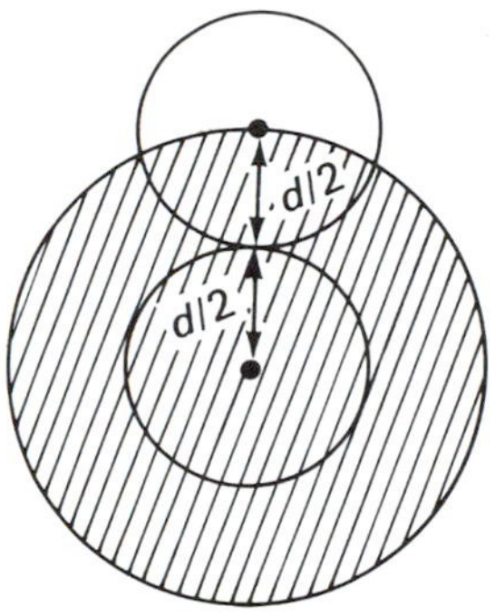

Figure 5.8 Excluded volume in a gas: the van der Waals radius of the 'hard-sphere' species is $d/2$ and the centres of two such species cannot approach more closely than twice $d/2$. Hence, proximal species cannot move freely in a sphere (shaded) of radius less than d.

5.3.11 Collision frequency

A collision occurs when two molecules approach to the distance d from each other; d is the *collision diameter* of the species and may be thought of as that distance of separation at which repulsion forces between two molecules become significant. A given molecule travelling a linear path sweeps out a cylinder of area πd^2 within which collisions occur; this area is called the *collision cross-section* σ. If the molecules, of mean speed $\bar{v}$, travel for a time t in the cylinder and there are N molecules in a volume V, then the number of molecules with their centres inside the volume swept out is $\sigma\bar{v}t(N/V)$. The collision frequency z is the number of collisions per unit time and is equal to $\sigma\bar{v}(N/V)$.

However, two molecules, each of average speed $\bar{v}$, moving in the same direction have a relative speed of zero. If they are moving away from each other along the same line then their relative speed would be $2\bar{v}$. An average situation between these two extremes would be motion at 90° to each other, in which case the relative speed would be $\bar{v}\sqrt{2}$. Hence,

$$z=\sqrt{2}\sigma\bar{v}(N/V) \tag{5.64}$$

where $\bar{v}$ is $[8k_BT/(\pi m)]^{1/2}$, from Section 5.3.5. If we require the total number of collisions per unit volume per unit time Z, then we have

$$Z=\tfrac{1}{2}z(N/V)=\sigma\bar{v}(N/V)^2/\sqrt{2} \tag{5.65}$$

where the factor of $\frac{1}{2}$ arises because collisions between any two given molecules must be counted once only.

Values of collision diameters d (obtained from measurements of gas viscosity) collision cross-sections σ for argon, nitrogen and carbon dioxide are given in Table 5.2, together with the van der Waals constant b calculated from them. The correspondence of the values of b to those in Table 5.1 indicates the approximate nature of b (similar to that of a).

5.3.12 Mean free path

A molecule of average speed $\bar{v}$ and collision frequency z spends a time $1/z$ between collisions and so travels a distance $\bar{v}/z$. The *mean free path* l of the molecule (between collisions) is given by

$$l=\bar{v}/z=1/(\sqrt{2}\sigma N/V) \tag{5.66}$$

Table 5.2 Collision diameter, collision cross-section and van der Waals' constant b for argon, nitrogen and carbon dioxide

	Ar	N_2	CO_2
d/nm	0.330	0.342	0.380
σ/nm^2	0.342	0.368	0.454
10^2b/dm^3 mol^{-1}	4.53	5.04	6.92

We illustrate the results (5.71) to (5.73) by continuing the study of sodium vapour, heated at 700 K in an oven of volume 200 dm^3; the collision diameter of sodium is 0.37 nm and its vapour pressure is 1 mmHg at the temperature of the experiment. We shall determine:

(a) the number of collisions of a single sodium atom in 1 s;
(b) the total number of collisions in 1 s;
(c) the mean free path of sodium atoms in the oven at 700 K.

Since $N/V=nN_A/V=p/(k_BT)$, $z=\sqrt{2}\sigma\bar{v}p/(k_BT)$, where σ is πd^2 and $\bar{v}$, from the example problem in Section 5.3.4, is 803 m s^{-1}. Thus,

(a)
$$z=\frac{\pi\sqrt{2}\times0.37^2\times10^{-18}\ \text{nm}^2\times803\ \text{m s}^{-1}\times(1/760)\times101325\ \text{N m}^{-2}}{1.3807\times10^{-23}\text{JK}^{-1}\times700\text{K}}$$
$$=6.74\times10^6\ \text{s}^{-1}$$

(b)
$$Z_{\text{NaNa}}=\tfrac{1}{2}zN/V=\tfrac{1}{2}zp/k_BT$$
$$=\frac{6.74\times10^6\ \text{s}^{-1}\times(1/760)\times101325\ \text{Nm}^{-2}}{2\times1.3807\times10^{-23}\text{JK}^{-1}\times700\text{K}}$$
$$=4.65\times10^{28}\ \text{s}^{-1}\ \text{m}^{-3}$$

Hence, in the oven of volume 200 dm^3, the total number of collisions is 0.93×10^{28} s^{-1}.

(c)
$$l=\bar{v}/z=803\ \text{m s}^{-1}/(6.74\times10^6\ \text{s}^{-1})=1.19\times10^{-4}\ \text{m}$$

We can see that the value of l is approximately $10^5\,d$, so we would expect the kinetic theory to be satisfactory in the several example problems with sodium vapour that we have studied.

5.3.13 The Joule–Thomson effect

Further experimental evidence for attractive forces between molecules of real gases derives from the porous-plug experiment of Joule and Thomson. A gas passed through a plug of a porous material, such as silk or unglazed porcelain, was generally found to become cooled; a similar effect is observed if the gas expands into a vacuum. Hydrogen and helium were found to increase in temperature in this experiment, unless the operating temperature was very low; there is, thus, a Joule–Thomson inversion temperature T_i.

The apparatus (Figure 5.9) is insulated from the surroundings, so that the porous-plug process is carried out adiabatically. A gas under conditions p_1, V_1, T_1 is compressed isothermally through the plug by means of a piston, such that V_1 is reduced to zero; the work

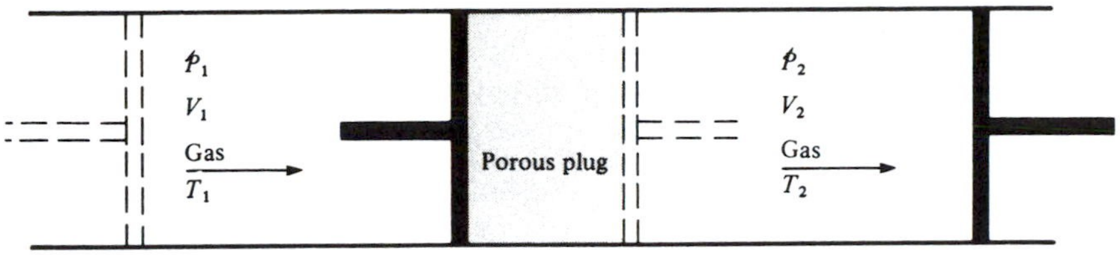

Figure 5.9 Schematic diagram of apparatus to demonstrate the Joule–Thomson effect; the apparatus is insulated from the surroundings. For $T_1 < T_i$, $\mu_{JT} > 0$ and T_2 is *less* than T_1 after streaming through the plug.

done on the gas is p_1V_1. On the exit side of the plug the gas expands isothermally against another piston, so that the volume on the exit side increases from zero to V_2. The work done on the gas is now p_2V_2, in total $p_1V_1 - p_2V_2$, and is equal to the change in internal energy. Hence, the expansion is isenthalpic ($\mathrm{d}H = 0$).

The observed property is $\Delta T/\Delta p$, and the Joule–Thomson coefficient μ_{JT} is given by

$$\mu_{JT} = \lim_{p \to 0} \Delta T/\Delta p = (\partial T/\partial p)_H \tag{5.67}$$

Following (4.14),

$$\mathrm{d}H = (\partial H/\partial T)_p \,\mathrm{d}T + (\partial H/\mathrm{d}p)_T \,\mathrm{d}p \tag{5.68}$$

From the experimental conditions $\mathrm{d}H = 0$, and

$$(\partial T/\partial p)_H = -(\partial H/\mathrm{d}p)_T/(\partial H/\partial T)_p \tag{5.69}$$

From (4.24), (4.94) and (4.100),

$$\mu_{JT} = (\partial H/\partial p)_T = T(\partial S/\partial p)_T + V = \frac{T(\partial V/\partial T)_p - V}{C_p} \tag{5.70}$$

For an ideal gas $T(\partial V/\partial T)_p = V$, so that μ_{JT} is zero. With a real gas, the cooling arises because the intermolecular forces are overcome as the gas streams through the porous plug.

For a gas that obeys the van der Waals equation, we may use (5.56) with $n=1$, ignore ab/V^2 because it is of relatively small magnitude and replace a/V by $\mathcal{R}T/p$ because it is a small quantity; then we obtain

$$pV = \mathcal{R}T + pb - ap/(\mathcal{R}T) \tag{5.71}$$

Dividing by p and differentiating with respect to T at constant pressure gives

$$(\partial V/\partial T)_p = \mathcal{R}/p + a/(\mathcal{R}T^2) \tag{5.72}$$

Substituting for $\mathcal{R}/p$ from (5.71), dividing by C_p and rearranging,

$$\frac{T(\partial V/\partial T)_p - V}{C_p} = \mu_{JT} = \frac{2a/(\mathcal{R}T) - b}{C_p} \tag{5.73}$$

The Joule–Thomson inversion temperature T_i corresponds to $\mu_{JT} = 0$, that is,

$$T_i = 2a/(\mathcal{R}b) \tag{5.74}$$

Table 5.3 lists some results for the Joule–Thomson effect. The low values of T_i for helium and hydrogen arise because their a values are very small, whereas their values of b are similar to those of other gases. Only at temperatures less than approximately 35 K and 223 K for

Table 5.3 Joule–Thomson data at 273 K

Gas	C_p/dm^3 atm K^{-1} mol^{-1}	μ_{obs}/K atm^{-1}	μ_{calc}/K atm^{-1}	T_i/K
Helium	0.205	−0.06	−0.10	35
Hydrogen	0.285	−0.03	−0.02	223
Nitrogen	0.287	0.27	0.30	860
Carbon	0.366	1.22	0.76	2051

helium and hydrogen, respectively, are the intermolecular forces sufficiently strong to lead to Joule–Thomson cooling.

5.4 Intermolecular forces

The general term *van der Waals' forces* relates to interactions between closed-shell species and may be divided into several classes, such as ion (monopole)–dipole, dipole–dipole, dipole– induced dipole and induced dipole–induced dipole. There are also interactions involving quadrupoles and, in general, multipoles, but they are of smaller magnitudes.

5.4.1 Electric moments

The first electric moment of a charge distribution is the dipole moment μ, which consists of numerical point charges $\pm qe$ separated by a distance r:

$$\mu = qer \tag{5.75}$$

We introduced this topic briefly in Section 2.11.12; the dipole moment is a vector quantity, directed along r, from the positive to the negative point charge.

The second moment is the quadrupole moment, which consists of four point charges with an overall charge of zero and, thus, a zero dipole moment. Carbon dioxide has a zero dipole moment but a significant quadrupole moment Θ equal to -14×10^{-40} C m^2. Generally, the moment of a 2^n-pole is given by the vector sum over all charges of each point charge multiplied by the nth power of its distance from the centroid of the charges.

5.4.2 Polarizability

In an electric field, the electron distribution in a chemical species becomes distorted. As a result, a dipole moment is created (in addition to one that may be present permanently) because the positions of the centroids of positive and negative charge are altered. The species is said to be polarized, and the induced dipole moment μ_{id} is proportional to the strength $\mathscr{E}$ of the applied electric field. For a material that is isotropic,

$$\mu_{id} = \alpha\mathscr{E} \tag{5.76}$$

where α is the *polarizability* of the species.

The hydrogen chloride molecule has a polarizability α of 2.93×10^{-40} F m^2. By dividing a value of α by $4\pi\epsilon_0$, where ϵ_0 is the permittivity of a vacuum (8.854×10^{-12} F m), we obtain the quantity α', equal to 2.63×10^{-24} cm^3, which has the dimensions of volume, and serves to emphasize the relationship between the polarizability and the size of the species. A *volume polarizability* α' may be defined therefore as

Table 5.4 Radii and volume polarizabilities

	He	Ar	F^-	Cl^-	Br^-	I^-
r/nm	0.09	0.191	0.136	0.181	0.195	0.216
$10^{24}\alpha'/\mathrm{cm}^3$	0.20	1.7	0.90	3.1	4.3	6.6

$$\alpha' = 10^{-6}\alpha/(4\pi\epsilon_0) \tag{5.77}$$

Generally, the larger the species or the more negative it is, the more easily it can be deformed by an external electric field, as the data in Table 5.4 show. In the larger species the nuclear charge holds the electrons less strongly, so that external fields have a greater polarizing effect on the electrons than they do in small species with few electrons.

5.4.3 Ion–dipole interaction

Of the several van der Waals interactions, we shall consider three and tabulate the others. Consider the interaction between an ion of charge $+Q_2$, where Q is a numerical charge q multiplied by the electron charge e, and a dipole of moment μ_1 (Figure 5.10a). The potential energy of interaction is based on the Coulombic formula $-Q_1Q_2/(4\pi\epsilon_0 r)$, and given by two pairwise additive terms:

$$V_{\mathrm{i,d}} = [1/(4\pi\epsilon_0)](Q_1Q_2/AC - Q_1Q_2/BC) \tag{5.78}$$

For AC, we have

$$\begin{aligned} AC &= [r^2 + \mathrm{r}r\cos(\theta) + \mathrm{r}^2/4]^{1/2} \\ &= r[1 + \mathrm{r}\cos(\theta)/r + \mathrm{r}^2/(4r^2)]^{1/2} \end{aligned} \tag{5.79}$$

Similarly for BC:

$$BC = r\{1 + [-\mathrm{r}\cos(\theta)/r + \mathrm{r}^2/(4r^2)]\}^{1/2} \tag{5.80}$$

whence

$$\begin{aligned} V_{\mathrm{i,d}} = &-[Q_1Q_2/(4\pi\epsilon_0 r)](\{1 + [-\mathrm{r}\cos(\theta)/r + \mathrm{r}^2/(4r^2)]\}^{-1/2} \\ &- \{1 + [\mathrm{r}\cos(\theta)/r + \mathrm{r}^2/(4r^2)]\}^{-1/2}) \end{aligned} \tag{5.81}$$

Since, in general, r is significantly larger than r, we can expand the terms of the form $(1+x)^{-1/2}$ in (5.81) by the binomial theorem to terms no larger than x^2 and, remembering that $\mu_1 = Q_1\mathrm{r}$, we obtain

$$V_{\mathrm{i,d}} = \frac{-\mu_1 Q_2 \cos(\theta)}{4\pi\epsilon_0 r^2}[1 - 3\mathrm{r}^2/(8r^2)] \tag{5.82}$$

Since $\mathrm{r}^2/r^2 \ll 1$, we may write

$$V_{\mathrm{i,d}} = -\mu_1 Q_2 \cos(\theta)/(4\pi\epsilon_0 r^2) \tag{5.83}$$

The maximum attraction arises when Q_2 is collinear with the dipole axis ($\theta = 0$), when

$$V_{\mathrm{i,d}} = -\mu_1 Q_2/(4\pi\epsilon_0 r^2) \tag{5.84}$$

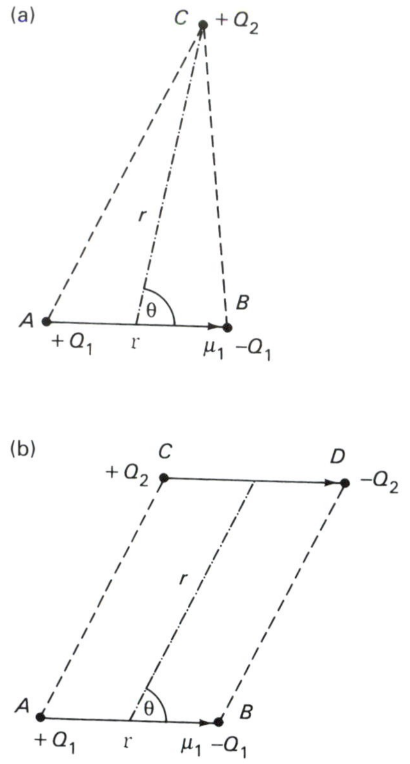

Figure 5.10 Dipolar interaction: (a) AB is a dipole vector of length r and charges $\pm Q_1$; Q_2 is a point charge at C distant r from the centre of the dipole of moment μ_1, such that the line of length r makes an angle θ with the dipole AB. (b) Dipoles AB and CD, each of length r, parallel and coplanar, with r as the distance between their centres. (In both diagrams, Q_i represents a numerical charge q_i multiplied by the electron charge e.)

5.4.4 Dipole–dipole interaction

The dipoles of two polar molecules attract each other and a force is set up between them. In a fluid, the molecules are able to rotate and the field of one dipole tends to orientate the dipole of a neighbouring molecule. Furthermore, the attractive forces dominate, being of longer range than the repulsive forces, and a net attractive potential exists.

Consider first two molecules with permanent dipole moments of magnitudes μ_1 and μ_2, separated by a distance r, fixed in orientation and lying in one and the same plane (not the most general case), as shown in Figure 5.10(b). For the purpose of calculation, we can ensure that the dipoles have the same length r by considering Q_2 modified as necessary with respect to Q_1. Proceeding as before, their potential energy $V_{\mathrm{d,d}}$ is given by

$$\begin{aligned} V_{\mathrm{d,d}} &= [1/(4\pi\epsilon_0)]\,(2Q_1Q_2/r - Q_1Q_2/AD - Q_1Q_2/BC) \\ &= [Q_1Q_2/(4\pi\epsilon_0)]\,(2/r - 1/AD - 1/BC) \\ &= [Q_1Q_2/(4\pi\epsilon_0 r)]\,(2 - \{1+[\mathrm{r}^2/r^2 + 2\mathrm{r}\cos(\theta)/r]\}^{-1/2} \\ &\quad - \{1+[\mathrm{r}^2/r^2 - 2\mathrm{r}\cos(\theta)/r]\}^{-1/2}) \end{aligned} \tag{5.85}$$

Making assumptions similar to those in ion–dipole interaction, we obtain

$$V_{\mathrm{d,d}}=-\mu_1\mu_2[3\cos^2(\theta)-1]/(4\pi\epsilon_0 r^3) \tag{5.86}$$

The general form of (5.86) is shown in Figure 5.11f; for the situation considered above, $\theta_1=\theta_2$ and $\phi=0$, whence (5.86) is obtained. Maximum interaction occurs in either a collinear, head-to-tail orientation ($\theta_1=\theta_2=0$), or an antiparallel orientation ($\theta_1=0$, $\theta_2=\pi$); in these cases, we have

$$V_{\mathrm{d,d}}=\mp 2\mu_1\mu_2/(4\pi\epsilon_0 r^3) \tag{5.87}$$

In a fluid, the molecules can rotate and we can write an average potential energy in the form

$$\overline{V}_{\mathrm{d,d}}=[-\mu_1\mu_2/(4\pi\epsilon_0 r^3)]\,\overline{f(\theta,\phi)P(\theta,\phi)} \tag{5.88}$$

If we take a constant (zero) value of ϕ, in accord with Figure 5.10(b), $f(\theta,\phi)$ is the expression $[3\cos^2(\theta)-1]$ developed above; for $P(\theta,\phi)$ we assume a Boltzmann distribution depending on θ, namely, $\exp[-V_\theta/(k_\mathrm{B}T)]$, where V is given by (5.86). Expanding the exponential to two terms because $V_\theta \ll k_\mathrm{B}T$, P may be approximated by $[1-V_\theta/(k_\mathrm{B}T)]$. Hence

$$\begin{aligned} V_{\mathrm{d,d}} &= -[\mu_1\mu_2/(4\pi\epsilon_0 r^3)]\,[(3\,\overline{\cos^2(\theta)}-1)]\,[(1-\overline{V_\theta}/(k_\mathrm{B}T)] \\ &= -[\mu_1\mu_2/(4\pi\epsilon_0 r^3)]([3\,\overline{\cos^2(\theta)}-1] \\ &\quad -\{-[\mu_1\mu_2/(4\pi\epsilon_0 r^3 k_\mathrm{B}T)]\,\overline{[3\cos^2(\theta)-1]^2}\} \end{aligned} \tag{5.89}$$

We can easily show through (5.43) and (5.49) that the average value of $\cos^2(\theta)$ is 1/3, whence $[3\,\overline{\cos^2(\theta)}-1]=0$. In a similar manner, $\overline{[3\cos^2(\theta)-1]^2}$ evaluates to 4/5, but a more detailed analysis, for the general case, shows that this factor is more correctly 2/3. Hence, we write

$$V_{\mathrm{d,d}}=\frac{-2\mu_1^2\mu_2^2}{3(4\pi\epsilon_0)^2 k_\mathrm{B}T r^6} \tag{5.90}$$

In this expression, we may note both the r^{-6} factor and the dependence upon temperature. The former shows that the dipole–dipole energy will in general be less in magnitude than that for ion–dipole interactions. The inverse dependence upon temperature means that the average dipole–dipole energy in a fluid decreases in magnitude as the temperature increases, because the tendency towards orientation of the dipoles is then randomized by the increased thermal motion of the molecules. For molecules of dipole moment approximately 1 D separated by 0.30 nm, (5.90) gives an energy of -1.3 kJ mol^{-1}; at 0.40 nm separation the energy is only -0.2 kJ mol^{-1}.

5.4.5 Induced dipole–induced dipole interaction

The form for the induced dipole–induced dipole interaction (the dispersion energy, or London energy) was deduced first by London in 1937; it is quantum mechanical in origin, but we can obtain an approximate result by the following argument. For a nonpolar system such as argon, a transient dipole of moment μ_1 exists at any instant in a given atom (1), because of fluctuations in the electron density, the strongest effect existing for a separation of charges equal to the Bohr radius a_0 (see Figure 2.8):

$$\mu_1=ea_0 \tag{5.91}$$

The field arising from this dipole will polarize a neighbouring species (2) and produce an interaction that is similar to the dipole–induced dipole interaction (the Debye energy – see Figure 5.11j). Hence,

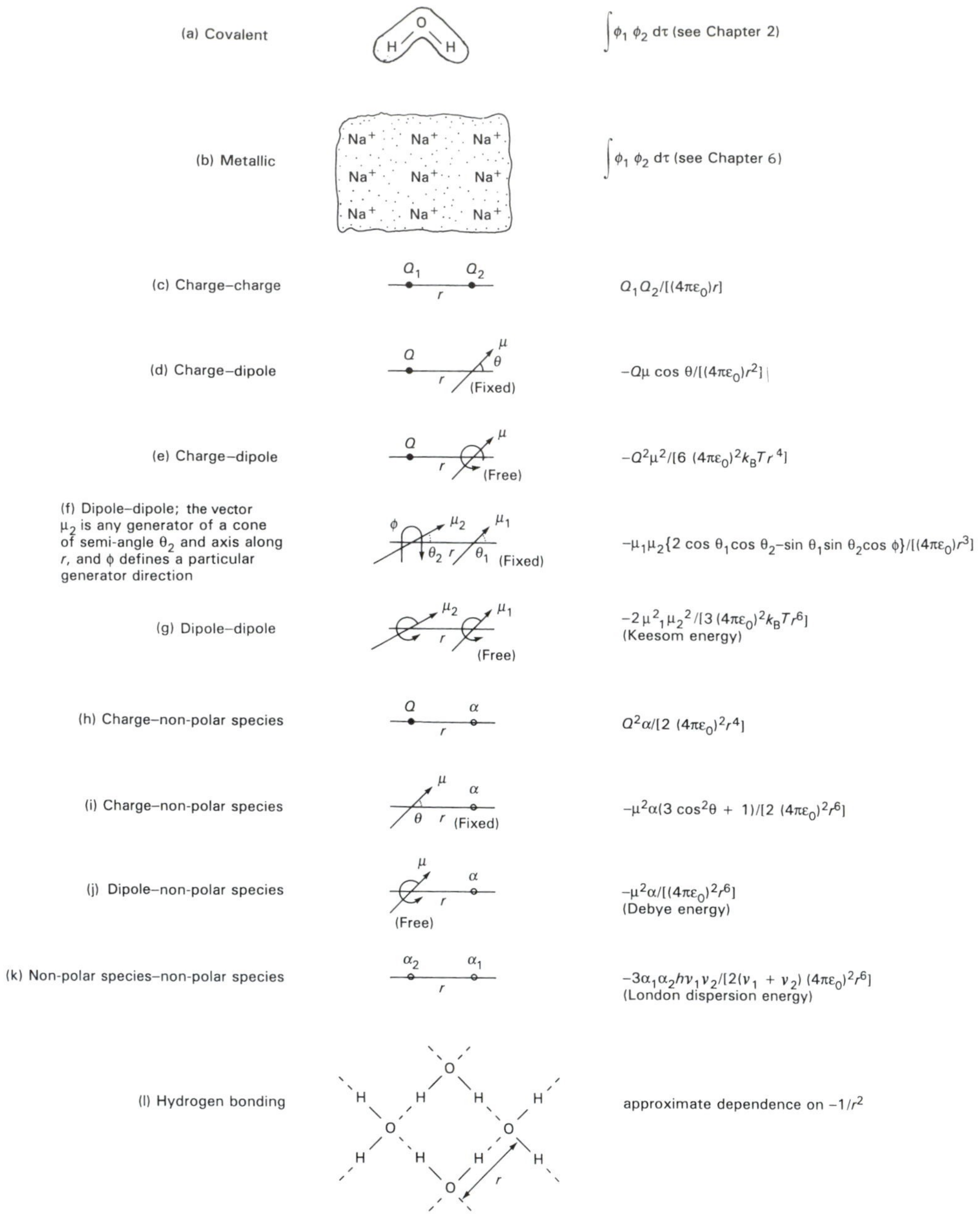

Figure 5.11 Types of intermolecular forces. (a) Covalent (Chapter 2). (b) Metallic (Chapter 6). (c) Charge–charge; r is the distance between the centres of the charges. (d) Charge–fixed dipole; r is the distance between the charge and the centre of the dipole axis. (e) Charge–free dipole; an inverse dependence upon temperature shows that the tendency towards orientation of the dipoles decreases as T increases. (f) Fixed dipole–fixed dipole; ϕ is the angle of rotation of μ_1 relative to μ_2 with respect to the dipole axis. (g) Free dipole–free dipole, also known as the Keesom energy; temperature dependent. (h) Charge–nonpolar species; α is the polarizability of the nonpolar species. (i) Fixed dipole–nonpolar species; α is the polarizability of the nonpolar species. (j) Free dipole–nonpolar species, also known as the Debye energy. (k) Nonpolar species–nonpolar species, also known as the London energy, or dispersion energy. (l) Hydrogen bonding; there is an approximate dependence on $-1/r^2$, where r is the closest nonbonded distance between two nonhydrogen atoms, e.g. O ... O in O−H ... O. It is common to refer to the strength of a hydrogen bond by this length −0.25 nm for O−H ... O would be termed 'strong' and 0.29 nm for the same liaison 'weak'.

$$V_{\mathrm{id,id}}=-\mu_1^2\alpha_2/[(4\pi\epsilon_0)^2r^6]=-(ea_0)^2\alpha_2/[(4\pi\epsilon_0)^2r^6] \tag{5.92}$$

As the first transient dipole changes its value and orientation, so the second (induced) dipole will follow it. It is because of this correlation that the multitude of transients do not average to zero, but give an overall attraction, which leads to an overall negative energetic result of induction. Since the polarizability α is a volume parameter, we write

$$\alpha\simeq4\pi\epsilon_0r^3 \tag{5.93}$$

In this analysis, (5.93) may be written

$$\alpha_2=4\pi\epsilon_0a_0^3 \tag{5.94}$$

and a_0 may be shown to be that distance at which the Coulomb energy $e^2/(4\pi\epsilon_0a_0)$ is equal to $2h\nu$, that is

$$a_0=e^2/(8\pi\epsilon_0h\nu) \tag{5.95}$$

Combining (5.92), (5.94) and (5.95) gives

$$V_{\mathrm{id,id}}=-\alpha_2^2h\nu/[(4\pi\epsilon_0)^2r^6] \tag{5.96}$$

This result has the correct structure, but London's formula for the dispersion interaction between two identical atoms of polarizability α gives the precise result

$$V_{\mathrm{id,id}}=-\tfrac{3}{4}\alpha^2h\nu/[(4\pi\epsilon_0)^2r^6] \tag{5.97}$$

and between dissimilar atoms

$$V_{\mathrm{id,id}}=-\tfrac{3}{2}\alpha_1\alpha_2h\nu_1\nu_2/[(\nu_1+\nu_2)(4\pi\epsilon_0)^2r^6] \tag{5.98}$$

Using ionization energies, with $I=h\nu$, (5.98) becomes

$$V_{\mathrm{id,id}}=-\tfrac{3}{2}\alpha_1\alpha_2I_1I_2/[(I_1+I_2)(4\pi\epsilon_0)^2r^6] \tag{5.99}$$

and with volume polarizabilities

$$V_{\mathrm{id,id}}=-\tfrac{3}{2}\alpha_1'\alpha_2'I_1I_2/[(I_1+I_2)r^6] \tag{5.100}$$

Again, the inverse sixth power of distance is in evidence. We may apply (5.100) to a nonpolar substance, such as methane. We take α' as 2.6×10^{-24} cm^3, I as 12.6 eV, and the distance of closest approach as 0.37 nm, whence $V_{\mathrm{id,id}}$ (CH_4)$=-8.5$ kJ mol^{-1}. A check on this value is given by the cohesive energy of the solid, which may be approximated by the sum of the enthalpies of melting and of vaporization; the resulting value of 8.9 kJ mol^{-1} is in good agreement with our calculation. Induction energies are negative because the induction process always tends to induce moments in the direction of the induction field; there is no antiparallel induction.

In any practical situation a range of intermolecular forces will operate. In a substance such as argon, the dispersion energy is of paramount importance. It is operative in other substances too, but it may be overshadowed in magnitude by other forces that are present. The range of possible intermolecular forces is shown diagrammatically by Figure 5.11, together with their formulations, some of which we have derived in the foregoing sections.

5.5 Intermolecular potentials

At large distances of separation r, two gas molecules do not interact, so that their joint potential energy is effectively zero. As they approach each other, an attraction develops between them which increases continuously as r decreases. However, at distances equal to or less than the collision diameter, strong repulsion between the electrons (and nuclei) of adjacent species causes the potential energy to rise very steeply. The superposition of these two potentials is a curve of the form of Figure 5.12a (see also Figure 2.20), with a minimum energy at the equilibrium distance r_e. Simpler potential functions are the hard-sphere and the square-well models, Figures 5.12b and c.

The potential energy function in Figure 5.12a is the Lennard-Jones m–n model, which has been used successfully in a wide range of applications. For $m=12$ and $n=6$, it takes the form

$$V(r)=4\epsilon_L[(\delta/r)^{12}-(\delta/r)^6] \tag{5.101}$$

The value of 12 for m in the repulsion function was chosen to reproduce the rapid increase (positive) in potential energy as r falls below the value of δ, and it is mathematically useful for m to be twice n.

If we differentiate (5.101) with respect to r and set the derivative to zero for $r=r_e$, we find that $r_e=2^{1/6}\delta$. Inserting this value into (5.101) shows that $V(r)$ is equal to $-\epsilon_L$ at the equilibrium distance r_e. Some values for the parameters ϵ_L and δ are listed in Table 5.5; ϵ is usually given in the form ϵ_L/k_B.

Approximate values for these parameters are given by $\epsilon_L/k_B=0.775T_c$ and $\delta=0.1(V_{m,c}/3)^{1/3}$, where the values of T_c and $V_{m,c}$ are as given in Table 5.1; the values above in parentheses were obtained in this way. The values of δ are very similar to the corresponding collision diameters d (Table 5.2). This is not unreasonable: the essential difference is that whereas the values of d are obtained through experimental measurements, such as on viscosity, δ (and ϵ_L) are chosen as best-fit parameters for (5.101) to reproduce molecular properties.

For argon and nitrogen, ϵ_L has the values 1.0 kJ mol^{-1} and 3.7 kJ mol^{-1} respectively. The corresponding classical thermal energies at 298 K are 3.7 kJ mol^{-1} ($\simeq\frac{3}{2}\mathcal{R}T$) and 6.2 kJ mol^{-1} ($\simeq\frac{5}{2}\mathcal{R}T$); the larger value for nitrogen arises because of extra (rotational and vibrational) degrees of freedom as well as those of translation. It is clear that, at normal temperatures, the thermal energy is four or more times greater than the intermolecular attractive energy, so that the kinetic theory postulates are sensibly upheld in these results.

We have considered aspects of both ideal and real gases, together with some of their properties, and the extent to which these properties are dependent upon intermolecular forces. Some of the results that we have obtained in this chapter will be transferred to our study of liquids in the ensuing section.

5.6 Liquids

In this section, the term 'liquid' will mean a pure substance. It may consist of single atomic species, such as argon, or of two or more bonded species, such as nitrogen or water. These substances are often called *classical liquids*, because their behaviour can be studied by the equations of classical physics. Certain liquids, among which helium is best known, exhibit abnormal behaviour in their flow and conduction properties. They are termed *quantum liquids*, or *superliquids*, but our concern will be with classical liquids.

5.6.1 Liquid–gas equilibrium

We may move into the subject of liquids by considering the equilibrium between a gas and the corresponding liquid.

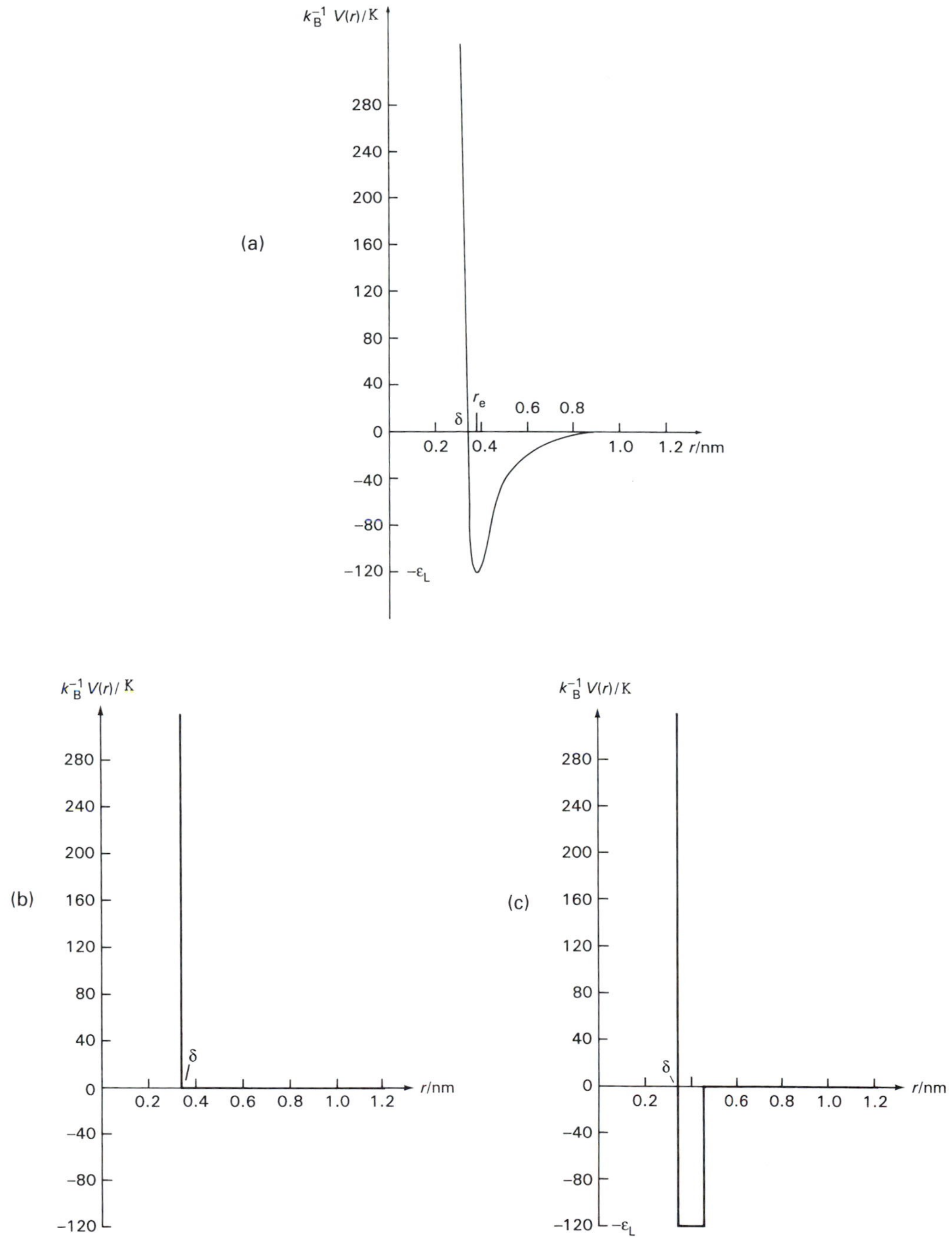

Figure 5.12 Intermolecular potential energy functions. (a) Lennard-Jones 12–6 intermolecular potential function: $V(r)=0$ at $r=\delta$, the Lennard-Jones collision diameter, and $-\epsilon_L$ at $r=r_e$; the equilibrium intermolecular distance r_e is equal to $2^{1/6}\delta$. If $\delta=0.34$ nm for argon, then $r_e=0.38$ nm. (b) Hard-sphere model: a discontinuity occurs at $r=\delta$, the hard-sphere diameter; this model cannot be used for equilibrium properties because there is no minimum in $V(r)$. (c) Square-well model: a minimum in $V(r)$ exists, but discontinuities are present.

Table 5.5 Parameters for the Lennard-Jones 12–6 potential function

	(ϵ_L/k_B)/K	δ/nm
Ar	124 (117)	0.34 (0.29)
N_2	92 (98)	0.37 (0.31)
CO_2	190 (235)	0.40 (0.32)

An individual molecule in a gas possesses a cohesive self-energy $E_s(r)$ which is the sum of its interactions with its neighbours. The self-energy is related to the *attractive* energy discussed in Section 5.1, which we may write as $-C/r^m$ $(m>3)$ for $r>\delta$, where δ is a 'hard-sphere' diameter of the molecule, and tends to infinity for $r<\delta$. The energy $E_s(r)$ is the sum of the pairwise additive interactions of a molecule with its neighbours, that is, those terms, such as Coulombic and dispersion, that do not involve self-cancelling induction interactions.

If the uniform number density in the gas be N, then the number of molecules lying in a spherical shell of radii between r and $r+\mathrm{d}r$ is $4\pi Nr^2\,\mathrm{d}r$. Hence,

$$E_s(r)=-4\pi NC\int_0^{\infty}(1/r^{m-2})\,\mathrm{d}r=-4\pi NC/[(m-3)r^{m-3}] \tag{5.102}$$

For the case with $m=3$, (5.102) illustrates the discontinuity of the hard-sphere model, Figure 5.12b. We write (5.102)

$$E_s(r)=-\alpha N \tag{5.103}$$

where $\alpha=4\pi C/[(m-3)r^{m-3}]$. Following Section 5.3.10, we replace v by $(v-\beta)$, to allow for the volume of the molecules, or in molar terms $(V_m-N_A\beta)$.

From (4.121) and Boyle's law, the chemical potential of the gas is

$$\mu_g=\mu_g^{-0-}+\mathcal{R}T\ln(V^{-0-}/V_m) \tag{5.104}$$

The standard chemical potential is proportional to the self-energy, so we may write μ_g^{-0-} as $-\gamma N$, where γ is a constant.

From (4.123), we develop

$$(\partial\mu_g/\partial p)_T=V_m=(\partial\mu_g/\partial N)_T(\partial N/\partial p)_T=N_A/N \tag{5.105}$$

whence

$$(\partial p/\partial N)_T=(N/N_A)(\partial\mu_g/\partial N)_T \tag{5.106}$$

$$p=\frac{1}{N_A}\int_0^N N(\partial\mu_g/\partial N)_T\,\mathrm{d}N \tag{5.107}$$

From (5.104), introducing $(N_A/N\simeq N_A\beta)$ in place of N_A/N we now have

$$\mu_g=-\gamma N+\mathcal{R}T\ln(V^{-0-})+\mathcal{R}T\ln[N/(N_A-N_ANB)] \tag{5.108}$$

whence

$$(\partial\mu_g/\partial N)_T=-\gamma+\mathcal{R}T/[N(1-\beta N)] \tag{5.109}$$

From (5.107),

$$p = \frac{1}{N_A}\int_0^N [-\gamma N + \mathcal{R}T/(1-\beta N)]\,dN \tag{5.110}$$

which integrates to

$$p = (1/N_A)[-\gamma N^2/2 - (\mathcal{R}T/\beta)\ln(1-\beta N)] \tag{5.111}$$

Since $\beta N/N \ll 1$, $\ln(1-\beta N/N_A)$ may be approximated through the following steps:

$$\ln(1-\beta N) \simeq -\beta N - (\beta N)^2/2 \simeq -\beta N(1+\beta N/2)$$
$$\simeq -(\beta N)/(1-\beta N/2) \simeq -\beta/(1/N - \beta/2)$$

whence

$$p = \frac{1}{N_A}[-\gamma N_A^2/(2V_m^2) + \mathcal{R}T/(1/N - \beta/2)] = -\gamma N_A/(2V_m^2) + \mathcal{R}T/(V_m - \beta N_A/2) \tag{5.112}$$

If we let $\gamma N_A^2/2 = a$ and $\beta N_A/2 = b$, we obtain

$$(p + a/V_m^2)(V_m - b) = \mathcal{R}T \tag{5.113}$$

which is the van der Waals equation (5.56), with $n=1$.

We have shown earlier that $b = \pi d^3/6$, where d is the collision diameter; here, we note that a is proportional to $C/[(m-3)r^{m-3}]$. Thus, we have demonstrated in a simple manner a relationship between the van der Waals constants a and b and the attractive and repulsive forces in a gas.

We have mentioned that the difference between a liquid and a gas in the critical region is indistinct. Thus, it not unreasonable to explore the properties of a simple liquid first by means of the van der Waals equation. From (4.105) and (5.113), we have

$$p = T(\partial p/\partial T)_V - (\partial U/\partial V)_T \tag{5.114}$$

$$p = \mathcal{R}T/(V_m - b) - a/V_m^2 \tag{5.115}$$

from which it is clear that

$$(\partial U/\partial V)_T = a/V^2 \tag{5.116}$$

Integrating (5.116) leads to

$$U_m = -a/V_m + f(T) \tag{5.117}$$

The limiting value of the function $f(T)$ as $a \to 0$ will be the same as that for an ideal gas and, for a monatomic species, is $\frac{3}{2}\mathcal{R}T$. The term $-a/V_m$ in (5.117) represents the contribution of the attractive forces to the internal energy of a gas that obeys the van der Waals equation of state.

The molar heat capacity at constant volume is given by

$$C_{v,m} = (\partial U/\partial T)_v \tag{5.118}$$

If we assume that the van der Waals equation were obeyed by a simple monatomic liquid such as argon, its molar heat capacity would be $3\mathcal{R}/2$, from (5.21). This value is the same as the ideal gas value, which cannot be correct. In fact, the molar heat capacity for liquid argon is close to $5\mathcal{R}/2$. It is evident that a better equation of state is needed for a liquid.

5.6.2 Radial distribution function

The internal energy of a simple monatomic liquid consists of a kinetic energy, which is $\frac{3}{2}k_BT$ per molecule, and a potential energy $V(r)$, which may be represented satisfactorily by the Lennard-Jones 12–6 equation (5.101), modified by a radial distribution function which we must consider before we can specify fully the internal energy of the liquid.

The radial distribution function characterizes the average structure of a liquid or, more precisely, the average distribution of its molecules relative to one another. For a completely uniform distribution of molecules, the number $N(r)$ of them lying within a spherical shell of radius r and thickness dr is

$$N(r)=4\pi r^2\mathcal{N}\,\mathrm{d}r \qquad (5.119)$$

where $\mathcal{N}$ is the number density[1] of the species.

The radial distribution function $g(r)$ measures the variation with r in the probability of observing one species at the distance r from another and is defined by

$$g(r)=N(r)/(4\pi r^2\mathcal{N})\,\mathrm{d}r \qquad (5.120)$$

More explicitly, consider a system of n atoms and take any one of them as an origin. For a sequence of values of r, count the number $N(r)$ of other atoms whose centres lie within spherical shells of volume $4\pi r^2\,\mathrm{d}r$. This procedure is repeated with each of the other $(n-1)$ atoms as centre, and finally the results for each value of r are averaged. Thus, we obtain a radial distribution function $g(r)\,\mathrm{d}r$ as a function of the parameter r.

The molar internal energy U_m for a simple liquid consists of the kinetic energy and the potential energy, and may be written in the form

$$U_m=\tfrac{3}{2}\mathcal{R}T+(4\pi\mathcal{N}N_A/2)\int_0^{\infty} r^2V(r)g(r)\,\mathrm{d}r \qquad (5.121)$$

where $V(r)$ is a pair potential function, such as (5.101), and the factor $\frac{1}{2}$ arises because each pair of interactions in $V(r)$ must be included only once. The determination of $g(r)$ can be achieved by diffraction studies, with X-ray or neutrons, or by computer simulation techniques. For a more detailed discussion of radial distribution functions, the reader is referred to a more specific treatment of this subject[2].

5.6.2.1 Diffraction studies

Evidence for a degree of order in liquids is provided by diffraction studies. Argon gas does not give a diffraction pattern because there is no characteristic spacing between the atoms.[3] In the liquid, however, diffraction peaks are obtained, thereby indicating order, although not anywhere near the extent that it exists in solids.

A diffraction pattern may be recorded on a photographic film (Figure 5.13) or, more conveniently, as a diffractometer trace. The intensity of scattering $I(S)$ from a liquid is given by the product of the scattering function for the radiation used and the radial distribution function, integrated over all space. Since the integrals over θ and ϕ evaluate to 4π, we have for X-ray diffraction

[1] In the units of m^{-3}.

[2] See, for example, Y Marcus, *Introduction to Liquid State Chemistry* (Wiley, 1977).

[3] Polyatomic gases give diffraction effects arising from the spacings between atoms in the molecule, rather than from intermolecular distances themselves.

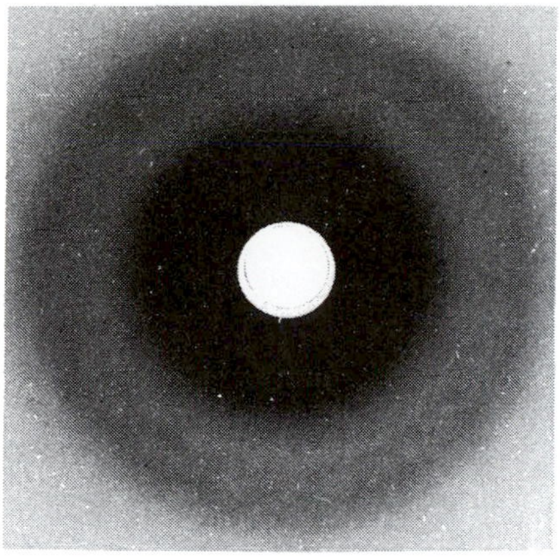

Figure 5.13 X-ray diffraction pattern from a liquid, showing characteristic concentric, diffuse maxima, decreasing in intensity as the angle of scattering increases (outwards). The white central spot arises because of a lead direct-beam trap.

$$I(S)=4\pi C\int_0^\infty r^2 g(r)\sin(Sr)/(Sr)\,\mathrm{d}r \tag{5.122}$$

where C is related to the X-ray atomic scattering factor and S is $4\pi\lambda^{-1}\sin(\theta)$. The Fourier transform of (5.122) may be written as

$$g(r)=[1/(2\pi^2 Cr^2)]\int_0^\infty I(S)S\sin(Sr)/(Sr)\,\mathrm{d}S \tag{5.123}$$

which represents a formal definition of the radial distribution function, and shows how it may be deduced from X-ray scattering data.

The X-ray diffractometer trace for argon gas is shown in Figure 5.14. In the gas there is no order; no particular value of r predominates. Hence, from our discussion $g(r)$ is unity and the number of atoms in spherical shells of radius r will be constant, proportional to the square of the given value of r, because of the uniform distribution within the gas. Thus, we expect the parabolic form shown by Figure 5.14; the proportionality constant is the number density $\mathcal{N}$, equal to N_A/V_m. If the temperature of the gas were increased the parabolic form would remain, but the value of $g(r)$ at any value of r would decrease because the spherical shell would then contain fewer molecules, that is $\mathcal{N}$ would be smaller in magnitude. There is a certain artificial nature about Figure 5.14: at a value of r less than the hard-sphere diameter δ, which is approximately 0.3 for argon, $g(r)$ would in fact tend to zero. Thus, more realistically, the curve should follow the dashed line for $r<\delta$.

With liquid argon, we obtain a trace for $g(r)$ as a function of r typified by Figure 5.15. The radial distribution function for the gas is now modified by the correlation between pairs of atoms arising from the localized order in the liquid. The first peak, at $r \simeq 0.37$ nm, represents nearest neighbours and the area A under the peak is the number of such neighbours:

$$A=4\pi\mathcal{N}_T\int_{0.3}^{r_{\max}} r^2 g(r)\,\mathrm{d}r \tag{5.124}$$

The upper limit $r_{\max}$ is not easy to determine, and an alternative calculation is to take the upper limit in (5.124) as the value of r at the peak and then double the result of the integration. The coordination number may be taken as the mean of the two determinations. The

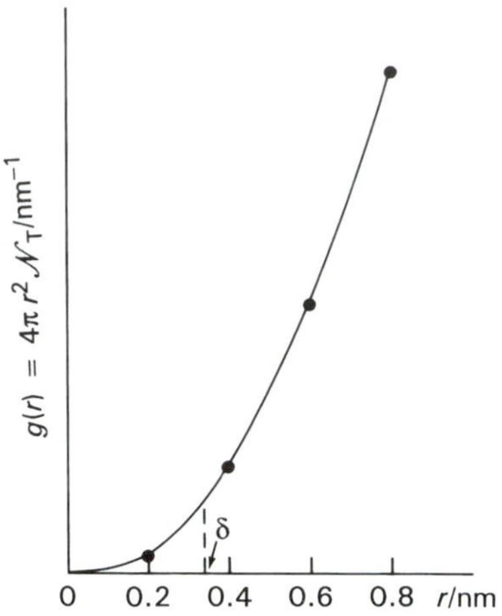

Figure 5.14 Radial distribution function for a gas: the parabolic form arises from the dependence of $g(r)$ solely on volume; realistically, the curve drops to zero at $r = \delta$, the collision diameter, because of strong repulsion forces for $r < \delta$.

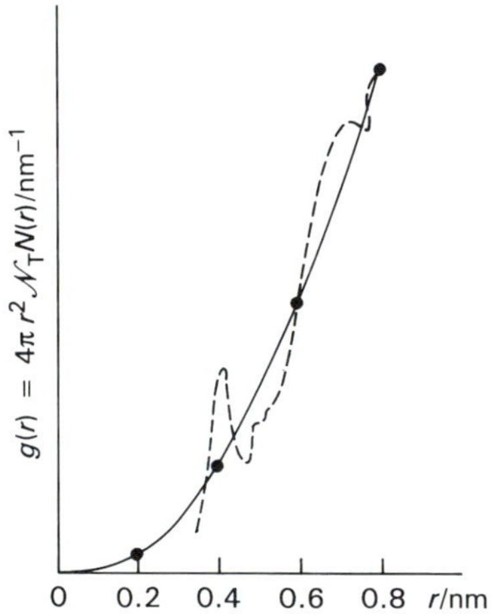

Figure 5.15 X-ray diffractometer trace (dashed line) of the radial distribution function for argon liquid near the triple-point temperature (83 K), superimposed on the curve for a gas (Figure 5.14). The area under the first peak, at $r \simeq 0.37$ nm, corresponds to an average coordination number of 10.4; a second enhanced distribution occurs at $r \simeq 0.70$ nm.

curve of Figure 5.15 tends to the parabolic form of Figure 5.14 as r increases, because the distribution in the liquid becomes uniform at large distances.

The curve shown in Figure 5.15 for the liquid near the triple-point temperature leads to a (first) coordination number of approximately 10.4; as the temperature increases, this value decreases continuously to 4.2. Near the triple-point temperature the high coordination number approaches the maximum (12) for a close-packed solid; the range and nonintegral nature of coordination numbers in liquids are structural features that distinguish them from solids, in which the integral coordination number does not change with temperature, unless a phase change occurs which leads to a different structure type.

The internal energy of a liquid (5.121) is not easy to evaluate because $g(r)$ is not a simple function. In order to show the quantities involved, we shall adopt a simplified procedure in examining argon as both gas and liquid at the critical temperature.

In argon gas at the critical point (see Table 5.1), $\mathcal{N}=N_A/V_{m,c}(g)$ which, from (5.63), is $N_A/(0.0966\times10^{-3}\ m^3\ mol^{-1})$, or $6.23\times10^{27}\ m^{-3}$. Since there is no correlation in the gas $g(r)=1$, and we will assume the Lennard-Jones potential energy function (5.93), with $\epsilon_L=124k_B$ J and $\delta=0.34$ nm. We now have for the configurational energy term of (5.121)

$$2\pi N_A \mathcal{N} \epsilon_L \int_{0.3\,\mathrm{nm}}^{\infty} r^2[(\delta\backslash 2/r)^{1/2}-(\delta/r)^6]\,dr$$

and the integral evaluates to

$$[-\delta^{12}/9r^9+\delta^6/3r^3]_{0.3\,\mathrm{nm}}^{\infty}=-5.60\times10^{-30}\ m^3$$

Introducing the other data from the text leads to a value of -0.23 kJ mol^{-1} for the configurational energy. The kinetic energy, $\frac{3}{2}\mathcal{R}T$, is 1.88 kJ mol^{-1}, so that the total internal energy $U_m(\text{gas})=1.65$ kJ mol^{-1}. We remarked earlier that, in the gaseous state, the kinetic energy is necessarily greater than the configurational energy; here, we have given that statement a more quantitative expression.

In the case of liquid argon, the differences from the calculation for the gas are first that the value for $\mathcal{N}$ is now (from Table 5.1) $N_A/(75.2\times10^{-6}\ m^3\ mol^{-1})$, or $8.01\times10^{27}\ m^{-3}$. Secondly, and more importantly, we need to integrate the complex function in the integrand of (5.121). In practice, it is carried out by simulation procedures. Here, however, we shall use an average value for $g(r)$ over the range 0 to 0.8 nm, from Figure 5.15. Introducing these data leads to values of -2.72 kJ mol^{-1} for the configurational energy in the liquid and -0.84 kJ mol^{-1} for the total energy. Refined calculations have given a value of -2.48 kJ mol^{-1} for the configurational energy at the critical temperature, compared with an experimental value of -2.47 kJ mol^{-1} but the results that we have obtained for the gas and liquid states of argon in the critical state may be deemed satisfactory for the approximations used.

5.6.3 Equation of state for a fluid

The intermolecular potential function and the macroscopic p, V, T properties of a fluid are connected through an equation of state. For a gas we may use the virial equation (5.54) which, if the interactions in the gas are small, may be truncated after the second term. The coefficient $B(T)$ may be related to the pair potential function $V(r)$ through the methods of statistical thermodynamics:

$$B(T)=2\pi N_A\int_0^{\infty}\{1-\exp[-V(r)/(k_BT)]r^2\}\,dr \tag{5.125}$$

For a full treatment of this subject the reader is referred to specialized texts on statistical mechanics.[4]

If the interatomic forces are zero, $B(T)$ and higher virial coefficients are zero, and (5.54) then reduces to the equation of state for the ideal gas. The integral in (5.125) can be evaluated readily for the hard-sphere model (Section 5.5). We have $V(r)=\infty$ for $r\leqslant\delta$, but zero otherwise; thus, (5.125) becomes

[4] See, for example, D Chandler, *Introduction to Statistical Mechanics* (OUP, 1987), and T L Hill, *Introduction to Statistical Thermodynamics* (Addison-Wesley, 1960).

$$B(T)=2\pi N_A \int_0^\delta r^2\,dr \tag{5.126}$$

which, on integration, gives

$$B(T)=2\pi N_A \delta^3/3 \tag{5.127}$$

With a value of δ as 0.34 nm (argon), $B(T)$ evaluates to 5.0×10^{-5} m^3 mol^{-1}, which is comparable to the values already given (page 209). Although the temperature dependence of (5.125) is lost in the simplicity of the hard-sphere model, we gain further evidence that b is related to the hard-sphere repulsion part of the intermolecular potential: if we write $p(V_m-b)\simeq\mathcal{R}T$, then $pV_m/(\mathcal{R}T)\approx1+b/V_m$, which is the equivalent to (5.54) to two terms, with $B(T)=b$.

The equation of state for a liquid is more complex, because the radial distribution function must be invoked to modify the interactional energy term. The statistical mechanics of a simple liquid leads to the equation of state

$$p=\mathcal{N}k_BT-(\mathcal{N}^2/6)\int_0^\infty g(r)\{r[dV(r)/dr]\}\,dr \tag{5.128}$$

or in terms of energy,

$$pV=Nk_BT-(\mathcal{N}N/6)\int_0^\infty g(r)\{r[dV(r)/dr]\}\,dr \tag{5.129}$$

where the second term on the right-hand side corresponds to the configurational energy $\mathscr{E}$ and N is the number of molecules in the volume V; the term $\{r[dV(r)/dr]\}$ is known as the *virial*. Equation (5.129) has been investigated by computer simulation techniques.

Two methods that have received most attention are the *Monte Carlo*[5] method, which leads to equilibrium thermodynamic parameters and the $g(r)$ function, given a form for the potential $V(r)$, and the *molecular dynamics*[6] (time-dependent) method, which provides also measures of transport properties for the liquid. Both methods examine small numbers of particles ($\simeq10^3$), being limited by the time length of computation, but include procedures which endeavour to ensure that the results are representative of a bulk sample. The success of the methods may be judged by calculating parameters, such as the internal energy, that depend upon the potential energy function. We shall look briefly at these two techniques for studying liquids and liquid structure.

5.6.4 Monte Carlo method for liquid structure

A model is set up in which an elementary cubic unit cell containing an initial configuration of approximately 10^3 atoms is repeated by side-by-side stacking in three dimensions, so that the macroscopic liquid is generated by lattice-type translations of the cubic cell. Figure 5.16 is a projection of the model after elapsed simulation time; the cell side a is chosen such that the desired number density, at a temperature T, $\mathcal{N}$ is achieved ($\mathcal{N}=N/V$, where $V=a^3$). The initial configuration $\mathbb{R}$ for N atoms of a monatomic liquid is based, typically, on a face-centred cubic lattice.

[5] See, for example, N A Metropolis, A W Rosenbluth, M N Rosenbluth, A H Teller and E Teller, *J. Chem. Phys.* **21**, 1087 (1953); Y Marcus, *loc. cit*.

[6] See, for example, B J Alder and T W Wainwright, *J. Chem. Phys.* **27**, 1208 (1957); *idem ibid*, **31**, 459 (1959); W W Wood and J D Jacobson, *ibid*, **27**, 1207 (1957); Y Marcus, *loc. cit*.

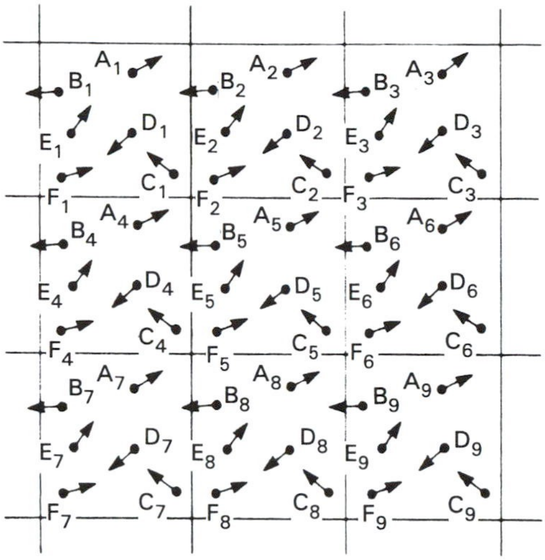

Figure 5.16 Projection of a configuration of atoms in a Monte Carlo simulation of a monatomic liquid. The positions of the atoms in any one cubic cell are random; the other cells are formed by equal translations in three dimensions. After sufficient simulation time, an equilibrium arrangement is obtained. As one molecule, say B_2, moves out of its cell, B_3 moves in from an adjacent cell so as to preserve a constant number density.

The model of Figure 5.16 contains a translational symmetry that is not present in the liquid. It is assumed that, provided the range of the potential energy of interaction between atoms is less than $a/2$, the potential experienced by any given atom is not affected by the symmetry of the model. It is clear that motion of the atoms will take them out of the unit cell. However, as B_1, say, moves out of its cell, the translation image B_2 moves in to take its place and so on. In this way, the atom density is conserved in each cell.

For a configuration of atoms, the potential energy ϵ_i of the ith atom is determined, using a potential function based on the Lennard-Jones 12–6 model. A given atom is then selected and random increments δx, δy and δz applied to its coordinates, and a new energy ϵ'_i calculated. If ϵ'_i is less than or equal to ϵ_i, over all N atoms, then the new configuration is accepted.

If ϵ'_i is greater than ϵ_i, a random number ζ is selected, such that $0<\zeta<1$. If $\zeta<\exp[-(\epsilon'_i-\epsilon_i)/(k_B T)]$, the new configuration is accepted; otherwise the configuration is rejected. This procedure is repeated many times, typically 10^5 to 10^6. The first 10^4 to 10^5 ($\simeq$10%) of the configurations depend strongly on the initial conditions and are discarded in achieving an equilibrium distribution.

The remaining $\mathcal{M}$ configurations are thus generated with relatively low energies that are distributed according to the Boltzmann probability $\Pi(\mathbb{R}^N)$, given by

$$\Pi(\mathbb{R}^N)=\frac{\exp[-\epsilon(\mathbb{R}^N)]/(k_B T)}{\int_{\mathbb{R}^N}\exp[-\epsilon(\mathbb{R})]/(k_B T)} \tag{5.130}$$

Finally, the configurational average $\overline{A}$ of a function of the coordinates $A(\mathbb{R}^N)$ is given by

$$\overline{A}=\frac{1}{\mathcal{M}}\sum_{\mathcal{M}} A(\mathbb{R}^N) \tag{5.131}$$

Table 5.6 Experimental and calculated values for pressure and internal energy for argon

		p/atm		$-U_m$/kJ mol^{-1}	
T/K	$10^6 V_m$/m^3 mol^{-1}	Calc	Expt	Calc	Expt
100.0	29.7	116	115	5.52	5.54
140.0	41.8	18	37	3.81	3.86
150.7	75.2	49	49	2.48	2.47

Some results for the pressure and internal energy of liquid argon are listed in Table 5.6, and it can be seen that the agreements with the corresponding experimental values are good.

5.6.5 Molecular dynamics method for liquid structure

Molecular dynamics treats the evolution with time of systems of particles that interact through conservative forces[7] operating under the laws of classical mechanics. In effect, it tracks the motion of molecules in condensed phases by solving Newton's equations of motion. Since solids are well represented by other theories, the majority of applications of molecular dynamics have been to liquids, particularly in elucidating transport and equilibrium parameters.

In applying molecular dynamics to a simulated liquid system, a set of initial coordinates is generated, usually from a face-centred Bravais lattice, at the required density. Initial momenta configurations $\mathbb{Q}$ can be assigned randomly, such that the system has the desired total energy. Periodic boundary conditions are imposed, as described for the Monte Carlo method.

Many molecular dynamics calculations have been carried out with the hard-sphere potential function. It is computationally simple, and also shows that the structure of simple liquids is almost independent of their chemical nature and may be approximated by the interaction of rigid spheres. This idea was present in earlier work by Bernal with physical models (see Section 5.6.7).

Figures 5.17a and b illustrate the results of molecular dynamics calculations of the liquid interface of argon, using 1500 atom sets. In Figure 5.17a, the atoms are vibrating about their mean positions in the solid state, the sites of a face-centred cubic lattice. Figure 5.17b shows the traces of the atoms, now in a typically liquid phase.[8] Detailed descriptions of various molecular dynamics algorithms have been given in the literature.[9]

5.6.6 Structure of liquid water

Many attempts have been made to simulate the properties of water by both Monte Carlo and molecular dynamics techniques. Care is needed in specifying the pair potential between water

[7] Forces that are derivable from the potential energy function through $F = -\mathrm{d}V/\mathrm{d}r$.

[8] A J C Ladd, *Computer Simulations of Liquids and Liquid Interfaces*, Ph D Thesis, Cambridge (1977).

[9] See, for example, M P Allen and D J Tildesley, *Computer Simulation of Liquids* (OUP, 1996) and references therein; J P Hansen and I R McDonald, *Theory of Simple Liquids* (Academic Press, 1986); G Ciccotti and W G Hoover (editors), *Molecular Dynamics Simulation of Statistical-Mechanical Systems* (North-Holland, 1986); G Ciccotti, D Frenkel and I R McDonald, *Simulation of Liquids and Solids* (North-Holland, 1987).

(a)

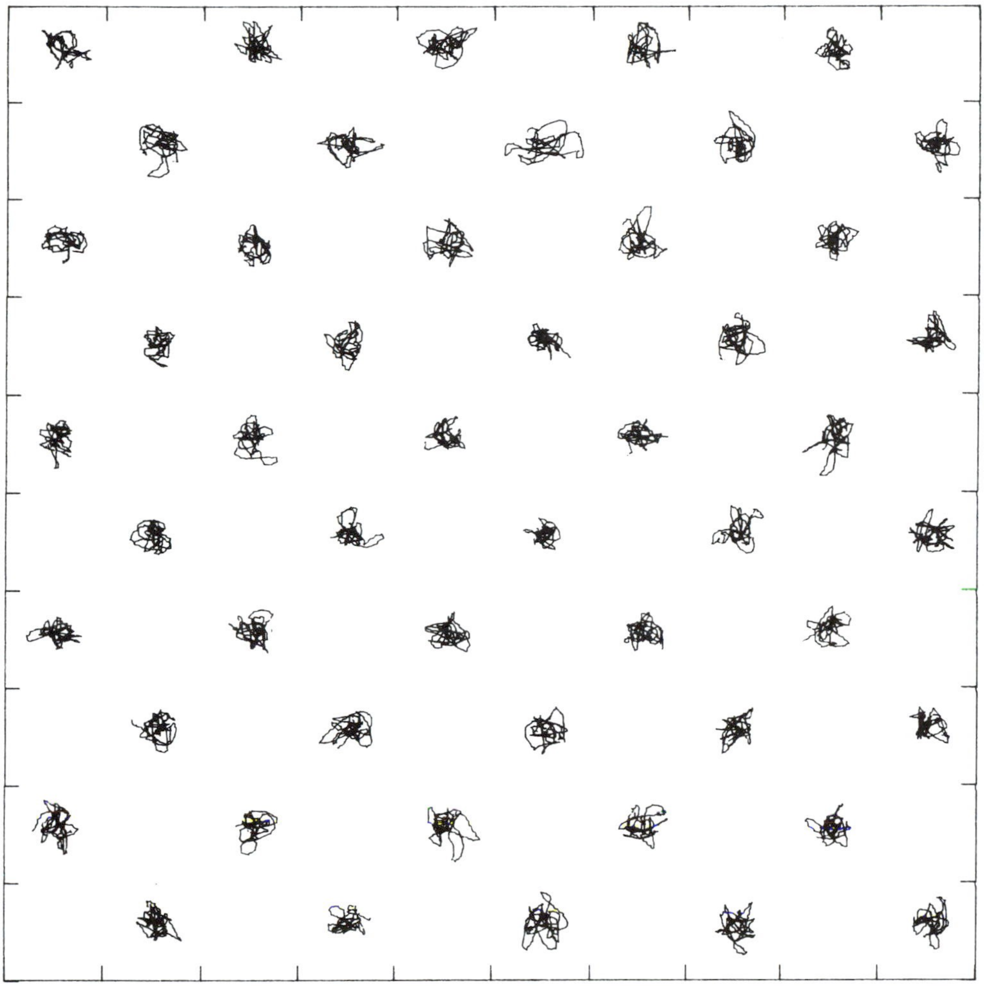

Figure 5.17 Models of argon from computer simulation solutions of Newton's equations of motion: (a) face-centred cubic (projection) solid state, showing clear evidence of the regular initial configuration; (b) equilibrium liquid state showing randomness, with only localized regions of order. (Reproduced with permission from A J C Ladd, *Computer Simulations of Liquids and Liquid Interfaces,* Ph D Thesis, Cambridge 1977.)

molecules, because of the relatively long range of the dipolar interactions.[10] Table 5.7 lists the configurational energy $\mathscr{E}$, the second term on the right-hand side of (5.129), an equation of state function and the constant-volume heat capacity, and compares the results with those from both molecular dynamics and experimental results.[11]

The structural properties of water were addressed by computing the radial distribution functions for O–O, O–H and H–H by sampling the pair distributions after every 250 configura-

[10] See M P Allen and D J Tildesley, *loc. cit.*

[11] F H Stillinger and A Rahman, *J. Chem. Phys.* **60**, 1545 (1974), MD – Tables 5.7 and 5.8 reference (a); A J C Ladd, *Molec. Phys.* **33**, 1039 (1977), MC – Tables 5.7 and 5.8 reference (b).

(b)

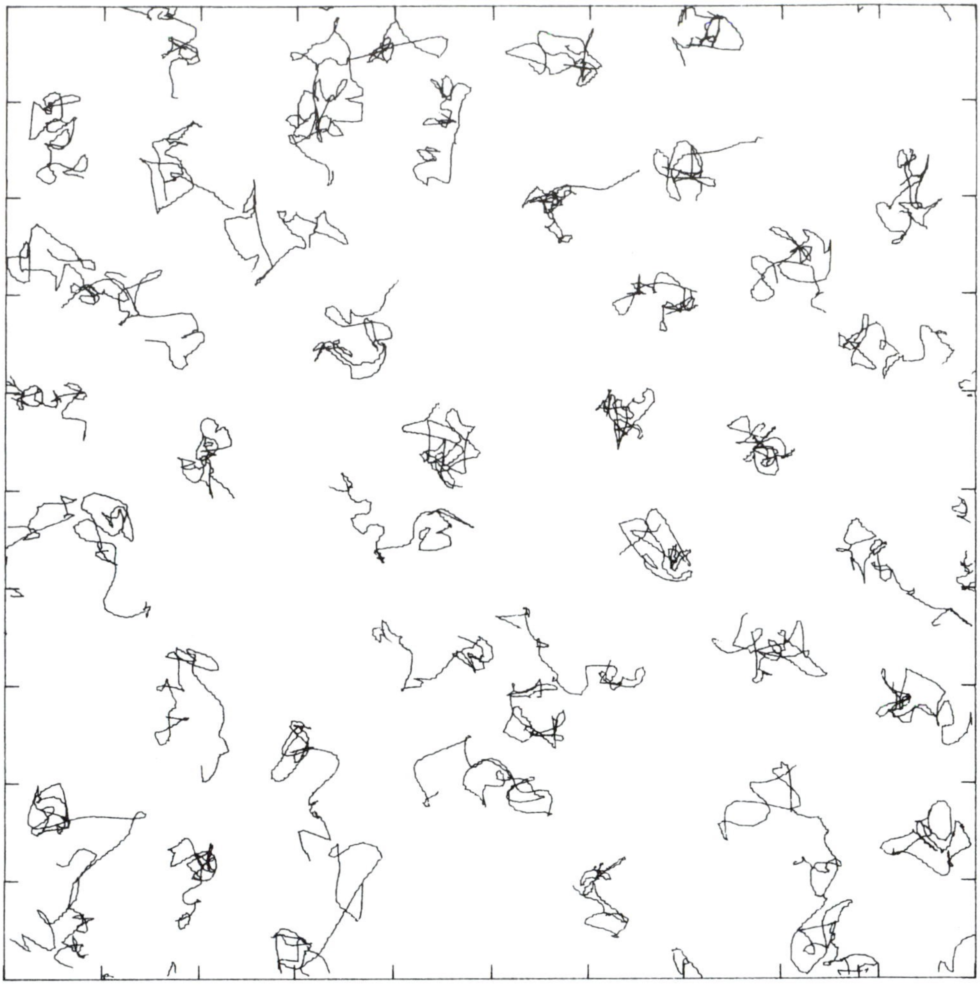

Figure 5.17 (*cont.*)

tions. Table 5.8 lists results for the positions and heights of the maxima for a 256-molecule system, and compares them with both molecular dynamics and experimental results.

The results of this simulation work can be seen to be very satisfactory in its representation of thermodynamic and structural properties of liquid water. Water has been subjected to many subsequent simulation calculations of this nature,[12] and the above general conclusions confirmed and refined.

5.6.7 Use of physical models

The configuration obtained by the simulation of a monatomic liquid shows that it is packed randomly. A hard-sphere model can be simulated by packing steel balls into an irregular-shaped rubber ball. If the centres of the balls in the model are joined by straight lines which

[12] See, for example, P Barnes, J L Finney, J D Nicholas and J E Quinn, *Nature*, **282**, 459 (1979); K Watanabe and M L Klein, *J. Chem. Phys.* **131**, 157 (1989); M P Allen and P J Tildesley, *loc. cit.*

Table 5.7 Thermodynamic properties of water by computer simulation

System	$\mathscr{E}$/kJ mol^{-1}	$pV/(\mathscr{R}T)$	C_v/J K^{-1} mol^{-1}
216[(a)]	−43.1	0.05	100
256[(b)]	−39.9±0.3	0.6±0.3	70
Experiment	−41.4	0.05	75

Table 5.8 Structural properties of water by computer simulation:, positions r and heights M of maxima in radial distribution functions

		r_1 /nm	M_1	r_2 /nm	M_2
g_{O-O}	216[(a)]	0.285	3.09	0.470	1.13
	256[(b)]	0.285	3.11	0.530	1.06
	Experiment	0.283	2.31	0.425	1.08
g_{O-H}	216[(a)]	0.190	1.38	0.340	1.60
	256[(b)]	0.191	1.24	0.332	1.53
	Experiment	0.190	0.80	0.335	1.70
g_{H-H}	216[(a)]	0.250	1.50	0.390	1.20
	256[(b)]	0.250	1.15	0.375	1.07
	Experiment	0.235	1.04	0.400	1.08

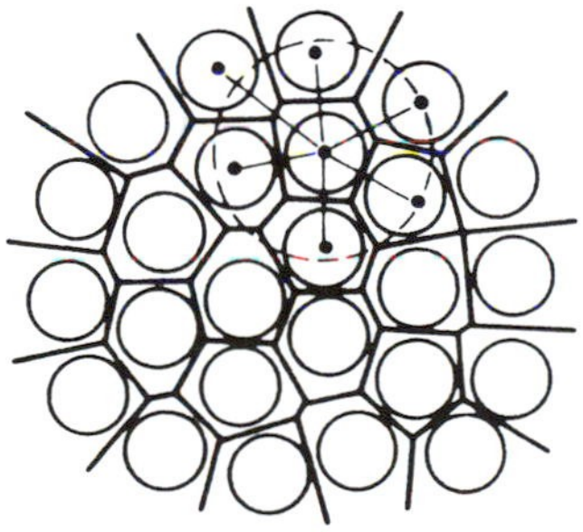

Figure 5.18 Voronoi polyhedra in projection (also known as Dirichlet polygons) around randomly packed atoms in a monatomic liquid; average coordination number 5.5.

are then bisected by perpendicular planes, the resulting polyhedra are Voronoi polyhedra (Figure 5.18), which fill space completely. Voronoi polyhedra can be simulated if plasticine balls are used in place of steel; individual balls from the practical simulation then have the shape of Voronoi polyhedra. The theoretical average numbers of faces and coordination number are 5.1 and 13.6 respectively, and the models show corresponding average values of 5 and 13.

The closest regular packing of equal spheres, as in many metals, has a coordination number of 12, with a volume per sphere of 1.351 (1/0.74). The close packing of irregular polyhedra leads to a volume per sphere of approximately 1.351×13.6/12, or 1.53, an increase of about 13% on regular close packing.

A model due to Bernal (1959) of a randomly packed liquid, or *heap,* is shown in Figure 5.19a; it may be contrasted with the regular array of identical spheres, or *pile* (Figure 5.19b), for which the coordination number is 12.

(a)

(b)

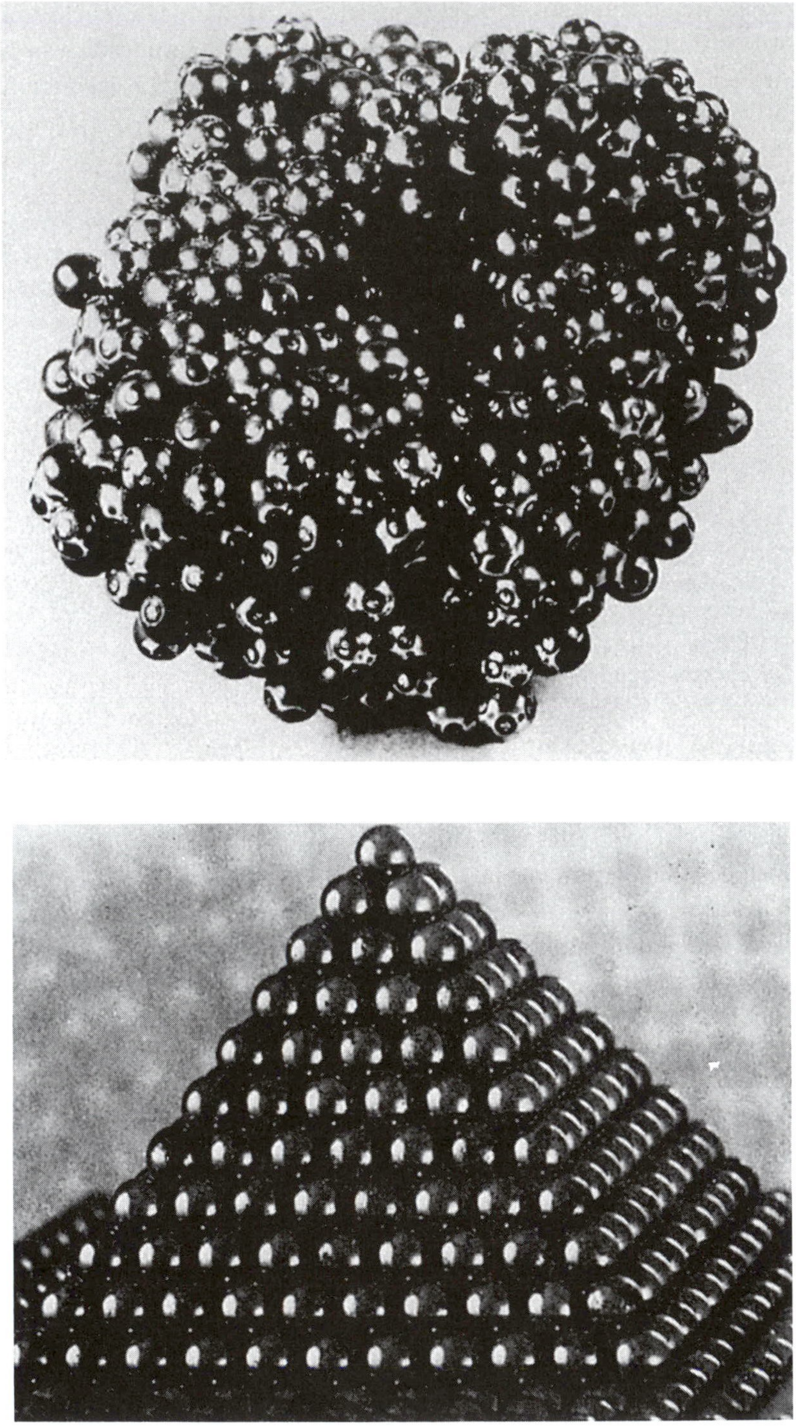

Figure 5.19 Stacking of identical spheres: (a) random close-packed *heap*, of average coordination number 5.5; (b) regular close-packed *pile* of coordination number 12. (After J D Bernal, in *Liquids: Structure, Properties and Solid Interactions*, edited by T J Hughel, (Elsevier, 1965).)

Problems 5

5.1 Given that, for any gas, $pV = \frac{1}{3}N m \overline{v^2}$, show how (a) Avogadro's, (b) Boyle's, (c) Charles' and (d) Graham's laws follow.

5.2 If the collision cross-section σ for argon at 298 K is 0.34 nm^2, calculate (a) the pressure at which the mean free path of argon is equal to its collision diameter, and (b) the number of collisions per second when the pressure is 5 atm.

5.3 The collision diameter of nitrogen gas is 0.34 nm at 298 K and 1 atm pressure. Assuming that the gas behaves ideally, calculate the mean free path for nitrogen at an altitude at which the temperature is 195 K and the pressure 0.045 atm.

5.4 Show that, for a gas that obeys the kinetic theory, the ratio of speeds $v_{max} : \overline{v} : (\overline{v^2})^{1/2}$ is $1 : 1.128 : 1.225$.

5.5 The probability that a molecule in a gas has an energy ϵ is given by the Boltzmann equation $\exp[-\epsilon/(2k_BT)]$. By invoking a normalization criterion, show that this equation can lead to the Maxwell–Boltzmann one-dimensional distribution equation.

5.6 Metallic potassium is heated at 900 K in an oven containing a minute hole in one wall. The atomic collision diameter is 0.45 nm, and at 900 K the vapour pressure of potassium is 115 mmHg. Calculate (a) the mean velocity of the emergent potassium atoms, (b) the mean speed of potassium atoms inside the oven, (c) the frequency of collisions made by a single potassium atom, (d) the frequency of collision for all the atoms in the oven, and (e) the mean free path of potassium atoms in the oven.

5.7 Convert the van der Waals equation of state into a virial equation in powers of $1/V$. Hence, obtain a value for the second virial coefficient $B(T)$ in terms of the van der Waals parameters a and b. Give an expression for the Boyle temperature in terms of a and b, and find its value for carbon dioxide. Comment on the result in the light of Table 5.1.

5.8 Determine the critical constants for oxygen, given that the van der Waals constants for this gas are $a = 1.36$ dm^6 atm mol^{-2} and $b = 3.18 \times 10^{-2}$ dm^3 mol^{-1}. Hence, evaluate a molecular radius for oxygen.

5.9 (a) The dipole moment for fluorobenzene is 1.70 D. What are the dipole moments for the three difluorobenzenes; which of the answers is the most certain, and why?

(b) The dipole moment for hydrogen fluoride is 1.83 D, and the H–F bond length is 0.0927 nm. Determine the effective point charges at the hydrogen and fluorine ends of the dipole.

5.10 Show that the equilibrium separation of atoms in a gas that is governed by a Lennard-Jones (12–6) potential is given by $r_e = 2^{1/6}\delta$. Assuming that r_e is 0.37 $\pm$0.02 nm, what is the value of δ?

5.11 The isothermal compressibility κ is given by $\kappa = -(1/V)(\partial V/\partial p)_T$. Assuming that liquid argon follows the van der Waals' equation of state, determine an expression for κ in terms of the van der Waals constants a and b. Find a value for κ at the critical temperature.

5.12 The following data were obtained for the radial distribution function of a simple liquid (the datum ar r =0.25 nm is an extrapolation):

$10^{-9}g(r)$/nm^{-1}	0	20	90	70	40	45	60	80	110	160
r/nm	0.25	0.30	0.35	0.40	0.45	0.50	0.55	0.60	0.65	0.70
							150	140	160	210
							0.75	0.80	0.85	0.90

Plot the radial distribution function, and determine the average number of nearest neighbours (the coordination number) and the average corresponding nonbonded distance. How may one explain that $g(r)$ tends to unity as r tends to ∞?

5.13 The following data have been obtained for ethane and ethene. At each pressure p, calculate by writing a program, or otherwise, each of the parameters p_r, pV_m, $\mathscr{Z}_m$, $\mathscr{Z}_m$ (virial equation) and pV_m (van der Waals' equation). Calculate $\mathscr{Z}_{m,c}$ for each gas. Plot $\mathscr{Z}_m$ against p for each gas, and plot, *on one graph*, $\mathscr{Z}_m$ against p_r for both gases. Comment on the total results.

	Ethane, C_2H_6	Ethene, C_2H_4
T/K	458.3	424.1
a/dm^6 mol^{-2}	5.49	4.47
b/dm^3 mol^{-1}	0.0638	0.0571
p_c/atm	48.50	49.98
V_c/dm^3 mol^{-1}	0.1417	0.1276
T_c/K	305.5	282.7
$B(T)$/dm^3 mol^{-1}	−0.0686	−0.0598
$C(T)$/dm^6 mol^{-1}	0.00575	0.00467
p/atm	V_m/dm^3 mol^{-1}	V_m/dm^3 mol^{-1}
1	37.59	34.90
25	1.4318	1.3220
50	0.6787	0.6263
100	0.3063	0.2841
150	0.1943	0.1796
200	0.1496	0.1378
250	0.1281	0.1167
300	0.1155	0.1050
350	0.1073	0.0973
400	0.0970	0.0919
500	0.0936	0.0860

5.14 Use the virial equation, to the third term, to determine expressions for $B(T)$ and $C(T)$ in terms of the van der Waals constants a and b. Hence, determine $B(T)$ and $C(T)$ for the gases in Problem 5.13, and note any discrepancy with the experimental values given.

CHAPTER 6

States of matter: solids

6.1 Introduction

A solid has a definite volume and shape, neither of which changes appreciably with changes in temperature or pressure. A study of the solid state is principally a study of crystalline materials, since almost all solids form crystals. Certain solids are described as amorphous: in some cases they are microcrystalline, but other examples, such as glass or many polymers, do not have the regularity in structure that is associated with true crystals. The term amorphous is best restricted to those solids in which order extends over only a few atomic dimensions. In crystals, the atoms or molecules are arranged on, or in a fixed relation to, the points of a Bravais lattice, where they vibrate about their mean positions. The mean positions are, normally, invariant with time, and the vibrational energy is a major contributory factor to the heat capacity of the solid.

The atomic vibrations are anharmonic (see also Section 3.6.5.1), so that an increase in temperature causes an increase in the distance between the mean positions of the atoms and the material expands. Even if the vibrations were harmonic, the increase in free energy with temperature would lead to an increase in volume,[1] although the expansion from this source is, normally, a second-order effect.

The time-invariance of mean atomic positions may be invalidated by disorder. In some solids, atoms may exhibit free rotation (dynamic disorder) in the solid state, or they may show static disorder. For example, sodium cyanide at room temperature has the sodium chloride structure type (Figure 1.2), whereas below 6 °C it transforms to the orthorhombic structure of lower symmetry shown in Figure 6.1. In the high-temperature form, the random orientations of the linear cyanide ion, averaged over many unit cells, simulate statistically the envelope of a sphere, and the higher symmetry of the cube is observed. The pseudo-spherical cyanide ions replace the Cl^- ions in the sodium chloride structure.

6.2 Amorphous solids

Solids such as glass, plastics and resins are described as amorphous. The amorphous condition of matter is a metastable phase, that is, it is not the thermodynamically lowest energy structure for the given substance. Glass can change extremely slowly from its normal, amorphous condition: ancient glass may be found in a devitrified condition, that is, some of its components have crystallized into stable structures. The slowness of the process implies a very high energy barrier ϵ_A to devitrification. Normal cooling of a liquid produces the crystalline state, which is the thermodynamically stable state.[2] If the liquid is cooled very quickly

[1] See, for example, E A Guggenheim, *Boltzmann's Distribution Law*, (North-Holland, 1963).

[2] There may be more than one crystalline state of a solid (polymorphs), but one of them will be of lower energy than the others at a given temperature.

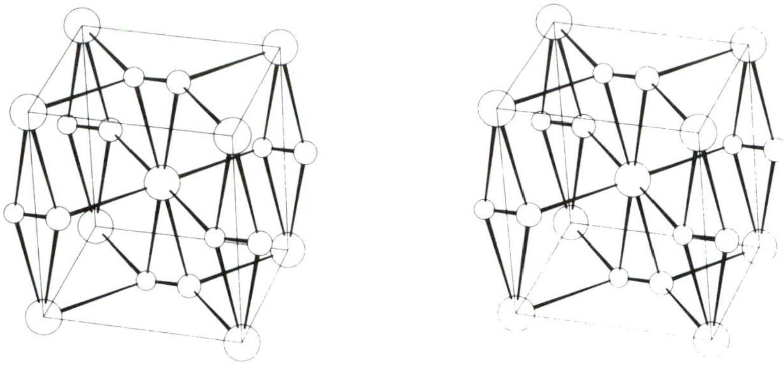

Figure 6.1 Stereoview of the unit cell and environs of the orthorhombic structure (below 279 K) of NaCN; circles in decreasing order of size represent Na^+, N and C (together CN^-). How is this structure related to the high-temperature cubic form (see Figure 1.2)?

and a solid is formed, the liquid structure may be locked in by the process, thus producing an amorphous solid (or supercooled liquid). The probability of the transition from the metastable state to the stable state is proportional to the Boltzmann factor $\exp[-\epsilon_A/(k_B T)]$. If $k_B T$ is small with respect to ϵ_A the probability factor is very small, and the rate of change may be then so slow as to be experimentally unobservable.

The criterion of crystallinity is the appearance of a discrete (spot) X-ray diffraction pattern (Figure 3.35). Amorphous solids give X-ray patterns very similar to those of liquids (see Figure 5.13), for which reason they may be considered as supercooled liquids. An amorphous solid may be characterized by a radial distribution function, rather like a liquid, and statistical information on the average coordination number and average nearest neighbour distances may be determined from it.

Figure 6.2 is an X-ray photograph of a drawn ethene–propadiene polymer fibre of about 70% crystallinity. The photograph shows evidence of a repeat distance, in the (vertical) fibre direction, superimposed on to the diffuse ring pattern that is characteristic of the amorphous state.

Glass is, perhaps, the most important amorphous solid. Figure 6.3a illustrates a model for silica glass, which may be compared with the structure of α-quartz in Figure 6.3b. Evidence for the basic structural unit SiO_4 in silica glass, which exists also in quartz and silicate minerals, has been obtained through a study of its radial distribution function.

6.3 Molecular solids

Crystalline solids are characterized by their three-dimensional regularity. Every crystalline solid has a Bravais lattice (see Section 3.5.1) as its geometrical basis, and a crystal is built up by the regular arrangement of its units of structure, atoms or molecules, in fixed vector relationships at or around each lattice point. It is the regularity of a crystal structure that enables a unit cell to be selected, representative of the whole structure. The macroscopic crystal may then be obtained, assuming no disorder, by a side-by-side stacking and sharing of these unit cells in three-dimensional space.

Figure 6.4 is an example of the molecular solid pentaerythritol, in which the molecules $C(CH_2OH)_4$ are disposed about the origin (0,0,0) and the centre $(\frac{1}{2},\frac{1}{2},\frac{1}{2})$ of a body-centred tetragonal unit cell. The atoms at the extremities of the unit cell, faces, edges and corners,

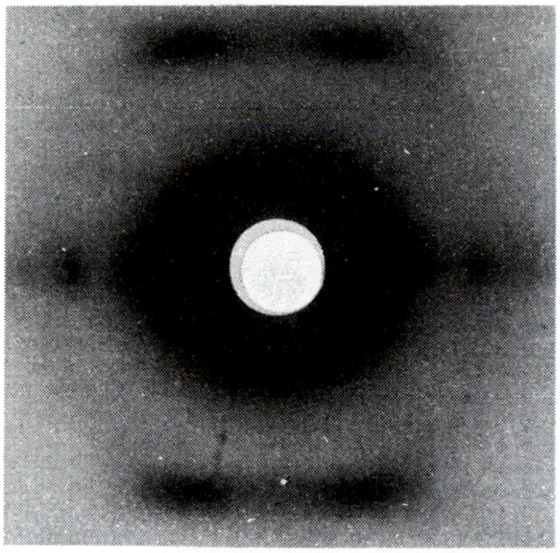

Figure 6.2 X-ray diffraction photograph of a drawn ethene–propadiene polymer fibre at 70% crystallinity. The spot pattern which is just discernible indicates a repeat period in the vertical (fibre) direction and is superimposed upon the diffuse ring pattern that is characteristic of a liquid (see Figure 5.13).

are shared among the immediate neighbouring unit cells. Intermolecular attraction arises through the permanent dipole of the molecule, the London forces and hydrogen-bonding; the shortest O–H...O hydrogen bond distance is 0.27 nm.

In this example, we need to give the coordinates of one molecule and the space group ($I\bar{4}$) in order to build up the complete crystal structure. The geometry and symmetry of crystals has been discussed in Section 3.5 and, with more detail, in other standard works on this subject.[3]

Molecular crystals contain discrete molecules, or in the case of the noble gases just individual atoms, that are stable in their own right. They are held together by van der Waals' forces, the nature of which we have discussed (see Section 5.4 *et seq.*). If the molecules are nonpolar the van der Waals energies are pairwise additive, and the energy of the crystal can be represented by the Lennard-Jones potential (5.101) or, better, by an equation of the type

$$V(r) = A\exp(-Br) - C/r^6 \tag{6.1}$$

where A and B are constants of the repulsive energy and C is a constant of the attractive potential energy which includes species-dependent parameters, such as are given in (5.98). The exponential function is in evidence in overlap integrals (Section 2.11.3) and so has a theoretical basis, whereas the r^{-12} function in (5.101) is empirical, as discussed in Section 5.5.

The repulsive potential is important in determining the packing in molecular crystals. Both the attractive and repulsive energies in solids include lattice sums that take into account the three-dimensional repeating nature of the unit cell and its contents. Results for benzene, for example, have given a cohesive (lattice) energy of -53 kJ mol^{-1} at 270 K. This value may be compared with the enthalpy of sublimation for benzene at the same temperature, which is 44 kJ mol^{-1}.

[3] See, for example, M F C Ladd, *Symmetry in Molecules and Crystals* (Horwood, 1992); *International Tables for Crystallography*, Volume A, edited T Hahn (Reidel, 1983) – formerly *International Tables for X-ray Crystallography*, Volume I, edited N F M Henry and K Lonsdale (Kynoch Press, 1965), which may be found to be simpler.

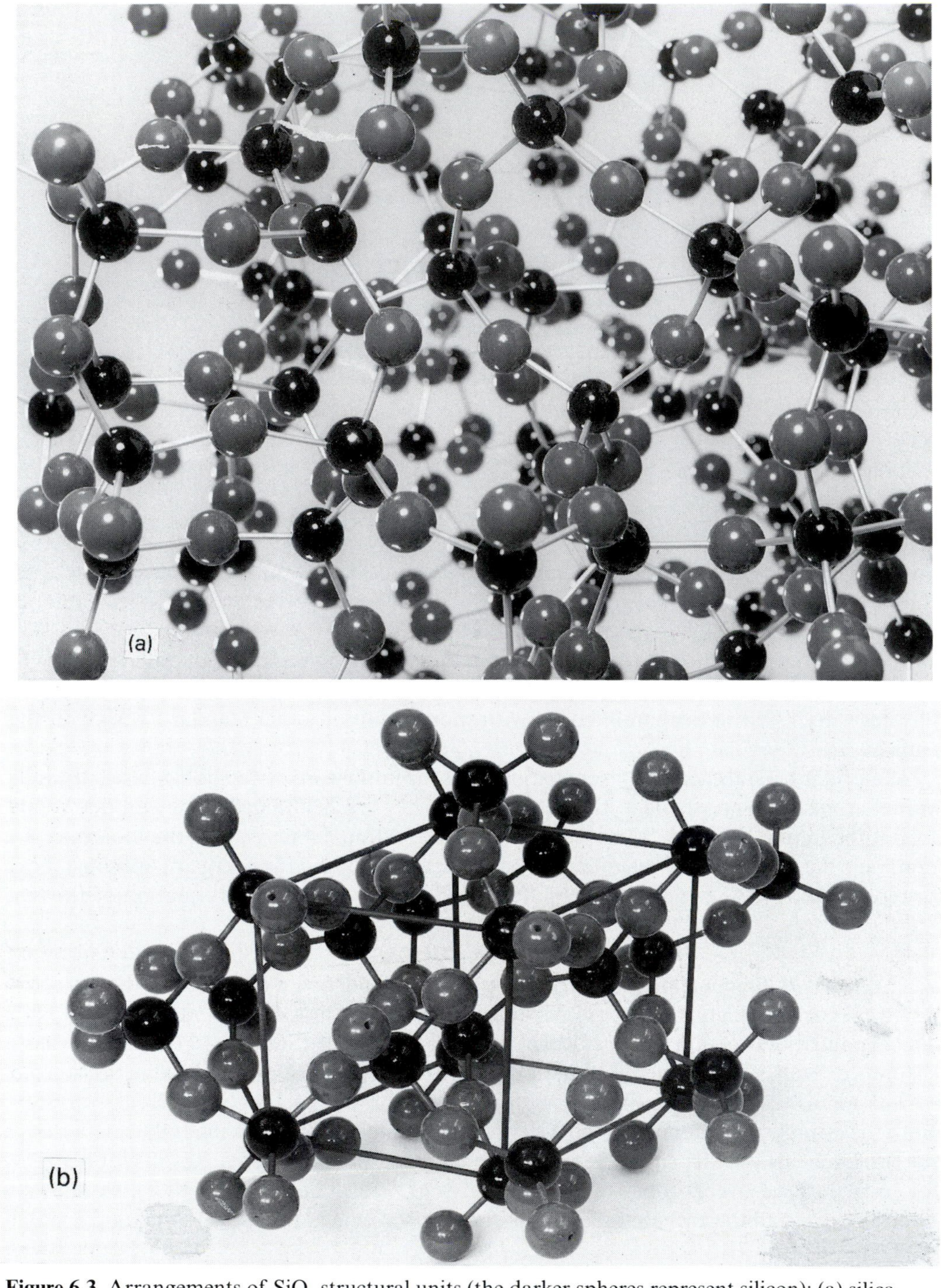

Figure 6.3 Arrangements of SiO_4 structural units (the darker spheres represent silicon): (a) silica glass; (b) α-quartz. The long-range regularity in quartz is absent in silica glass. (Crown copyright. Reproduced from NPL Mathematics Report Ma62 by R J Bell and P Dean, with the permission of the Director, National Physical Laboratory, Teddington.)

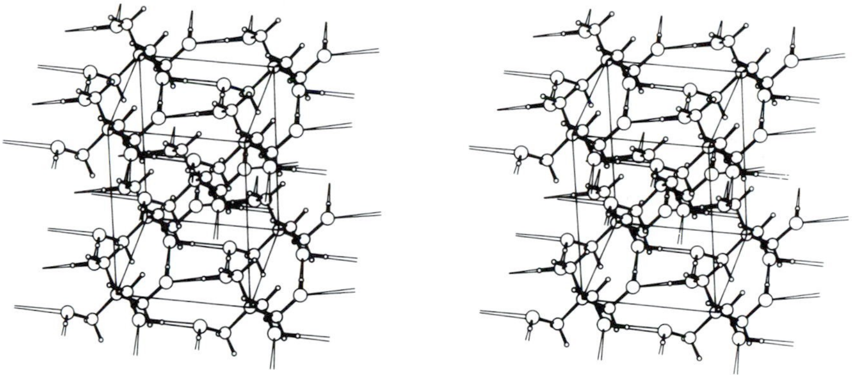

Figure 6.4 Stereoview of the unit cell and environs of the structure of pentaerythritol $C(CH_2OH)_4$; circles in decreasing order of size represent O, C and H. The double lines represent O–H...O hydrogen bonds of medium strength that assist in linking the molecules throughout the structure. The molecules at each corner of the unit cell are shared in the structure with another seven adjacent unit cells.

6.3.1 Packing of molecules

In the condensed state, chemical species pack together in order to make the most efficient use of the space available, which leads to a minimum in the cohesive energy. This principle was formulated first by Barlow in 1895 with respect to simple close-packed structures, such as metals and the alkali-metal halides, and it has been extended to other solids. Many organic molecules are irregular in shape, so that in order to achieve the maximum filling of space, the protrusion of a given molecule fits into a recess of an adjacent molecule, a sort of 'lock and key' mechanism. Thus, a particular molecular shape has a preferred mode of packing. Figures 6.5a and 6.5b illustrate good space-filling and less economical modes of packing with molecules of a given shape. This aspect of molecular organic crystals has been discussed at length by Kitaigorodskii.[4]

6.3.2 Classification of solids

The classification of solids is not, and cannot be, rigid. The basis of the classification adopted in this book is determined by the forces responsible for cohesion in the solid state. For example, although the forces between atoms in the molecule of anthracene are mainly covalent, the cohesion in the solid state arises through van der Waals' forces, and the properties of solid anthracene are determined by them. Thus, it is classified here as a molecular compound.

6.3.2.1 Noble gases

The noble gases except helium crystallize with the face-centred cubic structure (Figures 3.13c, and 6.16a later). The coordination number is 12, representative of the closest packing, with a packing efficiency of 74%. Helium is equally closely packed in the solid state but is hexagonal (see Figure 6.16b later), except for a small region of its phase diagram within which it is body-centred cubic with a coordination number of 8.

[4] See, for example, A I Kitaigorodskii, *Organic Chemical Crystallography* (Consultants Bureau, 1957).

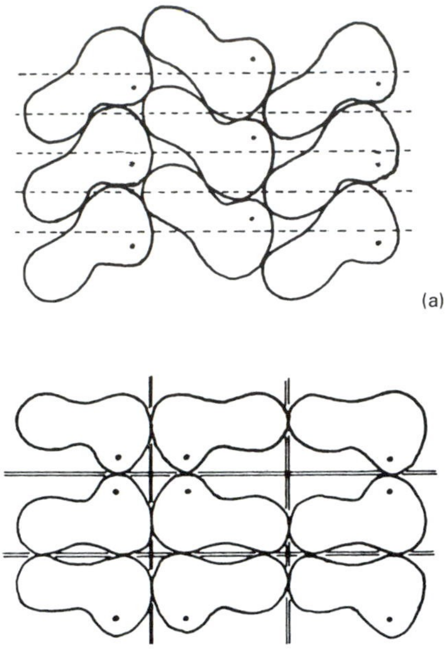

Figure 6.5 Packing of organic molecules: (a) economical packing, with the protrusion of any one molecule fitting into a recess in an adjacent molecule; (b) uneconomical packing, leading to an increase in the percentage of void space relative to (a).

The noble gases are nonpolar and cohesion arises through van der Waals' induced dipole–induced dipole energy (5.97). The weak nature of these forces is shown by the melting-point of the solids, 83 K for argon. At this temperature, the value of the translational energy, $\frac{3}{2}\mathcal{R}T$, is 1.035 kJ mol^{-1}. Taking ϵ/k_B from Table 5.5, the energy ϵ at the equilibrium interatomic distance is approximately -1.032 kJ mol^{-1}. Thus, the kinetic energy and configurational energy are balanced at this temperature, and hence a solid can exist. The temperature of 83 K is, in fact, the triple-point temperature for argon, the temperature at which gaseous, liquid and solid phases of this substance coexist in equilibrium.

6.3.2.2 Nonmetallic elements

Among the structures of the nonmetallic elements, there is a tendency for a species in periodic group n to form $18 - n$ covalent bonds with itself and its compounds.[5] However, although this rule may be explained in terms of the outermost electron configuration, it is only approximate; in group 14 it fails with lead and tin, and it has little applicability in group 13.

Nevertheless, we find that the noble gases are monatomic and that the elements in periodic groups 15 to 17 form molecular crystals. The diatomic molecules of the halogens are linked by London forces, with nonbonded distances in the range 0.33 nm to 0.35 nm (Figure 6.6). Fluorine has a close-packed cubic structure (see Figure 6.16a), because the molecules are in free rotation and attain time-averaged spherical envelopes of motion.

In group 16, the rule is obeyed through the formation of chains or rings; both structures

[5] This 'rule' was given originally as $8-n$, in relation to the earlier numbering of the periodic groups.

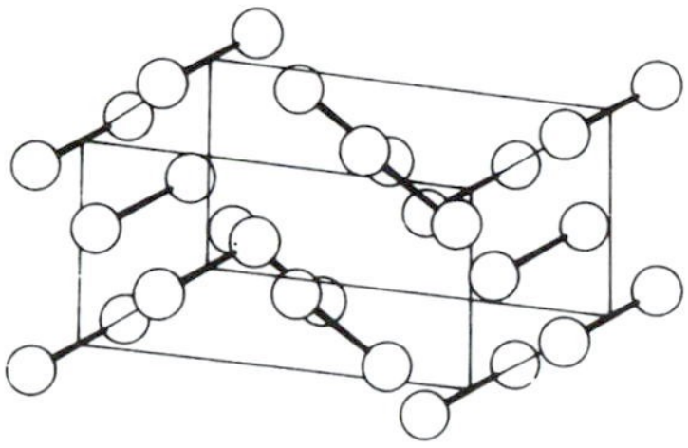

Figure 6.6 Stereoview of the unit cell and environs of the structure of iodine (shown), chlorine and bromine; each atom is linked to one other atom.

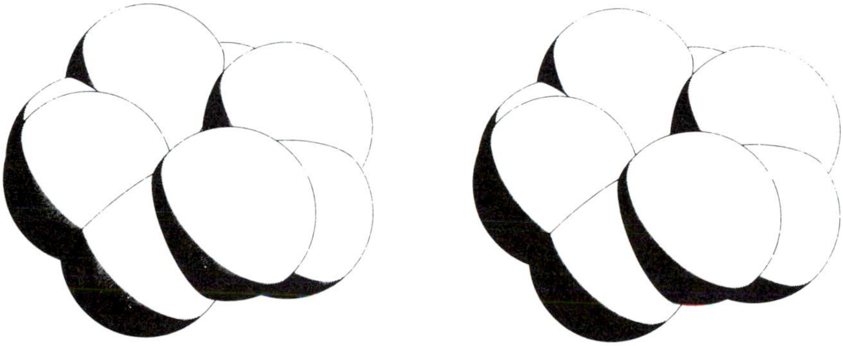

Figure 6.7 Stereoview of a space-filling model of the S_8 molecule of sulfur; each atom is linked to two other atoms by ring formation.

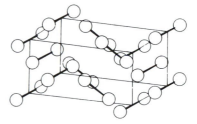

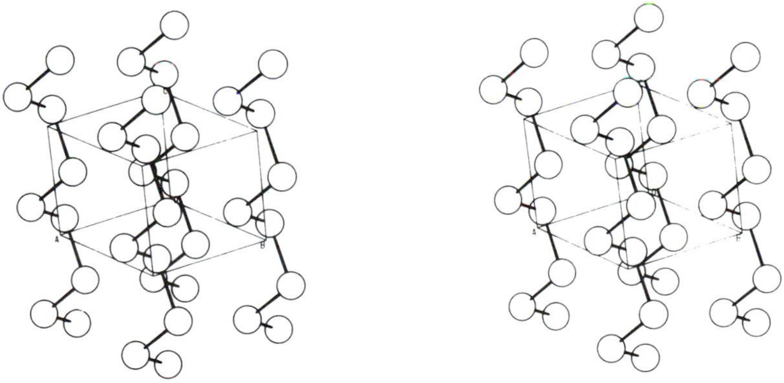

Figure 6.8 Stereoview of the unit cell and environs of the structure of selenium (shown) and tellurium; each atom is linked to two other atoms in infinite chains.

are known for sulfur, with the crown-shaped S_8 molecule[6] (Figure 6.7) occurring in the thermodynamically stable form at room temperature. Exceptions to the 18–*n* rule are not uncommon, and include the $[SiF_6]^{2-}$ anion (see Section 2.15).

Selenium and tellurium form infinite chains, as shown by Figure 6.8. Covalent bonds exist between the atoms in any chain, and the chains are linked by van der Waals' forces.

The group 15 elements arsenic, antimony and bismuth form puckered sheets, with each atom bonded to three others. The bonding between atoms in a sheet takes place through a

[6] The S_8 molecule possesses an S_8 roto-reflection symmetry axis.

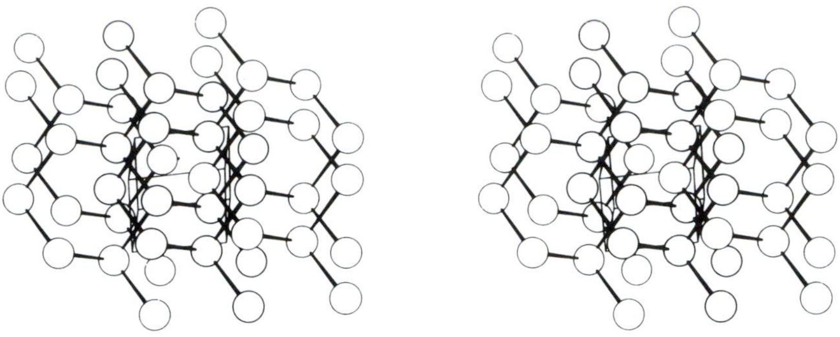

Figure 6.9 Stereoview of the unit cell and environs of the structure of arsenic (shown), antimony and bismuth; the bonding of any one atom to three others forms sheet structures with these elements.

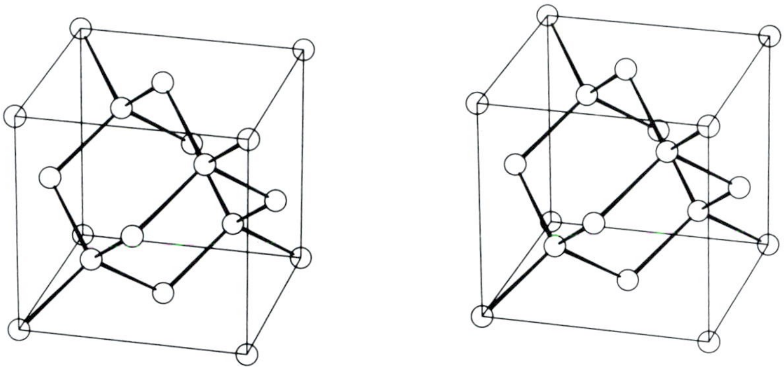

Figure 6.10 Stereoview of the unit cell and environs of the diamond structure of carbon. It is an almost wholly covalent solid, with the C−C single bonds forming a three-dimensional network. Each C_4 tetrahedral unit shares its corners with four other, adjacent tetrahedra.

hybrid of p and s orbitals, leading to bond angles greater than 90° (97° in the case of arsenic, slightly less for antimony and bismuth). The distances between the layers increase down the periodic group from 0.32 nm to 0.35 nm. The atoms have three close neighbours and three others at a greater distance, forming a distorted octahedral arrangement of nearest neighbours (Figure 6.9).

Group 14 elements tend to form covalent solids, with a tetrahedral arrangement of covalent bonds, as in the diamond structure of carbon (Figure 6.10), silicon, germanium and grey tin. However, white tin (Figure 6.11) and lead exhibit metallic properties.

Carbon exists in several forms including graphite (Figure 6.12). The bonds within the planar layers of the graphite structure arise through sp^2 hybrid orbitals, similar to the bonding in benzene, but the p orbitals normal to the plane of the ring overlap to form π molecular orbitals that are delocalized over entire layers. It is because of this structural effect of conjugated π bonds that graphite is such a good conductor of electricity along directions in the layers; normal to the layers, the electrical conductivity is very small. The σ bonds within the ring have a length of 0.142 nm, slightly longer than those in benzene, and the distance between the planes is approximately 0.34 nm.

In these structures the tendency of molecules to pack closely can be seen. In the halogens, the molecules are oriented so that one atom of a given molecule is opposite to the centre of

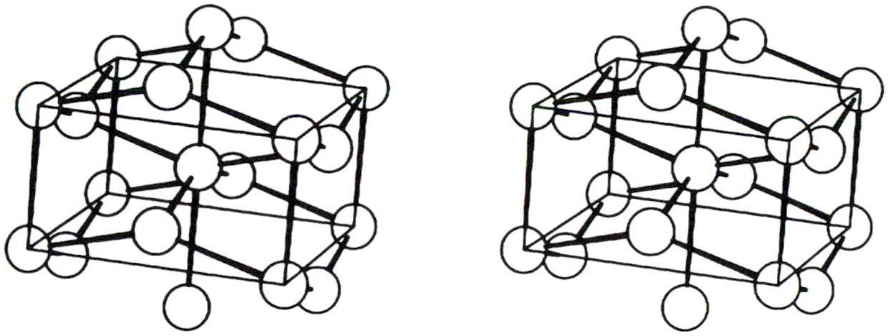

Figure 6.11 Stereoview of the unit cell and environs of the structure of white tin (β-Sn). White tin is metallic; each atom has four nearest neighbours and two others at a slightly larger distance. The transition in bond type to a true metal is completed in group 14 at lead. Grey tin (α-Sn) is intermediate between a metallic and a covalent solid, having the four-coordinate structure type shown by diamond (Figure 6.10), but exhibiting good electrical conduction (Section 6.2.3).

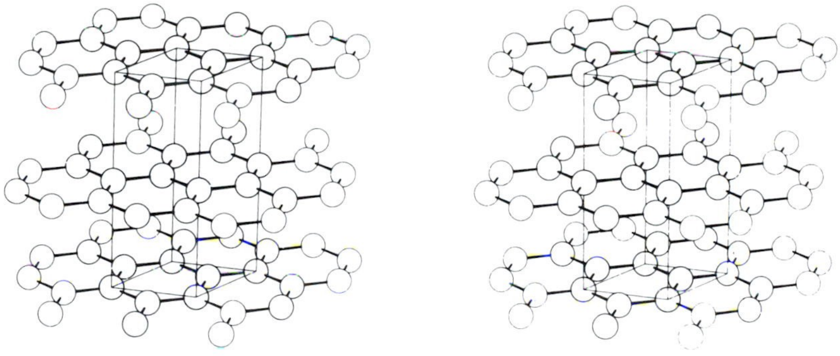

Figure 6.12 Stereoview of the unit cell and environs of the graphite structure of carbon. The π electrons are delocalized over the σ-bonded layers (C−C=0.140 nm) and this feature is responsible for the high electrical conductivity in the layer direction. The relative weakness of the van der Waals forces between the layers leads to the lubricant properties of graphite. The distance between the layers is 0.34 nm.

the bond of an adjacent molecule, so that a plane of molecules, in zig-zag formation, packs closely over an adjacent plane. In the helical chain structures, we see that the 'elbows' of one chain fit into recesses in an adjacent chain. In the ring structures, the centres of rings in one layer tend to lie above and below atoms in adjacent layers. All these modes of packing tend to minimize voids, thereby leading to good space-filling structures.

6.3.2.3 Small inorganic molecules

There are many inorganic solids that contain discrete molecules, linked by van der Waals' forces. Examples of these compounds are $HgCl_2$, SnI_4, SF_6, solid CO and so on. The structure of $HgCl_2$, for example, is that of linear molecules packed in a manner very similar to that shown by iodine. The coordination may be described as 6 : 3, reflecting the formula ratio of 2 to 1, which means that there are six nearest Cl neighbours for each Hg, and three nearest Hg neighbours for each Cl. Solid carbon monoxide has a close-packed hexagonal structure, like helium.

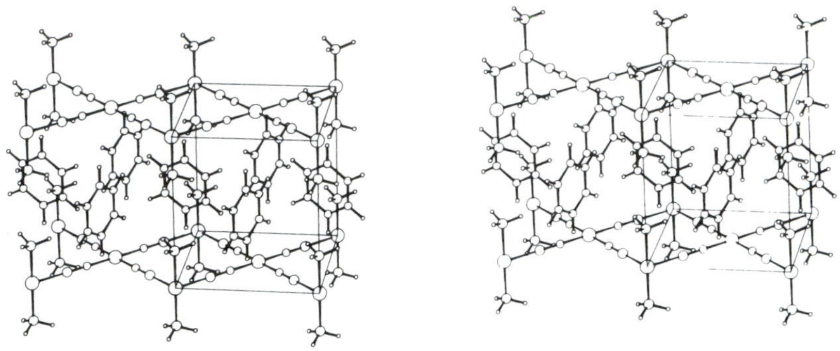

Figure 6.13 Stereoview of the unit cell and environs of the clathrate compound of nickel(II) cyanide, ammonia and benzene; circles in decreasing order of size are Ni, N, C and H. The benzene ring is trapped mechanically in the $Ni(CN)_2(NH_3)_2$ cage structure; the shortest contact distance $C_{(benzene)}-C_{(CN\ group)}$ is 0.36 nm.

The linear molecules of CO are in free rotation in the solid and simulate a close packing of spheres, thus attaining the higher observed symmetry. As liquid carbon monoxide is cooled to form a solid, the CO molecules will not all lie in the same orientation. Hence, a residual entropy arises, as discussed in Section 4.6.4.1.

6.3.2.4 Organic compounds

Under this heading, we include hydrogen-bonded compounds, clathrate compounds, charge-transfer compounds and π-electron overlap compounds, as well as organic compounds generally.

Clathrate compounds exist only in the solid state and may exhibit variable composition. There is often only a very small bonding interaction between the components, the host structure acting as a mechanical trap for the occluded molecule. A well-known example of a clathrate compound is illustrated in Figure 6.13. The Ni^{2+} cations are coordinated octahedrally to two NH_3 molecules and to four CN^- anions, which themselves form a square-planar arrangement. The groups link as shown in Figure 6.13 to form a cage that traps a benzene molecule. Its formula is $Ni(CN)_2(NH_3)_2C_6H_6$, and crystallization of this compound can be used to obtain benzene at a purity in excess of 99.995%.

A *charge-transfer* compound is exemplified by benzene–halogen complexes of the type $C_6H_6X_2$, where X_2 is a molecule of chlorine, bromine or iodine. The typical structure is shown for the chlorine complex in Figure 6.14. The Cl_2 molecule is oriented normal to the benzene ring plane, and the shortest distance from a Cl atom to the centre of the ring is 0.33 nm. The halogens behave as Lewis acids in these compounds, with the π electrons of benzene acting as donors to σ^* molecular orbitals on the halogens. The movement of electrons is called a charge-transfer transition and is revealed through the ultraviolet absorption spectra of these compounds.

Electron overlap is important among aromatic hydrocarbons, such as anthracene (see Figure 6.15). The molecular planes are oriented such that the delocalized electrons in the π orbitals of adjacent molecules can overlap, leading to enhanced stability. Table 6.1 shows an interesting relationship between melting-point T_m and relative molar mass M_r. Normally, an increase in molar mass leads to an increased melting-point. In aromatic compounds, however, π-electron overlap stabilizes them relative to their alicyclic analogues.

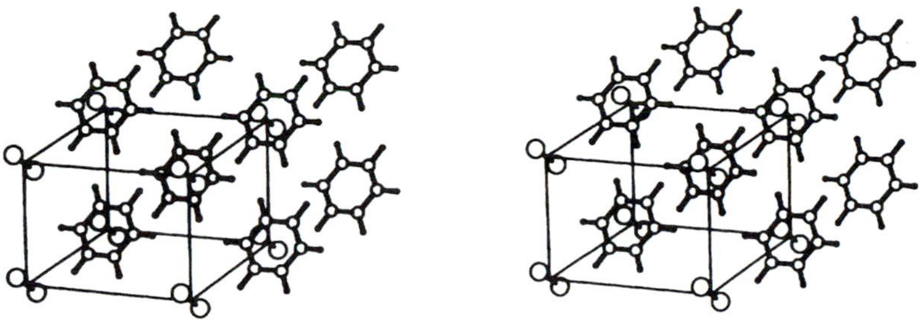

Figure 6.14 Stereoview of the unit cell and environs of the charge-transfer compound between benzene and chlorine (Cl_2) molecules in equimolar ratio; circles in decreasing order of size are Cl, C and H. The shortest contact distance between a Cl atom and the ring centre is 0.33 nm. Similar structures are formed between benzene and molecular bromine or iodine.

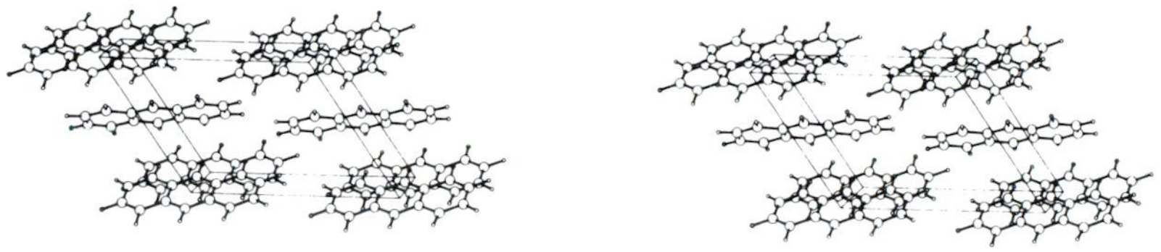

Figure 6.15 Stereoview of the unit cell and environs of the structure of anthracene $C_{14}H_{10}$; circles in decreasing order of size are C and H. The shortest intermolecular contact distance is 0.36 nm.

6.3.2.5 Standard values for bond lengths and angles

Among the numerous organic compounds that are now known, we can identify certain structural units that preserve their characteristic shapes, sizes and symmetries within quite small limits; they include bond lengths, bond angles, structural groups and even complete molecular entities. We have discussed some of the reasons for this constancy in Chapter 2. Table 6.2 lists standard values for a range of bond lengths and bond angles. The values vary with the environment, and Table 6.2 shows this variation by means of the notation C4, C3, N2, O1, and so on. The digit indicates the number of atoms directly bonded, the *connectivity*; thus, C4−C4 indicates a C(sp^3)−C(sp^3) bond.

As an example of a larger structural unit, consider the phenyl moiety: it is usually found to be planar, within experimental error, with C−C bonds of 0.140 nm, C−H bonds of 0.108 nm and C−C−C and H−C−C angles all close to 120°. According to the nature and position of substituents on the ring, there may be small deviations from these standard values, generally more in the bond lengths than in the bond angles.

It has been found experimentally that intermolecular nonbonded distances between pairs of atoms do not vary greatly. In the absence of hydrogen-bonding, carbon, nitrogen and oxygen atoms exhibit average nonbonded, or contact, distances of approximately 0.37 nm among a wide range of compounds. This feature led to the development of van der Waals' radii for atoms, which represent an average minimum distance of approach of atoms in neighbouring molecules. It is, perhaps, not surprising that this value is close to the δ parameter of the Lennard-Jones 12–6 potential function (5.101) for these species.

Table 6.3 lists the van der Waals radii for a number of common elements. The van der Waals radius of an atom can be correlated with the size of its outer orbital. Thus, a carbon

Table 6.1 Melting-points and relative molecular masses for some aromatic hydrocarbons and their fully hydrogenated counterparts

		T_m/K	M_r
Benzene	C_6H_6	279	78.1
Cyclohexane	C_6H_{12}	280	84.2
Naphthalene	$C_{10}H_8$	353	128.2
Decahydronaphthalene	$C_{10}H_{18}$	230, 241[a]	138.3
Anthracene	$C_{14}H_{10}$	490	178.2
Tetradecahydroanthracene	$C_{12}H_{24}$	335, 366[a]	192.4

Note:
[a] Polymorphs

2p orbital enclosing 99% of the 2p electron density extends from the nucleus to approximately 0.19 nm, which is very close to the van der Waals radius for this species.

6.3.2.6 Structural and physical characteristics of molecular compounds

Two atoms in neighbouring molecules may lie further apart than the sum of their van der Waals' radii, because steric effects may prevent the normal closest approach distances from being realized. In other compounds, nonbonded distances may be significantly less than the sum of the van der Waals radii. Thus, the distance between two nonbonded oxygen atoms may be as small as 0.24 nm if strong hydrogen-bonding exists between them. We may note that, for those atoms that can exist as clearly defined ions, the van der Waals radius is very close to the corresponding ionic radius. This result is in accord with the fact that repulsive forces increase very rapidly at short distances, because of their ($1/r^{12}$)-dependence. For example, the effective size of the bromine species is about the same whether the repulsion energy is balanced against the Coulombic $1/r$ potential function in KBr or against the relatively weaker $1/r^6$ attractive energy in C_6H_5Br.

Van der Waals' forces can bond an atom to an indefinite number of neighbours and they are spatially undirected. In the solid noble gases, van der Waals' interactions are the sole means of cohesion. In other molecular compounds, relatively short covalent bonds may exist between atoms, such as Cl—Cl (0.200 nm) and C—C (0.154 nm), but with characteristically longer nonbonded, or contact, distances between adjacent molecular entities.

The dependence of the van der Waals energy on polarizability is shown clearly by the trends in the melting-points of the silicon tetrahalides, SiX_4 (Table 6.4). In these compounds, the polarizability of the halogen increases more rapidly from fluorine to iodine than do the corresponding intermolecular distances. Thus, repulsion is progressively relatively less strong with a consequent enhancement of the lattice energy (enthalpy of sublimation), compared with a similar range of molar mass among the alkanes, for which the increase in melting-point is only about 140 °C.

Molecular compounds generally form soft, brittle crystals, with low melting-points and large thermal expansivities. The electrical and optical properties of molecular solids may be said to be the aggregate of those of the component molecules, since the electron systems of the molecules do not interact strongly in the solid state. These properties are, therefore, similar in the solid, melt and solution, whereas in the solid, anisotropy of physical properties is generally marked.

Table 6.2 Standard bond lengths and bond angles

Single bond	Length/nm	Single bond	Length/nm
H−H	0.074	C3−C2	0.145
C4−H	0.109	C3−N3	0.140
C3−H	0.108	C3−N2	0.140
C2−H	0.106	C3−O2	0.136
N3−H	0.101	C2−C2	0.138
N2−H	0.099	C2−N3	0.133
O2−H	0.096	C2−N2	0.133
C4−C4	0.154	C2−O2	0.136
C4−C3	0.152	N3−N3	0.145
C4−C2	0.146	N3−N2	0.145
C4−N3	0.147	N3−O2	0.136
C4−N2	0.147	N2−N2	0.145
C4−O2	0.143	N2−O2	0.141
C3−C3	0.146	O2−O2	0.148
Double bond	**Length/nm**	**Double bond**	**Length/nm**
C3−C3	0.134	C2−O1	0.116
C3−C2	0.131	N3−O1	0.124
C3−N2	0.132	N2−N2	0.125
C3−O1	0.122	N2−O1	0.122
C2−C2	0.128	O1−O1	0.121
C2−N2	0.132		
Triple bond	**Length/nm**	**Aromatic bond**	**Length/nm**
C2−C2	0.120	C3−C3	0.140
C2−N1	0.116	C2−N2	0.134
N1−N1	0.110	N2−N2	0.135
Apex atom	**Geometry**	**Angle/deg**	**Example**
C4	Tetrahedral	109.5	CH_4
C3	Planar	120	C_2H_4
C2	Bent	109.5	−CHO
	Linear	180	HCN
N4	Tetrahedral	109.5	NH_4^+
N3	Pyramidal	107.5	NH_3
	Planar	120	H_2N−CHO
N2	Bent	109.5	H_2CHN
	Linear	180	HNC
O3	Pyramidal	109.5	H_3O^+
	Bent	104.4	H_2O

Table 6.3 Van der Waals' radii of some common species

Species	Radius/nm	Species	Radius/nm
H	0.120	C	0.185
$-CH_3$	0.200	Si	0.210
N	0.150	P	0.190
As	0.200	Sb	0.220
O	0.140	S	0.185
Se	0.200	Te	0.220
F	0.135	Cl	0.180
Br	0.19	I	0.215
Half-thickness of phenyl ring[7]			0.185

Table 6.4 Melting-points, polarizabilities and van der Waals' radii sums for the silicon tetrahalides

	SiF_4	$SiCl_4$	$SiBr_4$	SiI_4
Melting point/K	183	203	278	394
$10^{40}\alpha$/F m^2	1.0	3.4	4.8	7.3
Σr/nm	0.35	0.39	0.41	0.43

6.4 Covalent solids

Within the form of classification described in Section 6.3.2 there are very few solids that will be termed covalent. The best example is diamond, Figure 6.10, in which each atom is bonded covalently to four others through sp^3 hybrid orbitals. Other elements of group 14 have the diamond structure, but tin is dimorphic (Figure 6.11). A continuous change in bond type from covalent to metallic is exemplified well by this group, and the change is paralleled in the electrical resistivities ρ of these elements:

	C (diamond)	Si	Ge	Sn(grey)	Sn(white)	Pb
ρ/Ω m	5×10^{12}	2×10^{3}	0.5	1×10^{-5}	1×10^{-7}	2×10^{-7}

Other essentially covalent solids include silicon carbide SiC, boron nitride $(BN)_n$, which is harder than diamond, and quartz (Figure 6.3b). The two common forms of zinc sulfide, blende and würtzite, although four-coordinated, have a high degree of ionic character and will be considered with ionic solids.

The covalent bond is strongly directional, and exists between an atom and a small number of neighbours. Covalent solids form strong, hard crystals with low compressibilities and expansivities, but with high melting-points. In these properties they are similar to ionic crystals, but they differ markedly from them by being electrical insulators in the solid and liquid states. Covalent solids are of low chemical reactivity and insoluble in all usual solvents.

[7] The closest approach distance of the planes of two phenyl rings would be expected to approximate to 0.37 nm.

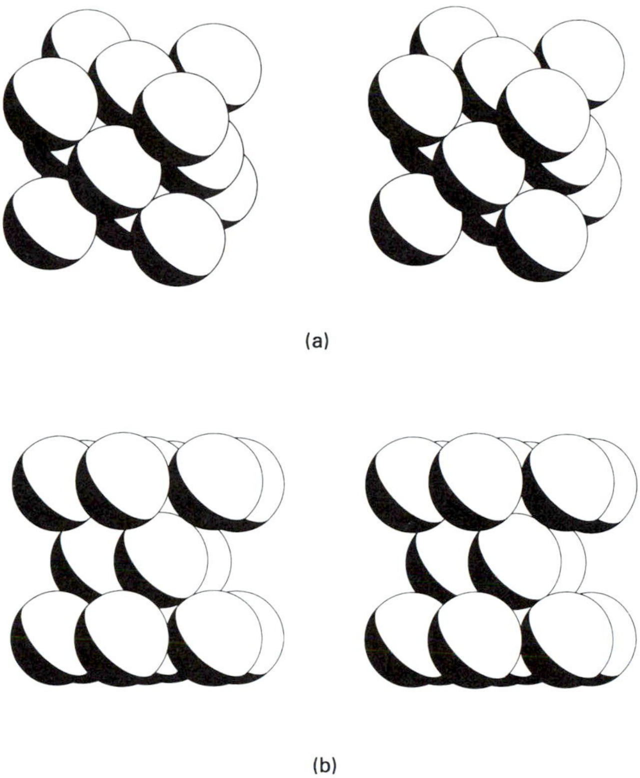

Figure 6.16 Closest packing of identical spheres: stereoviews of the unit cell and environs of (a) close-packed cubic CPC, also called CCP, FCC and A1, and (b) close-packed hexagonal CPH, also called HCP and A3. The coordination number is 12 and the packing efficiency 74% in each structure.

6.5 Metals

The metallic bond does not possess directional character, so that the structures adopted by metals are determined to a large extent by space-filling criteria. The metals of periodic groups 1, 2 and 13, together with the transition-type metals and certain others,[8] form relatively simple structures: the close-packed (face-centred) cubic (A1), the body-centred cubic (A2) and the close-packed (primitive) hexagonal (A3) arrangements. The close-packed structure types A1 and A3 represent the two modes of regular, closest packing of identical spheres (Figures 6.16a and b).

In both A1 and A3, a first layer is obtained by placing spheres in contact such that their centres form the apices of equilateral triangles (Figure 6.17a). A second similar layer is added such that the spheres of that layer rest in the depressions of the first layer (Figure 6.17b). A third layer can then be added in one of two ways. If it is arranged such that the spheres in the third layer lie above voids in *both* the first *and* second layers, the close-packed cubic (CPC) structure A1 is obtained (Figure 6.17c). If, however, the spheres of the third layer lie directly above those of the first layer, the close-packed hexagonal (CPH) structure A3 is obtained (Figure 6.17d). The sequence of close-packed layers in the two arrangements is the following:

[8] At approximately 3000 K and 10^6 atm, measurements of electrical resistivity on liquid hydrogen (and deuterium) reveal metallic character for these substances; the species remain in the diatomic molecular state.

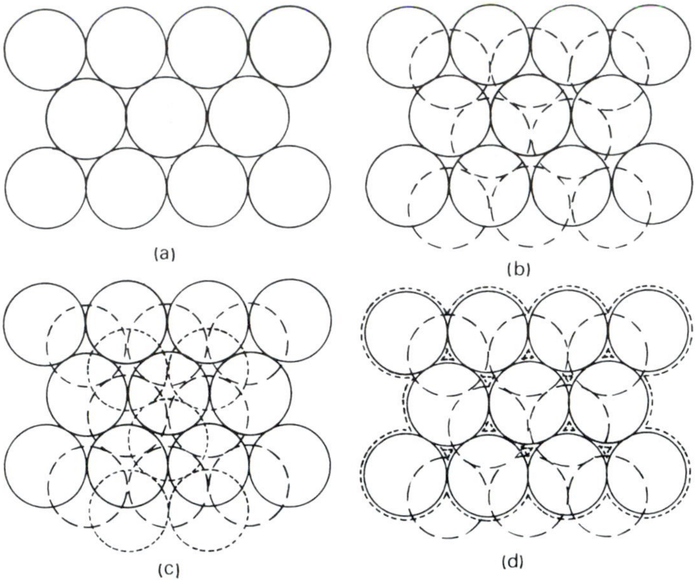

Figure 6.17 Closest packing of identical spheres. (a) A first layer A, full lines; the centres of the spheres form a succession of equilateral triangles. (b) A second layer B, long-dashed lines; there is only one way in which a second layer can be close-packed on to the first. (c) CPC, with the layer sequence ABCABCA. . .; the layer C is shown by short-dashed lines. (d) CPH, with the layer sequence ABABA. . .; again, the layer C is shown by short-dashed lines, but the C-layer circles have been drawn slightly larger, for clarity.

Type	Close-packed planes	Sequence
A1	(111) planes	1 2 3 1 2 3 1 2 ...
A2	(0001) planes	1 2 1 2 1 2 1 2 ...

In these two structure types, each sphere is in contact with 12 other spheres, the maximum coordination for identical spheres in regular packing in crystals. In the CPC structure, the volume of the unit cell is a^3, where a is the length of the side of the unit cell. The face diagonal is $a\surd 2$ and we have

$$a\surd 2=4r \tag{6.2}$$

where r is the radius of the sphere, because the spheres are in contact along that diagonal (Figure 6.16a). Hence the unit cell volume can be written as $(4r/\surd 2)^3$, and the volume *occupied* per sphere is $(4r/\surd 2)^3/4$, or $5.66r^3$. The actual volume of a sphere is $\frac{4}{3}\pi r^3$; hence, the fraction of space occupied, or the packing efficiency, is 0.74. The same value is obtained for the CPH structure type (Figure 6.16b).

In the body-centred cubic (BCC) structure type A2, the coordination number is eight; it is a less closely packed structure type. Reference to Figure 6.18 shows that the spheres are in contact along a body diagonal. Hence,

$$a\surd 3=4r \tag{6.3}$$

from which the packing efficiency may be shown to be 0.68.

Figure 6.18 Stereoview of the unit cell and environs of the body-centred cubic packing BCC, also called A2; the coordination number is 8 and the packing efficiency is 68%.

6.5.1 Metallic radii

We can develop the concept of a metallic radius r from equations (6.2) and (6.3), since a is an experimentally measurable parameter. Table 6.5 lists the radii for a selection of metals in structures of twelvefold cordination.

Certain metals exist in polymorphic modifications. From a study of their structures, the following empirical relationship between metallic radius and coordination number has been evolved, taking coordination number 12 as a reference standard:

Coordination number	12	8	6	4
Relative radius	1.00	0.97	0.96	0.88

6.5.2 Metallic bonding

Metals are distinguished from other substances by several physical properties, among which their high electrical and thermal conductivities, and their opacity to visible light are well known. About three-quarters of the elements are metals, yet their structure types are few in number and geometrically fairly simple, as we have seen.

The first theory of metals, the *Drude–Lorentz* (free-electron) theory, considered that electrons in a metal are of two kinds: those in closed inner shells, the core electrons belonging to a lattice of positive ions in the metal; and those free to move, the *valence* or *conduction* electrons that permeate the metal as a whole. The conduction electrons are influenced by an applied electrical or thermal field.

Thus, we have the model of a lattice of cations, with their core electrons intact, surrounded by a 'sea' of conduction electrons, or *electron gas*. The cohesive energy of a metal was considered to depend upon the attraction between the positive ions and the electron gas, without considering any interaction involving the electrons of the core.

The theory predated wave mechanics, and provided a satisfactory model with respect to certain properties. The high electrical conductivity of metals was explained in terms of the drifting of valence electrons under an applied potential gradient; their thermal conductivity was considered to arise from the redistribution of electrons in a thermal field, carrying thermal energy with them.

Table 6.5 Metallic radii r/nm

Li	0.152	Be	0.112	Cu	0.128
Na	0.186	Mg	0.160	Ag	0.144
K	0.227	Ca	0.197	Au	0.144
Rb	0.248	Sr	0.215	Fe	0.124
Cs	0.265	Ba	0.222	Co	0.125
Fr	0.293	Ra	0.229	Ni	0.125

6.5.3 Heat capacity

The atoms in a metal crystal may be considered to be oscillating harmonically, each about its mean position, with a common frequency ν. Each atom may then be treated as a simple harmonic oscillator, and its total energy is the sum of kinetic and potential energy components (2.1) and (2.58):

$$E=\tfrac{1}{2}mv^2+2\pi^2\nu^2mx^2 \tag{6.4}$$

where m is the mass of the vibrating atom, and v and x are, respectively, its speed and linear displacement from the equilibrium position, at any instant. The average thermal energy $\overline{E}$ for a classical oscillator is k_BT (Appendix 3). Generalizing to N oscillators with three mutually perpendicular directions of vibration, the mean total energy becomes

$$\overline{E}=3Nk_BT \tag{6.5}$$

For N equal to N_A, the average molar energy becomes

$$\overline{E}_m=3\mathcal{R}T \tag{6.6}$$

and from (4.19)

$$C_v=3\mathcal{R} \tag{6.7}$$

which is approximately 25 J K^{-1} mol^{-1}, a result given by the law of Dulong and Petit (1819): it is obeyed well at high temperatures but, as Figure 6.19 shows, it fails dramatically at low temperatures.

6.5.3.1 Einstein and Debye solids

Einstein (1906) treated the vibrations of N atoms in a lattice array of $3N$ independent oscillators, each of frequency ν, in one dimension x, but with the vibrational energy quantized according to Planck's equation (2.13) and distributed according to the Boltzmann equation. Thermal vibrations are thermally excited *phonons* and the phonon is the quantized unit of lattice vibration with energy $h\nu$, or $h\omega/(2\pi)$. The distribution of phonon energies is no longer continuous and the average energy is, following (5.43) but replacing the integrals with summations,

$$\overline{E}=3N\sum_{n=0}^{\infty} nh\nu\exp[-nh\nu/(k_BT)]\Big/\sum_{n=0}^{\infty}\exp[-nh\nu/(k_BT)] \tag{6.8}$$

Putting $-h\nu/(k_BT)=x$, we see that (6.8) can be recast in the form

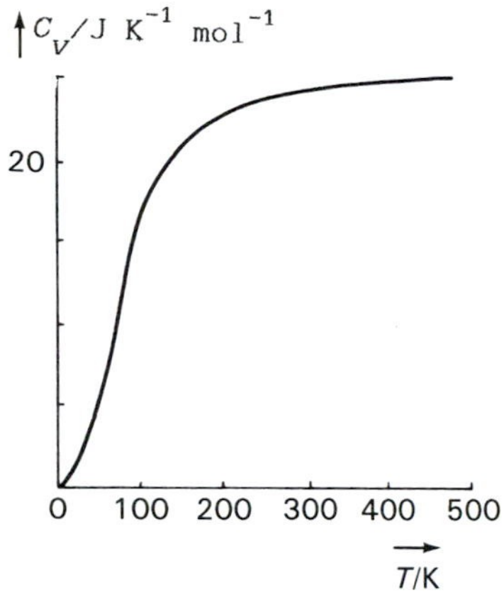

Figure 6.19 Variation with temperature of the molar heat capacity at constant volume of a monatomic solid that shows no phase changes over the temperature range given. The high-temperature limiting value of C_v is the Dulong and Petit value of $3\mathcal{R}$ (24.94 J K^{-1} mol^{-1}).

$$\overline{E}=3Nh\nu\frac{d}{dx}[\ln(1+e^x+e^{2x}+\ldots)]=3Nh\nu/(e^{-x}-1) \tag{6.9}$$

or, for N_A atoms,

$$\overline{E}=3N_Ah\nu/\{\exp[h\nu/(k_BT)]-1\} \tag{6.10}$$

whence from (4.19)

$$C_v=\frac{3\mathcal{R}[h\nu/(k_BT)]^2\exp[h\nu/(k_BT)]}{\{\exp[h\nu/(k_BT)]-1\}^2} \tag{6.11}$$

We can expand (6.11) to give

$$C_v=3\mathcal{R}[h\nu/(k_BT)]^2\frac{\{1+h\nu/(k_BT)+[h\nu/(k_BT)]^2/2+\ldots\}}{\{1+h\nu/(k_BT)+[h\nu/(k_BT)]^2/2+\ldots-1\}^2} \tag{6.12}$$

At high temperatures $h\nu/(k_BT)$ is small: thus, we may neglect both $h\nu/(k_BT)$ in relation to 1, and second-order and higher terms of the exponential expansion, whereupon (6.12) becomes

$$C_v=3\mathcal{R}[h\nu/(k_BT)]^2/[h\nu/(k_BT)]^2=3\mathcal{R} \tag{6.13}$$

which is the classical (Dulong and Petit) result. As $T\to 0$, $\exp[h\nu/(k_BT)]\to\infty$ and, from (6.10), E and, hence, C_v tend to zero exponentially. However, the Einstein equation does not give a good fit to the experimentally determined variation of heat capacity with temperature (Figure 6.19), particularly at low temperatures, where it varies experimentally as T^3.

Debye's correction (1912) assumed that not all oscillators had the same frequency. At wavelengths that are long relative to the interatomic spacings, large regions of crystal volume may be coupled in vibrational motion. At low temperatures some vibrations correspond to the condition $h\nu \ll k_BT$, and for them the long-wavelength vibrations are very important. They make a classical contribution to the energy, thereby modifying the exponential dependence on temperature shown by (6.10).

The entire treatment is complex and we state, without proof here, his formulation for the average energy and constant volume heat capacity:

Table 6.6 Debye temperatures Θ for some solids

Element	Θ/K
C (diamond)	2230
C (graphite)	420
Na	158
K	91
Al	428
Cu	343
Pb	105
NaCl	321
KCl	235

$$\bar{E}=9\mathcal{R}T/(\Theta/T)^3\int_0^{\Theta/T}\frac{x^3}{\exp(x)-1}\,dx \qquad (6.14)$$

$$C_v=9\mathcal{R}/(\Theta/T)^3\int_0^{\Theta/T}\frac{\exp(x)^4x}{[\exp(x)-1]^2}\,dx \qquad (6.15)$$

where $x=h\nu/(k_BT)$. The Debye temperature Θ is equal to $h\nu_{max}/k_B$, and the higher the value of Θ the lower the heat capacity, at a given temperature. The wavelength λ_{min} corresponding to ν_{max} is given by $\lambda_{min}=hv_s/(k_B\Theta)$, where v_s is the velocity of sound in the material. For copper v_s=4000 m s^{-1} and λ_{min}=0.56 nm, whence Θ(Cu)=343 K. Values for λ_{min} may be approximated by twice the metallic diameter, which is 0.52 nm for copper. The currently accepted values of Θ for a selection of solids are listed in Table 6.6.

At low temperatures the Θ/T integration limit in (6.14) can, for convenience and without sensible error, be set at infinity. Then, by integration, we have

$$\bar{E}_m=3\pi^4\mathcal{R}T^4/(5\Theta^3) \qquad (6.16)$$

whence

$$C_v=1944(T/\Theta)^3 \qquad (6.17)$$

Equation (6.17) establishes the T^3 law that we used in Section 4.6.4.

6.5.3.2 Heat capacity paradox

Although free-electron theory was capable of explaining conduction properties of metals it required all valence electrons to be used, so that the electron 'gas' possessed kinetic energy. If a mole of conduction electrons in a metal were to behave as free particles, its contribution to C_v would, from Appendix 3, be 1.5$\mathcal{R}$, so that the total heat capacity would be 4.5$\mathcal{R}$. Now, the molar heat capacities of monatomic solids do not deviate from the Dulong and Petit value of 3$\mathcal{R}$ by more than approximately 10% for Θ/T greater than 0.7. Since there can be no doubt about the nature of atomic vibrations and their contribution to the heat capacity, it would seem that the contribution from the electrons is grossly overestimated by the free-electron theory.

We write the molar heat capacity at low temperatures as the sum of the two terms

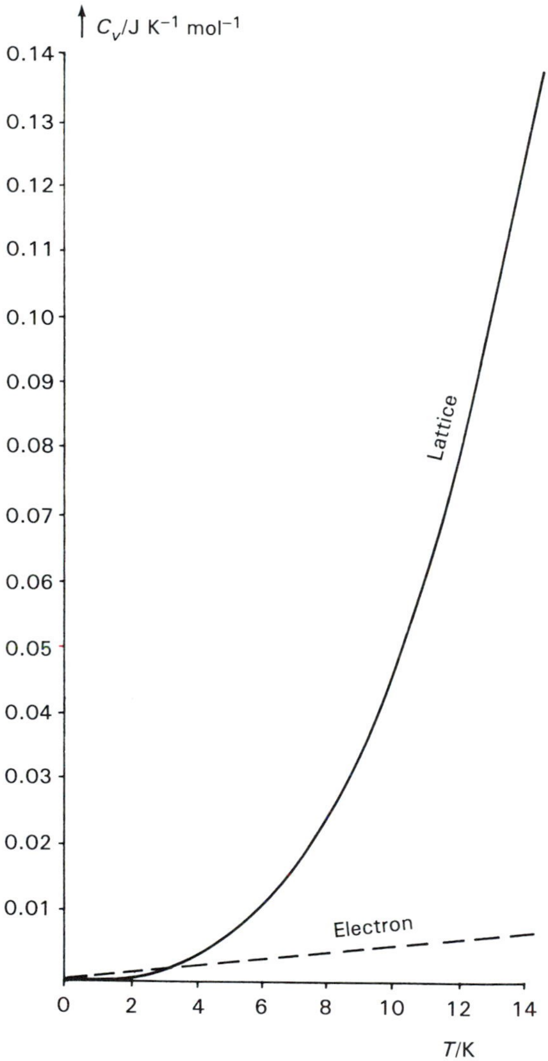

Figure 6.20 Contributions to the constant volume molar heat capacity of copper from lattice vibrations and electrons, between 0 K and 14 K; the electron contribution becomes a significant fraction of the total only at temperatures less than $\simeq$10 K.

$$C_v = aT^3 + bT \tag{6.18}$$

because it has been found experimentally that C_v/T varies linearly with T^2, as required by (6.17); the intercept at $T^2=0$ is b. Experiments with copper gave a value of 7×10^{-4} J K^{-2} mol^{-1}; hence, from (6.17) and Table 6.6, it follows that the electronic contribution to the molar heat capacity of copper exceeds 10% of the total only at temperatures less than approximately 10 K. Figure 6.20 is a plot of the two contributions to the molar heat capacity of copper between 0 K and 14 K. We may be satisfied that the apparent heat capacity paradox can be explained adequately.

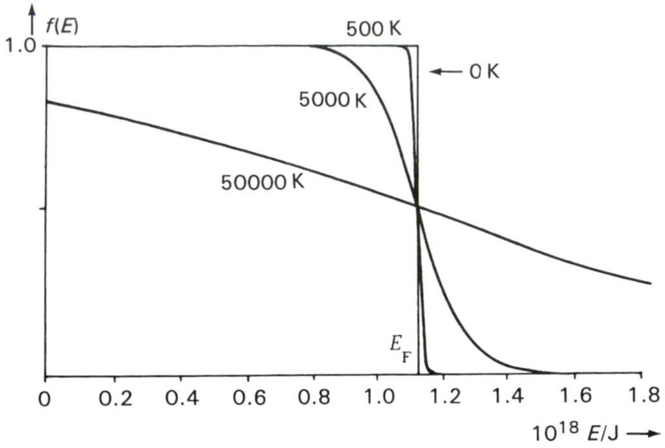

Figure 6.21 Fermi–Dirac distribution of energies $f(E)$ for copper at different temperatures; E_F is the Fermi energy. The curves for temperatures greater than zero intersect at $f(E)=\frac{1}{2}$, where $E=E_F$.

6.5.4 Wave-mechanical free-electron theory

In the three-dimensional particle in a box solution (Section 2.5.5), we were able to consider electrons occupying energy levels E_{n_x,n_y,n_z}, without reference to spin. Thus, if we assume a Boltzmann distribution of the energies of N electrons, the number N_i occupying an energy state ϵ_i of degeneracy g_i is

$$N_i = g_i N_0 \exp[-\epsilon_i/(k_B T)] \tag{6.19}$$

At $T=0$ K, $N_i=0$ for all i, that is, all electrons would occupy the state $\epsilon=0$. According to the Pauli exclusion principle (Section 2.8) this situation is forbidden; the maximum number of electrons in any given energy state ϵ_i is $2g_i$. At all temperatures, including 0 K, electrons are distributed among the energy states, and the distribution function $f(E)$ for the energies of particles with half-integral spin (Figure 6.21) is given by Fermi–Dirac statistics[9] (Appendix 15):

$$f(E) = 1/\{\exp[(E-E_F)/(k_B T)] - 1\} \tag{6.20}$$

At $T=0$, the energy states are built up according to the *Aufbau* principle until the Fermi energy E_F is reached, at which there is a discontinuity as $f(E)$ falls to zero; the Fermi energy acts as a cut-off level for allowed electron energies in solids.

For values of T greater than zero, $f(E)$ falls below unity for $E<E_F$ and is greater than zero for $E>E_F$. The high-energy 'Maxwellian tail' to the $f(E)$ distribution for $T>0$ corresponds to the situation $(E-E_F) \gg k_B T$, or $\exp[(E-E_F)/(k_B T)] \gg 1$. Then

$$f(E) \approx \exp[(E_F - E)/(k_B T)] \propto \exp[-E/(k_B T)] \tag{6.21}$$

which identifies with the classical Boltzmann distribution.

[9] Particles with zero or integral spin, such as photons or phonons, are governed by the Bose–Einstein distribution:
$$f(E) = 1/\{\exp[(E-\alpha)/(k_B T)] - 1\}$$
where α is a normalizing constant determined by the condition $\int f(E)\,dE = 1$.

6.5.5 Band theory

The theory used in the preceding sections has been shown to be capable of explaining many properties of metals. However, it does not treat the extreme variations in electrical resistivity shown by conductors, semiconductors and insulators.

The band theory of solids begins with the Schrödinger equation, and incorporates the periodic potential energy field $V(x)$ of a lattice of atoms. In this form, it is often called the Bloch equation:

$$\frac{-\hbar^2}{2m_e}\frac{d^2\psi}{dx^2}+V(x)=E\psi \tag{6.22}$$

Since the potential function is periodic, $V(x)=V(x+na)$, where a is the repeat distance along the x direction in the lattice and n is an integer. The solution of (6.22) has been given by Bloch as

$$\psi_k=E_k(x)\exp(ikx) \tag{6.23}$$

where $E_k(x)$ is an energy function with the periodicity a of the lattice. The dependence of the energy on k is quadratic, as in the free-electron theory but, for $k=\pm n\pi/a$, discontinuities appear in the energy that lead to a band structure (Figure 6.22).

The space in which k is measured is generally termed **k**-space. It is the reciprocal space of X-ray crystallography but, whereas crystallography is concerned with the (weighted) reciprocal lattice points, **k**-space is the *entire* space under investigation; as expected, k has the dimensions of reciprocal length. Each energy state specified by $\mathbf{k}$, or by[10] n_x, n_y and n_z, can accommodate two electrons with spins $\pm\frac{1}{2}$, and can be regarded as a point in **k**-space. The surface of constant energy E_F in **k**-space is known as the Fermi surface; it separates filled from unfilled orbitals.

For most values of k, the electrons behave very much like free electrons. However, at values of k equal to $\pm n\pi/a$, the condition for Bragg reflection of electron waves in one dimension is realized: $k=\pm n\pi/a$ is equivalent to the Bragg equation $2a\sin(\theta)=n\lambda$, where $k=2\pi/\lambda$, from (2.37), and $\sin(\theta)=1$. The first-order reflection at $k=\pm\pi/a$ arises because waves reflected from adjacent atoms interfere constructively, the phase difference being 2π. The region lying between $\pm\pi/a$ is called the first *Brillouin zone* (Figure 6.23). The energy is quasi-continuous within a zone, because the energy levels are very closely spaced in a solid, but discontinuous at the zone boundaries. As k increases towards $n\pi/a$, the eigenfunctions (6.23) contain increasing amounts of Bragg-reflected wave. At $k=\pi/a$, for example, the wave $\exp(i\pi x/a)$ reflects as $\exp(-i\pi x/a)$, and the resulting combinations are standing waves ψ_1 and ψ_2 of the forms $\cos(\pi x/a)$ and $\sin(\pi x/a)$ respectively. Brillouin zones are not normally encountered in X-ray crystal structure analysis, but they are essential to an analysis of electron energy levels in solids.

The probability densities of the two standing waves are $|\psi_1|^2$ and $|\psi_2|^2$, whereas that for the travelling wave is $\exp(i2kx)$. Figure 6.24 illustrates a one-dimensional periodic potential field and the wave probability functions just described. The travelling wave time-distributes charge uniformly along the x axis; ψ_1 has its peaks at na and ψ_2 has its peaks at $(n+\frac{1}{2})a$. The potential energies of the two distributions follow the order $|\psi_1|^2<\exp(i2kx)<|\psi_2|^2$. Hence, an energy gap ΔE arises, and the waves ψ_1 and ψ_2 correspond to the points A and B in Figure 6.22. The combination of the waves $\exp(\pm ikx)$ at $k=\pm n\pi/a$, that is, at the boundaries of the

[10] For a cubic crystal, $\mathbf{k}=\mathbf{i}(k_x+k_y+k_z)=\mathbf{i}(\pi/a)(n_x+n_y+n_z)$.

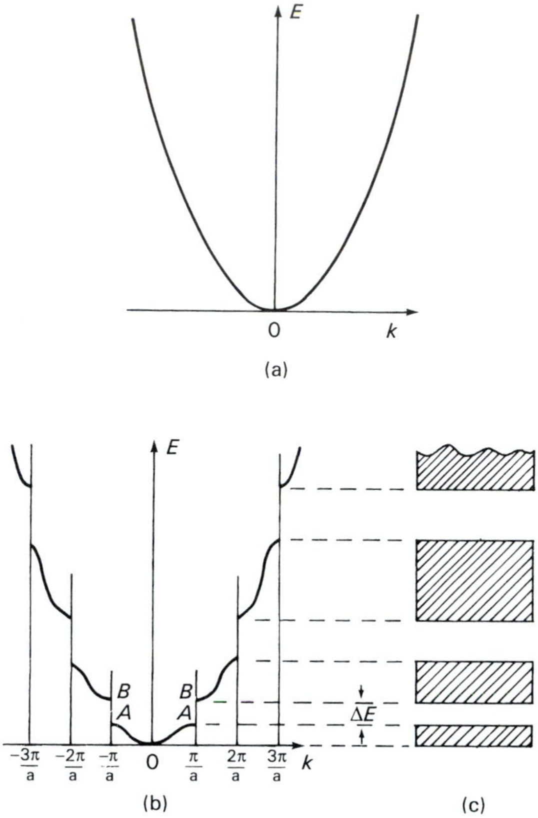

Figure 6.22 Electron energies in solids. (a) Free-electron theory, showing the quadratic dependence of E on k ($k\hbar$ is a quasi-momentum of the electron). (b) Energy bands, showing gaps ΔE at values of k for which the Bragg equation holds. The positions A and B correspond to waves ψ_1 ($\cos(\pi x/a)$) and ψ_2 ($\sin(\pi x/a)$), respectively; the values of k ($\pm n\pi/a$) define the boundaries of the one-dimensional Brillouin zones. (c) Structure of quasi-continuous energy bands separated by gaps ΔE_i ($i=1, 2, \ldots$).

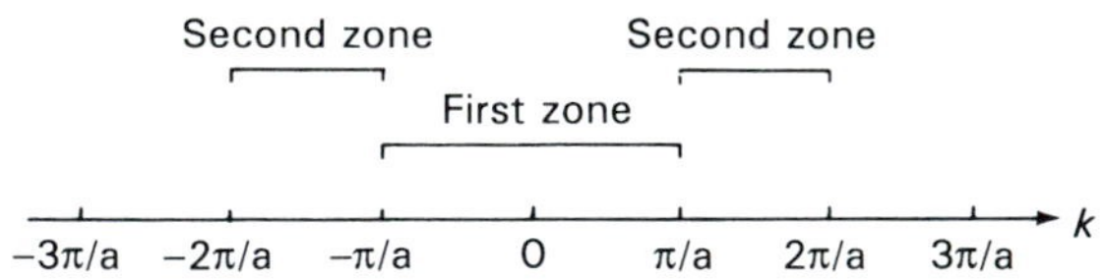

Figure 6.23 Boundaries for the first two Brillouin zones in a one-dimensional lattice of periodicity a.

Brillouin zones, leads to an energy gap ΔE of $2V_k$, where V_k is the potential energy function at the position in $\boldsymbol{k}$-space corresponding to k. This result may be compared with the bonding/antibonding situation in MO theory, which also depends on a core potential energy.

Brillouin zones may be extended to two and three dimensions. The zone boundaries are determined by the regions in $\boldsymbol{k}$-space where the Bragg equation is satisfied: they are governed by crystal structure rather than by chemical composition.

The first Brillouin zone is the smallest volume in $\boldsymbol{k}$-space that is entirely enclosed by planes that are normal to and bisect the shortest reciprocal lattice vectors drawn from the

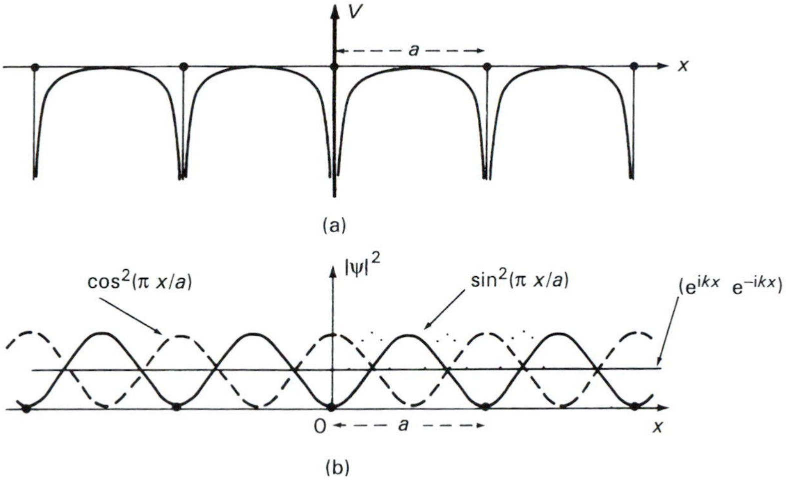

Figure 6.24 One-dimensional functions with a lattice periodicity a. (a) Periodic potential energy $V(x)$; $V(x)=V(x+na)$. (b) Probability densities $\cos^2(\pi x/a)$, $\sin^2(\pi x/a)$ and $[\exp(ikx)\exp(-ikx)]$.

origin of $\boldsymbol{k}$-space (see also Section 3.5.1.1). The Brillouin zones correspond to Wigner–Seitz cells[11] in $\boldsymbol{k}$-space. The Wigner–Seitz cell for the lattice based on a conventional primitive cubic unit cell is, itself, a cube; for the lattice based on the conventional face-centred cubic unit cell, it is a rhombic dodecahedron; the faces are the crystallographic *form*[12] of planes $\{011\}$.

A Wigner–Seitz cell is obtained by joining a point of the Bravais lattice to all the nearest lattice points and then bisecting these lines with perpendicular planes. (The same construction is encountered with Voronoi polyhedra.) The intersections of these planes contain the Wigner–Seitz cell.[13] The planes of larger area are those closer to the origin point; beyond a certain distance, lattice points will not contribute to a Wigner–Seitz cell because the bisecting planes lie outside the confines of the smallest polyhedron. Wigner–Seitz cells are unit cells in the sense that they stack to fill space completely, but not all of them may be generated by three noncoplanar translation vectors taken from any corner of the cell; it is preferable to use the term Wigner–Seitz *cell*.

If we let K in (3.18) take the value of 2π, and ϕ the value $(90-\theta)$ in Figure 3.33 (QA is the direction of the wave vector from the origin of $\boldsymbol{k}$-space), then the Bragg equation (3.55) may be written:

$$(2\pi/\lambda)\cos(\phi)=k\cos(\phi)=d^*/2 \tag{6.24}$$

Multiplying both sides by d^* gives

$$kd^*\cos(\phi)=d^{*2}/2 \tag{6.25}$$

[11] See, for example, E Wigner and F Seitz, *Phys. Rev.* **43**, 804 (1933); C Kittel, *Introduction to Solid State Physics*, 5th edition (Wiley, 1976); B K Vainshtein, *Fundamentals of Crystals*, Volume 1 (Springer-Verlag, 1994).

[12] The form of planes $\{hkl\}$ refers to those planes (hkl) related by the symmetry of the structure under consideration.

[13] This construction leads, generally, to a Voronoi polyhedron; see, for example, M F C Ladd, *J. Chem. Educ.* **74**, 461 (1997).

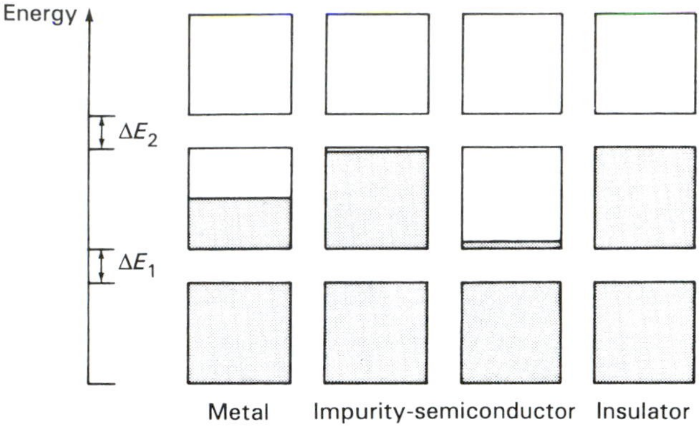

Figure 6.25 Schematic illustration of energy bands in solids; the shading indicates electron occupancy. The energy gaps ΔE_1 and ΔE_2 are forbidden ranges for electron energies.

or

$$\boldsymbol{k}\cdot\boldsymbol{d}^* = d^*/2 \tag{6.26}$$

which is the equation of a plane in $\boldsymbol{k}$-space, normal to d^* and at a perpendicular distance $d^*/2$ from the origin. Thus, the Bragg equation is satisfied for a wave vector $\boldsymbol{k}$ that terminates on a plane (line in two dimensions) normal to $\boldsymbol{d}^*$ and at the midpoint of d^*, and these terminations determine the boundaries of the Brillouin zones.

Figure 6.25 shows a schematic arrangement of energy bands for the three main classes of electrical conductors, metals (conductors), semiconductors and insulators. If a band is partially filled the solid behaves as a metal. If all bands are filled except for one or two bands which are either nearly filled or nearly empty, the solid is a semiconductor; and if all occupied bands are filled the solid is an insulator. If all occupied bands are filled but the energy gap between the uppermost filled band (*valence band*) and the band of next highest energy (*conduction band*) is small, then the solid can again show semiconduction. Clearly, there will be gradations throughout these classes, as measurements of electrical resistivity have shown (see Section 6.4).

6.5.6 Energy bands and molecular-orbital theory

The Bloch theory is essentially a molecular-orbital model of solids. The important feature of the MO theory as applied to a chemical species is that each electron moves in the potential field created by all other atoms in the molecule, the core field. It uses one-electron Hamiltonians and the solutions of the corresponding wave functions are obtained from an LCAO approximation. We saw that in a compound such as benzene, strong overlap of the π orbitals of the atoms led to electron delocalization over the whole molecule. In the case of a metal, the number of atoms is infinite, or at least very large, and we can envisage an extreme case of electron delocalization now over the whole solid, a situation often called the 'tight binding' approximation because the outer electrons are assumed initially to be associated with the atoms. The molecular orbitals are now the conduction orbitals and metallic properties depend upon their degree of overlap. In lithium, for example, the overlap integral $\int \psi_1(2s)\psi_2(2s)\,d\tau$ for adjacent atoms is approximately 0.5. An extension of this overlap to encompass all atoms in a metal crystal leads to the complete delocalization that establishes metallic character.

In order to highlight the important difference between the MOs in a metal and those that we discussed in Chapter 2, we will consider building up a crystal of the electronically simplest metal lithium; this element has the ground state electronic configuration $(1s)^2(2s)^1$. When we discussed the one-electron species H_2^+ (Section 2.11.7), we saw that, when the two hydrogen nuclei were brought together, two diatomic molecular orbitals $\Psi_\pm$ were obtained: from the *single* energy level of each atom, *two* energy levels arose in the molecule. If we add a third atom, its orbital overlaps strongly with that of its nearest neighbour (and only slightly with that of its next nearest) and three MOs are formed. The addition of a fourth atom leads to four MOs and so on. The general effect of adding more atoms is to spread out the range of energies spanned by the MOs, while filling in the range of energies with more and more MOs. When a large number N atoms have been added there are N MOs within a band of finite width. The Hückel determinant for this configuration is then of the form

$$\begin{vmatrix} \alpha-E & \beta & 0 & 0 & 0 \ldots 0 \\ \beta & \alpha-E & \beta & 0 & 0 \ldots 0 \\ 0 & \beta & \alpha-E & \beta & 0 \ldots 0 \\ 0 & 0 & \beta & \alpha-E & \beta \ldots 0 \\ \vdots & \vdots & \vdots & \vdots & \vdots \\ 0 & 0 & 0 & 0 & 0 \ldots \alpha-E \end{vmatrix} = 0 \tag{6.27}$$

and the solution of this determinant may be given as

$$E_n = \alpha + 2\beta \cos[n\pi/(N+1)] \tag{6.28}$$

where n is an integer ranging from 1 to N, and α and β have meanings as before. The reader may care to compute E_n for $N=4$ and to compare the results with those for butadiene (Section 2.12). Why does (6.28) not give the correct results for benzene ($N=6$)?

As N becomes very large, the difference between adjacent energy levels becomes vanishingly small, so that an effective band is obtained. The solutions of (6.28) for $n=1$ and $n=N$, as $N\to\infty$, are $\pm 2\beta$, so that the width of the band is 4β. The band consists of N molecular orbitals, and if it is formed from s orbitals, it is referred to as an s band; p orbitals lead to a p band. Figure 6.26 illustrates some aspects of this discussion on the MO model of a metal. In the case of lithium, a cubic crystal of side 1 mm contains approximately 5×10^{19} atoms and there are approximately 9.3×10^{19} 2s energy states present in this crystal of lithium, forming a quasi-continuous series of energy states; a continuum is the limiting situation for infinite degeneracy.

6.5.6.1 Occupation of orbitals

In lithium, each atom contributes one valence electron. At $T=0$ K the lowest $N/2$ MOs are filled and the Fermi level is the energy level corresponding to the highest occupied molecular orbital (HOMO). The important difference from the molecules that we discussed in Chapter 2, even with delocalization as great as that in benzene, is that there are unoccupied orbitals very close to the Fermi level in metals, so that very little energy is needed to excite electrons into higher states.

At $T>0$, the distinction between occupied and unoccupied MOs in a band is no longer distinct, and the thermal energy of the electrons near the Fermi level is sufficient to cause electron excitation. The energy states in a band occur in pairs (quantum numbers n_x, n_y, n_z and $\pm m_s$), and could represent waves travelling in opposite directions ($\pm m_s$) in the crystal. In the absence of an external electric field, the net momentum of the electrons is zero. However,

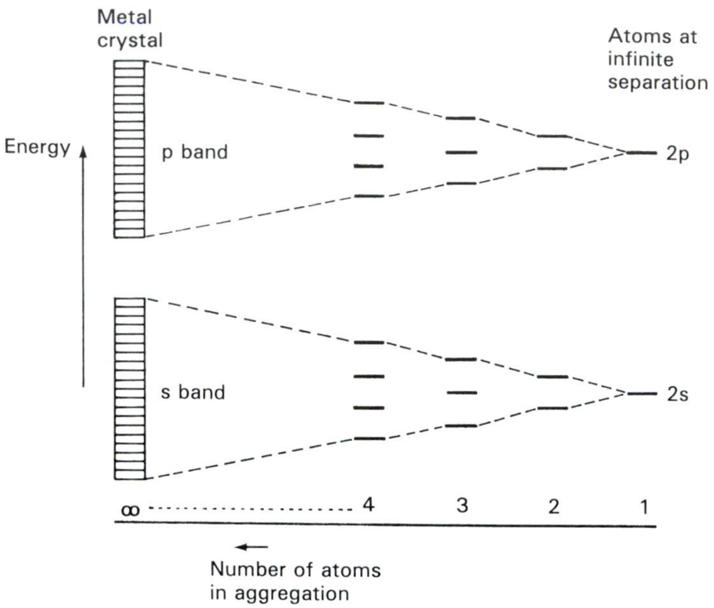

Figure 6.26 Schematic representation of the formation of s and p energy bands, consisting of very closely spaced energy levels. The overlap of s atomic orbitals gives rise to an s band, and the overlap of p atomic orbitals to a p band; the distance between the top of the s band and the bottom of the p band is the s–p energy band gap. The lowest and highest level of each band correspond to bonding and antibonding situations respectively.

if a potential difference were applied to the metal, there would be a net resultant electron flow and some electrons, those lying approximately k_BT below the Fermi energy, may be raised from the half-filled band to an empty, higher-level band in the process. Thus, for a metal with one valence electron, the band theory and the free-electron theory give similar results.

If the metal has two valence electrons per atom, like beryllium or magnesium for example, then the $2N$ valence electrons would fill the N orbitals of the s band. For $T=0$, the Fermi level now lies at the top of the band. Since there is normally a gap before the next highest band begins, these elements might be expected to be insulators. In order to explain the fact that they are good electrical conductors, it was postulated that the bands actually overlap for these metals. This apparent 'theory to fit the facts' has been adequately justified by experiment.

The electrical conductivity of a metal decreases with increasing temperature, even though the increased thermal energy of the electrons causes more electron excitation. However, that same thermal energy brings about an increase in the lattice vibrations of the positive ions, so that the electron-scattering, current-limiting process increases and the steady electron current decreases in magnitude. At $T=0$, the lattice vibrations cease, so that the electrical resistivity ρ (reciprocal of conductivity) should tail off to zero. The presence of impurities and defects lead to a small residual resistivity, even at absolute zero.

6.5.7 Semiconductors and insulators

At 0 K, a semiconductor has a completely filled valence band separated by a small energy gap, approximately 1 eV or less, from an empty conduction band. There is no electrical conduction because the electrons are unable to cross the gap to the conduction band. At temperatures greater than zero, some electrons may be excited to the conduction band. If an electrical field is then applied, it can act upon the energy states in both bands and a current will flow. The number of electrons that are promoted across the gap increases with increasing temperature. Hence, the electrical conductivity of a semiconductor increases with increasing temperature, in contradistinction to that of a metal.

If the gap is large, then only very few electrons may be promoted to the higher energy band. The effect again increases with increasing temperature, but it is so small in magnitude that the solid is classed as an insulator. Thus, the essential difference between a semiconductor and an insulator resides in the size of the energy gap between the valence and conduction bands, and does not have the absolute distinction possessed by a metal.

The most important semiconducting elements are silicon and germanium; the atoms of these elements have four outer electrons, $(1s)^2(2s)^2(2p)^2$. When the energies of the AOs are close, the bands formed from them may merge into a sort of hybrid band, rather than remain discrete. This situation arises for pure silicon and germanium, in which the s and p levels are hybridized to form a completely filled hybrid valence band. The valence band is separated from the conduction band by a gap of only 1 eV or less for silicon and germanium, and such pure materials are termed *intrinsic* semiconductors, and are prone to excitation by the presence of impurities.

Since the active electrons lie in the Maxwellian tail of the Fermi–Dirac distribution, the dependence on temperature of the electrical conductivity can be expected to follow a Boltzmann distribution, since the conductivity will depend on the population of charge carriers. If we let E_n be the energy of the uppermost level of the valence band, then we can write

$$E_F - E_n = \Delta E/2 \tag{6.29}$$

Then from (6.21)

$$f(E) \approx \exp[-\Delta E/(2k_B T)] \tag{6.30}$$

and the electrical conductivity may then be expressed as

$$\sigma = A \exp[-\Delta E/(2k_B T)] \tag{6.31}$$

which is an Arrhénius-type equation in which the energy of activation for electrical conductivity is one-half the band gap; the electrical conductivity σ is the reciprocal of the electrical resistivity ρ. The following example illustrates the application of (6.31) to germanium.

The resistance R of a sample of pure germanium has been found to vary with temperature as follows:

T/K	300	350	400
R/Ω	20.4	2.67	0.581

From (6.31) and the definition of conductivity, we have $R = A' \exp[\Delta E/(2k_B T)]$, where A' is a constant. Hence, the plot of ln (R/Ω) against 1/(T/K) will be a straight line of slope $[\Delta E/(2k_B)]$/K. By least squares, the slope is 4271.3/K^{-1}, so that the band gap $\Delta E = 5.90 \times 10^{-20}$ J, or 0.74 eV.

The number of charge carriers in silicon or germanium can be increased by *doping*, that is, by implanting impurity atoms in minute traces, approximately 1 in 10^9, into an otherwise highly pure material, thus forming *extrinsic* semiconductors. If the dopant atoms have fewer valence electrons than do the atoms in the bulk of the solid, such as gallium in silicon, they can extract electrons from the conduction band leaving 'holes' that allow movement of other electrons. Such materials are termed p-type semiconductors, because the hole is *positive* with respect to electrons. Alternatively, a dopant may introduce more valence electrons than those of the parent material, such as phosphorus in germanium. The additional electrons can occupy otherwise empty bands and give rise to n-type (*negative*) semiconductors.

6.5.8 Structural and physical characteristics of metallic compounds

The metallic bond is spatially undirected, and metal structures have high coordination numbers and high densities. Metals are opaque and possess high reflecting power. Electrons near the Fermi level in metals can absorb energy, in accordance with (2.13), and are raised in energy. If there is only slight scattering interaction between these electrons and the lattice of positive ions, the energy gained is radiated away without change of phase and the crystal is transparent to that radiation. Metals interact in this way with high-energy radiations, such as X-rays.

Metals have variable strength. Deformation by gliding is common in metals and it takes place most easily along close-packed planes of atoms in the crystal. In the CPC structure type there are four such planes, with Miller indices (111), ($\bar{1}11$), ($1\bar{1}1$) and ($\bar{1}\bar{1}1$), whereas there is only one such plane, (0001), in the CPH structure type and none of a similar degree of close packing in the BCC structure type. Consequently, we find that the more malleable and ductile metals, such as copper, silver, nickel and γ-iron, crystallize in the CPC structure type, whereas the harder and more brittle metals, such as beryllium, chromium, tungsten and α-iron, have the CPH or BCC structure types. The metallurgical importance of iron is related to its ability to adopt either the A1 or the A2 structure type according to heat treatment. Metals have sharp melting-points, but they vary widely (Hg, 234 K; W, 3683 K) and the liquid interval is long (Ga, 2370 K; Hf, 3400 K). Perhaps the most distinctive properties of metals are their high electrical and thermal conductivities, to which we referred earlier in this chapter.

6.6 Bond type among elements

A study of the elements of groups 12–17 in the periodic table reveals a continuous change in bond type towards increasing metallic character as the atomic number down any group increases. Thus, solid iodine (group 17) is a semiconductor; it has the lustre associated with a metal and under pressure becomes metallic. Again, selenium and tellurium in group 16 and, more especially, antimony and bismuth in group 15 show varying degrees of metallic character. Group 14 elements in the solid state show a clear gradation from covalent character in carbon (diamond) to metallic character in tin and lead. Lead has the CPC structure, whereas tin is dimorphic, grey tin (α-Sn) having the tetrahedrally coordinated structure type shown by diamond (Figure 6.10). White tin, β-Sn (Figure 6.11), has a flattened tetrahedral coordination with the equatorial bonds and two slightly longer axial bonds complete a distorted octahedral coordination; β-Sn is strongly metallic, with an electrical conductivity approximately 100 times that of α-Sn. All the members of group 12 are metallic. Mercury has a unique rhombohedral structure, but zinc and cadmium are close to the CPH structure type. Their unit cell axial ratios *c*/*a* are close to 1.9, whereas the ideal CPH ratio is $\surd(8/3)$, or 1.633.

6.7 Ionic solids

Ionic compounds are formed generally between species of widely varying electronegativity. One of them, typically a metal, becomes ionized by the loss of one or more electrons, and the other, typically a nonmetal, becomes ionized by acquiring an equivalent number of electrons. The charged species, or ions, then attract each other by Coulombic forces that are inversely proportional to the square of the distance between them and vary directly as the product of the charges on the ions.

6.8 Electrostatic model for lattice energy

If two ions have charges q_1e and q_2e, considered here as points, separated by a distance r, then the electrostatic, Coulombic energy between them is given by

$$U_E = -q_1q_2e^2/(4\pi\epsilon_0 r) \tag{6.32}$$

where the negative sign implies ions of opposite sign, leading to an attraction between them. In a crystal, it is necessary to include the effect of all ions present on the result (6.32). In the case of the sodium chloride structure type (Figure 1.2), for example, we write

$$U_E(r) = -\mathcal{A}q_1q_2e^2/(4\pi\epsilon_0 r) \tag{6.33}$$

and $\mathcal{A}$ is a constant for the given structure type, known as the Madelung constant (Madelung, 1918).

6.8.1 Madelung constant

The meaning of a Madelung constant may be considered first in one dimension. Figure 6.27 represents an infinite row of alternating positive and negative point charges of magnitude e, regularly spaced at a distance r. Let any ion be chosen as a reference origin. Its immediate neighbours give rise to an attractive potential energy of $-2e^2/(4\pi\epsilon_0 r)$. The next nearest neighbours set up a repulsive energy of $+2e^2/[4\pi\epsilon_0(2r)]$, and the next nearest an attraction of $-2e^2/[4\pi\epsilon_0(3r)]$, and so on. The continuation of this process leads to the series for the total electrostatic energy of the row of charges

$$U_E(r) = \frac{-2e^2}{4\pi\epsilon_0 r}(1 - 1/2 + 1/3 - 1/4 + \ldots) \tag{6.34}$$

The series of point charges has been treated in pairs $\pm 2\alpha/(nr)$, where $\alpha = e^2/(4\pi\epsilon_0)$ and $n = 1, 2, 3, \ldots$, which ensures its conditional convergence to ln (2). Hence, the effect of this infinite row of charges is to modify (6.32) by the factor 2 ln (2), or 1.3863, which number may be regarded as the Madelung constant for the one-dimensional structure under consideration.

In three dimensions the corresponding calculation is a little more difficult. Consider again

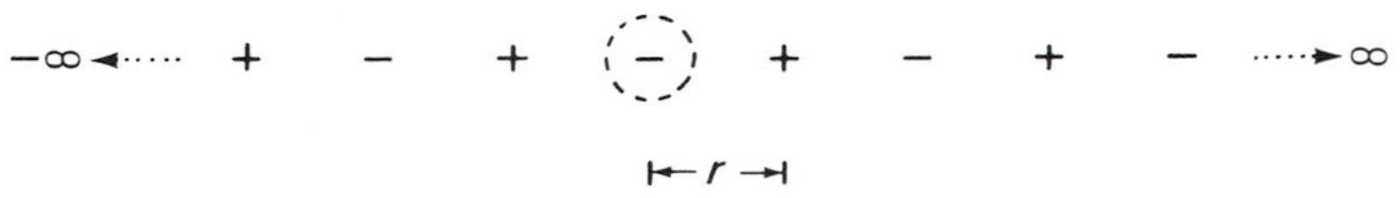

Figure 6.27 Infinite row of regularly spaced point charges alternating in sign, each of magnitude e; the dashed circle indicates an arbitrarily chosen origin for the calculation of the Madelung constant.

the sodium chloride structure type (Figure 1.2) and take the central Na^+ ion as a reference origin. The six nearest Cl^- neighbours give rise to an attractive energy of $-6e^2/(4\pi\epsilon_0 r)$, where r is the $Na^+ \ldots Cl^-$ distance, or one half of the cell side a for this structure type. The next nearest neighbours, at the centres of the cell edges, set up a repulsive energy of $+12e^2/(4\pi\epsilon_0 r\sqrt{2})$, and the next nearest, at the corners of the unit cell, an attraction of $-8e^2/(4\pi\epsilon_0 r\sqrt{3})$, and so on. Writing these terms as a series, we obtain

$$U_E(r) = \frac{-e^2}{4\pi\epsilon_0 r}(6 - 12/\sqrt{2} + 8/\sqrt{3} - \ldots) \tag{6.35}$$

This series has a very slow rate of convergence and is not suitable for computation of the Madelung constant of the structure. Evjen (1932) showed that conditional convergence can be achieved by working with nearly neutral blocks of structure: the potential energy falls off more rapidly with distance for a neutral group than it does with a group of excess charge.

Consider the unit cell of the sodium chloride structure type (Figure 1.2) as a nearly neutral block of structure. This method required each ion in the given unit cell to have a weighting factor, in order to achieve the neutrality condition. An ion at the corner of a unit cell is shared in the structure by eight adjacent unit cells, so that its contribution, or weight, in one unit cell is one eighth. Similarly, an ion at the centre of a cell edge has a weight of 1/4 and that at the centre of a face a weight of 1/2. If we now restate (6.35), with these weights included, we obtain

$$U_E(r) = \frac{-e^2}{4\pi\epsilon_0 r}(6/2 - 12/4\sqrt{2} + 8/8\sqrt{3}) \tag{6.36}$$

The sum of the terms in parentheses is 1.46, which is an approximation to the Madelung constant for this structure type. If we take a cube of twice the linear size, the value obtained is 1.75. The Madelung constant calculated by precise methods has the value 1.74756..., so that we come very close to the true value by using Evjen's method on a cube of side $2a$. A selection of Madelung constants is included in Table 6.7.

In Table 6.7, the values listed for $\mathscr{A}$ refer to the structure type with unit charges at the ion sites. The corresponding values including the ion charges are given in the next column; in some cases, they are, naturally, the same. Some sources quote the values of $q_1 q_1 \mathscr{A}$ given here as $\mathscr{A}$ itself, so that it is necessary to identify correctly the term given in any reference work. There is no unique value of r_e in some structures; in TiO_2 (rutile), for example, the value given here for $\mathscr{A}$ was calculated in terms of the shortest interionic distance in the structure. The sodium chloride structure type is common to many compounds; when magnesium oxide, for example, is being considered, $q_1 q_1 \mathscr{A}$ becomes 6.9904. The final column of Table 6.7 gives the electrostatic energy for each structure type when divided by the appropriate value of r_e in nanometres.

6.8.2 Lattice energy equation

As with molecular compounds, so in ionic crystals the attractive energy is balanced by a repulsive energy that is important at small values of r. The Madelung constant incorporates the electrostatic ionic repulsions, but a second term is required to represent the electron–electron repulsion of closed shells, and to achieve an equilibrium state in terms of energy.[14] The earliest repulsion potential function was given in the form of $1/r^n$, where n varied between 9

[14] The reader may wish to review Earnshaw's theorem in electrostatics.

Table 6.7 Madelung constants for some simple structure types

Structure type	q_1	q_2	$\mathscr{A}$	$q_1q_2\mathscr{A}$	r_eU_E/kJ mol^{-1} nm
CsCl	1	1	1.7627	1.7627	244.90
NaCl	1	1	1.7476	1.7476	242.80
α-ZnS	2	2	1.6407	6.5628	911.80
β-ZnS	2	2	1.6381	6.5524	910.36
CaF_2	2	1	2.5194	5.0388	700.07
TiO_2	4	2	2.3851	19.0808	2650.99
β-SiO_2	4	2	2.2011	17.6088	2446.48

and 13. As we have noted, theroretical studies have shown that an exponential form of repulsion potential is more satisfactory, and the cohesive energy, usually called the lattice energy,[15] may be given as

$$U(r) = -\mathscr{A}q_1q_2e^2/(4\pi\epsilon_0 r) + B\exp(-r/\rho) \tag{6.37}$$

where B is a constant for the structure type and ρ is a constant of the structure itself. For simplicity of manipulation, we write

$$U = \mathscr{A}'/r + B\exp(ar) \tag{6.38}$$

where $\mathscr{A}' = -\mathscr{A}q_1q_2e^2/(4\pi\epsilon_0)$ and $a = -1/\rho$. The two terms in (6.38) lead to a curve similar in form to that of Figure 5.12a, except that in a crystal the actual situation is more complex than that of a diatomic molecule. Nevertheless, we can identify an energy minimum at the equilibrium interionic distance r_e. From (6.38),

$$\mathrm{d}U/\mathrm{d}r = -\mathscr{A}'/r^2 + Ba\exp(ar) \tag{6.39}$$

At $r = r_e$, $\mathrm{d}U/\mathrm{d}r = 0$, and we obtain

$$B = \mathscr{A}'/[r_e^2 a\exp(ar_e)] \tag{6.40}$$

Hence, from (6.37) and (6.40), inserting the values for $\mathscr{A}'$ and a and multiplying by N_A to obtain the energy in molar terms, we have

$$U(r_e) = -N_A\mathscr{A}q_1q_2e^2(1-\rho/r_e)/(4\pi\epsilon_0 r_e) \tag{6.41}$$

A similar equation is still sometimes quoted in which $1/n$ replaces ρ/r_e, but it is less satisfactory: ρ/r_e is remarkably close to 0.1 in many 1 : 1 ionic compounds, although both r_e and ρ themselves show considerable variation, over a similar range of structures.

The parameter ρ is related, not surprisingly, to the compressibility of the crystal, because the repulsive energy becomes very large as r is reduced below r_e, as would be the case for a crystal under compressive stress. From (4.105) at 0 K, $(\partial U/\partial V)_T = -p$; at 298.15 K, neglect of the term $T\alpha/\kappa$ leads to an underestimate of the magnitude of the lattice energy by approximately 1%; we may correct the energy from 0 K to T K through the integral $\int_0^T C_v\,\mathrm{d}T$. At 0 K, we have

$$(\partial p/\partial V)_T = -(\partial^2 U/\partial V^2)_T \tag{6.42}$$

[15] This name is strictly a misuse of the term *lattice* but is, nevertheless, traditional usage.

Since the isothermal compressibility κ is $-(1/V)(\partial V/\partial p)_T$ (Section 4.8.1), the constant temperature condition leads to

$$1/(\kappa V)=\mathrm{d}^2U/\mathrm{d}V^2 \tag{6.43}$$

For isotropic crystal structures in which there is a unique interionic distance r, we can write

$$v=\mathcal{K}r^3 \tag{6.44}$$

where v is the volume occupied by a pair of oppositely charged ions and $\mathcal{K}$ is a constant. The first and second differentials of (6.44) are

$$\left.\begin{aligned}\mathrm{d}v/\mathrm{d}r&=3\mathcal{K}r^2=3v/r\\ \mathrm{d}^2v/\mathrm{d}r^2&=6\mathcal{K}r=6v/r^2\end{aligned}\right\} \tag{6.45}$$

By the rules of differentiation, we have

$$\left.\begin{aligned}\mathrm{d}U/\mathrm{d}v&=(\mathrm{d}U/\mathrm{d}r)/(\mathrm{d}v/\mathrm{d}r)\\ \mathrm{d}^2U/\mathrm{d}v^2&=\frac{1}{(\mathrm{d}v/\mathrm{d}r)}\frac{\mathrm{d}}{\mathrm{d}r}[(\mathrm{d}U/\mathrm{d}r)/(\mathrm{d}v/\mathrm{d}r)]\\ &=\frac{(\mathrm{d}v/\mathrm{d}r)(\mathrm{d}^2U/\mathrm{d}r^2)-(\mathrm{d}U/\mathrm{d}r)(\mathrm{d}^2v/\mathrm{d}r^2)}{(\mathrm{d}v/\mathrm{d}r)^3}\end{aligned}\right\} \tag{6.46}$$

From (6.39), we obtain

$$\mathrm{d}^2U/\mathrm{d}r^2=2\mathcal{A}'/r^3+Ba^2\exp(ar) \tag{6.47}$$

and the totality of these results leads to

$$\frac{1}{\kappa v}=\frac{(3v/r)[2\mathcal{A}'/r^3+Ba^2\exp(ar)]-(6v/r^2)[-\mathcal{A}'/r+Ba\exp(ar)]}{(3v/r)^3} \tag{6.48}$$

Eliminating B through (6.40) at the equilibrium distance r_e and rearranging, we obtain

$$9v/\kappa-2\mathcal{A}'/r_\mathrm{e}=a\mathcal{A}'=a\mathcal{A}'r_\mathrm{e}/r_\mathrm{e} \tag{6.49}$$

Finally, introducing the values for $\mathcal{A}'$ and a, we obtain

$$\rho/r_\mathrm{e}=\frac{\mathcal{A}q_1q_2e^2/(4\pi\epsilon_0r_\mathrm{e})}{9v/\kappa+2\mathcal{A}q_1q_2e^2/(4\pi\epsilon_0r_\mathrm{e})} \tag{6.50}$$

and this value of ρ/r_e is used in (6.41) to calculate the lattice energy. We will apply these equations to sodium chloride, given that $r_\mathrm{e}=0.282$ nm and $\kappa=4.1\times10^{-11}$ N^{-1} m, with $v=2r_\mathrm{e}^3$ for this structure type. Executing the calculation gives $\rho/r_\mathrm{e}=0.113$ and U(NaCl)$=-764$ kJ mol^{-1}. We shall return to the precision of this value and, indeed, of the model that we have used, but it is desirable first to be able to set up a test model with experimentally determined parameters.

6.9 Thermodynamic model for lattice energy

We shall consider the stages in the formation of an ionic compound in terms of enthalpy, using the sodium chloride structure as an example, and then convert the lattice enthalpy to lattice energy. Figure 6.28 illustrates a thermochemical cycle for lattice enthalpy, put forward

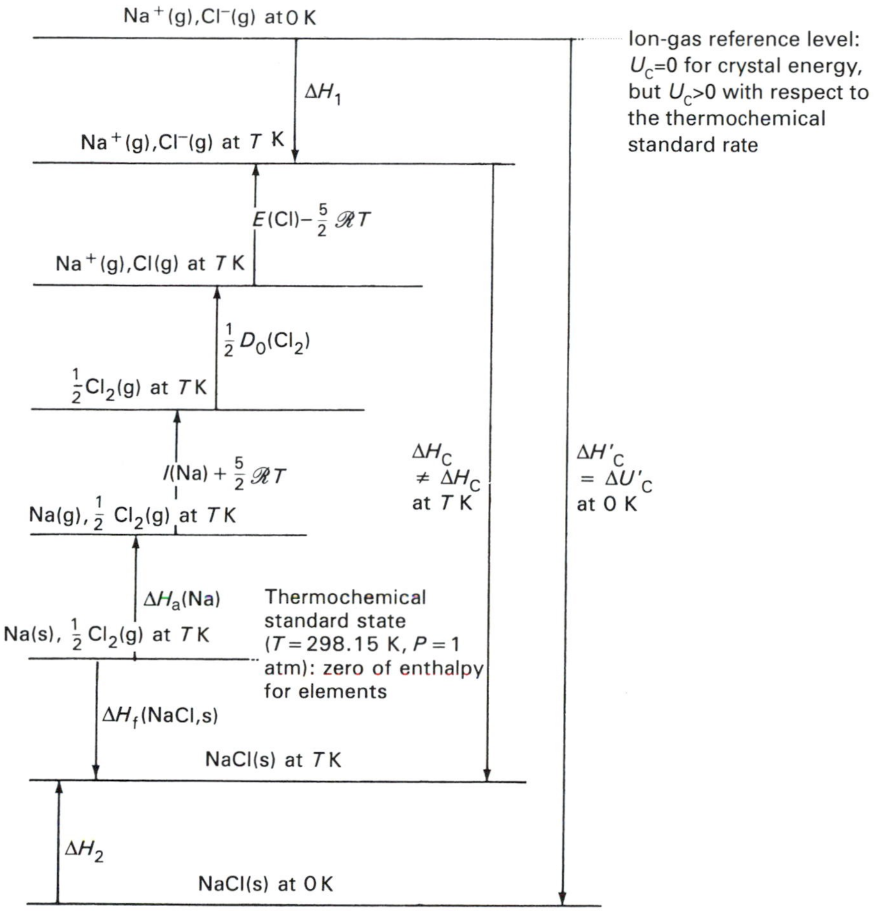

Figure 6.28 Born–Haber–Fajans thermochemical cycle for the lattice energy ΔU_C (and enthalpy ΔH_C) of an ionic solid, illustrated here for NaCl; care should be taken to distinguish between the two reference levels, one for elements in their standard state – for *formation*, and the other for the ion-gas in its standard state – for *lattice energy*.

independently by Born, Haber and Fajans. We have discussed already most of the processes involved, in Sections 3.6.2.1, 3.6.2.2, 3.6.5.2, 4.5, 4.5.2 and 4.5.3 with some summaries in Table 4.3.

The lattice energy is defined as *that energy liberated when one mole of crystal is formed from its component gaseous ions in the reference state*. The lattice enthalpy is the corresponding enthalpy change, and at 0 K these two quantities are equal. As defined they are negative quantities, and refer to the difference between two thermodynamic states, the ion-gas and the crystal. Thus, for a substance AX,

$$\Delta U_C(AX) = U_C(AX) - [U(A^+,g) + U(X^-,g)] \qquad (6.51)$$

where the subscript C refers to the crystal in the thermodynamic discussion of lattice energetics.

The standard state of zero lattice energy is defined as the ideal ion-gas at 0 K and 1 atm.

There is no energy of interaction between the ions in the (hypothetical) standard state; hence, for any crystal at 0 K

$$\Delta U_C = U_C \tag{6.52}$$

Thus, although we are concerned with a change between two states, the lattice energy is usually given the symbol U_C, referring implicitly to 0 K; at any other temperature the symbol ΔU_C is used for the energy and ΔH_C for the lattice enthalpy.

We note that in calculating ΔH_C from the cycle (Figure 6.28), the correction of $\frac{5}{2}\mathcal{R}T$ per mole ($\int_0^T C_p(\mathrm{g})\,\mathrm{d}T = \int_0^T C\frac{5}{2}\mathcal{R}\,\mathrm{d}T$) has to be added to the ionization energy I at any temperature other than 0 K (see also Sections 2.10 and 4.5.2), but a similar amount is deducted from the electron affinity, in both cases to take account of the enthalpy of 1 mole of ideal gas (electrons) at constant pressure.

The quantity ΔH_1 is the enthalpy change for the ideal ion-gas between the standard state at 0 K and another temperature T, and from the above is equal to $\frac{5}{2}\mathcal{R}T$ per mole. For two moles of gaseous ions at 298.15 K, $\Delta H_1 = 12.4$ kJ mol^{-1}. The term ΔH_2 is the enthalpy change for the crystal between 0 K and a temperature T, given by $\int_0^T C_p(\mathrm{s})\,\mathrm{d}T$, which may be evaluated from heat capacity measurements.

We may equate the lattice enthalpy and lattice energy at 0 K, but at another temperature T, we have

$$\Delta U_C = \Delta H_C - p\,\Delta V = \Delta H_C - p[V(\mathrm{s}) - V(\mathrm{g})] \tag{6.53}$$

At a constant pressure, the volume of a solid is very small in comparison with that of the same mass of gas. Hence, in (6.53), we can neglect V(s) in comparison with V(g) and, since the ion-gas is ideal, we replace pV(g) by $n\mathcal{R}T$, where n is the number of moles of gaseous species.

From the cycle (Figure 6.28), we can write generally, for the standard lattice energy of a compound AX at a temperature T,

$$\Delta U_C^{\ominus} = \Delta H_f^{\ominus}(AX,\mathrm{s}) - \Delta H_a^{\ominus}(A) - I(A) - \tfrac{1}{2}D_0^{\ominus}(X_2)$$
$$- E(X) + n\mathcal{R}T + \Delta H_1 \tag{6.54}$$

We see that the energy needed to form the gaseous ions Na^+ and Cl^- from the elements in their standard state is 378.9 kJ mol^{-1}, which does not suggest spontaneity. However, the driving force for the formation of sodium chloride may be said to lie in the energy released, the lattice energy, when the gaseous ions condense together to form the solid.

Completing the evaluation of the lattice energy of sodium chloride from (6.54), we obtain $\Delta U_C = -774$ (± 2) kJ mol^{-1}. We may note that, in common with other thermodynamic calculations, that for the lattice energy involves no knowledge of the particular type of bonding present in the given substance. This information is locked away practically in the enthalpy of formation, and theoretically in the lattice energy itself, each with respect to its own standard state.

6.9.1 Precision of the thermodynamic lattice energy

The precision of the thermodynamic lattice energy is governed by the total precision in the quantities set out in (6.54). Tables 6.8 and 6.9 list data on the enthalpy of sublimation of the alkali metals and the enthalpy of formation of the alkali-metal halides. Data on the other parameters in the Born–Haber–Fajans cycle have been presented earlier (see Sections 3.6.2 and 3.6.5.2).

Table 6.8 Standard enthalpies of sublimation of the alkali metals A

A	$\Delta H_a^{-0-}(A)$/kJ mol^{-1}
Li	161.5±5.5
Na	110.2±1.5
K	90.0±1.9
Rb	85.8±2.1
Cs	78.7±2.1

Table 6.9 Standard enthalpies of formation for the alkali-metal halides; all values are in kJ/mol^{-1} and the precisions are given in parentheses

	Li	Na	K	Rb	Cs
F	−612.1(8.4)	−571.1(2.1)	−562.7(4.2)	−549.4(8.4)	−530.9(16.7)
Cl	−405.4(8.4)	−412.5(0.8)	−436.0(0.8)	−430.5(8.4)	−433.0(8.4)
Br	−348.9(8.4)	−361.7(1.7)	−392.0(2.1)	−389.1(8.4)	−394.6(12.6)
I	−271.1(8.4)	−290.0(2.1)	−327.6(1.3)	−328.4(8.4)	−336.8(10.5)

The precision of ±2 kJ mol^{-1} that we quoted earlier for the lattice energy of sodium chloride derives from the uncertainties in its component quantities. The precision attainable in ΔU_C for sodium chloride and potassium chloride is high, but rarely is it matched with other compounds.

6.10 Polarization in ionic compounds: precision of the electrostatic model for lattice energy

The lattice energy equation (6.37) was set up in terms of a point-charge model for ions. The ions in the alkali-metal halides are subject to symmetrical polarization and, although ionic compounds have no permanent dipole moment, there are dipolar-type interactions to consider. Calculation shows that *induced* dipole–dipole and dipole–quadrupole terms have a significance at the level of precision of the thermodynamic calculations. We extend (6.37) to a form applicable at a finite temperature T:

$$U(r) = -\mathcal{A}q_1q_2e^2/(4\pi\epsilon_0 r) + B\exp(-r/\rho) - C/r^6 - D/r^8 + \phi(T\alpha/\kappa) \tag{6.55}$$

where C and D are constants of the induced dipole–dipole (see Section 5.4.5) and dipole–quadrupole energies, and the final term arises from the equation of state (4.105) for a solid at a temperature T other than zero.

Following the analysis of Section 6.8.2 leads to the equations

$$\rho/r_e = \frac{\mathcal{A}q_1q_2e^2/(4\pi\epsilon_0 r_e) + 6C/r_e^6 + 8D/r_e^8 - 3VT\alpha/\kappa}{9V\Phi(T,p)/\kappa + 2\mathcal{A}q_1q_2e^2/(4\pi\epsilon_0 r_e) + 42C/r_e^6 + 72D/r_e^8} \tag{6.56}$$

$$\Phi(T,p) = 1 + T(\partial\kappa/\partial T)_p/\kappa + \alpha(\partial\kappa/\partial p)_T + 2\alpha/3 \tag{6.57}$$

$$U(r_e) = -N_A[\mathcal{A}q_1q_2e^2(1-\rho/r_e)/(4\pi\epsilon_0/r_e) + C(1-6\rho/r_e)/r_e^6 + D(1-8\rho/r_e)/r_e^8 + 3VT\alpha(\rho/r_e)/\kappa] \tag{6.58}$$

Fuller descriptions of these equations have been given elsewhere.[16] Their general application is limited because there is only a small number of data available on temperature and pressure coefficients of compressibility, $(\partial\kappa/\partial T)_p$ and $(\partial\kappa/\partial p)_T$, respectively. Furthermore, once we move outside the realm of cubic structures, properties such as expansivity and compressibility become anisotropic.

When the full analysis given by (6.56) to (6.58) is applied to sodium chloride, the value obtained for $U(r_e)$ is -774 kJ mol^{-1}, in excellent agreement with the result from the thermodynamic (experimental) cycle. Not only is this result satisfactory in itself, but also it confirms the applicability of the electrostatic model for ionic compounds.

We note also that although our analysis in Section 6.4.2 used the form of (4.105) appropriate to 0 K for simplicity, the values of r_e and κ apply to normal temperatures. These errors are relatively small, but the best precision demands the use of (6.58), wherever possible, and with data appropriate to one and the same temperature. We shall consider polarization in ionic compounds again in a later section of this chapter.

6.11 Approximate calculation of lattice energy

Notwithstanding approximations exist in the lattice energy calculations that we have discussed, they are of value in obtaining information on a range of parameters for which experimental determinations are either difficult or not possible. However, in view of the paucity of the physical data needed for precise calculations of lattice energies, it is pertinent to consider the use of approximate equations.

One equation that is often put forward is that discussed uncritically by Kapustinskii[17] (1956). He gave an equation for the lattice energy $U(r_e)$ in kJ mol^{-1} as

$$U(r_e) = -(120.2\ \text{kJ mol}^{-1}\ \text{nm})\,\frac{nq_1q_2}{r_+ + r_-}\left(1 - \frac{0.0345}{r_+ + r_-}\right) \tag{6.59}$$

where n is the number of ions in the formula-weight and r_e is replaced by $(r_+ + r_-)$, the sum of the individual ionic radii. Effectively, this equation reduces all crystals to the sodium chloride type, and inserting the values in nanometres for the radii of Na^+ and Cl^- leads to the value of -748 kJ mol^{-1} which, not surprisingly, is a reasonable approximation to the best known value. There is, however, little to commend this equation for general use.

It is unnecessary to approximate r_e by the sum of the corresponding ionic radii, since so many reliable data are available readily from X-ray structure analyses. Furthermore, the calculation of a Madelung constant is not a problem, given even a small personal computer (see Appendix 16). It has been shown[18] that ρ/r_e is very close to 0.1 for a wide range of 1 : 1 ionic compounds. Hence, it is possible to approximate (6.41) by

$$U(r_e) = \frac{-(125.0\ \text{kJ mol}^{-1})\mathscr{A}q_1q_2}{r_e/\text{nm}} \tag{6.60}$$

Inserting the data for sodium chloride gives -775 kJ mol^{-1} for $U(r_e)$. This calculation can be very useful because, in the absence of sufficient other data, it may be the only approach available for the estimation of lattice energies.

[16] See, for example, M F C Ladd, *J. Chem. Soc. Dalton*, 220 (1977); M P Tosi, in *Solid State Physics*, Volume 16, 1ff, (Academic Press, 1964).

[17] A F Kapustinskii, *Quart. Rev. Chem. Soc.* **10**, 283 (1956).

[18] M F C Ladd and W H Lee, *J. Inorg. Nucl. Chem.* **11**, 264 (1959).

6.12 Uses of lattice energies

We consider here three applications of lattice energies that lead to information not always obtainable by other means.

6.12.1 Electron affinities and thermodynamic parameters

The first use of the electrostatic model with the alkali-metal halides, together with terms in (6.54), led to the quantities $\frac{1}{2}D_0(X_2)+E(X)$ for the halogens. The constancy of these composite terms for each halogen was regarded as good evidence for the model. As D_0 became measurable, so E was obtained from the average value of $(\frac{1}{2}D_0+E)$ for each halogen.

Single-electron affinities have been measured for the halogens (see Section 3.6.2.2) and for some other species, such as oxygen, sulfur and the hydroxyl group. The affinities of species such as oxygen or sulfur for two electrons must be deduced by combining the result of the electrostatic calculation of the appropriate lattice energy with the corresponding Born–Haber–Fajans cycle. In the case of oxygen, a value of 607 ± 25 kJ mol^{-1} has been found for $E(O\rightarrow O^{2-})$. It should be noted that, whereas the addition of a single electron to oxygen (and to other species) is a thermodynamically spontaneous process, the addition of a second electron requires an expenditure of energy on the system; the driving force for attaching the second electron lies in the combination of the gaseous ions to form the ionic solid; the lattice energy of magnesium oxide, for example,[19] is -3784 kJ mol^{-1}.

The separate terms D_0 and E do not have a clear meaning for polyatomic ions, such as $[NO_3]^-$. They may be addressed collectively by the term $\Delta H_f^{-0-}(NO_3^-,g)$, for example, representing the process *elements in the standard state*→ion-gas. If we set up an electrostatic calculation of $U(r_e)$ for a suitable nitrate, the enthalpy of formation of the ion-gas can be deduced; for the nitrate ion, a value of -351 kJ mol^{-1} for $\Delta H_f^{-0-}(NO_3^-,g)$ has been calculated.[20] The precision of these derived parameters will generally be less than that of the corresponding quantities obtained by experiment.

6.12.2 Compound stability

If an ionic compound NeCl were to exist, it would be not unreasonable for it to have the sodium chloride structure type. From (6.54), we may write

$$\Delta H_f(\mathrm{NeCl,s})\approx U_c(\mathrm{NeCl})+\Delta H_a(\mathrm{Ne})+I(\mathrm{Ne})+\tfrac{1}{2}D_0(\mathrm{Cl_2})+E(\mathrm{Cl})-n\mathcal{R}T-\Delta H_1 \qquad (6.61)$$

From (6.60), $U(r_e)=-218/(r_e/\mathrm{nm})$, and inserting the known quantities into (6.61) gives $\Delta H_f(\mathrm{NeCl,s})=U_c(\mathrm{NeCl})+1836$ kJ mol^{-1}. If the enthalpy of formation of NeCl is to be negative, which is a reasonable requirement, then $U_C(\mathrm{NeCl})$ must be less than -1836 kJ mol^{-1}, that is, $-218/(r_e/\mathrm{nm})<-1836$, or $r_e<0.119$ nm. Since the radius of the Cl^- ion alone is approximately 0.18 nm (see Table 6.12) the required situation cannot be achieved, and we conclude that NeCl cannot form a stable ionic compound. Because the ionization energy of neon is so large ($\simeq$2080 kJ mol^{-1}), the amount of energy that has to be expended in forming the ion-gas cannot be recovered in forming a crystalline ionic solid of the type AX.

Interesting compounds are formed by the noble gases under suitable conditions. The gas-phase reaction

$$O_2(g)+PtF_6(g)\rightarrow O_2[PtF_6](s) \qquad (6.62)$$

[19] M F C Ladd, *J. Chem. Phys.* **62**, 4583 (1974).
[20] M F C Ladd and W H Lee, *J. Inorg. Nucl. Chem.* **13**, 218 (1960).

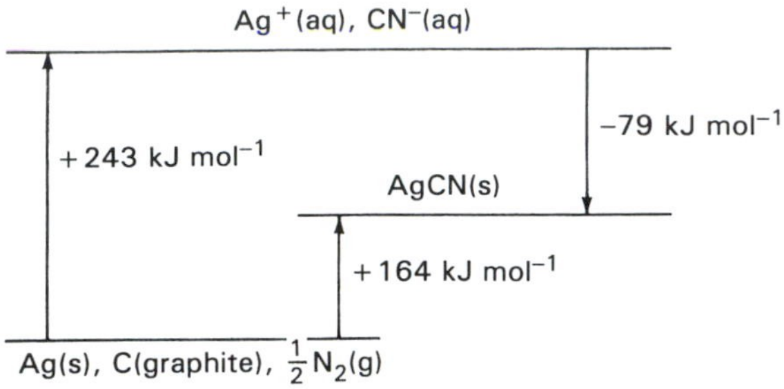

Figure 6.29 Free-energy-level diagram for silver cyanide; the solid is unstable with respect to the elements by 164 kJ mol^{-1}, but it is stable with respect to the aqueous ions by 79 kJ mol^{-1} and can be prepared by precipitation from aqueous solution.

led to a crystalline compound $O_2[PtF_6]$, and X-ray analysis showed that it consisted of the species O_2^+ and $[PtF_6]^-$; the oxygen molecule is oxidized by PtF_6. The lattice energy was calculated to be approximately -502 kJ mol^{-1} from an equation like (6.60), taking the experimental r_e of 0.435 nm and assuming the Madelung constant of 1.7476 for an NaCl-like structure.

The ionization energies for $O_2(g) \rightarrow O_2^+(g)$ and $Xe(g) \rightarrow Xe^+(g)$ are 1182 kJ mol^{-1} and 1175 kJ mol^{-1}, respectively. Thus, it is feasible to conduct the process of (6.62) with xenon in place of oxygen. It was found that xenon was oxidized in a similar manner and the crystalline compound $Xe[PtF_6]$ with a lattice energy of -460 kJ mol^{-1} was formed. These experiments heralded the chemistry of the noble gases, and several compounds, such as KrF_2, XeF_n (n=2, 4 and 6), XeO_3 and $XeNF(SO_2F)_2$, have been prepared. A summary of the chemistry of the noble gases has been given by Holloway.[21]

The existence of Cu(I)F has been challenged on the grounds that its free energy of formation is positive. This criterion refers to thermodynamic stability with respect to the elements in their standard states and does not preclude the existence of the compound. Another example is silver cyanide, AgCN. This compound, too, has a positive free energy of formation (+164 kJ mol^{-1}), but it can be prepared by precipitation from aqueous solutions of silver nitrate and potassium cyanide. Silver cyanide is thermodynamically unstable with respect to its elements in their standard state, but not with respect to the aqueous ions (Figure 6.29). Caution should be exercised in referring to 'stability'; the reference level should be precisely defined.

6.12.3 Charge distribution on polyatomic ions

Lattice energies may be used to estimate the distribution of charge on polyatomic ions; we shall illustrate the procedure with respect to the cyanide ion. At 298.15 K, KCN and NaCN, but not LiCN, have the sodium chloride structure type (Figure 1.2), with the cyanide ions in orientational disorder (see Section 6.1). At 279 K and 233 K, NaCN and KCN, respectively, transform to an orthorhombic structure (Figure 6.1). The cyanide ions are locked into fixed positions in the orthorhombic structure of the sodium salt at the higher temperature because of their stronger interaction with the Na^+ cation.

At the transition temperature, two polymorphs of a species have the same free energy; hence, at constant pressure, the free energy change at the transition temperature is given by

[21] J Holloway, *Chemistry in Britain*, 658 (1987).

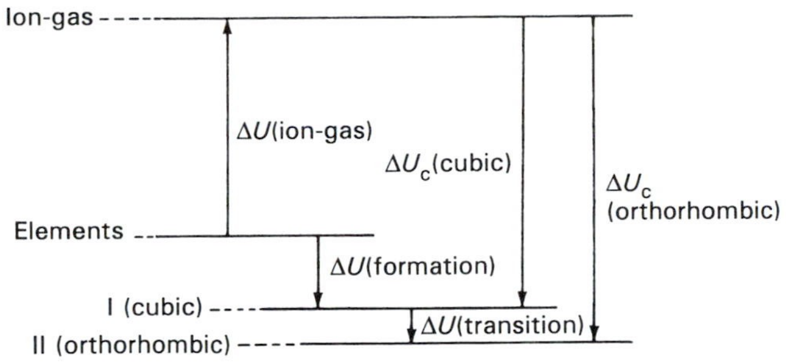

Figure 6.30 Energy-level diagram for the polymorphs I (cubic) and II (orthorhombic) of NaCN and KCN at a transition temperature; the two polymorphs may be related in energy through the value of $\Delta U\,(T_t\,\Delta S_t)$ at the transition temperature.

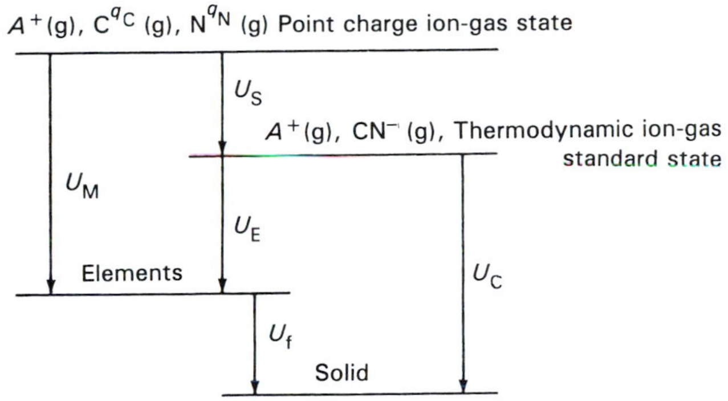

Figure 6.31 Energy-level diagram showing the relationship between electrostatic energies U_M (Madelung energy for point charges), U_S (self-energy of the CN^- ion) and U_E (Madelung energy for Na^+ and CN^- species) and the lattice energy U_C, for NaCN and KCN; U_f relates to the formation of the crystal from its elements and completes the cycle, although it is not used in the discussion.

$$\Delta G = \Delta U - T\Delta S + p\,\Delta V \tag{6.63}$$

The volume change from one solid polymorph to the other is negligible, and since $\Delta G=0$ at equilibrium we have

$$\Delta U = T\Delta S \tag{6.64}$$

In other words, the difference in the lattice energies of the two polymorphs may be determined from the entropy[22] change at the transition temperature (Figure 6.30). The values of ΔU are 2.9 kJ mol^{-1} and 1.3 kJ mol^{-1} for NaCN and KCN, respectively. The lattice energies for NaCN and KCN have been calculated in their cubic form, at the transition temperatures, and the corrections ΔU from (6.64) applied.

In the orthorhombic structures the CN^- ions are in fixed positions and the electrostatic energy must be calculated in two parts. The negative charge on the cyanide ion may be considered to be divided as $[C^{q_C}N^{q_N}]^-$, where $q_C + q_N = -1$. In Figure 6.31, the components of the electrostatic energy for a cyanide ACN are shown: U_M is the electrostatic energy obtained with a Madelung constant calculated on the lattice array of charged species A^+, C^{q_C} and N^{q_N}

[22] At the transition temperature, $\Delta H_t/T_t = \Delta S_t$.

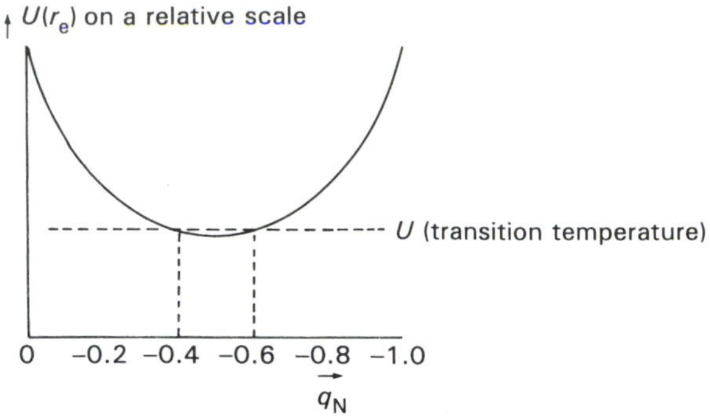

Figure 6.32 Variation of $U(r_e)$ for KCN, in its orthorhombic structure, as a function of q_N, the charge on the nitrogen atom; the symmetry of the structure is reflected in the shape of the curve, and the value of U_C (cubic) at the transition temperature intersects the curve of $U(r_e)$ at $q_N=-0.4$ and -0.6. A similarly symmetrical curve was obtained for NaCN but not for LiCN, which allowed the choice of -0.6 for q_N to be made.

and N; U_E is the electrostatic energy of the array of species A^+ and CN^- actually present in the crystal; the difference between them is U_S, the electrostatic self-energy of the cyanide ion. This term arises from the hypothetical process

$$C^{q_C}(g)+N^{q_N}(g)\rightarrow CN^-(g) \qquad (6.65)$$

which is included in the cycle of Figure 6.31.

The term U may be defined formally by

$$U_S=\sum_{i,j} q_iq_je^2/(4\pi\epsilon_0r_{ij}) \qquad (6.66)$$

where r_{ij} is the distance between the ith and jth species of charges q_i and q_j, respectively, in the complex ion. For the cyanide ion it reduces to

$$U_S=q_Cq_Ne^2/(4\pi\epsilon_0r_{e,CN^-}) \qquad (6.67)$$

Thus, we are able to determine $U(r_e)$ as a function of q_N. Then, by plotting $U(r_e)$ against q_N (Figure 6.32), the equivalent value of U from the cubic structure is satisfied by q_N equal to either -0.4 or -0.6. The two values arise because the orthorhombic structure is symmetrical with respect to spatial interchange of the C and N species for both NaCN and KCN, as far as has been determined by X-ray methods. This situation does not occur with LiCN, and the following results were obtained for the three cases:

	KCN	NaCN	LiCN
$U\ (q_N=-0.4)$	−669	−732	−682
$U\ (q_N=-0.6)$	−669	−732	−791

It is unreasonable that the lattice energy value for LiCN should lie between those for NaCN and KCN, because of the important proportionality to $1/r_e$; thus, the value $q_N=-0.6$ is preferred. Subsequently, this result was supported by quantum mechanical calulations on the CN^- ion,[23] which gave $q=-0.60$.

[23] M F C Ladd, *Trans. Faraday Soc.* **65**, 2712 (1969); *idem, J. Chem. Soc. Dalton*, 220 (1977).

Table 6.10 Equilibrium interionic distances/nm for the alkali-metal halides

	Li	Δ	Na	Δ	K	Δ	Rb	Δ	Cs	$\bar{\Delta}^{b}$
F	0.201	*0.030*	0.231	*0.036*	0.267	*0.015*	0.282	*0.018*	0.300	
Δ	*0.056*		*0.050*		*0.047*		*0.046*		*0.056*	*0.051*
Cl	0.257	*0.024*	0.281	*0.033*	0.314	*0.014*	0.328	*0.028*	0.356	
Δ	*0.018*		*0.017*		*0.015*		*0.015*		*0.015*	*0.016*
Br	0.275	*0.023*	0.298	*0.031*	0.329	*0.014*	0.343	*0.028*	0.371	
Δ	*0.025*		*0.025*		*0.024*		*0.023*		*0.024*	*0.024*
I	0.300	*0.023*	0.323	*0.030*	0.353	*0.013*	0.366	*0.029*	0.395	
$\bar{\Delta}^{b}$		*0.025*		*0.033*		*0.014*		*0.026*[a]		

Notes:
[a] It should be noted that CsCl, CsBr and CsI have the cesium chloride structure type under normal conditions of temperature and pressure; all others have the sodium chloride structure type.
[b] The symbol $\bar{\Delta}$ denotes the mean value of Δ for the row or column.

Several detailed reviews of lattice energy calculations and allied topics have been given[24] in the chemical literature.

6.13 Crystal chemistry

Crystal chemistry is concerned with structures themselves, and with the relationship between the properties of solids and their internal structure. Systematic crystal chemistry may be said to have begun in about 1920, after the publication of measurements of the interionic distances r_e for all the alkali-metal halides (Table 6.10).

The values of Δ show that the differences between the r_e values for two halides of a given cation are almost independent of the nature of the cation. Similarly, the differences between the r_e values for two halides of a given anion are almost independent of the nature of the anion. These results suggest that ions have a characteristic radius, whereby the sums of the radii are equivalent to the corresponding interionic distances. Thus

$$r_e(\mathrm{KCl}) = r(\mathrm{K}^+) + r(\mathrm{Cl}^-) \tag{6.68}$$

$$r_e(\mathrm{NaCl}) = r(\mathrm{Na}^+) + r(\mathrm{Cl}^-) \tag{6.69}$$

whence the difference Δ for the two chlorides becomes

$$\Delta(\mathrm{KCl} - \mathrm{NaCl}) = r(\mathrm{K}^+) + r(\mathrm{Na}^+) \tag{6.70}$$

and this relationship would apply whichever halogen were involved.

6.13.1 Ionic radii

The experimental interionic distance r_e must be divided into its ionic components. In the earliest method (Landé, 1920), six compounds that have the sodium chloride structure type, listed in Table 6.11, were studied. The constancy of r_e for the selenides and sulfides,

[24] M F C Ladd and W H Lee, *Progress in Solid State Chemistry*, Volume 1 (Pergamon, 1964); *idem ibid.* Volume 2 (Pergamon, 1965); *idem ibid.* Volume 3 (Pergamon, 1966), and references therein; T C Waddington, *Advances in Inorganic Chemistry and Radiochemistry*, Volume 1 (Academic Press, 1959).

Table 6.11 Interionic distances for some compounds with the sodium chloride structure type

	r_e/nm		r_e/nm
MgO	0.210	MnO	0.222
MgS	0.260	MnS	0.261
MgSe	0.273	MnSe	0.273

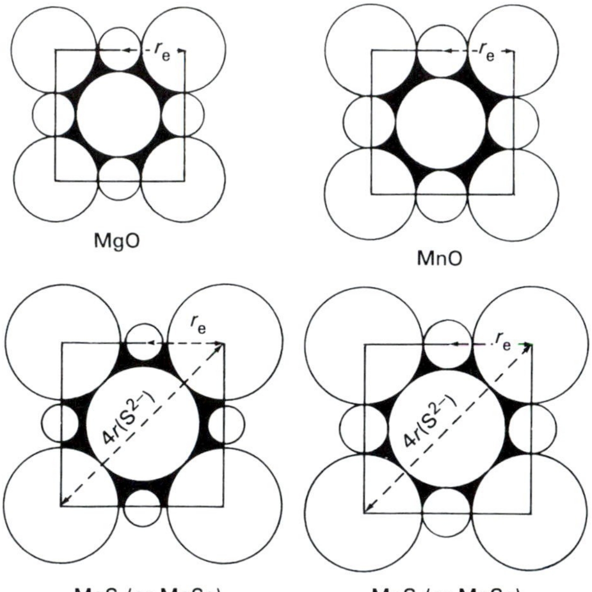

Figure 6.33 Close-packed arrays of ions, as seen in projection on to a cube face; the sodium chloride structure type exists for all of the examples here. Close contact of the anions across the face of a cube diagonal is obtained for MgS (MgSe) and MnS (MnSe), but not for MgO (MnO); thus, for the sulfides and selenides, $2r_e\sqrt{2}=4r_-$.

irrespective of the change in cation, was taken to indicate that the anions were in close contact, with the smaller cations occupying the interstices among the close-packed anions (Figure 6.33).

In the close-packed arrangement of anions, it is evident that $2r_e\sqrt{2}=4r_-$. Hence, $r(Se^{2-})=0.193$ nm and $r(S^{2-})=0.184$ nm. The oxides cannot be treated in this way because their interionic distances indicate a lack of, or uncertainty about, close packing in these compounds. If we accept the additivity of ionic radii, that is, $r_e=r_++r_-$, as implied by Table 6.10, then other radii can be deduced.

Pauling (1930) showed that the radius of an atom is governed mainly by the configuration of the outermost electrons in the corresponding atom. For an isoelectronic series of ions, he gave the relationship

$$r_i=c/(Z_i-\sigma) \tag{6.71}$$

Figure 6.34 Idealized electron density contour map for LiF (sodium chloride structure type), as seen in projection on to a cube face; it may be noted that the anions are not in contact in this structure whereas they are in LiI and LiBr. The dashed lines represent the contours of zero electron density, within experimental error, and so define the spatial limits of the ions.

where c is a constant for an isoelectronic series of ions, Z_i is the atomic number of the ith ion of radius r_i and σ is the screening constant given by Slater's rules (see Appendix 7).

Considering NaF, for example, we have from (6.71)

$$r(\mathrm{Na}^+)=c/(11-4.15) \tag{6.72}$$

$$r(\mathrm{F}^-)=c/(9-4.15) \tag{6.73}$$

whence

$$(\mathrm{Na}^+)/r(\mathrm{F}^-)=0.708 \tag{6.74}$$

Since r_e (NaF)=0.231 nm (Table 6.10), then using $r_e=r_++r_-$, it follows that $r(\mathrm{Na}^+)=0.096$ nm and $r(\mathrm{F}^-)=0.135$ nm.

Ladd[25] showed that by applying Landé's method to LiI (and to LiBr), for which, among the alkali-metal halides, a close-packed array of anions was most likely to exist, the same set of radii was obtained as was deduced from direct measurements on electron density contour maps that were obtained by X-ray crystallographic studies on the alkali-metal halides[26] (Figure 6.34). Table 6.12 lists Ladd's radii and Pauling's radii for several species, together with values for cations from Shannon and Prewitt.[27] It may be seen that some uncertainty is still attached to the precise value of an ionic radius. However, the close correspondence between the results given by Ladd and by Shannon and Prewitt, obtained by different procedures, would tend to indicate that a greater reliability may be attached to these values.

With the exception noted, the radii in Table 6.12 refer to six-coordination. Measurements on polymorphic species show that radii are dependent on the coordination number. With six-coordination as the standard, the following scheme applies:

four-coordination	standard−5%
six-coordination	standard
eight-coordination	standard+3%

[25] M F C Ladd, *Theoret. Chim. Acta* **12**, 333 (1968).

[26] H Witte and H Wîlfel, *Z. phys. Chem.* (Frankfurt), **3**, 296, (1955); *idem, Rev Modern Phys.* **30**, 51 (1958).

[27] R D Shannon and C T Prewitt, *Acta Crystallog.* B **25**, 925 (1969); *idem ibid.* B **26**, 1046 (1970).

Table 6.12 Radii/nm of some ionic species

	Ladd	Pauling	Shannon and Prewitt
Li^+	0.086	0.060	0.088
Na^+	0.112	0.095	0.116
K^+	0.144	0.133	0.152
Rb^+	0.158	0.148	0.163
Cs^+	0.184	0.169	0.184
NH_4^+	0.166	0.148	—
Ag^+	0.127	0.126	0.129
Tl^+	0.154	0.140	—
Be^{2+}	0.048[a]	0.031[a]	0.041[a]
Mg^{2+}	0.087	0.065	0.086
Ca^{2+}	0.118	0.099	0.114
Sr^{2+}	0.132	0.113	0.130
Ba^{2+}	0.149	0.135	0.150
H^-	0.139	0.208	—
F^-	0.119	0.136	—
Cl^-	0.170	0.181	—
Br^-	0.187	0.195	—
I^-	0.212	0.216	—
O^{2-}	0.125	0.140	—
S^{2-}	0.170	0.184	—
Se^{2-}	0.181	0.198	—
Te^{2-}	0.197	0.221	—

Note:
[a] Four-coordination; otherwise six-coordination.

6.13.2 Radius ratio and AX structure types

The radius ratio $\mathbb{R}$ is defined as the ratio of the cationic radius to the anionic radius, r_+/r_-. It is a guide to probable structure type, but it is not followed in all compounds. We review the concept first in relation to simple structures of the type AX.

Consider the cesium chloride structure type (Figure 6.35), in which the coordination pattern is 8 : 8. Let the ions be of such a size that adjacent anions at the corners of the unit cell are in contact with one another and with the central cation. Then the unit-cell side a is equal to $2r_-$ and the body-diagonal $a\sqrt{3}$ is equal to $2(r_+ + r_-)\sqrt{3}$. Hence, it follows that

$$r_-\sqrt{3} = r_+ + r_- \tag{6.76}$$

whence

$$\mathbb{R}_8 = r_+/r_- = 0.732 \tag{6.77}$$

On reducing the cation radius while the anion radius remains constant, contact between the cation and the anions is lost. Thus, there is a separation of charged regions, although the distance between the ion centres has not changed, and one may consider whether or no a more stable structure might be attained by a change in the coordination pattern. In fact, the sodium chloride structure type, with a 6 : 6 pattern, offers an energetically more stable configuration at the smaller value of $\mathbb{R}$ now under consideration. Referring to Figure 6.33

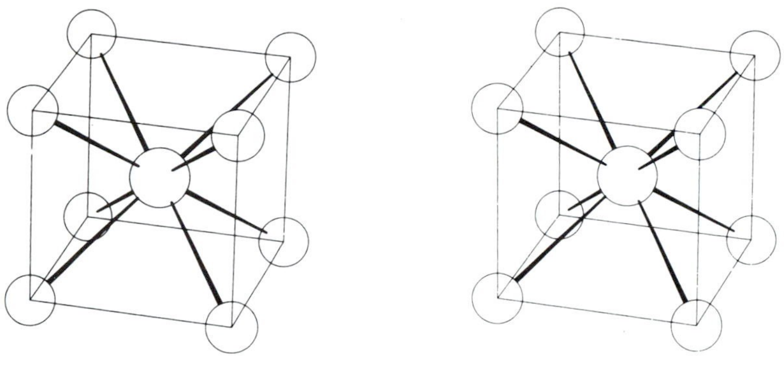

Figure 6.35 Stereoview of the unit cell and environs of the cesium chloride structure type; the circles represent, in decreasing order of size, Cl^- and Cs^+. It is a primitive (P) cubic unit cell, with one ion of each type per unit cell volume.

for MnS, we can see that the maximum contact situation corresponds to the new value for $\mathbb{R}$ given by

$$\mathbb{R}_6=\sqrt{2}-1=0.414 \tag{6.78}$$

Continuing in this manner, we obtain $\mathbb{R}_4=0.225$ for the 4 : 4 coordination in the würtzite and blende structures of zinc sulfide (Figures 6.36a and b).

We can interpret these results by means of the lattice energy equation. Let (6.41) be written in the form

$$U(r_e)=\alpha/r_e \tag{6.79}$$

where $\alpha=-10^{-3}N_A\mathcal{A}q_1q_2e^2(1-\rho/r_e)/(4\pi\epsilon_0)$. Let $q_1=q_2=1$ and $\rho/r_e=0.1$; then, assuming that $r_e=r_++r_-$ and keeping a fixed value of r_-, we can write (6.79) as

$$U(r_e)=\beta/(\mathbb{R}+1) \tag{6.80}$$

where $\beta=\alpha/r_-$ and $\mathbb{R}$ is the radius ratio. The graph of $U(r_e)$ against $\mathbb{R}$ is shown in Figure 6.37. Starting at $\mathbb{R}=1$, the cesium chloride structure type is the most stable. As $\mathbb{R}$ is decreased, keeping r_- constant, $U(r_e)$ decreases, as the graph shows. When $\mathbb{R}=0.732$ the ions are in maximum contact and cannot become closer packed, even though the central cation may become smaller than its surrounding hole.

However mobile we may consider the cation in its central hole, r_e remains constant and the curve for the cesium chloride structure type becomes horizontal at $\mathbb{R}=0.732$, in terms of this model. At this value of $\mathbb{R}$, the energy of the structure may be decreased further by its adopting the sodium chloride arrangement of component ions. Similar arguments can be applied at $\mathbb{R}=0.414$. We may summarize the results as follows:

Structure type	$\mathbb{R}$
CsCl	≥ 0.732
NaCl	0.414–0.732
α-ZnS/β-ZnS	0.225–0.414

Values of $\mathbb{R}$ for the alkali-metal halides are listed in Table 6.13.

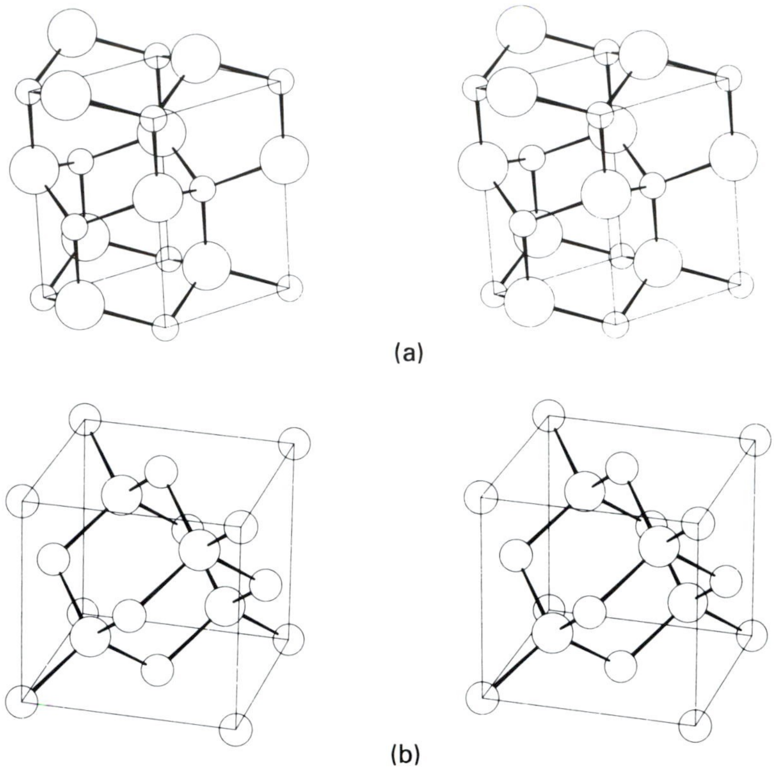

Figure 6.36 Stereoviews of the unit cells and environs of the zinc sulfide structures; circles in decreasing order of size represent S and Zn: (a) α-ZnS, würtzite, (b) β-ZnS, blende. Tetrahedral coordination (4 : 4) is present in both of these structures (compare Figure 1.2). Blende transforms to würtzite at 1023 °C; both forms occur in nature, so that a return to the equilibrium state at normal temperatures is infinitely slow.

Eight values of $\mathbb{R}$ greater than 0.732, enclosed by the broken boundary in Table 6.13, are found for halides that have the sodium chloride structure under normal conditions. The radius ratio is a geometrical concept based on the packing of spheres, and we must not be too disturbed to find that such a simple approach has its limitations. There are three factors to consider.

The energy difference between the cesium chloride and sodium chloride structure types for a given compound is only about 8 kJ mol^{-1}; it depends, in a first analysis, upon the difference in the Madelung constants (Table 6.7). Remembering that this value is based upon the point-charge model (6.41), we may expect that the dipole and quadrupole terms in (6.58) will take on particular significance when considering small differences in electrostatic (Madelung) energy. It has been shown by calculating the lattice energy of potassium chloride, by (6.58), in both the cesium chloride and sodium chloride structure types that the sodium chloride form is more stable by approximately 8 kJ mol^{-1}.

There is also a small covalent contribution to the lattice energy, even in these highly ionic compounds. In the sodium chloride structure, the p orbitals of adjacent ions are directed towards one another, thus facilitating a tendency towards overlap. In the cesium chloride

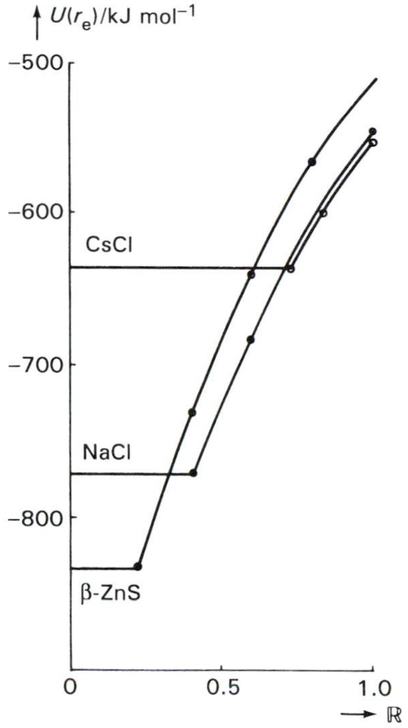

Figure 6.37 Variation of lattice energy $U(r_e)$ with radius ratio ℝ, at a constant value of r_-, according to (6.80); the small difference in electrostatic energy between the CsCl and NaCl structure types is very evident. Thus, contributions to the lattice energy from polarization become important in determining the structure type that is obtained.

structure type this situation does not obtain, because of the different coordination pattern in relation to the p orbital directions.

A change in external conditions that would lead to closer packing, that is, an increase in pressure or a decrease in temperature, often brings about a transformation in the solid state. For example, rubidium chloride transforms at 83 K to the cesium chloride structure type, without appreciable change in the radius ratio.

The difference in energy between α-ZnS and β-ZnS is only 0.2% (Table 6.7). Figure 6.37 indicates that, at ℝ<0.414, the zinc sulfide structure type is more stable than that of sodium

Table 6.13 Experimental radius ratios ℝ for the alkali metal halides

	Li	Na	K	Rb	Cs
F	0.72	0.94	0.83[a]	0.75[a]	0.65[a]
Cl	0.51	0.66	0.85	0.93[a]	0.92
Br	0.46	0.60	0.77	0.84	0.98
I	0.41	0.53	0.68	0.75	0.87

Note:
[a] For ℝ>1, 1/ℝ is quoted.

Table 6.14 Interionic distances and radii sums in the silver halides

	Structure type	r_e/nm	Σr_i/nm	Δ/nm
AgF	NaCl	0.246	0.246	0.00
AgCl	NaCl	0.277	0.297	0.20
AgBr	NaCl	0.288	0.314	0.26
AgI	β-ZnS	0.281	0.322[a]	0.41

Note:
[a] Including a −5% adjustment to take into account the change in coordination from six to four.

chloride; nonetheless, it does not occur, even in lithium iodide. Phillips[28] has pointed out empirically that four-coordinated structures tend to arise when the degree of ionic character is less than about 0.78. Calculations for the alkali-metal halides indicate a range of ionic character from 0.89 in LiI to 0.96 in RbF, so that a four-coordinated structure is unlikely among these compounds.

Other attempts to set up demarcation lines between four- and six-coordinated ionic structures based, for example, on different scales of electronegativity simply restate in other terms that the structural and physical data needed to establish good energy calculations have yet to be measured with precision. When such data are known, a refined electrostatic model such as (6.58) will provide a satisfactory account of the energetics of ionic compounds.

6.13.3 Polarization in ionic structures

Polarization, which is expressed in (6.58) through the C and D terms, may be likened to a distortion of the electron density of the ions by the electric field of neighbouring ions, with consequent induced dipole–dipole and dipole–quadrupole interactions. Separately, there is also a tendency towards covalent overlap where the orbitals of the ions have favourable relative orientations. The presence of these enhancements of the simple electrostatic energy may be indicated by a comparison of sums of ionic radii with the corresponding experimental r_e distances; the silver halides provide a good example (Table 6.14).

Among ions of similar radii, it is known that polarization effects are larger when an outermost d electron configuration exists, probably because of the screening effect of the d electrons on the outermost electrons. If we compare this screening in Ag^+ with that in the comparably sized Na^+ and K^+, we find the following results:

	Ag^+	Na^+	K^+
Z_{eff}	8	17%	10
Z_{eff}/Z	17%	64%	53%

The extent of screening of the nuclear charge in the silver ion results in its electron distribution being much less strongly held than would have been expected for an ion of its size; thus, the Ag^+ electron distribution is more susceptible to distortion than those of Na^+ and K^+.

[28] J C Phillips, *Rev. Modern Phys.* **42**, 317 (1970).

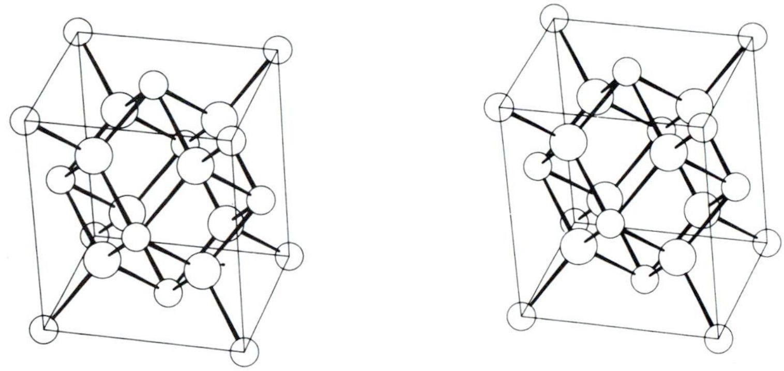

Figure 6.38 Stereoview of the unit cell and environs of the fluorite (CaF_2) structure type; 8 : 4 coordination. Circles in decreasing order of size represent F and Ca (compare Figure 6.35).

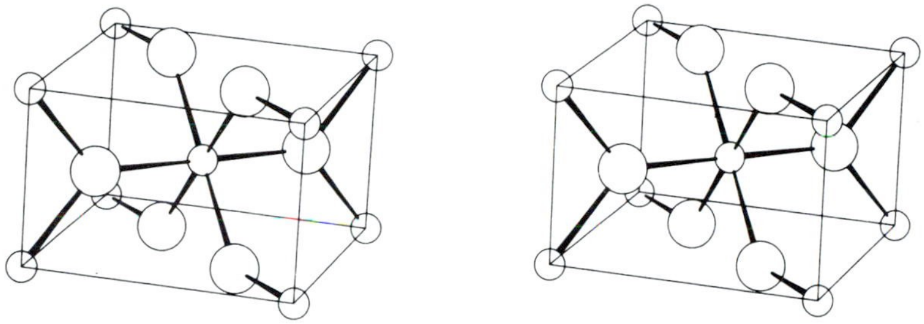

Figure 6.39 Stereoview of the unit cell and environs of the rutile (TiO_2) structure type; 6 : 3 coordination. Circles in decreasing order of size represent O and Ti (compare Figure 1.2).

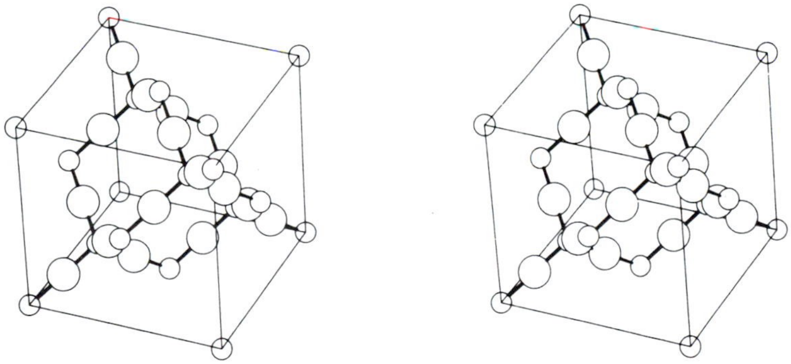

Figure 6.40 Stereoview of the unit cell and environs of the β-cristobalite (SiO_2) structure type; 4 : 2 coordination. Circles in decreasing order of size represent O and Si. β-Cristobalite is related to blende (Figure 6.36b) as tridymite, another form of SiO_2, is to würtzite (Figure 6.36a).

The result of this effect is further manifested in solubility relationships, as we shall discuss later.

6.13.4 Radius ratio and AX_2 structure types

The most common AX_2 structure types are fluorite (CaF_2) and rutile (TiO_2); β-cristobalite (SiO_2) is found less frequently. These structures are illustrated in Figures 6.38 to 6.40. Since

Table 6.15 Radius ratios ℝ for some AX_2 structure types

Fluorite		Rutile		β-Cristobalite	
BaF_2	1.25	$CaCl_2$	0.69	β-SiO_2	0.30
SrF_2	1.11	$CaBr_2$	0.63	BeF_2	0.23
$BaCl_2$	0.88	MgF_2	0.73		
CaF_2	0.99	MnF_2	0.68		
$SrCl_2$	0.78	ZnF_2	0.62		

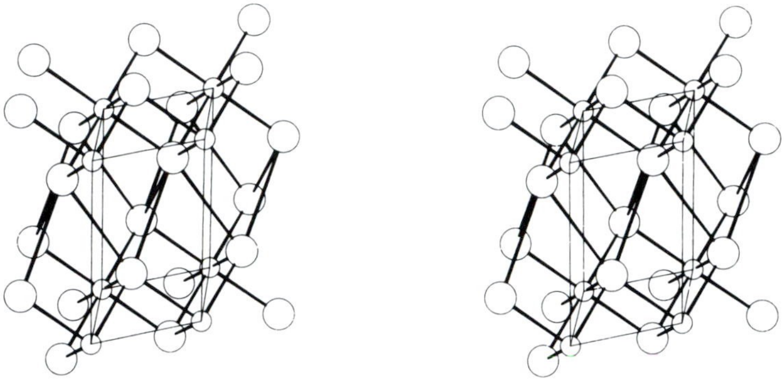

Figure 6.41 Stereoview of the unit cell and environs of a typical layer structure, CdI_2; circles in decreasing order of size represent I and Cd. The 6 : 3 coordination consists of a slightly distorted octahedron of iodine around cadmium, with the three cadmium nearest neighbours of any iodine atom lying to one side of it in the composite layer, forming a trigonal pyramid with iodine at the apex.

their coordination patterns are 8 : 4, 6 : 3 and 4 : 2, respectively, the radius ratio limits are the same as those for the cesium chloride, sodium chloride and blende (β-ZnS) structure types, respectively. Table 6.15 lists the radius ratios for some structures in this group.

Among these compounds, the radius ratio is obeyed without exception. From Table 6.7, it is evident that the electrostatic energy differences among these structures are very much greater than are those among the *AX* structures. Although polarization effects are larger with the smaller, more highly charged species, the electrostatic component of their lattice energies remains dominant.

Among AX_2 structures, a decrease in the radius ratio, which can be accompanied by an increase in polarization, may lead to layer structures, of which cadmium iodide is typical (Figure 6.41). As polarization increases or ℝ decreases further, discrete groups of atoms may segregate into molecules, characteristic of molecular solids. Mercury(II) chloride is such a case, and we referred to this compound in Section 6.3.2.3. It provides an example of the inevitable merging within the arbitrary classification of compounds that we defined. The sequence of changes among AX_2 structure types with changes in radius ratio and polarization is shown diagrammatically in Figure 6.42.

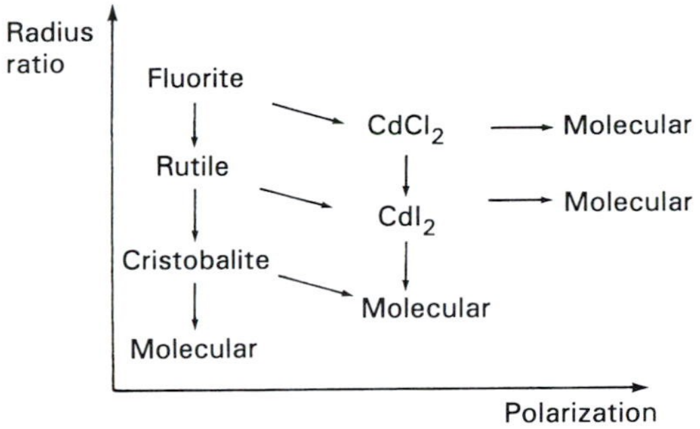

Figure 6.42 Schematic illustration of transformations among AX_2 structure types according to radius ratio and polarization; $CdCl_2$ and CdI_2 are typical layer structures, intermediate in type between the ionic and molecular AX_2 compounds.

6.14 Structural and physical characteristics of ionic compounds

Ionic bonds link an ion to an indefinite number of neighbours. The bond has no particular directionality in space, so that ionic structures tend to be governed, to a first approximation, by geometrical considerations, subject to electrical neutrality of the structure as a whole. The Na^+ ion is sufficiently small that 12 of them could be packed around a Cl^- ion. However, the Cl^- ion cannot reciprocate this behaviour and, under electrical neutrality, sixfold coordination is obtained. Polarization and partial covalency enhance the value of the lattice energy of ionic compounds, and may modify the predictions of the radius ratio, particularly when the difference in the electrostatic energies of two polymorphs is small.

A refined electrostatic model gives satisfactory lattice energy values. It should be remembered that agreement between the electrostatic equation (6.58) and that based on thermodynamic terms (6.54) need not necessarily imply the correctness of the electrostatic model, but rather that it forms a satisfactory basis for the calculation of lattice energetics.

Molecules do not occur in ionic structures, because electrons are almost totally localized in the atomic orbitals of ions and could not overlap with one another sufficiently to form molecules without a change in status of the compound. In solids containing polyatomic ions such as nitrates, carbonates and sulfates, covalent bonding is predominant within the complex ion, with ionic bonding taking place between the polyatomic ions and those species of complementary charge in the structure. Figure 6.43 illustrates the structure of calcite. The planes of the CO_3^{2-} ions are all normal to the vertical direction, a threefold symmetry axis of the rhombohedral unit cell, and the Ca^{2+} ions are each coordinated by six oxygen atoms from different carbonate ions.

Ionic compounds form hard crystals of low compressibility and expansivity, but of high melting-point. Generally, they are electrical insulators in the solid, but when molten, or in solution, they conduct electricity by ion transport; this property distinguishes ionic solids most clearly from covalent and van der Waals' solids. Certain unusual solids, such as MAg_4I_5 where M includes K, NH_4 or Rb, have high electrical conductances, in the region of 10 Ω^{-1} m^{-1}; they exhibit disorder in the solid state.[29]

Trends in physical properties can be related to lattice energies through the dominant $1/r_e$ proportionality. Table 6.16 lists some results on hardness and melting-point.

[29] See, for example, M F C Ladd and W H Lee, *Z. Kristallog.* **129**, 157 (1969).

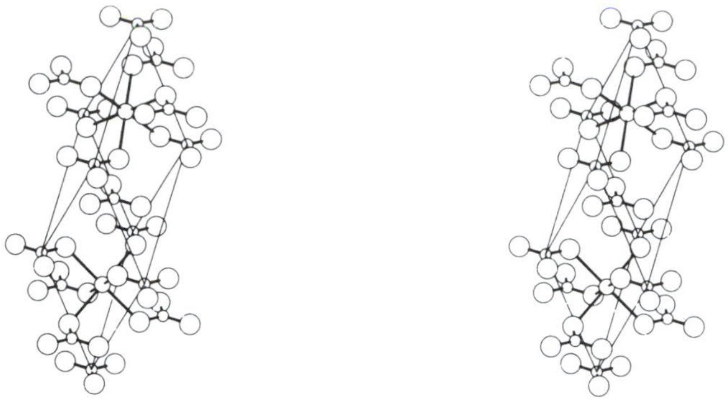

Figure 6.43 Stereoview of the unit cell and environs of the calcite $CaCO_3$ structure; circles in decreasing order of size represent O, Ca and C. The calcium ion is in sixfold coordination with oxygen atoms of different carbonate ions. Each carbonate ion has a trigonal planar structure and lies normal to the direction of the threefold symmetry axis in the rhombohedral unit cell. Sodium nitrate exhibits the same structure type.

6.15 Vibrations and defects in ionic compounds

The first model that we discussed for ionic crystals considered them to be perfectly regular, three-dimensional arrays of point charges. This picture was modified by introducing terms to allow for van der Waals' interactions and also admitted the existence of a covalent contribution to the lattice energy. It retained, nevertheless, a static system of particles, each on its geometrically correct lattice site. Real ionic solids possess defects in their structure; they undergo transitions that may result in the appearance of colour; and the ions execute vibrational motion about their lattice sites, which makes the major contribution to the heat capacity of solids (see Section 6.5.3).

6.15.1 Absorption spectra

All ionic solids absorb in the ultraviolet region of the spectrum; some compounds absorb also in the visible region. Unlike metals, the absorption does not produce opacity; often, apparently opaque crystals are simply intensely coloured, like permanganates, and transmission of light can be observed through thin sections. Ionic solids are coloured for two main reasons. Some of them may contain ions that give rise to a characteristic colour through transitions involving their d electrons, as with salts of $[Ti(H_2O)_6]^{3+}$, $[Fe(H_2O)_6]^{2+}$ or $[Cr_2O_7]^{2-}$. In Chapter 2, we saw that d electron states in a free atom are degenerate. In a polyatomic ion, however, the spherical symmetry of the environment of the atom is lost through combination with ligands, and electrons can absorb energy by making transitions to the allowed d energy states, t_{2g} and e_g of the central atom (see Section 2.14). The electron travels a relatively large distance in such a transition, so that the transient dipole moment is large and the absorption correspondingly intense. The ligand-field splitting energy parameter Δ is such ($\approx$2.5 eV) that transitions occur in the visible range of the spectrum, so that the compound appears coloured. The colour in the $[MnO_4]^-$ ion arises from a charge-transfer transition, in which electrons are transferred from the ligand to the d orbitals of the central atom; in this example, Δ is approximately 2.3 eV, so that λ is approximately 540 nm.

In other coloured solids, the ions that impart no colour when in solution undergo polarization in the solid state and colour arises from transitions among partially delocalized

Table 6.16 Hardness and melting-point for some ionic solids

	BeO	MgO	CaO	SrO	BaO
r_e/nm	0.165	0.210	0.240	0.257	0.276
Hardness (Moh's scale)[30]	9.0	6.5	5.5	4.1	3.3
	NaF	NaCl	NaBr	NaI	
r_e/nm	0.231	0.282	0.298	0.323	
T_m/K	1266	1074	1020	934	

electrons. If the absorption were to shift from the ultraviolet just into the visible, the wavelengths absorbed would lie at the blue end of the spectrum and the compound would appear yellow to red. Examples of this situation are found with the compounds AgI, Ag_3PO_4 and Pb_3O_4.

6.15.2 Heat capacity

As the temperature of a solid is increased its vibrational energy increases. If there are no translational energy modes available, then, since rotation of a species about its own axis does not constitute a degree of freedom, all of the energy imparted to the solid in the form of heat enhances its vibrational energy. A vibration can be resolved along Cartesian axes into three mutually perpendicular squared terms (those that depend upon the square of a velocity or of a coordinate), so that there are $6N_A$ degrees of freedom per mole of a solid AX and each squared term contributes an amount of energy kT (see Appendix 3).

Sodium chloride contains $2N_A$ ions per mole of substance; if we assume that each ion vibrates independently of the others present, the total vibrational energy is $6N_A k_B T$, or $6\mathcal{R}T$ per mole. Thus, C_v is $6\mathcal{R}$, or approximately 50 kJ mol^{-1} (twice the Dulong and Petit value for monatomic species; see Section 6.5.3).

The principal molar heat capacities are related by $C_p - C_v = \alpha^2\, TV/\kappa$ (see the solution to Problem 4.8). For sodium chloride at 298 K, $\alpha = 1.1\times10^{-4}$, $\kappa = 4.1\times10^{-11}$ N^{-1} m and $V = 2.7\times10^{-5}$ m^3 mol^{-1}. Hence, $C_p - C_v = 2.4$ J mol^{-1} K^{-1}. For the ideal gas, $C_p - C_v = \mathcal{R}$; the smaller value for the solid arises from the fact that its heat capacity is determined mainly by the vibrational energy, and this parameter does not vary appreciably between constant pressure and constant volume conditions.

A vibrating atom acquires energy in quanta of $h\nu$, and the (Boltzmann) probability that a vibrating atom receives this amount of energy is proportional to $\exp[-h\nu/(k_B T)]$. At room temperature, almost all of the vibrational degrees of freedom are active, so that C_v tends to its limiting value of approximately 50 J K^{-1} mol^{-1} . As the temperature is decreased vibrations of increasingly lower frequency cease to be excited. Thus, C_v decreases with decreasing temperature, because the average energy of the solid decreases. In the limit as $T\to0$, $C_v\to0$ for a pure crystalline solid, in accordance with the third law of thermodynamics.

The effect of a change in temperature on the atomic vibrations of a solid can be gleaned from the following example. Sodium chloride gives a single absorption band at 164 cm^{-1}, and the ratio of the probabilities that this oscillator will acquire the corresponding energy $hc\bar{\nu}$ at 500 K and at 50 K is approximately 70, from the Boltzmann law. The energy absorbed is used mainly in increasing the amplitudes of the vibrating atoms.

[30] On Moh's scale of scratch hardness, diamond is 10 and talc 1, on a scale of 1 to 10.

The requirement that a vibration shall be active in the infrared region of the spectrum is that the species shall contain an oscillating dipole (see Section 3.6.5.3), and the rate of change of dipole moment with time determines the ability of a species to absorb infrared radiation. In ionic solids, each pair of vibrating ions of opposite sign is equivalent to an oscillating dipole and its vibrations are excited by infrared radiation; the heavier the species the smaller the wavenumber (energy) of the vibration.

6.15.3 Defects in crystals

From perfect crystals of vibrating atoms, we treat the fully realistic situation of such crystals containing imperfections. Almost all crystalline substances contain defects. Defects may be classed as *intrinsic*, when they occur in an otherwise perfect substance; or they may be termed *extrinsic*, when they derive from the presence of impurities (cf. Section 6.5.7). The subject of crystal defects is extensive and we shall investigate here only those imperfections known as *point defects,* which can exist in both intrinsic and extrinsic forms. Point defects may be limited to random single sites, or they may be linked in one or more dimensions to form *extended* defects.

The introduction of defects into a perfect solid requires an expenditure of energy on the system, but the creation of defects brings about an increase in disorder (entropy) that reduces the free energy for the system with defects. Provided that the change in the Helmholtz free energy (4.85) for the creation of defects (at constant volume) is negative, a system with defects is thermodynamically the more stable.

6.15.3.1 Schottky defect

The simplest intrinsic point defect in an otherwise perfect crystal is the Schottky defect, for which an ion (or atom) is transferred from the interior of the crystal to its surface (Figure 6.44a). In a crystal at thermal equilibrium with its surroundings, a certain number of vacancies at atom sites throughout the lattice array always exists and contributes to the entropy of the crystal, at any temperature above absolute zero.

Consider a fixed volume of crystal and let n Schottky defects be distributed over N lattice sites. The first defect can be distributed in N ways, the second defect in $(N-1)$ ways, and the nth in $[N-(n-1)]$ ways. Thus, the total number of arrangements w of the n defects is given by

$$w=N(N-1)(N-2)\ldots[N-(n-1)]=N!/(N-n)! \tag{6.81}$$

However, defects are indistinguishable from one another: we cannot differentiate between defects i and j on a given site. There are n indistinguishable ways of obtaining the first defect, $(n-1)$ ways of obtaining the second, $(n-2)$ ways for the third, and so on, and in all $n!$. Then, the total number of arrangements, or probability W of the system, that we require is

$$W=N!/[(N-n)!\ n!] \tag{6.82}$$

Let Δu represent the energy change for the formation of a single defect; then

$$\Delta A=n\,\Delta u-T\,\Delta S \tag{6.83}$$

for the formation of the n defects. Using (4.71),

$$\Delta A=n\,\Delta u-k_{\mathrm{B}}T\ln\{N!/[(N-n)!\ n!]\} \tag{6.84}$$

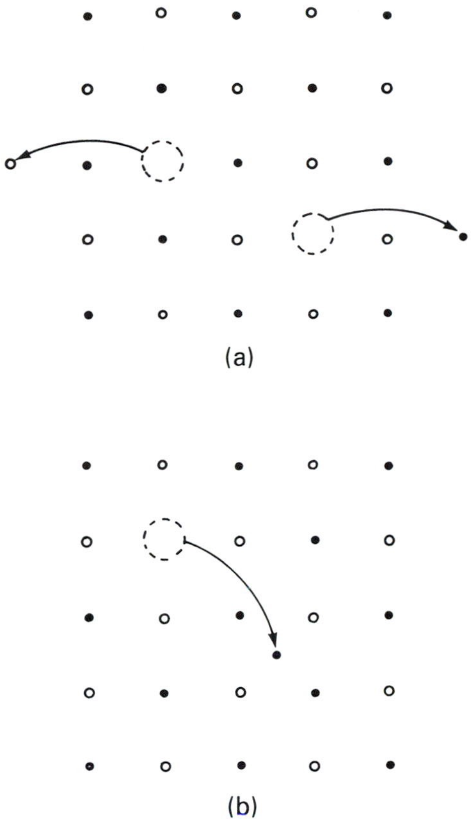

Figure 6.44 Intrinsic point defects; cations are the filled circles, anions the open circles. (a) Pair of Schottky defects in KCl; the ions are removed from the bulk of the substance to its surface. (b) Frenkel defect in AgCl; the cation moves to an interstitial position and so comes under a compressive stress.

At equilibrium, from Sections 4.7 and 4.9,

$$(\partial \Delta A/\partial n)_T = 0 \tag{6.85}$$

and using Stirling's approximation for factorials

$$\ln(X!) = X \ln(X) - X \tag{6.86}$$

we obtain

$$\Delta u - k_B T \ln[(N-n)/n] = 0 \tag{6.87}$$

If we assume that N is very large, then $n \ll N$ and (6.87) rearranges to

$$n = N \exp[-\Delta u/(k_B T)] \tag{6.88}$$

In ionic crystals, it is energetically favourable to form Schottky defects in pairs, one of each sign, in order to maintain the best electrical balance. Then W is squared and we obtain

$$n_{\pm} = N \exp[-\Delta u/(2k_B T)] \tag{6.89}$$

The average nearest neighbour bond energy in a solid is approximately 1 eV, and this amount of energy has to be expended on the system to create a vacancy. The structure then readjusts around the vacancy, and approximately two-thirds of the energy expended is recovered in this process. The entropy change is positive and contributes to the driving force for the creation of defects, according to (6.83).

In potassium chloride at 300 K, the value of Δu for a pair of K^+ Cl^- defects is approximately 1 eV, or 96 kJ mol^{-1}. Hence, from (6.89)

$$n_{\pm}/N = \exp[-96000/(600\mathscr{R})] = 4.4\times10^{-9}$$

If N is set equal to the Avogadro constant, $n_{\pm}$ becomes 2.6×10^{15}, or approximately 1 defect pair per 2.3×10^{8} sites. At 600 K, this ratio is increased to 1 in 15000.

6.15.3.2 Frenkel defect

Another type of intrinsic point defect, found in silver chloride, for example, is the Frenkel defect. It implies the existence of a vacant cation site, the Ag^+ ion having been transferred to an interstitial position in the structure (Figure 6.44b).

We now need to consider N lattice sites and N' interstices, as well as the n defects. The probability can be evaluated as for the Schottky defect and the result for W is now

$$W = N!\, N'!/\{[(N-n)!\, n!\, (N'-n)!\, n!]\} \tag{6.90}$$

Following out the analysis as before leads to the expression

$$n = (NN')^{1/2}\exp[(-\Delta u/(2k_B T)] \tag{6.91}$$

for the creation of n Frenkel defects consisting of the interstitial ion and its hole. The reason that silver chloride exhibits, preferentially, Frenkel defects is related to its greater tendency towards covalent character in comparison with sodium or potassium chlorides. Moreover, the silver cation is under compressive stress in an interstitial position, and this situation would increase the tendency towards covalent character.

6.15.4 Defects and ion mobility

Notwithstanding we have indicated that ionic solids are, in general, electrical insulators, some of them exhibit electrical conduction, dependent upon the nature and concentration of defects present in an ionic crystal. A defect concentration can be increased by heating the solid to a high temperature and then quenching it rapidly. The defects at the high temperature then become locked into the structure and, because ion mobility is related to defect concentration, an enhanced electrical conductivity results. In the alkali-metal halides, the electrical conductivity is proportional to the ion mobilities and to an exponential factor, as indicated by (6.88). The mobility of an ion may be defined as its drift speed under unit applied field strength. It is often quoted in the units cm^2 s^{-1} V^{-1}, which implies the rate of movement of a unit surface under an applied potential difference of 1 V.

In a perfect crystal structure it would be difficult to envisage a mechanism for ion transport. Studies have been carried out by radioactive tracer techniques to determine solid-state conductivities. In a typical experiment, a thin slice of ^{24}NaCl was contained between two plates of nonradioactive sodium chloride and the composite maintained at a constant temperature. The defect concentrations throughout the sample were determined by cutting thin sections of the composite after given times and then counting the radioactivity in each section. The diffusion of the tracer is related to the defect concentration c and, at a distance d from the surface after a time t, it has been shown that

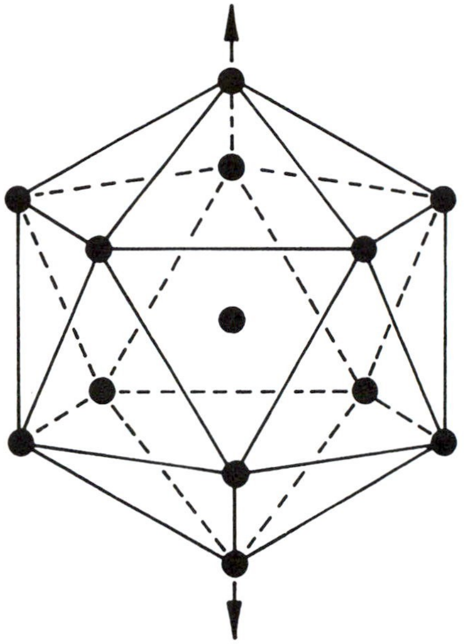

Figure 6.45 An icosahedron, point-group symmetry $m\bar{5}m$ (I_h). The polyhedron is centrosymmetric, and the vertical direction is one of the six fivefold inversion axes present.

$$\ln(c) = \alpha - d^2/(Dt) \qquad (6.92)$$

where α is a constant and D is the diffusion coefficient for the substance under examination. Further experiments showed that D itself varied with temperature according to the equation

$$D = D_0 \exp[-E_d/(\mathcal{R}T)] \qquad (6.93)$$

The activation energy E_d for diffusion in sodium chloride is approximately 173 kJ mol^{-1}, which includes the energy needed both to create the vacancy and to move an ion into the vacant site. Thus, for a single ion, using our datum in Section 6.15.3.1,

$$E_d \simeq E_\mu + 96 \text{ kJ mol}^{-1} \qquad (6.94)$$

where E_μ, the energy needed to induce migration, is approximately 77 kJ mol^{-1}.

6.16 Quasicrystals

It has been found[31] that a rapidly cooled alloy of composition $Al_{86}\,Mn_{14}$ gives a pattern of discrete spots by electron diffraction that shows the icosahedral symmetry $m\bar{5}m$. It follows that, although the structure is not based on a translation unit cell, it has to possess a high degree of three-dimensional regularity in order to exhibit Bragg reflection. Such materials are termed *quasicrystalline*.

Figure 6.45 shows an icosahedron; the vertical direction is one of the six $\bar{5}$ axes. An icosahedron may be formed by allowing 20 regular tetrahedra to share a common apex, the central dot in Figure 6.45, each tetrahedron distorting slightly in the process. Other species of icosahedral symmetry are known. For example, the turnip yellow mosaic virus has crystals of

[31] D Schechtman, I Blech, D Gratias and J W Cahn, *Phys. Rev. Lett.* **53**, 1951 (1984).

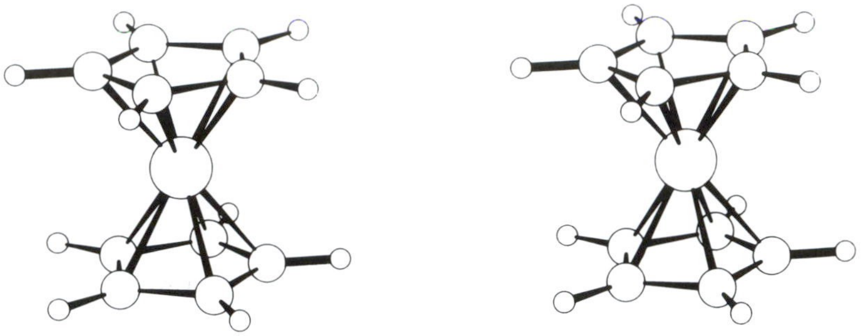

Figure 6.46 Stereoview of the molecular structure of biscyclopentadienyl iron, showing the centrosymmetric point-group symmetry $\bar{5}m$ for the molecules; the two rings are in the staggered position.

point group 532 (I). In this case, the inherent icosahedral strain is relieved by incorporating water molecules into the structure.

Molecules are known that show fivefold symmetry, as in the metal biscyclopentadienyls $(C_5H_5)M$. For M=Fe, with the cyclopentadienyl rings staggered (Figure 6.46), the point-group symmetry is $\bar{5}m$ ($\mathscr{D}_{5d}$), whereas for M=Ru, with the rings in the eclipsed position, the symmetry is $\overline{10}m2$ ($\mathscr{D}_{5h}$)[32].

A quasicrystal may be considered as a three-dimensional generalization of the two-dimensional Penrose tiling pattern (Figure 6.47), which consists of rhombi of angles 144° and 108°; the former occurs $(1+\sqrt{5})/2$ times more frequently than the latter; this number is referred to as the *golden mean* τ. A pentagon of unit side has a diagonal of $2\cos(\pi/5)$, or τ; the same value occurs for the distance from the vertex of a regular decagon of unit length sides to its centre. If three rectangular cards with sides in the ratio of $\tau:1$ are slotted centrally such that they can be interlinked to form three mutually perpendicular planes, the corners of the card become the vertices of an icosahedron. The value of τ (1.618033989...) is also the limit of the ratio of two adjacent terms f_{n+1}/f_n in the Fibonacci series, as n tends to infinity. For example, for $f_{n+1}=10^6$, $f_n=618034$, and $f_{n+1}/f_n=1.618033959\ldots$.

Although there is no pure translational symmetry in the tiling pattern, the regular decagons are similarly oriented, and lines can be drawn through the corners (atom positions) at 72° (360°/5) to each other. In three dimensions, two different rhombohedral cells (compressed cubes) can be packed together to give an icosahedral structure. For further discussions on quasicrystals the reader is referred to the literature[33] and references therein.

6.17 Liquid crystals

Certain organic crystals, on being heated slowly, pass into a form intermediate between the solid and liquid states and known as liquid crystals. Cholesteryl benzoate, $C_{34}H_{50}O_2$, melts sharply at 419 K to form an opaque liquid crystal which clears to a normal liquid at 452 K. Liquid crystals consist of large, elongated molecules that possess one or more polar groups, such as $=NH_2$ or $=CO$. In the crystalline state, these substances pack with their molecules

[32] $\overline{10}=5/m$.

[33] R Penrose, *Math. Intell.* **2**, 32 (1979); D L Nelson and B I Halperin, *Science*, **229**, 233 (1985); A L Mackay, *Physica* A, **114**, 609 (1982); V E Dmitrienko, *J. Physique*, **51**, 2717 (1990).

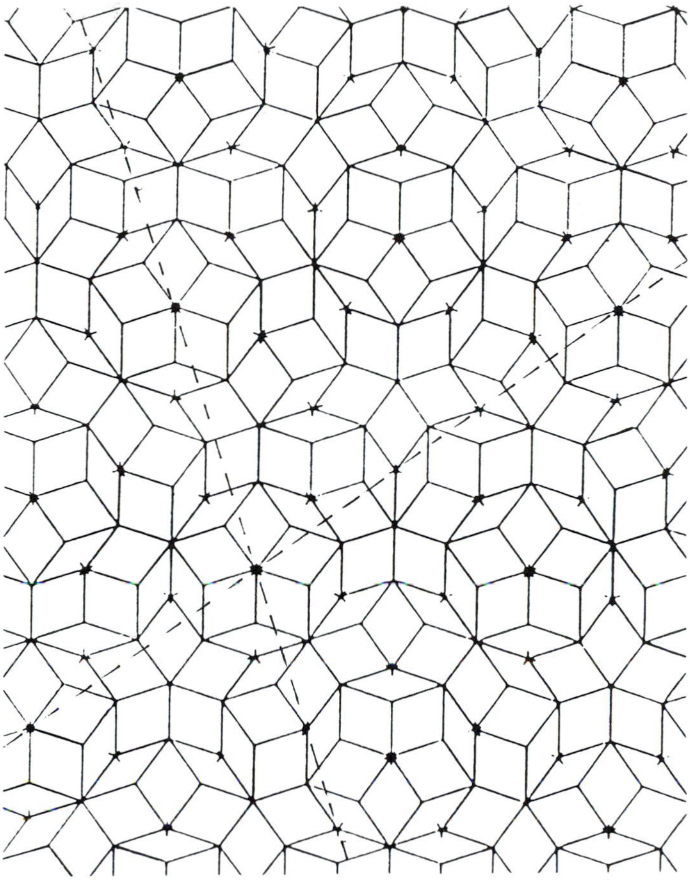

Figure 6.47 Two-dimensional Penrose tiling pattern; the rhombi have angles of 144° and 108°. The similar orientations of the decagons are highlighted by the dashed lines, at 72° to each other, such lines permeating the whole pattern. It is this type of regularity in three dimensions that enables a quasicrystal to develop Bragg reflections.

aligned parallel to one another. Forces of attraction exist between the polar groups, in addition to the van der Waals forces described above. On heating, the weaker van der Waals forces are overcome first, by the thermal energy supplied to the crystal, and relative movement occurs. Increased heating breaks the links between the polar groups as well, whereupon the substance then passes into the true liquid state.

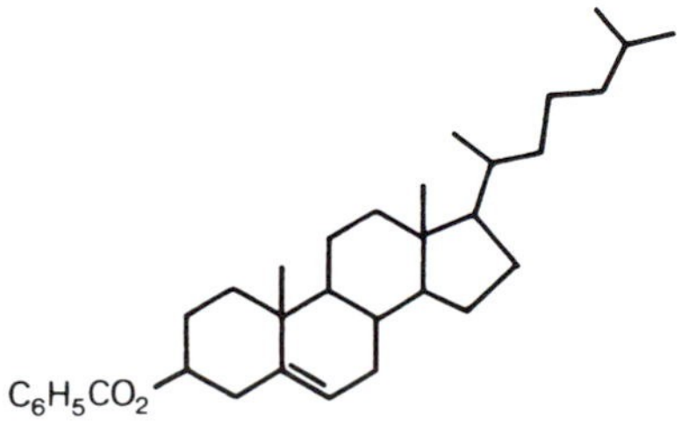

Cholesteryl benzoate

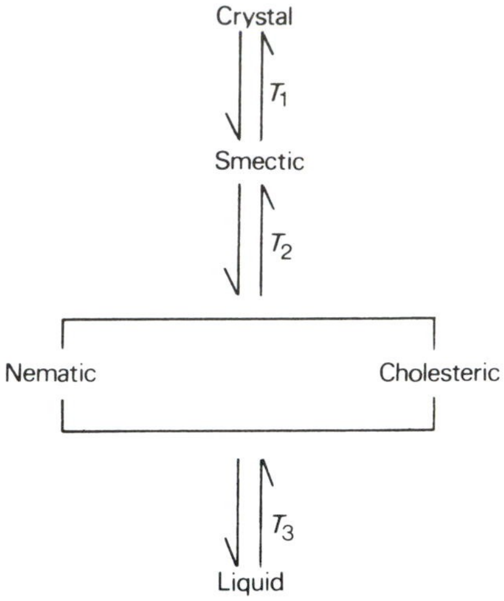

Figure 6.48 Possible phases for liquid crystals, although a given substance would probably not show all of them; T_1, T_2 and T_3 represent transition temperatures.

Several phases of liquid crystal are recognized (Figure 6.48), but not all liquid crystals exhibit every phase. In the *nematic* phase, shown by ammonium oleate, $CH_3(CH_2)_7CH=CH(CH_2)_7CO_2NH_4$, for example, the long molecules are arranged with their lengths parallel, rather like an army of descending parachutists. In the *smectic* phase, shown by 4-azoxyanisole

$$H_3CO-\langle\bigcirc\rangle-N(O)=N-\langle\bigcirc\rangle-OCH_3$$

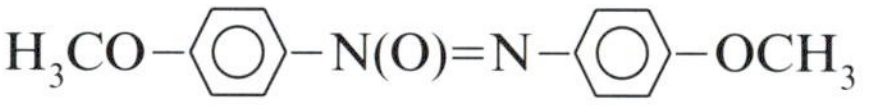

for example, the molecules are arranged on equally spaced planes, but without any lateral periodicity, rather like a crowd of shoppers in a departmental store.

The transitions between the phases are reversible, and occur at definite transition temperatures that vary with pressure according to the Clapeyron equation (Section 7.5):

$$dp/dT = \Delta H_t/[T(V_2-V_1)] \tag{6.95}$$

where V_1 and V_2 are volume quantities in two phases and ΔH_t is the enthalpy of transition at the temperature T.

The *cholesteric* liquid crystals are optically anisotropic and exhibit colours that depend upon temperature. This property is very marked in nematic liquid crystals, and forms the basis of digital display functions in instruments and watches[34].

6.18 Molten salts

Pure inorganic salts melt normally at temperatures between 500 K and 1500 K. The melts possess high electrical conductivity because the ionic state is preserved in the liquid. On the other hand, inorganic molecular compounds, such as $HgCl_2$, have very small electrical conductivities in the liquid, showing that the molecule remains intact on melting.

X-ray diffraction experiments lead to radial distribution functions for melts, of the kind discussed in Section 5.6.2.1. For molten alkali-metal halides of the NaCl structure type, the coordination number ranges from 4.0 to 5.8 for unlike ions and from 8.0 to 13.0 for like ions.

[34] A brief general account of liquid crystals has been given by C Booth, *Physics World*, **10**, 33 (1977).

These values compare with 6 and 12 for the NaCl structure type in the solid state. It may be noted that, whereas the coordination number for unlike ions is less than that in the crystalline state, that for like ions can exceed the value of 12, so that very close packing can occur in a melt.

Computer simulation procedures with ionic melts are more complicated than with other liquids, because Coulombic forces, proportional to $1/r$, are of much longer range than are dispersion forces, or even dipolar forces, and many more particles are needed to achieve good simulation.

The alkali-metal halides increase in volume on melting by 10–15%. Computer simulation studies suggest that transient voids are present in the melt structure, which are filled by concerted migration of neighbouring ions. This is both equivalent and preferable to the concept of moving 'holes' (missing ions), as an explanation of the mechanism of electrical conduction in these materials.

There is an entropy increase on melting of 20–25 J K^{-1} mol^{-1}, indicating the expected decrease in order accompanying fusion. The entropy change on vaporization has a mean value of about 98 J K^{-1} mol^{-1}, which is comparable to the Trouton value of 85 J K^{-1} mol^{-1} for other liquids (Section 4.6.3). It arises because again the gaseous states involved are not fundamentally different in the two cases and the gas phase makes the major contribution to the entropy of the system.

Problems 6

6.1 Calculate the electronic and lattice contributions to the constant volume molar heat capacity of aluminium at 10 K, given that $\Theta(\mathrm{Al})=428$ K and that the lattice contribution is bT, where $b=9\times10^{-1}$ J K^{-2} mol^{-1} . At what temperature are the two contributions equal?
6.2 Calculate the packing efficiency for the closest packing of identical spheres in the CPH (A3) structure type.
6.3 CdS is a photoconducting material with a band gap of 2.42 eV. What is the largest wavelength of (visible) radiation that can excite a valence electron to the conduction band? What is the ratio of the resistivities of this material at 300 K and 400 K?
6.4 The anharmonic vibrations of atoms about their mean positions in a metal crystal may be represented approximately by the potential energy function $V(r)=ar^2-br^3$, where a and b are constants. By determining an expression for the mean displacement $\bar{r}$, show that $\bar{r}$ is directly proportional to temperature, that is, it is consistent with the solid expanding on heating. Note that, since r is small, $\exp(br^3)\approx(1+br^3)$.
6.5 Use the appropriate Debye equation to calculate the molar heat capacity at constant volume of copper at 298.15 K.
6.6 How might one describe the crystal structure of gold (A1) in concise crystallographic terms?
6.7 (a) Draw a two-dimensional square lattice in ***k***-space and mark in the boundaries of the first three Brillouin zones. (b) What is the first Brillouin zone for a primitive cubic lattice of side a?
6.8 From the *standard* data below (I and E refer to 0 K), show how the formation of $MgCl_2$(s) is preferred to that of MgCl(s):

I_1(Mg,g)	6.09 eV	I_2(Mg,g)	11.82 eV
ΔH_a(Mg,s)	149.0 kJ mol^{-1}	$D_0(Cl_2$,g)	243.0 kJ mol^{-1}
E(Cl,g)	−348.6 kJ mol^{-1}	ΔH_f(MgCl,s)	−221.8 kJ mol^{-1}
$\Delta H_f(MgCl_2$,s)	−641.8 kJ mol^{-1}		

6.9 Use the following data to construct a thermochemical cycle for the enthalpy of formation of NH_4Cl, and find the standard value of this parameter:

	ΔH^{-0-}/kJ mol^{-1}
$\frac{1}{2}N_2(g)+\frac{3}{2}H_2(g)\rightarrow NH_3(g)$	−46.0
$NH_3(g)+aq\rightarrow NH_4^+(aq)+OH^-(aq)$	−34.7
$\frac{1}{2}H_2(g)+\frac{1}{2}Cl_2(g)\rightarrow HCl(g)$	−92.5
$HCl(g)+aq\rightarrow H^+(aq)+Cl^-(aq)$	−74.9
$NH_4^+(aq)+OH^-(aq)+H^+(aq)+Cl^-(aq)\rightarrow NH_4^+(aq)+Cl^-(aq)$	−52.3
$NH_4Cl(s)+aq\rightarrow NH_4^+(aq)+Cl^-(aq)$	+15.1

6.10 From the following structural and standard thermodynamic data (I and E refer to 0 K) on calcium oxide, calculate the affinity of oxygen for two electrons:

Structure type	NaCl
Unit cell side a	0.4811 nm
Madelung constant $\mathcal{A}$	1.7476 (remember the ion charges)
Compressibility κ	$0.895\times10^{-11}N^{-1}$ m
I_1(Ca,g)	589.5 kJ mol^{-1}
I_2(Ca,g)	1145.0 kJ mol^{-1}
ΔH_a(Ca,s)	176.6 kJ mol^{-1}
$D_0(O_2$,g)	489.9 kJ mol^{-1}
ΔH_f(CaO,s)	−635.5 kJ mol^{-1}

6.11 The sulfate ion may be considered as a regular tetrahedral arrangement of oxygen atoms around sulfur. If the charge on oxygen is −1 and the S−O distance 0.130 nm, calculate the electrostatic self-energy of the sulfate ion.

6.12 If $r(Cs^+)$ and $r(I^-)$ are 0.184 and 0.212 nm respectively, and the departure of r_e from the additivity of radii is −0.001 nm, calculate the density of crystalline CsI.

6.13 Refer to Figure 6.39 for rutile (TiO_2): the coordination around Ti is distorted octahedral, whereas that around O is an isosceles triangle. Use the following X-ray crystallographic data to calculate the Ti−O bond lengths (two different values) and O−Ti−O bond angles (two different values).

Tetragonal crystal system: $a=b=0.4593$ nm, $c=0.2959$ nm. Two formula-entities per unit cell at fractional coordinates

2	Ti	0, 0, 0;	$\frac{1}{2}, \frac{1}{2}, \frac{1}{2}$
4	0	$x, x, 0$;	$\bar{x}, \bar{x}, 0$;
		$\frac{1}{2}+x, \frac{1}{2}-x, \frac{1}{2}$;	$\frac{1}{2}-x, \frac{1}{2}+x, \frac{1}{2}$

with $x=0.3056$. It may be found helpful first to make sketches of the structure in plan, from which the unique distances and angles to be calculated should be clear.

6.14 Show that the radius ratio for the würtzite structure type with atoms in maximum contact is 0.225.

6.15 Derive an expression for the number n of Frenkel defects in a crystal of silver chloride containing N lattice sites and N' interstices. If the energy needed to set up a single defect in silver chloride is $\simeq$1 eV, calculate the fraction of Frenkel defects in this substance at 500 K.

6.16 Radioactive silver was allowed to diffuse through a silver–indium alloy at 1000 K. The penetration depth x was determined after time intervals t of 6×10^4 s by measuring the radioactivity β_t. Show that the diffusion process follows the equation

$$\beta_t = A \exp[-x^2/(Dt)]$$

and find the value of the diffusion constant D. The experimental data follow; β_t is dimensionless.

x/mm	β_t	x/mm	β_t
0.000	600	0.329	88
0.084	540	0.376	50
0.132	450	0.425	25
0.183	360	0.470	12
0.230	250	0.520	5
0.279	160	0.568	2

6.17 In a further experiment, values of the diffusion coefficient D were obtained as a function of temperature. Given that D is related to T by the Boltzmann equation

$$D = D_0 \exp[-E_d/(\mathcal{R}T)]$$

determine the value of the activation energy E_d for the diffusion process; the relevant data follow:

T/K	878	1007	1176	1253	1322
D/m^{-2} s	1.6×10^{-18}	4.0×10^{-17}	1.1×10^{-15}	4.0×10^{-15}	1.0×10^{-14}

What is the value of D_0 and what is its significance?

6.18 Calculate the standard molar entropy of nickel from the following data (use the Debye approximation below 15.05 K):

T/K	15.05	25.20	47.10	67.13	82.11
C_p/J K^{-1} mol^{-1}	0.1945	0.5994	3.532	7.639	10.10
T/K	133.4	204.1	256.5	283.0	
C_p/J K^{-1} mol^{-1}	17.88	22.72	24.81	26.09	

6.19 Consider a face of the cubic unit cell, side a, of the sodium chloride structure type (Figure 1.2) as a two-dimensional square array of unit charges of alternating sign. Calculate the Madelung constant for this array by Evjen's method. Repeat the calculation for squares of side $2a$, $3a$ and $4a$. By a method of constant second differences, or by a suitable extrapolation, report a probable value of the Madelung constant for the two-dimensional array.

6.20 Draw the Wigner–Seitz unit cells for (a) a two-dimensional hexagonal p unit cell ($a=b$; $\gamma=120°$), and (b) a three-dimensional P tetragonal unit cell in which $c>a$.

Phase rule and properties of solutions

7.1 Introduction

Under certain conditions of temperature and pressure the states of matter, gas, liquid and solid, transform from one to another. Each state has a range of temperature and pressure over which it is the thermodynamically stable form of the given substance. The application of heat to ice, at 1 atm, brings about the following changes:

$$\underset{\text{Ice}}{H_2O} \underset{273.15\text{ K}}{\rightleftharpoons} \underset{\text{Water}}{H_2O} \underset{373.15\text{ K}}{\rightleftharpoons} \underset{\text{Steam}}{H_2O} \tag{7.1}$$

The transition temperatures may be varied if the pressure is altered, and the results are summarized conveniently by a phase diagram, Figure 7.1, which we shall study.

7.2 Phase rule

A general relation, the *phase rule*, exists between the number of phases P, the number of components C and the number of degrees of freedom F in a system; thus,

$$P+F=C+2 \tag{7.2}$$

A *phase* is a distinct and uniform state of matter. Ice is a single phase, ice and water together constitute two phases. A single phase is a *homogeneous* system; a system of two or more phases is *heterogeneous.*

The number of *components* refers to those species needed to specify the composition of *all* phases in a given system. Thus, for the equilibrium (7.1) C is unity. In (7.3) C is 2; although three substances (and three phases) are present, they are related by stoichiometry:

$$CaCO_3(s) \rightleftharpoons CaO(s)+CO_2(g) \tag{7.3}$$

The number of *degrees of freedom* refers to those intensive properties, such as temperature and pressure, that can be varied independently without altering the number of phases present in equilibrium in the given system.

The ice–water–steam system (Figure 7.1) is an example of a one-component system. In general, each phase has two degrees of freedom, from (7.2). A position such as m requires both the temperature and pressure to be stated in order to specify it completely. At points such as m' and m'', water is in equilibrium with ice and water vapour (steam) respectively. Two phases exist at these points and the system is then univariant, as it is along the lines OA, OB and OC. At the composition represented by the triple point O (see Section 4.6.1.2) the system is invariant, which, together with the ease of obtaining very pure water, enables the triple point of water (273.16 K) to be used as a fundamental temperature standard (see

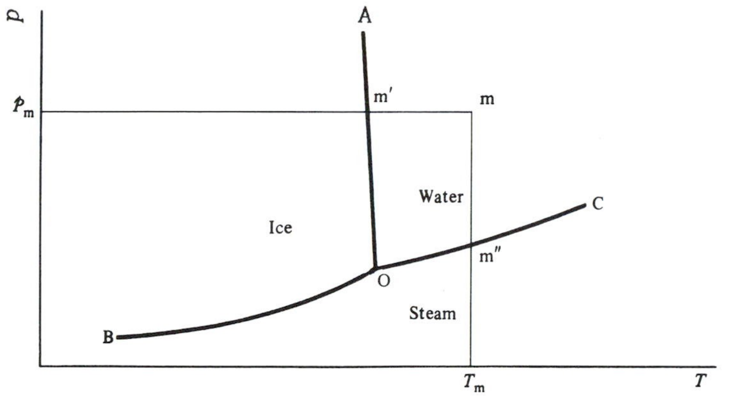

Figure 7.1 Phase diagram for the one-component ice–water–steam system; O is the invariant triple point, at which ice, water and water vapour are in equilibrium.

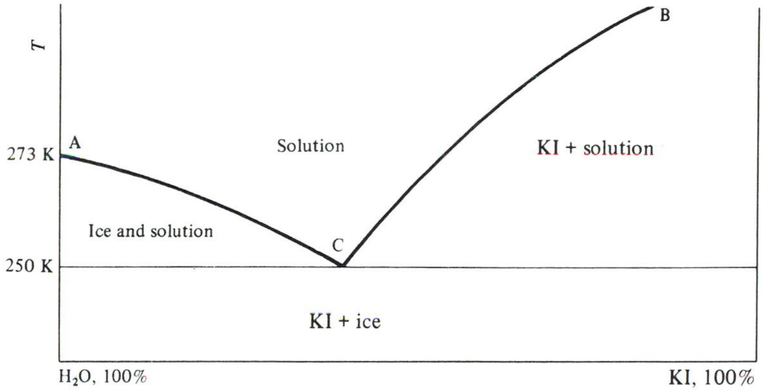

Figure 7.2 Portion of the phase diagram for the two-component water–potassium iodide system; A is the (univariant) melting-point of ice, and C is the (invariant) eutectic point. The curve CB intersects the right-hand ordinate at 954 K, the (univariant) melting-point of potassium iodide.

Section 5.2.2). The melting-point at 1 atm is 273.15 K and defines the zero of the Celsius scale. The slope of OA is negative, indicating that the melting-point of ice decreases with increasing pressure.

7.3 Two-component systems

The system potassium iodide–water is an example of a two-component system and is illustrated at a constant pressure of 1 atm by Figure 7.2. No hydrates are formed in this system at normal pressures. On cooling a solution represented by a point in the neighbourhood of A, ice separates from the solution when the curve AC is reached. Further cooling causes a continued separation of ice and the composition of the solution follows the curve AC. Along the line BC potassium iodide separates out on cooling, and the composition proceeds along the line BC. The curve BC represents the variation with temperature of the solubility of potassium iodide in water, because the solid is in equilibrium with its saturated solution. The curve meets the right-hand pressure ordinate at 954 K, the melting-point of potassium iodide.

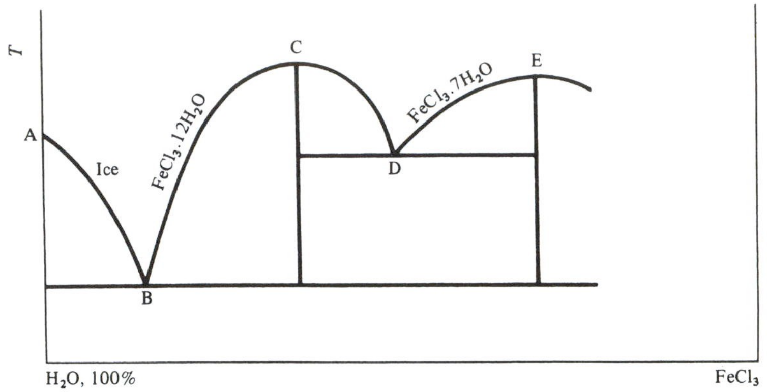

Figure 7.3 Portion of the phase diagram for the two-component system water–iron(III) chloride; A is the melting-point of ice, B and D are eutectic points, and C and E are (invariant) congruent melting-points.

Similarly, the curve AC represents the variation in the freezing-point of ice in the presence of dissolved potassium iodide. The point C is a *eutectic* point; at this composition and temperature, a mixture of ice and potassium iodide crystallizes together and the temperature remains constant until the liquid phase disappears completely.

The eutectic is not a compound but a mixture of two phases. Since the pressure is constant we have a *condensed* system of phases and (7.2) takes the modified form $F=C+1-P$. At the point C there are two components and three phases in equilibrium, so that F is zero. When the liquid at C finally disappears P is reduced by unity and F is 1; then, there is no restriction to a continued fall in temperature.

Figure 7.3 is a portion of the phase diagram for the two-component system iron(III) chloride–water at a constant pressure of 1 atm. In this system, compounds, that is, hydrates, are formed. The line AB represents ice in equilibrium with solution, forming a eutectic at B of ice and iron(III) chloride dodecahydrate. At the invariant point C the liquid and solid phases in equilibrium have the same composition; C is termed a *congruent* melting-point. (When the two compositions are not the same, the invariant point is an *incongruent* melting-point, or *peritectic* point.) At the point D a eutectic is formed between iron(III) chloride dodecahydrate and iron(III) chloride heptahydrate, and another congruent melting-point occurs at E.

The variation of the vapour pressure of water at a constant temperature of 25 °C above the hydrates of copper(II) sulfate is illustrated in Figure 7.4. If a saturated solution of copper(II) sulfate is evaporated isothermally, the vapour pressure falls along the line BA. At the point A the solution is saturated with respect to the pentahydrate, which then crystallizes. The vapour pressure remains constant at $p=23.0$ mmHg until all the saturated solution has crystallized. The phase rule shows that this situation is to be expected: since there are three phases, the system is univariant along any horizontal line and is specified by the composition.

Continued dehydration at the pentahydrate composition causes the vapour pressure to fall discontinuously to 7.8 mmHg, at which pressure the pentahydrate and trihydrate are in equilibrium. Further loss of water of crystallization results in the stepwise curve being followed, until the material is fully anhydrous.

This diagram shows how a hydrate may be dried. In order to obtain a pure sample of

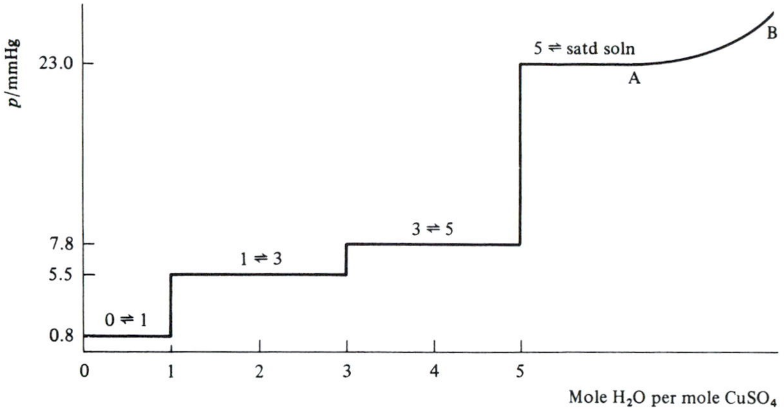

Figure 7.4 Isothermal evaporation of a saturated solution of copper(II) sulfate, at 25 °C. Any horizontal line represents an equilibrium between the pentahydrate and the saturated solution, or between two hydrates, or between the lowest hydrate and the anhydrous compound. The pentahydrate and trihydrate coexist at a partial pressure of water of 7.8 mmHg.

copper(II) sulfate pentahydrate, of the stoichiometric composition $CuSO_4{\cdot}5H_2O$, a sample of the pentahydrate containing a small amount of the saturated solution is confined over a mixture of trihydrate and pentahydrate, as a 'drying' agent. The excess moisture in the sample is taken up by the trihydrate and the vapour pressure will equilibrate at 7.8 mmHg, at which value the composition $CuSO_4{\cdot}5H_2O$ will be attained by the sample, as long as some trihydrate remains in the trihydrate– pentahydrate drying mixture.

We will now consider some of the thermodynamics associated with the processes that we have just described.

7.4 Thermodynamics of the phase rule

Consider a system of two phases α and β. From (4.81), at constant composition and energy, $dS=0$. If S_α and S_β are the entropies of the two phases, then

$$dS=dS_\alpha+dS_\beta=0 \tag{7.4}$$

and from (4.59), it follows that

$$T_\alpha=T_\beta \tag{7.5}$$

Thus, the temperatures of two phases in equilibrium are equal, a result that we might expect intuitively. A similar condition holds for the pressures of the two phases.

The requirement for chemical equilibrium at constant temperature and pressure is that $dG=0$. Thus,

$$dG=dG_\alpha+dG_\beta=0 \tag{7.6}$$

Consider a process in which dn_i mole of component i is taken from phase α and added to phase β. Then, from (4.119)

$$dG=-\mu_{i,\alpha}+\mu_{i,\beta}=0 \tag{7.7}$$

whence

$$\mu_{i,\alpha}=\mu_{i,\beta} \tag{7.8}$$

Thus, for a closed system at constant temperature and pressure, equilibrium between two phases requires that their chemical potentials be equal. This result may be generalized to a system of P phases, as follows.

The mole fraction x_i of the ith component in any phase of a system is, by definition

$$x_i = n_i / \sum_i n_i \tag{7.9}$$

and it follows that for each phase present

$$\sum_i x_i = 1 \tag{7.10}$$

Because of this relation, only $(C-1)$ components need be specified in any phase. Thus, for a total of P phases, the number of variables to be specified is $P(C-1)+2$, where the added two variables represent temperature and pressure.

At equilibrium, an extension of (7.8) leads to

$$\mu_{i,j} = \mu_{i,k} \qquad (i=1 \text{ to } C; j,k=1 \text{ to } P; j \neq \text{k}) \tag{7.10}$$

As with (7.9), each equality sign in this set of C equations indicates a restriction on the number of variables. The total number of such restrictions is $C(P-1)$, because for any two phases (7.8) holds. The number of degrees of freedom F is now given by

$$F = P(C-1)+2-C(P-1) = C+2-P \tag{7.11}$$

which is equivalent to the phase rule (7.2).

7.5 Thermodynamics of the p–T phase diagram

The equilibrium boundary between any two phases may be discussed in terms of its slope, $\mathrm{d}p/\mathrm{d}T$. Consider any two phases α and β in equilibrium. The temperature and pressure are T and p respectively, and their chemical potentials μ_α and μ_β are equal. Let T and p be changed by amounts $\mathrm{d}T$ and $\mathrm{d}p$, but with the α and β phases remaining in equilibrium. The chemical potentials μ' for the α and β phases at $T+\mathrm{d}T$ and $p+\mathrm{d}p$ may be given by Taylor expansions to the first-order terms:

$$\left.\begin{aligned} \mu'_\alpha &= \mu_\alpha + (\partial\mu_\alpha/\partial T)_p \mathrm{d}T + (\partial\mu_\alpha/\mathrm{d}p)_T \mathrm{d}p \\ \mu'_\beta &= \mu_\beta + (\partial\mu_\beta/\partial T)_p \mathrm{d}T + (\partial\mu_\beta/\mathrm{d}p)_T \mathrm{d}p \end{aligned}\right\} \tag{7.12}$$

We know that $\mu'_\alpha = \mu'_\beta$, and $\mu_\alpha = \mu_\beta$. Hence, (7.12) may be arranged to give

$$\mathrm{d}p/\mathrm{d}T = -\frac{-[(\partial\mu_\beta/\partial T)_p - (\partial\mu_\alpha/\partial_T)_p]}{(\partial\mu_\beta/\partial p)_T - (\partial\mu_\alpha/\partial p)_T} \tag{7.13}$$

At equilibrium, from (4.97) and (4.117), $(\partial\mu/\partial T)_p = -S_\mathrm{m}$ and $(\partial\mu/\partial p)_T = V_\mathrm{m}$, where the subscript m indicates a molar property. Hence,

$$\mathrm{d}p/\mathrm{d}T = \Delta S_\mathrm{m}/\Delta V_\mathrm{m} \tag{7.14}$$

which is the Clapeyron equation, and it applies to any phase transition of a pure compound.

7.5.1 Liquid–vapour equilibrium

At a transition temperature T, the molar entropy of vaporization is $\Delta H_\mathrm{v}/T$, where ΔH_v is the molar enthalpy of vaporization. The change ΔV_m is equal to $V_{\mathrm{m,g}} - V_{\mathrm{m},\ell}$. Since $V_{\mathrm{m,g}} > V_{\mathrm{m},\ell}$, $\Delta V_\mathrm{m} \simeq V_{\mathrm{m,g}}$; if further we assume ideal behaviour by the vapour, $V_{\mathrm{m,g}} = \mathcal{R}T/p$. Then, from (7.14),

$$d\ln(p)/dT=\Delta H_v/(\mathcal{R}T^2) \tag{7.15}$$

which is the Clausius–Clapeyron equation.[1] We may write this equation as

$$d\ln(p)/d(1/T)=-\Delta H_v/\mathcal{R} \tag{7.16}$$

so that a plot of the logarithm of vapour pressure against the reciprocal of the absolute temperature is a straight line of slope $-\Delta H_v/\mathcal{R}$, within the limits of the approximations used in (7.15).

7.5.1.1 *Vapour under applied pressure*

Consider an idealized system at a fixed temperature T, in which a pure liquid is confined within a cylinder by a piston, rather like Figure 4.3 except that here the piston is assumed permeable to the vapour of the liquid. Following Section 7.5, we obtain

$$(\partial\mu_\ell/\partial p)_T\,dp_\ell=(\partial\mu_g/\partial p)_T\,dp_g \tag{7.17}$$

From (4.97) and (4.117), $(\partial\mu/\partial p)=V_m$; hence,

$$V_{m,\ell}\,dp_\ell=V_{m,g}\,dp_g \tag{7.18}$$

If we assume ideal behaviour of the vapour, $V_{m,g}=\mathcal{R}T/p_g$ and

$$d\ln(p_g)=[V_{m,\ell}/(\mathcal{R}T)]\,dp_\ell \tag{7.19}$$

Over a small range of pressure the change in $V_{m,\ell}$ is negligible, and integration of (7.19) over the range p_1 to p_2 then gives

$$\ln(p_{g,2}/p_{g,\ell})=[V_{m,\ell}/(\mathcal{R}T)](p_{\ell,2}-p_{\ell,1}) \tag{7.20}$$

We may illustrate the use of (7.20) by calculating the vapour pressure of bromine under an applied pressure of 10 atm at 58.78 °C (the boiling-point); the density is 2928 kg m^{-3}. $V_{m,\ell}=0.15981$ kg mol^{-1}/2928 kg m$^{-3}=5.458\times10^{-5}$ m^{-3} mol^{-1}: $\ln(p_{\ell,2}/p_{\ell,1})=5.458\times10^{-5}$ m^{-3} mol$^{-1}\times$(10 atm$-$1 atm)/(0.08206$\times10^{-3}$ m^3 atm K^{-1} mol$^{-1}\times$331.93 K)$=1.803\times10^{-2}$. Because $p_{\ell,1}$ at 58.78 °C is 1 atm, $p_{\ell,2}$ (at 10 atm) is 1.018 atm.

7.6 Vapour pressure: Raoult's and Henry's laws

We defined chemical potential when considering gases in Sections 4.9.2 and 4.9.3. We can apply (4.121) to a liquid in relation to its dilute vapour, with which the liquid is in equilibrium. Thus, for a pure liquid A, we write

$$\mu_{A,\ell}=\mu_A^{-0-}+\mathcal{R}T\ln(p_{A,\ell}/p^{-0-}) \tag{7.21}$$

In the presence of a second substance we have a *binary* solution and we write the chemical potential of A as

$$\mu_A=\mu_A^{-0-}+\mathcal{R}T\ln(p_A/p^{-0-}) \tag{7.22}$$

where p is the vapour pressure of A in the solution. Eliminating μ_A^{-0-} between (7.21) and (7.22), we obtain

[1] $d\ln(p)$ is equal to $(1/p)\,dp$, and is dimensionless.

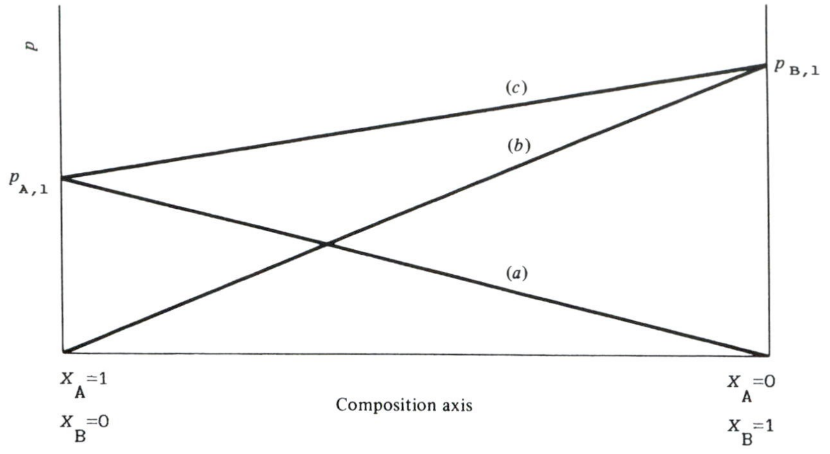

Figure 7.5 Vapour pressure–composition diagram for an ideal binary solution; (*a*) and (*b*) are the vapour pressure curves for pure A and pure B components respectively; (*c*) is the curve of the total pressure and, symbolically, (*c*)=(*a*)+(*b*).

$$\mu_A=\mu_{A,\ell}+\mathcal{R}T\ln(p_A/p_{A,\ell}) \quad (7.23)$$

Raoult's law states that, *for dilute solutions of a solute* B *in a solvent* A *at a given temperature, the vapour pressure of* A *is proportional to its mole fraction in the solution.* Thus,

$$p_A=x_A p_{A,\ell} \quad (7.24)$$

where x_A is the mole fraction (7.9), and the chemical potential for a solvent in a solution behaving ideally may thus be given as

$$\mu_A=\mu_{A,\ell}+\mathcal{R}T\ln(x_A) \quad (7.25)$$

The vapour pressure of ethanol at 292 K is 40.00 mmHg and that of water at the same temperature is 16.48 mmHg. Consider a solution containing 2 g water in 50 g ethanol. Assuming the solution to be ideal, that is, to obey Raoult's law, then $x_{EtOH}=0.9072$ and $x_{H2O}=0.0928$, so that $p_{EtOH}=36.29$ mmHg and $p_{H_2O}=1.529$ mmHg. Thus, the decrease in the vapour pressure of ethanol on addition of water means that its effective concentration is decreased.

The vapour pressure–composition isotherms for an ideal binary solution are shown in Figure 7.5, and (7.24) is obeyed for both solvent (a) and solute (b). From Dalton's law (5.28), the total pressure in a binary solution of A and B is given by $p=p_{A,\ell}+p_{B,\ell}$, represented by the line (c). A few liquids behave in this ideal manner, of which the pairs bromoethane and iodoethane at 303 K, and benzene and dichloroethane at 323 K are good examples.

Henry's law applies to real solutions at low concentrations. It states that, *at a given temperature, the vapour pressure of the solute is proportional to its mole fraction.* Thus,

$$p_B=H_B x_B \quad (7.26)$$

where H_B is the constant of Henry's law, with the dimension of pressure, for the solute B; it is the limit as x_B tends to 1 of the tangent to the experimental vapour pressure curve (see Figure 7.27 later and accompanying text), for the solution of A and B at $x_B=0$. Henry's law may be used to determine the solubilities of gases in liquids.

The Henry's law constant for nitrogen in water at 25 °C is 8.57×10^4 atm and its partial pressure in the presence of air at that temperature is 0.79 atm. For 1 kg of water, $n_{H2O}=55.51$

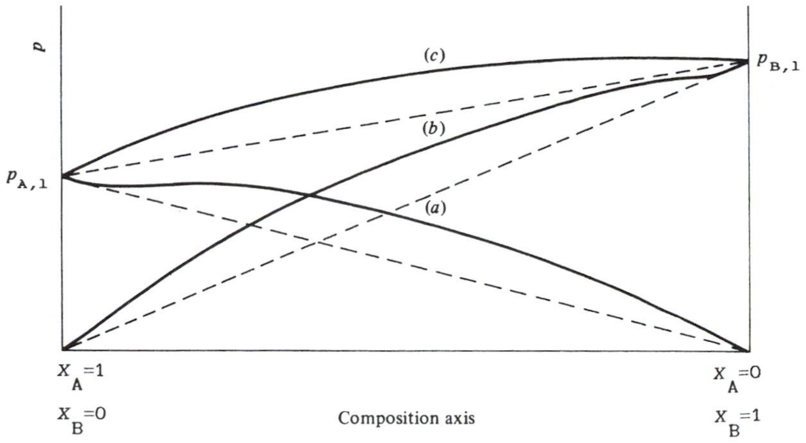

Figure 7.6 Vapour pressure–composition diagram for a binary solution that shows positive deviations from Raoult's law; the lines (*a*), (*b*) and (*c*) have the meanings as before. The dashed lines correspond to ideality.

mol and $x_{N_2}=n_{N_2}/(n_{N_2}+55.51 \text{ mol})\simeq n_{N_2}/55.51$ mol. From Henry's law, 0.79 atm$=8.57\times10^4$ atm$(n_{N_2}/55.51$ mol), whence $n_{N_2}=5.12\times10^{-4}$ mol, that is, the solubility is 5.12×10^{-4} molal.

Henry's law is used also in defining the standard state for the activity of a solute in a solution, as we discuss later.

7.6.1 Nonideal solutions

Most pairs of liquids form nonideal solutions. Sometimes the partial pressure of component A is enhanced when the second component B is added, as shown by line (a) in Figure 7.6; a similar effect is then observed for the partial pressure of B when A is added to it (b). The total vapour pressure (c) at all compositions is greater than that corresponding to ideality, except at $x_A=1$ or $x_B=1$. Curves (a) and (b) represent *positive* deviations from Raoult's law, and are exhibited by liquid mixtures in which the molecules repel one another and so enter the vapour phase more readily than they do for the pure liquids. Examples of this behaviour are the pairs pentane–ethanol and heptane– tetrachloromethane.

If the molecules in a binary liquid tend to attract one another, their escape into the vapour phase will be inhibited and *negative* deviations result, Figure 7.7. This behaviour is exemplified by the pairs ethanone–trichloromethane and pyridine–ethanoic acid. It is evident that two liquids in admixture both exhibit the same type of departure from ideality, over part of the composition range at least, or else none at all.

The deviation from Raoult's law may be large enough for the total vapour pressure curve to exhibit a maximum or minimum. Figure 7.8 shows a maximum at a composition *c*; the total vapour pressure rises above that of pure A or pure B.

We may consider this situation from another point of view. The vapour pressure at each composition of the mixture will rise with increase in temperature until it attains the same pressures as the surroundings, say 1 atm; the mixture then boils. Of all the mixtures of A and B, that with the composition *c* will attain a pressure of 1 atm at the lowest temperature. The component B attains the ambient pressure at a higher temperature, and component A does so at a higher temperature still. Hence, the boiling-point–composition curve for mixtures of

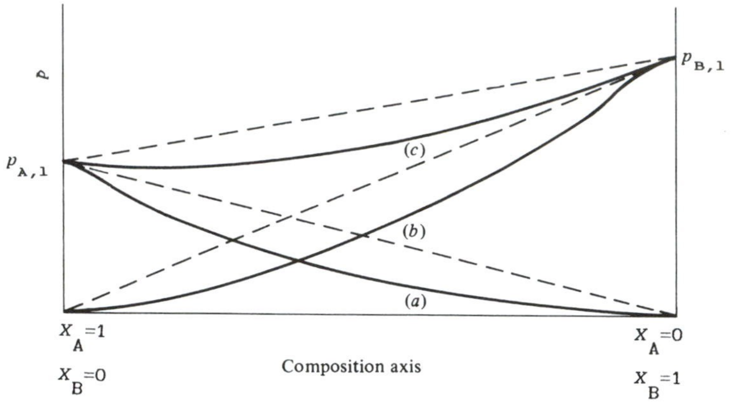

Figure 7.7 Vapour pressure–composition diagram for a binary solution that shows negative deviations from Raoult's law; the lines (*a*), (*b*) and (*c*) have the meanings as before.

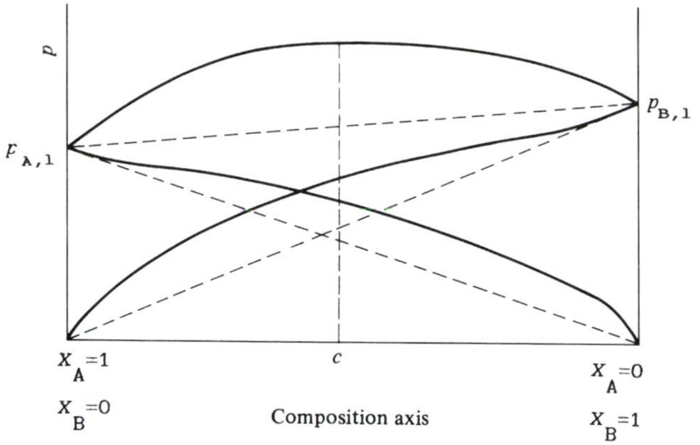

Figure 7.8 Vapour pressure–composition diagram for a binary solution showing positive deviations from Raoult's law that are sufficiently large to lead to a maximum in the vapour pressure, at the composition *c*.

compounds such as A and B will exhibit a minimum boiling-point at *c*, Figure 7.9. Pairs of liquids that show this behaviour are ethanol–water and ethyl ethanoate–water, for example.

Similar enhancements of negative deviations lead to a minimum vapour pressure, Figure 7.10, at a composition *d*, with a maximum in the boiling-point–composition curve at the corresponding concentration. Examples of this behaviour are the systems hydrochloric acid–water, and nitric acid–water.

Measurements of the lowering in vapour pressure of a solvent A of known relative molar mass on addition of small amounts of a solute may be used to determine the relative molar mass of the solute. The procedure is conducted most easily when the solute B is nonvolatile.

An aqueous solution of 15 g of glucose $C_6H_{12}O_6$ in 100 g of water has a vapour pressure of 17.28 mmHg at 20 °C. At the same temperature the vapour pressure of pure water is 17.54 mmHg. From (7.24), 17.28 mmHg/17.54 mmHg$=x_{H_2O}=$(100/18.015)/[(100/18.015)+(15/M_r)]. Thus, $M_r=179.6$. ($C_6H_{12}O_6=180.15$.)

Small changes in vapour pressure are not easy to measure accurately, and it is preferable to employ related properties, such as freezing-point depression, as we shall discuss shortly.

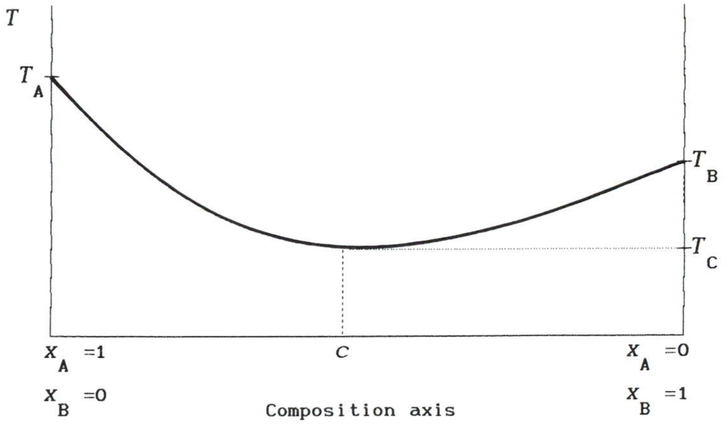

Figure 7.9 A maximum in the vapour pressure–composition curve (Figure 7.8) leads to a minimum boiling-point at the same composition, *c*.

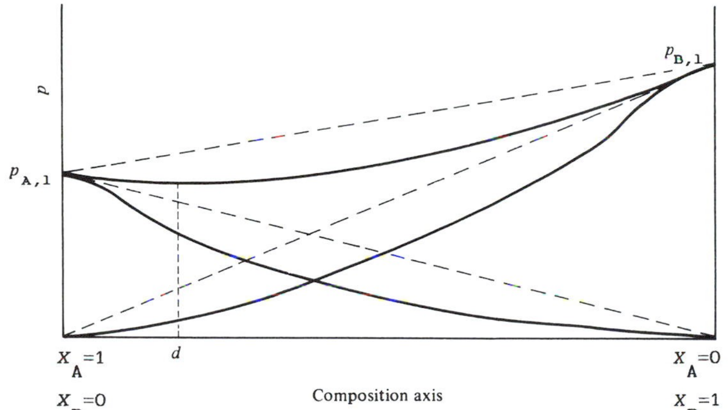

Figure 7.10 Binary solution showing a minimum in the vapour pressure–composition curve, at the composition *d*. This system would have a maximum in the boiling-point at the same composition.

7.6.2 Distillation

If the distillation of a binary liquid system is carried out isothermally, by reduction of the external pressure, then the compositions of both the liquid and the vapour remain unchanged as distillation proceeds. Usually, however, distillation is carried out at constant pressure, and the boiling-point of the mixture changes with change in composition, Figure 7.11. The upper curve is the *condensation curve*; the lower curve is the *boiling-point curve.* In general, the vapour above the mixture has a higher mole fraction of the more volatile component than does the liquid mixture, as we shall show.

Let a binary liquid mixture of mole fraction x_A be in equilibrium with vapour of mole fraction y_A. From Dalton's law, (7.9) and (7.24) $y_A = p_A/p$, where p is the total vapour pressure. Furthermore, $y_A = x_A p_{A,\ell}/p$, and $p = x_A p_{A,\ell} + (1-x_A)p_{B,\ell}$. Thus, $y_A = x_A\ p_{A,\ell}/[x_A (p_{A,\ell} - p_{B,\ell}) + p_{B,\ell}]$, from which it is clear that, if $p_{A,\ell} > p_{B,\ell}$ then $y_A > x_A$, a condition that would be found for the system in Figure 7.11.

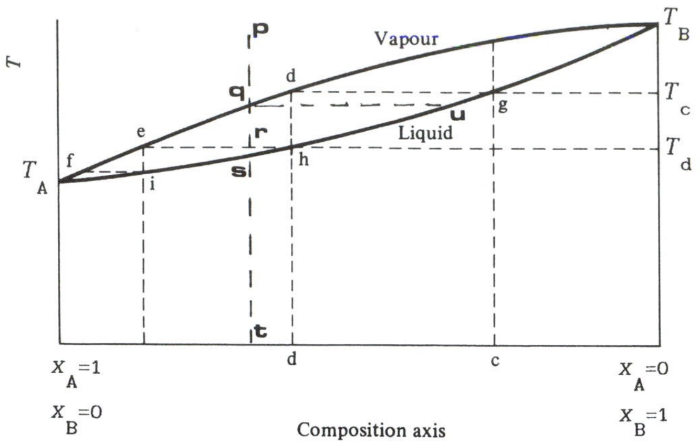

Figure 7.11 Temperature–composition diagram for a binary solution that shows a regular change in boiling-point. The upper curve is the condensation curve for the vapour, or *vaporous*, and the lower curve is the boiling-point curve for the liquid, or *liquidus*. The diagram is described fully in the text.

Consider the liquid mixture of composition c, Figure 7.11. On heating to the temperature T_C, point g on the liquidus, the liquid will boil, producing a vapour of composition d that is richer in the lower boiling-point component A. This vapour condenses at h to a liquid of composition d which, upon distillation, yields the vapour e at temperature T_d. Horizontal lines, such as gd, that connect the compositions of liquid and vapour that are in equilibrium are called *tie-lines*. By careful distillation, highly pure components can be obtained, A as the final distillate and B as the final residual liquid in the distillation vessel.

Fractional distillation implies the progression c, g, d, h, e, ..., leading to the pure component A. A fractionating column carries out the successive vaporizations and condensations without the need for separating each distillate. In the bubble cap column, Figure 7.12, ascending vapour bubbles through the liquid films on each plate and, because the liquid is cooler than the vapour, some condensation takes place. As the vapour passes up the column, enrichment of the lower boiling-point compound takes place at each plate, thus moving through Figure 7.11 in the stepwise manner indicated.

7.6.2.1 Lever rule

Consider a vapour at the point p (Figure 7.11), of composition represented by t. In the one-phase vapour region the system is bivariant, because the pressure is constant. As the temperature is decreased, the first observable change occurs at q, when the liquid in equilibrium, represented by the point u on the tie-line qu, condenses. The temperature is further decreased in the two-phase region, and we consider a typical position r. Here, the vapour has the composition e and the liquid has composition d, and the proportions of each may be calculated as follows.

Let (n_v+n_ℓ) be the sum of the numbers of moles of components A and B in the vapour and liquid phases, respectively. From a mass balance on B, $(n_v+n_\ell)\mathrm{t}=n_v\mathrm{e}+n_\ell\mathrm{d}$. Rearranging, we obtain

$$n_v/n_\ell=(\mathrm{d}-\mathrm{t})/(\mathrm{t}-\mathrm{e})=\mathrm{rh}/\mathrm{re} \tag{7.27}$$

where rh and re are lengths on the tie-line eh. This relation is called the *lever rule*, because a

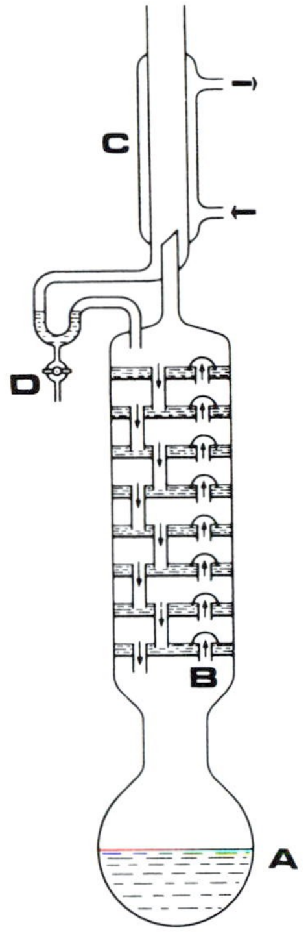

Figure 7.12 Fractionating column: A boiling vessel, B first plate with bubble-cap, C reflux condenser, D condensate of lower boiling-point component. The ascending vapour bubbles through the liquid traps; some condenses, but a vapour richer in the more volatile component passes upwards through the column.

similar rule relates the masses at two ends of a mechanical lever to their distances from the fulcrum.

7.6.3 Maximum and minimum boiling-point systems

If a mixture shows a maximum or a minimum in the boiling-point–composition curve, separation into *two* pure components by distillation is not possible. Figure 7.13 shows a maximum boiling-point T_m at a composition m. Distillation of a mixture of composition c produces a vapour e at a temperature T_d and so on, until the distillate is pure A and the residue has the composition m, at which it distils unchanged in composition. The vapour is always richer in the more volatile of the two species A and m.

If the starting composition were c′, distillation would ultimately produce component B and the mixture m. No one mixture of A and B can be separated into its pure components. A similar argument applies to a mixture with a minimum boiling-point, Figure 7.14. Mixtures such as m that distil unchanged at a fixed temperature are called *azeotropes* or *constant boiling-point* mixtures. Hydrochloric acid and water form a constant boiling-point mixture, which can be used as a primary standard acid in volumetric analysis, according to Table 7.1.

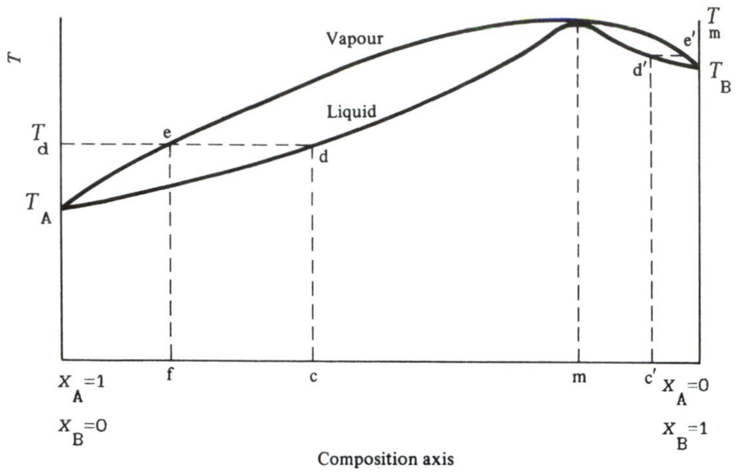

Figure 7.13 Temperature–composition diagram for a binary system that forms a maximum boiling-point mixture. A mixture at composition *c* distils to pure A and the azeotrope of composition *m*, whereas a mixture at *c′* distils to pure B and the same azeotrope composition *m*.

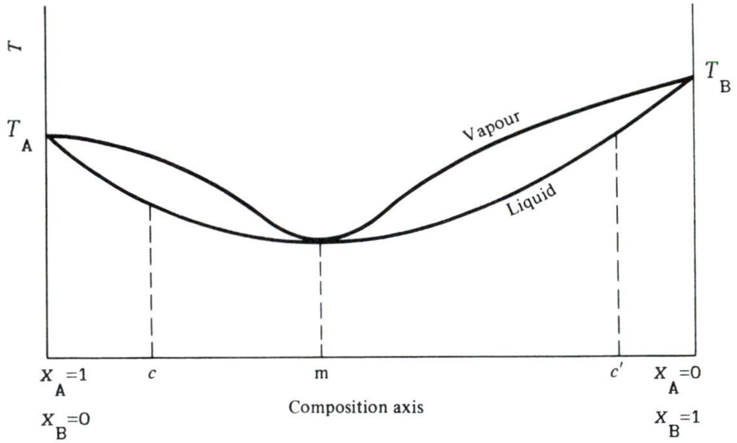

Figure 7.14 Binary system with a minimum boiling-point. Again, compositions *c* or *c′* distil to A or B respectively, each with the azeotrope composition *m*.

7.7 Partially miscible liquids

If a little phenol is added to an excess of water it will dissolve completely. Further addition of the organic compound leads to the formation of two liquid layers. One layer is water saturated with phenol, and the other layer is phenol saturated with water.

At a given temperature these two layers are in equilibrium and are called *conjugate solutions.* At atmospheric pressure, there are two components and, in general, three phases; thus, the system is univariant. At a given temperature, the compositions of the two conjugate solutions are fixed; addition of water or phenol to an equilibrium system changes only the relative *amounts* of the two layers.

Figure 7.15 shows a portion of the phenol–water phase diagram; above the curve is a one-

Table 7.1 Composition of constant boiling-point hydrochloric acid–water mixtures

p/mmHg	Bp/K	%wt HCl
400	365.23	21.235
500	370.36	20.916
600	375.36	20.638
700	379.57	20.360
760	381.57	20.222
800	383.16	20.155

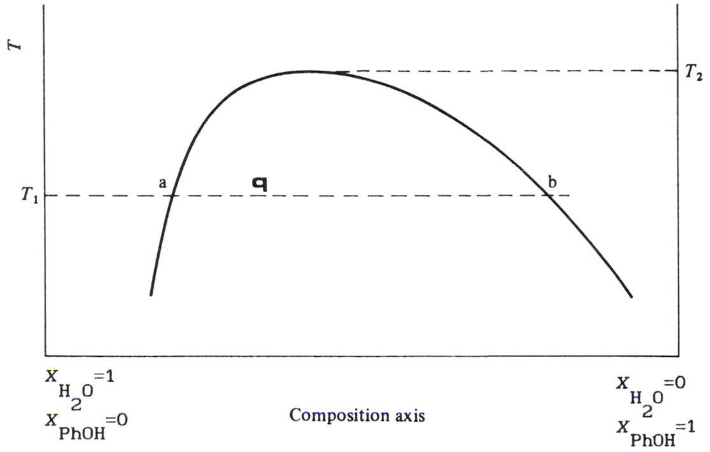

Figure 7.15 Portion of the phenol–water phase diagram; the region within the curve is two-phase, outside the region is one-phase, whereas above the consolute temperature T_2 there is complete miscibility. The ratio of compositions *a*/*b* at a typical point *q* is given by *qb*/*qa*.

phase system and below the curve there are two phases present. At the temperature T_1 the conjugate solutions, of compositions represented by the points a and b, are in equilibrium; ab is a tie-line. Mutual solubilities increase with increasing temperature and the system becomes homogeneous above the temperature T_2, the *consolute temperature*, also known as the critical temperature.

Figure 7.16 shows the nicotine–water system. Upper (481 K) and lower (334 K) consolute temperatures are shown by this system. Outside the temperature region 334 K to 481 K and above the curve the two components are completely miscible.

7.7.1 Steam-distillation

When the mutual solubilities of two liquids are very small, as with aniline and water, they may be considered immiscible. At equilibrium, two liquid phases (almost pure components) and a vapour coexist. The phase rule (7.2) shows that this system has one degree of freedom. The total vapour pressure is the sum of the vapour pressures of the two components and depends on temperature. It is independent of the composition of the mixture; addition of more aniline only changes the relative amounts of the two layers.

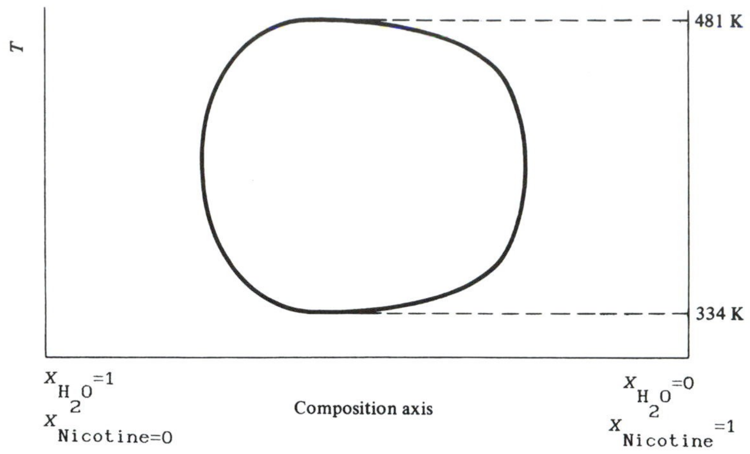

Figure 7.16 The nicotine–water phase diagram; the region inside the closed curve is two-phase; the upper and lower consolute (critical) temperatures are 481 K and 334 K, respectively.

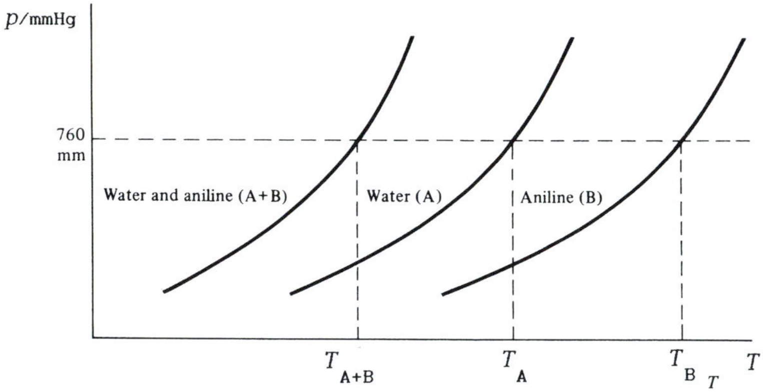

Figure 7.17 Vapour pressure–temperature curves for the aniline–water system (not to scale).

Vapour pressure–temperature curves for the aniline–water system are shown in Figure 7.17. The total vapour pressure is 760 mmHg at T_{A+B}=371.6 K; T_A and T_B are the boiling-points of the components water and aniline respectively, at 760 mmHg. At 371.6 K, the partial vapour pressures are p_{H_2O}=717.6 mmHg and $p_{C_6H_6NH_2}$=42.4 mmHg. The sum of these partial pressures is 760 mmHg, and the mixture boils at this temperature.

The proportions of aniline and water in the vapour at this temperature are in the ratio 42.4 : 717.6, because both chemical species possess the same kinetic energy at a given temperature. From (5.20) it follows that, in the vapour, $p \propto n$, whence

$$\frac{n_{H_2O}}{n_{C_6H_5NH_2}} = \frac{p_{H_2O}}{p_{C_6H_5NH_2}} \tag{7.28}$$

Since the relative molar masses of water and aniline are approximately 18 and 93 respectively, the weights of aniline and water in the vapour are in the ratio of (42.4×93) : (717.6×18), or 0.305. Thus, the condensed vapour contains approximately 23.4 wt% of aniline. By using these principles, organic liquids such as aniline or nitrobenzene can be separated from water

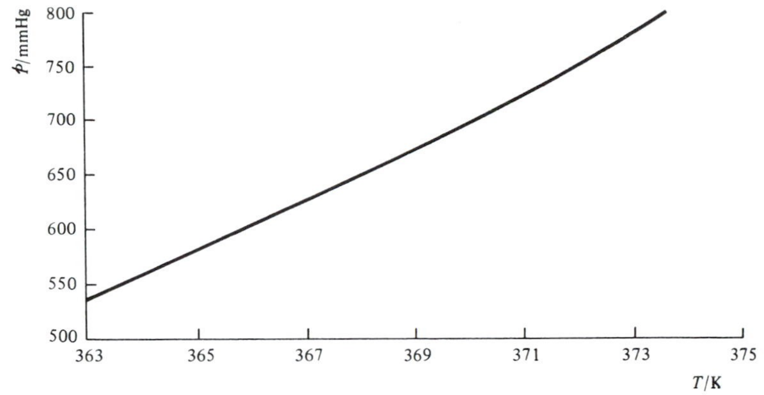

Figure 7.18 Vapour pressure–temperature curve for the nitrobenzene–water system; at 760 mmHg (ambient pressure) the boiling-point is 372.4 K.

by distillation. The process is normally carried out by passing steam into the reaction mixture, from which the term *steam-distillation* arises.

The vapour pressures of water and nitrobenzene vary with temperature as follows:

T/K	363	368	371	373
$p_{C_6H_5NO_2}$/mmHg	12.9	16.7	19.2	20.9
p_{H_2O}/mmHg	525.8	633.9	707.3	760.0
p/mmHg	538.7	650.6	726.5	780.9

From Figure 7.18 the boiling-point of nitrobenzene–water mixtures at 760 mmHg is 372.4 K, and the relative molar masses of nitrobenzene and water are approximately 123 and 18 respectively. Hence, by interpolating the above data for a temperature of 372.4 K

$$w_{H_2O}/w_{C_6H_5NO_2} = (741 \times 18)/(20.25 \times 123) = 5.36$$

Thus, the steam-distillate contains 100[1−(5.36/6.36)], or 15.7, wt% nitrobenzene.

7.7.2 Solvent extraction

Consider a solute that is soluble in two immiscible solvents, such as benzoic acid in an ether–water mixture; both the ethereal and the aqueous layers will contain some solute, benzoic acid. At equilibrium, at a given temperature, there is a fixed ratio between the concentrations of the solute in the two liquid layers, independent of the total amount present, provided that the solute is in the same molecular form in both solvents.

Let w_1 g of solute of relative molar mass M_r be contained in V_1 cm^3 of solution; the molarity is $1000w_1/(M_rV_1)$. If this solution is shaken with V_2 cm^3 of an immiscible solvent and w_2 g of solute passes into it; then, the molarities in the two layers are $1000(w_1-w_2)/(M_rV_1)$ in solvent 1, and $1000w_2/(M_rV_2)$ in solvent 2. The ratio of molarities is $w_2V_1/[(w_1-w_2)V_2]$ and is a constant D, the *distribution coefficient* for the system at the given temperature. These principles form the basis for the solvent-extraction process.

Let an aqueous solution of 5 g (w_1) of solute in 100 cm^3 of aqueous solution be extracted with ether, and let the distribution coefficient D for the system be 10. The solution is shaken with 15 cm^3 of ether and w_2 g of solute are extracted; then $100w_2/(5-w_2)=10$, whence $w_2=3.0$ g.

Table 7.2 Distribution of iodine between tetrachloromethane and water

c_1/mol dm^{-3}	c_2/mol dm^{-3}	D
0.020	0.000235	85.1
0.040	0.000469	85.2
0.060	0.000703	85.4
0.080	0.000930	86.0
0.100	0.001140	87.5

It is easy to see that, in this example, each such extraction will remove 3/5 of the solute remaining from the previous extraction. Thus, three extractions remove 93.6% of the solute.

A single extraction with the total of 45 cm^3 of solvent would remove 81.8%, so that it is always advantageous to extract with several small portions of solvent rather than with one portion of the same total volume.

7.7.3 Distribution law

Solvent extraction is a particular example of the general distribution law. If to a system of two immiscible liquids a solute, soluble in both liquids, is added, then the distribution law

$$c_1/c_2 = D \tag{7.29}$$

holds at a given temperature; c_1 and c_2 are the molar concentrations of the solute in the two solvent layers. The data in Table 7.2 have been obtained for the distribution of iodine between tetrachloromethane (1) and water (2). The value of D is very nearly constant for a range of concentrations c_1 and c_2.

The gradual increase in D with increasing concentration implies a small inexactness in the theory. In fact, concentrations should be replaced by activities, a topic that we consider in detail shortly; suffice it here to say that the activity tends to the concentration as the concentration decreases, so that the values of D at the lower concentrations are more constant and closer to the limiting value.

We can use the distribution law to study equilibria in solution. We use the triiodide equilibrium as an example:

$$I_2(s) + I^-(aq) \rightleftharpoons I_3^-(aq) \tag{7.30}$$

This system may be studied by using the immiscible solvents water (1) and tetrachloromethane (2). The equilibrium (7.30) exists only in the aqueous solution, whereas iodine is soluble in both solvents. The ratio

$$\frac{[I_2 \text{ in aqueous layer}]}{[I_2 \text{ in } CCl_4 \text{ layer}]} = D \tag{7.31}$$

may be found by equilibriating a solution of iodine in tetrachloromethane with water at a given temperature and then analysing the two layers. (The notation [X] is often used in solution chemistry to represent the molar concentration of the species X.)

Let iodine be distributed between aqueous potassium iodide of molarity c and tetrachloromethane; the iodine will be present as I_2 in both the aqueous and the organic layers,

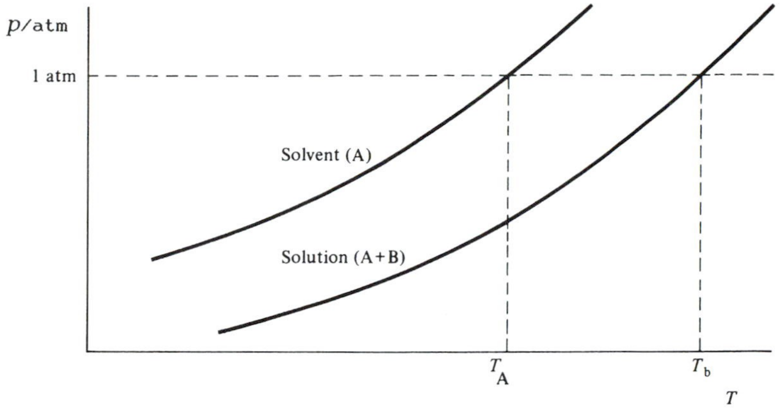

Figure 7.19 Boiling-point elevation diagram, showing the vapour pressure–temperature curves for a binary system of solvent and nonvolatile solute.

but only as I_3^- ions in the aqueous layer. If the total iodine concentrations in layers 1 and 2 are c_1 and c_2 respectively, then the *free* iodine in layer 1 will be Dc_2, and the concentration of triiodide ions is $c_1 - Dc_2$. The equilibrium constant K for the triiodide equilibrium is given by

$$K = \frac{[I_3^-]}{[I_2][I^-]} = \frac{c_1 - Dc_2}{Dc_2[c-(c_1-Dc_2)]} \tag{7.32}$$

At 298 K, the value of K, also called the *stability constant* of the triiodide ion, is 85.0.

7.8 Elevation of the boiling-point

We consider in these next sections three more properties that depend upon the number of *particles* present in the solution, referred to collectively as *colligative properties*, namely, elevation of the boiling-point, depression of the freezing-point and osmosis. The driving force in each case is the decrease in the chemical potential of the solvent on addition of a small quantity of solute and the consequent increase in the entropy of the system (see Section 4.9.2).

Vapour-pressure–temperature curves for dilute solutions of an involatile solute B in a solvent A are shown in Figure 7.19. The solvent boils under atmospheric pressure at T_A; the solution has a lower vapour pressure and, hence, an increased boiling-point T_b, compared with pure solvent. The increase $\Delta T_{bp} = T_b - T_A$ is the boiling-point elevation.

For this equilibrium we use (7.25), here applied to the solvent A in the vapour phase, where the chemical potential is $\mu_{A,v}$, and in solution, where the chemical potential is $\mu_{A,\ell} + \mathcal{R}T \ln(x_A)$; the term $\mathcal{R}T \ln(x_A)$ represents the decrease ($x_A < 1$) in $\mu_{A,\ell}$ owing to the presence of the nonvolatile solute B. The difference of these two chemical potentials is the free energy of vaporization ΔG_v; hence,

$$\frac{\mu_{A,v} - \mu_{A,\ell}}{\mathcal{R}T_b} = \frac{\Delta G_v}{\mathcal{R}T_b} = \ln(1-x_B) = \frac{\Delta H_v}{\mathcal{R}T_b} - \frac{\Delta S_v}{\mathcal{R}} \tag{7.33}$$

or

$$\ln(1-x_B) = \frac{\Delta H_v}{\mathcal{R}T_b} - \frac{\Delta S_v}{\mathcal{R}} \tag{7.34}$$

Table 7.3 Molal ebullioscopic constants

	K_e/kg K mol^{-1}
Benzene	2.65
Tetrachloromethane	5.00
Ethanol	1.20
Water	0.51

Following Section 4.6.3, we know that $\Delta H_v/T_A=\Delta S_v$; a similar conclusion is reached by setting $x_B=0$ (pure solvent) in (7.34). If x_B is small, then $\ln(1-x_B)\simeq -x_B$ and (7.34) may now be rearranged to give

$$x_B=(\Delta H_v/\mathcal{R})(1/T_A-1/T_b)=(\Delta H_v/\mathcal{R})\,\Delta T_{bp}/T_A^2 \tag{7.35}$$

where ΔT_{bp} is the boiling-point elevation and $T_A^2\simeq T_A T_b$. A further approximation that is usually applied is

$$x_B=n_B/(n_A+n_B)\simeq n_B/n_A \tag{7.36}$$

since n_B is much smaller than n_A, because the solution is *dilute*. For 1 kg of solvent of molar mass M_A, $n_A=1$ kg$/M_A$ and $x_B=n_B M_A/1\text{kg}=m_B M_A$. Thus, we obtain

$$\Delta T_{bp}=\mathcal{R}T_A^2 M_A m_B/\Delta H_v=K_e m_B \tag{7.37}$$

where m_B is the molality of the solute B and K_e is the experimentally determined *ebullioscopic constant*, given by

$$K_e=\mathcal{R}T_A^2 M_A/\Delta H_v \tag{7.38}$$

Experimental values of K_e for some solvents are listed in Table 7.3.

An ebulliometer, based on measurements of the resistance of a temperature-sensitive resistor (thermistor), is illustrated in Figure 7.20. Solvent is boiled in the borosilicate glass bulb; a cold-finger reflux condenser prevents the loss of solvent by evaporation. The boiling liquid is pumped over the tip of the thermistor probe P, and the resistance of the thermistor measured by a Wheatstone bridge device. A weighed pellet of solute is then introduced, and the resistance found at the elevated temperature; this resistance will be lower, because the thermistor, a semiconductor, has a negative temperature coefficient of resistivity (see Section 6.5.7). Further pellets may be added, but the theory developed depends upon the solution being dilute, and extrapolation to $m_B=0$ may be desirable.

A solution of 2.001 g urea $CO(NH_2)_2$ in 0.125 kg water boiled at 373.136 K; at the same pressure, water boiled at 373.00 K; $\Delta H_v=40.656$ kJ mol^{-1}. The calculated K_e is 0.513 kg K mol^{-1}. It is preferable to obtain K_e by experiment, taking the relative molar mass of urea as 60.056, whence K_e, from (7.38), is 0.510 kg K mol^{-1}.

7.9 Depression of the freezing-point

One result of the lowering of the vapour pressure of a solvent by the addition of a non-volatile solute is that the freezing-point is also lowered, and we need to investigate the equilibrium between a given solvent as a solid and in a dilute solution of a solute. Blagden's law states that *the lowering of the freezing-point is proportional to the solute concentration, for*

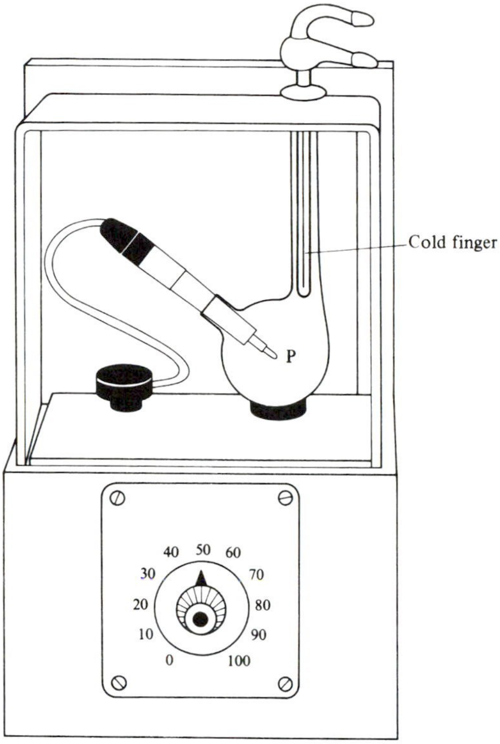

Figure 7.20 Small-scale ebulliometer, for the practical determination of the elevation of the boiling-point; ΔT_{bp} is the length $T_A T_b$ in Figure 7.19.

dilute solutions, and Raoult established that *equimolar solutions of different substances depressed the freezing-point by equal amounts, provided that the solutions were dilute, and that the solute did not associate or dissociate in the solvent.* A further requirement is that only pure solvent separates out on freezing.

Figure 7.21 shows the vapour pressure–temperature curves for a solvent A and solution (A+B) containing a nonvolatile solute B, in the neighbourhood of the freezing-point; the curve for the solution lies below that for the solvent. The freezing-point of the pure solvent is T_A and its vapour pressure is $p_{A,s}$. The freezing-point of the solution is T_f and the corresponding vapour pressure is $p_{A,\ell}$. At T_f, the solvent in solution is in equilibrium with the solid solvent; the freezing-point depression ΔT_{fp} is then $T_A - T_f$, and the distance XZ is the lowering of the vapour pressure, $p_{A,s} - p_{A,\ell}$. In practice, ΔT_{fp} is a small quantity, so that Y and Z on the diagram are very close.

For this equilibrium, the chemical potentials are $\mu_{A,s}$ for the solid solvent, and $\mu_{A,\ell} + \mathcal{R}T\ln(x_A)$ for the solvent in solution, from (7.25). At equilibrium, we have

$$\mu_{A,s} = \mu_{A,\ell} + \mathcal{R}T\ln(x_A) \tag{7.39}$$

We could now follow the procedure developed in Section 7.8, but we choose a slightly different route. From (7.39), it follows that $(\mu_{A,s} - \mu_{A,\ell})/T = -\Delta G_{fu}/T = \mathcal{R}\ln(1-x_B)$, the negative sign arising because the process of fusion would involve the opposite signs on the chemical potentials. From (4.109),

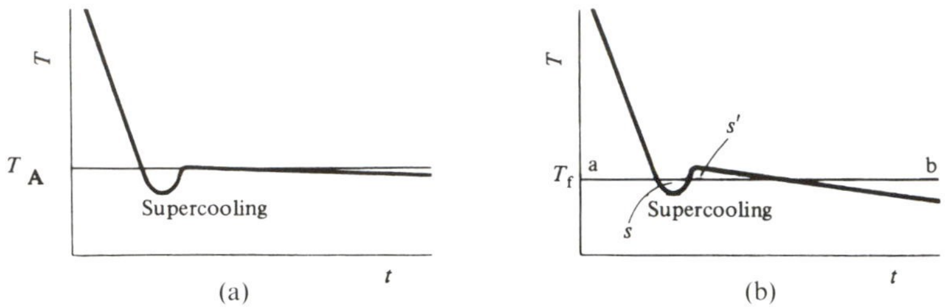

Figure 7.23 Curves of temperature T against time t for the freezing of solvent from pure solvent (a) and from solution (b); in (b) the areas s and s' are made equal in defining T_f.

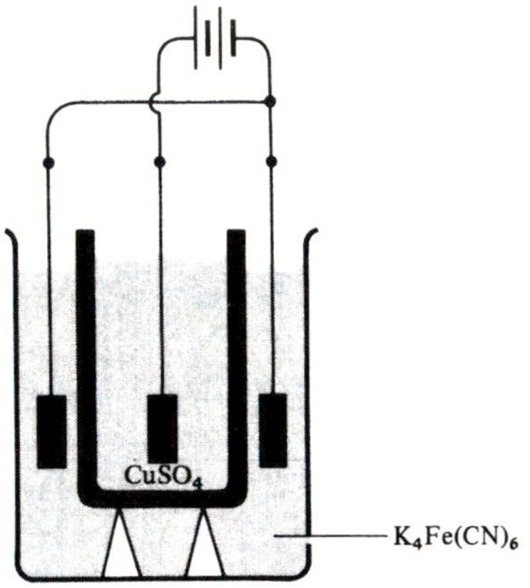

Figure 7.24 Formation of a semipermeable membrane of $Cu_2[Fe(CN)_6]$ in the wall of a porous pot. By electrolysis, Cu^{2+} and $[Fe(CN)_6]^{4-}$ ions move in opposite directions and meet to form $Cu_2[Fe(CN)_6]$ in the pores of the pot.

ponent; if the pressure is constant (ambient) the system is invariant (F=0) and the temperature varies with composition. Hence, the initial temperature rise may continue above the freezing-point. In this case, Figure 7.23b, T_f is fixed by drawing the horizontal line *ab* on the cooling graph such that the areas s and s' are equal.

7.10 Osmosis

It was discovered in the middle of the eighteenth century that if alcohol and water were separated from each other by an animal membrane, such as a pig's bladder, water passed through the membrane into the alcohol, but alcohol did not pass through into the water. Subsequently, this property was demonstrated for solutions generally. The essential property of the membrane is that it allows the passage of solvent molecules, e.g. H_2O, but not those of the solute, e.g. C_2H_5OH; such membranes are called *semipermeable.*

A semipermeable membrane may be formed in the wall of a porous pot by electrolyzing solutions of copper(II) sulfate and potassium ferrocyanide that are separated by the pot, Figure 7.24. Copper(II) ferrocyanide is deposited in the pores of the pot, which then acts as a semipermeable membrane.

Osmosis refers to the passage of solvent from a dilute solution through a semipermeable membrane to a more concentrated solution. It may be *demonstrated* with the apparatus of

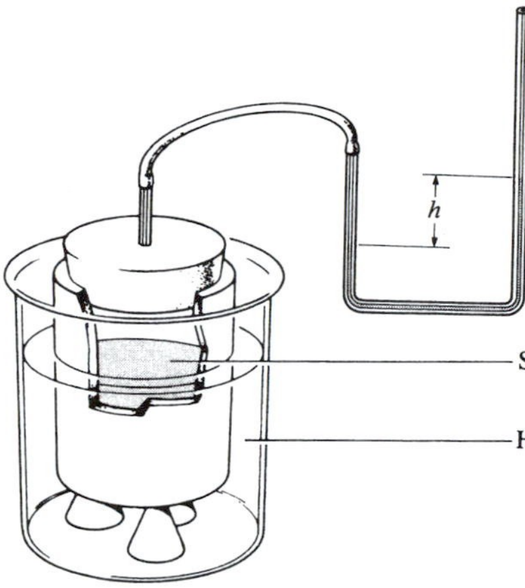

Figure 7.25 Demonstration of osmosis: water passes through the semipermeable membrane (porous pot) into the sugar solution; the pressure above the sugar solution increases, as is indicated by the height h on the manometer M. Osmotic pressure cannot be measured in this way because the concentration of the solution changes during the process.

Figure 7.25. As water enters the sugar solution through the walls of the pot, the pressure increases and is indicated by the height h in the manometer M. Osmotic pressure cannot be measured with this apparatus because the solution changes in concentration during the experiment. In the absence of an external applied pressure, osmosis would continue until the solutions on each side of the membrane were *isotonic*, that is, of the same osmotic pressure.

Osmotic pressure is defined as *that external pressure, applied to a concentrated solution separated from a more dilute solution by a semipermeable membrane, which just prevents osmosis from taking place*. Its measurement is particularly important for determining the relative molar masses of polymers and other macromolecules. The amount of boiling-point elevation or freezing-point depression decreases as the relative molar mass increases. Polymers are often sparingly soluble, so that the solutions are very dilute. However, a small osmotic pressure can produce a significant capillary rise in a modern osmometer, Figure 7.26.

In this apparatus the capillary tubes A and B are identical in bore: A is the measuring capillary, in contact with the solution, and B is the reference capillary, dipping into the solvent. The osmotic head is the difference in the meniscus heights in A and B. The apparatus is calibrated with solutions of known osmotic pressure.

At equilibrium under an applied pressure p, the chemical potential of the solvent on the pure solvent side is $\mu_{A,\ell}$; on the solution side the chemical potential μ_A is increased because the pressure is enhanced by the osmotic pressure Π, but decreased because the mole fraction x_A is less than unity. At equilibrium $\mu_{A,\ell}=\mu_A$, and from Section 7.5.1.1 and (4.97),

$$\mu_A=\mu_{A,\ell}+\int_p^{p+\pi} V_m \, dp \tag{7.45}$$

As $\mu_{A,\ell}=\mu_A+\mathcal{R}T\ln(x_A)$, and replacing x_A by $1-x_B$ with the earlier assumption that $\ln(1-x_B)\simeq -x_B$, it follows that

$$\mathcal{R}Tx_B=\Pi V_m \tag{7.46}$$

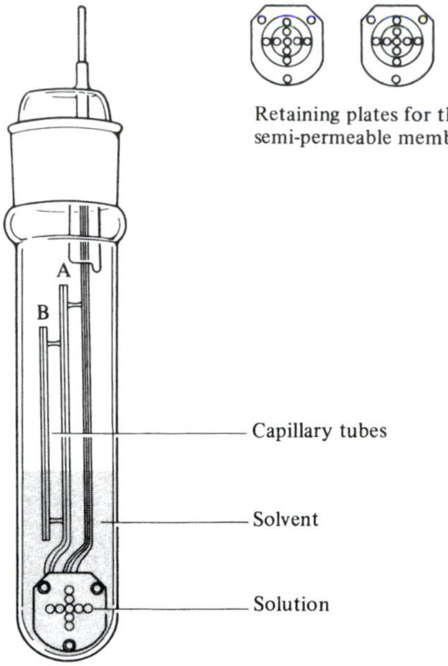

Figure 7.26 Osmometer: A and B are capillary tubes of identical bore; A is the measuring capillary dipping into the solution, and B is the reference capillary in contact with the solvent. The osmotic pressure is proportional to the difference in heights of the menisci in A and B. The arrangement compensates for surface tension effects in the capillaries, but the apparatus must be calibrated.

Under the dilute solution assumption, we can, as we have seen, approximate x_B to n_B/n_A, whereupon we obtain the van 't Hoff equation

$$\Pi = c\mathcal{R}T \tag{7.47}$$

where c is the molarity of the solution, n_B/V_m. It is evident that (7.47) is comparable with the equation of state for an ideal gas. However, solutions of polymers are far from ideal, and comparison of (5.4) with $n=1$ and (5.54) suggests that the nonideality of the polymer solution requires an equation of the type

$$\Pi = c\mathcal{R}T(1 + Bc + Cc^2 + \ldots) \tag{7.48}$$

In practice, the osmotic pressure is measured at several concentrations, and Π/c is extrapolated to $c=0$ in order to find M_B.

The following data have been measured at 25 °C for solutions of a polymer; the pressure is represented by the manometric height h of a liquid of density D equal to 0.9840 g cm^{-3}:

w/g dm^{-3}	1.000	2.000	3.000	5.000	7.500	10.000
h/dm	0.0281	0.0726	0.1305	0.2945	0.5850	0.9730
hw^{-1}/dm^4 g^{-3}	0.0281	0.0363	0.0435	0.0589	0.0780	0.0973

A graph shows that the plot of hw^{-1} against w is linear and, by least squares, the intercept at w/g dm^{-3}=0 is 0.02063 mol kg^{-1} with an estimated standard deviation of 0.00016; the correlation coefficient is 0.99997, indicating an excellent linear fit. Since $w/M=c$ and $\Pi=hDg$, where g is the gravitational acceleration (980.7 cm s^{-2}), we have $M=\mathcal{R}T/(0.02063\ \text{dm}^4\ \text{g}^{-1}\ Dg)=$

Table 7.5 Van 't Hoff *i*-factors and apparent degrees of dissociation α for three ionic chlorides

	i	α
NaCl	1.90	0.90
$MgCl_2$	2.10	0.55
$LaCl_3$	1.24	0.08

124.5 kg mol^{-1}. The estimated standard deviation is 1.0 kg mol^{-1}, and we may write the relative molar mass as 124500 ±1000.

7.11 Anomalous behaviour

A solution containing 0.35 g of potassium chloride in 100 g water has a freezing-point depression of 0.143 K. Since M_{KCl} is 0.07455 kg mol^{-1} and the cryoscopic constant for water is 1.86 kg K mol^{-1}, we expect, from (7.44), that ΔT_{fp} would be 0.087 K, or only 0.6 times the observed value. The parameter i, given for dissociation by

$$i = \text{observed property/calculated property} \tag{7.49}$$

was introduced by van 't Hoff to represent the apparent anomalies, or departures from ideal behaviour, and is known as the van 't Hoff *i*-factor.

Table 7.5 lists some measurements of the *i*-factors for a given molar concentration c of three ionic chlorides that dissociate into ions in aqueous solution:

$$\begin{array}{lll} MCl_x + aq \rightarrow & M^{x+} + & xCl^- \\ c(1-\alpha) & c\alpha & xc\alpha \end{array} \tag{7.50}$$

If we *assume* a degree α of dissociation then, at equilibrium, the concentrations of the three species are as shown above, and the total concentration is $c(1+x\alpha)$. Hence, the *i*-factor is, from (7.49), $c(1+x\alpha)/c = 1+x\alpha$, so that α may be calculated from the *i*-factor; it is listed in Table 7.5.

We saw in Chapter 5 that ionic compounds are ionized in the solid state, and that the ions separate upon dissolution in water. The *i*-factors in Table 7.5 do not have the values of 2, 3 and 4 that might, at first, have been expected. Although the values of α suggest incomplete dissociation, this explanation is not correct for these compounds; there are no 'molecules' of NaCl in solution. An ion in aqueous solution develops a hydration sphere, or 'atmosphere', of ions around it, and oppositely charged atmospheres attract one another electrostatically. As a result, the total concentration of ions is less than that expected from the stoichiometric concentration (see also Sections 9.2.6 and 9.2.7). Since colligative properties depend on the number of particles in solution, it follows that the extent of vapour pressure lowering by a dissolved solute will be less than that calculated from the stoichiometric concentration, and that the *i*-factor will be less than that given by (7.49).

Let us reconsider the solution of potassium chloride above. Its molality is 0.0469, and each ion molality is also of this value. The activity coefficient (see Section 7.12) for potassium chloride in this solution has been measured as 0.820; hence, the effective concentration, or *activity*, of the solution is actually 0.82×0.0469, or 0.0385. Thus, ΔT_{fp} would become 0.0715 K, and i=0.143/0.0715=2.0.

Some solutes associate in solution. Aliphatic carboxylic acids tend to dimerize in aqueous solution, through intermolecular hydrogen-bonding:

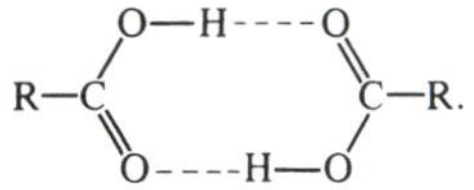

*R*CO$_2$ H dimer

Association leads to *i*-factors of less than unity:

$$\underset{c(1-\beta)}{2\,R\mathrm{CO_2H}} \rightleftharpoons \underset{c\beta/2}{(R\mathrm{CO_2H})_2}$$

From arguments similar to those used above, the concentration at equilibrium, from an initial concentration c, are readily obtained; β is a degree of association, and the i-factor is $1-\beta/2$.

A solution of 1.425 g of ethanoic acid in 100 g benzene has a freezing-point depression of 0.608 K. From (7.44) m_{B}=0.1188, so that M_{r}=119.9. Since M_{r} for ethanoic acid is 60.05, it is almost completely dimerized in this solution; β=0.998.

7.12 Activities

We conclude this chapter with a brief study of the thermodynamic concept of activity, to which we have referred from time to time, and we shall consider the meaning of activity in relation to a solvent liquid and a solute. The corresponding quantity for a gas is the fugacity; we considered this topic in Section 4.9.3, and some of the material there will be relevant to this discussion.

An activity a is related to the corresponding concentration function θ by an *activity coefficient* γ. Generally, we have that $a=\gamma\theta$, where θ will normally be a molarity, molality, mole fraction or partial pressure. The properties of the activity coefficient will depend upon the particular conditions, and may be different for solvent and solute. For example, if the solvent A is present in excess, then because it obeys Raoult's law more closely as it approaches purity, the activity $a_{\mathrm{A}} \to x_{\mathrm{A}}$ as $x_{\mathrm{A}} \to 1$. Then,

$$a_{\mathrm{A}}=\gamma_{\mathrm{A}} x_{\mathrm{A}} \tag{7.51}$$

which satisfies the criterion that as $x_{\mathrm{A}} \to 1$ $\gamma_{\mathrm{A}} \to 1$, too.

7.12.1 Activity of a solvent

If we measure the vapour pressure of the solvent in a binary system of water A and a non-volatile solute B, such as sucrose, we can obtain the activity of the solvent from vapour pressure measurements:

$$a_{\mathrm{A}}=p_{\mathrm{A}}/p_{\mathrm{A},\ell} \tag{7.52}$$

where the subscript A,l indicates the pure liquid A as before, and

$$\mu_{\mathrm{A}}=\mu_{\mathrm{A},\ell}+\mathcal{R}T\ln(p_{\mathrm{A}}/p_{\mathrm{A},\ell}) \tag{7.53}$$

If the system were to behave ideally then we would use (7.23), repeated here for convenience,

Table 7.6 Activity data for aqueous sucrose at 50 °C; p_{H_2O}=92.51 mmHg

x_A	p_A	a_A	$\gamma_{x,A}$	x_B	a_B	$\gamma_{x,B}$
0.9940	91.95	0.9939	1.000	0.0060	0.0060	1.000
0.9864	91.16	0.9854	0.999	0.0136	0.0145	0.897
0.9826	90.65	0.9799	0.997	0.0174	0.0209	0.833
0.9762	89.95	0.9723	0.996	0.0238	0.0318	0.748
0.9665	88.97	0.9617	0.995	0.0335	0.0443	0.756
0.9559	87.67	0.9477	0.991	0.0441	0.0656	0.672
0.9439	86.03	0.9300	0.985	0.0561	0.0938	0.598
0.9323	83.66	0.9043	0.970	0.0677	0.1401	0.483
0.9098	81.02	0.8758	0.963	0.0902	0.2108	0.428
0.8911	75.30	0.8140	0.913	0.1089	0.3414	0.319

where we could replace $p_A/p_{A,\ell}$ by the mole fraction x_A. If the system is nonideal then, following (7.51), we have

$$\mu_A=\mu_{A,\ell}+\mathcal{R}T\ln(a_A) \tag{7.54}$$

Table 7.6 lists data for sucrose solutions at 50 °C; γ_x refers to the activity coefficient in terms of the mole fraction concentration function.

It is clear from Table 7.6 that $\gamma_{x,A}$ tends to unity as x_A approaches unity. We can expand (7.54) to the form

$$\mu_A=\mu_{A,\ell}+\mathcal{R}T\ln(x_A)+\mathcal{R}T\ln(\gamma_{x,A}) \tag{7.55}$$

from which it is clear that the standard state, which is the pure liquid at 1 atm, is established when $x_A=1$ (and $\gamma_{x,A}=1$, from (7.51)); the deviation from ideality is represented by the term $\mathcal{R}T\ln(\gamma_{x,A})$ in (7.55).

7.12.2 Activity of a solute

If the solute B were also volatile, then its standard state could be chosen as the pure compound, like the solvent A, so that the activities of B would be obtained from vapour pressure measurements. This situation holds particularly when the solute concentration has a wide mole fraction range.

When the concentration range is small, say from $x=0$ to $x=0.1$, the standard state for the solute is defined by extrapolating the Henry's law line, the tangent to the real solution curve at $x_B=0$, to $x=1$, Figure 7.27. The activity of the solute B in the standard state is equal to H_B, the Henry's law constant:

$$a_B=p_B/H_B=\gamma_B x_B \tag{7.56}$$

The difference between the two conventional standard states for the solute is that, on the basis of Raoult's law, $\gamma\to1$ as $x\to1$ (as with component A in Table 7.6), whereas on the basis of Henry's law, $\gamma\to1$ as $x\to0$ (as with component B in Table 7.6). When the solute is nonvolatile, as in the example of sucrose, the standard state is based on Henry's law. The activity of the sucrose in aqueous solution can be calculated from the vapour pressure of the

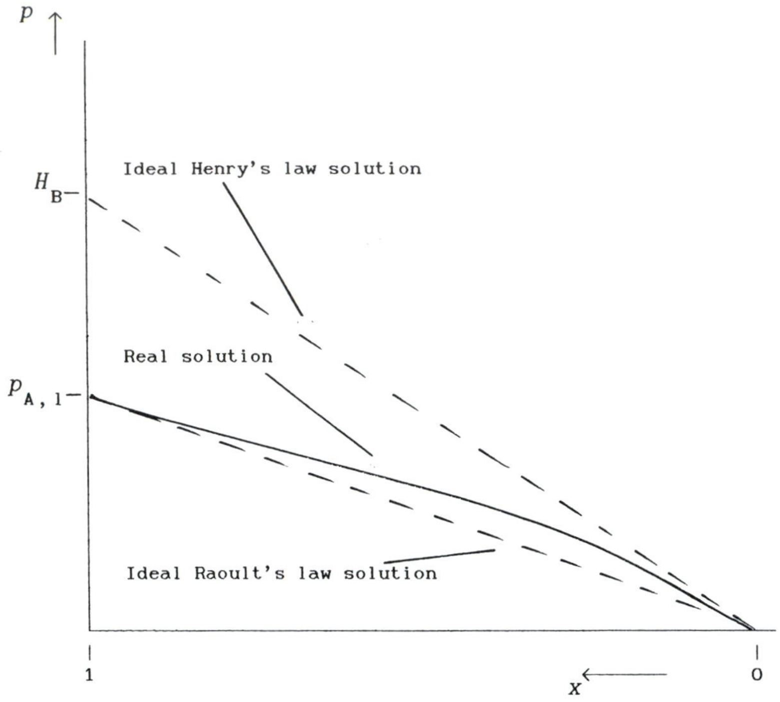

Figure 7.27 Vapour pressure p as a function of mole fraction x: standard states for solvent and solute in a binary liquid. Close to $x=1$, the solvent is nearly pure and Raoult's law is obeyed, that is, $p \propto x$, and the proportionality constant is the vapour pressure of the pure solvent $p_{A,1}$. The standard state for the solvent is, thus, the pure liquid at 1 atm, at the vapour pressure $p_{A,1}$. Close to $x=0$, Henry's law is obeyed for the solute, that is, $p \propto x$, but the proportionality constant now is H_B, the constant of Henry's law for the solute. The standard state for the solute is, thus, the value of p at $x=1$, as though Henry's law were obeyed for the solute close to $x=1$, that is, it is at $p=H_B$, at 1 atm. The fact that this state is hypothetical does not debar its use as a standard state.

solvent over the solute concentration range from x_B to the highest value of interest. However, we need first to consider partial molar properties and the Gibbs–Duhem equation.

7.12.2.1 Partial molar volume: Gibbs–Duhem equation

Consider a solution of n_A mole of A and n_B mole of B in a volume V that is sufficiently large that the addition of 1 mole of A or B would not change the concentration significantly. The increase in volume per mole of A added under these conditions is the *partial molar volume* of A, defined by

$$\overline{V}_A = (\partial V/\partial n_A)_{T,p,n_B} \tag{7.57}$$

At constant temperature and pressure, and following (4.14),

$$dV = (\partial V/\partial n_A)_{n_B} dn_A + (\partial V/\partial n_B)_{n_A} dn_B \tag{7.58}$$

Introducing the partial molar volume from (7.57),

$$dV = \overline{V}_A\, dn_A + \overline{V}_B\, dn_B \tag{7.59}$$

and by integrating (7.59), we obtain

$$V = n_A \overline{V}_A + n_B \overline{V}_B \tag{7.60}$$

If we form the complete differential of (7.60), still under the conditions of constant temperature and pressure, we have

$$dV = V_A\, dn_A + V_B\, dn_B + n_A\, dV_A + n_B\, dV_B \tag{7.61}$$

and by comparing (7.59) and (7.61), it follows that

$$n_A\, dV_A + n_B\, dV_B = 0 \tag{7.62}$$

The same argument may be applied to the partial molar Gibbs free energy, or chemical potential from (4.116); hence, we may derive

$$n_A\, d\mu_A + n_B\, d\mu_B = 0 \tag{7.63}$$

which is the Gibbs–Duhem equation.

Consider the addition of 60 cm^3 of water (A) to 40 cm^3 of ethanol (B). Is the final volume 100 cm^3? The numbers n_A and n_B are 3.330 mol and 0.682 mol respectively, and the mole fractions in the solution are $x_A=0.830$ and $x_B=0.170$. From graphs of $\overline{V}$ against x for each substance, $\overline{V}_A$ at x_A is 17.8 cm^3 and $\overline{V}_B$ at x_B is 54.7 cm^3. Hence, from (7.60), $V=96.6$ cm^3.

The reason that volume is not an additive property lies in the interaction between the molecules of the different species. In the case of polar species, like water and ethanol, hydrogen-bonding acts between water and ethanol molecules, as well as in the pure components themselves, drawing them closer together; thus, the final volume is less than the stoichiometric volume.

To return now to the activity of the solute, we combine (7.54) and (7.63) to give

$$n_A\, d\ln(a_A) + n_B\, d\ln(a_B) = 0 \tag{7.64}$$

Dividing by (n_A+n_B) leads to

$$x_A\, d\ln(a_A) + x_B\, d\ln(a_B) = 0 \tag{7.65}$$

Hence,

$$\int d\ln(a_B) = -\int x_A/(1-x_A)\, d\ln(a_A) \tag{7.66}$$

In order to avoid the indeterminacy with $x_B=0$, that is, $x_A=1$, the lower limit of integration is the value of x_A at which the solvent follows Raoult's law, that is, where $x_A=a_A$. The integral in (7.66) is readily computed by numerical methods (see Appendix 14), and the results a_B and, hence, $\gamma_{x,B}$, for the sucrose solutions are listed in Table 7.6.

As an example of the application of (7.66), we will calculate a_B *at* $x_B=0.1089$ (Table 7.6). The integral *in extenso* becomes

$$\ln(a_{0.1089}/0.0060) = -\int_{0.0060}^{0.1089} x_A/(1-x_A)\, d\ln(a_A)$$

The function $x_A/(1-x_A)$ is plotted against $\ln(a_A)$ and the area under the curve calculated. The area, by Gaussian quadrature, was 4.0414; hence, a $(x_B=0.1089)=0.0060/\exp(-4.0414)$ $=0.3414$.

7.12.2.2 Isopiestic method

We consider water as a solvent A with a solute B. Following (7.65) with molality as the concentration function,

$$m_{\mathrm{A}}\,\mathrm{d}\ln(a_{\mathrm{B}})+m_{\mathrm{A}}\,\mathrm{d}\ln(a_{\mathrm{A}})=0 \tag{7.67}$$

or

$$\mathrm{d}\ln(m_{\mathrm{B}})+\mathrm{d}\ln(\gamma_{\mathrm{B}})=-55.51\,\mathrm{d}\ln(a_{\mathrm{A}})/m_{\mathrm{B}} \tag{7.68}$$

The molal osmotic coefficient ϕ is defined by

$$\phi=-55.51\ln(a_{\mathrm{A}})/m_{\mathrm{B}} \tag{7.69}$$

and from (7.52) $\mathrm{d}\ln(a_{\mathrm{A}})=\mathrm{d}\ln(p_{\mathrm{A}}/p_{\mathrm{A,l}})$, so that ϕ is determined by the vapour pressure of the solvent and the molality of the solute. Rearranging (7.69) and differentiating:

$$\mathrm{d}(\phi m_{\mathrm{B}})=-55.51\,\mathrm{d}\ln(a_{\mathrm{A}})=\phi\,\mathrm{d}m_{\mathrm{B}}+m_{\mathrm{B}}\,\mathrm{d}\phi \tag{7.70}$$

Manipulation of (7.68) and (7.70) leads to the result

$$\mathrm{d}\ln(\gamma_{\mathrm{B}})=\mathrm{d}\phi-\frac{(1-\phi)}{m_{\mathrm{B}}}\,\mathrm{d}m_{\mathrm{B}} \tag{7.71}$$

whence

$$\int_0^{m_{\mathrm{B}}}\mathrm{d}\ln(\gamma_{\mathrm{B}})=\int_0^{m_{\mathrm{B}}}\mathrm{d}\phi-\int_0^{m_{\mathrm{B}}}\frac{(1-\phi)}{m_{\mathrm{B}}}\,\mathrm{d}m_{\mathrm{B}} \tag{7.72}$$

Since the activity of the pure water is unity, and $m_{\mathrm{B}}\to 0$ as $\phi\to 1$, we obtain

$$\ln(\gamma_{\mathrm{B}})=(\phi-1)-\int_0^{m_{\mathrm{B}}}\frac{(1-\phi)}{m_{\mathrm{B}}}\,\mathrm{d}m_{\mathrm{B}} \tag{7.73}$$

where the integration is carried out numerically.

In practice, the activity of water is determined by placing a reference sucrose solution and the solution of unknown activity in a vacuum chamber. Water migrates from the solution of higher vapour pressure to that of lower vapour pressure. At equilibrium, the *isopiestic point*, the vapour pressures become equal, and the activity of water in each solution is the same. The compositions of the solutions are analysed, and the method repeated to obtain the function ϕ over the desired range of molality. Further details on this procedure may be found in standard reference works.[3] The method is of particular use in determining the activity coefficients of very soluble salts, such as lithium bromide and potassium iodide, in their saturated solutions.

7.13 Activity and molality

We have seen that, because molality is independent of temperature, elevation of the boiling-point and depression of the freezing-point are discussed in terms of that parameter. Hence, we need to define activity also in terms of molality.

In Section 7.8, for example, we used the approximation $x_{\mathrm{B}}\simeq n_{\mathrm{B}}/n_{\mathrm{A}}$, because the solution was dilute. Since $n_{\mathrm{B}}\propto m_{\mathrm{B}}$, we may write

$$x_{\mathrm{B}}=km_{\mathrm{B}}/m^{-0-} \tag{7.74}$$

[3] See, for example, K G Denbigh, *The Principles of Chemical Equilibrium*, (CUP, 1981); H S Harned and B B Owen, *The Physical Chemistry of Electrolytic Solutions* (Reinhold, 1967); G N Lewis and M Randall, *Thermodynamics* (McGraw-Hill, 1923).

where k is a constant. We choose $m^{-0-}=1$ mol kg^{-1}, so that both activity and activity coefficient become dimensionless:

$$a_B = \gamma_B m_B/m^{-0-} \tag{7.75}$$

where $\gamma_B \rightarrow 1$ as $m_B \rightarrow 0$. Thus, using (7.25) with (7.74) the chemical potential of the solute may be given as

$$\mu_B = \mu_B^{-0-} + \mathcal{R}T \ln(m_B/m^{-0-}) \tag{7.76}$$

The standard chemical potential μ_B^{-0-} incorporates the constant k in (7.74), and corresponds to the chemical potential of a solution in which the molality $m_B = m^{-0-} = 1$ mol kg^{-1}.

We may summarize the discussion of standard states as follows:

Solvent A

Raoult's law basis: $a_A = p_A/p_{A,l} = \gamma_A x_A$; $\gamma_A \rightarrow 1$ as $x_A \rightarrow 1$. The standard state is the pure solvent at 1 atm.

Solute B

Wide range of concentration. Raoult's law basis: as for a solvent.

Small range of concentration. Henry's law basis: $a_B = p_B/H_B = \gamma_B x_B$; $\gamma_B \rightarrow 1$ as $x_B \rightarrow 0$. The standard state (hypothetical) is the pure solute. In *molality* terms, $a_B = \gamma_B m_B/m^{-0-}$; $\gamma_B \rightarrow 1$ as $m_B \rightarrow 0$. The standard state (hypothetical) is the 1 molal solution, in which the environment of each species is the same as that at infinite dilution (see Section 8.6.1.2).

Problems 7

7.1 How many phases, components and degrees of freedom are present in the following systems?

(a) $NH_4Cl(s) \rightleftharpoons NH_3(g) + HCl(g)$
(b) $NH_4Cl(s) + HCl(g) \rightleftharpoons NH_3(g) + 2HCl(g)$
(c) $S(l) \rightleftharpoons S(s)$ (constant pressure)
(d) $He_I(l) \rightleftharpoons He_{II}(l) \rightleftharpoons He(s)$

7.2 Show that the pressures of two phases in equilibrium are equal.

7.3 Show that for two phases α and β in equilibrium, the Clapeyron equation may be derived from (4.97) and (4.117).

7.4 (a) The following data for the vapour pressure of molten lead were obtained from measurements of the rate of effusion of the vapour through a small hole into a vacuum:

T/K	895.4	922.1	964.5	1009.7	1045.5
p/N m^{-2}	0.0783	0.205	0.539	1.40	3.40

Determine the enthalpy of vaporization of lead for the experimental temperature range and its estimated standard deviation.

(b) The molar heat capacity of lead may be represented by the following equations:

$$298\text{–}600\text{ K } C_p = 23.56\text{ J K}^{-1}\text{ mol}^{-1} + (0.00975\text{ J K}^{-2}\text{ mol}^{-1})T$$

$$600\text{–}1200\text{ K } C_p = 32.43\text{ J K}^{-1}\text{ mol}^{-1} - (0.00310\text{ J K}^{-2}\text{ mol}^{-1})T$$

The enthalpy of fusion of lead at the melting-point of 600 K is 4.81 kJ mol^{-1}. If the result for ΔH_v from (a) applies at the mean temperature of 970 K, determine the standard enthalpy

of vaporization of lead, assuming that the vapour behaves ideally. It may be helpful first to construct a thermochemical cycle in order to show the processes involved in the calculation.
7.5 Determine the melting-point of ice under an applied pressure of 500 atm. At 0 °C the density of ice is 0.91680 g cm^{-3} and that of water is 0.99987 g cm^{-3}; the enthalpy of fusion is 6.008 kJ mol^{-1}.
7.6 From the following data on the vapour pressure of solid iodine, determine the enthalpy of sublimation over the given temperature range.

T/°C	38.7	73.2	97.5
p/mmHg	1.0	10	40

7.7 The vapour pressures of water and of propanone at 312.5 K are 54.30 and 400.0 Torr respectively. Calculate the partial vapour pressures of the two components above a solution containing 10 g water and 20 g propanone at this temperature, assuming the system to be ideal.
7.8 The vapour pressure of ethanol at 292 K is 40.00 mmHg. If 1.032 g of a nonvolatile substance B is dissolved in 98.7 g ethanol, the lowering of the vapour pressure is 0.432 mmHg. Calculate the molar mass of B.
7.9 An aqueous solution of a weak monobasic acid containing 0.100 g in 21.7 g water depresses the freezing-point of the water by 0.187 K. Given that the cryoscopic constant for water is 1.86 kg K mol^{-1}, determine the relative molar mass of the acid.
7.10 Interpret fully the fact that a 0.1 mol dm^{-3} aqueous solution of glucose ($C_6H_{12}O_6$) freezes at the same temperature as does a solution of calcium chloride ($CaCl_2$) containing 0.44 g in 100 g water.
7.11 A solution of 0.125 g urea in 170 g water gave a boiling-point elevation of 0.63 K. Calculate the relative molar mass of urea, taking the ebullioscopic constant as 0.51 kg K mol^{-1}. Derive any formula used.
7.12 The following data refer to the freezing-point depression of benzene containing concentrations c of tetrachloromethane (RMM=153.82):

c/mol dm^{-3}	0.1184	0.3499	0.8166
T_f/K	0.603	1.761	4.005

Calculate the cryoscopic constant at each concentration, and comment upon the results.
7.13 The following osmotic pressures were obtained for a solution of a polymer at 25 °C, in terms of a manometric height h:

w/kg m^{-3}	3.201	4.798	5.702	6.898	7.797
h/m	0.0291	0.0583	0.0787	0.1099	0.1397

The osmotic pressure may be given as $\Pi = c\mathcal{R}T(1+Bc+\ldots) = w(\mathcal{R}T/M_m)[1+(B/M_m)w+\ldots]$, where $c=w/M_m$. Use the data to find M_m and B (second osmotic virial coefficient); the density of the manometric liquid was 925 kg m^{-3}, and g may be taken as 9.807 m s^{-2}. Find also the estimated standard deviations in M and B.
7.14 The activity coefficient of 0.05 molal aqueous sodium chloride is 0.821. What would be the freezing-point depression for this solution? (K_c=1.86 kg K mol^{-1})
7.15 A nonvolatile solute B of relative molar mass 500 is dissolved in a solvent of cryoscopic

constant 6.9 kg K mol^{-1} to form a solution of concentration 0.01 mol kg^{-1}. The density D of the solution is 1.022 g cm^{-3}. Calculate

(a) the freezing-point depression for the solvent;
(b) the osmotic pressure of the solution; first obtain an equation relating m (molality) to c (molarity).

Verify that, for a sparingly soluble compound of high relative molar mass, it is more practicable to determine the relative molar mass from osmotic pressure measurements than from freezing-point depression measurements. It may be assumed that a temperature change can be measured with a probable error of $\pm$0.002 K, and an osmotic pressure to $\pm$2 mmHg.

7.16 The freezing-point depressions for 4-hydroxytoluene (*p*-cresol) solutions in benzene (K_c=5.12 kg K mol^{-1}) are 0.420 K and 5.002 K for molalities 0.0860 and 1.850 respectively. Calculate the degree of dimerization at each concentration.

7.17 100 cm^3 of an aqueous solution of phenol containing 15 g dm^{-3} are shaken with (a) one 50 cm^3 portion of pentanol, (b) two 25 cm^3 portion of pentanol. If the ratio of the concentration of phenol in water to that in pentanol is 0.0625, calculate the total weights of phenol extracted by the two processes.

7.18 In the steam-distillation of bromobenzene, the boiling-point was 369 K at 770 Torr. If the partial vapour pressure of water is 657.6 Torr at 369 K, calculate the percentage by weight of bromobenzene in the steam-distillate.

7.19 The antibiotic gliotoxin $C_{13}H_{14}O_4N_2S_2 \cdot xH_2O$ crystallizes with a unit-cell volume, determined by X-ray diffraction, of 1.451 $\pm$0.002 nm^3; its experimental density is 1543 $\pm$10 kg m^{-3}. Chemical evidence suggests a relative molar mass M_r of approximately 330. Determine

(a) the number of molecules in the unit cell;
(b) a precise value for M_r, and its estimated standard deviation.

7.20 The following data were obtained for the vapour pressures of mixtures of ethyl ethanoate A and iodoethane B at 323 K at the mole fractions listed. At each mole fraction, determine (a) the activity coefficients of both components on the basis of Raoult's law, and (b) the activity coefficients of component B as a solute on the basis of Henry's law.

x_A	1.000	0.9421	0.8905	0.8082	0.7647	0.6282
p_A/Torr	280.4	266.1	252.3	231.4	220.8	187.9
p_B/Torr	0.000	28.0	52.7	87.7	105.4	155.4
x_A	0.4522	0.3651	0.1747	0.0707	0.000	
p_A/Torr	144.2	122.9	66.6	38.2	0.000	
p_B/Torr	213.3	239.1	296.9	322.5	353.4	

CHAPTER 8

Chemical equilibrium

8.1 Introduction

So far, we have considered the physical chemistry of a number of different systems for which conditions of equilibrium were invoked. Here, we discuss equilibrium itself in a more general manner and apply the results in several different situations. Chemical reactions tend towards an equilibrium, in which all reactants and products are present in varying amounts. The equilibrium is dynamic, that is, there is a continuous interchange of individual reactant and product species but with no net change, at a given temperature.

For all practical purposes, some reactions may be said to go to completion. If we mix an aqueous solution of silver nitrate with the stoichiometric amount of hydrochloric acid, then silver chloride is precipitated to the extent that only 10^{-6} mol dm^3 (0.01 mg in 100 cm^3) remains in solution:

$$Ag^+(aq)+Cl^-(aq) \rightleftharpoons AgCl(s) \tag{8.1}$$

Such a reaction would normally be said to be complete, although an equilibrium still exists.

In contrast, the esterification of ethanoic acid with ethanol at 25 °C reaches an equilibrium in which about two thirds of the acid is esterified:

$$CH_3CO_2H(l)+C_2H_5OH(l) \rightleftharpoons CH_3CO_2C_2H_5(l)+H_2O(l) \tag{8.2}$$

In the reactions (8.1) and (8.2), the existence of a dynamic equilibrium means that the forwards and backwards reactions in each system are taking place at the same rate; there is no net change in the equilibrium concentrations, provided that the temperature remains constant.

8.2 Mass action

The chemical equilibrium equation was deduced first from a study of the rate of the acid-catalysed hydrolysis of sucrose. It was found that the rate of hydrolysis was proportional to the amount of sucrose unreacted. Similar studies were carried out on the esterification of acids; some results for the reaction (8.2) at 100 °C are listed in Table 8.1.

Guldberg and Waage expressed chemical equilibrium in a more general form, using a reaction such as

$$A+B \rightleftharpoons C+D \tag{8.3}$$

and their law of mass action states that *the rate of the forwards reaction is proportional to the product* [C][D], where [C] refers to the molar concentration[1] of the species C and so on. The

[1] Guldberg and Waage used the term *active mass* rather than concentration, which is, fortuitously, closer to the correct descriptor, *activity*.

Table 8.1 Esterification of ethanoic acid with ethanol at 100 °C

n_{HAc}/mol (init.)	n_{EtOH}/mol (init.)	n_{EtAc}/mol (equil.)	[Products]/[Reactants]	
1.00	0.19	0.18	4.0	
1.00	0.32	0.29	3.9	
1.00	0.48	0.41	4.1	
1.00	1.00	0.67	4.1	
1.00	2.00	0.85	4.2	
1.00	8.00	0.97	4.5	Mean 4.1

rate of the backwards reaction is proportional to [A][B] and at equilibrium, at a given temperature, the two rates are equal and their ratio is the *equilibrium constant K*:

$$K=[C][D]/[A][B] \tag{8.4}$$

We note that, although (8.4) is a correct statement of the chemical equilibrium equation, it is not a proof because it is based upon the assumption of particular rate equations. While this assumption is, in fact, valid for (8.2), it is not true for the reaction

$$H_2(g)+I_2(g) \rightleftharpoons 2HI(g) \tag{8.5}$$

although the equilibrium constant would still be formulated according to (8.4), so that it is necessary to consider a more general approach to equilibrium.

8.3 Equilibrium and free energy

We explained in Chapter 4 that a thermodynamically spontaneous reaction at constant temperature and pressure is accompanied by a decrease in the Gibbs free energy of the system. Here we shall direct our attention to this parameter.

Consider a simple chemical reaction

$$A \rightleftharpoons B \tag{8.6}$$

as is met in the isomerization of cyclopropane to propene, for example:

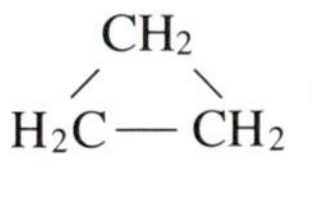

$$\begin{array}{c} CH_2 \\ / \quad \backslash \\ H_2C \!-\! CH_2 \end{array} \rightleftharpoons CH_3\!-\!CH\!=\!CH_2 \tag{8.7}$$

Initially, let there be n_A mol of A and none of B. For a small decrease dn_A in the amount of A, an equivalent amount dn_B of B is formed. It is convenient to define an *extent* ξ of reaction, such that $dn_A=-d\xi$ and $dn_B=d\xi$. At constant temperature and pressure, applying (4.118) leads to

$$dG=-\mu_A d\xi+\mu_B d\xi \tag{8.8}$$

or

$$(\partial G/\partial\xi)_{T,p}=\mu_B-\mu_A=\Delta G \tag{8.9}$$

where ΔG is the free energy change accompanying a finite change (8.6). If ΔG is positive, the tendency is B→A, whereas if ΔG is negative, the tendency is A→B. If we were to plot

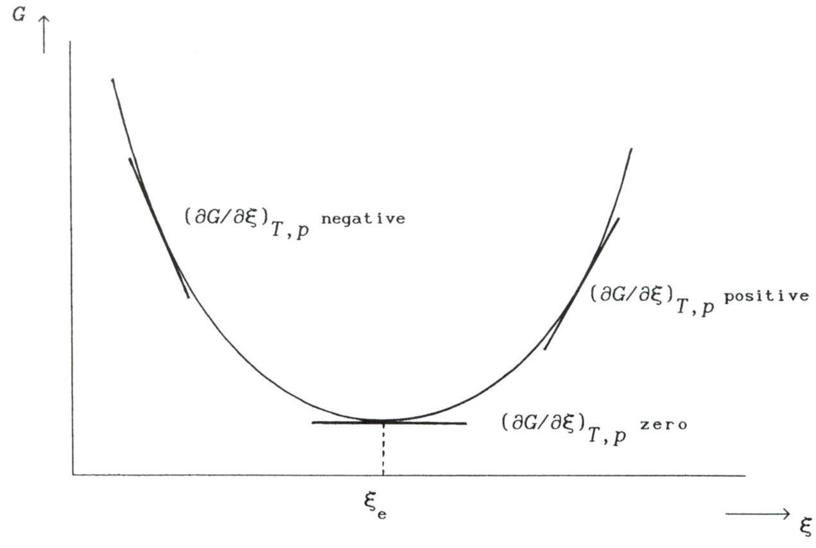

Figure 8.1 Variation in the free energy G of a reacting system with the extent ξ of reaction; at an extent ξ_e, corresponding to dynamic equilbrium in the system, $(\partial G/\partial \xi)_{T,p}=0$.

G for the system as a function of ξ, we would obtain a graph of the form shown by Figure 8.1.

At any point on the curve, the gradient is $(\partial G/\partial \xi)_{T,p}$: where the gradient is negative, the reaction is $A \rightarrow B$; where it is positive, the reaction is $B \rightarrow A$. At the minimum, $\Delta G=0$, $\mu_A=\mu_B$ and the system is at equilibrium.

For the example reaction (8.6), $\Delta G=G_B-G_A$, but since A and B are individual substances, G_A and G_B will be $\Delta G_{f,A}$ and $\Delta G_{f,B}$ respectively, the free energies of formation; under standard conditions, $\Delta G^{\ominus}=\Delta G^{\ominus}_{f,B}-\Delta G^{\ominus}_{f,A}$. If we use (4.121) with (8.9), assuming that A and B are gases that behave ideally, we obtain

$$\Delta G=\Delta G^{\ominus}+\mathcal{R}T\ln(p_B/p_A) \tag{8.10}$$

where $\Delta G^{\ominus}=\mu^{\ominus}_B-\mu^{\ominus}_A$. At equilibrium $\Delta G=0$; hence,

$$\Delta G^{\ominus}=-\mathcal{R}T\ln(K_p) \tag{8.11}$$

K_p is the equilibrium constant (p_B/p_A) for reaction (8.6) and depends upon the temperature; the subscript p indicates that the concentrations of the reacting species are given in terms of partial pressure.

At 350 K, equilibrium in the isomerization of *cis*-but-2-ene to *trans*-but-2-ene is established when 0.67 mol of *trans*-but-2-ene is formed from 1 mol of *cis*-but-2-ene. Since partial pressure is proportional to mole fraction,

$$K_p=p_{trans}/p_{cis}=2.03$$

$$\Delta G^{\ominus}=-\mathcal{R}T\ln(K_p)=-2.06\ \text{kJ mol}^{-1}$$

We consider a more general case of (8.6), which may be written as

$$\nu_A A+\nu_B B \rightleftharpoons \nu_C C+\nu_D D \tag{8.12}$$

where ν_A is the number of molecules of A and so on. Following the previous argument, an extent of reaction $d\xi$ implies that $dn_A = -\nu_A\, d\xi$, $dn_B = -\nu_B\, d\xi$, $dn_C = \nu_C\, d\xi$ and $dn_D = \nu_D\, d\xi$, whence

$$dG = \sum_j \nu_j \mu_j\, d\xi \tag{8.13}$$

where the chemical potentials μ_j of the product terms have a positive sign and those of the reactants a negative sign, and from (8.9)

$$\Delta G = (\partial G/\partial \xi)_{T,p} = \sum_j \nu_j \mu_j \tag{8.14}$$

Following the previous argument, using activity in place of partial pressure, we now write

$$\Delta G = \Delta G^{\ominus} + \mathcal{R}T \ln\left(\frac{a_C^{\nu_C}\, a_D^{\nu_D}}{a_A^{\nu_A}\, a_B^{\nu_B}} \right) \tag{8.15}$$

This equation is sometimes referred to as the van 't Hoff isotherm.

At equilibrium $\Delta G = 0$, the activities are then the equilibrium values and this term is identified with the equilibrium constant K, whence (8.15) becomes

$$\Delta G^{\ominus} = -\mathcal{R}T \ln(K) \tag{8.16}$$

which is similar to (8.11), where K is given now by

$$K = \frac{a_C^{\nu_C}\, a_D^{\nu_D}}{a_A^{\nu_A}\, a_A^{\nu_B}} \tag{8.17}$$

In terms of partial pressures, we write K_p in place of K, where

$$K_p = \frac{p_C^{\nu_C}\, p_D^{\nu_D}}{p_A^{\nu_A}\, p_A^{\nu_B}} (1/p^{\ominus})^{\Delta\nu} \tag{8.18}$$

where $p^{\ominus}$ is a standard pressure of 1 atm and $\Delta\nu = \Sigma\nu(\text{Products}) - \Sigma\nu(\text{Reactants})$; like K, K_p is dimensionless.

We may wish to specify concentration other than by partial pressure. For example, we could write (8.18) as

$$K_p = \frac{[C]^{\nu_C}[D]^{\nu_D}}{[A]^{\nu_A}[B]^{\nu_B}} (\mathcal{R}T)^{\Delta\nu} = K_c (\mathcal{R}T)^{\Delta\nu} \tag{8.19}$$

where K_c is an equilibrium constant in terms of molarity. The composition of a mixture may also be expressed in terms of mole fraction: $x_j = p_j/p$, where p is the total pressure, so that we obtain

$$K_p = \frac{x_C^{\nu_C}\, x_D^{\nu_D}}{x_A^{\nu_A}\, x_B^{\nu_B}} (p/p^{\ominus})^{\Delta\nu} = K_x (p/p^{\ominus})^{\Delta\nu} \tag{8.20}$$

where K_X is the equilibrium constant in terms of mole fraction.

For gases, we may use the fugacity f (see Section 4.9.3) in place of activity and, with $a_j = f_j / p^{-0-}$, K_p is again dimensionless. In the case of liquids and solutions the appropriate activity is used, as discussed in Section 7.12ff. Thus, with (8.2) and the data in Table 8.1, we use mole fraction, where $a_j = \gamma_j x_j$, or molality, where $a_j = \gamma_j m_j / m^{-0-}$, with m^{-0-} equal to a standard molality of 1 mol kg^{-1}.

8.4 Temperature and pressure effects on equilibrium

It is evident from (8.10) that equilibrium constants depend on temperature. Furthermore, K_p is independent of the total pressure, because it is determined by ΔG^{-0-}, which is itself defined at 1 atm. However, in using mole fraction, as in (8.20), an additional dependence arises, on the total pressure p. Similar considerations apply to the use of K_c, (8.19).

It is important to distinguish between the effect of temperature and pressure on an equilibrium *constant* and on an equilibrium *composition*, or position of equilibrium. The Le Chatelier–Braun principle was given in Section 1.3; here, we shall investigate this principle a little more fully, through the synthesis of ammonia:

$$\tfrac{1}{2}N_2(g) + \tfrac{3}{2}H_2(g) \rightleftharpoons NH_3(g) \qquad (8.21)$$

Let there be initially a mol of nitrogen and b mol of hydrogen. At equilibrium, let the extent of the reaction be ξ, when there will be $a - \xi/2$ mol nitrogen, $b - 3\xi/2$ mol hydrogen and ξ mol ammonia. Then, from (8.20) with $\Delta\nu = -1$,

$$K_p = \frac{\xi(a+b-\xi)}{(a-\xi/2)^{1/2}(b-3\xi/2)^{3/2}}(p^{-0-}/p) \qquad (8.22)$$

where p is the total pressure. It may be noted that, had (8.21) been written without fractions, by multiplying throughout by 2, then the values for K_p in (8.22) and Table 1.3 would have been squared. This result is in accord with the fact that free energy is an extensive property.

Table 1.3 (Chapter 1) shows that K_p is sensibly independent of pressure. However, as the temperature is increased K_p decreases, according to the Le Chatelier–Braun principle: because the formation of ammonia is exothermic, the effect of an increase in temperature tends to be annulled by the backwards (endothermic) reaction.

The product of the fugacity coefficients $\Pi(\gamma)^{\Delta\nu} = \gamma(\text{Products})/\gamma(\text{Reactants})$ applicable with (8.18) has the following values at 723 K:

	p/atm		
	10	50	100
$10^2 K_p$	0.659	0.690	0.725
$\Pi(\gamma)^{\Delta\nu}$	0.990	0.945	0.898
$10^2 K$	0.652	0.652	0.651

It is evident that the small drift of K_p with p in Table 1.3 reflects the departure from ideal behaviour of the system (8.21).

An increase in temperature also has the effect of increasing the speed at which equilibrium is attained in any reaction (see also Section 10.8.2). The rates of both the forwards and backwards reactions increase with an increase in temperature. Hence, the rate at which equilibrium is attained also increases, but the effect on the position of equilibrium depends upon the sign of the enthalpy change for the direction of reaction under consideration.

Table 1.2 shows that an increase in temperature decreases the amount of ammonia at equilibrium, for the reason just discussed. However, in these data we see also a pressure effect, namely, that an increase in pressure increases significantly the proportion of ammonia at equilibrium. Again, this result is explained by the Le Chatelier–Braun principle: since the forwards reaction is accompanied by a decrease in volume, the effect of an increase in pressure tends to be annulled by an increased extent of the forwards reaction. It may be noted that a pressure effect arises only when there is a change in volume (number of molecules) for the reaction.

The reaction between nitrogen and oxygen to form nitrogen monoxide is endothermic:

$$\tfrac{1}{2}N_2(g)+\tfrac{1}{2}O_2(g) \rightleftharpoons NO(g) \tag{8.23}$$

and the equilibrium constant is given by

$$K_p=\frac{p_{NO}}{p_{N_2}^{1/2}\,p_{O_2}^{1/2}}=K_x=K_C \tag{8.24}$$

from which the pressure independence is evident; an increase in temperature promotes the formation of nitrogen monoxide.

8.5 Van 't Hoff equation

The effect of pressure on the equilibrium constant has been discussed sufficiently in Section 8.4. Examination of (8.20) shows how a change in pressure affects only the *position* of equilibrium, that is, the composition at equilibrium, when there is a change in the total number of molecules in the course of the reaction.

From (8.11) at constant temperature and pressure, we can write

$$\frac{d}{dT}\ln(K_p)=(-1/\mathcal{R})\frac{d}{dT}(\Delta G^{-0-}/T) \tag{8.25}$$

Using the Gibbs–Helmholtz equation (4.109), we obtain

$$\frac{d}{dT}\ln(K_p)=\frac{\Delta H}{\mathcal{R}T^2} \tag{8.26}$$

which is the van 't Hoff equation. It shows that the rate of change of K_p with temperature for a reaction depends on the standard enthalpy change for that reaction, at the given temperature T. Furthermore, an exothermic reaction, for example, will have a negative value for $d\ln(K_p)/dT$; thus, an increase in T leads to a decrease in K_p, as we have discussed in Section 8.4.

It is a straightforward matter to use (8.26) to calculate K_p at another temperature, given the value at one temperature. Assuming that the variation of ΔH^{-0-} with temperature may be neglected, integration leads to

$$\ln(K_{p_2})-\ln(K_{p_1})=(-\Delta H^{-0-}/\mathcal{R})(1/T_2-1/T_1) \tag{8.27}$$

Consider next the exothermic reaction

$$CO(g)+H_2O(g) \rightleftharpoons H_2(g)+CO_2(g) \tag{8.28}$$

for which the following data are available:

T/K	400	600	800	1000
K_p	1540	28.10	4.155	1.397

We can recast (8.26) as

$$\ln(K_p) = -\Delta H^{-0-}/\mathcal{R}T + C \tag{8.29}$$

Where C is a constant. Assuming that ΔH^{-0-} does not vary with temperature, a least-squares plot of $\ln(K_p)$ against $1/T$ gives the equation

$$\ln(K_p) = 4685.9/(T/\mathrm{K}) - 4.4085 \tag{8.30}$$

The least-squares correlation coefficient r is 0.9998, indicating a good linear fit to the data. If we calculate K_p at 298.15 K from this equation, the value obtained is 1.759×10^5. From the free energies of formation of the species in (8.28), $\Delta G^{-0-} = -28.62$ kJ mol^{-1}. Using (8.11), K_p at 298.15 K is 1.033×10^5. Thus, it would appear that the assumption that the variation in ΔH^{-0-} is negligible becomes invalid when extrapolated as far as 298.15 K.

From (4.48), (4.50) and (4.51),

$$d(\Delta H_T) = \Delta C_p\, dT = (\Delta a + \Delta b\, T + \Delta c/T^2)\, dT \tag{8.31}$$

where Δa represents the contributions to ΔC_p in a, summed as discussed in Section (4.5.2), and similarly for b and c. On integration

$$\Delta H_T = \Delta a\, T + \Delta b\, T^2/2 - \Delta c/T + \Delta H_0 \tag{8.32}$$

where ΔH_0 is an integration constant that may be determined from ΔH_T at a chosen temperature: since $\Delta H_{298}^{-0-} = -41.16$ kJ mol^{-1}, $\Delta H_0 = -47.367$ kJ mol^{-1}. If we now substitute (8.32) in (8.26) and integrate, we obtain

$$\ln(K_p) = (\Delta a/\mathcal{R})\ln(T) + \Delta b\, T/(2\mathcal{R}) - \Delta c\, T^2/(2\mathcal{R}) + \Delta H_0/(\mathcal{R}T) + \mathcal{I} \tag{8.33}$$

where $\mathcal{I}$ is the constant for this integration. To obtain $\mathcal{I}$, we need the value of K_p at a chosen temperature, say 400 K, whence $\mathcal{I} = -156020$. We may now calculate K_p at any temperature with (8.33), and a comparison of results is given in Table 8.2; the agreement between (8.33) and the experimental results is highly satisfactory.

8.5.1 Heterogeneous equilibria

So far the systems that we have studied have been in either the liquid phase or the gas phase and, therefore, homogeneous. A typical heterogeneous reaction is the decomposition of calcium carbonate:

$$CaCO_3(s) \rightleftharpoons CaO(s) + CO_2(g) \tag{8.34}$$

From (8.17), we may write the equilibrium constant as

$$K = a_{CaO}a_{CO_2}/a_{CaCO_3} \tag{8.35}$$

It is conventional to choose the activity of a pure solid to be unity. The activity, or fugacity, of a gas is γp, from (4.122), where p is the partial pressure of the gas. Furthermore, if we are in a range of temperature and pressure within which carbon dioxide may be assumed to behave ideally, then $\gamma = 1$, whence

$$K_p = p_{CO_2}/p^{-0-} \tag{8.36}$$

Table 8.2 Calculation of K_p values for (8.28)

T/K	K_p(expt)	K_p(8.30)	K_p(8.33)
298.15	$(1.033\times10^5)^a$	1.759×10^5	1.033×10^5
400	1540	2926	1540
600	28.10	53.72	28.10
800	4.155	7.280	4.155
1000	1.397	2.194	1.398

Note:
[a] From ΔG^{-0-}

The equilibrium dissociation pressure of carbon dioxide in this system varies from 1 mmHg at 800 K to 3000 mmHg at 1300 K. Since ΔH for the decomposition is positive, an increase in temperature increases the percentage decomposition at equilibrium.

Another heterogeneous system is provided by the thermal decomposition of ammonium chloride:

$$NH_4Cl(s) \rightleftharpoons NH_3(g) + HCl(g) \tag{8.37}$$

The equilibrium vapour pressures at 700 K and 730 K are 6.00 atm and 10.70 atm, respectively. We will calculate K_p at each temperature and the entropy of decomposition at 700 K.

From (8.18),

$$K_p = p_{NH_3}p_{HCl}/(p^{-0-})^2 = (p_{NH_3}/p^{-0-})^2 = p^2/(p^{-0-})^2.$$

Thus, K_p is 9.00 at 700 K and 28.62 at 730 K. From (8.27), ΔH at 700 K is 163.84 kJ mol^{-1}. Using (8.11), ΔG at 700 K$=-12.79$ kJ mol^{-1}; hence, ΔS for the decomposition reaction is 252 J K^{-1} mol^{-1}. The large difference between ΔG and ΔH for this reaction arises from the very large increase in entropy when the system changes from one mole of solid to two mole of *gases*.

8.6 Solubility of ionic compounds

We consider next the first of two topics of considerable importance in the study of chemical equilibrium; equilibria between acids and bases in solution will be discussed subsequently. Acids, bases and salts whose ions dissociate in solution are termed *electrolytes.*

The solubility of ionic compounds in water can be treated quantitatively by thermodynamics, which shows at the same time the relationship between solubility and lattice energy. Ionic solids are said to be soluble in water, but covalent solids are not. The latter part of this statement is true, particularly with covalent solids defined as in this book, but the first part of the statement is only partly correct. Again, it is said that solubility decreases with increasing polarization or covalent character, but the following series of compounds shows that this statement is too general:

AgF	AgCl	AgBr	AgI
→	Decreasing solubility		
→	Increasing polarization/covalent character		
CaF_2	$CaCl_2$	$CaBr_2$	CaI_2
→	Increasing solubility		
→	Increasing polarization/covalent character		

A thermodynamic treatment of solubility requires first a definition of the standard states involved.

8.6.1 Standard states for solubility

Consider the equilibrium

$$AX(s) \rightleftharpoons A^{+}(aq)+X^{-}(aq) \tag{8.38}$$

The equilibrium constant K is given from (8.17) as

$$K=a_{A+}a_{X-}/a_{AX} \tag{8.39}$$

and we recall that it depends only on temperature. We have defined the activity of a pure, crystalline solid in the standard state as unity (Section 8.5.1); hence,

$$K=a_{A+}a_{A-} \tag{8.40}$$

or

$$K=a_{\pm}^{2}=(m/m^{-0-})^{2}\gamma_{\pm}^{2} \tag{8.41}$$

where $a_{\pm}$ is the *mean activity*, m is the *solubility* (molality at saturation) and $\gamma_{\pm}$ is the *mean activity coefficient* for the electrolyte at molality m; we shall assume a temperature of 298.15 K (25 °C) unless otherwise stated.

8.6.1.1 Mean activity and activity coefficient

The requirement of overall electrical neutrality of a solution means that there is no way to measure separately the activity or activity coefficient of a single ion, so that mean properties must be sought. Consider an electrolyte of general formula

$$A_{\nu+}X_{\nu-}(s)+aq \rightleftharpoons \nu_{+}A^{|\nu_{-}|+}(aq)+\nu_{-}X^{|\nu_{+}|-}(aq) \tag{8.42}$$

The numerical charges carried by the ionic species are governed formally by the formula-composition; the total number of ions ν is $(\nu_{+}+\nu_{-})$, and activity is defined in terms of a mean parameter $a_{\pm}$:

$$a_{\pm}^{\nu}=a_{+}^{\nu+}a_{-}^{\nu-} \tag{8.43}$$

In the example of lead phosphate

$$Pb_3(PO_4)_2(s)+aq \rightleftharpoons 3Pb^{2+}(aq)+2(PO_4)^{3-}(aq) \tag{8.44}$$

$a_{\pm}^{5}=a_{Pb2+}^{3}\times a_{PO_4^{3-}}^{2}$. We shall define individual activities and activity coefficients by the equations

$$a_{+}=\gamma_{+}m_{+} \qquad a_{-}=\gamma_{-}m_{-} \tag{8.45}$$

Mean molalities and mean activity coefficients may now be defined:

$$a_{\pm}^{\nu}=m_{+}^{\nu+}m_{-}^{\nu-}\gamma_{+}^{\nu+}\gamma_{-}^{\nu-} \tag{8.46}$$

Writing the mean activity coefficient in the form of (8.43), we obtain

$$\gamma_{\pm}^{\nu}=\gamma_{+}^{\nu+}\gamma_{-}^{\nu-} \tag{8.47}$$

For a stoichiometric molality m, $m_{+}=\nu_{+}\,m$ and $m_{-}=\nu_{-}\,m$. Hence,

$$\gamma_{\pm}=a_{\pm}/m(\nu_{+}^{\nu+}\nu_{-}^{\nu-})^{1/\nu}=a_{\pm}/m_{\pm} \tag{8.48}$$

In the case of lead phosphate, of molality m,

$$\gamma_{\pm}=a_{\pm}/(108^{1/5}m) \tag{8.49}$$

8.6.1.2 Standard state for solution

The standard state for solution is the infinitely dilute solution, in which the ratio of the activity of the solute to its molality is unity. The fact that this standard state is hypothetical does not invalidate the arguments that follow. We may regard the standard state as a solution of mean molality 1 mol kg^{-1} and unit mean activity coefficient, in which the partial molar heat content of the solute is the same as that at infinite dilution, as we now show (see also Section 7.13).

In an equilibrium between a solute and its solution, the solid and the solute in the saturated solution are at the same chemical potential. For any species i we have, following (7.54)

$$\mu_i/T-\mu_i^{-0-}/T=\mathcal{R}\ln(a_i) \tag{8.50}$$

From (4.88), but using partial molar quantities:

$$\overline{G}_i=\overline{H}_i-T\overline{S}_i=\mu_i \tag{8.51}$$

Using (4.118) with $\mathrm{d}\overline{G}=\mathrm{d}\mu_i$, $(\partial\mu_i/\partial T)_{p,n_i}=-\overline{S}_i$; rearranging and dividing by $-T^2$

$$-\mu_i/T^2+(\partial\mu_i/\partial T)_{p,n_j}/T=\left\{\frac{\partial(\mu_i/T)}{\partial T}\right\}=-\overline{H}_i/T^2 \tag{8.52}$$

If we now differentiate (8.50) with respect to T at constant p and n_j, we obtain

$$\mathcal{R}\left\{\frac{\partial\ln(a_i)}{\partial T}\right\}_{p,n_j}=\left\{\frac{\partial(\mu_i/T)}{\partial T}\right\}_{p,n_j}-\left\{\frac{\partial(\mu_i^{-0-}/T)}{\partial T}\right\}_{p,n_j} \tag{8.53}$$

whence

$$\left\{\frac{\partial\ln(a_i)}{\partial T}\right\}_{p,n_j}=\frac{H_i^{-0-}-\overline{H}_i}{\mathcal{R}T^2} \tag{8.54}$$

where $\overline{H}_i$ is the partial molar enthalpy of the ith constituent in the solution and H_i^{-0-}, equivalent here to $\overline{H}_i^{-0-}$, is the corresponding value in the pure state.

Thus, the standard state for a solution is the infinitely dilute solution, the activity a being defined as already discussed. This state then corresponds to a (hypothetical) solution of unit activity coefficient and unit activity, in which, from (8.54), the partial molar enthalpy of the solute in the standard state $\overline{H}_i^{-0-}$ has the same value as that in the infinitely dilute solution $\overline{H}_i^{-0-}$. Thus, in discussing solubility, it is convenient to refer enthalpy to the infinitely dilute solution and free energy to unit activity, knowing that they both relate to one and the same standard state.

Consider the equilibrium

$$\mathrm{NaCl(s)+aq \rightleftharpoons Na^+(aq)+Cl^-(aq).}$$

The standard state requires $m_{\pm}=1$ and $\gamma_{\pm}=1$. This choice has the required property that $a_{\pm}(\mathrm{NaCl})=1=m^2\gamma_{\pm}^2$. In NaCl, $m=m_{\pm}$. Since each ionic species also has unit molality $\gamma_{\pm}=\gamma_{+}+\gamma_{-}$, and $\mu^{-0-}(\mathrm{NaCl})=\mu^{-0-}(\mathrm{Na^+})+\mu^{-0-}(\mathrm{Cl^-})$.

Next, consider

$$\mathrm{MgCl_2(s)+aq \rightleftharpoons Mg^{2+}(aq)+2Cl^-(aq).}$$

The equation $\mu(MgCl_2) = \mu^{-0-}(MgCl_2) + \mathcal{R}T\ln[a(MgCl_2)]$ requires that, in the standard state, $a_{\pm}(MgCl_2) = 1 = 4m^3\gamma_{\pm}^3$. The standard state refers to a unit mean molality ($4m^3$) and unit mean activity coefficient. The molality of $MgCl_2$ in the standard state is $4^{-1/3}$, that of Mg^{2+} being $4^{-1/3}$ and that of Cl^- being $2(4^{-1/3})$. Since $4^{-1/3}(2\times4^{-1/3})^2 = 1$, it appears that the standard state for Cl^- in $MgCl_2$ is different from that in NaCl.

The following counter argument can be applied. Let 1 mol Mg^{2+} be concentrated from the hypothetical solution of $m = 4^{-1/3}$ to $m = 1$, while 2 mol Cl^- are diluted from the hypothetical solution of $(2\times4^{-1/3})$ to $m = 1$. Both new solutions will be deemed to obey the requirement $\gamma_{Mg^{2+}} = \gamma_{Cl^-}$. Hence, $\Delta G_{Mg^{2+}} = -\mathcal{R}T\ln(4^{-1/3}) = \frac{1}{3}\mathcal{R}T\ln(4)$. Again, $\Delta G_{Cl^-} = -2\mathcal{R}T\ln(2\times4^{-1/3}) = \frac{2}{3}\mathcal{R}T\ln(4) - 2\mathcal{R}T\ln(2)$. The total free energy change is zero, and we can write $\mu_{MgCl_2}^{-0-} = \mu_{Mg^{2+}}^{-0-} + 2\mu_{Cl^-}^{-0-}$, which compares with the equation for NaCl. Provided that we refer to the standard state of unit molality *and* unit activity coefficient, electrolyte solutions may be compared on a common thermodynamic basis.

8.6.2 Solubility relationships

For convenience, let us simplify (8.38) to

$$A(s) \rightleftharpoons B(\text{saturated solution}) \tag{8.55}$$

The solid A is in equilibrium with the saturated solution B, and

$$K = a_B/a_A \tag{8.56}$$

The chemical potentials of the components in the system (8.55) are given by

$$\mu_A = \mu_A^{-0-} + \mathcal{R}T\ln(a_A) \tag{8.57}$$

$$\mu_B = \mu_B^{-0-} + \mathcal{R}T\ln(a_B) \tag{8.58}$$

At equilibrium, the solid A and the solute in the saturated solution B are at the same chemical potential. Hence, $\mu_A = \mu_B$, and the standard free energy change of dissolution for (8.55) is given by

$$\Delta G_d^{-0-} = \mu_B^{-0-} - \mu_A^{-0-} = -\mathcal{R}T\ln(a_B/a_A) = -\mathcal{R}T\ln(K) \tag{8.59}$$

By analogy, ΔG_d^{-0-} for (8.38) is

$$\Delta G_d^{-0-} = -\mathcal{R}T\ln(K) = -\mathcal{R}T\ln(a_{\pm}^2) = -\mathcal{R}T\ln(m_{\pm}^2\gamma^2) \tag{8.60}$$

where m is the stoichiometric molality of AX in its saturated solution and $\gamma_{\pm}$ is the mean activity coefficient of the ions at saturation. This equation represents the standard free energy change of dissolution and governs solubility.

8.6.2.1 Two example calculations

We shall consider two calculations based on (8.60) to determine the quantities involved and to see what further analysis might be desirable.

Silver iodide

The solubility of silver iodide in water at 25 °C is 1.02×10^{-8} mol dm^{-3}. At this concentration $\gamma_{\pm}$ is sensibly unity, m may be replaced by c and, from (8.60), ΔG_d^{-0-} is +91.2 kJ mol^{-1}; silver iodide is highly insoluble in water.

Lithium fluoride

The solubility of lithium fluoride at 25 °C is 0.09 mol dm^{-3}. Using the same procedure as with silver iodide, $\Delta G_d^{\ominus}$ is +11.9 kJ mol^{-1}. However, we are not justified here in taking $\gamma_\pm$ as unity, because the saturated solution is not sufficiently dilute (although we may still use c in place of m). From the Debye limiting equation (Section 8.6.4.1) $\gamma_\pm$ is approximately 0.70 and $\Delta G_d^{\ominus}$ now becomes +13.7 kJ mol^{-1}. If we use the Debye limiting equation with silver iodide, $\gamma_\pm$ is unity to within 0.01%.

Do we need any further discussion on solubility? These calculations require activity coefficients for saturated solutions. When solubilities are less than approximately 0.1 mol dm^{-3}, $\gamma_\pm$ may be calculated with sufficient accuracy. In more concentrated solutions the calculation of $\gamma_\pm$ is not reliable. For example, potassium chloride is saturated at 4.8 mol dm^{-3}. At this concentration $\gamma_\pm$ would be given as 0.077 from the Debye limiting equation, whereas it is 0.574. There is, unfortunately, a paucity of data on activity coefficients for saturated solutions. Furthermore, from an analysis of $\Delta G_d^{\ominus}$ we can obtain a clearer understanding of solubility.

8.6.3 Solubility and energy

The solubility of a substance is the molality m of its saturated solution. However, it is not unusual, particularly with sparingly soluble substances, for the solubility to be quoted in molarity c, that is, mol dm^{-3}; the conversion of the two concentration units requires knowledge of the density of the saturated solution:

$$c = mD/(1 + mM) \tag{8.61}$$

where D is the density of the solution in kg dm^{-3} (g cm^{-3}) and M is the molar mass of the solute in kg mol^{-1}. It is clear from (8.61) that as $m \to 0$, $D \to 1$ and $mM \ll 1$; then, $c \simeq mD \simeq m$.

A sucrose solution of 1% by weight has a density of 1.0021 kg dm^{-3}. The molality m is 0.02951 and the molarity c, from (8.61), is 0.02928, or 0.296 if mM is ignored.

The important quantity $\Delta G_d^{\ominus}$ may be expanded in the usual way, all the terms involved being molar properties:

$$\Delta G_d^{\ominus} = \Delta H_d^{\ominus} - T\Delta S_d^{\ominus} \tag{8.62}$$

where $\Delta H_d^{\ominus}$ is the standard enthalpy change for the dissolution process[2], referred to its standard state of infinite dilution, and $\Delta S_d^{\ominus}$ is the corresponding change in entropy. We may write

$$\Delta S_d^{\ominus} = \Sigma \Delta \overline{S}_i^{\ominus} - S_C^{\ominus} \tag{8.63}$$

where $\Sigma \Delta \overline{S}_i^{\ominus}$ is the sum of the standard relative partial molar entropies of the hydrated ions in their standard state (Section 9.3.3.4), the sum extending over the pair of ions under consideration, and $S_C^{\ominus}$ is the standard molar entropy of the crystal (Section 4.6.4); these data are readily available in the chemical literature.[3]

The relationship between solubility and lattice energy is indicated by Figure 8.2, which is drawn in terms of enthalpy changes:

$$\Delta H_d^{\ominus} = \Delta H_h^{\ominus} - \Delta H_C^{\ominus} \tag{8.64}$$

[2] Sometimes called loosely the 'heat of solution'.

[3] See, for example, Appendix 18 herein; F D Rossini *et al.* Circular No 500 and later Supplements (National Bureau of Standards, 1952).

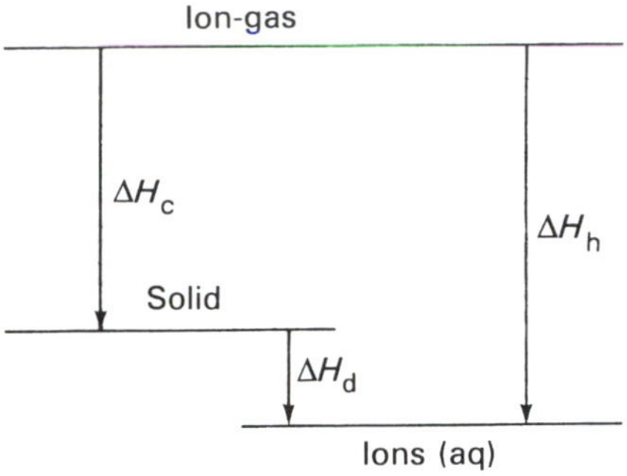

Figure 8.2 Enthalpy-level diagram relating parameters of solubility: ΔH_C lattice enthalpy, ΔH_h enthalpy of hydration, ΔH_d enthalpy of dissolution; solubility is calculated in terms of the corresponding Gibbs free energies, through $\Delta G_d^{\ominus} = \Delta G_h^{\ominus} - \Delta G_C^{\ominus}$.

where $\Delta H_h^{\ominus}$ is the standard enthalpy change for the hydration of the ion-gas to the standard state of infinite dilution and $\Delta H_C^{\ominus}$ is the lattice enthalpy (Section 6.9). This equation shows how, from lattice enthalpy and enthalpy of dissolution data, we can obtain hydration enthalpies that may not be otherwise available.

The experimental measurements of $\Delta H_d^{\ominus}$ must be extrapolated to infinite dilution, because the process of dilution itself can produce significant enthalpy changes. For example, ΔH_d for cadmium sulfate in 200 mole of water is -43.9 kJ mol^{-1}, but at infinite dilution $\Delta H_d^{\ominus}$ becomes -53.6 kJ mol^{-1}. The difference between two such values may be commensurate with, or even larger than, $\Delta H_d^{\ominus}$ itself.

Table 8.3 lists thermodynamic data related to solubility for a range of ionic compounds. The interaction between ions and water molecules on the one hand, and that between ions in the crystal on the other, both increase as the ions become smaller or more highly charged, because the dominant Coulombic energy is proportional to[4] q_+q_-/r_e. Both ΔH_d and ΔG_d depend on the difference between two quantities, generally of large magnitudes, one concerned with the solid and the other with the hydrated ions, both being referred to the ion-gas zero of lattice energy.

A decrease in the value of ΔG_h, that is to a more negative value, tends to stabilize the hydrated state with respect to the ion-gas and so promote solubility. However, a decrease in either ΔG_C or ΔS_d tends to decrease solubility, the former by stabilizing the crystal with respect to the ion-gas, and the latter by making the hydrated state relatively less probable. The term ΔS_d becomes very important with small or highly charged ions, such as in LiF or MgF_2, as Table 8.3 shows.

On transferring a gaseous ion isothermally into water, two processes occur. There is a structure-breaking reaction on the water itself because of the interaction between ions and water molecules, including a disruption of some hydrogen bonds in the water. Then there is a structure-making reaction arising from the coordination of ions by water molecules in hydration shells around the ions. The first of these reactions is important with large ions because it tends to increase the $S_i^{\ominus}$ values, as may be seen by comparing SrF_2 and BaF_2 for example. The second process is significant with small ions, since it acts to decrease $S_i^{\ominus}$, as exemplified by a comparison of CaF_2 and MgF_2. Any given case involves an interplay of

[4] In aqueous solution the sum of the effective radii of a pair of ions would be different from r_e, but the strong dependence on $1/r$ remains true.

Table 8.3. Thermodynamic data relating to the solubilities of some ionic halides at 298.15 K[a].

Halide	$\Delta H_d^{\ominus}$ /kJ mol^{-1}	$\Sigma\overline{S}_i^{\ominus}$ /J mol^{-1}K^{-1}	$S_C^{\ominus}$ /J mol^{-1}K^{-1}	$T\Delta S_d^{\ominus}$ /kJ mol^{-1}	$\Delta G_d^{\ominus}$ /kJ mol^{-1}
LiF	4.6	4.6	36.0	−9.4	14.0
LiCl	−37.2	69.5	55.2	4.3	−41.5
LiBr	−49.0	95.0	69.0	7.8	−56.8
LiI	−63.2	123.4	75.7	14.2	−77.4
NaF	0.4	50.6	58.6	−2.4	2.8
NaCl	3.8	115.5	72.4	12.9	−9.1
NaBr	−0.8	141.0	85.8	16.5	−17.3
NaI	−7.5	169.5	92.5	23.0	−30.5
KF	−17.6	92.9	66.5	7.9	−25.5
KCl	17.2	157.7	82.8	22.3	−5.1
KBr	20.1	183.3	96.7	25.8	−5.7
KI	20.5	211.7	104.2	32.1	−11.6
RbF	−26.4	114.6	72.8	12.5	−38.9
RbCl	16.7	179.5	94.6	25.3	−8.6
RbBr	21.8	205.0	108.4	28.8	−7.0
RbI	25.9	233.5	118.0	34.4	−8.5
CsF	−37.7	123.4	79.9	13.0	−50.7
CsCl	18.0	188.3	97.5	27.0	−9.0
CsBr	25.9	213.8	121.3	27.6	−1.7
CsI	33.1	242.3	129.7	33.6	−0.5
AgF	−20.5	64.4	83.7	−5.8	−14.7
AgCl	66.5	129.3	96.2	9.9	56.6
AgBr	84.1	154.8	107.1	14.2	69.9
AgI	111.7	183.3	114.2	20.6	91.1
TlF	−2.5	111.7	87.9	7.1	−9.6
TlCl	43.5	182.4	108.4	22.1	21.4
TlBr	57.3	207.9	119.7	26.3	31.0
TlI	74.1	236.4	123.0	33.8	40.3
MgF_2	−18.4	−137.2	57.3	−58.0	39.6
$MgCl_2$	−155.2	−7.5	89.5	−28.9	−126.3
$MgBr_2$	−186.2	43.5	123.0	−23.7	−162.5
MgI_2	−214.2	100.4	145.6	−13.5	−200.7
CaF_2	13.4	−74.5	69.0	−42.8	56.2
$CaCl_2$	−82.8	55.2	113.8	−17.5	−65.3
$CaBr_2$	−110.0	106.3	129.7	−7.0	−103.0
CaI_2	−120.1	163.2	142.3	6.2	−126.3
SrF_2	10.5	−58.6	89.5	−44.2	54.7
$SrCl_2$	−51.9	71.1	117.2	−13.7	−38.2
$SrBr_2$	−71.5	122.2	141.4	−5.7	−65.8
SrI_2	−90.4	179.1	164.0	4.5	−94.9
BaF_2	3.8	−6.7	96.7	−30.8	34.6
$BaCl_2$	−13.0	123.0	125.5	−0.7	−12.3
$BaBr_2$	−25.5	174.1	148.5	7.6	−33.1
BaI_2	−47.7	230.9	171.1	17.8	−65.5

[a] M F C Ladd and W H Lee *Trans. Farad. Soc.* **54**, 34 (1958).

these factors and, while solubility may be less easy to explain in molecular terms, the thermodynamic analysis is quantitative and the results precise, within the limits of experimental error.

Writing (8.62) in the form

$$\Delta G_d^{-0-}/T = \Delta H_d^{-0-}/T - \Delta S_d^{-0-} \tag{8.65}$$

we can see that, when a dissolution is exothermic, $\Delta H_d^{-0-}/T$ acts in the same sense as does ΔS_d^{-0-}, promoting solubility. Sodium nitrate dissolves endothermically, $\Delta H_d^{-0-} = 4.9$ kJ mol^{-1}, but $T\Delta S_d^{-0-}$ is 6.4 kJ mol^{-1}, so that $\Delta G_d^{-0-} = -1.5$ kJ mol^{-1}. This example illustrates the inadvisability of considering solubility solely in terms of enthalpy.

We can now explain the results with which we introduced this discussion on solubility. In the series of calcium halides, ΔG_h increases (becomes more positive) less rapidly than does ΔG_C from fluoride to iodide. As the anionic radius increases, the demands of regular packing in the solid determine the smallest r_e distance. In solution, however, the same packing demands do not exist and the total ion–solvent interaction remains strong, despite the increase in anionic radius. Hence, ΔG_d becomes more negative in this direction, despite some large, negative ΔS_d terms, and solubility increases.

In the series of silver halides, however, ΔG_C decreases from fluoride to iodide more rapidly than would be expected for halides containing an ion the size of r_{Ag^+}. A full calculation of the lattice energies of the silver halides using (6.58) gives results that agree with the thermodynamic model (6.54) for silver fluoride, but show increasing discrepancies from chloride to iodide. A partial covalent character in these compounds (see also Section 6.13.3), in addition to the polarization terms, is responsible for the greater stability of the crystal with respect to the ion-gas and, hence, for the decrease in solubility along this series.

Compounds such as MgO and CaO, for example, are highly ionic but insoluble in water. The lattice energies are numerically large because the electrostatic energy includes the product of the ionic charges and is proportional to four times the Madulung constant for NaCl. The O^{2-} ion in these structures hydrates to a pair of OH^- ions and ΔH_h (OH^-,g) is approximately -1000 kJ mol^{-1}. Nevertheless, the hydration energy does not overcome the very large, negative lattice energy ($\simeq -3300$ kJ mol^{-1}), and these oxides are not soluble in water.

In those cases for which activity coefficient data at saturation are available, the above analysis of solubility is confirmed. For example, sodium chloride has a solubility of 6.10 mol kg^{-1}. At this molality, the mean activity coefficient has been measured as 1.02. Hence, from (8.60), $\Delta G_d^{-0-} = -9.06$ kJ mol^{-1}, which is in excellent agreement with the value reported in Table 8.3.

8.6.4 Solubility product

For a sparingly soluble compound, the equilibrium constant in (8.40) is often called the solubility product and given the symbol K_S. For the general electrolyte in (8.42), let the solubility be s; then

$$K_S = a_\pm^\nu = (s/m^{-0-})^\nu \gamma_\pm^\nu \tag{8.66}$$

and if the solubility is small, $\gamma_\pm \to 1$.

The symmetrical electrolyte silver chloride has a solubility product of 1.6×10^{-10}. From (8.66),

$$s=(1.6\times10^{-10})^{1/2}m^{-0-}=1.26\times10^{-5}\text{ mol kg}^{-1}$$

Lead phosphate has a solubility product of 1.7×10^{-32}. In this case, the dissociation of the compounds gives Pb^{2+} ions of total molality $3s$ and PO_4^{3-} ions of total molality $2s$; from (8.48) $m_\pm=(2^23^2)^{1/5}$; hence,

$$s=(K_S/108)^{1/5}m^{-0-}=1.74\times10^{-7}\text{ mol kg}^{-1}$$

8.6.4.1 Common ion effect and ionic strength

The precipitation of silver chloride (8.1) may be made even more complete by the presence of excess of a common ion, Cl^- or Ag^+. Since the solution is very dilute, $\gamma_\pm\simeq1$ and $m\simeq c$. Thus, from (8.66), we may write

$$s=[Ag^+]=[Cl^-]=K_S(c^{-0-})^2/[Cl^-] \qquad (8.67)$$

where the unit molality c^{-0-}, for dilute solutions, serves a purpose similar to that of m^{-0-} in (8.66). Hence, if the concentration of chloride ion is increased, say to $[Cl^-]=0.001\text{ mol dm}^{-3}$, then $[Ag^+]$ and, hence, the solubility of silver chloride in solution, decreases to approximately $K_S(c^{-0-})^2/(0.001\text{ mol dm}^{-3})$, or $1.6\times10^{-7}\text{ mol dm}^{-3}$.

The complete argument is, however, a little more complex. Addition of excess chloride ion must be accompanied by an equivalent amount of positive ion and the total ion concentration has a bearing upon the activity of the solution. The dependence is expressed by the *ionic strength I* of the solution. The ionic strength derives from the Debye–Hückel theory of electrolyte solutions,[5] and is defined by

$$I=\tfrac{1}{2}\sum_i q_i^2 m_i \qquad (8.68)$$

where q_i is the numerical charge on the ith ion of molality m_i and the sum is taken over all ionic species present in the solution. In dilute solution, from (8.61) we may write

$$I\simeq\tfrac{1}{2}\sum_i q_i^2 c_i \qquad (8.69)$$

The low-concentration approximation in the Debye–Hückel theory leads to the *limiting equation* for activity coefficients:

$$\ln(\gamma_\pm)=-[1.330\times10^5\,D^{1/2}/(\epsilon_r T)^{3/2}]|q_+q_-|\sqrt{I} \qquad (8.70)$$

For water at 298.15 K, $D=997.07\text{ kg m}^{-3}$ and $\epsilon_r=78.54$, whence

$$\ln(\gamma_\pm)=-(1.172\text{ mol}^{-1/2}\text{ kg}^{1/2})|q_+q_-|\sqrt{I} \qquad (8.71)$$

This equation is satisfactory for ionic strengths up to approximately 0.01 molal; the calculation may be extended to 0.1 molal with Davies' empirical equation:

$$\ln(\gamma_\pm)=-(1.172\text{ mol}^{-1/2}\text{ dm}^{3/2})|q_+q_-|\left(\frac{\sqrt{I}}{1+\sqrt{I}}-0.30I\right) \qquad (8.72)$$

Returning to our problem with silver chloride, the saturated solution has an activity coefficient of 0.996, which has negligible effect on (8.67). In the presence of chloride ion of

[5] See, for example, R A Robinson and R H Stokes, *Electrolyte Solutions*, 2nd edition (Academic Press, 1959); H S Harned and B B Owen, *The Physical Chemistry of Electrolytic Solutions* (Academic Press, 1967).

Table 8.4 Activity coefficients for silver bromate in aqueous potassium nitrate

Limiting equation				Davies's equation			
$[KNO_3]$	$\gamma_\pm$	s_{AgBrO_3}/mol dm^{-3}	I/mol$^{1/2}$dm$^{-3/2}$	$[KNO_3]$	$\gamma_\pm$	s_{AgBrO_3}/mol dm^{-3}	I/mol$^{1/2}$dm$^{-3/2}$
0.000	0.898	0.0085	0.0085	0.000	0.909	0.0084	0.0084
0.001	0.892	0.0085	0.0095	0.001	0.904	0.0084	0.0094
0.010	0.851	0.0090	0.0190	0.010	0.874	0.0087	0.0187
0.050	0.750	0.0102	0.0602	0.050	0.812	0.0094	0.0594
0.100	0.676	0.0113	0.1113	0.100	0.776	0.0098	0.1098

concentration 0.001 mol dm^{-3} (and another ion, say Na^+, of the same concentration), the total ionic strength is 0.001 mol dm^{-3}, and $\gamma_\pm$ is now 0.964. Hence, $s_{AgCl}=1.66\times10^{-7}$ mol dm^{-3}. If, say, sodium nitrate had been added at the same concentration, in place of sodium chloride, then $\gamma_\pm$ would still be 0.964 but s_{AgCl} would become $K_S^{1/2}(c^{-0-})^2/0.964=1.66\times10^{-5}$ mol dm^{-3}. Thus, the addition of an electrolyte without an ion in common increases the solubility of the solute, sometimes called a *salting-in* effect; the decrease in solubility from the common ion effect may be called *salting-out*. In these calculations, we have, justifiably, ignored the very small contribution to the ionic strength by the silver chloride in solution.

Finally in this section, we study silver bromate ($K_S=5.8\times10^{-5}$) in aqueous potassium nitrate of varying concentrations. Table 8.4 lists activity coefficients up to 0.1 M, calculated by the limiting equation (8.71) and Davies' equation (8.72):

With these 1 : 1 electrolytes agreement between the results from the two equations is satisfactory up to approximately 0.01 M, after which Davies' equation is to be preferred. For concentrations of 1 molal or more no satisfactory theory has yet been developed, probably because the species in electrolyte solutions have been treated as uniform distributions at all concentrations (see Section 5.6.2ff).

8.7 Acid–base equilibria

We consider now the equilibria between acids and bases, the second of the two important topics to which we alluded earlier.

Water is ionized to a small but finite extent:

$$H_2O+H_2O \rightleftharpoons H_3O^++OH^- \tag{8.73}$$

The hydrogen ion, a bare proton, is always hydrated in the presence of water, so that H_3O^+, the *hydroxonium ion*,[6] is a more precise designation than H^+. The Brønsted–Lowry definition of an acid is a *proton donor*, whereas a base is a *proton acceptor.* Thus, in (8.73) the water molecule acts both as an acid and as a base, which is an example of *autoprotolysis*. Substances that can act as both an acid and a base are termed *amphiprotic*.

If the first water molecule in (8.73) were thought of as an acid, then the second water molecule would be a base. The H_3O^+ ion is the *conjugate acid* of the 'base' H_2O, and the OH^- ion is the *conjugate base* of the 'acid' H_2O. The equilibrium could be viewed in the opposite sense, whereupon the definitions are interchanged. Thus, the species in an acid–base equilibrium may be said to exist in acid and base conjugate pairs.

[6] The hydroxonium ion will normally be further hydrated, $H_3O^+(aq)$.

The equilibrium constant for (8.73) may be written following (8.17). However, since the extent of dissociation is very small, the activity of water may be taken as unity. The equilibrium constant for water is given the special symbol K_w:

$$K_w = a_{H_3O^+} a_{OH^-} \tag{8.74}$$

The value of K_w at 25 °C is 1.008×10^{-14}, or $pK_w = -\log_{10}(K_w) = 14.00$.

As in earlier sections, we may replace activities by concentrations when the ion concentrations are small. For (8.73), the mean activity coefficient is, from the Debye–Hückel limiting equation, 0.9996, so that we may write $a_{H_3O^+} = m_{H_3O^+}/m^{\ominus} \simeq [H_3O^+]/c^{\ominus}$, whence (8.74) becomes

$$K_w = [H_3O^+][OH^-]/(c^{\ominus})^2 \tag{8.75}$$

and is known as the *ionic product for water*; it is a function of temperature, in accordance with (8.26).

Another example of autoprotolysis is shown by pure sulfuric acid:

$$2H_2SO_4 \rightleftharpoons H_3SO_4^+ + HSO_4^- \tag{8.76}$$

The ionic product $K_{H_2SO_4}$ is 2.4×10^{-4} at 25 °C and $pK_{H_2SO_4} = 3.62$.

Pure nitric acid does not set up this type of equilibrium. The solvated proton $H_2NO_3^+$ is unstable and decomposes:[7]

$$2HNO_3 \rightleftharpoons H_2NO_3^+ + NO_3^- \rightarrow NO_2^+ + NO_3^- + H_2O \tag{8.77}$$

8.7.1 Strong acid–strong base equilibria

Acidity is often expressed on the pH scale:

$$pH = -\log_{10}(a_{H^+}) \simeq \log_{10}([H_3O^+]/c^{\ominus}) \tag{8.78}$$

Insofar as it is not possible to measure the single-ion entity a_{H^+}, (8.78) is a formal, but useful, definition of pH; we will consider an operational definition in the next chapter. When the ion concentration is sufficiently small, we may write $pH = -\log_{10}([H_3O^+]/c^{\ominus})$. A pH of 7 represents *neutrality* which, as we shall see, may be different from a stoichiometric *equivalence* point. Pure water has a pH of 7.00 at 25 °C, but in equilibrium with the atmosphere it absorbs carbon dioxide and an equilibrium is set up, leading to a solution of pH$\simeq$5:

$$2H_2O(l) + CO_2(g) \rightleftharpoons H_3O^+(aq) + HCO_3^-(aq) \tag{8.79}$$

An electrolyte, acid, base or salt, is considered to be *strong* if it is fully dissociated into its ions in solution; otherwise it is said to be *weak*. The role of the solvent is important: thus, hydrogen chloride is a strong acid in water, but a weak acid in pure ethanoic acid. The difference arises from the differing abilities of the two solvents to accept a proton. The terms 'strong' and 'weak' should not be confused with *concentrated* and *dilute*, which refer to the stoichiometric amount of electrolyte present.

We consider next the relationship between the pH of a strong acid or strong base and its stoichiometric concentration. Let a strong acid be represented by HX; then,

$$HX(aq) + H_2O(l) \rightleftharpoons H_3O^+(aq) + X^-(aq) \tag{8.80}$$

[7] See, for example, W H Lee in *The Chemistry of Nonaqueous Solvents*, edited J J Lagowski, Volume 4 (Academic Press, 1967).

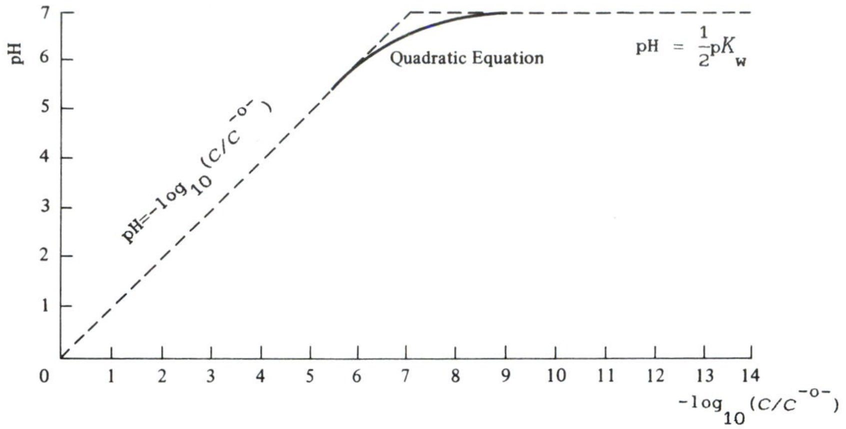

Figure 8.3 Variation of pH with concentration c for the strong acid–strong base combination: for concentrations down to approximately 10^{-6} M, both the quadratic (8.83) and the linear (8.84) equations are satisfactory; between 10^{-6} M and 10^{-9} M the quadratic equation is essential; below $c=10^{-9}$ M, the pH is effectively pK_w.

Hereinafter, we shall presume the hydrated state for ionic species, unless stated otherwise, *and for convenience in this chapter, we shall write a concentration term in molarity as* $[X]$, *with the understanding that it implies* $[X]/c^{-0-}$, *unless otherwise specified.*

If the stoichiometric concentration of the acid is c mol dm^{-3}, the following equations can be set up:

Ionic product $$K_w=[H_3O^+][OH^-] \tag{8.81}$$

Electroneutrality $$[H_3O^+]_{total}=[X^-]+[OH^-]=c+[OH^-] \tag{8.82}$$

Simple manipulation leads to

$$[H_3O^+]=c/2\pm\{(c/2)^2+K_w\}^{1/2} \tag{8.83}$$

If K_w is negligible with respect to $([c]/2)^2$ and with the positive root,

$$[H_3O^+]=c \tag{8.84}$$

$$pH=-\log_{10}(c) \tag{8.85}$$

If $c \ll K_w$, then

$$[H_3O^+]=K_w^{1/2} \tag{8.86}$$

$$pH=\tfrac{1}{2}\,pK_w \tag{8.87}$$

Figure 8.3 shows the variation of pH with concentration for a strong acid. It is clear that the contribution to $[H_3O^+]_{total}$ from the dissociation of water is significant only for $c<5\times10^{-6}$ mol dm^{-3}. A strong base can be treated in a similar manner, and we note that at 25 °C

$$pH+pOH=14.00 \tag{8.88}$$

As examples, we calculate here the pH for (a) 0.01 M sodium hydroxide and (b) 10^{-7} M nitric acid.

(a) $[OH^-]=0.01$ mol dm^{-3}; $pH=14.00-2.00=12.00$.

(b) $[H_3O^+]=(0.5\times10^{-7}\ \text{mol dm}^{-3})\pm[(0.25\times10^{-14}\ \text{mol}^2\ \text{dm}^{-6}+1.00\times10^{-14})]^{1/2}=1.62\times10^{-7}$ mol dm^{-3}; pH=6.79.

8.7.2 Weak/strong acid–strong/weak base equilibria

We consider the example of a weak acid HA and a strong base, such as sodium hydroxide. Because the acid is weak, an equilibrium is set up:

$$\text{H}A+\text{H}_2\text{O} \rightleftharpoons \text{H}_3\text{O}^+ + A^- \tag{8.89}$$

and we write the equilibrium constant K_a (the dissociation constant for the acid) as

$$K_a=\frac{[\text{H}_3\text{O}^+][A^-]}{[\text{H}A]} \tag{8.90}$$

We have also the following equations in this equilibrium:

Ionic product $$K_w=[\text{H}_3\text{O}^+][\text{OH}^-] \tag{8.91}$$

Electroneutrality $$[\text{H}_3\text{O}^+]=[A^-]+[\text{OH}^-] \tag{8.92}$$

Mass balance on A $$c=[\text{H}A]+[A^-] \tag{8.93}$$

The general solution of (8.90) to (8.93) is readily found to be

$$[\text{H}_3\text{O}^+]^3+K_a[\text{H}_3\text{O}^+]^2-(cK_a+K_w)[\text{H}_3\text{O}^+]-K_aK_w=0 \tag{8.94}$$

We may make certain simplifying assumptions: since the product K_aK_w is very small and $K_w \ll cK_a$, we may write

$$[\text{H}_3\text{O}^+]^2+K_a[\text{H}_3\text{O}^+]-cK_a \tag{8.95}$$

If, furthermore, we let $[\text{H}_3\text{O}^+]$ be negligible with respect to c, then

$$[\text{H}_3\text{O}^+]=(K_ac)^{1/2} \tag{8.96}$$

$$\text{pH}=\tfrac{1}{2}\,\text{p}K_a-\tfrac{1}{2}\log_{10}(c) \tag{8.97}$$

Figure 8.4 shows the variation of pH with c, calculated from (8.94), (8.95) and (8.96/8.97). For concentrations down to approximately 0.01 M (8.96) is satisfactory; to approximately 10^{-6} M (8.95) may be used; for lower concentration it is essential to use the cubic equation (see Appendix A1.4).

The salts of weak acids or bases are hydrolysed in solution; we consider sodium methanoate, the sodium salt of methanoic acid ($K_a=1.77\times10^{-4}$), and determine an equation for $[\text{H}_3\text{O}^+]$ at a concentration c. The methanoate ion HCO_2^-, which we shall symbolize as Meth$^-$, is hydrolysed:

$$\text{Meth}^-+\text{H}_2\text{O} \rightleftharpoons \text{HMeth}+{}^-\text{OH} \tag{8.98}$$

and the hydrolysis constant K_h for this equilibrium is given by

$$K_h=\frac{[\text{HMeth}][\text{OH}^-]}{[\text{Meth}^-]} \tag{8.99}$$

This equation and that for the dissociation of the acid in water are not independent of the ionic product for water, as may be seen by multiplying the numerator and denominator of (8.99) by $[\text{H}_3\text{O}^+]$. Conveniently, we may use the following five equations for the methanoate equilibrium:

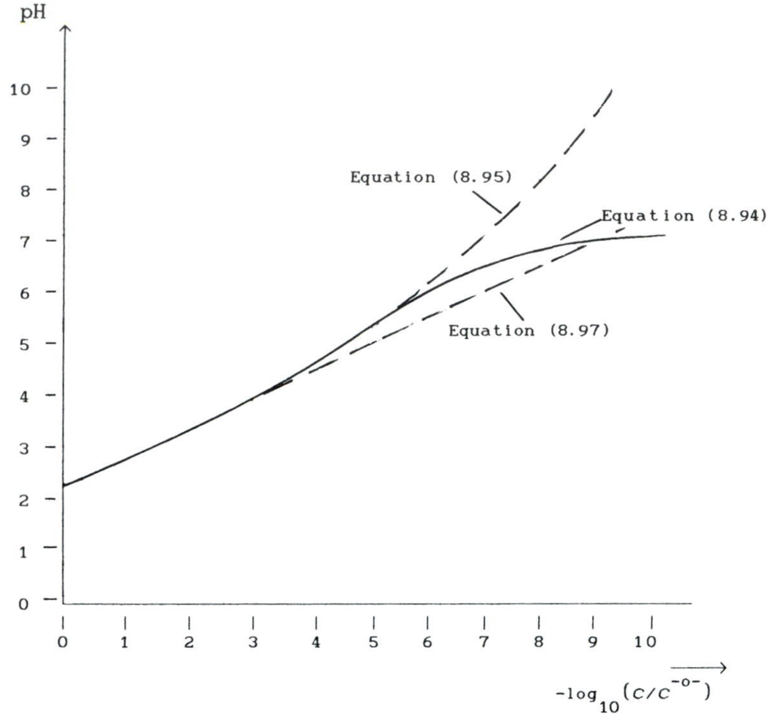

Figure 8.4 Variation of pH with concentration c for the weak acid–strong base combination ($K_a = 1.76 \times 10^{-5}$); the ranges of applicability of the three equations developed are as follow:

c/mol dm^{-3}	1 to 10^{-2}	10^{-2} to 10^{-6}	$<10^{-6}$
Equations	(8.94)–(8.96)	(8.94)–(8.95)	(8.94)

If K_a is reduced to 1.76×10^{-8} these ranges change as follow, because K_w is of more significance in relation to K_a:

c/mol dm^{-3}	1 to 10^{-4}	$<10^{-4}$
Equations	(8.94)–(8.96)	(8.94)

Dissociation $$K_a = \frac{[H_3O^+][Meth^-]}{[HMeth]} \tag{8.100}$$

Ionic product $$K_w = [H_3O^+][OH^-] \tag{8.101}$$

Salt concentration $$[Na^+] = c \tag{8.102}$$

Electroneutrality $$[H_3O^+] + [Na^+] = [Meth^-] + [OH^-] \tag{8.103}$$

Mass balance on Meth $$c = [Meth^-] + [HMeth] \tag{8.104}$$

The rearrangement of these equations leads to

$$[H_3O]^3 + (c + K_a)[H_3^+O]^2 - K_w[H_3O^+] - K_aK_w = 0 \tag{8.105}$$

A solution of sodium methanoate is basic: if we neglect the small term $[H_3O^+]^3$, the equation reduces to the quadratic

$$(c+K_a)[H_3O^+]^2-K_w[H_3O^+]-K_aK_w=0 \quad (8.106)$$

which is easily solved for $[H_3O^+]$. Thus, if $c=0.001$ mol dm^{-3}, pH=7.41, which is basic.

A simpler approach begins with (8.98), from which $[\text{HMeth}]=[OH^-]$. If we assume that the amount of hydrolysis is so small that $[\text{Meth}^-]\simeq c$, then, from (8.99), $[OH^-]=(K_hc)^{1/2}=(K_wc/K_a)^{1/2}$, so that pOH=6.62 and pH=7.38 (ΔpH against (8.106)=0.03, or 0.4%). The higher the concentration c the smaller is the error in this approximation.

Similar equations can be constructed for the strong acid–weak base combination. The validity of the approximations discussed depends upon both the concentration and the value of the appropriate equilibrium constant.

8.7.3 Weak acid–weak base equilibria

As an example we may consider ammonium ethanoate. Both the acid NH_4^+ and the conjugate base $CH_3CO_2^-$, which we shall symbolize as Eth^-, are hydrolysed by water:

$$NH_4^+ + H_2O \rightleftharpoons H_3O^+ + NH_3 \quad (8.107)$$

$$\text{Eth}^- + H_2O \rightleftharpoons OH^- + \text{HEth} \quad (8.108)$$

If the stoichiometric concentration of ammonium ethanoate is c, then the initial concentrations of NH_4^+ and Eth^- are both c. Thus, the mass balance may be given as $c=[NH_4^+]+[NH_3]=[\text{Eth}^-]+[\text{HEth}]$. In all, six equations are available for this equilibrium, which may be solved to give

$$[H_3O^+]^4+(c+K_a+K_w/K_b)[H_3O^+]^3+(K_aK_w/K_b-K_w)[H_3O^+]^2$$
$$-(K_aK_wc/K_b+K_aK_w+K_w^2/K_b)[H_3O^+]-K_aK_w^2/K_b=0 \quad (8.109)$$

The final term is very small and the equation may be treated as cubic. In the case of ammonium ethanoate $K_a=1.76\times10^{-5}$ and $K_b=1.79\times10^{-5}$, and the similarity of these values results in the pH being very close to 7.00 and almost independent of c. In other cases (8.109) shows that pH→7 as $c\rightarrow0$.

If we assume that the concentrations of both $[H_3O^+]$ and $[OH^-]$ are small, then we may use the following equilibrium in an approximate solution for pH:

$$K_a[\text{HEth}]=[H_3O^+][\text{Eth}^-] \quad (8.110)$$

$$K_b[NH_3]=[NH_4^+][OH^-] \quad (8.111)$$

$$K_w=[H_3O^+][OH^-] \quad (8.112)$$

$$c=[\text{HEth}][\text{Eth}^-]=[NH_3]+[NH_4^+] \quad (8.113)$$

$$[H_3O^+]+[NH_4^+]=[OH]+[\text{Eth}^-] \quad (8.114)$$

Combining (8.113) and (8.114) gives

$$[\text{HEth}]=[OH^-]+[NH_3]-[H_3O^+] \quad (8.115)$$

and with the approximation in (8.114) and (8.115)

$$[NH_4^+]=[Eth^-] \tag{8.116}$$

$$[NH_3]=[HEth] \tag{8.117}$$

From (8.110) to (8.112)

$$K_wK_a/K_b=[H_3O^+]^2[Eth^-][NH_3]/([HEth][NH_4^+]) \tag{8.118}$$

which, with (8.116) and (8.117) gives

$$pH=\tfrac{1}{2}pK_w+\tfrac{1}{2}pK_a-\tfrac{1}{2}pK_b \tag{8.119}$$

Under this approximation pH is independent of c, and for $K_a \simeq K_b$ pH$\simeq$7 at 25 °C.

If K_a and K_b are large with respect to $\sqrt{K_w}$ and commensurate with c, $[H_3O^+]$ and $[OH^-]$ are negligible in (8.114) but not in (8.115). The solution of the set of equations then gives the better approximation

$$[H_3O^+]=\frac{K_wK_a(c+K_b)}{K_b(c+K_a)} \tag{8.120}$$

The dissociation constants for HF and NH_3 are 6.75×10^{-4} and 1.79×10^{-4}, respectively. A solution of NH_4F of concentration 0.1 M has a pH of 6.21, from (8.120). Under these conditions, (8.119) gives pH=3.04, a value seriously in error and demonstrating the limitations of the gross approximation.

8.7.4 Buffer solutions

In many chemical and biological investigations, it may be necessary to maintain a medium of nearly constant pH during a reaction which produces or utilizes hydrogen ions. The reaction is carried out in a *buffer solution*, which is a mixture of a weak acid and its salt (conjugate base) or a weak base and its salt (conjugate acid). An example of a buffer solution is ethanoic acid HEth ($K_a=1.76\times10^5$) and sodium ethanoate NaEth.

As an example, let the buffer solution have the concentration 0.05 M in both HEth and NaEth. The complete equation for this system can be set up in the manner discussed above. However, we may make the approximation that [HEth] is also 0.05 M, since its small dissociation would be repressed even further by the concentration of Eth^-. Hence, from (8.100)

$$[H_3O^+]=K_a=1.76\times10^{-5}\ \text{mol dm}^{-3} \tag{8.121}$$

$$pK_a=4.754 \tag{8.122}$$

Let 1 cm^3 of 1 M hydrochloric acid be added to 100 cm^3 of the buffer solution. The added hydrogen ions react with ethanoate to give ethanoic acid. Since 0.001 mol of $[H_3O^+]$ has been added, $[Eth^-]$ and [HEth] are decreased and increased, respectively, by 0.001 M, and the volume of solution becomes 101 cm^3. Thus, we have

$$[Eth^-]=(0.05-0.001)100/101=0.0485\ \text{M}$$

$$[HEth]=(0.05+0.001)100/101=0.0505\ \text{M}$$

Using (8.100) again, pH$=pK_a-\log_{10}(0.0505/0.0485)=4.736$ and ΔpH$=-0.018$. If there had been no buffer action, then $[H_3O^+]$ would have been 0.0099 M, with pH=2.00.

To generalize this procedure a little further, we write the equation for the equilibria in this buffer system:

Acid dissociation	$K_a[\mathrm{HEth}]=[\mathrm{H_3O^+}][\mathrm{Eth^-}]$	(8.123)
Ionic product	$K_w=[\mathrm{H_3O^+}][\mathrm{OH^-}]$	(8.124)
Electroneutrality	$[\mathrm{H_3O^+}]+[\mathrm{Na^+}]=[\mathrm{OH^-}]+[\mathrm{Eth^-}]$	(8.125)
Mass balance, Eth	$[\mathrm{HEth}]+[\mathrm{Eth^-}]=c_{\mathrm{HEth}}+c_{\mathrm{Eth^-}}$	(8.126)
Mass balance, Na	$[\mathrm{Na^+}]=c_{\mathrm{Eth^-}}$	(8.127)

Thus, it follows that

$$[\mathrm{H_3O^+}]=c_{\mathrm{HEth}}+[\mathrm{Eth^-}]+[\mathrm{OH^-}]-[\mathrm{HEth}] \tag{8.128}$$

If we assume that $[\mathrm{H_3O^+}]$ and $[\mathrm{OH^-}]$ are both less than c, which is frequently the case with buffer solutions, then

$$[\mathrm{HEth}]=c_{\mathrm{HEth}} \tag{8.129}$$

$$[\mathrm{Eth^-}]=c_{\mathrm{Eth^-}} \tag{8.130}$$

Hence, from (8.123), we obtain

$$\mathrm{pH}=\mathrm{p}K_a+\log_{10}([\mathrm{salt}]/[\mathrm{acid}]) \tag{8.131}$$

which is often called Henderson's equation. For the comparable system of a weak base and its salt, such as ammonia and ammonium chloride, we have

$$\mathrm{pOH}=\mathrm{p}K_b+\log_{10}([\mathrm{salt}]/[\mathrm{base}]) \tag{8.132}$$

or

$$\mathrm{pH}=\mathrm{p}K_w-\mathrm{p}K_b-\log_{10}([\mathrm{salt}]/[\mathrm{base}]) \tag{8.133}$$

Equations (8.131) and (8.133) are satisfactory for the concentration ranges 10^{-3} M to 10^{-6} M and 10^{-8} M to 10^{-11} M, respectively. The limitations are clear: for example, (8.131) shows that for a constant ratio of [salt]/[acid] the pH is independent of the actual concentrations, but pH→7 as $c\to 0$, as the exact solution shows.

If equations (8.123) to (8.127) are solved for c_A at a given ratio of c_{HA}/c_A a linear equation in c_A is obtained:

$$c_A=\frac{(K_a/[\mathrm{H_3O^+}]+1)([\mathrm{H_3O^+}]-K_w/[\mathrm{H_3O^+}])}{(c_{HA}K_a/c_A[\mathrm{H_3O^+}]-1)} \tag{8.134}$$

8.7.4.1 Van Slyke buffer index

The capacity of a buffer to resist the addition of strong acid or strong base may be measured by a useful practical parameter, the van Slyke buffer index β:

$$\beta=-\mathrm{d}(c_a/c^{-0-})/\mathrm{d}(\mathrm{pH})=\mathrm{d}(c_b/c^{-0-})/\mathrm{d}(\mathrm{pH}) \tag{8.135}$$

The addition of $\mathrm{d}c_b$ mol of strong base to 1 dm^3 of buffer solution increases the pH and increases the stoichiometric concentration of the basic component by that amount at the expense of the acid component. The equation for a weak acid and its salt may be derived as follows.

Let a solution contain c_a mol of hydrochloric acid, c_b mol of sodium hydroxide, and a weak acid HA such that $[\mathrm{HA}]+[\mathrm{A^-}]=c$. Then, we have the equilibria:

Acid dissociation $$K_a[HA]=[H_3O^+][A^-] \tag{8.136}$$

Ionic product $$K_w=[H_3O^+][OH^-] \tag{8.137}$$

Mass balance, Na^+ $$c_b=[Na^+] \tag{8.138}$$

Mass balance, Cl^- $$c_a=[Cl^-] \tag{8.139}$$

Mass balance, A $$c=[HA]+[A^-] \tag{8.140}$$

Electroneutrality $$[H_3O^+]+[Na^+]=[OH^-]+[Cl^-]+[A^-] \tag{8.141}$$

From (8.136) and (8.140)

$$[A^-]=cK_a/(K_a+[H_3O^+]) \tag{8.142}$$

Substituting (8.137), (8.138) and (8.139) in (8.142)

$$c_b=K_w/[H_3O^+]+c_a-[H_3O^+]+cK_a/(K_a+[H_3O^+]) \tag{8.143}$$

Since $dc_b/d(pH)$ is $-\ln(10)\,[H_3O^+]\,dc_b/d([H_3O^+])$, β is obtained by differentiating (8.143) with respect to $[H_3O^+]$ at a constant c_a:

$$\beta=\ln(10)\left(\frac{K_w}{[H_3O^+]}+[H_3O^+]+\frac{cK_a[H_3O^+]}{(K_a+[H_3O^+])^2}\right) \tag{8.144}$$

For the above example with ethanoic acid and sodium ethanoate, the approximate pH was 4.736, so that $[H_3O^+]\simeq1.84\times10^{-5}$ M. From (8.144), $\beta=0.0576$ and, since $\Delta c_a=0.001$ mol, $\Delta(pH)=-0.017$, which is very close to the approximate value (-0.018) deduced from Henderson's equation.

The variation of β with pH is illustrated by Figures 8.5 to 8.7, in which ammonia, ethanoic acid and phenol, with their respective salts, are represented at different values of the concentration c. The maximum useful buffer capacity occurs at pK_a, or pK_b, for the system under consideration. At these values of pH the ratios of [salt]/[acid] or [salt]/[base] are unity. A tenfold decrease in the salt concentration produces an approximately tenfold decrease in β. Minimum buffer action occurs in the pH regions corresponding to the acid or the conjugate base. For example, with 0.05 M ethanoic acid and 0.05 M sodium ethanoate, the minima in β occur at pH approximately 3 for the acid and at pH approximately 8 for the conjugate base.

The buffer index of 0.05 M sodium ethanoate is only 2×10^{-5}; thus, this solution is sensitive to the presence of atmospheric carbon dioxide, so that accurate measurement of its pH requires considerable care. The useful range of buffer action for the ethanoic acid and sodium ethanoate system is seen from Figure 8.5 to be pH 3 to 6.5. Comparable deductions may be made about the other systems discussed, and an equation for β similar to (8.144) can be drawn up for the addition of strong base to a weak base/salt buffer, such as ammonia/ammonium chloride.

8.7.5 Polyprotic acids

We will consider a diprotic acid H_2A, such as carbonic acid or oxalic acid. There are two stages of dissociation:

$$H_2A+H_2O \rightleftharpoons H_3O^++HA^- \tag{8.145}$$

$$HA^-+H_2O \rightleftharpoons H_3O^++A^{2-} \tag{8.146}$$

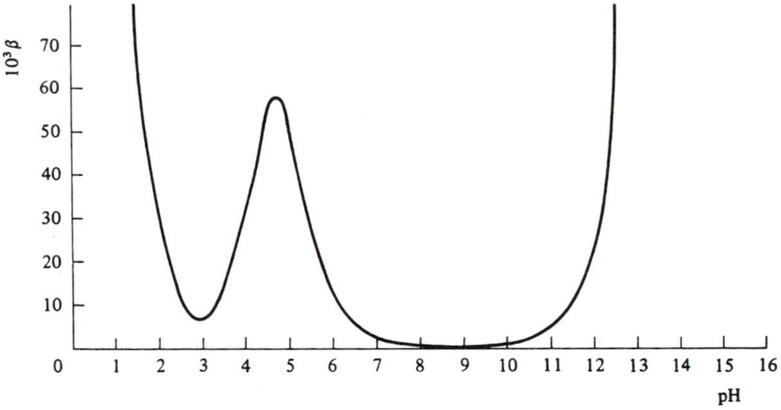

Figure 8.5 Buffer index β as a function of pH for the 0.05 M ethanoic acid–0.05 M sodium hydroxide system, total concentration c=0.1 M; the useful buffer range is pH 3 to 6.5.

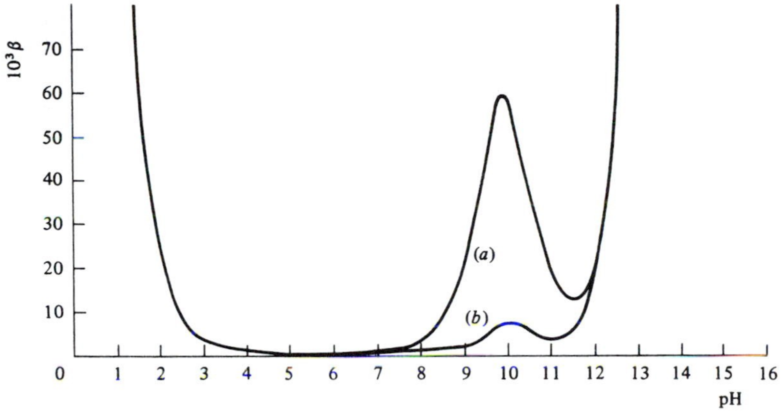

Figure 8.6 Buffer index β as a function of pH for phenol–sodium hydroxide systems: (*a*) c=0.05 M; (*b*) c=0.005 M. The useful buffer range is pH 8.0 to 11.0; the buffer capacity decreases tenfold for the dilution from (*a*) to (*b*).

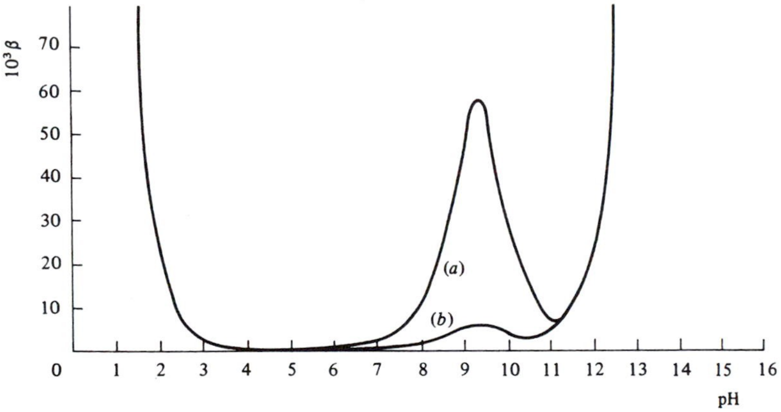

Figure 8.7 Buffer index β as a function of pH for ammonia–hydrochloric acid systems: (*a*) c=0.05 M; (*b*) c=0.005 M. The useful buffer range is pH 7.5 to 10.5; the tenfold decrease in buffer capacity on dilution is again evident.

Table 8.5. Evaluation of pH as a function of concentration c for carbonic acid ($K_1=4.3\times10^{-7}$, $K_2=5.6\times10^{-11}$) and oxalic acid ($K_1=5.9\times10^{-2}$, $K_2=6.4\times10^{-5}$)

c/mol dm^{-3}	Carbonic acid		Oxalic acid	
	(8.152)	(8.153)	(8.152)	(8.153)
10^0	3.18	3.18	0.67	0.67
10^{-1}	3.68	3.68	1.28	1.28
10^{-2}	4.19	4.18	2.06	2.05
10^{-3}	4.69	4.68	2.98	3.01
10^{-4}	5.20	5.18	3.88	(4.00)
10^{-5}	6.31	(5.73)	4.75	
10^{-6}	6.83		5.70	
0	7.00		0	7.00

Five equations can be set up to represent these equilibria:

First dissociation $$K_1[H_2A]=[H_3O^+][HA^-] \tag{8.147}$$

Second dissociation $$K_2[HA^-]=[H_3O^+][A^{2-}] \tag{8.148}$$

Ionic product $$K_w=[H_3O^+][OH^-] \tag{8.149}$$

Mass balance on A $$c=[H_2A]+[HA^-]+[A^{2-}] \tag{8.150}$$

Electroneutrality $$[H_3O^+]=[HA^-]+2[A^{2-}]+[OH^-] \tag{8.151}$$

Solution of this set of equations by the usual procedure leads to:

$$[H_3O^+]=\frac{c-[H_3O^+]+\dfrac{K_w}{[H_3O^+]}}{\dfrac{[H_3O^+]}{K_1}-\dfrac{K_2}{[H_3O^+]}}+\frac{2K_2\left(c-[H_3O^+]+\dfrac{K_w}{[H_3O^+]}\right)}{[H_3O^+]\left(\dfrac{[H_3O^+]}{K_1}-\dfrac{K_2}{[H_3O^+]}\right)}+\frac{K_w}{[H_3O^+]} \tag{8.152}$$

This equation may be solved for $[H_3O^+]$ by iteration, or rearranged into a quartic and solved systematically, as with equations discussed earlier. If K_2 is much smaller than K_1 and the pH is less than approximately 6, then by setting $K_2=0$ in (8.152) and neglecting terms in K_w, we obtain

$$[H_3O^+]^2+K_1[H_3O^+]-cK_1=0 \tag{8.153}$$

which may be compared to (8.95) with $K_a=K_1$. That this simplification may be applied satisfactorily to carbonic acid but less so to oxalic acid is shown by the data listed in Table 8.5.

8.7.6 Acid–base indicators

An acid–base indicator is a weak acid or weak base and, therefore, only slightly dissociated in solution. The acid and its conjugate base, or the base and its conjugate acid, are generally differently coloured. The two dissociations may be represented by

$$HIn \rightleftharpoons [H_3O^+]+In^- \tag{8.154}$$

$$InOH \rightleftharpoons OH^-+In^+ \tag{8.155}$$

Table 8.6. Data for acid–base indicators

	pH range	$pK_{In} \pm 1$	Colour change acidic→alkaline
Methyl orange	2.9–4.0	2.7–4.7	Red→orange
Bromophenol blue	3.0–3.6	3.0–5.0	Yellow→blue
Bromocresol green	3.8–5.4	3.7–5.7	Yellow→blue
Methyl red	4.4–6.0	4.1–6.1	Red→yellow
Bromothymol blue	6.0–7.6	6.0–8.0	Yellow→blue
Phenol red	6.8–8.4	6.6–8.6	Yellow→red
Phenolphthalein	8.3–10.0	8.4–10.4	Colourless→red
Thymolphthalein	9.3–10.5	8.9–10.9	Colourless→blue

where HIn represents a weak-acid indicator, such as methyl red, and InOH represents a weak-base indicator, such as phenolphthalein.

We shall consider methyl red as a typical example of a weak-acid indicator. The equilibrium constant for (8.154) is

$$K_{In} = [H_3O^+][In^-]/[HIn] \qquad (8.156)$$

and the ratio $[In^-]/[HIn]$ is governed by the pH of the system. The red form of methyl red is HIn (acidic) and the yellow form is In^- (basic). If a fraction α of the total indicator present is dissociated, then from (8.156)

$$[H_3O^+] = K_{In}(1-\alpha)/\alpha \qquad (8.157)$$

We assume that when α is less than 0.09 the yellow In^- colour is just not visible to the eye, so that the highest hydrogen-ion concentration at which In^- can be detected is

$$[H_3O^+] = K_{In}(0.91/0.09) \simeq 10K_{In} \qquad (8.158)$$

and $pH = pK_{In} - 1$. Furthermore, when 91% of the indicator is dissociated, the red colour of HIn is just not visible to the eye. Thus, we obtain a working range for the indicator of $pK_{In} \pm 1$. Table 8.6 lists a number of indicators with the experimental working ranges that have been determined with buffer solutions; they compare favourably with $pK_{In} \pm 1$.

In the phthalein indicators, the colour change is not due to ionization alone. Phenolphthalein (I) exists in two tautomeric forms, the quinonoid forms (II) and (III) predominating in alkaline solution; here, $K_{In} = K_1 K_2$.

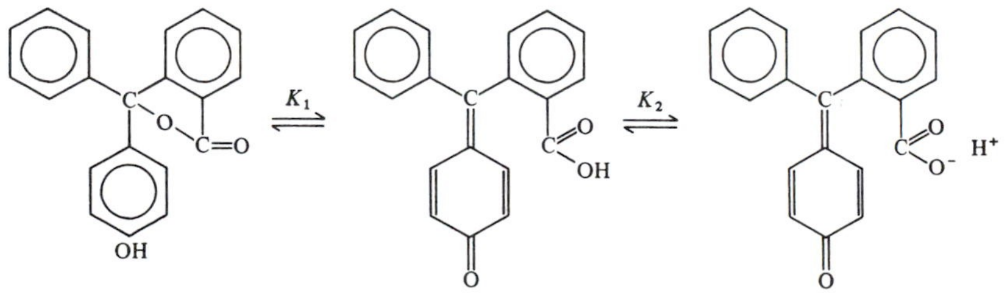

Phenolphthalein formulae

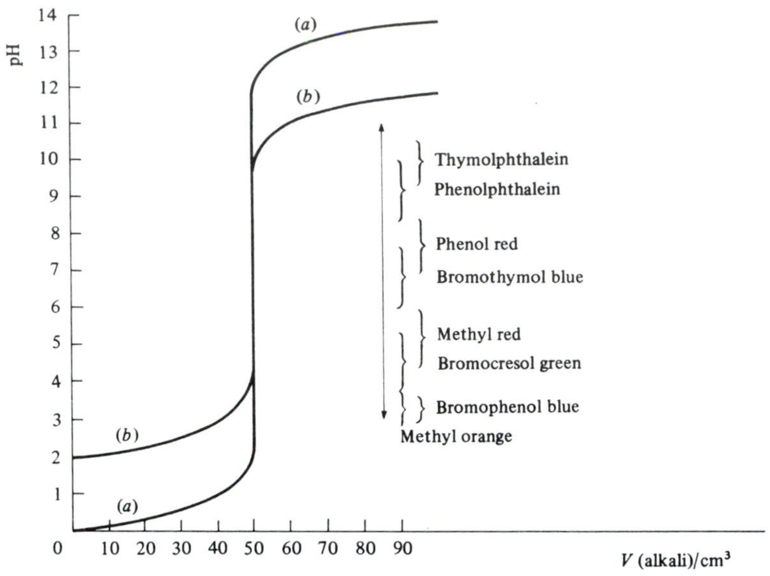

Figure 8.8 Variation of pH with volume V of sodium hydroxide added to 50 cm^3 hydrochloric acid: (b) 1 M solutions; (b) 0.01 M solutions. The indicator working-ranges from Table 8.6 are shown, from which satisfactory indicators may be chosen.

8.7.7 Acid–base titrations

The changes in pH during the titration of an acid with a base can be calculated by means of the equations already developed, and suitable indicators chosen.

8.7.7.1 Titrations with strong acids and strong bases

The titration curves of 50 cm^3 of 1 M hydrochloric acid with 1 M sodium hydroxide are shown in Figure 8.8a; Figure 8.8b is the corresponding curve for 0.01 M solutions. The pH changes rapidly in the neighbourhood of the *equivalence point*, which is the same as neutrality for these systems. Suitable indicators can be chosen in accordance with their working ranges; in particular, the curves show that methyl orange would be unsatisfactory with 0.01 M solutions, because the colour change would occur before the equivalence point was reached.

8.7.7.2 Titrations involving weak acids or weak bases

We consider the pairs ethanoic acid ($K_a = 1.76 \times 10^{-5}$) with sodium hydroxide, and pyridine ($K_b = 1.70 \times 10^{-9}$) with hydrochloric acid, all in 0.1 M solutions. On adding sodium hydroxide to ethanoic acid, the pH increases slowly at first, but flattens off as the system passes through the buffer working-range (Figure 8.9a). The pH changes rapidly between 7.5 and 9.5, and phenolphthalein indicator would be suitable to determine the equivalence point. The buffer–index curve (Figure 8.5) may be reviewed.

The system pyridine–hydrochloric acid shows a rapid decrease from pH 9.3 on addition

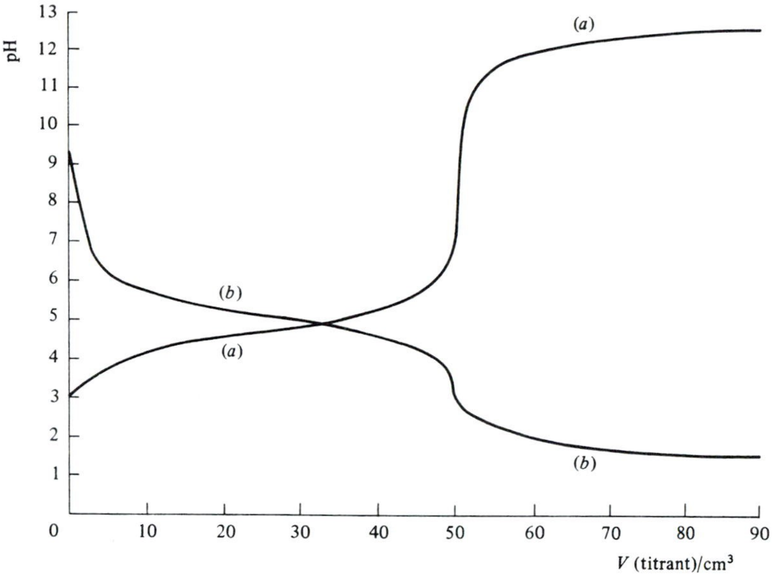

Figure 8.9 Variation of pH with volume V of titrant added: (*a*) 0.1 M sodium hydroxide added to 50 cm^3 of 0.1 M ethanoic acid; (*b*) 0.1 M hydrochloric acid added to 50 cm^3 of 0.1 M pyridine. Phenolphthalein indicator is suitable for (*a*), but the equivalence point in (*b*) is not easily located with precision.

of the acid. The decrease slows down through the buffer working-range of pH 4 to 6 (Figure 8.9b). The change at the equivalence point is not very marked; no indicator is wholly satisfactory, the best is probably benzylaniline-azobenzene.

8.7.7.3 Differential titration curves

The equivalence point is better defined if $\Delta(\mathrm{pH})/\Delta V$ is plotted aginst V, where V is the volume of titrant. Figures 8.10a and b show, respectively, the differential curves for the strong acid−strong base titration in Figure 8.8b and the weak acid−strong base titration in Figure 8.9b. It is evident that the equivalence point is more precisely defined, but the method requires the measurement of pH during the titration; we shall consider the necessary technique in the next chapter.

8.7.8 Activities in acid–base equilibria

We have used concentration terms in the discussion of the previous sections. In solutions of ionic strength of approximately 0.001 mol dm^{-3} at equivalence (for 1 : 1 electrolytes), the error arising by neglecting activities is approximately 4%; at an ionic strength of 0.005 mol dm^{-3} the error is approximately 8%. The arguments developed above are not affected in principle by the use of concentrations and in practice the equivalence points are obtained correctly.

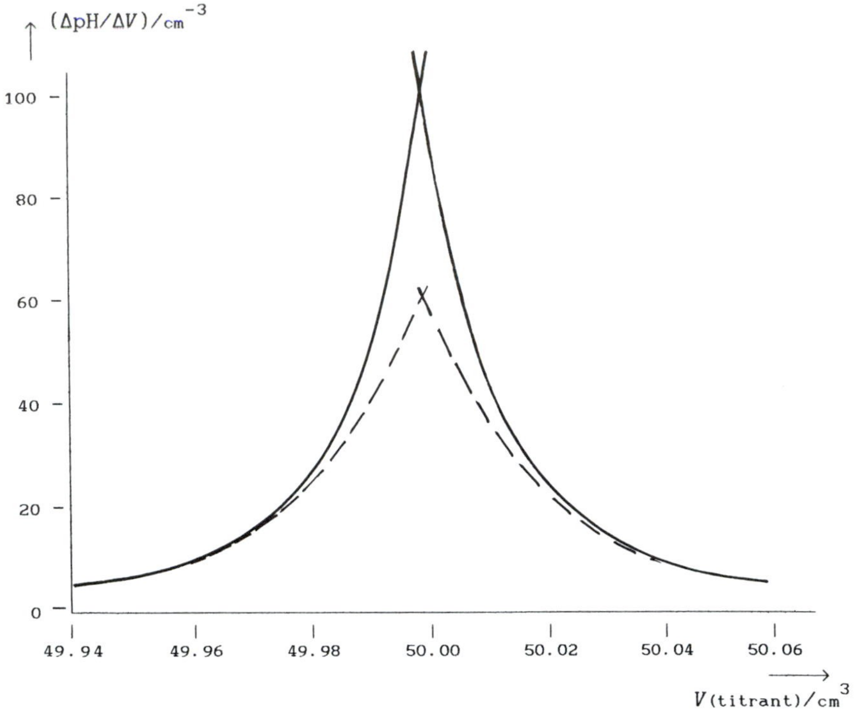

Figure 8.10 Differential curves: Δ(pH)/Δ*V* plotted against the volume *V* of titrant in the neighbourhood of the equivalence point: (*a*) strong acid–strong base (Figure 8.8b); (*b*) weak base–strong acid (Figure 8.9b). The improvement in the precision of the equivalence point for the latter titration is particularly notable.

Problems 8

8.1 Derive an expression for an equilibrium constant in the homogeneous liquid system of equation (8.2) and show that it leads to the results in Table 8.1. How is it that no mention of pressure arises, although the concentrations are given in mole fraction?

8.2(a) The values of $\Delta G^{\ominus}$ for cyclopropane and propene are 104.5 kJ mol^{-1} and 62.8 kJ mol^{-1} respectively. Calculate $\Delta G^{\ominus}$ and K_p for the isomerization reaction of equation (8.7).
(b) The standard entropies for cyclopropane and propene are 53.3 J K^{-1} mol^{-1} and 20.4 J K^{-1} mol^{-1} respectively. Calculate ΔH_{298} for the isomerization.
(c) Calculate K_p at 400 K, assuming that ΔH_{298} applies satisfactorily up to this temperature. Is the change in K_p in accordance with Le Chatelier's principle?

8.3 The following data refer to the equilibrium $N_2O_4 \rightleftharpoons 2NO_2$ at two temperatures; *p* is the total pressure. Evaluate K_p at both temperatures. Is the reaction exothermic or endothermic? Confirm your deduction by calculating ΔH for the dimerization process, over the given temperature interval.

	298 K		308 K	
Initial N_2O_4/mol	N_2O_4/mol at equilibrium	p/atm	N_2O_4/mol at equilibrium	p/atm
6.28×10^{-3}	3.90×10^{-3}	0.212	3.17×10^{-3}	0.238
1.26×10^{-2}	8.98×10^{-3}	0.394	7.73×10^{-3}	0.440
1.98×10^{-2}	1.514×10^{-2}	0.600	1.34×10^{-2}	0.662

8.4 The following data were obtained in a study of the equilibrium in (8.34). Calculate ΔH in the five successive temperature intervals.

T/K	p_{CO_2}/mmHg
773	7.73×10^{-2}
873	1.84
973	22.2
1073	167
1173	793
1273	2940

8.5 The dissociation of ammonium carbamate may be represented by the equation $NH_4CO_2NH_2(s) \rightleftharpoons 2NH_3(g)+CO_2(g)$; ΔH for the forwards reaction is negative. What changes in (a) the equilibrium concentrations and (b) the rate of attainment of equilibrium would be expected for

(i) decrease in pressure;
(ii) increase in temperature;
(iii) addition of ammonia to the system;
(iv) addition of an inert gas to the system?

8.6 A sample of 25 cm^3 of 0.1 M aqueous pyridine ($K_b=1.70\times10^{-9}$) is titrated with 0.1 M hydrochloric acid. Calculate the pH initially, at the equivalence point and after a total of 30 cm^3 of hydrochloric acid has been added.
8.7 The solubility product of barium sulfate is 1.1×10^{-10} at 25 °C. Calculate the solubility of this salt in (a) water, (b) 10^{-3} M sodium sulfate and (c) 10^{-2} M sodium chloride, at the given temperature.
8.8 Calculate the pH of the following aqueous solutions at 25 °C: (a) 0.01 M NaOH, (b) 0.005 mol dm^{-3} H_2SO_4, (c) 1 cm^3 2M HCl in 500 cm^3 water and (d) 5×10^{-9} M barium hydroxide.
8.9 The values of K_w for water at 273 K, 298 K and 333 K are 1.2×10^{-15}, 1.0×10^{-14} and 9.6×10^{-14}, respectively. Calculate the pH of water at each temperature, and determine an average enthalpy of dissociation of water over the given temperature range.
8.10 Determine the pH of the following aqueous solutions:

(a) 0.005 M ethanoic acid ($K_a=1.76\times10^{-5}$)
(b) 0.02 mol dm^{-3} pyridine ($K_a=1.70\times10^{-9}$)
(c) 0.1 mol dm^{-3} ammonium nitrate ($K_h=5.59\times10^{-10}$)
(d) 0.001 M sodium benzoate ($K_a=6.50\times10^{-5}$)

8.11 Calculate the pH of a solution that is 0.1 M in methanoic acid and 0.2 M in potassium methanoate ($K_a=1.80\times10^{-4}$). What is the change in pH on addition of 2 cm^3 of 1.0 mol

dm^{-3} sodium hydroxide to 100 cm^3 of the solution? In the second part of this problem, use each of the procedures documented in the text.

8.12 Calculate the pH of a 0.05 M solution of phosphoric acid H_3PO_4 (K_1=7.5×10^{-3}, K_2=6.2×10^{-8}, K_3=2.2×10^{-13}). The full solution involves a quintic equation in [H_3O^+], but it may be assumed here that the third dissociation is negligible at the given concentration.

8.13 By differentiating (8.144) with respect to [H_3O^+], find the conditions for the maximum and minimum values for β. By setting the differential in a suitable form, find the pH values for the extrema by iteration and compare them with Figure 8.5. Use K_a=1.76×10^{-5} (ethanoic acid) and c=0.05 M.

8.14 Calculate the temperature at which calcium carbonate begins to undergo thermal decomposition, given the following data:

	$CaCO_3(s)$	$CaO(s)$	$CO_2(g)$
$\Delta G_f^{\ominus}$/kJ mol^{-1}	−1128.8	−604.0	−394.4
$\Delta H_f^{\ominus}$/kJ mol^{-1}	−1206.9	−635.1	−393.5

8.15 Use the data below on the variation of compression factor $\mathcal{Z}$, $pV/(\mathcal{R}T)$, with pressure for oxygen gas at −70 °C to determine its fugacity at 100 atm:

p/atm	10	20	30	40	50
$\mathcal{Z}$	0.9696	0.9385	0.9060	0.8730	0.8405
p/atm	60	70	80	90	100
$\mathcal{Z}$	0.8085	0.7765	0.7455	0.7155	0.6870

CHAPTER 9

Electrochemistry

9.1 Introduction

In this chapter, we shall consider mainly electrical conduction and equilibrium electrochemistry, but with a brief reference to certain nonequilibrium topics. We begin with a brief résumé of electrolysis which, although strictly a nonequilibrium process, is a well-known topic with which to introduce our study of electrochemistry.

9.2 Electrical conduction

Two simple examples of electrical conduction are illustrated in Figures 9.1a and b. In the first of them, the electric current consists of a stream of electrons through the wires and the metal filament of the bulb, and may be called *electronic conduction*. The electrons are part of the structure of the conductor (Section 6.5ff.); the battery supplies the electromotive force required to drive them through the circuit, from a region of high electron potential, the 'negative' terminal of the battery, to one of low electron potential, the 'positive' terminal of the battery. We note that the flow of electrons is in the *opposite* direction $(- \rightarrow +)$ to that of conventional electric current $(+ \rightarrow -)$.

In Figure 9.1b, the electric current is carried through a dilute sulfuric acid solution by positive hydroxonium ions and negative sulfate ions. At the negative electrode the hydroxonium *cations* gain electrons and are discharged from the system as hydrogen gas:

$$H_3O^+(aq) + e^- \rightarrow \tfrac{1}{2}H_2(g) + H_2O(l) \tag{9.1}$$

The sulfate *anions* are not discharged in this way: they travel towards the positive electrode, but the electron involved in the reduction process (9.1) arises from the oxidation of a water molecule:

$$\tfrac{3}{2}H_2O(l) \rightarrow H_3O^+(aq) + \tfrac{1}{4}O_2(g) + e^- \tag{9.2}$$

This process is energetically favourable in comparison with

$$\tfrac{1}{2}SO_4^{2-}(aq) \rightarrow \tfrac{1}{2}SO_4(aq) + e^- \tag{9.3}$$

because the sulfate radical SO_4 is not a stable species.

Solutions that contain ions are able to conduct an electric current and are called *electrolyte* solutions. Generally, acids, bases and salts in solutions or in the melt are good electrical conductors.

The velocity of movement of ions towards the anode and cathode, under an applied potential gradient, is small. For example, potassium ions in 0.1 mol dm^{-3} aqueous potassium chloride solution at 25 °C under a potential gradient of 1 V m^{-1} have a velocity of

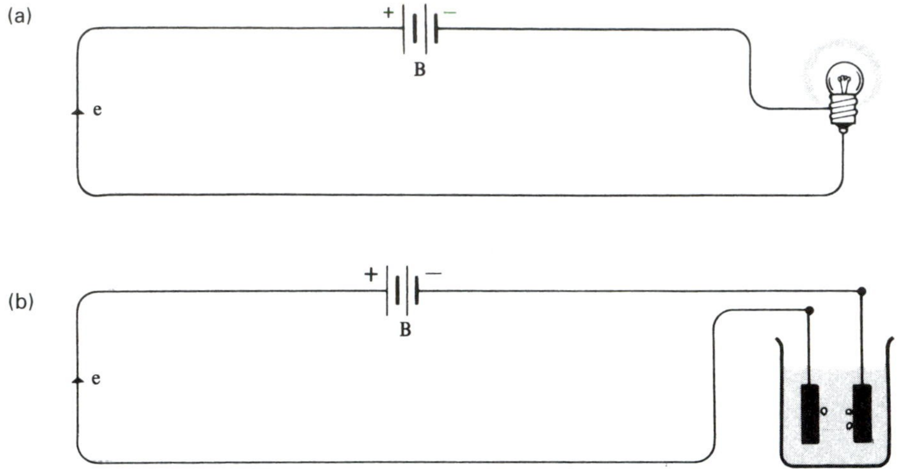

Figure 9.1 Electrical conduction: (*a*) electrons flow through the solid conductors, wire and metal filament; (*b*) electrons flow through the solid conductors, but transport in the solution takes place through ions. In both cases electron flow is from − to +.

approximately 10^{-7} m s^{-1} (2.2×10^{-7} mph); random thermal motions of the ions are much more energetic.

9.2.1 Laws of electrolysis

The decomposition of an electrolyte by the passage of an electric current through it is termed *electrolysis*. Its quantitative aspect is governed by Faraday's laws, which may be summarized as

$$m_i = M_i I\tau / (\mathcal{F}|q_i|) \tag{9.4}$$

where m_i is the mass of a species i of numerical charge $|q_i|$ that is discharged by a current of I ampere flowing for τ second, and M_i is the molar mass of the species. If $I\tau=\mathcal{F}$, $m_i=M_i/|q_i|$, that is, $\mathcal{F}$ coulomb is the amount of electricity required to discharge $1/|q_i|$ mole of any ionic species, and is known as the Faraday constant. The total charge on 1 mole of ions is $N_A|q_i|e$; hence, $N_A|q_i|e=|q_i|\mathcal{F}$, or

$$e=\mathcal{F}/N_A \tag{9.5}$$

9.2.2 Electrical conductivity[1]

The motion of ions in a solution may be studied through the measurement of the electrical resistance of the solution. In the case of a solid conductor, the electrical resistivity ρ is given by

$$\rho=RA/l \tag{9.6}$$

where R is the resistance and A and l are the physical dimensions, cross-sectional area and length, of the conductor; ρ has the units of Ω m. The *conductivity* κ is the reciprocal of the resistivity:

[1] Conductance G is the reciprocal of resistance R; conductivity κ is the reciprocal of resistivity ρ.

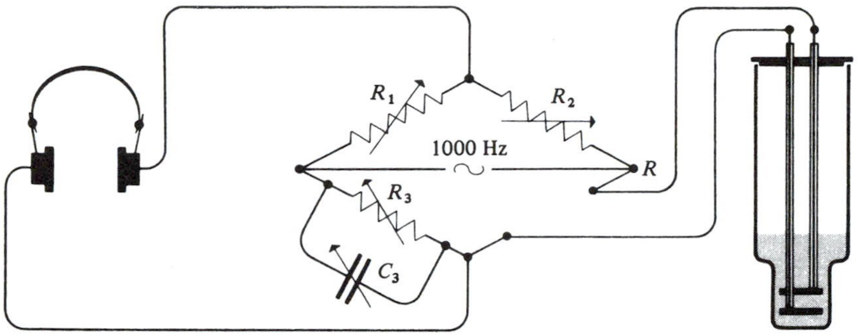

Figure 9.2 Wheatstone bridge network for the measurement of the resistance of an electrolyte solution of resistance R. A balance, zero signal in the headphones, is obtained by adjusting both R_3 and C_3 for an appropriate ratio R_1/R_2. The cell is calibrated with potassium chloride solution, preferably of similar conductivity to that of the solution under examination (see Table 9.1).

$$\kappa = 1/\rho = l/(RA) \tag{9.7}$$

The same definitions apply to electrolyte solutions. However, it is difficult to measure precisely the physical dimensions of a solution in a conductivity cell (Figure 9.2) because the conductance path is not normal to the electrodes. A *cell constant* χ, equal to l/A is determined by calibration with a solution of known conductivity, usually potassium chloride (Table 9.1), the conductivity of which has been determined against pure mercury in a capillary cell of precisely known dimensions.

If the resistance of the conductivity cell and solution is R, the conductivity of the solution is given by

$$\kappa = \chi/R \tag{9.8}$$

In order to compare different electrolytes on a common basis, a *molar conductivity* is specified. Consider one mole of an electrolyte solution of concentration c mol m^{-3} , that is, one mole of electrolyte is present in $1/c$ m^3. If this volume of solution were to be enclosed by two parallel, plane electrodes separated by a distance of 1 m, the area of each electrode would be $1/c$ m^2. Each cubic metre of solution would have a conductivity κ, so that the conductivity of one mole would be κ/c; this quantity is the molar conductivity Λ_c, measured in Ω^{-1} m^2 mol^{-1}:

$$\Lambda_c = \kappa/c \tag{9.9}$$

Two electrolytes, say MCl and MCl$_2$, may then be compared in terms of $\Lambda_c(M\mathrm{Cl})$ and $\frac{1}{2}$ $\Lambda_c(M\mathrm{Cl}_2)$.

9.2.2.1 Measurement of conductivity

The electrical property of a solution that is directly measurable is its resistance R. In the Wheatstone bridge circuit of Figure 9.2, the resistance of the cell and solution is measured by balancing the bridge, at an appropriate ratio R_1/R_2, with resistance R_3 and capacitance C_3 for a sharply defined minimum in the headphones detector. An alternating EMF is used to operate the bridge in order to prevent polarization at the electrodes; it is because of the alternating EMF that the capacitance of the cell must be balanced, too.

A more refined method employs a transformer ratio-arm bridge, Figure 9.3. A balance

Table 9.1. Electrical conductivity[2] $\kappa/\Omega^{-1}\,m^{-1}$ of standard aqueous potassium chloride

c/mol dm^3	18 °C	25 °C
0.010	0.12227	0.14115
0.100	1.1192	1.2886
1.000	9.820	11.173

is obtained by switching in the appropriate components from a range of standard conductors G_S and capacitors C_S. Fine control is obtained by adjusting the ratio of the numbers of turns T_1/T_2 on the tapped transformers for zero output, with maximum gain, at the detector across the secondary winding T_2. The required conductance G_X is given by G_ST_1/T_2, whence

$$\kappa = \chi G_X \tag{9.10}$$

and χ is the cell constant. Temperature control of the conductivity cell to within 0.1 K is essential for reliable measurements.

A solution of potassium chloride of concentration (c) 0.1 M has a molar conductance of 0.012886 $\Omega^{-1}\,m^2\,mol^{-1}$ at 25 °C. In a given conductivity cell this solution had a measured resistance of 30.05 Ω. Thus, $\kappa=0.012886\,c\times10^3=1.2886\,\Omega^{-1}\,m^{-1}$, and the cell constant $\chi=30.05\,\Omega\times\kappa=38.72\,m^{-1}$.

A solution of 0.0005 M HCl had a resistance of 1832 Ω when measured in the same cell. Thus, $\kappa=38.72\,m^{-1}/1832\,\Omega=0.021135\,\Omega^{-1}\,m^{-1}$, and $\Lambda=0.021135\,\Omega^{-1}\,m^{-1}/(10^3c)=0.04227\,\Omega^{-1}\,m^2\,mol^{-1}$. We note that the factor of 10^3 appears in these calculations because concentrations are usually reported in mol dm^{-3} in this context.

9.2.3 Independent conductivities of ions

It was noted first by Kohlrausch that plots of Λ_c against $\sqrt{c}$ for strong electrolytes were linear at low concentrations. The molar conductivities of weak electrolytes, such as ethanoic acid or ammonia, showed marked curvature on similar plots, Figure 9.4. When the plots are linear, they may be extrapolated to limiting molar conductivities at $c=0$, also called the molar conductivity at infinite dilution. The limiting equation for molar conductivity may be given as

$$\Lambda_c = \Lambda_0 - \sqrt{c} \tag{9.11}$$

where c extends over the range of linearity. At infinite dilution, there is no interaction between the ions and they move with characteristic maximum velocities. We define the limiting value Λ_0 by the equation

$$\Lambda_0 = \lim_{c\to0} \kappa/c \tag{9.12}$$

As $c\to0$ $\kappa\to0$, and the ratio κ/c has a finite limit, characteristic of the electrolyte.

[2] The SI unit of reciprocal resistance is the siemens, with conductivity measured in S m^{-1}. However, much of the literature on electrical conductivity uses Ω^{-1} rather than S, and this practice is adopted here (1 S=1 Ω^{-1}).

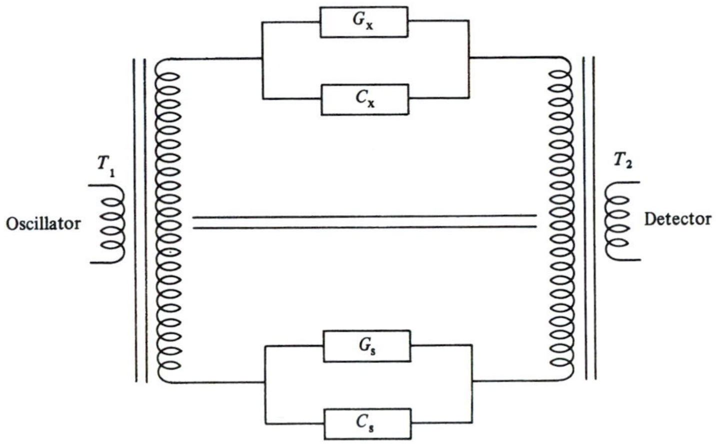

Figure 9.3 Transformer ratio-arm bridge circuit; conductive G_S and reactive C_S components are adjusted for a balance in the circuit, and the unknown conductance G_X is measured in terms of the conductive component G_X and the cell constant ($R=1/G$).

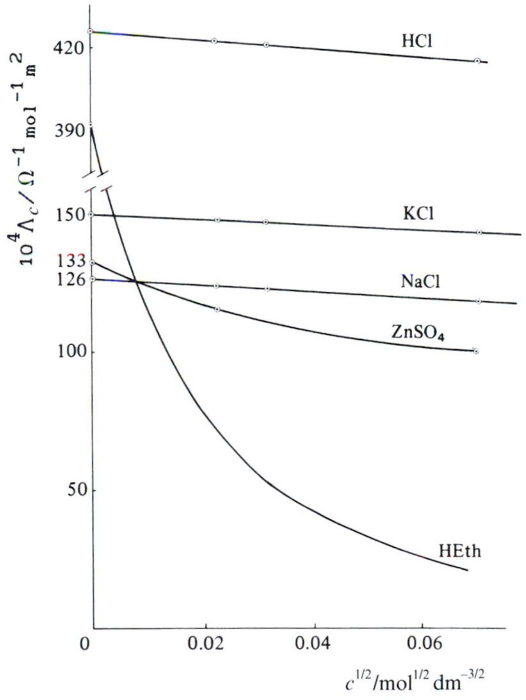

Figure 9.4 Plots of molar conductivity as a function of $\sqrt{c}$; at low concentrations the plots are linear for strong electrolytes, so that extrapolation to $c=0$ is practicable. For weak electrolytes, the curve rises rapidly at low concentrations and reliable extrapolation is not possible, although Λ_0 is comparable to the values for strong electrolytes.

Table 9.2. Equivalent conductivities Λ_0 at 25 °C and their differences Δ (in italics)

KA	$\Lambda_0/\Omega^{-1}\ m^2\ mol^{-1}$	$\Lambda/\Omega^{-1}\ m^2\ mol^{-1}$	NaA	$\Lambda_0/\Omega^{-1}\ m^2\ mol^{-1}$
KCl	0.01500	*0.00236*	NaCl	0.01264
$\Delta/\Omega^{-1}\ m^2\ mol^{-1}$	*0.00049*			*0.00049*
KNO_3	0.01451	*0.00236*	$NaNO_3$	0.01215
$\Delta/\Omega^{-1}\ m^2\ mol^{-1}$	*0.00298*			*0.00298*
KEth	0.01153	*0.00236*	NaEth	0.00917
$\Delta/\Omega^{-1}\ m^2\ mol^{-1}$	*−0.00382*			*−0.00382*
$\frac{1}{2}K_2SO_4$	0.01535	*0.00236*	$\frac{1}{2}NaSO_4$	0.01299

The concept of molar conductivity at infinite dilution leads to Kohlrausch's law of independent ion conductivities. In Table 9.2, the difference between Λ_0 for KA and NaA is a constant (0.00236), independent of the nature of the species A. Similarly, the difference between Λ_0 for MCl and MNO_3 is a constant (0.00049), independent of the nature of M (cf. Section 6.13). Similar remarks apply to all other pairs of electrolytes and we may write

$$\Lambda_0(MA)=\lambda_0(M^+)+\lambda_0(A^-) \tag{9.13}$$

and from Table 9.2 we can extract relationships such as

$$\lambda_0(K^+)-\lambda_0(Na^+)=0.00236\ \Omega^{-1}\ m^2\ mol^{-1} \tag{9.14}$$

These relations are strictly true only at infinite dilution, where there are no ionic interactions.

We can use Kohlrausch's law to determine Λ_0 for a weak electrolyte, for which extrapolation is not possible (Figure 9.4). Consider a weak acid HA. We need measurements of Λ_c against c for the salt NaA, a strong acid HY, and the sodium salt NaY, all of which are strong electrolytes. The data are extrapolated to give Λ_0 for each substance. Then, from the law of independent conductivities, we have

$$\Lambda_0(NaA)+\Lambda_0(HY)-\Lambda_0(NaY)=\lambda_0(H^+)+\lambda_0(A^-)=\Lambda_0(HA) \tag{9.15}$$

In this way, Λ_0 for ethanoic acid at 25 °C has been found to be 0.0396 Ω^{-1} m^2 mol^{-1}. Even at the low concentration of 10^{-3} mol dm^{-3}, Λ_c for ethanoic acid is only 0.00515 Ω^{-1} m^2 mol^{-1}, so that extrapolation is clearly impracticable.

9.2.4 Dissociation of electrolytes

Arrhénius suggested that the ratio Λ_c/Λ_0 was a measure of the extent of dissociation of an electrolyte into its ions:

$$\Lambda_c/\Lambda_0=\xi \tag{9.16}$$

We consider ethanoic acid, HEth. In (8.89), we write the extent of the dissociation as ξ; then, for an initial concentration c of acid, we obtain the concentrations at equilibrium [HEth]$=c(1-\xi)$, $[H_3O^+]=[Eth^-]=c\xi$, and (8.90) becomes

$$K_a=\xi^2c/(1-\xi) \tag{9.17}$$

which can be solved for ξ at any concentration that is not so dilute that the dissociation of water need be considered. If we now substitute $\Lambda_c/\Lambda_0=\xi$, we obtain

Table 9.3. Dissociation of ethanoic acid at 25 °C

c/mol dm^{-3}	Λ_c/Ω^{-1} m^2 mol^{-1}	$\Lambda_c/\Lambda_0(\xi)$	K_a
0.100	0.000523	0.0132	1.77×10^{-5}
0.020	0.001156	0.0292	1.76×10^{-5}
0.001	0.00495	0.125	1.79×10^{-5}
Limit	0.0396	1.000	–

Table 9.4. Dissociation of hydrochloric acid at 25 °C

c/mol dm^{-3}	Λ_c/Ω^{-1} m^2 mol^{-1}	$\Lambda_c/\Lambda_0(\xi)$	Apparent K
0.100	0.03913	0.918	1.03
0.010	0.04120	0.967	0.28
0.001	0.04214	0.989	0.089
Limit	0.04262	1.000	–

$$K_a = \frac{c\Lambda^2_c}{\Lambda_0(\Lambda_0 - \Lambda_c)} \tag{9.18}$$

which shows that Λ_c is not a linear function of c for a weak electrolyte. This equation is known as Ostwald's dilution law. Table 9.3 shows values of K_a obtained from (9.18) that are satisfactory. Results for a similar calculation with hydrochloric acid, listed in Table 9.4, show that this theory is not valid for a strong electrolyte, and another explanation will be required for their smaller variation in Λ_c and its linearity with $\sqrt{c}$. We shall return to a consideration of strong electrolytes in a later section of this chapter.

9.2.5 Transport properties

If two plane electrodes are placed at a distance d apart in an electrolyte solution and maintained at a potential difference V, the ions come under an electric field of strength $\mathcal{E}$ given by

$$\mathcal{E} = V/d \tag{9.19}$$

An ion of charge $|q|$ experiences a force of magnitude

$$F_e = |q|e\mathcal{E} = |q|eV/d \tag{9.20}$$

As the ions move through the solution to their appropriate electrodes, they are hindered by frictional forces in solution. According to Stokes' law[3], the frictional force F_f on a particle of radius a is given by

$$F_f = 6\pi\eta a v_d \tag{9.21}$$

where η is the coefficient of viscosity of the solution and v_d is the drift speed of the ions when the forces F_e and F_f are balanced. Thus, the drift speed is given by

[3] See, for example, H S Harned and B B Owen *The Physical Chemistry of Electrolytic Solutions* (Reinhold, 1967).

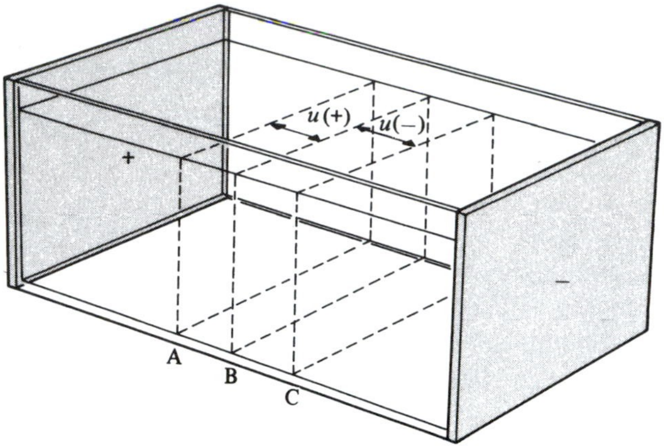

Figure 9.5 Schematic diagram to illustrate the calculation of ion mobility. The shaded ends of the cell represent electrodes of area 1 m² each, separated by a length of 1 m; *A*, *B* and *C* are imaginary planes in the central region of the cell, separated by distances $|u_+|$ and $|u_-|$, as shown. At a concentration of c mol m^{-3}, the cell contains c mol of electrolyte. The number ν_i of ions i in a volume u_i m³ is, thus, cu_i.

$$v_d = |q| e V / (6\pi a \eta d) \tag{9.22}$$

For an Na^+ ion, $q=1$, and a is the radius sum for the ionic radius and its hydration sphere. With $V/d = 1$ V m^{-1}, v_d is approximately 5.2×10^{-8} m s^{-1}; since $\eta = 0.00089$ kg m^{-1} s^{-1}, (9.22) gives $a = 1.602\times10^{-19}$ C$\times$1V m^{-1}/($6\pi\times5.2\times10^{-8}$ m s$^{-1}\times0.00089$ kg m^{-1} s^{-1}), or 0.18 nm.

The value of a just obtained for the *hydrodynamic radius* of the hydrated sodium ion is not unreasonable, since $r_{cryst} = 0.112$ nm. However, if we carry out the same calculation for the chloride ion the resulting value is 0.12 nm, which is smaller than r_{cryst} (0.170 nm). It is clear that a degree of uncertainty attaches to the estimation of a in this manner.

9.2.5.1 Ion mobilities

The mobility u of an ion is its velocity towards an electrode under an applied field gradient of 1 V m^{-1}. The calculation above actually used the value of the mobility of the sodium ion, since the field strength was unity; the units of ion mobility are m² s^{-1} V^{-1}.

The conductivity of an ion is related to its mobility. Consider the cell of Figure 9.5, in which c mol of an electrolyte $M_{\nu_+}A_{\nu_-}$ are enclosed between plane electrodes of 1 m² cross-sectional area, separated by a distance of 1 m; the molarity of the solution is, thus, c. If a current of I A flows under a potential difference of 1 V, then from Ohm's law and the cell geometry

$$I = 1/R = \kappa \tag{9.23}$$

The conductivity arises from the flow of ions. We imagine three planes parallel to the electrodes: B is a central reference plane, and planes A and C lie one on each side of it, at distances from it numerically equal to u_+ and u_-, as shown. In 1 s, all positive ions ν_+ in the volume u_+ m³ reach or cross plane B, and all negative ions ν_- in the volume u_- m³ reach or cross B from the opposite direction. Let ξ be the extent of dissociation of the electrolyte.

Since $v_+ = cu_+$ and $v_- = cu_-$, the charge Q crossing plane B in τ s is given by

$$Q = \mathscr{F}\xi\tau(q_+v_+ + q_-v_-) = \mathscr{F}\xi\tau c(q_+u_+ + q_-u_-) \tag{9.24}$$

Since $Q = I\tau$, and using (9.9), (9.16) and (9.23), we obtain

$$\Lambda_c = \mathscr{F}\xi(q_+u_+ + q_-u_-) \tag{9.25}$$

As $c \to 0$ $\xi \to 1$, so that the limiting value of (9.25) becomes

$$\Lambda_0 = \mathscr{F}(q_+u_+ + q_-u_-) \tag{9.26}$$

We may divide (9.26) into individual ionic conductivities, since they are additive (Section 9.2.3), such that $\lambda_{0,+} = \mathscr{F}q_+u_+$ and so on. In general,

$$\lambda_i = \mathscr{F}q_iu_i \qquad \lambda_{0,i} = \mathscr{F}q_iu_{0,i} \tag{9.27}$$

9.2.5.2 Transport numbers

In using the law of independent conductivities to obtain Λ_0 for a weak electrolyte, we did not require the individual conductivities $\lambda_{0,i}$. They could be obtained from measurements of their mobilities (9.27), but are usually determined indirectly from their *transport numbers*.

The transport number t_i is that fraction of the charge passed through an electrolyte that is carried by the *i*th ion. Thus, if q_{Na^+} is the charge carried by the sodium ion in sodium chloride,

$$t_{Na^+} = q_{Na^+}/(q_{Na^+} + q_{Cl^-}) \tag{9.28}$$

A similar equation for t_{Cl^-} shows that $t_{Na^+} + t_{Cl^-} = 1$, and generally

$$\sum_i t_{i,+} + \sum_j t_{j,-} = 1 \tag{9.29}$$

where the sums are taken over all ionic species present in the given electrolyte solution.

The total charge passed through an electrolyte is not, in general, equally divided between the anions and cations. We shall consider *limiting* transport numbers, in order to avoid the effects of ionic interaction. The fractional charge carried by an ion is proportional to its limiting molar conductivity. Thus, we have from Section 9.2.3 and (9.27)

$$t_+ = \lambda_{0,+}/(\lambda_{0,+} + \lambda_{0,-}) = u_{0,+}/(u_{0,+} + u_{0,-}) \tag{9.30}$$

$$t_- = \lambda_{0,-}/(\lambda_{0,+} + \lambda_{0,-}) = u_{0,-}/(u_{0,+} + u_{0,-}) \tag{9.31}$$

A solution in which u_+ and u_- are numerically different remains electrically neutral when a current is passed through it, but the concentration does not remain uniform throughout the solution. This fact enables the transport numbers of ions to be measured.

9.2.5.3 Measurement of transport numbers

Consider an aqueous copper sulfate solution of known concentration, between two copper electrodes in a Hittorf transport cell, Figure 9.6. If the transport number of the copper ion is t_+, the charge transported by the copper ions is $t_+\mathscr{F}$. Thus, for the passage of one $\mathscr{F}$, $\frac{1}{2}t_+$ mol Cu^{2+} move from the anode compartment to the centre compartment. A similar quantity

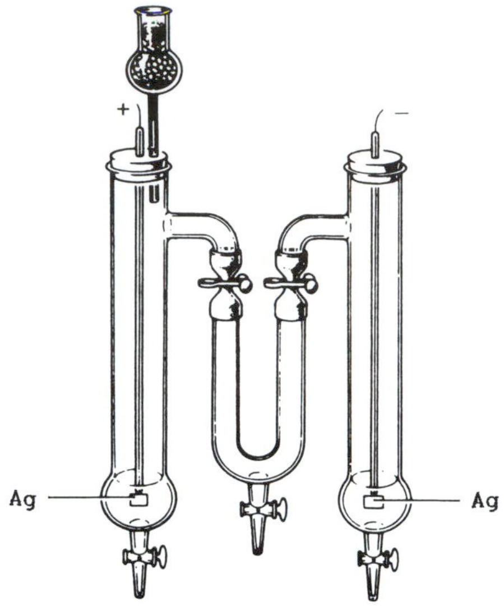

Figure 9.6 Hittorf cell for the determination of transport numbers. The two clips are open during the experiment and closed while the contents of the other compartments are removed for analysis, so that no mechanical mixing of the solutions occurs. The whole of the compartments are analysed.

of copper ions move from the centre compartment to the cathode compartment, and at the cathode $\frac{1}{2}$mol of copper ions is discharged and deposited on the cathode:

$$\tfrac{1}{2}Cu^{2+}(aq)+e^{-}\rightarrow\tfrac{1}{2}Cu(s) \tag{9.32}$$

The net loss of copper ions from the cathode compartment is $(\frac{1}{2}-\frac{1}{2}t_+)$ mol, or $\frac{1}{2}t_-$ mol. The cathode compartment loses $\frac{1}{2}t_-$ mol of SO_4^{2-} ions, so that $\frac{1}{2}$ mol of $CuSO_4$ is removed from this compartment.

In the anode compartment, $\frac{1}{2}t_+$ mol Cu^{2+} migrate out, as we have discussed, and $\frac{1}{2}t_-$ mol of SO_4^{2-} ions enter, having effectively passed through the centre compartment. The sulfate ions are not discharged and the reaction at the anode is

$$\tfrac{1}{2}Cu(s)\rightarrow\tfrac{1}{2}Cu^{2+}(aq)+e^{-} \tag{9.33}$$

The anode compartment gains $(\frac{1}{2}-\frac{1}{2}t_+)$, or $\frac{1}{2}t_-$, mol $CuSO_4$.

In practice, only a very small charge is passed, so that the changes at the electrodes are confined to their respective compartments; the charge passed is measured by a coulometer in series with the Hittorf cell. The *whole* of each compartment is analysed: the total number of moles in each compartment is needed, not just the concentrations, as the problem below shows; the centre compartment should remain unchanged in composition throughout the experiment.

If platinum electrodes were used, the cathode compartment would again lose $\frac{1}{2}t_-$ mol $CuSO_4$. At the anode, the process would be

$$\tfrac{3}{2}H_2O(l)\rightarrow H_3O^{+}(aq)+\tfrac{1}{4}O_2(g)+e^{-} \tag{9.34}$$

If this compartment were analysed, there would be a loss of $\frac{1}{2}$ mol Cu^{2+} and a gain of $\frac{1}{2}t_-$ mol SO_4^{2+}, per $\mathcal{F}$ C. The difference is $\frac{1}{2}t_- -(-\frac{1}{2}t_+)$, or $\frac{1}{2}$ mol $CuSO_4$, and represents the hydroxonium ions discharged at the anode during the oxidation process. Thus, we must know the electrode reactions before the changes in electrolyte content of the compartments can be interpreted in terms of transport numbers.

A solution containing 10.850 g silver nitrate in 1000 g water was electrolysed between silver electrodes in a Hittorf cell. A series coulometer recorded the passage of 143.0 C. The anode and centre compartments were analysed totally after the experiment.

Mass of anode compartment	54.900 g
Mass of $AgNO_3$ present	0.7220 g
Mass of centre compartment	46.200 g
Mass of $AgNO_3$ present	0.4960 g
Mass of water in centre compartment	45.704 g
Mass of $AgNO_3$ present originally in this amount of water (45.714×10.850/1000)	0.4960 g

Thus, the centre compartment remains unchanged in amount of $AgNO_3$.

Mass of water in the anode compartment	54.178 g
Mass of $AgNO_3$ present originally in this amount of water (54.178×10.850/1000)	0.5878 g
Mass of $AgNO_3$ after the experiment	0.7220 g

The anode reaction is $Ag(s) \rightarrow Ag^+(aq) + e^-$, so that the anode compartment will gain t_- mol of anions. Since the measured gain is 0.1342 g $AgNO_3$ for the passage of 142 C, the gain would be $(0.1342\ g/169.87\ g\ mol^{-1}) \times (96485\ C\ mol^{-1}/143\ C) = 0.533$, which is the value of t_-; thus, $t_+ = 0.467$.

Table 9.5 lists some results of transport number measurements at 298 K and several concentrations. The variation of t_+ with concentration arises from ionic interaction and the limiting values were obtained by extrapolation.

The change in transport number with electrolyte concentration is much less than the corresponding variation of ionic conductivity because it involves a ratio, (9.30). The individual conductivities are obtained from the transport numbers and the conductivity of the electrolyte; thus,

$$\lambda_i = t_i \Lambda_i \qquad (9.35)$$

and limiting values for some species are listed in Table 9.6.

9.2.5.4 Conduction and hydration

The relatively large conductivities of the hydroxonium and hydroxyl ions are related through the self-dissociation of water. A chain process of conduction is envisaged for these ions, as illustrated by Figure 9.7; the movement of hydroxyl ions may be considered as the movement of protons in the opposite direction:

$$2H_2O \rightleftharpoons H_3O^+ + OH^-$$

Another example of chain conduction is provided by the autoprotolysis of pure sulfuric acid. In this viscous medium, chain conduction is the only feasible mechanism; the transport number of the HSO_4^- ion lies in the range 0.96 to 0.99. Water-insoluble salts such as barium sulfate and lead sulfate dissolve readily in pure sulfuric acid:

$$2H_2SO_4 \rightleftharpoons H_3SO_4^+ + HSO_4^-$$

Table 9.5. Cation transport numbers t_+ at 298 K

	c/mol dm^{-3}				
	0.20	0.10	0.05	0.01	Limit
HCl	0.834	0.831	0.829	0.825	0.821
KCl	0.489	0.490	0.490	0.490	0.491
NaCl	0.382	0.385	0.388	0.392	0.396
$AgNO_3$	0.470	0.468	0.466	0.465	0.464

Table 9.6. Limiting molar conductivities of some ions at 298 K

	$\lambda_0/\Omega^{-1}m^2\,mol^{-1}$		$\lambda/\Omega^{-1}\,m^2\,mol^{-1}$
H_3O^+	0.03500	$\frac{1}{3}La^{3+}$	0.00720
K^+	0.00745	OH^-	0.01920
NH_4^+	0.00745	Cl^-	0.00755
Ag^+	0.00635	NO_3^-	0.00706
Na^+	0.00509	$CH_3CO_2^-$	0.00408
Li^+	0.00387	$\frac{1}{2}SO_4^{2-}$	0.00790
$\frac{1}{2}Ca^{2+}$	0.00600	$\frac{1}{4}Fe(CN)_6^{4-}$	0.01110

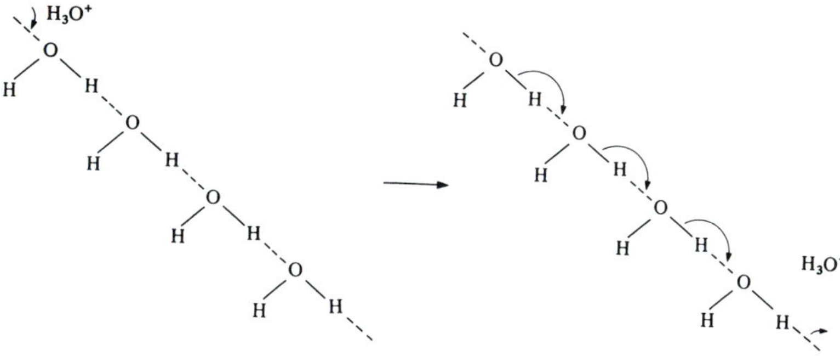

Figure 9.7 Schematic illustration of a chain mechanism for proton transfer in an aqueous electrolyte solution. The transfer of protons between water molecules simulates a movement of H_3O^+ ions and leads to a rapid current transport. The movement of OH^- ions may be thought of as a movement of hydroxonium ions in the opposite sense.

$$BaSO_4(s) \xrightleftharpoons{H_2SO_4\,(l)} Ba^{2+} + SO_4^{2-} \quad \Big\downarrow +H_3SO_4^+ \quad H_2SO_4(l) + 2HSO_4^- \tag{9.36}$$

The transport number $t_{Ba^{2+}}$ in this medium is approximately 0.009, compared with 0.45 in aqueous barium chloride.

Table 9.7. Crystal radii[4] r, hydration numbers n and hydrodynamic radii a of some cations and anions

	r/nm	n	a/nm		r/nm	n	a/nm
Li^+	0.086	6.2	0.31	Cl^-	0.170	3.7	0.19
Na^+	0.112	4.4	0.26	Br^-	0.187	3.1	0.21
K^+	0.144	3.7	0.20	NO_3^-	0.114	2.5	0.15
Mg^{2+}	0.087	7.5	0.36	SO_4^{2-}	0.138	5.4	0.20

The Stokes equation (9.21) would tend to suggest that in a series of ions such as that of the alkali metals, the smaller ions would move more rapidly and, therefore, have the higher conductivities, since $\lambda \propto u$, from (9.30). The existence of the reverse order (Table 9.6) is explained by the fact that the parameter a in (9.21) refers to a hydrated ion. The degree of hydration depends on the electric field strength of the ion, $|q|e/(4\pi\epsilon_0 r^2)$, and so is greater for the smaller and more highly charged ions. Thus, among the hydrated alkali-metal cations, lithium is the largest. Table 9.7 lists parameters of some common ions. The values of a have been determined through Gorin's equation:

$$\Lambda = \frac{\Lambda_0 + (a_- \lambda_{0,+} + a_+ \lambda_{0,-})}{1 + (a_+ + a_-)\kappa} \tag{9.37}$$

The results appear to be more acceptable in relation to the crystal radii than are the values obtained through the Stokes equation. In the light of these data, the trends for λ_0 listed in Table 9.6 are understandable.

9.2.6 Strong electrolytes

A strong electrolyte, such as sodium chloride, is considered to be completely dissociated into its ions at all concentrations in aqueous solution. The molar conductivity decreases with increasing concentration because interactions between the anions and cations increase. A similar effect was considered in Section 7.11. The effect of interionic attraction is to reduce the velocities of the ions and, hence, their conductivities, from (9.27). Table 9.8 presents data for ethanoic acid and sodium chloride for the concentration range 0.1 mol dm^{-3} to infinite dilution. The velocities of the ions in ethanoic acid remain nearly constant, because so few ions are present in the ethanoic acid solution, whereas those in sodium chloride decrease by approximately 15%.

The Debye–Hückel–Onsager theory, the development of which has been well documented,[5] leads to an equation for dilute solutions of strong electrolytes in the form deduced empirically by Kohlrausch:

$$\Lambda_c = \Lambda_0 - \mathcal{S}\sqrt{c} \tag{9.38}$$

[4] See Table 6.12.

[5] See, for example, R A Robinson and R H Stokes, *Electrolyte Solutions* (Academic Press, 1959); H S Harned and B B Owen, *The Physical Chemistry of Electrolytic Solutions* (Reinhold, 1967).

Table 9.8. Variation of molar conductivity Λ_c and mobility u with concentration at 298 K in aqueous ethanoic acid HEth and sodium chloride; ξ is the extent of dissociation

	c/mol dm^{-3}	Λ_c/Ω^{-1} m^2 mol^{-1}	ξ	$10^6\, u_+$/m s^{-1}	$10^6\, u_-$/m s^{-1}
HEth	0.100	0.000417	0.012	0.359	0.421
	0.010	0.00182	0.047	0.362	0.423
	Limit	0.0396	1.00	0.363	0.425
NaCl	0.100	0.01067	1	0.428	0.680
	0.010	0.01185	1	0.481	0.747
	Limit	0.01265	1	0.517	0.795

where $\mathcal{S}$, the Onsager slope, refers to the slope of the lines shown in Figure 9.4 for strong electrolytes at low values of $\sqrt{c}$. The slope may be interpreted as the sum of two terms:

$$\mathcal{S}=A\Lambda_0+B \tag{9.39}$$

The ionic atmosphere that is set up around each ion in solution does not respond instantaneously to the movement of the central ion. Thus, the atmosphere ahead of the ion is incomplete, whereas that behind it is in excess. The effect is that the central ion is ahead of the centre of charge of the atmosphere, so that the mobility of the moving ion is reduced. A relaxation effect is created, which is the term $A\Lambda_0$ in (9.39). Furthermore, the frictional effect on the moving ion is enhanced because the ionic atmosphere, being of opposite sign to the central ion, moves in an opposing direction, thus creating an electrophoretic effect[6] represented by the term B in (9.39). The constant A is inversely proportional to $(\epsilon_r T)^{-3/2}$ and B is inversely proportional to $(\epsilon_r T)^{-1/2}\eta$. We shall not be concerned with the actual values of A and B.

The molar conductivity may now be written as

$$\Lambda_c=\Lambda_0-(A\Lambda_0+B)\sqrt{c} \tag{9.40}$$

and results for aqueous sodium chloride are listed in Table 9.9; the agreement is very good, and is similar for most 1 : 1 electrolytes at these concentrations.

9.2.7 Ion association

In solvents of relative permittivity less than 50, 1 : 1 electrolytes exhibit lower conductivities than would be expected from (9.40) because of ion association. Even in water, where $\epsilon_r\simeq 80$, 2 : 2 electrolytes show this effect.

The force of attraction F between ions M^{q+} and A^{q-} separated by a distance d in a medium of relative permittivity ϵ_r is given by

$$F=|q_+q_-|e^2/(4\pi\epsilon_0\epsilon_r d) \tag{9.41}$$

where ϵ_0 is the permittivity of a vacuum. Thus, attraction between oppositely charged ions increases as the charges increase, but decreases as the relative permittivity increases. An equilibrium is set up:

$$M^{q+}(\text{aq})+A^{q-}(\text{aq}) \rightleftharpoons (M^{q+}A^{q-})^{|q_+q_-|}(\text{aq}) \tag{9.42}$$

[6] Electrophoresis refers to the motion of a particle caused by a potential drop.

Table 9.9 Calculated and experimental molar conductivities for aqueous sodium chloride at 298 K

c/mol dm^{-3}	Λ_c/Ω m^2 mol^{-1} (expt)	Λ_c/Ω^{-1} m^2 mol^{-1} (calc)
Limit	0.01265	–
0.0005	0.01245	0.01245
0.001	0.01237	0.01237
0.005	0.01202	0.01207
0.01	0.01176	0.01185

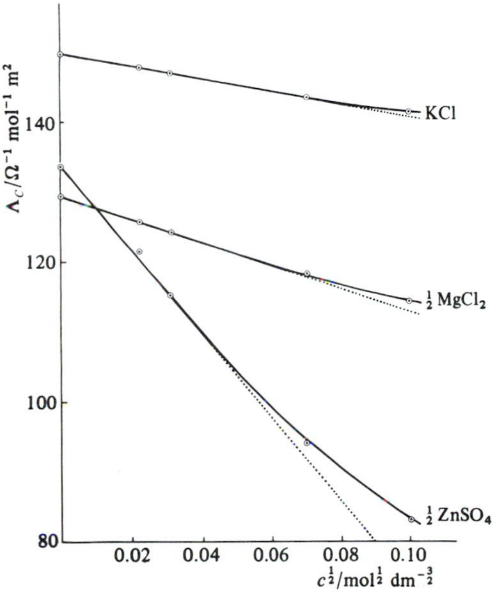

Figure 9.8 Variation of molar conductivity with $\sqrt{c}$ for 1 : 1, 2 : 1 and 2 : 2 electrolytes; the slope and the departure from linearity both increase as the ionic charges increase.

With symmetrical electrolytes, the ion-pair, on the right-hand side of (9.42), is uncharged, and does not contribute to the conductivity of the solution. Figure 9.8 illustrates the variation of molar conductivity with $\sqrt{c}$ for a 1 : 1, a 2 : 1 and a 2 : 2 electrolyte.

For the given solvent, water, the slope of each plot increases as the force of attraction between anions and cations increases, because of the product $|q^+q^-|$. In the case of zinc sulfate, this attraction is sufficient for an equilibrium to be established:

$$Zn^{2+}(aq)+SO_4^{2-}(aq) \rightleftharpoons [Zn^{2+}SO_4^{2-}](aq) \qquad (9.43)$$

The ion-association constant K_A is 4.9×10^{-3} at 298 K.

Ion association may be pairwise or more extensive, but we shall consider only the pairwise case. The ion-pair formation may be treated in the manner of weak electrolytes. Thus, if ξ is the extent of ion-pair formation in a symmetrical electrolyte of concentration c, we have

$$K_A=\xi^2 c\gamma_{\pm}^2/(1-\xi) \qquad (9.44)$$

Table 9.10. Ion-association constants at 298 K

	Solvent	ϵ_r	$\log_{10}(K_A)$
$Ca(OH)_2$	H_2O	78.5	1.30
$MgSO_4$	H_2O	78.5	2.23
$CaSO_4$	H_2O	78.5	2.28
$Ca_2Fe(CN)_6$	H_2O	78.5	3.77
CuMal[a]	H_2O	78.5	5.70
$Cu(ClO_4)_2$	CH_3CN	36.2	2.28
$(CH_3)_4NClO_4$	CH_3OH	32.6	1.75
CsI	C_2H_5OH	24.3	2.15

Note:
[a] Mal is the malonate (propane dicarboxylate) ion, $CH_2(CO_2)_2^-$.

Both ξ and K_A may be determined from conductivity measurements and, thus, the concentrations of the species present in solution. The concentration of an ion pair may be of importance when these species show important chemical and biological activity compared with that of the free solvated ions.

In order to include the effect of ion pairing in (9.40), we write

$$\Lambda_c = \xi\Lambda_0 - \xi(A\Lambda_0 + B)(\xi c)^{1/2} \tag{9.45}$$

This equation may be solved by iteration: a starting value of $\xi=1$ in the term $(\xi c)^{1/2}$ leads to a better approximation ξ', and so on, until two successive values of ξ agree to within the precision of the experimental data. Some data on ion-association constants are listed in Table 9.10 and further information may be found in the standard literature.[7]

9.2.8 Applications of conductivity measurements

In this section we shall consider how conductivity measurements may be used in determining solubility products and charges on complex ions and in monitoring acid–base titrations.

9.2.8.1 Determination of solubility products

As an example, let the sparingly soluble silver bromate have a solubility s; then, the molar conductivity of the saturated solution is given by

$$\Lambda_s = \kappa_s / s \tag{9.46}$$

Since the solution is very dilute, $\Lambda_s \simeq \Lambda_0 = \lambda_{0,+} + \lambda_{0,-}$; the latter data may be found from the literature. Thus, from a measurement of κ_s the solubility product K_s may be calculated. A correction must be applied for the conductivity of the water.

The conductivity κ_s of saturated silver chloride at 298 K is $1.7574\times10^{-4}\ \Omega^{-1}\ m^{-1}$; the conductivity of the water is $2.01\times10^{-5}\ \Omega^{-1}\ m^{-1}$. The limiting ion conductivities are $\lambda_{0,+}=0.00543\ \Omega^{-1}\ m^2\ mol^{-1}$ and $\lambda_{0,-}=0.00655\ \Omega^{-1}\ m^2\ mol^{-1}$, so that $\Lambda_0=0.01198\ \Omega^{-1}\ m^2\ mol^{-1}$. Thus, $s=1.5564\times10^{-4}\ \Omega^{-1}\ m^{-1}/0.01198\ \Omega^{-1}\ m^2\ mol^{-1}=0.01299\ mol\ m^{-3}$, or $1.299\times10^{-5}\ mol\ dm^{-3}$. At this concentration $\gamma_\pm=0.990$, so that $K_s=1.65\times10^{-10}$.

[7] See, for example, C W Davies, *Ion Association* (Butterworth, 1962).

Table 9.11. Ionization data for complex-ion electrolytes at 298 K

	$\Lambda_{1024}/\Omega^{-1}$ m^2 mol^{-1}	Ionization mode
Electrolyte type		
MA	0.0120–0.0140	M^+A^-
MA_2, M_2A	0.0240–0.0280	$M^{2+}2A^-$, $2M^+A^{2-}$
MA_3, M_3A	0.0350–0.0430	$M^{3+}3A^-$, $3M^+A^{3-}$
MA_4	0.0450–0.0520	$M^{4+}4A^-$
Examples		
$Pt(NH_3)_6Cl_4$	0.0500	$Pt(NH_3)_6^{4+}$, $4Cl^-$
$K_3Co(NO_2)_6$	0.0418	$3K^+$, $Co(NO_2)_6^{3-}$
$Co(NH_3)_3(NO_2)_3$	0.00015	Nonionic

9.2.8.2 Determination of the charge on a complex ion

Singly charged ions, excluding H_3O^+ and OH^-, in water at 298 K have an average limiting conductivity of approximately 0.0065 Ω^{-1} m^2 mol^{-1}, with correspondingly larger values for more highly charged species. Thus, a comparison of Λ_c for a complex-ion electrolyte with the sum of the values for the individual species allows the charge on the complex ion to be predicted. For this work, c is usually 2^{-10} mol dm^{-3}, or 1 mol in 1024 dm^3 of solution. Table 9.11 illustrates the procedure sufficiently clearly.

9.2.8.3 Acid–base titrations

Acid–base titrations may be monitored by measuring the electrical conductivity of the solution during the course of the titration. The shape of the curve and, hence, the precision of the equivalence point, is determined by the relative strengths – not concentrations – of the acid and base.

If a strong base is added to a strong acid, say sodium hydroxide to hydrochloric acid, the highly conducting hydroxonium ions are replaced by sodium ions and the conductivity of the solution falls rapidly to the equivalence point. Addition of base after equivalence has been reached introduces excess hydroxyl ions and the conductivity rises rapidly. If the volume change is kept small, by titrating 0.01 M acid with 1 M base, the changes in conductivity are linear and the equivalence point is the intersection of the two lines, Figure 9.9a.

If the hydrochloric acid is replaced by a weak acid, such as salicylic acid, there is an initial decrease in conductivity, Figure 9.9b. The anion formed then represses the dissocation of the acid, so that the fall in conductivity is less rapid. The conductivity then begins to rise because of the equilibrium

$$\mathrm{HSal(aq)+Na^+(aq)+OH^-(aq)} \rightleftharpoons \mathrm{H_2O(l)+Sal^-(aq)+Na^+(aq)} \quad (9.47)$$

Beyond the equivalence point conductivity increases rapidly because of the excess hydroxyl ions present in the solution.

We saw in Section 8.7.7.2 that the location of the equivalence point in the titrations of weak acids and weak bases by means of indicators was imprecise, particularly so if both electrolytes were weak. However, it is possible to titrate ethanoic acid with ammonia to good precision by following the conductivity of the solution, Figure 9.10.

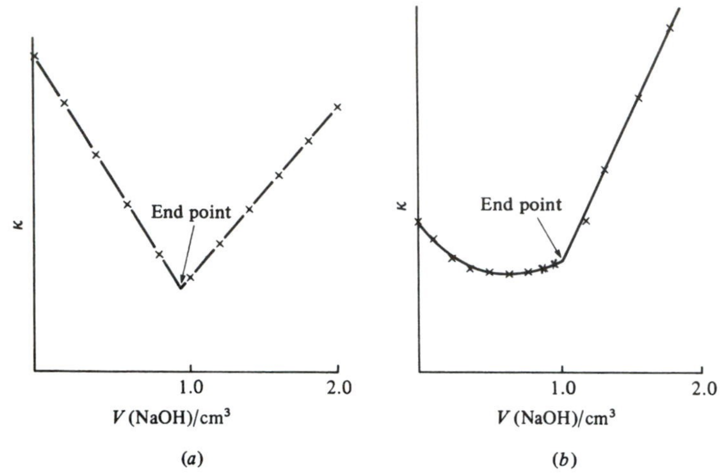

Figure 9.9 Monitoring acid–base titrations by conductivity measurements. (*a*) 100 cm^3 0.01 M hydrochloric acid titrated with 1.038 M sodium hydroxide; the equivalence point is readily located. (*b*) 100 cm^3 0.01 M salicylic acid titrated with 1 M sodium hydroxide; the relatively flat portion of the curve before the equivalence point arises from the equilibrium (9.47); as more Na^+ is formed the curve rises to the equivalence point, and then more rapidly as OH^- ions are added in excess.

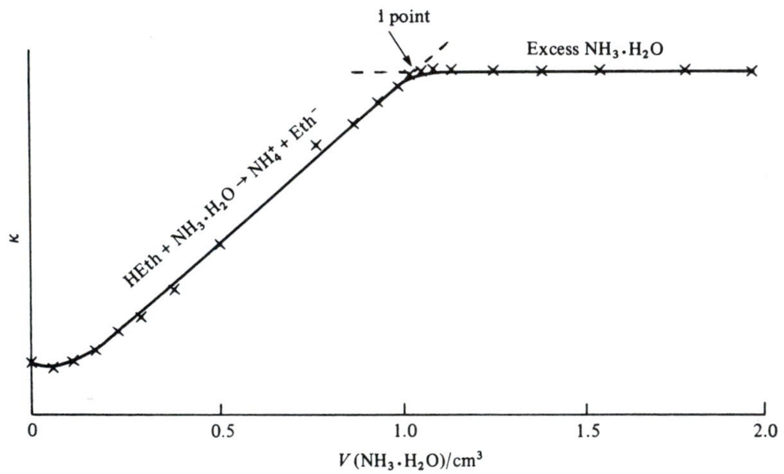

Figure 9.10 Conductivity changes during the titration of 100 cm^3 0.01 M ethanoic acid with 1 M ammonia; there is almost no decrease in conductivity, because of the repression of the dissociation of ethanoic acid by the ethanoate ion. The conductivity increases to the equivalence point because of the formation of ammonium ions; after the equivalence point there is negligible change in conductivity because ammonia is a weak base. Nevertheless, the equivalence point can be located with precision by extrapolating the straight line portions of the curve.

Figure 9.11 shows the conductivity curve for the titration of a mixture of hydrochloric acid and ethanoic acid by sodium hydroxide. The dissociation of the weak acid is repressed by the hydroxonium ions until the strong acid has been nearly neutralized. The extrapolation of the straight lines to obtain the equivalence point is necessary because dissociation of the weak acid begins just before the strong acid has been completely neutralized. The first

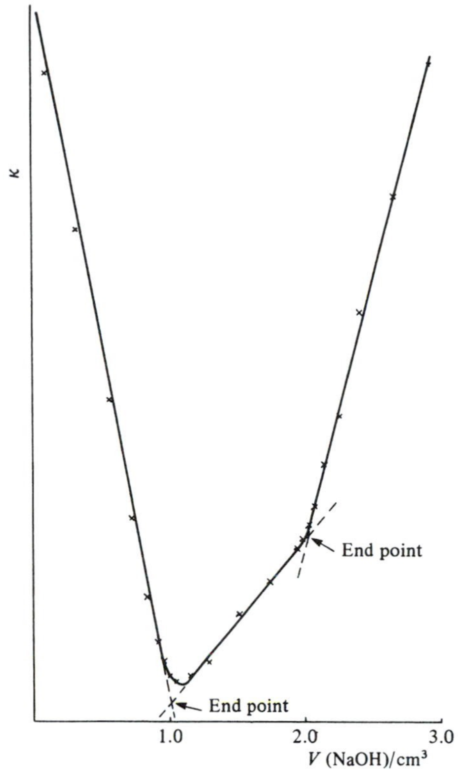

Figure 9.11 Titration of a mixture of 50 cm^3 0.02 M hydrochloric acid and 50 cm^3 0.02 M ethanoic acid with 1 M sodium hydroxide. The first equivalence point is that for the strong acid; the disociation of the weak acid is negligible almost to the equivalence point, but extrapolation of the straight lines locates that point precisely. After the second equivalence point, the excess hydroxyl ions cause a sharp rise in conductivity.

equivalence point relates to the strong acid and the second equivalence point to the weak acid. Precipitation reactions, such as the titration of a chloride solution with silver nitrate, in which there is a conductivity change at the equivalence point may be monitored in a similar manner.

9.3 Equilibrium electrochemistry

Metals have a small but finite tendency to dissolve in water producing cations, leaving their valence electrons on the metal. The metal acquires a negative potential, which prevents the release of further cations, and an equilibrium is established:

$$M(\mathrm{s}) \rightleftharpoons M^{q+}(\mathrm{aq}) + q\mathrm{e}^- \qquad (9.48)$$

The tendency to dissolve is diminished, and may be reversed, if there are already cations of the metal in solution. For zinc, the reversal occurs at an activity of approximately 2.6×10^{25}; for copper, the corresponding concentration would be 5.5×10^{-12} mol dm^{-3}. Thus, at all practical concentrations, zinc tends to oxidize, or donate electrons, whereas copper tends to be reduced, or gain electrons (Figure 9.12):

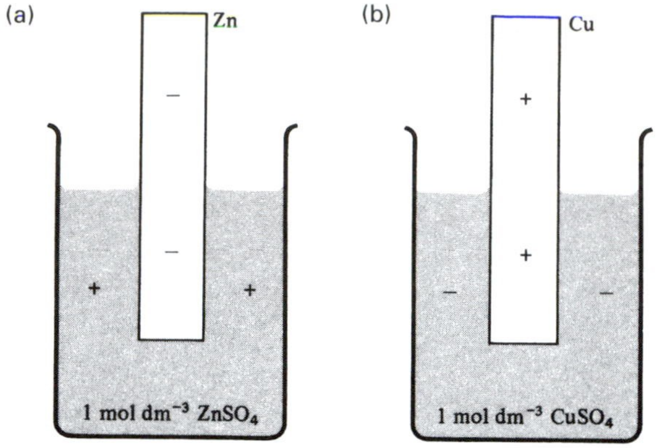

Figure 9.12 Metallic zinc and copper in contact with their own ions: (*a*) zinc tends to dissolve and to be charged negatively with respect to its neighbourhood solution; (*b*) copper tends to deposit from its ions, leaving the neighbourhood solution relatively negative.

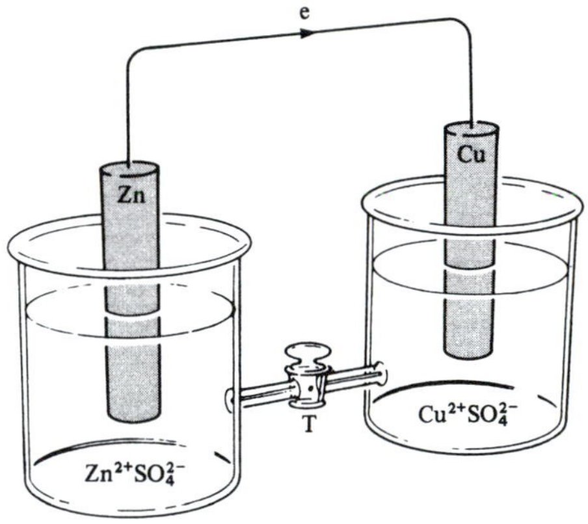

Figure 9.13 A Daniell cell; when contact is established by opening the tap, electrons flow through the external circuit from zinc to copper. The current flow in the cell takes place by ion transport.

$$\tfrac{1}{2}Zn(s) \rightleftharpoons \tfrac{1}{2}Zn^{2+}(aq)+e^- \tag{9.49}$$

$$\tfrac{1}{2}Cu^{2+}(aq)+e^- \rightleftharpoons \tfrac{1}{2}Cu(s) \tag{9.50}$$

If the two systems Zn^{2+}/Zn and Cu^{2+}/Cu are joined appropriately, as in Figure 9.13 with the tap open, electrons flow *from zinc to copper through the external circuit*. The flow of current is maintained by the discharge of copper at its electrode with anions being transferred to the zinc solution; zinc cations flow into the copper solution. The assembly constitutes a galvanic, or electrochemical, cell; it is, in fact, the Daniell cell in chemical dress.

In all galvanic cells, similar oxidation and reduction processes take place in the operation of the cell. In the Leclanché dry cell, the following reactions occur at the electrodes:

$$\tfrac{1}{2}Zn(s) \rightleftharpoons \tfrac{1}{2}Zn^{2+}(aq)+e^- \tag{9.51}$$

$$MnO_2(s)+H_3O^+(aq)+e^- \rightleftharpoons \tfrac{1}{2}Mn_2O_3(s)+\tfrac{3}{2}H_2O(l) \tag{9.52}$$

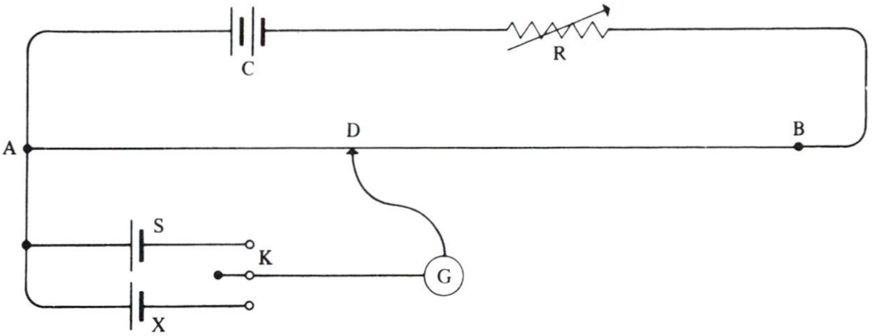

Figure 9.14 Basic potentiometer circuit for measurement of EMF under thermodynamically reversible conditions: C is an external battery opposing the EMFs of the cells; the standard Weston cell S and the cell under investigation X are switched by the key K; R is a variable resistor; AB is a calibrated, uniform resistance wire; a balance point D is established for zero current through the galvanometer G. In a modern potentiometer, the calibrated wire is replaced by high-quality standard resistors, with fine tuning controls.

ganese(III) in dimanganese trioxide. The hydroxonium ions are supplied by a paste of ammonium chloride:

$$NH_4^+(aq)+2H_2O(l) \rightleftharpoons NH_3{\cdot}H_2O(aq)+H_3^+O(aq) \qquad (9.53)$$

The sum of the potential differences between each electrode and its solution represents the electromotive force (EMF) of the cell. Here, EMF has its usual meaning of the potential difference between two electrodes in the absence of an IR voltage drop, that is, when the current I is zero. A conventional voltmeter draws an appreciable current from the circuit, so the EMF must be measured under balanced conditions, as with a potentiometer.

9.3.1 Measurement of EMF

Figure 9.14 illustrates the basic potentiometer circuit. The measurement of EMF is performed under reversible conditions because, at balance, no current flows through the cell. Then, we can use equilibrium thermodynamics to discuss the system involved.

The external storage battery C is connected via a variable resistor R to a uniform, calibrated resistance bridge wire AB. The moving contact D taps a known resistance from the wire, and is linked through the galvanometer G to either a standard cell S or an unknown cell X, connected in opposition to the battery. The bridge is calibrated by the standard cell, giving a balance point D_S on the wire for zero current through the galvanometer. Then a similar procedure is carried out for the cell X, giving a balance at D_X. Then, $E_X=E_S$ (AD_S/AD_X).

The standard cell S is normally a Weston cadmium cell; its EMF at 293 K is 1.0183 V, and the variation of EMF with temperature is given by

$$E_T=1.0183\text{ V}-(4.06\times10^{-5}\text{ V K}^{-1})(T-293\text{ K})$$
$$-(9.5\times10^{-7}\text{ V K}^{-2})(T-293\text{ K})^2 \qquad (9.54)$$

The Weston cell has a very reproducible EMF, and measurements may be carried out in this way with high precision. The basic circuit can be enhanced by equipment that includes standard resistors with fine control and a temperature-compensation device.

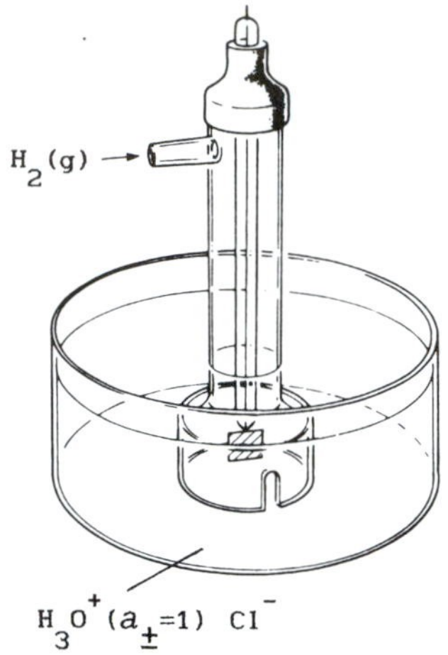

Figure 9.15 Reference standard hydrogen electrode (SHE): at unit activity of hydroxonium ions, 1 atm partial pressure of hydrogen gas and 298.15 K, the EMF is defined to be zero.

9.3.2 Electrode (reduction) potentials

It is evident that the measurement of the potential difference between an electrode and its ions requires a second electrode in the solution to complete an electrical circuit. A reference electrode, the standard hydrogen electrode (SHE) is chosen, Figure 9.15. It consists of a platinum electrode coated with electrolytically deposited, finely divided platinum (platinum black) in an aqueous solution containing hydroxonium ions at unit activity; hydrogen gas at a partial pressure of 1 atm is bubbled through the solution, over the electrode. An equilibrium is set up:

$$\tfrac{1}{2}H_2(g)+H_2O(l) \rightleftharpoons H_3O^+(aq,\ a_\pm=1)+e^- \tag{9.55}$$

At 298.15 K, the potential of this electrode is defined as 0 V.

The potential of any other electrode system can be determined by measuring the EMF of a cell in which that electrode is combined with the SHE. In Figure 9.16a, the SHE is the cathode, attracting electrons from the zinc anode. The overall cell reaction is the sum of the electrode reactions:

Anode
$$\tfrac{1}{2}Zn(s) \rightleftharpoons \tfrac{1}{2}Zn^{2+}(aq)+e^- \tag{9.56}$$

Cathode
$$H_3O^+(aq)+e^- \rightleftharpoons \tfrac{1}{2}H_2(g)+H_2O(l) \tag{9.57}$$

Overall
$$\tfrac{1}{2}Zn(s)+H_3O^+ \rightleftharpoons \tfrac{1}{2}Zn^{2+}(aq)+\tfrac{1}{2}H_2(g)+H_2O(l) \tag{9.58}$$

In Figure 9.16b, the SHE is the anode, and the overall reaction may be established as above, resulting in

$$\tfrac{1}{2}H_2(g)+\tfrac{1}{2}Cu^{2+}(aq)+H_2O(l) \rightleftharpoons H_3O^+(aq)+\tfrac{1}{2}Cu(s) \tag{9.59}$$

At 298.15 K and unit activity of zinc ions, the EMF of cell (a) is 0.763 V.

We adopt the following conventions with cells and electrode potentials.

(a) A cell is set up such that the electrode on the left-hand side involves an oxidation process. Thus, we write

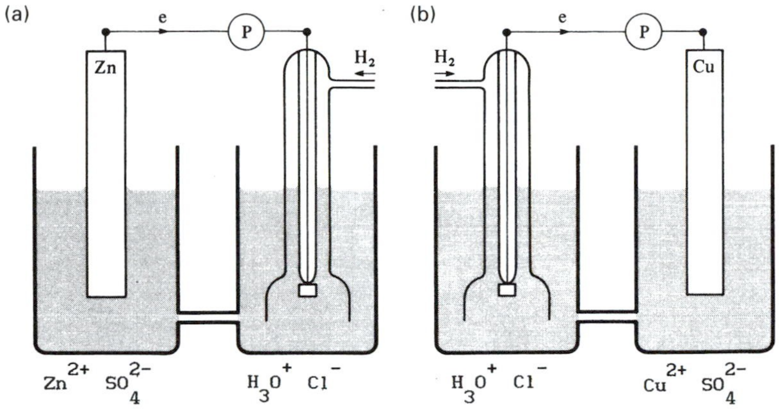

Figure 9.16 The SHE in combination with zinc and copper half-cells: (*a*) reduction is taking place at the SHE, which is behaving as the cathode; (*b*) oxidation is taking place at the SHE, which is behaving as the anode. The EMF of a cell is always positive for a spontaneous cell reaction, so that the metal half-cell is the left-hand electrode in (*a*) and the right-hand electrode in (*b*). It is, perhaps, best not to think of half-cells in terms of anode and cathode, because a given half-cell can assume both identities, as we have shown.

$$\mathrm{Zn}|\mathrm{Zn}^{2+}(a_{\pm}=1)|\mathrm{H_3O^+}(a_{\pm}=1)|\mathrm{H_2,Pt} \qquad (9.60)$$

(b) Electrode potentials are quoted as *reduction potentials*; thus, the zinc reduction potential or *half-cell* is written in terms of the reaction

$$\tfrac{1}{2}\mathrm{Zn^{2+}(aq)}+\mathrm{e^-} \rightleftharpoons \tfrac{1}{2}\mathrm{Zn(s)} \qquad (9.61)$$

(c) The EMF of a cell is the potential of the right-hand electrode *minus* the potential of the left-hand electrode. Thus, from the cell (9.60), Figure 9.16a, with an EMF of 0.763 V, it follows that the standard reduction potential for zinc may be written as $\pi^{-0-}(\mathrm{Zn^{2+}/Zn})=-0.763$ V.

A simple way to remember the sign conventions is (i) *zinc is negative and on the left in the Daniell cell representation*; (ii) *reduction potentials above hydrogen in the electrochemical series (Section 9.3.7.4) are negative*; (iii) *electrons flow from left to right in the external circuit of a cell.*

The EMF for the cell in Figure 9.16b is 0.339 V and the reduction potential for the copper system is written as $\pi^{-0-}(\mathrm{Cu^{2+}/Cu})=0.339$ V. Thus, when the spontaneous half-cell reaction is an oxidation, as in (9.49), the potential is negative; when it corresponds to a reduction, as in (9.50) reversed, the potential is positive. We shall discuss these results more fully shortly.

In constructing a cell we place two half-cells back-to-back, but we must place them in the correct order, namely, with the *more negative* reduction potential at the *left-hand* electrode.

9.3.2.1 Subsidiary reference electrode

The SHE is a rather cumbersome device, and a subsidiary reference electrode is generally incorporated into an electrochemical cell. Figure 9.17 shows a calomel (mercury(I) chloride) reference electrode. A paste of mercury and mercury(I) chloride is in contact with a saturated solution of potassium chloride. Electrical contact is made through a sheathed platinum

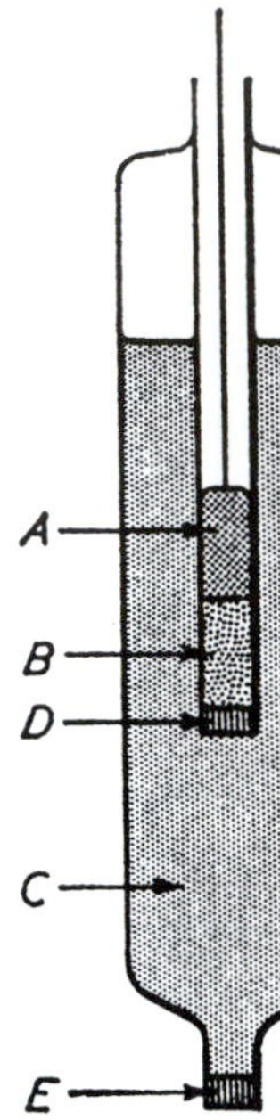

Figure 9.17 Calomel reference electrode: A mercury, carrying a platinum wire for external connection; B paste of mercury, mercury(I) chloride (calomel) and potassium chloride; C saturated potassium chloride solution, in contact with excess solid potassium chloride to ensure saturation; D sintered glass plug, to provide electrical contact between B and C; E sintered glass disc, to provide electrical contact between the electrode and the solution under investigation. The electrode is reversible to chloride ions and its potential is dependent upon their activity; $\pi^{-0-}=0.242$ V at 298.15 K.

wire assembly, the tip of which passes into mercury that is in contact with the calomel paste. Electric contact between the electrode and the solution is made through a fine sintered glass plug.

The electrode reduction reaction for the calomel half-cell is

$$\tfrac{1}{2}Hg_2Cl_2(s)+e^- \rightleftharpoons Hg(l)+Cl^-(aq) \qquad (9.62)$$

$\pi^{-0-}(Hg_2Cl_2/Hg)=0.267$ V, with chloride ions at unit activity, but with saturated potassium chloride $\pi^{-0-}=0.242$ V at 298.15 K; the temperature variation of EMF for this electrode is given by

$$E_T=0.242\ \text{V}-(7.0\times10^{-4}\text{V K}^{-1})(T-298) \qquad (9.63)$$

A second practical reference electrode is the silver–silver chloride electrode. It consists of a silver wire (coated with silver chloride by anodic deposition from a solution of potassium chloride) in contact with potassium chloride solution. The electrode reaction is

$$Ag(s)+Cl^-(aq) \rightleftharpoons AgCl(s)+e^- \qquad (9.64)$$

In construction it is similar to the electrode assembly in Figure 9.17, and its standard reduction potential at unit activity of chloride ions, $\pi^{-0-}(AgCl/Cl^-)$, is 0.2224 V. Both this electrode and the calomel half-cell are reversible to chloride ions.

We may note that a practical indication of the reversible nature of a cell is that a balance point with a potentiometer may be approached from either side of the zero position of the galvanometer. Thus the cell (9.60) is reversible, but the cell

$$Zn|Zn^{2+}(a_{\pm})||H_3O^+(a_{\pm})|Pt \tag{9.65}$$

is reversible only at the left-hand electrode. The reaction at the platinum electrode is

$$H_3O^+(aq)+e^- \rightarrow H_2O(l)+\tfrac{1}{2}H_2(g) \tag{9.66}$$

which is not reversible in the absence of a supply of hydrogen gas.

9.3.2.2 Liquid junction potential

The central bar in the representation (9.60) implies a junction between the two solutions. In general, the junction of two different solutions gives rise to a junction potential because the ions crossing the liquid interface have different velocities. The liquid junction potential is measurable only with difficulty, but it can be effectively eliminated by employing a *salt-bridge* junction.

A salt bridge is made by stirring a small quantity of agar-agar into a heated, concentrated electrolyte solution, which then forms a gel on cooling. It may be contained in a U-tube; then, instead of linking the two half-cells in Figure 9.13 by a capillary and tap, they are brought into electrical contact by means of the salt bridge dipping into both solutions. For a reason that will be apparent when we discuss concentration cells with transference, an electrolyte is selected for which the transport numbers are approximately equal. Thus, potassium chloride or ammonium nitrate are suitable materials for making a salt bridge. Representation (9.60) would then be written as

$$Zn|Zn^{2+}(a_{\pm}=1)||H_3O^+(a_{\pm}=1)|H_2,Pt \tag{9.67}$$

where the double bar implies a salt-bridge junction, as in (9.65), too.

9.3.3 Thermodynamics of galvanic cells

From the discussion in Section 4.7, the maximum electrical work w_e (nonexpansion work) at constant temperature and pressure is given by ΔG for the reaction under consideration. We are able to determine this value because we have arranged to work under reversible conditions. When the potentiometer bridge-circuit is balanced, we measure the *tendency* for a reaction to occur.

In order to perform a measurement on a reaction, it has to advance by a minute amount. Let the forwards reaction advance infinitesimally by an amount $d\xi$ mol; then from (8.9)

$$đw_e = \Delta G\, d\xi \tag{9.68}$$

In an advancement $d\xi$, the total absolute electronic charge transferred between the electrodes is $(-e)qN_A\, d\xi$, or $-q\mathcal{F}\, d\xi$. The work done $đw_e$ is, therefore, $-q\mathcal{F}E\, d\xi$, where E is the cell EMF, or the potential difference between the electrodes on open circuit. Thus, from (9.68), we now obtain $\Delta G = -q\mathcal{F}E$. Because we have arranged the cell and half-cell reactions to correspond to a single electron transfer, $q=1$, and we have

$$\Delta G = -\mathcal{F}E \tag{9.69}$$

It is now evident that a cell reaction with a positive EMF is a thermodynamically spontaneous process.

From (8.15) and (9.69), we obtain

$$E = -\Delta G^{-0-}/\mathcal{F} - (\mathcal{R}T/\mathcal{F})\ln(\prod_i a_i^{\nu_i}) \tag{9.70}$$

where the term $\Pi a_i^{\nu_i}$ is a convenient notation for the activity expression in (8.17); ν is negative for the denominator. Following (9.69), $-\Delta G^{-0-}/\mathcal{F}$ is the standard EMF E^{-0-} corresponding to unit activity and $\Pi_i a_i^{\nu_i}=1$; hence, we write

$$E = E^{-0-} - (\mathcal{R}T/\mathcal{F})\ln(\prod_i a_i^{\nu_i}) \tag{9.71}$$

which is the Nernst equation, relating the EMF of a cell to the activities of its components. If a reaction is written such that a transfer of n mole of electrons takes place, then $(\mathcal{R}T/\mathcal{F})$ is replaced by $[\mathcal{R}T/(n\mathcal{F})]$: thus, if (9.72) were multiplied throughout by 2, n would take this value. We note that the EMF of a cell, for a given set of conditions, has a fixed value, but that ΔG for the same cell is directly proportional to n: E is an intensive and G an extensive property.

The overall reaction for the Daniell cell is

$$\tfrac{1}{2}Zn(s) + \tfrac{1}{2}Cu^{2+}(aq) \rightleftharpoons \tfrac{1}{2}Zn^{2+}(aq) + \tfrac{1}{2}Cu(s) \tag{9.72}$$

If this reaction is performed by adding finely powdered zinc to copper sulfate at 298 K, the energy available from the reaction is dissipated as heat ($\Delta H^{-0-} = -109.3$ kJ mol^{-1}), which may be measured calorimetrically. If the system is set up as the electrochemical cell

$$Zn|Zn^{2+}(a_\pm=1)||Cu^{2+}(a_\pm=1)|Cu \tag{9.73}$$

the energy available is harnessed to perform electrical work ($E^{-0-}=+1.102$ V; $\Delta G^{-0-} = -106.3$ kJ mol^{-1}).

From (9.71), it is evident that a cell standard EMF E^{-0-} requires that $\Pi_i a_i^{\nu_i}$ is unity. If, instead, the zinc sulfate concentration were 0.005 mol kg^{-1} and that of copper sulfate 0.001 mol kg^{-1}, the corresponding activity coefficients would be $\gamma_{Zn^{2+}}=0.575$ and $\gamma_{Cu^{2+}}=0.761$, whence $\Pi_i a_i^{\nu_i}=3.778$ and $E=1.068$ V, at 298.15 K.

When a cell has reached equilibrium $\Delta G=0$ and $\Pi_i a_i^{\nu}=K$, whence

$$E^{-0-} = (\mathcal{R}T/\mathcal{F})\ln(K) \tag{9.74}$$

For the Daniell cell $K=\exp$ [1.102 V$\times$96485 C mol^{-1}/(8.3145 J K^{-1} mol$^{-1}\times$298.15 K)]$=4.24\times10^{18}$, which implies a high degree of completion of the forwards reaction in (9.72).

9.3.3.1 Measurement of standard reduction potentials

We illustrate the determination of E^{-0-} with reference to a cell consisting of a hydrogen electrode at an activity $a_\pm$ of hydrochloric acid and a silver–silver chloride electrode, set up in a thermostat at 25 °C; $p_{H_2}=1$ atm.

$$Pt,H_2(g)|HCl(aq)(a_\pm)|AgCl,Ag \tag{9.75}$$

The electrode reactions involved are

Left-hand electrode $\quad \tfrac{1}{2}H_2(g) + H_2O(l) \rightleftharpoons H_3O^+(aq) + e^- \quad$ (9.76)

Right-hand electrode $\quad AgCl(s) + e^- \rightleftharpoons Ag(s) + Cl^-(aq) \quad$ (9.77)

which sum to the overall cell reaction

$$\tfrac{1}{2}H_2(g) + AgCl(s) + H_3O(l) \rightleftharpoons H_3O^+(aq) + Ag(s) + Cl^-(aq) \tag{9.78}$$

Table 9.12. EMF of cell (9.75) as a function of molality, and mean activity coefficients, at 298.15 K

$10^2\,m$/mol kg^{-1}	E/mV	$\gamma_{\pm}$(expt)	$\gamma_{\pm}$(D-H)	$\gamma_{\pm}$(Davies)
0.3222	520.5	0.939	0.936	0.938
0.5628	492.6	0.929	0.916	0.925
0.9146	468.6	0.910	0.894	0.906
1.3406	449.7	0.894	0.873	0.891

Since the activities of solids are unity and the amount of water is constant, the EMF is, from (9.71)

$$E=E^{\ominus}-(\mathcal{R}T/\mathcal{F})\ln[(a_{\pm,\mathrm{HCl}})^2/(f_{\mathrm{H_2}}/p^{\ominus})^{1/2}] \quad (9.79)$$

where $E^{\ominus}$ is the standard reduction potential of the silver–silver chloride electrode. Since we shall be working with dilute solutions and at 1 atm partial pressure of hydrogen, $f_{\mathrm{H_2}}=p^{\ominus}$; hence,

$$E=E^{\ominus}-(\mathcal{R}T/\mathcal{F})\ln(m/m^{\ominus})^2-(\mathcal{R}T/\mathcal{F})\ln(\gamma_{\pm})^2 \quad (9.80)$$

From (8.71) $\ln(\gamma_{\pm})=-1.172\sqrt{m}$, and (9.80) becomes

$$E=E^{\ominus}-2(\mathcal{R}T/\mathcal{F})\ln(m/m^{\ominus})+2.344(\mathcal{R}T/\mathcal{F})(m/m^{\ominus})^{1/2} \quad (9.81)$$

We rearrange this equation to give

$$E+2(\mathcal{R}T/\mathcal{F})\ln(m/m^{\ominus})=E^{\ominus}+2.344(\mathcal{R}T/\mathcal{F})(m/m^{\ominus})^{1/2} \quad (9.82)$$

The EMF E is measured for varying molalities of hydrochloric acid (Table 9.12); then, E is plotted against $(m/m^{\ominus})^{1/2}$ and extrapolated to $m=0$ to obtain $E^{\ominus}$. The least-squares fit of these data to (9.82) leads to $\pi^{\ominus}_{(\mathrm{AgCl/Ag,Cl})}=0.223$ V. The use of better activity data than those provided by the Debye–Hückel limiting equation (D–H) leads to the accepted value of 0.2224 V.

9.3.3.2 Measurement of activity coefficients

Once the value of $E^{\ominus}$ has been determined for a cell, the analysis given in the previous section permits calculation of activity coefficients. By rearranging (9.80), we obtain

$$\ln(\gamma_{\pm})=\mathcal{F}(E^{\ominus}-E)/(2\mathcal{R}T)-\ln(m/m^{\ominus}) \quad (9.83)$$

and from measurements of cell EMFs at different molalities m, the activity coefficient $\gamma_{\pm}$ may be calculated at each molality (Table 9.12).

9.3.3.3 Variation of EMF with temperature

From (4.97) it follows that

$$(\partial\Delta G/\partial T)_p=-\Delta S \quad (9.84)$$

If we substitute in (9.69) and impose standard conditions,

$$(\partial E^{\ominus}/\partial T)_p=\Delta S^{\ominus}/\mathcal{F} \quad (9.85)$$

and it follows that the standard entropy change for a cell reaction may be obtained by measuring $E^{\ominus}$ at two or more temperatures. Thus, $E^{\ominus}$ values for the Daniell cell were determined to be 1.102 V and 1.095 V at 25 °C and 90 °C, respectively. Hence, the temperature coefficient of EMF over this range of temperature is -1.077×10^{-4} V K^{-1}, and $\Delta S^{\ominus}$ is -10.4 J K^{-1} mol^{-1}.

The values of $\Delta H^{\ominus}$ (-109.3 kJ mol^{-1} and $\Delta G^{\ominus}$ (-106.3 kJ mol^{-1}) for the Daniell cell reaction are so similar that it was thought originally[8] that the electrical work could be equated to the enthalpy change. As we have just shown in the above calculation, both the temperature coefficient of EMF for this cell and the entropy change are small.

In the cell

$$\mathrm{Pb}|\mathrm{Pb}^{2+}(a_{\pm}=1)\|\mathrm{Cu}^{2+}(a_{\pm}=1)|\mathrm{Cu} \tag{9.86}$$

$E^{\ominus}=0.465$ V, $(\partial E^{\ominus}/\partial T)_p=8.129\times10^{-4}$ and $\Delta S^{\ominus}=78.4$ J K^{-1} mol^{-1}. The fact that the temperature coefficient can be positive or negative shows that a spontaneous reaction is not always accompanied by an increase in the entropy *of the system*; it is always accompanied by an increase in the entropy of the system (cell) plus the surroundings and by a decrease in the Gibbs free energy of the system. Some of the factors that determine the signs of the entropies of hydrated ions have been discussed in Section 8.6.3.

9.3.3.4 Measurement of partial molar entropies of ions

The partial molar entropy of a hydrated ion may be obtained from the temperature coefficient of EMF for an appropriate cell. Consider zinc chloride in the following cell:

$$\mathrm{Zn}|\mathrm{Zn}^{2+}(m)\|\mathrm{H_3O}^{+}(a_{\pm}=1)|\mathrm{H_2,Pt} \tag{9.87}$$

for which the spontaneous cell reaction is

$$\tfrac{1}{2}\mathrm{Zn(s)}+\mathrm{H_3O^+(aq)} \rightleftharpoons \tfrac{1}{2}\mathrm{Zn^{+}(aq)}+\tfrac{1}{2}\mathrm{H_2(g)} \tag{9.88}$$

The EMF is measured for solutions of zinc chloride of different molalities m and extrapolated to find $E^{\ominus}$ as discussed in Section 9.3.3.1. The measurements are repeated at three or four temperatures between, say, 288 K and 308 K, from which $(\partial E^{\ominus}/\partial T)_p$ and $\Delta S^{\ominus}$ may be obtained.

For (9.88) the standard entropy change is given by

$$\Delta S^{\ominus}=[\tfrac{1}{2}\overline{S}^{\ominus}(\mathrm{Zn^{2+},aq})+\tfrac{1}{2}S^{\ominus}(\mathrm{H_2,g})]$$
$$[\tfrac{1}{2}S^{\ominus}(\mathrm{Zn,s})+\overline{S}^{\ominus}(\mathrm{H_3O^+,aq})] \tag{9.89}$$

By convention $\overline{S}^{\ominus}(\mathrm{H_3O^+,aq})$ is defined as zero; hence,

$$\overline{S}^{\ominus}(\mathrm{Zn^{2+},aq})=2\Delta S^{\ominus}+S^{\ominus}(\mathrm{Zn,s})-S^{\ominus}(\mathrm{H_2,g}) \tag{9.90}$$

In a typical experiment, $(\partial E^{\ominus}/\partial T)_p$ at 298 K was found to be -1.00×10^{-4} V, $\Delta S^{\ominus}=-9.65$ J K^{-1} mol^{-1}, $\overline{S}^{\ominus}$ (Zn,s)$=41.6$ J K^{-1} mol^{-1} and $S^{\ominus}(\mathrm{H_2,g})=130.7$ J K^{-1} mol^{-1}. Hence, $\overline{S}^{\ominus}=-108.4$ J K^{-1} mol^{-1}; partial molar entropy values are relative to the zero value for the hydroxonium ion.

9.3.4 Concentration cells

In the cells that we have considered so far, we have discussed the tendency for a given spontaneous reaction to occur, and the properties that may be derived from such a cell operating under

[8] At that time the calorie was used, so they looked even closer (-26.1 and -25.4 kcal mol^{-1}).

reversible conditions. If an electrolyte is transferred from a solution of mean activity a_1 to a solution of final activity a_2, then the free energy change will be the difference of the chemical potentials of the two solutions. Following (7.54), we write for one mole transferred

$$\Delta G = \mathcal{R}T \ln (a_2/a_1) \quad (9.91)$$

If a_2 is less than a_1 ΔG is negative, corresponding to the spontaneous change on allowing the solution to mix. In a galvanic cell this free energy change can be linked with the EMF of the cell, through (9.69). Consider the cell

$$\text{Pt,H}_2\text{(g)|HCl}(a_1)\text{|HCl}(a_2)\text{|H}_2\text{,Pt} \quad (9.92)$$

Anode (LHS) $\quad \frac{1}{2}H_2(g) + H_2O(l) \rightleftharpoons H_3O^+(aq, a_2) + e^- \quad (9.93)$

Cathode (RHS) $\quad H_3O^+(aq, a_1) + e^- \rightleftharpoons H_2O(l) + \frac{1}{2}H_2(g) \quad (9.94)$

Thus, 1 mol H_3O^+ ions passes out of solution into the gas phase at the cathode, but this loss is partially compensated by the transport of t_+ mol of H_3O^+ ions towards the cathode, across the common junctions of the solutions. At the anode, 1 mol H_3O^+ ions is formed, but t_+ mol of them is transported away, across the liquid junction. At the same time, t_- mol Cl^- ions is transported from the cathode to the anode. Note that hydroxonium ions travel from anode to cathode and chloride ions in the reverse direction, *inside the cell.*

The net result is the transfer of $1-t_+$, or t_-, mol of both H_3O^+ and Cl^- ions, that is, t_- mol hydrochloric acid, from the solution of activity a_1 to that of final activity a_2. Hence, ΔG in (9.91) is modified to

$$\Delta G = t_- \mathcal{R}T \ln (a_2/a_1) \quad (9.95)$$

setting up an EMF

$$E = -t_-(\mathcal{R}T/\mathcal{F}) \ln (a_2/a_1) \quad (9.96)$$

In practice, a_1 and a_2 are mean values $a_{\pm,1}$ and $a_{\pm,2}$, so that we obtain finally

$$E = -2t_-(\mathcal{R}T/\mathcal{F}) \ln (a_{\pm,2}/a_{\pm,1}) \quad (9.97)$$

Note that the transport number that appears in the equation refers to that ion for which the half-cells are *not* reversible. This type of cell is a *concentration cell with transference* (or transport); its source of EMF lies in the differing activities $a_1 > a_2$ in the two half-cells. It follows that we can use such cells to find transport numbers or activities.

Suppose that we know $a_{\pm,1}$ and need to find $a_{\pm,2}$. We could conveniently use the cell

$$\text{Ag,AgCl|HCl}(a_1)\text{|HCl}(a_2)\text{|AgCl,Ag} \quad (9.98)$$

It is left as an exercise to the reader to show that

$$E = -2t_+(\mathcal{R}T/\mathcal{F}) \ln (a_{\pm,2}/a_{\pm,1}) \quad (9.99)$$

Precise transport data are not always available and, instead of (9.98), we could set up the cell

$$\text{Ag,AgCl|HCl}(a_1)\text{|H}_2\text{(g),Pt}-\text{Pt,H}_2\text{(g)|HCl}(a_2)\text{|H}_2\text{(g),Pt} \quad (9.100)$$

with $a_1 > a_2$ again. There is now no common liquid junction, so that no transport of ions can take place between the half-cells. The reaction of the left-hand electrode is

$$Cl^-(aq,a_1) + Ag(s) + H_3O^+(aq,a_1) \rightleftharpoons AgCl(s) + \tfrac{1}{2}H_2(g) \quad (9.101)$$

The right-hand electrode operates in the opposite sense, but at an activity a_2. Hence, the passage of 96485 C would result in an effective transfer of 1 mol of hydrochloric acid from mean activity $a_{\pm,1}$ to mean activity $a_{\pm,2}$, so that

$$E=-2(\mathcal{R}T/\mathcal{F})\ln(a_{\pm,2}/a_{\pm,1}) \tag{9.102}$$

and the transport numbers are not invoked; this cell is an example of a *concentration cell without transference.* When it is not possible to find electrodes that are reversible to both ions in solution, the cells may be joined by a salt bridge (Section 9.3.2.2). Ideally, $t_+=t_-=\frac{1}{2}$, so that (9.97) and (9.99) merge as

$$E=2(\mathcal{R}T/\mathcal{F})\ln(a_{\pm,\mathrm{conc}}/a_{\pm,\mathrm{dil}}) \tag{9.103}$$

9.3.5 Redox equilibria

Another type of reversible electrode consists of an inert material, often platinum or carbon, in a solution containing metal ions in two oxidation states; two examples and their standard potentials are

$$Fe^{3+}(aq)+e^- \rightleftharpoons Fe^{2+}(aq) \qquad \pi^{\ominus}(Fe^{3+}/Fe^{2+})=0.771\ \mathrm{V} \tag{9.104}$$

$$\tfrac{1}{5}MnO_4^-(aq)+\tfrac{8}{5}H_3O^+(aq)+e^- \rightleftharpoons \tfrac{1}{5}Mn^{2+}(aq)+\tfrac{12}{5}H_2O(l)\ \pi^{\ominus}(MnO_4^-Mn^{2+})=1.51\ \mathrm{V} \tag{9.105}$$

The combination of these two systems in the cell

$$Pt|Fe^{3+}(aq),Fe^{2+}(aq)||MnO_4^-(aq),Mn^{2+}(aq)|Pt \tag{9.106}$$

provides information about the reaction

$$\tfrac{1}{5}MnO_4^-(aq)+\tfrac{8}{5}H_3O^+(aq)+Fe^{2+}(aq) \rightleftharpoons \tfrac{1}{5}Mn^{2+}(aq)+Fe^{3+}(aq) + \tfrac{12}{5}H_2O(l) \tag{9.107}$$

The standard EMF is 1.51 V$-$0.771 V$=$0.739 V; hence $\Delta G^{\ominus}=-71.3$ kJ mol^{-1} and $K=3.1\times10^{12}$, so that the reaction is complete for all practical purposes.

A knowledge of half-cell reactions is useful in constructing equations for more complex redox processes. Consider the oxidation of Mn(II) to Mn(VII) by Bi(V) in acid solution. The Bi(V) half-cell reduction reaction is

$$\tfrac{1}{2}BiO_3^-(aq)+3H_3O^+(aq)+e^- \rightleftharpoons \tfrac{1}{2}Bi^{3+}(aq)+\tfrac{9}{2}H_2O(l) \tag{9.108}$$

and $\pi^{\ominus}(BiO_3^-/Bi^{3+})=1.71$ V. It is evident that the spontaneous reaction involves (9.105) now in the opposite sense; hence, the cell and complete reaction are

$$Pt|Mn^{2+}(aq),MnO_4^-(aq)||BiO_3^-(aq),Bi^{3+}(aq)|Pt \tag{9.109}$$

$$\tfrac{1}{2}BiO_3^-(aq)+\tfrac{1}{5}Mn^{2+}(aq)+\tfrac{7}{5}H_3O^+(aq) \rightleftharpoons \tfrac{1}{2}Bi^{3+}(aq) +\tfrac{1}{5}MnO_4^-(aq)+\tfrac{21}{10}H_2O(l) \tag{9.110}$$

In this example, $E^{\ominus}=0.20$ V, $\Delta G^{\ominus}=-19.3$ kJ mol^{-1} and $K=2.40\times10^3$.

9.3.6 Measurement of pH

The inability to measure the activity of a single ion leads to the desirability of an operational definition of pH, as well as that provided by (8.78). Consider a cell consisting of the SHE and a reference electrode such as the calomel half-cell:

$$Pt,H_2(g)|H_3O^+(aq)||\mathrm{calomel} \tag{9.111}$$

The solution containing hydroxonium ions is first a buffer of known pH, say pH_s. Then, we have the cell

$$E_S = \pi_{calomel} - [\pi^{-0-}_{(H_3O^+,H_2)} + (\mathcal{R}T/\mathcal{F})\ln(a_{H_3O^+})_S] \tag{9.112}$$

since $p_{H_2}=1$ atm with the SHE. For the cell with a solution of unknown pH, say pH_X, we have

$$E_X = \pi_{calomel} - [\pi^{-0-}_{(H_3O^+,{}_xH_2)} + (\mathcal{R}T/\mathcal{F})\ln(a_{H_3O^+})_X] \tag{9.113}$$

Thus, at 298.15 K

$$E_S - E_X = (0.05916 \text{ V})(pH_S - pH_X) \tag{9.114}$$

or

$$pH_X = pH_S + (E_S - E_X)/(0.05916 \text{ V}) \tag{9.115}$$

The measurement of pH is, thus, a measurement of EMF. A pH reference standard is a 0.05 mol dm^{-3} solution of potassium hydrogen phthalate, the pH of which is defined by

$$pH = 4.000 + 0.05[(T-288)/100]^2 \tag{9.116}$$

Other standards are saturated potassium hydrogen tartrate[9] (pH=3.557 at 298.15 K) and 0.01 mol kg^{-1} disodium tetraborate (pH=9.180 at 298.15 K).

9.3.6.1 Glass electrode

The pH of a solution H*X* may be measured directly by the cell

$$\text{Ag,AgCl}|\text{HCl(aq)}\blacksquare\text{H}X\text{(aq)}|\text{KCl(satd)}|\text{Hg}_2\text{Cl}_2\text{,Hg} \tag{9.117}$$

The left-hand half-cell represents the *glass electrode*, Figure 9.18, and the symbol ■ indicates the glass membrane separating the hydrochloric acid solution inside the bulb of the electrode from the solution H*X* of unknown pH. The glass electrode is a concentration half-cell; the bulb is permeable to H_3O^+ ions, so that the potential difference set up depends on the difference in the hydroxonium ion activities in the bulb and the solution of H*X*. Alternatively, the glass electrode may contain a phosphate buffer solution and a solution of hydrochloric acid: conveniently, a cell with this electrode on the left-hand side of (9.117) has zero EMF when the solution H*X* has pH 7.00 at 25 °C.

A circuit containing a glass electrode has an impedance of several mega-ohms, so that the conventional potentiometer is not sufficiently sensitive for pH measurement. The glass electrode is used in conjunction with an electrometer of high input impedance. A pH meter is such an instrument, and its basis is the cell (9.117). It is calibrated against standard buffer solutions, and incorporates temperature-compensation circuits.

The pH of a solution is measured by first calibrating the pH meter, at the known temperature of operation, with a standard buffer solution. Then the buffer solution is replaced by the solution under investigation, and the pH read directly from the meter.

9.3.6.2 Specific-ion electrodes

The glass electrode is one of a series of electrodes that are specific to a particular ion, in which a membrane separates the solution under test from a medium with a known

[9] 0.0341 molal solution.

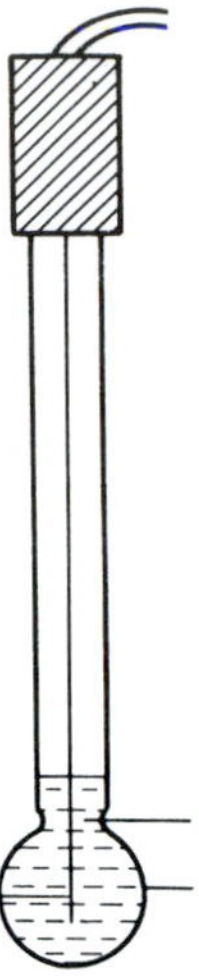

Figure 9.18 Glass electrode: the bulb encloses either hydrochloric acid or a phosphate buffer–hydrochloric acid solution, and is reversible to hydroxonium ions; the high impedance of the glass electrode requires the concomitant use of an electrometer to monitor EMF in pH measurement, as in a pH meter.

concentration of that ion. These *specific-ion*, or ion-selective, electrodes measure the mean ionic activity of the given species at the total ionic strength of the solution. Data for some of these electrodes are given in Table 9.13.

The extent of interference of other ions is important: Table 9.13 shows the concentration range for the species Y being measured over which the cell EMF is a linear function of pY; $[X]$ is the concentration of an ion X that could lead to an error of $\simeq$10% in $[Y]$ at $a_y = 10^{-3}$. Further details on specific-ion electrodes are available in dedicated texts.[10]

9.3.7 Potentiometric titrations

We will consider briefy acid–base, precipitation and redox systems, and how they may be monitored by EMF measurements.

9.3.7.1 Acid–base reactions

If we set up a cell of the type

$$\text{glass electrode}|\mathrm{H}X(\mathrm{aq})|\text{calomel electrode} \tag{9.118}$$

the acid solution can be titrated with a base and the course of the reaction followed by monitoring the EMF of the cell:

$$E = \pi_{\text{calomel}} - [\pi^{-0-}_{\text{glass}} + \mathcal{R}T\ln(a_{\mathrm{H_3O^+}})] \tag{9.119}$$

Thus, at a given temperature, $E \propto$pH, and Figure 9.19 is the curve of E against volume V of 0.5 M sodium hydroxide in the titration of 25 cm^3 of 0.1 M hydrochloric acid; the equiva-

[10] See, for example, J Koryta, *Ion-selective Electrodes* (CUP, 1975).

Table 9.13. Data for some specific-ion electrodes

Species Y	Range of $[Y]$/mol dm^{-3}	Species X	$[X]$/mol dm^{-3}
Na^+	3 to 1×10^{-6}	K^+	1×10^{-1}
Ca^{2+}	1 to 1×10^{-5}	Li^+	5×10^{-5}
Pb^{2+}	1 to 1×10^{-7}	Mg^{2+}, Sr^{2+}	8×10^{-3}
F^-	Saturated to 1×10^{-6}	OH^-	1×10^{-1}
Br^-	1 to 1×10^{-6}	I^-	2×10^{-4}
NO_3^-	1 to 5×10^{-6}	ClO_4^-, CN^-, NO_2^-	3×10^{-3}

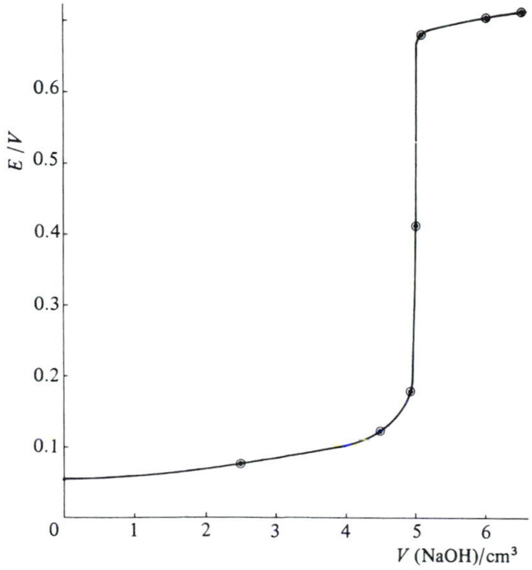

Figure 9.19 Variation of cell EMF with volume V of sodium hydroxide in the pH (acid–base) titration of hydrochloric acid. Weak acids and bases can be estimated by this technique, and differential titration (Section 8.7.7.3) is readily carried out in this way.

lence point is very clearly indicated. Similar results could be obtained in terms of pH directly, and more conveniently by means of a pH meter.

Other pH electrodes include the quinhydrone system, which is useful in nonaqueous solutions. In practice, a few milligrams of sparingly soluble quinhydrone, a 1 : 1 molecular complex of quinone Q and hydroquinone HQ, are added to the unknown solution, and the equilibrium established at a platinum electrode, with reference to the calomel half-cell. The standard reduction potential for the quinhydrone electrode is 0.699 V. We can write

$$\pi = \pi^{-0-}_{(Q.HQ)} + (\mathcal{R}T/\mathcal{F}) \ln (a_Q a^2_{H_3O^+}/a_{HQ}) \qquad (9.120)$$

The ratio a_Q/a_{HQ} is unity, because Q and HQ are solids, and we obtain

$$\pi = 0.699 \text{ V} - (0.05916 \text{ V})\text{pH} \qquad (9.121)$$

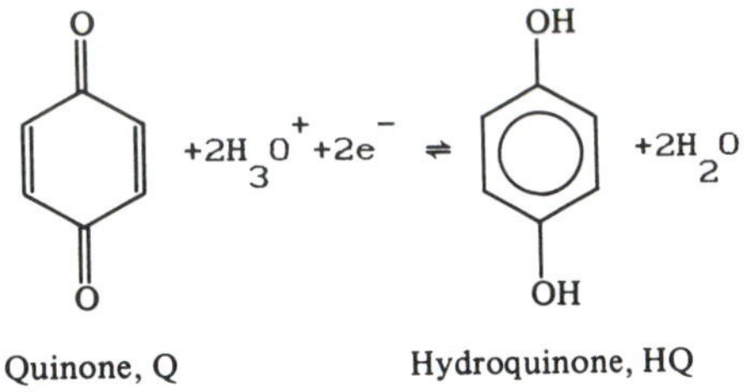

Quinone–hydroquinone equilibrium

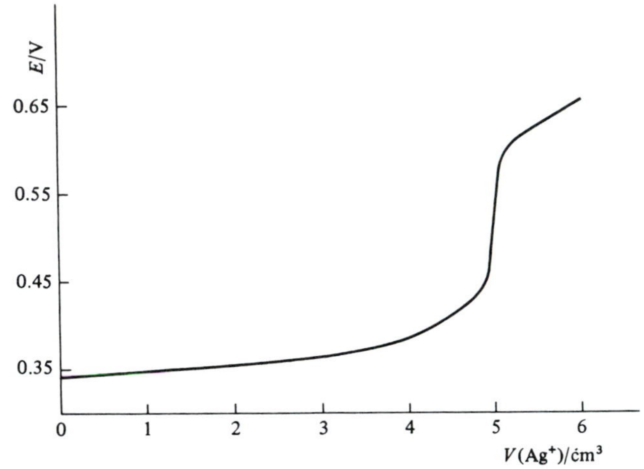

Figure 9.20 Variation in cell EMF with volume V of silver nitrate in the precipitation titration of chloride ions; a clearly defined equivalence point is shown.

9.3.7.2 Precipitation reactions

The concentration of a halide ion X^- may be determined by titration with silver nitrate solution. The process may be carried out potentiometrically in the cell

$$\text{Ag}X\text{,Ag}||M^+\text{(aq)}, X^-\text{(aq)}|\text{calomel} \tag{9.122}$$

The salt bridge used here would contain ammonium nitrate, and the EMF of the cell is given by

$$E = \pi_{\text{calomel}} - [\pi^{-0-}_{(\text{Ag,AgX})} - (\mathcal{R}T/\mathcal{F})\ln(a_{\text{X}^-})] \tag{9.123}$$

or

$$E = \alpha - (0.05916\text{ V})\text{p}X \tag{9.124}$$

where α is a constant. A typical titration curve is shown in Figure 9.20, for 0.01 M potassium chloride solution with 0.05 M silver nitrate. Before the equivalence point, a_{Cl^-} is the activity of the unreacted chloride ion in the total volume of solution. At the equivalence point, the solution is saturated with silver chloride, and a_{Cl^-} is $\sqrt{K_s}$, where K_s is the solubility product of silver chloride. Beyond the equivalence point $a_{\text{Cl}^-} = K_s/a_{\text{Ag}^+}$, where a_{Ag^+} is the activity of silver ions at the total ionic strength of the solution.

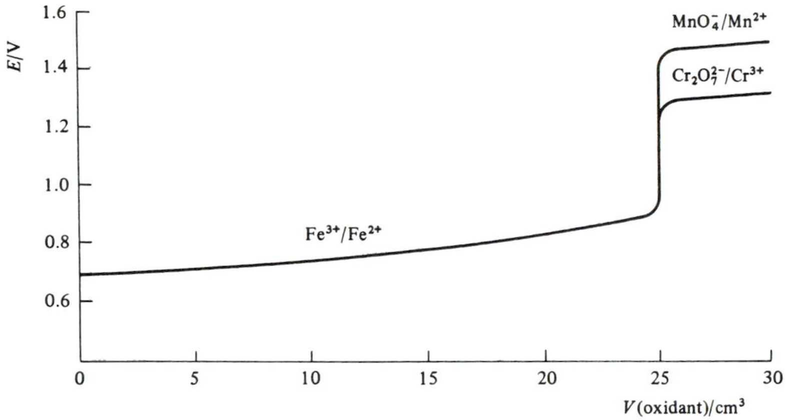

Figure 9.21 Variation in cell EMF with volume *V* of oxidant in the titration of Fe(II) with Mn(VII) or Cr(VI). The equivalence point is very clearly shown in both titrations; the height of the break is less with dichromate than with permanganate because it has the smaller reduction potential, 1.33 V compared with 1.51 V.

9.3.7.3 Redox reactions

The reaction (9.107) may be followed in terms of the half-cell

$$\text{calomel}|Fe^{3+}(aq),\ Fe^{2+}(aq), H_3O^+(aq)|Pt \tag{9.125}$$

where

$$E=[\pi^{\ominus}_{(Fe^{3+},Fe^{2+})}+(\mathcal{R}T/\mathcal{F})\ln(a_{Fe^{3+}}/a_{Fe^{2+}})]-\pi_{\text{calomel}} \tag{9.126}$$

As permanganate ions are added to the solution the ratio $a_{Fe^{3+}}/a_{Fe^{2+}}$, which is initially very small, increases to its maximum value at the equivalence point, which is indicated clearly (Figure 9.21). Beyond this point, the EMF is determined by the permanganate equilibrium and follows the equation

$$E=\pi^{\ominus}_{(MnO_4^-,Mn^{2+})}+(\mathcal{R}T/\mathcal{F})\ln\left(\frac{(a_{MnO_4^-})^{1/5}(a_{H_3O^+})^{8/5}}{(a_{Mn^{2+}})^{2/5}(a_{H_2O})^{12/5}}\right) \tag{9.127}$$

9.3.7.4 Electrochemical series

Chemical reactions may be considered as a combination of oxidation (O) and reduction (R) tions, and may be written generally as a cell

$$R_1, O_1 \| O_2, R_2 \tag{9.128}$$

the example (9.73), zinc is R_1, zinc ions O_1, copper ions O_2 and copper R_2. Thus, in the tation for the cell, the reduction half-cell reaction (9.50) is reversed at the right-hand elecrode.

If we arrange the reduction potentials in order of increasing positivity we obtain the *electrochemical series of the elements*. Any element will, thermodynamically, reduce ions of an element below it in the series, but the conditions must be correct. For example, lithium will not displace calcium from aqueous calcium chloride, because it will preferentially discharge hydroxonium ions from the water as hydrogen gas. It will, however, discharge metal-

Table 9.14. Portion of the electrochemical series of the elements

Element	Reduction reaction	Reduction potential π^{-0-}/V
Li	$Li^+(aq)+e^- \rightleftharpoons Li(s)$	−3.05
Ca	$\frac{1}{2}Ca^{2+}(aq)+e^- \rightleftharpoons \frac{1}{2}Ca(s)$	−2.87
Na	$Na^+(aq)+e^- \rightleftharpoons Na(s)$	−2.71
Mg	$\frac{1}{2}Mg^{2+}(aq)+e^- \rightleftharpoons \frac{1}{2}Mg(s)$	−2.36
Al	$\frac{1}{3}Al^{3+}(aq)+e^- \rightleftharpoons \frac{1}{3}Al(s)$	−1.71
Zn	$\frac{1}{2}Zn^{2+}(aq)+e^- \rightleftharpoons \frac{1}{2}Zn(s)$	−0.763
Fe	$\frac{1}{2}Fe^{2+}(aq)+e^- \rightleftharpoons \frac{1}{2}Fe(s)$	−0.440
Sn	$\frac{1}{2}Sn^{2+}(aq)+e^- \rightleftharpoons \frac{1}{2}Sn(s)$	−0.136
Pb	$\frac{1}{2}Pb^{2+}(aq)+e^- \rightleftharpoons \frac{1}{2}Pb(s)$	−0.126
H	$H_3O^+(aq)+e^- \rightleftharpoons \frac{1}{2}H_2(g)+H_2O(l)$	Zero
Cu	$\frac{1}{2}Cu^{2+}(aq)+e^- \rightleftharpoons \frac{1}{2}Cu^+(aq)$	+0.158
Cu	$\frac{1}{2}Cu^{2+}(aq)+e^- \rightleftharpoons \frac{1}{2}Cu(s)$	+0.339
O	$\frac{1}{4}O_2(g)+\frac{1}{2}H_2O(l)+e^- \rightleftharpoons OH^-(aq)$	+0.401
Fe	$Fe^{3+}(aq)+e^- \rightleftharpoons Fe^{2+}(aq)$	+0.771
Ag	$Ag^+(aq)+e^- \rightleftharpoons Ag(s)$	+0.799
Pt	$\frac{1}{2}Pt^{2+}(aq)+e^- \rightleftharpoons -\frac{1}{2}Pt(s)$	+1.20
O	$\frac{1}{4}O_2(g)+H_3O^+(aq)+e^- \rightleftharpoons \frac{3}{2}H_2O(l)$	+1.23
Cl	$\frac{1}{2}Cl_2(g)+e^- \rightleftharpoons \frac{1}{2}Cl^-(aq)$	+1.36
Au	$Au^+(aq)+e^- \rightleftharpoons Au(s)$	+1.69
F	$\frac{1}{2}F_2(g)+e^- \rightleftharpoons \frac{1}{2}F^-(aq)$	+2.87

lic calcium from molten calcium chloride (see also Section 9.4.1). A selection of the electrochemical series is listed in Table 9.14.

9.4 Nonequilibrium electrochemistry

We shall conclude this chapter with an outline of some aspects of nonequilibrium electrochemistry.

9.4.1 Overpotential

In a practical application of a galvanic cell, in which a current flow occurs, the relative positions of two electrode systems in the electrochemical series might not describe correctly their behaviour in combination. For example, hydrogen gas does not reduce copper ions in aqueous solution

$$\tfrac{1}{2}Cu^{2+}(aq)+\tfrac{1}{2}H_2(g)+H_2O(l) \rightleftharpoons Cu(s)+H_3O^+(aq) \qquad (9.129)$$

although the standard Gibbs free energy change for (9.129) is, from Table 9.14, −32.7 kJ mol^{-1}. This situation arises because there is a large *overpotential* for the oxidation of hydrogen at a copper surface.

Table 9.15. Hydrogen overpotentials in dilute sulfuric acid

Electrode	η/V	Electrode	η/V
Pt (platinized)	0.005	Sn	0.53
Pt (smooth)	0.09	Pb	0.64
Ag	0.15	Zn	0.70
Cu	0.23	Hg	0.78

The voltage of a galvanic cell, that is, the difference in potential between its electrodes when a current is flowing, is less than its EMF. The difference between the EMF, corresponding to zero current, and the voltage at which the current is finite and measurable is the overpotential of the cell. Theoretically, a voltage of approximately 1.2 V should cause the decomposition of an aqueous solution of an acid or a base. In practice, the voltage required is appreciably larger, at about 1.7 V. Table 9.15 lists some hydrogen overpotentials at different metal surfaces.

The high overpotential of hydrogen at mercury is of importance in polarographic analysis, by which process cations such as zinc and sodium are reduced at a mercury cathode in preference to hydroxonium ions.

9.4.2 Corrosion

Corrosion is the partial or total transformation of a metal or an alloy to a combined state by its interaction with the environment. It is costly to businesses and the protection of materials from corrosion is necessary. Some materials, such as aluminium, are protected from degradation by an oxide film.

9.4.2.1 Thermodynamics of corrosion

The electrochemical series in Table 9.14 provides a useful guide to the energetics of corrosion. If for two half-cells A and B $\pi_{(A,A^+)} < \pi_{(B,B^+)}$, then we expect that the forwards reaction of the equilibrium

$$A(s) + B^+(aq) \rightleftharpoons A^+(aq) + B(s) \tag{9.130}$$

will predominate. In the particular and important example that A is iron and B^+ is the hydroxonium ion, two processes may be formulated:

$$\tfrac{1}{2}Fe(s) + H_3O^+(aq) \rightleftharpoons \tfrac{1}{2}Fe^{2+}(aq) + \tfrac{1}{2}H_2(g) + H_2O(l) \tag{9.131}$$

which has a standard EMF of +0.440 V, and

$$\tfrac{1}{2}Fe(s) + H_3O^+(aq) + \tfrac{1}{4}O_2(g) \rightleftharpoons \tfrac{1}{2}Fe^{2+}(aq) + \tfrac{3}{2}H_2O(l) \tag{9.132}$$

with a standard EMF of +1.69 V. Under alkaline conditions, we have

$$\tfrac{1}{2}Fe(s) + \tfrac{1}{2}H_2O(l) + \tfrac{1}{4}O_2(g) \rightleftharpoons \tfrac{1}{2}Fe^{2+}(aq) + OH^-(aq) \tag{9.133}$$

where $E^{\ominus} = +0.841$ V. In each case, iron is a left-hand half-cell, at which oxidation takes place, and the free energy changes are negative, so that all three processes are thermodynamically spontaneous.

Gold is resistant to these degradative reactions because its reduction potential is very high, 1.69 V. Its dissolution in the presence of oxidizing agents and chloride ions arises from the formation of the $AuCl_4^-$ complex ion. Platinum is also highly resistant to corrosion at almost all pH levels. There is, however, a small zone of corrosion at high acidity ($pH<-1$) and galvanic potential of approximately 1 V.

9.4.2.2 Corrosion inhibition

The reduction of corrosion by lowering the potential in the system is the basis of cathodic protection of metals. The galvanizing of iron-based materials consists in coating the metal with zinc (−0.763 V). This metal lies above iron (−0.440 V) in the electrochemical series; hence, zinc corrodes and the iron is preserved. The zinc is itself protected by an oxide film, so that its corrosion is minimized. Iron plated by tin is only weakly protected, because the tin potential is −0.136 V and tin is, therefore, reduced by iron:

$$\tfrac{1}{2}Sn^{2+}(aq)+\tfrac{1}{2}Fe(s) \rightleftharpoons \tfrac{1}{2}Sn(s)+\tfrac{1}{2}Fe^{2+}(aq) \tag{9.134}$$

$\Delta G^{-0-}=-29.3$ kJ mol^{-1}, and $K=1.4\times10^5$ at 25 °C.

Another method of inhibiting corrosion is by the sacrificial action of another metal. If an iron object is connected to a metal of more negative potential such as magnesium (−2.36 V), the magnesium is able to supply electrons to the iron while itself being oxidized to Mg^{2+} ions; this process has been of importance in reducing the corrosion of the hulls of ships.

9.4.3 Fuel cells

As Figure 9.22 shows, a fuel cell has an inherent simplicity; the problems that exist are linked to its power output and cost of operation. The Bacon cell in the diagram operates at about 200 °C and 30 atm, with a flowing electrolyte of approximately 35% potassium hydroxide. The overall cell reaction is the catalytic reaction of hydrogen and oxygen gases to form liquid water. Under the normal operating conditions, the open-circuit EMF is less than the theoretical value, and the voltage obtainable depends on the current density, Table 9.16.

The lead–acid accumulator used in the motor car is the best known fuel cell. It consists of two lead grids, one filled with a paste of lead dioxide PbO_2 and the other with finely divided lead, both immersed in 30% sulfuric acid (4.4 M). When the cell is supplying current, the following reactions occur:

$$\text{at the anode } \tfrac{1}{2}PbO_2(s)+2H_3O^+(aq)+\tfrac{1}{2}SO_4^{2-}(aq)+e^- \rightleftharpoons \tfrac{1}{2}PbSO_4(s)+H_2O(l) \tag{9.135}$$

$$\text{at the cathode } \tfrac{1}{2}Pb(s)+\tfrac{1}{2}SO_4^{2-}(aq) \rightleftharpoons \tfrac{1}{2}PbSO_4(s)+e^- \tag{9.136}$$

The standard reduction potentials for these half-cells are −0.36 V and 1.46 V, respectively, so the supply of an electric current is a spontaneous process. The standard EMF is 1.82 V, but is marginally greater in the presence of the 4.4 M sulfuric acid electrolyte. When the accumulator is charged, a voltage is applied in the reverse sense, whereupon the reactions (9.135) and (9.136) are reversed.

Many systems have been investigated that rely on hydrogen–oxygen or methanol–oxygen reactions, but their kinetics are relatively sluggish. A cell that attempts to overcome this deficiency is based on the reaction of hydrogen and oxygen in the presence of molten potassium

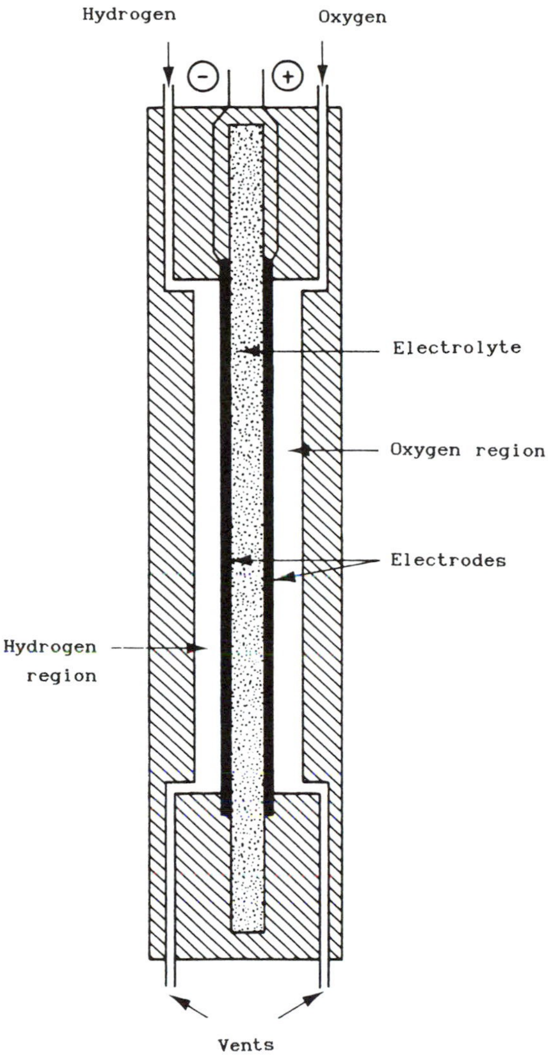

Figure 9.22 Bacon-type (hydrogen–oxygen) fuel cell. The electrodes are of titanium coated with platinum and the electrolyte is a cation-exchange resin; the external load is connected at the ± terminals. The electrode reactions are

$$\tfrac{1}{2}H_2(g)+H_2O(l)\rightarrow H_3O^+(aq)+e^-$$

$$\tfrac{1}{4}O_2(g)+H_3O^+(aq)+e^-\rightarrow\tfrac{3}{2}H_2O(l)$$

The standard EMF is 1.23 V (Table 9.14), but the voltage depends upon the load (Table 9.15).

and lithium carbonates at approximately 650 °C, between silver electrodes. Because of the good kinetics, high-purity hydrogen and catalysts are both unnecessary. The electrode reactions are as follow (M^+(l) implies a solvated state of the cation):

$$\tfrac{1}{2}H_2(g)+\tfrac{1}{2}M_2CO_3(l)\rightleftharpoons\tfrac{1}{2}CO_2(g)+\tfrac{1}{2}H_2O(l)+M^+(l)+e^- \qquad (9.137)$$

Table 9.16. Voltage–current density relationships in a Bacon cell

σ/mA cm^{-2}	0	10	50	100	300
V/V	1.04	1.01	0.95	0.89	0.78

$$\tfrac{1}{2}CO_2(g)+\tfrac{1}{4}O_2(g)+M^+(l)+e^- \rightleftharpoons \tfrac{1}{2}M_2CO_3(l) \tag{9.138}$$

which, overall, is the hydrogen–oxygen reaction:

$$\tfrac{1}{2}H_2(g)+\tfrac{1}{4}O_2(g) \rightleftharpoons \tfrac{1}{2}H_2O(l) \tag{9.139}$$

Electricity production relies on the conversion of thermal energy to electrical energy. A version of this concept uses two electrochemical cells at different temperatures, the forwards reaction at one temperature and the reverse reaction at another temperature, constituting a thermally regenerative electrochemical system (TRES). A reaction with a large temperature variation of EMF (and ΔG) implies a large entropy change for the reaction (Section 9.3.3.3), and one system, taking place at approximately 1200 K, is

$$\tfrac{1}{6}CH_4(g)+\tfrac{1}{6}H_2O(g) \rightleftharpoons \tfrac{1}{2}H_2(g)+\tfrac{1}{6}CO(g) \tag{9.140}$$

with the reverse reaction taking place at approximately 400 K. The Gibbs free energy changes are -12.7 kJ mol^{-1} forwards, and -21.0 kJ mol^{-1} reverse. The cell efficiency is given by the Carnot equation (Section 4.6.1.1):

$$\eta=(1200\ \text{K}-400\ \text{K})/1200\ \text{K}=0.67 \tag{9.141}$$

The free energies given above show that the methane cell has a potential of only 0.35 V, and the search for more powerful cells continues.

Problems 9

9.1 A water coulometer and a silver coulometer are connected in series. Calculate the mass of silver that is discharged during the time that 80 cm^3 of hydrogen gas is liberated at 293 K and 755 mmHg from a hydrogen coulometer in series; the density of hydrogen at 273.15 K and 760 mmHg is 0.0890 g dm^{-3}.
9.2 Write equations for the electrode reactions that take place in the electrolyses of the following aqueous solutions:

(a) copper sulfate with platinum electrodes;
(b) copper sulfate with copper electrodes;
(c) dilute sodium chloride with carbon electrodes;
(d) concentrated sodium chloride with carbon electrodes.

9.3 At 298.15 K the resistance of a conductivity cell containing 0.100 mol dm^{-3} potassium chloride is 307.62 Ω. The same cell containing 0.100 mol dm^3 silver nitrate has a resistance of 362.65 Ω. Refer to the text for the conductivity of the potassium chloride solution, and calculate

(a) the cell constant;
(b) the molar conductivity of the silver nitrate solution.

9.4 The dissociation constant of propanoic acid is 1.34×10^{-5} at 298.15 K, and the limiting conductivities of its ions are $\lambda_{0,\pm}=0.0350\ \Omega^{-1}\ m^2\ mol^{-1}$, and $\lambda_{0,-}=0.00358\ \Omega^{-1}\ m^2\ mol^{-1}$. Calculate the conductivity of a 0.100 mol dm^{-3} solution of propanoic acid.

9.5 From the precise data below, determine the limiting molar conductivities for hydrochloric acid, ethanoic acid, sodium chloride and sodium ethanoate.

	$\Lambda/\Omega^{-1}\ m^2\ mol^{-1}$		
c/mmol dm^{-3}	NaCl	CH_3CO_2Na	HCl
2	0.01125	0.00895	0.04197
10	0.01176	0.00867	0.04125
36	0.01096	0.00822	0.04008
80	0.01013	0.00776	0.03887
100	0.00830	0.00759	0.03844

9.6 The conductivity of a saturated aqueous solution of barium sulfate at 298.15 K is $3.59\times10^{-4}\ \Omega^{-1}\ m^{-1}$. and that of the water is $6.180\times10^{-5}\ \Omega^{-1}\ m^{-1}$. The limiting conductivities of the ions are $\frac{1}{2}\lambda_{0,+}=0.0065\ \Omega^{-1}\ m^2\ mol^{-1}$, and $\frac{1}{2}\lambda_{0,-}=0.0079\ \Omega^{-1}\ m^2\ mol^{-1}$. Calculate the solubility product for barium sulfate at the given temperature.

9.7 From the data in the text on limiting molar conductivities of ions, calculate the transport number of the chloride ion in the following infinitely dilute solutions, at 298.15 K: (a) HCl, (b) NaCl, (c) KCl, (d) $CaCl_2$ and (e) $LaCl_3$. Why is $t_{0,-}$ so much smaller in the hydrochloric acid solution?

9.8 Using the following values of π^{-0-}, set up three electrochemical cells: $\pi^{-0-}_{(Zn^{2+},Zn)}=-0.763$ V, $\pi^{-0-}_{(H_3O^+,H_2)}=0$ V and $\pi^{-0-}_{(Ag^+,Ag)}=0.799$ V. For each cell, write

(a) the *electrode* reactions;
(b) the overall reaction;
(c) E^{-0-};
(d) ΔG^{-0-};
(e) the direction of electron flow in an *external* circuit containing the cell.

9.9 Calculate the standard EMF and temperature coefficient of EMF for the cell Pt,H_2(1 atm)|HCl($a_\pm=1$)|AgCl,Ag, using the data below; the junction potential may be ignored.

	ΔG_f^{-0-}/kJ mol^{-1}	ΔH_f^{-0-}/kJ mol^{-1}
AgCl(s)	−127.2	−109.6
HCl($a_\pm=1$)	−167.4	−131.4

9.10 The standard reduction potentials for reaction of iron in acid medium are as follow:

$$\tfrac{1}{2}Fe^{2+}(aq)+e^- \rightleftharpoons \tfrac{1}{2}Fe(s) \qquad \pi^{-0-}=-0.440\ V$$

$$Fe^{3+}(aq)+e^- \rightleftharpoons Fe^{2+}(aq) \qquad \pi^{-0-}=0.771\ V$$

(a) Calculate ΔG^{-0-} and the equilibrium constant for the reaction $\frac{1}{2}Fe(s)+Fe^{3+}(aq) \rightleftharpoons \frac{3}{2}Fe^{2+}(aq)$.

(b) For the reaction $\frac{1}{4}O_2(g)+H_3O^+(aq)+e^- \rightleftharpoons \frac{3}{2}H_2O(l)$, $\pi^{\ominus}=1.23$ V. Is iron(II) inherently unstable with respect to oxidation by oxygen gas in acid solution? What is $E^{\ominus}$ for this reaction?

9.11 The EMF of the following cell at a given activity a of the acid HA is 219 mV at 25 °C:

$$\text{Hg,Hg}_2\text{Cl}_2|\text{KCl(saturated)}|1\,\text{H}A(a),\ \text{quinhydrone(saturated)}|\text{Pt}$$

Given that the potentials of the calomel and quinhydrone half-cells are 0.2442 V and 0.699 V, respectively, determine the pH of the solution of HA.
9.12 The EMF of the cell

$$\text{Ag}|\text{HCl(aq),AgCl(saturated),Hg}_2\text{Cl}_2\text{(saturated)}|\text{Hg}$$

is 0.0455 V at 298.15 K. (a) What is the cell reaction? (b) If the temperature coefficient of EMF is 3.4×10^{-4} V K^{-1}, calculate the standard molar enthalpy change of the reaction.
9.13 The EMF of the cell Pt,H_2(1 atm)|HCl(m)|Hg_2Cl_2,Hg has been determined at 25 °C, with the following precise results:

10^2m/mol kg^{-1}	0.16077	0.30769	0.50403	0.76938	1.09474
E/mV	600.80	568.25	543.66	522.67	505.32

Determine $E^{\ominus}$ and the mean activity coefficient of HCl at each molality. Compare the latter results with the direct use of the Debye–Hückel limiting equation and the Davies equation.
9.14 The EMF of the cell Pt,H_2(1 atm)|NaOH(m),NaCl(m')|AgCl,Ag has been measured precisely for m=0.0100 mol kg^{-1} and m'=0.01125 mol kg^{-1}:

T/°C	20	25	30
E/V	1.04774	1.04864	1.04942

Write the cell reaction. Determine pK_w at each temperature, and the standard enthalpy and entropy changes for the ionization of water.

Chemical kinetics and mechanisms of chemical reactions

10.1 Introduction

We complete our introduction to physical chemistry with a study of the actual chemical processes by which reactions take place. Normally, we need to investigate the rates of chemical reactions, because they tell us about how quickly reactions proceed, and lead to an understanding of how the reaction is initiated and carried on subsequently. Thus, a full analysis of a reaction mechanism involves both kinetic and dynamic aspects of the interaction of chemical species.

The usual stoichiometric equation cannot necessarily be taken to imply the process by which a reaction occurs. The three equations

$$H_2(g)+Cl_2(g)\rightarrow 2HCl(g) \qquad (10.1)$$

$$H_2(g)+Br_2(g)\rightarrow 2HBr(g) \qquad (10.2)$$

$$H_2(g)+I_2(g)\rightarrow 2HI(g) \qquad (10.3)$$

are formally similar, but the details of these reactions are very different, as is shown by their greatly differing rates under comparable conditions.

The rates of reactions are influenced by various external factors, the most important of which are temperature, pressure, concentrations of reagents and the presence of catalysts. In general, reaction rates increase with an increase in temperature, and we shall study this and other effects in ensuing sections of this chapter. First, however, we consider some experimental methods by which reactions rates are determined.

10.2 Experimental methods in chemical kinetics

The basic requirement in experimental kinetics is the ability to follow the course of a reaction with time, by measuring concentration, or some function of it, in the reacting system. We consider here some of the processes by which this goal may be achieved.

The time taken for reactions to reach completion varies widely, from nanoseconds to days. The methods that are employed most frequently to determine reaction rates involve monitoring pressure, as with the reaction

$$N_2O_5(g)\rightarrow 2NO_2(g)+\tfrac{1}{2}O_2(g) \qquad (10.4)$$

or monitoring optical activity, as with the mutarotation of glucose

$$\alpha\text{-D-}(+)\text{-}C_6H_{12}O_6(aq) \rightleftharpoons \beta\text{-D-}(+)\text{-}C_6H_{12}O_6(aq) \qquad (10.5)$$

The specific rotation $[\alpha]_D^{20}$ of an optically active solution is given by

$$[\alpha]_D^{20} = \frac{\text{observed rotation/deg}}{\text{length/dm} \times \text{density/g cm}^{-3}} \qquad (10.6)$$

at 20 °C in sodium D light. For α-D-(+)-glucose, $[\alpha]_D^{20} = +112°$, for β-D-(+)-glucose, $[\alpha]_D^{20} = +19°$, and at equilibrium, after mutarotation, it is +52.7°.

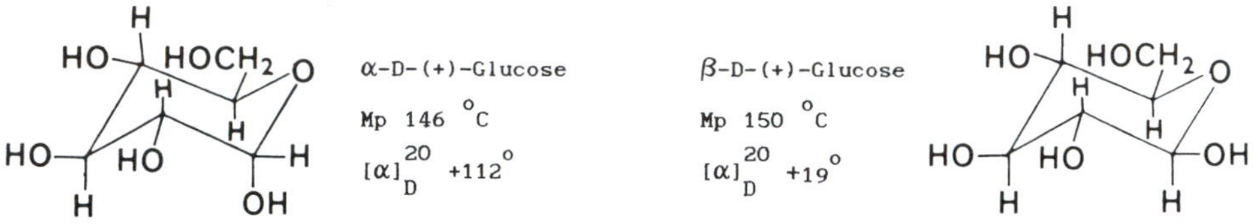

Diastereoisomers (anomers) of glucose[1]

A process may be followed also by a monitoring the absorption of white light, as with the reaction

$$H_2(g) + Br_2(g) \rightarrow 2HBr(g) \qquad (10.7)$$

or by acidimetry, as in the hydrolysis of an ester:

$$CH_3CO_2CH_3(aq) + H_2O(l) \rightleftharpoons CH_3CO_2H(aq) + CH_3OH(aq) \qquad (10.8)$$

In each case, we are following a concentration change, either directly or through a concentration-dependent physical process.

Reaction rates are very dependent upon temperature, and generally are found to be proportional to the Boltzmann factor $\exp[-E_a/(\mathcal{R}T)]$; E_a is an activation energy for the reaction, which we shall consider later. Evidently, strict temperature control is essential in obtaining precise reaction rates.

10.3 Rate and order of reaction

If we consider a simple notional reaction

$$A + B \rightarrow C \qquad (10.9)$$

where C represents the products of the reaction, then the rate R at which the reaction proceeds may be given as the rate of disappearance of a reactant, or the rate of appearance of a product:

$$R = -d[A]_t/dt = -d[B]_t/dt = d[C]_t/dt \qquad (10.10)$$

where $[A]_t$ represents the concentration of the species A at the time t and so on.[2] A more general reaction might be given as

$$A + 2B \rightarrow 3C + 4D \qquad (10.11)$$

whence we write for the rate

$$R = -d[A]_t/dt = -\tfrac{1}{2}d[B]_t/dt = \tfrac{1}{3}d[C]_t/dt = \tfrac{1}{4}d[D]_t/dt \qquad (10.12)$$

[1] Diastereoisomers are stereoisomers that are not mirror images; anomers are diastereoisomers that differ in configuration at C1.

[2] The notation [A] for the concentration of species A is conventional.

It is evident that the rate of a reaction must be given in terms of a specific reactant.

In Section 8.1 we considered the esterification of ethanoic acid. In this reaction the rate is given by

$$R=\frac{-\mathrm{d}[CH_3CO_2H]_t}{\mathrm{d}t}=k_2[CH_3CO_2H]_t[C_2H_5OH]_t \qquad (10.13)$$

since it has been found, by experiment, to depend upon the concentrations of both reactants.

The rate of isomerization of cyclopropane (Section 8.3) is given by

$$R=\frac{\mathrm{d}[CH_3CH\uparrow CH_2]_t}{\mathrm{d}t}= k_1\left[\begin{array}{ccc} & CH_2 & \\ / & & \backslash \\ CH_2 & — & CH_2 \end{array}\right]_t \qquad (10.14)$$

whereas in the gas-phase reaction

$$2NO(g)+Cl_2(g)\rightarrow 2NOCl(g) \qquad (10.15)$$

the rate of reaction has been found to be

$$R=\mathrm{d}[NOCl]_t/\mathrm{d}t=k_3[NO]_t^2[Cl_2]_t \qquad (10.16)$$

The dependence of the rate of a reaction upon a particular reactant is expressed by the *order* of the reaction with respect to that component. Thus, (10.13) describes a reaction that is first order in ethanoic acid, first order in ethanol and second order *overall*; (10.14) is an example of first-order kinetics. In (10.16), we have an example that is second order in nitrogen monoxide, first order in chlorine and third order overall. We shall see, however, that the order of a reaction is not always an integer constant.

It may be noted that by writing (10.15) as

$$NO(g)+\tfrac{1}{2}Cl_2(g)\rightarrow NOCl(g) \qquad (10.17)$$

we give a different impression of the order of reaction, although it is still correct stoichiometrically: it is essential to determine the order experimentally.

The constants k_i in the above rate equations are the *rate constants*, independent of concentration but dependent on temperature; i indicates the overall order of the given reaction.

The nitration of an aromatic molecule, such as toluene, is formally the reaction

$$C_6H_5CH_3+HNO_3\rightarrow C_6H_4CH_3NO_2+H_2O \qquad (10.18)$$

Under certain conditions, the concentration of toluene may be increased up to fourfold without appreciable change in the rate of nitration. The *rate-determining step* is the reaction

$$2HNO_3\rightarrow NO_2^+ + NO_3^- + H_2O \qquad (10.19)$$

and is followed by the *rapid* reaction

$$C_6H_5CH_3+NO_2^+ + H_2O\rightarrow C_6H_4CH_3NO_2+H_3O^+ \qquad (10.20)$$

The nitration is second order with respect to nitric acid, but *zero*th order in toluene.

10.4 Integrated rate equations

In order to obtain a best fit to a set of experimental rate data, it is necessary to obtain an integrated rate equation. Rate equations are differential equations and in many commonly

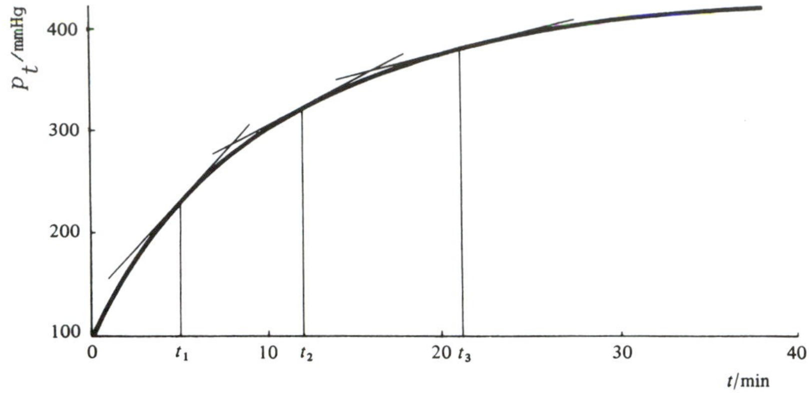

Figure 10.1 Variation of the partial pressure p_t of dinitrogen pentoxide with time t during its thermal decomposition at 329 K. At any time t_i, the rate of reaction is proportional to the tangent to the curve at the point $t=t_i$.

occurring reactions analytical integration is straightforward. In some complex reactions, computer-simulation studies can yield useful results, and we shall examine this approach for consecutive reactions.

10.4.1 First-order reactions

We begin this section by considering the reaction (10.4); it is accompanied by an increase in pressure and the reaction may be followed by monitoring the partial pressure of the reactant. The equation indicates that the rate of decomposition is proportional to the square of the concentration of dinitrogen pentoxide. However, experiments have shown that an increase in the concentration of reactant caused a proportionate increase in the rate; thus, the reaction is first order, with the rate equation

$$-\mathrm{d}[N_2O_5]_t/\mathrm{d}t=k_1[N_2O_5]_t \tag{10.21}$$

Suppose that the initial concentration of dinitrogen pentoxide is $[A]_0$, and that after a time t_1 the concentration remaining is $[A]_1$, then $-\mathrm{d}[A]_1/\mathrm{d}t=k_1[A]_1$. Similar results may be obtained after times t_2, t_3 and so on (Figure 10.1). The general result is $\mathrm{d}[A]_t/\mathrm{d}t=-k_1[A]_t$, which may be integrated to give

$$\ln([A]_t)=-k_1t+\mathcal{I} \tag{10.22}$$

For $t=0$, it is evident that the integration constant $\mathcal{I}$ must be $\ln([A]_0)$; hence,

$$\ln([A]_t/[A]_0)=-k_1t \tag{10.23}$$

which is the *integrated form* of the first-order rate equation.

To apply this result to the decomposition of dinitrogen pentoxide, let the partial pressure of N_2O_5 be p_0 initially, p_t after a time t and p_∞ after such a time that no further reaction can be detected. Then the initial concentration of N_2O_5 is proportional to the total change in pressure $(p_\infty-p_0)$, and that after a time t is proportional to the change in pressure $(p_\infty-p_t)$. Then, from (10.23), we obtain

$$\ln\{(p_\infty-p_t)/(p_\infty-p_0)\}=-k_1t \tag{10.24}$$

Table 10.1. Kinetics of decomposition of N_2O_5 at 329 K

t/min	p_t/mmHg	$(p_\infty - p_t)$/mmHg
0	100	331
3	178	253
5	220	211
10	297	134
15	345	86
20	377	54
30	409	22
40	421	10
∞	431	0

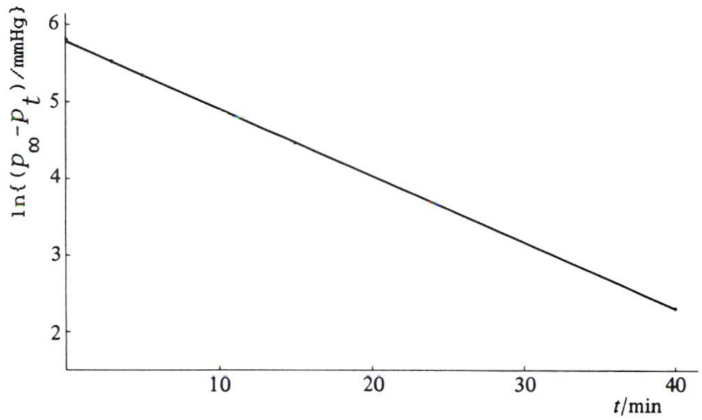

Figure 10.2 Variation of $\ln(p_\infty - p_t)$ as a function of t for the thermal decomposition of dinitrogen pentoxide at 329 K; the slope of the line is $-k_1$. In a typical experiment, $k_1 = 0.0884\ \text{min}^{-1}$ at 329 K, with a linear least-squares correlation r of 0.9997.

A plot of $\ln(p_\infty - p_t)$ against t is expected to be linear, with a slope equal to $-k_1$. The data in Table 10.1 were obtained in a typical experiment, and the plot is shown in Figure 10.2. By least squares the slope of the line is -0.08835 min (the correlation coefficient $r = 0.9997$); hence, the rate constant is $8.84 \times 10^{-2}\ \text{min}^{-1}$.

10.4.2 Second-order reactions

We shall consider two types of second-order reaction, exemplified by the thermal decomposition of hydrogen iodide

$$2HI(g) \rightarrow H_2(g) + I_2(g) \qquad (10.25)$$

and the alkaline hydrolysis of the 2,2,3,3-tetrachlorobutandicarboxylate (tetrachlorosuccinate) ion

$$\begin{matrix} CCl_2CO_2^- \\ | \\ CCl_2CO_2^- \end{matrix}(aq) + OH^-(aq) \rightarrow \begin{matrix} CCl(OH)CO_2^- \\ | \\ CCl_2CO_2^- \end{matrix}(aq) + Cl^-(aq) \qquad (10.26)$$

In the decomposition of hydrogen iodide, the rate equation has been determined as

$$-\mathrm{d[HI]}_t/\mathrm{d}t=k_2[\mathrm{HI}]_t^2 \tag{10.27}$$

If the initial concentration of hydrogen iodide is $[\mathrm{A}]_0$ and that after a time t is $[\mathrm{A}]_t$, then

$$-\mathrm{d[A]}_t/\mathrm{d}t=k_2[\mathrm{A}]_t^2 \tag{10.28}$$

which integrates readily to give

$$1/[\mathrm{A}]_t=k_2t+\mathcal{I} \tag{10.29}$$

For $t=0$, $\mathcal{I}=1/[\mathrm{A}]_0$, whence the integrated equation becomes

$$1/[\mathrm{A}]_t-1/[\mathrm{A}]_0=k_2t \tag{10.30}$$

Thus, a plot of $1/[\mathrm{A}]_t$, the concentration of hydrogen iodide at any time t, against t should be linear, with a slope equal to the second-order rate constant k_2. For this reaction, the half-life (q.v.) is

$$t_{1/2}=1/(k_2[\mathrm{A}]_0) \tag{10.31}$$

In the hydrolysis of the tetrachlorobutandicarboxylate ion (10.26), we are able to relate the concentrations of the reactants through the equation of the reaction. Thus, formally, we write for the rate of hydrolysis

$$-\mathrm{d[TCB]}_t/\mathrm{d}t=k_2[\mathrm{TCB}]_t[\mathrm{OH}^-]_t \tag{10.32}$$

where TCB represents the tetrachlorobutandicarboxylate ion. If the initial concentrations of TCB and alkali are $[\mathrm{A}]_0$ and $[\mathrm{B}]_0$, respectively, then as the concentration of TCB decreases to $[\mathrm{A}]_0-x$, the stoichiometry of (10.26) shows that the concentration of alkali decreases to $[\mathrm{B}]_0-x$; thus, we have related the two concentrations, and (10.32) becomes

$$\mathrm{d}x/\mathrm{d}t=k_2([\mathrm{A}]_0-x)([\mathrm{B}]_0-x) \tag{10.33}$$

or

$$\frac{\mathrm{d}x}{([\mathrm{A}]_0-x)([\mathrm{B}]_0-x)}=k_2\,\mathrm{d}t \tag{10.34}$$

By the method of partial fractions, $1/\{([\mathrm{A}]_0-x)([\mathrm{B}]_0-x)\}$ may be written as

$$\frac{1}{([\mathrm{B}]_0-[\mathrm{A}]_0)}\left(\frac{1}{([\mathrm{A}]_0-x)}-\frac{1}{([\mathrm{B}]_0-x)}\right)$$

and integration gives

$$\frac{1}{([\mathrm{B}]_0-[\mathrm{A}]_0)}\{-\ln([\mathrm{A}]_0-x)+\ln([\mathrm{B}]_0-x)\}=k_2t+\mathcal{I} \tag{10.35}$$

For $t=0$ $x=0$, and $\mathcal{I}=\ln\{([\mathrm{B}]_0/[\mathrm{A}]_0)/([\mathrm{B}]_0-[\mathrm{A}]_0)\}$, whence

$$\frac{1}{([\mathrm{B}]_0-[\mathrm{A}]_0)}\ln\left(\frac{[\mathrm{A}]_0([\mathrm{B}]_0-x)}{[\mathrm{B}]_0([\mathrm{A}]_0-x)}\right)=k_2t \tag{10.36}$$

from which it is clear that a plot of $\ln\{([\mathrm{B}]_0-x)/([\mathrm{A}]_0-x)\}$ against t should be linear, with a slope of $k_2([\mathrm{B}]_0-[\mathrm{A}]_0)$.

The data in Table 10.2 were obtained for the hydrolysis of the sodium tetrachlorobutandicarboxylate by aqueous sodium hydroxide at 298 K, over a period of 3 months.

Table 10.2. Alkaline hydrolysis of the tetrachlorobutandicarboxylate ion at 298 K with $[\text{TCB}]_0=0.0200$ mol dm^{-3} and $[\text{OH}^-]_0=0.0250$ mol dm^{-3}

t/hour	10^2x/mol dm^{-3}	$([\text{B}]_0-x)/([\text{A}]_0-x)$
500	0.339	1.3010
1000	0.588	1.3541
1500	0.779	1.4095
2000	0.929	1.4669
2500	1.051	1.5269

By least squares ($r=1.0000$) the slope of the plot of $\ln\{([\text{B}]_0-x)/([\text{A}]_0-x)\}$ against t was 8.002×10^{-5}/hour, and division by $([\text{B}]_0-[\text{A}]_0)$/mol dm^{-3} gives k_2 as 1.60×10^{-2} dm^3 mol^{-1} hour^{-1}.

There is no unique expression for the half-life of this reaction, since it depends upon both $[\text{A}]_0$ and $[\text{B}]_0$. However, if the reaction were carried out with equal initial values of $[\text{TCB}]_0$ and $[\text{OH}^-]_0$, the kinetics would reduce to (10.30) with $[\text{A}]_t=[\text{A}]_0-x$ and, for $[\text{A}]_0=0.020$ mol dm^{-3}, the half-life *under those* conditions would be given by (10.31) as 3125 hours.

10.4.3 Half-life and order of reaction

We consider again the decomposition of dinitrogen pentoxide (Section 10.4.1). Let the concentration of dinitrogen pentoxide remaining after a time t_β be the fraction β of the initial concentration $[\text{A}]_0$. Then, from (10.23)

$$\ln(\beta[\text{A}]_0/[\text{A}]_0)=-k_1t_\beta \tag{10.37}$$

or

$$t_\beta=\frac{\ln(1/\beta)}{k_1} \tag{10.38}$$

In particular, if $\beta=\frac{1}{2}$

$$t_{1/2}=\ln(2)/k_1 \tag{10.39}$$

where $t_{1/2}$ is the *half-life* of the reaction; it is the time taken for a reactant concentration at any given time to be decreased to half of its value at that time. Thus, in the decomposition reaction for dinitrogen pentoxide, the half-life is ln (2)/0.0884, or 7.8 min. It is evident that the half-life of a first-order reaction is independent of the initial concentration a, and the reaction is approximately 99% complete after seven half-lives. More generally, a half-life is dependent upon concentration, as we shall see.

The half-life may be used to indicate the order n of a reaction. For a rate law $-\text{d}[\text{A}]_t/\text{d}t=k_n[\text{A}]_t^n$, with $n\geq2$ or 0, $t_{1/2}$ is given (Problem 10.19) by

$$k_nt_{1/2}=\frac{2^{n-1}-1}{(n-1)[\text{A}]_0^{n-1}} \tag{10.40}$$

The special case of $n=1$ is covered by (10.39).

We illustrate a practical method for using the half-life to determine the order of reaction by the following data for the hydrolysis of an ester E:

t/s	100	200	300	400	500
$[E]_t$/mol dm^{-3}	0.0552	0.0382	0.0292	0.0236	0.0198

A series i of successive 'starting' values for $[E]_{0,i}$ is chosen and, by analytical (see Appendix A1.4) or graphical interpolation, values of t at $[E]_{0,i}/2$ are determined for each value of $[E]_{0,i}$, as follows:

i	1	2	3	4
$[E]_{0,i}$/mol dm^{-3}	0.050	0.046	0.042	0.038
$\frac{1}{2}[E]_{0,i}$/mol dm^{-3}	0.025	0.023	0.021	0.019
$t_{1/2}$/s	252	274	298	325

Figure 10.3 is a plot of $[E]_t$ against t; the chosen values of $[E]_{0,i}$ and $\frac{1}{2}[E]_{0,i}$ (i=1–4) are indicated as horizontal dashed lines. The corresponding four values for Δt, labelled 1–4, which are $t_{1/2}$ test-values, are clearly unequal. Since $t_{1/2}$ is not independent of $[E]_{0,i}$ the reaction is not of first order. Following (10.40), we can next plot $t_{1/2}$ against $1/[E]_{0,i}$, which would be linear for a second-order reaction, or we could plot $\ln(t_{1/2}/\text{s})$ against $\ln([E]_{0,i}/\text{mol dm}^{-3})$, which would be linear, of slope of $(1-n)$/mol dm^{-3} s.

By the first of these procedures, a least-squares straight line is obtained (r=0.9994) with a slope of 11.53/mol dm^{-3} s; hence, $k_2=8.66\times10^{-2}$ dm^3 mol^{-1} s^{-1}. By the second procedure a straight line is obtained (r=0.9993) with slope -0.925; thus, n=1.925, which we may take to be second order. This method is of particular value when the reaction under investigation may be of nonintegral order.

10.5 Reactions tending to equilibrium

Not all chemical reactions proceed to a stage at which the concentrations of the reactants become vanishingly small. We discussed this matter briefly in Section 8.1 and here we consider the kinetics of such reactions.

Let a reaction be represented in general terms by the scheme

$$\mathrm{A} \underset{k_{-1}}{\overset{k_1}{\rightleftharpoons}} \mathrm{B} \tag{10.41}$$

where k_1 and k_{-1} represent the rate constants for the forwards and reverse reactions, respectively. The equilibrium constant for this reaction may be written as

$$K=[B]_\infty/[A]_\infty=k_1/k_{-1} \tag{10.42}$$

where the subscript ∞ refers to a time t, sufficiently long to establish equilibrium at the given temperature.

The initial concentration of species A is $[A]_0$, and that of B is $[B]_0$. After a time t, let the concentration of species A be $[A]_t$ and that of B be $[B]_t$. The total rate of change of $[A]_t$ is given by

$$d[A]_t/dt=-k_1[A]_t+k_{-1}[B]_t \tag{10.43}$$

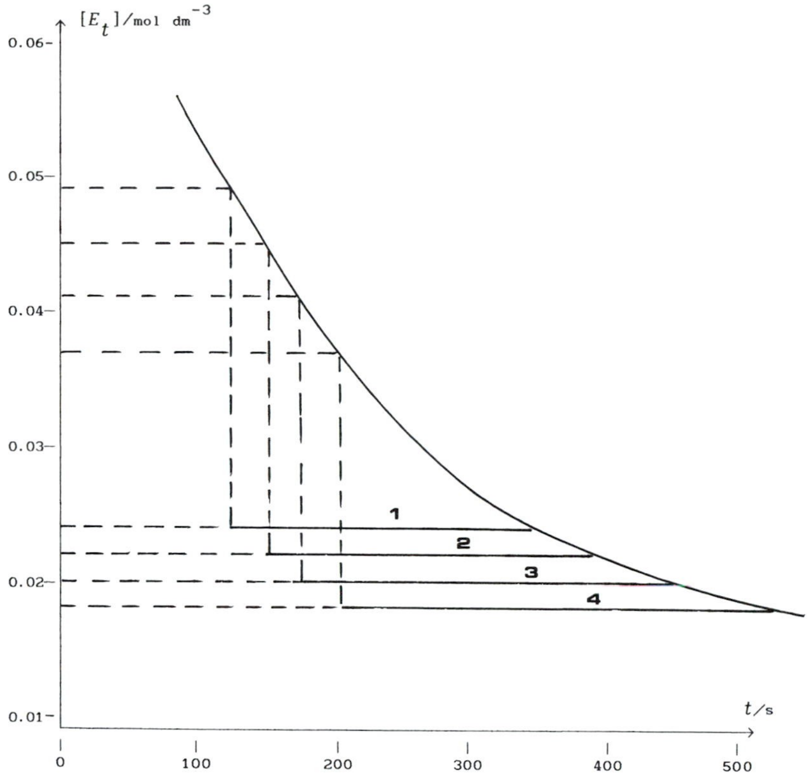

Figure 10.3 Plot of the concentration $[E]_t$ at times t in the alkaline hydrolysis of an ester E at 298 K. The upper four horizontal dashed lines are equidistant starting values $[E]_0$, with the similar lower four lines as the corresponding values $\frac{1}{2}[E]_0$. The lengths Δt (the horizontal full lines), which are the test-values of $t_{1/2}$, are clearly unequal in length, indicating a departure from first-order kinetics.

If, as is usual, $[B]_0$ is initially zero, it follows from a mass balance that at any time t, $[B]_t=[A]_0-[A]_t$, whence

$$d[A]_t/dt=-k_1[A]_t+k_{-1}([A]_0-[A]_t) \tag{10.44}$$

or

$$d[A]_t/dt=-(k_1+k_{-1})\left([A]_t-\frac{k_{-1}}{k_1+k_{-1}}[A]_0\right) \tag{10.45}$$

Now, from (10.42) we have $[B]_\infty/[A]_\infty=([A]_0-[A]_\infty)/[A]_\infty=k_1/k_{-1}$, or $[A]_\infty=[A]_0k_{-1}/(k_1+k_{-1})$. Introducing this result into (10.45), we obtain

$$d[A]_t/dt=-(k_1+k_{-1})([A]_t-[A]_\infty) \tag{10.46}$$

Integrating

$$\ln([A]_t-[A]_\infty)=-(k_1+k_{-1})t+\mathcal{I} \tag{10.47}$$

For $t=0$, $\mathcal{I}=\ln([A]_0-[A]_\infty)$; hence,

$$\ln\left(\frac{[A]_t-[A]_\infty}{[A]_0-[A]_\infty}\right)=-(k_1+k_{-1})t \tag{10.48}$$

Table 10.3. Mutarotation of an optically active ketone at 298 K

t/hour	α_t/deg	$(\alpha_t-\alpha_\infty)$/deg
0	189.0	157.7
3	169.3	138.0
5	156.2	124.9
7	145.9	114.6
11	124.6	93.3
15	110.4	79.1
24	84.5	53.2
∞	31.3	0

We may note that in the absence of a back reaction, that is, when $k_{-1}=0$, (10.48) reduces to a form comparable with (10.24).

Reaction (10.5) is one example of a process that tends to an equilibrium; another example is given by the data in Table 10.3, for the mutarotation of an optically active ketone K. The conversion of this ketone to its anomer K′ may be monitored by measuring the change in optical rotation of a solution with time. The optical rotation α_t at any time t is proportional to the concentration $[A]_t$; hence,

$$\ln\left(\frac{[\alpha]_t-[\alpha]_\infty}{[\alpha]_0-[\alpha]_\infty}\right)=-(k_1+k_{-1})t \tag{10.49}$$

Figure 10.4 shows a plot of ln $\{(\alpha_t-\alpha_\infty)/\text{deg}\}$ against time t. The slope of the least-squares line (r=0.9997) was $-0.0455/\text{hour}^{-1}$, whence $(k_1+k_{-1})=0.0455\ \text{hour}^{-1}$. From the above argument, it follows that $(\alpha_0-\alpha_\infty)/\alpha_\infty=k_1/k_{-1}=5.038$; hence, $k_1=3.80\times10^{-2}\ \text{hour}^{-1}$ and $k_{-1}=7.54\times10^{-3}\ \text{hour}^{-1}$.

10.6 Consecutive reactions

If the product of a reaction undergoes a further process in the same reacting system, consective reactions are set up. Two consecutive first-order reactions may be represented by the scheme

$$A \xrightarrow{k_1} B \xrightarrow{k_1'} C \tag{10.50}$$

An example of such a process is the thermal decomposition of propanone:

$$(CH_3)_2CO \rightarrow CH_2=CO+CH_4 \tag{10.51}$$
$$\downarrow$$
$$\tfrac{1}{2}C_2H_4+CO$$

Following procedures already described, we write

$$d[A]_t/dt=-k_1[A]_t \tag{10.52}$$

$$d[B]_t/dt=k_1[A]_t-k_1'[B]_t \tag{10.53}$$

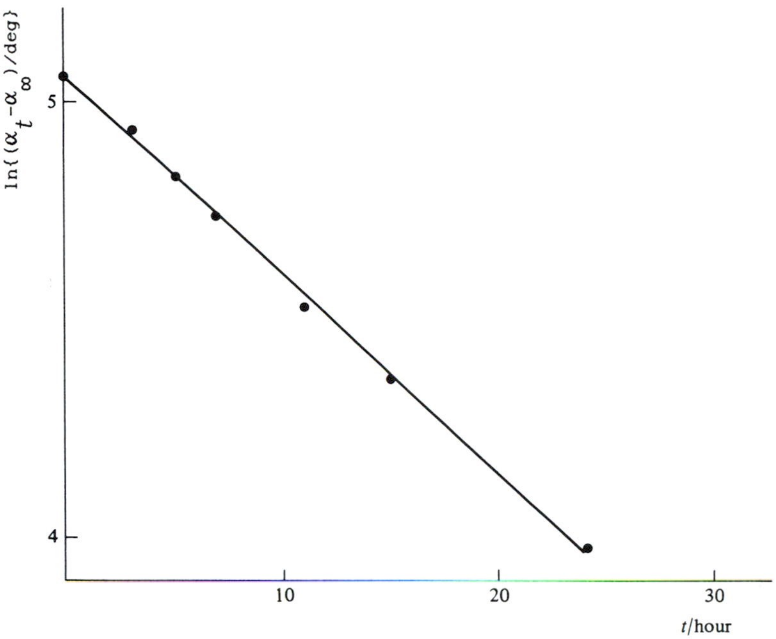

Figure 10.4 Plot of the variation of the optical rotation function $\ln(\alpha_t - \alpha_\infty)$ with time t in the mutarotation of an optically active ketone at 298 K. The least-squares line had a slope of $-0.0455/\text{hour}^{-1}$ ($r=0.9997$); $k_1=3.80\times10^{-2}$ hour^{-1} and $k_{-1}=7.54\times10^{-3}$ hour^{-1}.

$$d[C]_t/dt = k_1'[B]_t \qquad (10.54)$$

If we make the reasonable assumption that, at the time $t=0$, the concentrations of the species A, B and C are $[A]_0$, 0 and 0, respectively, then (10.52) may be integrated to

$$[A]_t = [A]_0 \exp(-k_1 t) \qquad (10.55)$$

Substituting (10.55) in (10.53) leads to

$$d[B]_t/dt = k_1[A]_0 \exp(-k_1 t) - k_1'[B]_t \qquad (10.56)$$

This equation may be solved by first multiplying throughout by the *integrating factor* $\exp(\int k_1'\, dt)$:

$$\exp(\int k_1'\, dt)\, d[B]/dt + \exp(\int k_1'\, dt) k_1'[B]_t = \exp(\int k_1'\, dt) k_1[A]_0 \exp(-k_1\, dt)$$

The left-hand side is the differential of $[B]_t \exp(\int k_1'\, dt)$ with respect to t, and $\exp(\int k_1'\, dt)$ is $\exp(k_1' t)/k_1'$ (the integration constant is common to both sides): hence,

$$[B]_t \exp(k_1' t)/k_1' = \int k_1[A]_0 \exp(-k_1 t) \exp(k_1' t)/k_1'\, dt$$

Multiplying throughout by k_1', and integrating:

$$[B]_t \exp(k_1' t) = \frac{k_1[A]_0 \exp\{(k_1' - k_1)t\}}{k_1' - k_1} + \mathcal{I}$$

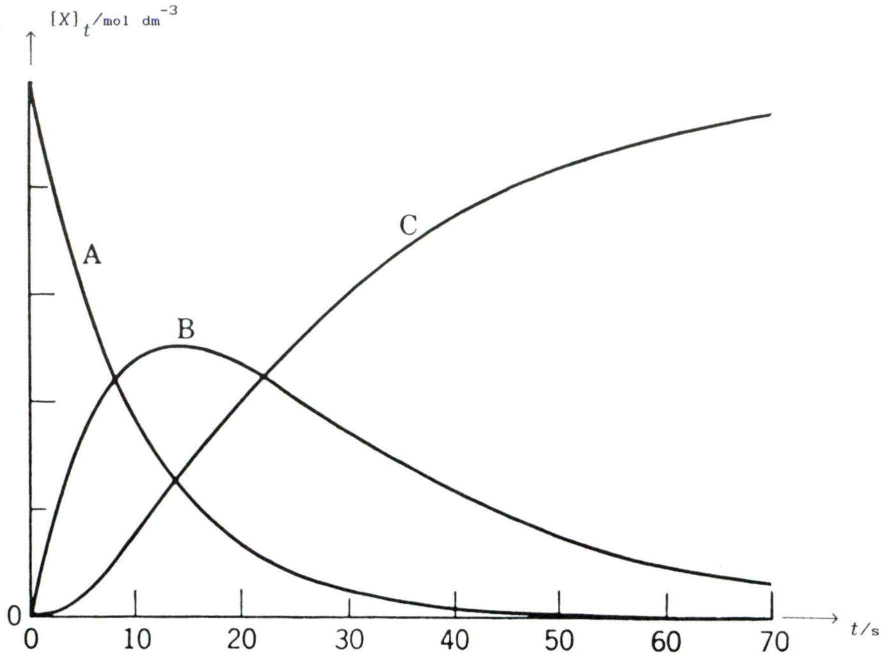

Figure 10.5 Variation of concentration $[X]_t$ (X=A, B, C) with time t for the consecutive reaction

$$A \xrightarrow{k_1} B \xrightarrow{k_1'} C,$$

for $[A]_0=1$ mol dm^{-3}, $[B]_0=[C]_0=0$, $k_1=0.1$ s^{-1} and $k'=0.05$ s^{-1}. A computer-simulation fit to the reaction (Section 10.7) reproduced the curves within their experimental error.

For $t=0$, $\mathcal{I}=-k_1[A]_0/(k_1'-k_1)$, whence by rearrangement

$$[B]_t=\frac{k_1[A]_0}{k_1'-k_1}\{\exp(-k_1t)-\exp(-k_1't)\} \tag{10.57}$$

From a mass balance $[C]_t=[A]_0-[A]_t-[B]_t$, and simple manipulation leads to

$$[C]_t=[A]_0\left(1-\frac{k_1'\exp(-k_1t)+k_1\exp(-k_1't)}{k_1'-k_1}\right) \tag{10.58}$$

In Figure 10.5, the curves for $[A]_t$, $[B]_t$ and $[C]_t$ have been plotted, assuming that $[A]_0=1$ mol dm^{-3}, $[B]_0=[C]_0=0.0$, $k_1=0.1$ s^{-1} and $k_1'=0.05$ s^{-1}.

The curves show that $[A]_t$ falls and $[B]_t$ rises rapidly. There is a time lag in the formation of species C, but $[C]_t$ subsequently increases and $[B]_t$ passes through a maximum. The ratio $[A]_t:[B]_t:[C]_t$ is 0.25 : 0.5 : 0.25 at $t=13.86$ s, the maximum in $[B]_t$, as differentiation of (10.57) confirms. Many examples of consecutive reactions may be found in the radioactive disintegration series and these reactions obey first-order, consecutive kinetics.

10.7 Computer-simulation studies of kinetics

It is evident that, even for the simple sequence of reactions in (10.50), the manipulation of the equations for the kinetic processes is moderately complex. Much of physical chemistry is concerned with *stochastic processes*, that is, reactions which are characterized by a sequence of random events that are governed by probability laws. The reactions between molecules are just such random processes and, although it is not possible to predict the behaviour of individual molecules, the statistical behaviour of a very large number of molecules ($\simeq N_A$) can be determined with good precision.

We have referred to computer-simulation studies in Sections 5.6.3ff. and the programs outlined in Appendix A1.4 contain some simple applications of the Monte Carlo method. Here, we consider the application of this technique to chemical kinetics.

Let the process A→B take place in a solvent, and let these species be identified by the integers 1, 2 and 0, respectively. A model is set up in the form of a grid of a very large number of cells and a set of molecules A is distributed at random among them. Next, a set of random numbers is generated, spanning the grid of cells and representing their location. The simulation is then started: if a '1' is found in a given cell, it is replaced by '2', indicating that the reaction has formed one of the species B; if '2' or '0' is encountered, no action is taken. As this process is repeated, it is clear that the population '1' decreases, and the number of correspondences at any 'visit' is a measure of the progress of the reaction. The time factor is somewhat arbitrary, but the shape of the concentration–time curve approximates closely to the true state of the reaction.

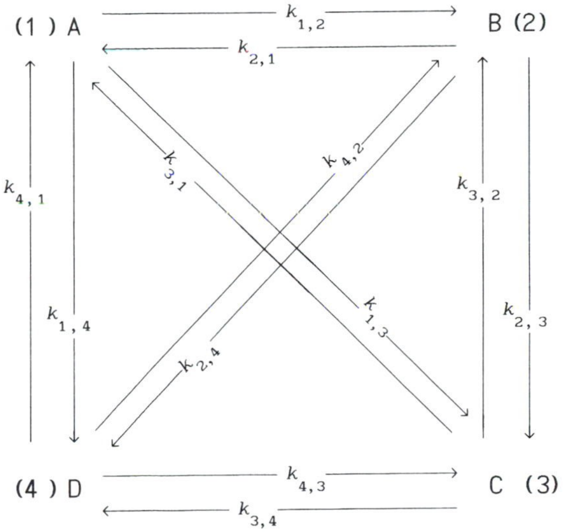

In the program outlined in Appendix A1.4, a grid of 2000 cells is defined and up to four reacting species are permitted, so that up to 12 (4^2-4) possible reaction paths can be identified, as the scheme above indicates.

If the species A to D are identified by the numbers 1 to 4, then the rate constants are of the form $k_{i,j}$, as shown in the diagram. A number of first-order processes can be studied in this way. The reaction

$$\begin{array}{ccc} & k_{1,2} & k_{2,3} \\ A & \longrightarrow B & \longrightarrow C \\ & k_{4,2} \Big\Updownarrow k_{2,4} & \\ & D & \end{array}$$

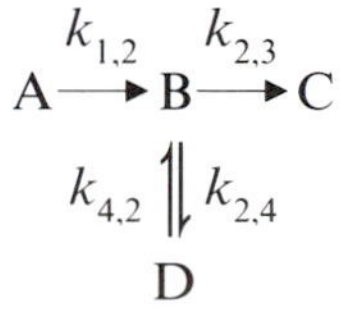

is typical of the systems that can be handled in this manner.

10.8 Reaction mechanism

After the rate law and the rate constant for a reaction have been determined, it is necessary to consider the actual course of the reaction in molecular terms, that is, its *mechanism*.

10.8.1 Molecularity

Reactions can usually be classified as *unimolecular*, in which a single molecules breaks up under the influence of heat or radiation, or *bimolecular*, in which two molecules react through a collision process. We define the *molecularity* of a reaction as the number of reacting species defining its rate-determining step. As we shall see, a reaction may involve several steps, some of which may be unimolecular and others bimolecular. Thus, while the order is an experimental quantity obtained from the rate equation, the molecularity depends upon the mechanism of the reaction.

The esterification of ethanoic acid is a second-order reaction, because its rate is proportional to the concentrations of both ethanoic acid and ethanol (10.13). It is also bimolecular, because it proceeds by collisions between two molecules of the reacting species:

$$CH_3CO_2H(aq)+C_2H_5OH(aq) \rightleftharpoons CH_3CO_2C_2H_5(aq)+H_2O(l) \qquad (10.59)$$

There are many chemical reactions that proceed by a bimolecular collision process; they are generally termed *simple* reactions, and we shall consider them first.

10.8.2 Dependence of rates of simple reactions upon temperature

The rates of most chemical reactions are sensitive to changes in temperature. It has been long known that the rate of a simple reaction at room temperature is approximately doubled for a rise in temperature of 10 K.

Arrhénius proposed the equation

$$k_2=A\exp\{-E_a/(\mathcal{R}T)\} \qquad (10.60)$$

where A is a *frequency factor*, which is only slightly dependent upon temperature, and E_a is the *activation energy* for the reaction, that is, the energy that reacting species must acquire in order to engage in a reaction to form products. It is evident that we can obtain both A and E_a if the rate constant is known at two or more temperatures.

The base-catalysed hydrolysis of bromoethane is a second-order, bimolecular reaction:

$$C_2H_5Br(l)+H_2O(l) \xrightarrow{OH^-(aq)} C_2H_5OH(aq)+HBr(aq) \qquad (10.61)$$

The data in Table 10.4 were obtained for reaction (10.61) at temperatures between 288 K and 333 K. It is evident from (10.60) that a plot of $\ln(k_2)$ against $1/T$ should be linear, with a slope of $-E_a/\mathcal{R}$ and an intercept of $\ln(A)$. A least-squares fit to these data (r=0.9997) leads to E_a=89.3 kJ mol^{-1} and A=1.85×10^{27} dm^3 mol^{-1} s^{-1}. We should note that, because E_a is not independent of temperature over a wide range of this variable, the best-fit value of A obtained here applies over the temperature range of the experiment.

The Arrhénius equation may be justified by the following argument. In order for two molecules A and B in the gas phase to react they must collide. We have discussed the collision frequency Z for gases in Section 5.3.11; its magnitude for ordinary gases is approximately 10^{34} s^{-1} m^{-3} at 298 K and 1 atm. If every collision resulted in reaction, the rate would be extremely fast. Again, kinetic theory shows that Z is inversely proportional to $\sqrt{T}$; thus, the ratio Z_{308}/Z_{298} should be 0.98, which is not in harmony with observed results.

Table 10.4. Variation of rate constant with temperature for the alkaline hydrolysis of bromoethane

T/°C	$10^3/(T/\mathrm{K})$	$k_2/\mathrm{dm^3\ mol^{-1}\ s^{-1}}$	$\ln(k_2/\mathrm{dm^3\ mol^{-1}\ s^{-1}})$
25	3.354	4.18×10^{11}	26.76
30	3.299	7.59×10^{11}	27.35
35	3.245	13.50×10^{11}	27.93
40	3.193	23.58×10^{11}	28.49

In order for a collision to result in reaction, the colliding molecules must possess sufficient energy, the activation energy E_a. Thus, the number of *effective* collisions may be written as PZ, where P is the fraction of collisions for which the energy is greater than or equal to the activation energy. The distribution in Figure 5.5 (Section 5.3.5) has been plotted with respect to mean speed. It is easy to show ($E=\frac{1}{2}mv^2$; $\mathrm{d}E=(2mE)^{1/2}\mathrm{d}v$) that (5.51) can be formulated in terms of energy, which has the same form of distribution:

$$\Phi(E)\,\mathrm{d}E=2\pi(\pi k_BT)^{-3/2}E^{1/2}\exp[-E/(k_BT)]\,\mathrm{d}E \qquad (10.62)$$

where $\Phi(E)\,\mathrm{d}E$ represents the probability that a molecule will have an energy lying between E and $E+\mathrm{d}E$. Thus, by analogy, an increase in temperature produces a broader distribution of energies, but the fraction P above the threshold value E_a for reaction is greatly increased. The fraction P is given by the Boltzmann distribution; thus,

$$P\propto\exp[-E_a/(\mathcal{R}T)] \qquad (10.63)$$

Clearly, P is more sensitive to changes in temperature than is Z; for a typical activation energy of 50 kJ mol^{-1}, the ratio $P_{308}/P_{298}=1.92$.

Values of the frequency factor A in (10.60) from experiment and calculation vary widely for certain reactions. For example, in the thermal decomposition of nitrosyl chloride, NOCl, to nitrogen monoxide and chlorine, A_{expt} and A_{calc} are in fairly good agreement, but for the hydrogenation of ethene they vary by approximately 10^5. A steric factor, the relative orientation of the reacting species at the moment of collision, is also of importance in promoting the reaction.

A model path of reaction is traced in Figure 10.6, for a gaseous bimolecular process of the type $\mathrm{A+B \rightleftharpoons C+D}$. The parameters E_1 and E_2 are the activation energies (strictly, enthalpies), for the forwards and reverse reactions, respectively. The energy E_1 is the minimum energy of the reactants A and B that would lead to reaction. Since $E_2<E_1$, the forwards reaction is thermodynamically spontaneous (assuming only a small entropy change), provided that sufficient energy can be supplied to raise the reactants to the energy threshold. Often, this energy is thermal, supplied simply by heating the reactants; it may be supplied also by irradiation. Thus, the reaction between hydrogen and chlorine *in the dark* is a very slow, bimolecular process; under irradiation, however, the reaction becomes very fast because its activation energy is greatly decreased.

The energy position at the maximum of the curve represents an activated complex formed from the reactants in their energized, transition state. From this position, the reaction can go forwards to form the products or the activated complex can return to the original state of the reactants; the path taken will be dependent upon the Gibbs free energy change for the process. This model of reaction is termed the *transition state theory* and the state of the energized reactants is the *activated complex*. A full discussion of transition state theory requires

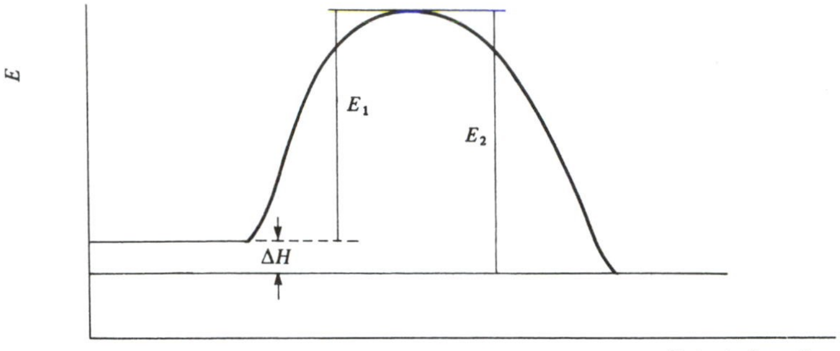

Figure 10.6 Model path of reaction for the bimolecular process $A+B \rightleftharpoons C+D$; E_1 and E_2 are the activation enthalpies ('energies') for the forwards and reverse processes, respectively. The forward reaction is favoured thermodynamically by the amount ΔH, assuming the entropy change to be small.

the use of statistical thermodynamics, for which reference should be made to more advanced texts.[3]

When more than one reaction can occur, the relative activation energies may determine the main product of reaction. Suppose that a compound A can react in two ways, with the free energy changes as indicated.

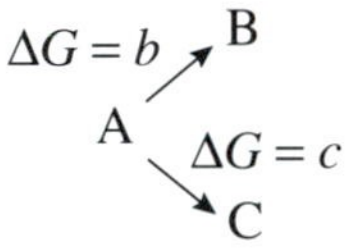

If $b<c$, the decomposition of A to form B is favoured thermodynamically over the reaction $A \rightarrow C$. However, suppose that the reaction paths are those traced in Figure 10.7. The activation energy for the formation of C is significantly the smaller, so that the formation of C is favoured kinetically. This situation is often encountered in organic reactions, whereby a number of different products could be formed; that produced in the greatest yield is often determined kinetically rather than thermodynamically.

10.9 Unimolecular decay in first-order reactions

Many gaseous dissociation and isomerization reactions follow first-order kinetics at normal pressures. If the activation energy is attained by collisions between the molecules, it might be expected that such reactions would be of second order; experiment shows that, at low pressures, the reactions do take place by second-order kinetics.

The Lindemann–Hinshelwood theory of these reactions proposes that a species A becomes energized by collision. The excited molecule A* may then collide with another molecule, thus losing its energy, or it may form a new product by unimolecular decay. These pathways may be shown as follow:

$$A+A \xrightarrow{k_2} A^*+A \tag{10.64}$$

[3] See, for example, T L Hill, *An Introduction to Statistical Thermodynamics* (Dover, 1986); A Ben-Naim, *Statistical Thermodynamics for Chemists and Biologists* (Plenum, 1992); K J Laidler, *Chemical Kinetics* (Harper and Row, 1987).

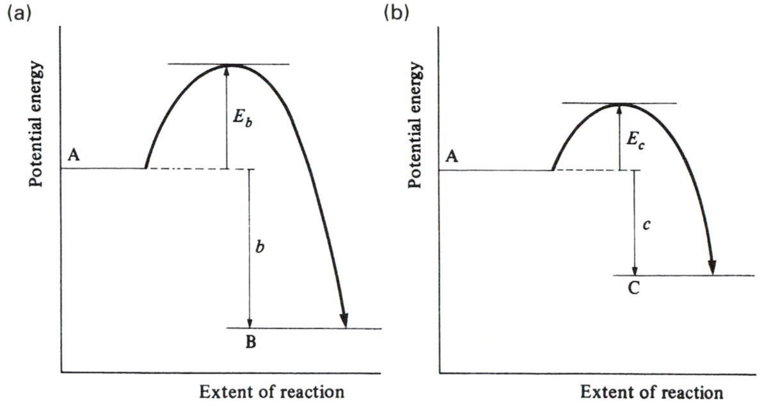

Figure 10.7 Model paths for the parallel reactions A→B and A→C; the former is favoured thermodynamically because the energy of B is lower than the energy of C, but the latter may occur in higher yield because of its lower activation energy ($E_C < E_B$).

$$A^* + A \xrightarrow{k_{-2}} A + A \tag{10.65}$$

$$A^* \xrightarrow{k_1} \text{Products} \tag{10.66}$$

If the formation of A* is fast, the rate-determining step for products is the unimolecular reaction (10.66), so that first-order kinetics ensue.

The rate of formation of A* is given by

$$d[A^*]_t/dt = k_2[A]_t^2 - k_{-2}[A^*]_t[A]_t - k_1[A^*]_t \tag{10.67}$$

The theory next adopts the *steady state approximation*, that is, that after a period of time the rates of formation and deactivation of the intermediate species A* are equal; thus, $d[A^*]_{eq}/dt = 0$ and (10.67) can be solved to give

$$[A^*]_{eq} = \frac{k_2[A]_t^2}{k_1 + k_{-2}[A]_t} \tag{10.68}$$

so that the rate of formation of the products is given by

$$d[P]_t/dt = \frac{k_1 k_2[A]_t^2}{k_1 + k_{-2}[A]_t} \tag{10.69}$$

If reaction (10.65) is fast with respect to (10.66), then $k_{-2}[A]_t \gg k_1$ and

$$d[P]/dt = \frac{k_1 k_2[A]_t}{k_{-2}} \tag{10.70}$$

which expresses first-order kinetics. However, if the pressure of the reacting species A is made very small, that is, $[A]_t$ is small, then $k_{-2} < k_1$, whereupon (10.69) becomes a second-order process:

$$d[P]/dt = k_2[A]_t^2 \tag{10.71}$$

The rate-determining step at low pressure is, thus, the bimolecular reaction (10.64).

A refinement of the Lindemann theory is the additional step in (10.66):

$$A^* \xrightarrow{k_1'} A^{\dagger} \xrightarrow{k_1''} P \tag{10.72}$$

where a distinction is made between the *energized species* A* and the *activated state* $A^{\dagger}$, and $(k_1'+k_1'')=k_1$; the difference depends upon the distribution of energy within the species A*. The theory has been tested and confirmed with such reactions as the isomerization of cyclopropane and the thermal decomposition of dinitrogen pentoxide (see also Problem 10.20).

10.10 Chain reactions

Many chemical reactions occur through a series of relatively simple steps, some of which may be repeated many times. Free radicals are involved in the processes, that is, transient species possessing an unpaired electron, and the order of the reaction is very different from that implied by its stoichiometry. We shall examine several such reactions, in order to elaborate the mechanisms that can be involved.

10.10.1 Hydrogen–chlorine reaction

This reaction may be initiated photochemically. Exposure to light of wavelength 500 nm or less causes a rapid and essentially complete reaction. In the dark, the reaction may be initiated by traces of sodium vapour. The surface : volume ratio of the reaction vessel and the nature of its surface affect the reaction once it has been initiated. There is also a small increase in pressure just before the first traces of hydrogen chloride are formed, because the total number of species is increased transiently.

The following steps are involved in the formation of hydrogen chloride:

Initiation $$Cl_2+h\nu \rightarrow Cl^{\bullet}+Cl^{\bullet} \tag{10.73}$$

Propagation $$\left.\begin{array}{l} H_2+Cl^{\bullet} \rightarrow HCl+H^{\bullet} \\ Cl_2+H^{\bullet} \rightarrow HCl+Cl^{\bullet} \end{array}\right\} \tag{10.74}$$

Termination $$\left.\begin{array}{r} Cl^{\bullet}+Cl^{\bullet} \rightarrow Cl_2 \\ H^{\bullet}+H^{\bullet} \rightarrow H_2 \\ H^{\bullet}+Cl^{\bullet} \rightarrow HCl \end{array}\right\} \tag{10.75}$$

Reaction (10.73), the initiation of the chain, must carry sufficient energy ($\simeq$243 kJ mol^{-1}, or $\lambda=493$ nm) to dissociate the chlorine molecules (D_0(Cl$_2$=243 kJ mol^{-1}; $D_0(H_2)=436$ kJ mol^{-1}). The chlorine atoms propagate the reaction by the steps (10.74), and it has been estimated that approximately 10^6 repetitions of these steps occur per chlorine radical before termination through steps (10.75) takes place.

The termination steps are relatively slow, since each involves a form of three-body collision:

$$Cl^{\bullet}+Cl^{\bullet}+M \rightarrow Cl_2+M^* \tag{10.76}$$

where the excited species M^* carries away the enthalpy of reaction, thus preventing the reverse reaction; M may be the reaction vessel wall, or an added molecule inert to the reaction, such as nitrogen monoxide or argon.

The 'dark' initiation of the reaction between hydrogen and chlorine may be brought about by sodium vapour. This vapour is largely diatomic, but under the reaction conditions of 10^{-4} mmHg and approximately 350 K, the initiating agent is the sodium atom[4] and the initiation may be represented by

[4] M Polanyi, *Trans. Faraday Soc.* **24** 606 (1928).

$$Na^{\bullet}+Cl_2 \rightarrow NaCl+Cl^{\bullet} \quad (10.77)$$

whereupon the propagation reactions (10.74) come into play. In this chain reaction, approximately 10^4 molecules of hydrogen chloride are formed per molecule of sodium chloride.

10.10.2 Hydrogen–bromine reaction

The thermal, 'dark' reaction between hydrogen and bromine in the gas phase at approximately 500 K involves several stages:

Initiation $$Br_2 \xrightarrow{k_1} Br^{\bullet}+Br^{\bullet} \quad (10.78)$$

Propagation $$\left.\begin{array}{l} Br^{\bullet}+H_2 \xrightarrow{k_2} HBr+H^{\bullet} \\ Br_2+H^{\bullet} \xrightarrow{k_3} HBr+Br^{\bullet} \end{array}\right\} \quad (10.79)$$

Inhibition $$HBr^{\bullet}+H \xrightarrow{k_4} H_2+Br \quad (10.80)$$

Termination $$Br^{\bullet}+Br^{\bullet} \xrightarrow{k_5} Br_2 \quad (10.81)$$

From these equations, we have

$$d[HBr]_t/dt=k_2[Br^{\bullet}]_t[H_2]_t+k_3[Br_2]_t[H^{\bullet}]_t-k_4[HBr]_t[H^{\bullet}]_t \quad (10.82)$$

We write next equations for the steady-state concentrations of the hydrogen and bromine free radicals:

$$d[H]_{eq}/dt=k_2[Br^{\bullet}]_t[H_2]_t-k_3[H^{\bullet}]_t[Br_2\,]_t-k_4[HBr]_t[H^{\bullet}]_t=0 \quad (10.83)$$

$$d[Br]_{eq}/dt=2k_1[Br_2]_t-k_2[Br^{\bullet}]_t[H_2]_t+k_3[Br_2]_t[H^{\bullet}]_t$$
$$+k_4[HBr]_t[H^{\bullet}]_t-2k_5[Br^{\bullet}]_t^2=0 \quad (10.84)$$

Solving for $[H^{\bullet}]_t$ and $[Br^{\bullet}]_t$, somewhat laboriously, gives

$$[Br^{\bullet}]_t=(k_1/k_5)[Br_2]_t \quad (10.85)$$

$$[H^{\bullet}]_t=k_2(k_1/k_5)^{1/2}[H_2]_t[Br_2]_t^{\ 1/2} \quad (10.86)$$

and substitution into (10.82) leads to

$$d[HBr]_t/dt=\frac{2k_2k_3k_4^{-1}(k_1/k_5)^{1/2}[H_2]_t[Br_3]_t^{1/2}}{k_3k_4^{-1}+[HBr]_t[Br_2]_t^{-1}} \quad (10.87)$$

In the early stages of the reaction, $[HBr]_t/[Br_2]_t$ is a small fraction, so that $d[HBr]_t/dt=2k_2(k_1/k_5)^{1/2}[H_2]_t[Br_2]_t^{1/2}$, and the reaction is of overall order 1.5. The term $(k_1/k_5)^{1/2}$ is the square root of the equilibrium constant for the dissociation of the bromine molecule.

10.10.3 Hydrogen–iodine reaction

The process indicated by (10.3) is of second order, with a rate constant k_2 of 0.024 $dm^3\ mol^{-1}\ s^{-1}$, but it does not proceed in the bimolecular manner that the equation would seem to indicate. The essential propagation reaction is the free radical mechanism

$$H_2+I^{\bullet}\rightarrow HI+H^{\bullet} \tag{10.88}$$

in which the iodine atoms are supplied by the dissociation of the iodine molecule. Reaction (10.88) is slow below 700 K, because it has a large activation energy.

10.10.4 Thermal decomposition of ethanal

We have seen how the apparently similar reactions (10.1) to (10.3) are very different when their mechanisms are considered in detail. The gas-phase decompositions of many organic compounds proceed by chain mechanisms; one example is the thermal decomposition of ethanal:

$$CH_3CHO\rightarrow CH_4+CO \tag{10.89}$$

The following stages in the process have been recognized:

Initiation
$$CH_3CHO\xrightarrow{k_1}CH_3^{\bullet}+CHO^{\bullet} \tag{10.90}$$

Propagation
$$\left.\begin{aligned}CH_3^{\bullet}+CH_3CHO&\xrightarrow{k_2}CH_4+CH_3CHO^{\bullet}\\ CH_3CO^{\bullet}&\xrightarrow{k_3}CH_3^{\bullet}+CO\end{aligned}\right\} \tag{10.91}$$

Termination
$$CH_3^{\bullet}+CH_3^{\bullet}\xrightarrow{k_4}C_2H_6 \tag{10.92}$$

By applying the steady-state treatment to $[CH_3^{\bullet}]$, we obtain

$$k_1[CH_3CHO]_t+k_3[CH_3CO^{\bullet}]_{eq}=k_2[CH_3^{\bullet}]_{eq}[CH_3CHO]_t+k_4[CH_3^{\bullet}]_{eq}^2 \tag{10.93}$$

Similarly, for $[CH_3CO^{\bullet}]$:

$$k_3[CH_3CO^{\bullet}]_{eq}=k_2[CH_3^{\bullet}]_{eq}[CH_3CHO]_t \tag{10.94}$$

On substituting (10.94) in (10.93) we obtain

$$[CH_3^{\bullet}]_{eq}^2=(k_1/k_4)[CH_3CHO]_t \tag{10.95}$$

The rate of disappearance of ethanal is equivalent to the rate of production of methane, from (10.89); hence,

$$d[CH_4]_t/dt=k_2[CH_3^{\bullet}]_{eq}[CH_3CHO]_t=k_2(k_1/k_4)^{1/2}[CH_3CHO]_t^{3/2} \tag{10.96}$$

and the order of reaction is 3/2, rather than unity suggested by (10.89).

10.10.5 Hydrogen–oxygen reaction

As a final example of chain reactions we consider the well-known reaction

$$2H_2+O_2\rightarrow 2H_2O \tag{10.97}$$

The chain reactions that we have discussed so far have been characterized by the fact that no step increased the number of free radicals present. The hydrogen–oxygen reaction is an example of a *branched-chain* reaction, that is, the number of radicals is increased in some steps of the process. A branched-chain mechanism is one way in which a reaction may proceed at an explosive rate. An explosive rate may also be induced thermally with an

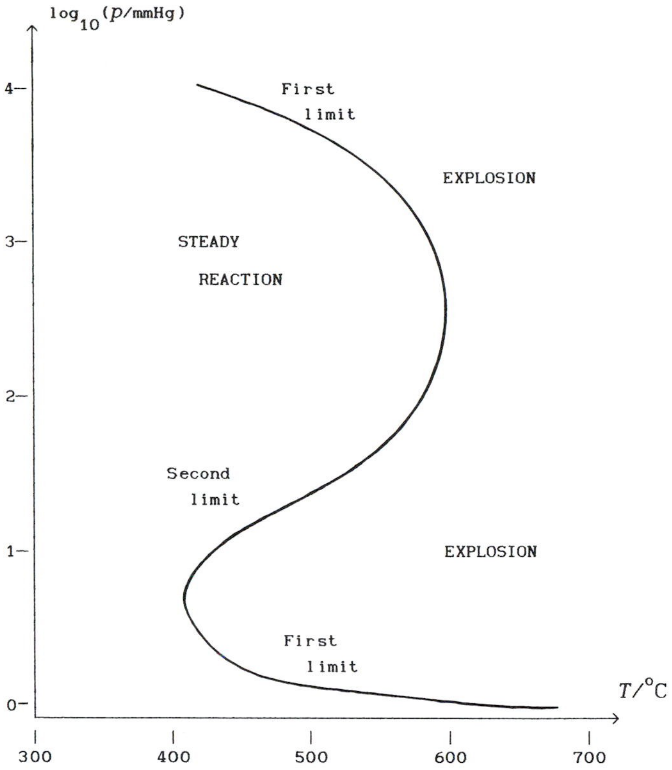

Figure 10.8 Steady-rate and explosive-rate regions for the hydrogen–oxygen reaction, in terms of temperature and total pressure. At a typical temperature, say 500 °C, three boundary limits between the steady- and explosive-rate reactions are encountered as the pressure is increased.

exothermic reaction, because of the exponential dependence of rate upon temperature. The hydrogen–oxygen reaction is complex and the following free-radical steps have been proposed for it:

Initiation $$H_2 + O_2 \rightarrow HO_2^{\bullet} + H^{\bullet} \tag{10.98}$$

Propagation $$\left.\begin{array}{l} H_2 + HO_2^{\bullet} \rightarrow HO^{\bullet} + H_2O \\ H_2 + HO^{\bullet} \rightarrow H + H_2O \end{array}\right\} \tag{10.99}$$

Branching $$\left.\begin{array}{l} O_2 + H^{\bullet} \rightarrow HO^{\bullet} + O^{\bullet} \\ H_2 + O^{\bullet} \rightarrow HO^{\bullet} + H^{\bullet} \end{array}\right\} \tag{10.100}$$

Termination $$HO_2^{\bullet}/HO^{\bullet}/H^{\bullet} + \text{Wall} \rightarrow \text{Removal} \tag{10.101}$$

The existence of an explosive rate of reaction depends upon the temperature and pressure of the gases, as shown by Figure 10.8.

If the mixture is sparked at low pressures and below approximately 400 °C, it undergoes steady reaction; the conditions lie outside the explosion limits. The radicals from the branching processes (10.100) reach the walls of the containing vessel and give up their energy (10.101) before generating more radicals.

Increasing the pressure at a temperature greater than 400 °C takes the mixture through the first limit and explosion occurs, because the branching reactions are outrunning the terminating reactions. A further increase in pressure takes the mixture into the region of steady combination; a termination step, logically different from that operating first, has again taken preference over the branching reactions. The existence of a third limit at high pressures serves to indicate the complexity of this well-known reaction, which is not yet understood fully. At temperatures greater than 600 °C, the reaction proceeds at an explosive rate at all pressures.

10.11 Photochemical reactions

In the hydrogen–chlorine reaction (Section 10.10.1), the initiation (10.73) was brought about by exposure to radiation; it is a photochemically induced reaction. Photochemical reactions underlie many chemical and biochemical processes and they are governed by two basic laws.

The Grotthus–Draper law states that *only radiation that is absorbed can lead to chemical reaction*. This statement seems self-evident, but it may be distinguished from the stimulated emission effect whereby a molecule in an excited state is caused to radiate even though it has not absorbed the stimulating radiation.[5] The second law is that of Stark and Einstein, which states that *a molecule absorbs a single quantum of radiation in the primary step of a photochemical reaction*:

$$M+h\nu \rightarrow M^* \tag{10.102}$$

The photochemical reaction, or *photolysis*, leading to the decomposition of hydrogen iodide (Section 10.10.3) may be activated with radiation of approximately 254 nm. Several steps are involved: $HI+h\nu \rightarrow H^{\bullet}+I^{\bullet}$, $H^{\bullet}+HI \rightarrow H_2+I^{\bullet}$ and $I^{\bullet}+I^{\bullet} \rightarrow I_2$. Thus, *two* molecules have been decomposed for the absorption of *one* quantum of radiation. The *quantum efficiency* Φ of a photochemical process is defined by

$$\Phi=\frac{\text{Number of molecules reacted}}{\text{Number of quanta absorbed}} \tag{10.103}$$

In a typical experiment, 300 J of radiant energy (λ=254 nm) decomposed 0.00128 mol HI. The number of photons absorbed by the HI is $(300\ \text{J})/(hc/254\ \text{nm})=3.84\times10^{20}$. An *einstein* of photons is N_A photons: hence, the hydrogen iodide absorbed 6.38×10^{-4} einstein, and $\Phi=0.00128/6.38\times10^{-4}$, or 2.01, which is in agreement with the above steps of the reaction.

The quantum efficiency may be very large, 10^5 or more, for a reaction that involves a chain mechanism. We refer again to the hydrogen–chlorine reaction stages (10.73) to (10.75). The rate of the initiation step (10.73) is proportional to the rate of absorption of photons, from the Grotthus–Draper and Stark–Einstein laws. Thus, we write $d[Cl^{\bullet}]_{init}/dt=2I_a$, where I_a is the radiant energy absorbed, including the concentration term $[Cl_2]$ arising from (10.73). Using the propagation reactions, with rate constants k_2 and k_3 respectively, the first of the termination reactions (k_4), and setting up steady-state expressions of $[H]_{eq}$ and $[Cl]_{eq}$, we obtain the result

$$d[HCl]_t/dt=2\sqrt{2}k_2[H_2]_t(I_a/k_4)^{1/2} \tag{10.104}$$

confirming an experimentally determined dependence of the rate of formation of hydrogen chloride upon the square root of the intensity of irradiation.

[5] Laser action depends upon light amplification by stimulated emission of radiation. See, for example, D L Andrews, *Lasers in Chemistry* (Springer-Verlag, 1990).

10.12 Catalysis

The rate of a chemical reaction can be altered by the presence of small amounts of substances, *catalysts*, that are foreign to the reacting system. Normally, the purpose of a catalyst is to increase the rate of reaction. The so-called 'negative catalysts', or inhibitors, such as traces of propan-1,2,3-triol (glycerol) that retard the decomposition of hydrogen peroxide, do not act in the opposite way to a normal catalyst. Here, we shall give the term catalyst its normal meaning.

As we have seen, even when the thermodynamics of a reaction indicate spontaneity, it may be very slow because the activation energy for the rate-determining step is high. A catalyst increases the rate of a reaction by lowering the activation energy. Two classes of catalyst are recognized, namely, *homogeneous* catalysts, which are in the same phase as the reactants, as with the base-catalysed hydrolysis of an ester, and *heterogeneous* catalysts, which are in a different phase from the reactants, as with the platinum-catalysed oxidation of ammonia to nitrogen monoxide. In both cases, however, certain features are common.

(a) The catalyst is unchanged in chemical constitution and amount at the end of the reaction, although its physical appearance may have altered. Thus, granular manganese dioxide added to potassium chlorate to catalyse its decomposition is returned in a very finely divided form. It is evident that a catalyst takes part in a reaction.
(b) Very small amounts of catalyst have a profound effect; for example, finely divided (colloidal) platinum at a concentration of 1 μg dm^{-3} is a very effective catalyst for the decomposition of hydrogen peroxide.
(c) A catalyst does not alter the position of equilibrium in a reaction (see Section 8.4). It follows that for a reaction symbolized by $A \rightleftharpoons B$, the rate constants of both the forwards and reverse reactions are increased to the same extent.

10.12.1 Homogeneous catalysis

A homogeneous gas-phase catalysed reaction is the aerial oxidation of sulfur dioxide, in the presence of nitrogen monoxide:

$$NO + \tfrac{1}{2}O_2 \rightarrow NO_2 \tag{10.105}$$

$$NO_2 + SO_2 \rightarrow SO_3 + NO \tag{10.106}$$

The mutarotation of glucose in aqueous solution is catalysed by acids and bases, an example of *acid–base* catalysis. Table 10.5 shows the varying efficiencies of different catalysts, all at the same concentration, revealed in the rate constants for the reaction; the hydroxyl ion is clearly the most effective catalyst for this reaction.

A homogeneous catalyst provides an alternative path, or mechanism, that involves a lower value of E_a for a reaction, compared with the activation energy for the uncatalysed reaction. Thus, if C is a catalyst for the unimolecular reaction of a species A, we can represent two reaction paths as

$$A \xrightarrow{k_1} \text{Products} \tag{10.107}$$

$$C + A \xrightarrow{c} \text{Products} \tag{10.108}$$

with the rate equations

Table 10.5. Efficiencies of catalysts, at the same concentration, in the mutarotation of aqueous glucose

Acid	k/s^{-1}	Base	k/s^{-1}
CH_3CO_2H	2.0×10^{-3}	$CH_3CO_2^-$	2.7×10^{-2}
$C_6H_5CH(OH)CO_2H$	6.0×10^{-3}	$C_6H_5CH(OH)CO_2^-$	6.1×10^{-2}
H_3O^+	1.4×10^{-1}	OH^-	6.0×10^{3}

Table 10.6. Acid-catalysed hydrolysis of ethyl diazoethanoate, at 298 K:

$$N_2CHCO_2C_2H_5+H_2O \xrightarrow{H_3O+} HOCH_2CO_2C_2H_5+N_2$$

$10^3[H_3O^+]$/mol dm^3	pH	k_C/dm^3 mol^{-1} min^{-1}	$10^{-3}k'_C$/min^{-1}
0.36	3.44	2.4	6.67
0.90	3.05	5.8	6.44
1.82	2.74	11.7	6.43
3.25	2.49	20.8	6.40

$$-d[A]_t/dt=k_1[A]_t \qquad (10.109)$$

$$-d[A]_t/dt=k_C[A]_t[C] \qquad (10.110)$$

Since C is a catalyst, [C] is unchanged, so that with $k'_C=k_C[C]$, we have

$$-d[A]_t/dt=k'_C[A]_t \qquad (10.111)$$

and the total rate becomes

$$-d[A]_t/dt=(k_1+k'_C)[A]_t \qquad (10.112)$$

If the catalysed reaction is much the faster, $k'_C \gg k_1$ and

$$-d[A]_t/dt \simeq k'_C[A]_t \qquad (10.113)$$

Thus, the rate would be expected to increase in proportion to [C], since $k'_C=k_C[C]$, while k'_C itself remains constant. This prediction is realized in the acid-catalysed hydrolysis of ethyl diazoethanoate, Table 10.6.

10.12.1.1 Kinetic salt effects

The rate of a reaction in solution may be increased or decreased by the addition of a salt that does not take part directly in the reaction. The added salt increases the ionic strength (Section 8.6.4.1) of the solution and, thus, the activities of the species present; its effect upon the reaction rate is termed the *primary kinetic salt effect.*

Consider a bimolecular reaction between species A and B, in which the activation process leads to the formation of an intermediate, or activated complex (Section 10.8.2), AB*. The rate of reaction is then the rate of decomposition of the activated complex to form the products P:

$$d[P]_t/dt = k^*[AB^*]_t \tag{10.114}$$

A treatment of the equilibrium

$$A + B \overset{k^*}{\rightleftharpoons} AB^* \tag{10.115}$$

by statistical thermodynamics shows that the rate constant k^* is given by

$$k^* = k_B T/h \tag{10.116}$$

where k_B is the Boltzmann constant and h is the Planck constant. The equilibrium constant K^* for (10.115) is given, in the usual way, by

$$K^* = \frac{[AB^*]\,\gamma_t^*}{[A]_t[B]_t\gamma_A\gamma_B} \tag{10.117}$$

where the activity coefficients will be important if some or all of the species are charged, that is, are present as ions. From the foregoing, we now write

$$d[P]_t/dt = k^*[AB^*] = k[A]_t[B]_t \tag{10.118}$$

where k is $k^*K^*\gamma_A\gamma_B/\gamma_t^*$. We let k_0 be the value of k for which all activity coefficients are unity, that is, $k_0 = k^*K^*$. Then,

$$\ln(k) = \ln(k_0) + \ln(\gamma_A) + \ln(\gamma_B) - \ln(\gamma_t^*) \tag{10.119}$$

The activity coefficients are influenced by the ionic strength of the solution which, in turn, affect the values of k. If the solutions are dilute, the Debye–Hückel limiting equation (8.71) may be used in the single-ion form, replacing the numerical value 1.172, applicable to aqueous solutions at 25 °C, by the temperature-dependent constant $\mathscr{A}_T$, for aqueous solutions at a temperature T (Section 8.6.4.1); thus

$$\ln(\gamma_i) = -\mathscr{A}_T q_i^2 I^{1/2} \tag{10.120}$$

where q_i is the charge on the given ion i. Then (10.119) becomes

$$\ln(k) = \ln(k_0) - \mathscr{A}_T I^{1/2}(q_A^2 + q_B^2) + \mathscr{A}_T I^{1/2}(q_A + q_B)^2 \tag{10.121}$$

where $(q_A + q_B)$ is the charge on the activated complex AB^*.

Simplifying (10.121) leads to

$$\ln(k) = \ln(k_0) + 2\mathscr{A}_T q_A q_B I^{1/2} \tag{10.122}$$

which is known as the Brønsted–Bjerrum equation; it shows that $\ln(k/k_0)$ will vary with $I^{1/2}$ according to the charges q_A and q_B. Figure 10.9 illustrates this equation for varying ionic charges. Since the (corrected) rate constant k_0 is a constant it would be sufficient to plot with $\ln(k)$.

For the reaction

$$S_2O_8^{2-}(aq) + 3I^-(aq) \rightarrow 2SO_4^{2-}(aq) + I_3^-(aq) \tag{10.123}$$

q_Aq_B is 2, and the rate of reaction increases with increasing ionic strength. In the case of ester hydrolysis

$$RCO_2R'(aq) + OH^-(aq) \rightarrow RCO_2^-(aq) + R'OH(aq) \tag{10.124}$$

q_Aq_B is zero and the rate is invariant with ionic strength. With the reaction

$$[Co(NH_3)_5Br]^{2+}(aq) + OH^-(aq) \rightarrow [Co(NH_3)_5OH]^{2+}(aq) + Br^-(aq) \tag{10.125}$$

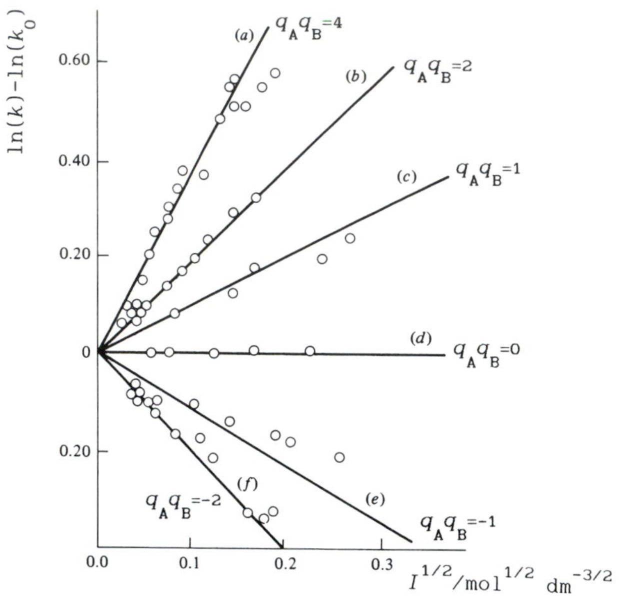

Figure 10.9 Variation of the rate constant function $\ln(k/k_0)$ with the square root of the ionic strength of the solution $I^{1/2}$, for typical ionic reactions in aqueous media:

(*a*) $2[Co(NH_3)_5Br]^{2+} + Hg^{2+} \rightarrow P$
(*b*) $S_2O_8^{2+} + 3I^- \rightarrow P$
(*c*) $(NO_2NCO_2C_2H_5)^- + OH^- \rightarrow P$
(*d*) $C_{12}H_{22}O_{11} + OH^- \rightarrow P$
(*e*) $H_2O_2 + H_3O^+ + Br^- \rightarrow P$
(*f*) $[Co(NH_3)_5Br]^{2+} + OH^- \rightarrow P$.

$q_A q_B$ is -2, and the rate decreases with increasing ionic strength. This dependence of the reaction rate upon the ionic strength of the solution is the *primary kinetic salt effect.*

The rate of a reaction in solution may depend also upon the ionic strength of the solution when the reaction is catalysed by an ion provided by the dissociation of a weak electrolyte; for example, hydroxonium ions from the dissociation of a weak acid HA. If the dissociation constant of the weak acid is K_a, then the amount of catalyst formed is given by

$$[H_3O^+] = \frac{K_a[HA]\gamma_{HA}}{[A^-]\gamma_{A^-}\gamma_{H_3O^+}} \tag{10.126}$$

and depends upon the ionic strength, because of the effect of ionic strength upon activity coefficients. This dependence is termed the *secondary kinetic salt effect.*

10.12.2 Heterogeneous catalysis: adsorption

A solution of hydrogen peroxide decomposes only very slowly under ambient conditions, but in the presence of platinum or platinum-black, a steady stream of oxygen is evolved. Since the platinum is chemically inert under these conditions, it is evident that the surface of the metal must play an important part in the reaction. The actual material and the physical state of its surface are both relevant to reaction. For example, if the vapour of methanoic acid is passed through a heated glass tube, two reactions occur to approximately equal extents:

$$HCO_2H \rightarrow H_2O + CO \tag{10.127}$$

$$HCO_2H \rightarrow H_2 + CO_2 \tag{10.128}$$

If the tube is packed with aluminium oxide Al_2O_3, the first of these reactions occurs; if zinc oxide ZnO is used, the second reaction takes place. A study of heterogeneous catalysis brings us into contact with the physical chemistry of surfaces.

The catalytic activity of a surface depends upon the reactants in contact with the surface being *adsorbed* on it. Adsorption should be regarded as a form of chemical adhesion, wholly different from absorption. A reaction then takes place on the surface, after which the products must be desorbed from the surface and carried away. Generally, the chemical reaction at a surface is the rate-determining step in heterogeneous catalysis.

10.12.2.1 Langmuir adsorption isotherm

A model is set up that depends upon there being a fixed number of adsorption sites on a catalyst surface, each site capable of holding one molecule of adsorbate. The energy of interaction of the adsorbate is assumed to be the same for each site, so that there is no interaction between adsorbed molecules. Langmuir adsorption is monomolecular; it does not consider the case that further adsorption may take place on the adsorbate already present on the surface.

If θ_t is the fraction of a catalyst surface that is occupied at a time t, the rate of desorption is proportional to the number of adsorbed species, $k_d\theta_t$:

$$d\theta_t/dt = k_d\theta_t \qquad (10.129)$$

where k_d is the rate constant for desorption. The rate of adsorption at a time t will be proportional to the unoccupied surface $1-\theta_t$, and to the rate at which the molecules strike the surface, that is, to the pressure p, so that we have also $d(1-\theta_t)/dt = k_a p(1-\theta_t)$. After a period of time, a steady state is attained and the two rates are equal; hence,

$$k_d\theta = k_a p(1-\theta) \qquad (10.130)$$

where θ is the equilibrium value of the fraction adsorbed at a given temperature and k_a is the rate constant for adsorption; with $K_L = k_a/k_d$, (10.130) may be rearranged to give

$$\theta = K_L p/(1+K_L p) \qquad (10.131)$$

which is the Langmuir adsorption isotherm; K_L has the dimensions of $(\text{pressure})^{-1}$ and is dependent upon temperature.

The data in Table 10.7 were obtained for the adsorption of carbon monoxide on a fixed amount of charcoal at 290 K; they can be analysed in terms of the Langmuir isotherm.

Since θ is proportional to the volume adsorbed we write $\theta = \alpha V_{CO_2}$. Simple manipulation shows that a plot of $1/V_{CO}$ against $1/p$ should be linear, with a slope of α/K_L and an intercept of α. By least squares ($r = 1.0000$) the intercept is 9.48/dm^{-3} and the slope is 8860/mmHg dm^{-3}. We identify α with $1/V$ when $1/p$ is zero, that is, $\alpha = 1/V_{max}$, the volume of CO_2 that would be adsorbed if all surface sites were occupied. Thus, $V_{max} = 0.105$ dm^3 and the constant K_L is 1.07×10^{-3} mmHg^{-1}.

Experimental results from Langmuir isotherms for several substances have shown that the equation is less precise at high values of θ, that is at higher values of pressure, and that, for a given pressure p, θ increases with K_L; θ tends to unity only at very high pressures.

Table 10.7. Adsorption of CO on charcoal at 290 K

p/mmHg	100	200	300	400	500	600
$10^2 V_{CO}$/dm^3	1.38	2.51	3.45	4.24	4.93	5.52

10.12.2.2 Other isotherms

The observed behaviour at a surface may be represented by different isotherms under differing conditions. The Langmuir assumption of the equivalence of all sites is an ideal that may not always be realized. For example, adsorption at a particular site may depend upon the occupancy of neighbouring sites, with the most energetically favourable sites being occupied first. An attempt to take this situation into account is represented by the Freundlich isotherm, which is often useful in studying adsorption from solution:

$$\theta = K_F p^{1/n} \tag{10.132}$$

where K_F and n are constants. This isotherm is readily studied by plotting $\ln(\theta)$ against $\ln(p)$, whereupon both K_F and n can be determined.

10.12.2.3 Chemical reaction at a surface

Adsorption at a surface may take place through van der Waals' forces, in which case an average enthalpy of adsorption is approximately -40 kJ mol^{-1}, or it may be governed by covalent forces, in which case the average enthalpy of adsorption is nearly ten times greater. The latter process, known as *chemisorption*, is of importance in the acceleration of reaction rates, whereby the chemisorbed layer acts as a transition state towards the product compound.

As an example of the application of the Langmuir isotherm to surface catalysis, we can consider the case of the decomposition of arsine AsH_3 on a tungsten surface. We use

$$\theta = k_a p/(k_d + k_a p) \tag{10.133}$$

If adsorption is weak $k_d \gg k_a p$ and

$$\theta \simeq k_a/k_d = K_L p \tag{10.134}$$

and the fraction of surface covered depends solely on the pressure. Normally, the surface area of the catalyst is sufficiently large that a relatively small amount of adsorbed arsenic does not significantly hinder its catalytic function. Since the rate of adsorption $-\mathrm{d}p_t/\mathrm{d}t$ is proportional to θ, we have the rate law

$$-\mathrm{d}p_t/\mathrm{d}t = k'\theta = k'K_L \mathrm{p}_t = kp_t \tag{10.135}$$

where $k = k'K_L$; hence,

$$\ln(p_t/p_0) = -kt \tag{10.136}$$

Experimental data on the decomposition of dinitrogen monoxide on gold at 1170 K are listed in Table 10.8. The data have been fitted by linear least squares ($r=0.9999$), with a slope of -0.0132/min^{-1}; thus, the kinetics are first order and $k=1.32\times10^{-2}$ min^{-1}.

If the adsorption is very strong $k_a p \gg k_d$, and

$$\theta \simeq 1 \tag{10.137}$$

Table 10.8. Decomposition of N_2O on gold at 1170 K

t/min	0	15	30	65	80	100
p_t/Torr	200	167	136	86	70	54

Table 10.9. Decomposition of NH_3 on tungsten at 1100 K; p_0=200 Torr

t/s	100	200	300	400	500	1000
p_t/Torr	186	173	162	152	141	88

Table 10.10. Decomposition of SbH_3 on antimony at 300 K

t/min	0	5	10	15	20	25
p_t/Torr	760	556	387	249	144	71

so that the fraction of surface covered is independent of the pressure. The reaction under these conditions is of zeroth order, but it may differ according to the catalyst employed. Thus, the decomposition of hydrogen iodide is a first-order reaction on platinum, but zeroth order on a gold surface.

From (10.137), since pressure is proportional to θ, $-\mathrm{d}p_t/\mathrm{d}t=k$, where k is a constant; thence

$$p_t-p_0=-kt \tag{10.138}$$

The decomposition of ammonia on tungsten has been found to be almost independent of the initial pressure. At 1100 K the data in Table 10.9 were obtained. By least squares, a linear plot (r=0.9997) confirmed a zeroth-order reaction, with k=0.108 Torr s^{-1}.

It is reasonable that, in the region of intermediate adsorption, θ is proportional to a fractional power of the pressure, that is, it follows the Freundlich isotherm:

$$-\mathrm{d}p_t/\mathrm{d}t=k'\theta=k'K_\mathrm{F}p^{1/n}=kp^{1/n} \tag{10.139}$$

The decomposition of stibine SbH_3 on an antimony surface at 300 K follows this law, and the data in Table 10.10 have been recorded.

Integration of (10.139) gives

$$1/p_0^{n-1}-1/p_t^{n-1}=(1-n)kt \tag{10.140}$$

but this equation is not easy to handle analytically. Given that $0<n<1$, the solution by iteration converges at n=0.550 and k =1.155 Torr$^{0.55}$ min^{-1}.

Once reaction has occurred on a surface, the products need to be removed, since they, too, can be adsorbed. An important case arises when the products of the catalytic reaction are themselves strongly adsorbed.

Table 10.11. Effect of hydrogen on the decomposition of NH_3 on platinum at 1400 K; $p_{0,A}=100$ Torr; $t=120$ s

p_{t,NH_3}/Torr	p_{H_2}/Torr (excess)	(p_{0,NH_3}/Torr$-p_{t,NH_3}$/Torr)
67	50	33
73	75	27
84	100	16
90	150	10

Consider a reaction of type A→B+C, for which the product B is strongly adsorbed. Let $(1-\theta_B)$ be the fraction of the surface free from the adsorbate B, and let $p_{t,A}$ and $p_{t,B}$ be the pressures of A and B, respectively, after a time t. The rate of the reaction will be proportional to both the free surface fraction $(1-\theta_B)$ and the pressure of A. From (10.131) generally,

$$1-\theta=1/(1+K_L p) \tag{10.141}$$

so that the rate of reaction is given by

$$-\mathrm{d}p_{t,A}/\mathrm{d}t=k(1-\theta_B)p_{t,A}=kp_{t,A}/(1+K_L p_{t,B}) \tag{10.142}$$

If the adsorption is strong $K_L p_{t,B} \gg 1$ and, applying this result to the reaction under consideration,

$$-\mathrm{d}p_{t,A}/\mathrm{d}t=k' p_{t,A}/p_{t,B} \tag{10.143}$$

This behaviour is observed in the catalytic decomposition of ammonia on platinum at 1400 K, where the reaction is inhibited by excess hydrogen adsorbed on the surface of the platinum catalyst, Table 10.11.

A threefold increase in the pressure of the excess hydrogen gas decreases the decomposition from 33% of the total to 10% over a constant time period. If the adsorption of the product is less strong, (10.142) applies without further approximation.

10.12.2.4 Enzyme catalysis

Many biological reactions are made possible by the catalytic activity of enzymes. If an enzyme be denoted by E and the substrate on which it acts by S, then the overall reaction is

$$\mathrm{S+E \rightarrow P+E} \tag{10.144}$$

where P represents the products of enzymolysis and the enzyme remains unchanged. The rate of reaction has been found to be proportional to the enzyme concentration [E] and the Michaelis mechanism may be written as

$$\mathrm{S+E} \underset{k_{-1}}{\overset{k_2}{\rightleftharpoons}} \mathrm{+SE} \xrightarrow{k_1} \mathrm{P+E} \tag{10.145}$$

where SE is an intermediate, bound state of substrate and enzyme that can either decompose to give the products or revert to the original substrate and free enzyme. The rate of formation of products is given by

$$\mathrm{d[P]}_t/\mathrm{d}t=k_1\mathrm{[SE]}_t \tag{10.146}$$

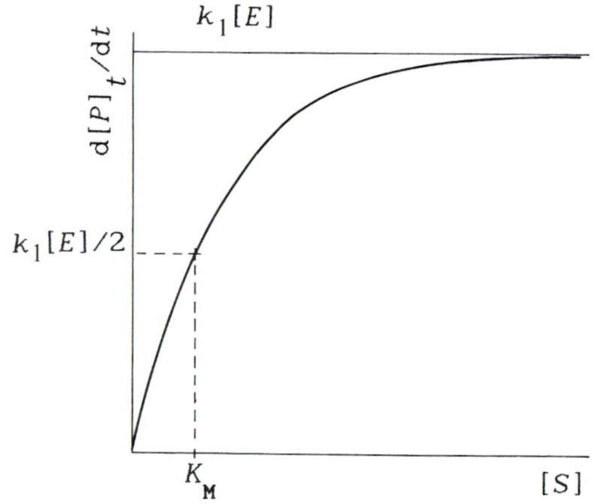

Figure 10.10 Rate of enzymolysis $d[P]_t/dt$ as a function of substrate concentration [S] for a given total enzyme concentration [E]. The higher [S] is, the closer the rate approaches the limiting value k_1[E]; the Michaelis constant expresses that value of [S] at which the rate of enzymolysis is half the limiting value.

For $[SE]_t$ itself, we have

$$d[SE]_t/dt = k_2[S]_t[E]_t \tag{10.147}$$

$$-d[SE]_t/dt = k_{-1}[SE]_t + k_1[SE]_t \tag{10.148}$$

Application of the steady-state hypothesis gives

$$k_2[S]_{eq}[E]_{eq} = (k_{-1} + k_1)[SE]_{eq} \tag{10.149}$$

If [E] is the total concentration of enzyme,

$$[E] = [E]_{eq} + [SE]_{eq} \tag{10.150}$$

Because the concentration of enzyme is much less than that of the substrate, $[S]_{eq} \simeq [S]$, the total substrate concentration; thus,

$$[SE]_{eq} = \frac{k_2([E] - [SE]_{eq})[S]}{k_{-1} + k_1} = \frac{k_2[E][S]}{k_{-1} + k_1 + k_2[S]} \tag{10.151}$$

whence, from (10.146)

$$d[P]_t/dt = \frac{k_1 k_2[E][S]}{k_{-1} + k_1 + k_2[S]} = \frac{k_1[E][S]}{K_M + [S]} \tag{10.152}$$

Thus, the enzymolysis is first order in both [E] and [S], and K_M is the Michaelis constant $(k_{-1} + k_1)/k_2$, which has the units of concentration.

From (10.152), we see that the rate of the enzymolysis reaction increases with [S] to the limit of $k_1[E]/(1 + K_M/[S])$ as [S] tends to a large value, that is, to the maximum k_1[E], Figure 10.10: K_M may be defined as that substrate concentration at which the rate attains half its maximum value.

The rates of the reactions of a nonspecific enzyme with two substrates are compared in Figure 10.11. The enzyme has the greater affinity, that is, it has the larger rate of reaction for a given concentration, for the substrate of lower K_M; the lower the value of K_M the higher

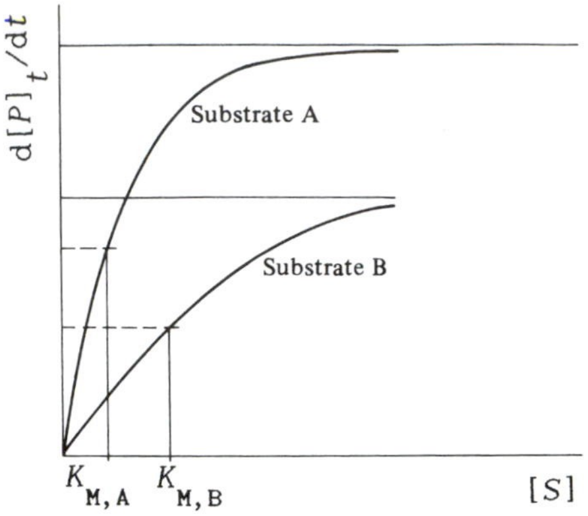

Figure 10.11 Rate of enzymolysis d[P]$_t$/dt as a function of substrate concentration [S] for substrates A and B; the smaller the Michaelis constant the larger the rate of enzymolysis.

the affinity for a substrate. When a very small amount of the substrate is present the condition $[S] \ll K_M$ holds and the rate of reaction becomes

$$d[P]_t/dt = (k_1/K_M)[E][S] \tag{10.153}$$

so that the enzymolysis is then second order overall.

The Michaelis constant can be determined through (10.152) if we write $d[P]_t/dt = k[E]$, where $k = k_1/(1 + K_m/[S])$ and is determined experimentally; then,

$$1/k = 1/k_1 + K_M/k_1[S] \tag{10.154}$$

and a plot of $1/k$ against $1/[S]$ will be linear with an intercept of $1/k_1$ and a slope of K_M/k_1.

10.13 Third-order reactions

A very small number of gas-phase reactions shows apparent third-order kinetics; they all involve nitrogen monoxide and another species, such as chlorine, oxygen or hydrogen.

Consider a general reaction of the form

$$2A + B \rightarrow P \tag{10.155}$$

Let the initial concentrations of A and B be a and b, respectively, and let the extent of the reaction at a time t be ξ; then $[A]_t = a - 2\xi$ and $[B]_t = b - \xi$; the rate equation is

$$d\xi/dt = k_3(a - 2\xi)(b - \xi) \tag{10.156}$$

which may be integrated to give

$$\frac{1}{(2b-a)^2}\left\{\frac{2\xi(2b-a)}{a(a-2\xi)} + \ln\left(\frac{b(a-2\xi)}{a(b-\xi)}\right)\right\} = k_3 t \tag{10.157}$$

This equation has been applied to the reaction

$$2NO(g) + H_2(g) \rightarrow N_2O(g) + H_2O(g) \tag{10.158}$$

and data at 1100 K are listed in Table 10.12. At a time τ when half the nitrogen monoxide is reacted $2\xi = a/2$, whence

Table 10.12. Apparent third-order kinetics for the reaction $2NO(g)+H_2(g)\rightarrow N_2O(g)+H_2O(g)$ at 1100 K

p_{NO}/Torr	p_{H_2}/Torr	τ/s	10^8k_3/Torr^{-2} s^{-1}
110	316	270	5.6
144	324	227	5.1
152	404	204	4.2
181	210	264	5.8
232	313	152	5.1
300	404	100	4.7
359	400	89	4.6
370	376	92	4.6

$$\frac{1}{(2b-a)^2}+\ln\left\{\frac{2b-a}{a}+\ln\left(\frac{2b}{4b-a}\right)\right\}=k_3\tau \tag{10.159}$$

In Table 10.2, $p_{NO}=a$ and $p_{H_2}=b$, and the values calculated for k_3 are in fair agreement.

An alternative treatment has been put forward for the similar reaction of nitrogen monoxide with oxygen

$$2NO(g)+O_2(g)\rightarrow 2NO_2(g) \tag{10.160}$$

that leads to *overall* third-order kinetics, but without the need to postulate three-body collisions. It has been found also that the rate of this reaction apparently decreases with temperature, which is contrary to the behaviour of other reactions. The experimental observations can be explained by the following reaction scheme:

$$\begin{array}{c} 2NO(g) \underset{k_b}{\overset{k_f}{\rightleftharpoons}} N_2O_2(g) \\ +O_2(g)\ \downarrow k_2 \\ 2NO_2(g) \end{array} \tag{10.161}$$

The rate of formation of nitrogen dioxide is given by the second-order reaction

$$d[NO_2]_t/dt=k_2[N_2O_2]_t[O_2]_t \tag{10.162}$$

We eliminate the term $[N_2O_2]_t$ through the equilibrium constant K (k_f/k_b), equal to $[N_2O_2]/[NO_2]^2$, giving

$$d[NO_2]_t/dt=k_2K[NO_2]_t^2[O_2]_t \tag{10.163}$$

The dimerization of nitrogen monoxide is exothermic; hence, K will decrease with increasing temperature (Section 8.4). This change will oppose the normal increase in k_2 with temperature and, if the temperature coefficient of k_2 is small, the product k_2K will decrease with an increase in temperature, as observed. It follows from (10.60) that E_a is given by

$$E_a=\mathcal{R}T^2\frac{d}{dt}\ln(k_2K) \tag{10.164}$$

This may be broken down to give

$$E_a = \mathcal{R}T^2\left(\frac{d}{dt}\ln(k_2) + \frac{d}{dt}\ln(k_f) - \frac{d}{dt}\ln(k_b)\right) \quad (10.165)$$

or

$$E_a = E_{a,2} + E_{a,f} - E_{a,b} \quad (10.166)$$

It is evident that the sign of the temperature coefficient of the apparent third-order rate constant (k_2K) is governed by the relative magnitudes of the three activation energies involved in the reaction.

Problems 10

10.1 The kinetics of the acid-catalysed hydrolysis of methyl ethanoate at 298 K

$$CH_3CO_2CH_3(aq) + H_2O(l) \xrightarrow{H_3O+} CH_3CO_2H(aq) + CH_3OH(aq)$$

were followed by withdrawing 2 cm^3 portions of the reaction mixture at times t after the initial mixing, adding them to 50 cm^3 ice-cold water, and then titrating with standard barium hydroxide solution. The following titres V were obtained:

t/min	0	10	21	40	115	∞
V/cm^3	18.5	19.1	19.7	20.7	23.6	34.8

Determine the order, the rate constant and the half-life for the reaction.

10.2 (a) The half-life of the krypton isotope ^{85}Kr is 10.6 year. How long is required for 99% of a sample of ^{85}Kr to disintegrate? (b) Calculate the mass of radon ^{222}Rn, in equilibrium with 1 g ^{226}Ra, if the half-lives are 3.83 day and 1622 year, respectively.

10.3 The acid-catalysed conversion of sucrose into glucose and fructose is a first-order reaction, which leads to a reversal of the sign of the optical rotation α. The following data were obtained with a polarimeter, measuring the rotation α as a function of time t:

t/min	5.0	20	44	90	140	175	∞
α_t/deg	12.2	9.95	6.95	2.70	0.10	−1.30	−4.00

Determine the rate constant and the half-life for the reaction.

10.4 The following data refer to the decomposition of ammonia on a tungsten surface at 1100 K: $2NH_3(g) \rightarrow N_2(g) + 3H_2(g)$. Show that the reaction follows zeroth-order kinetics, and determine the rate constant.

$p_{init}(NH_3)$/mmHg	265	130	58	16
$t_{1/2}$/min	7.6	3.7	1.7	0.5

10.5 In the decomposition of nitrogen monoxide, it has been shown that, at constant temperature, the time $t_{1/2}$ required for half the reaction to be completed is inversely proportional to the initial pressure p_0. The following results were obtained at the temperatures shown. Deduce the order of reaction, and calculate the rate constant at 967 K and the activation energy over the given temperature range.

T/K	967	1030	1085
p_0/mmHg	294	360	345
$t_{1/2}$/s	1520	212	53

10.6 The rearrangement of *N*-chloroacetanilide to 4-chloroacetanilide is catalysed by hydroxonium ions:

$$C_6H_5N(Cl)COCH_3 \xrightarrow{H_3O^+} ClC_6H_4N(H)COCH_3$$

N-chloroacetanilide liberates iodine from aqueous potassium iodide, and the reaction can be followed by titration with sodium thiosulfate. From the following data on titre volumes *V* at times *t*, deduce the order of reaction and determine the rate constant for the rearrangement reaction.

t/min	0	15	30	45	60	75
V/cm³	24.5	18.1	13.3	9.7	7.1	5.2

10.7 The mutarotation of α-D-(+)-glucose has been followed by polarimetric measurements of the optical rotation α_t at times *t* from the start of the experiment:

t/min	0	10	20	30	40	50	60	∞
α_t	130.7	112.3	98.1	87.0	78.3	71.9	66.3	47.5

Assuming the forwards and reverse reactions to be of first order, calculate the two rate constants.

10.8 It has been shown that the observed rate of the gas-phase decomposition of dinitrogen pentoxide, $N_2O_5(g) \rightarrow 2NO_2(g) + \frac{1}{2}O_2(g)$, may be interpreted in terms of the following sequence of reactions:

(i) $N_2O_5(g) \underset{k_b}{\overset{k_a}{\rightleftharpoons}} NO_2(g) + NO_3(g)$

(ii) $NO_2(g) + NO_3(g) \xrightarrow{k_c} NO_2(g) + NO(g) + O_2(g)$

(iii) $NO(g) + N_2O_5(g) \xrightarrow{k_d} 3NO_2(g)$

in which the bimolecular reaction (ii) is the rate-determining step. Verify that this mechanism leads to the first-order rate equation

$$-d[N_2O_5]_t/dt = 2k_a k_c [N_2O_5]_t/(k_b + k_c).$$

10.9 For a reaction A + 2B → Products, the rate equation is given by $d\xi/dt = k_2[A]_t[B]_t$, where ξ represents the extent of the reaction. Obtain an integrated rate equation and indicate the nature of the graphical plot that would be used to determine k_2.

10.10 By a computer-simulation technique, or otherwise, obtain a graphical representation of the concentrations as a function of time for the species in the reaction scheme

$$A \xrightarrow{k_1} B \xrightarrow{k_2} C$$
$$k_4 \upharpoonleft \downharpoonright k_3$$
$$D$$

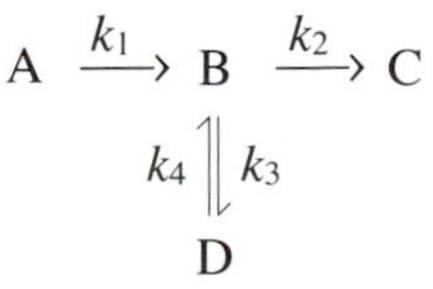

where $[A]_0$=0.250 mol dm^{-3} and $[B]_0$=$[C]_0$=$[D]_0$=0; k_1=0.8 min^{-1}, k_2=0.3 min^{-1}, k_3=0.2 min^{-1} and k_4=0.3 min^{-1}. Rationalize the simulated time scale t' with the value of 0.5 min for $t_{1/2}$ in the reaction A→B. (For computation, $[X]$ may be scaled by 10^3 and k by 10.)

10.11 A series of experiments on the auto-oxidation of hydroxylamine led to the following results:

T/K	273	283	288	298	313
$10^4 k_2$/mol^{-1} dm^3 s^{-1}	0.237	0.680	1.02	2.64	9.04

Determine the activation energy and the Arrhénius frequency factor for the auto-oxidation.

10.12 Gaseous hydrogen iodide is irradiated at 253 nm, and 0.196 mol decomposes per 250 J of radiant energy absorbed. What is the quantum efficiency of the decomposition reaction?

10.13 The quantum efficiency of the gas-phase combination of H_2 and Cl_2 is $\simeq 10^6$, when irradiated at 480 nm. How many moles of hydrogen chloride are formed per joule of radiant energy absorbed?

10.14 The catalytic decomposition of aqueous hydrogen peroxide solution was monitored by titrating aliquots at times t with aqueous potassium permanganate, leading to the following results:

t/min	5	10	20	30	50
$KMnO_4$/cm^3	37.1	29.8	19.6	12.3	5.0

Determine (a) the order of the reaction, (b) the rate constant for the reaction and (c) the permanganate titration at t=0 min.

10.15 The hydrolytic decomposition of hydroxylamine at 80 °C varies with ionic strength as shown by the following data:

$10^3 I$/mol dm^{-3}	5.06	11.21	15.85	22.94
k/mol^{-1} dm^3 hour^{-1}	1.07	1.02	0.98	0.89

Determine (a) the charges on the reacting species and (b) the rate constant corrected for the primary kinetic salt effect; the Debye–Hückel limiting equation constant at 80 °C is 1.322.

10.16 The following data refer to the adsorption of carbon monoxide on mica at 85 K:

p/mmHg	100	200	300	400	500	600
V/cm^3	0.126	0.155	0.167	0.174	0.179	0.182

(a) Deduce whether the Freundlich or Langmuir isotherm is the better representation of the data, (b) what is the maximum volume of CO adsorbed, (c) determine K_F (or K_L) and (d) what volume would be adsorbed at 1 atm?

10.17 Using results from Table 10.8, calculate the pressure of N_2O in contact with a gold surface at 1170 K after 2.5 hour, if the initial pressure was 350 Torr. After what time will the decomposition be 95% complete?

10.18 The following data refer to the adsorption of acetic acid, from varying initial concentrations c, on wood charcoal at 25 °C; m is the mass adsorbed per g of charcoal:

c/mol dm^{-3}	0.05	0.10	0.50	1.00	1.50
m/g	0.045	0.061	0.110	0.148	0.178

Deduce whether the Freundlich or Langmuir isotherm is the better representation of the data, and find the constants of the chosen isotherm.

10.19 The rate R of nitrogen elimination by an enzyme on a substrate was measured for a series of substrate concentrations [S]. Determine the Michaelis constant for the reaction, from the following data:

10^2[S]/mol dm^{-3}	5.0	3.0	1.0	0.50	0.20
R/mm^3 min^{-1}	16.7	15.0	10.0	6.67	3.33

10.20 For a reaction A→ Products, the rate is given by $-d[A]_t/dt=k_n[A]_t^n$. Show that, for $n \geq 2$,

$$t_{1/2}=\frac{2^{n-1}-1}{(n-1)_n k\,[A]^{n-1}}$$

Does this expression for $t_{1/2}$ hold at (a) $n=1$, and (b) $n=0$?

10.21 The Lindemann theory may be explored for the isomerization of cyclopropane to propene by means of the following data, obtained by Pritchard, Sowden and Trotman-Dickenson (1953):

(a) Moderate pressures

p_0/mmHg	200	200	400	400	600	600
p_t/mmHg	186	173	373	347	559	520
t/s	100	200	100	200	100	200

(b) Low pressures

p/mmHg	84.1	11.0	2.89	0.569	0.120	0.067
$10^4k'$/s^{-1}	2.98	2.23	1.54	0.857	0.392	0.303

Test the applicability of the Lindemann theory of unimolecular reactions through these data sets.

Problem-solving with personal computers

An important part of the study of physical chemistry is problem-solving. This activity may range from the evaluation of a parameter from a given equation to the detailed interpretation of a set of experimental data. An involvement with numerical work conveys an understanding of the magnitudes of physical quantities. This facility is important, because computers and hand calculators produce sensible results only if they are supplied correctly with good data.

A1.1 Solving numerical problems

Numerical problems give practice in relating experimental observations to theoretical models. The insertion of magnitudes into a given equation is a common scientific activity that should be mastered.

The solving of problems leads to an appreciation of several important features:

(a) the orders of magnitude of physico-chemical quantities;
(b) the need for an understanding of units;
(c) the value of checking dimensional homogeneity;
(d) the sources of physico-chemical data;
(e) the precision of the data and its transmission to the result.

Most problems involve algebraic manipulation. However, it is essential to obtain a clear picture of the physical chemistry involved in a problem before embarking on a series of mathematical processes, and it is useful to obtain an explicit algebraic expression before inserting numerical values. There are several advantages in so doing:

(f) the expression can be checked dimensionally;
(g) the possible cancellation of terms would simplify the arithmetic;
(h) similar problems with other magnitudes can be solved with little additional effort;
(i) it is good examination practice.

If the data are formed into an expression as numbers 1 to 9 multiplied by the appropriate powers of 10, it is easy to estimate an approximate answer. Suppose that we have for a relative permittivity ϵ_r

$$(\epsilon_r - 1) = \mathcal{N}\mu^2/(9\epsilon_0 k_B T) \tag{A1.1}$$

$\mathcal{N}$ (number of molecules per unit volume)	$=2.461 \times 10^{25}$	m^{-3}
μ (dipole moment)	$=5.11 \times 10^{-30}$	C m
ϵ_0 (permittivity of a vacuum)	$=8.8542 \times 10^{-12}$	$F\ m^{-1}$
k_B (Boltzmann constant)	$=1.3807 \times 10^{-23}$	$J\ K^{-1}$
T (absolute temperature)	$=298.15$	K

Inserting approximate parameters into (A1.1) gives

$$(\epsilon_r - 1) = \frac{2.5\times10^{25}\ \text{m}^{-3}\times25\times10^{-60}\ \text{C}^2\ \text{m}^2}{10\times10\times10^{-12}\ \text{F m}^{-1}\times1\times10-^{23}\ \text{J K}^{-1}\times300\ \text{K}} \tag{A1.2}$$

Equation (A1.2) is dimensionally correct, insofar as both sides are dimensionless. We can see that $(\epsilon_r - 1) \simeq 2\times10^{-3}$. Thus, when the expression is evaluated, we may write with confidence $(\epsilon_r - 1) \simeq 1.959\times10^{-3}$.

A1.2 Suggested procedure

There are different ways of tackling problems, so that these notes are offered only as a guide. Sometimes a recommended stage may be changed or bypassed. Elegant derivations are often concise: the converse in not necessarily true, and failure to justify a stage in a derivation may indicate a lack either of judgement or of confidence. Over-elaboration of trivial detail and of arithmetic manipulation are equally unacceptable in a polished answer to a problem. Some degree of subjective judgement is involved in the process of problem solving, but it is expected that a correct numerical answer should include adequate evidence of the method used. Thus,

(a) Read the problem carefully. If you think that it contains an ambiguity, assume the simplest interpretation of the ambiguity and comment on it.
(b) Summarize the information given by appropriate means, such as
   (i) labelled drawings;
   (ii) energy-level diagrams;
   (iii) sketch-graphs, correctly labelled;
   (iv) defining symbols used in diagrams and formulae;
   (v) listing numerical values with units.
(c) Indicate relevant laws and equations that are employed in developing the problem, at least initially;
(d) State methods to be used; for example, 'take $\log_e$ of each side of equation (1)';
(e) When appropriate, formulate an explicit equation before inserting numerical data. Look for cancellations of terms, and indicate any physical or functional approximations used;
(f) Do not make needless numerical approximations, but state any approximations that are made and include an estimate of the probable error, as far as is possible from the given data;
(g) For convenience, substitute a new symbol for a group of symbols in deriving an expression;
(h) Check the dimensions of equations for consistency. Remember that exponential, logarithmic and other functional arguments are dimensionless;
(i) Insert numerical values into expressions carefully and determine an approximate result, as with (A1.1);
(j) Think about the answer in terms of the physical chemistry of the problem. Doubtful results should be checked;
(k) Keep a neat format in the answer to any problem.

A1.3 Example problem and solution

The diffusion coefficient D of carbon in α-iron, as a function of temperature, is assumed to follow the equation:

$$D = D_0 \exp[-E_d/(\mathcal{R}T)] \tag{A1.3}$$

Values have been obtained for D as a function of the temperature T:

T/K	300	500	700	900	1100
$D/\text{m}^2\ \text{s}^{-1}$	4.73×10^{-21}	3.35×10^{-15}	1.80×10^{-12}	2.66×10^{-11}	2.05×10^{-10}

It is required to verify the above equation, to find values both for the activation energy for diffusion E_d and for D_0, and to comment briefly on the results (R=8.3145 J K^{-1} mol^{-1}). The following solution attempts to illustrate some of the points discussed above.

The presence of $\mathcal{R}$ in the exponent, and its units, indicate that E_d is expected in J mol^{-1}; D_0 is the limiting value of D as $T\to\infty$. Taking the natural logarithm of each side of (A1.3) gives

$$\ln(D) = \ln(D_0) - E_d/(RT) \tag{A1.4}$$

If the graph of $\ln(D)$ against $1/T$ is linear (A1.3) would be verified. The slope of the line, $\Delta\ln(D)/\Delta(1/t)$, is $(-E_d/\mathcal{R})$, and the intercept at $1/T=0$ is $\ln(D_0)$. Setting out the required data, we have the following:

$D/\text{m}^2\ \text{s}^{-1}$	$\ln(D/\text{m}^2\ \text{s}^{-1})$	T/K	$10^3/(T/\text{K})$
4.73×10^{-21}	−46.800	300	3.3333
3.35×10^{-15}	−33.330	500	2.0000
1.08×10^{-12}	−27.554	700	1.4286
2.66×10^{-11}	−24.350	900	1.1111
2.05×10^{-10}	−22.308	1100	0.90909

The straight-line graph (Figure A1.1) verifies (A1.3) for the diffusion of carbon in α-iron over the experimental temperature range. By least squares (see Appendix 8) the slope of the line is −10103 K, giving E_d=84.001 kJ mol^{-1}; D_0=1.999$\times10^{-6}$ m^2 s^{-1}. It is clear from the graph, and the least-squares correlation coefficient (r=1.0000), that no datum merits exclusion from the calculation.

The sign and magnitude of E_d are reasonable, since work must be done on the system to bring about diffusion, and the energies of such processes are usually of the order of 1 eV per atom ($\simeq$96 kJ mol^{-1}). The coefficient D increases with increasing temperature and, if (A1.3) continued to hold, would tend to the limiting value 1.999$\times10^{-6}$ m^2 s^{-1}. However, at very high temperatures the material would melt, or even vaporize. The usefulness of D_0 here is in calculating D from (A1.3).

The estimated standard deviation of the slope is (see Appendix 8) 0.75 K, which is transmitted to E_d as 0.006 kJ mol^{-1}; thus, the final value for E_d is (84.001±0.006) kJ mol^{-1}, usually written as 84.001(6) kJ mol^{-1}.

A1.4 Computer methods in problem-solving

Throughout this book, the reader is referred to procedures, such as linear least squares and numerical integration, that often need to be invoked while gaining the desired familiarity with the subject matter. To this end a number of computer programs has been written that meet the requirements of problem-solving in the context of the book.

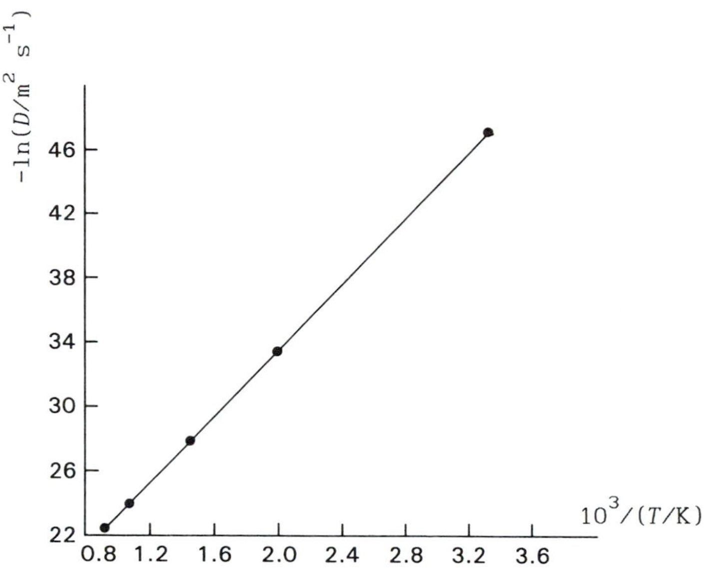

Figure A1.1 Graph of $-\ln(D/\mathrm{m^2\,s^{-1}})$ against $10^3/(T/\mathrm{K})$.

A1.4.1 Program design and availability

The programs have been written in a self-contained and self-explanatory manner. They may be executed on any IBM-compatible computer, and need only a monitor and a printer as peripherals. The word in parentheses after the title of each program listed is its calling name; thus, the linear least squares program is called by LSLI. The information on the monitor guides the user through each program. Access to the programs is through the Internet (Worldwide Web) www.cup.cam.ac.uk, from where further instructions are available.

A1.4.2 Programs

The programs available are listed below; in most cases, test data are supplied so as to allow the correct working of each program to be checked.

Linear least squares (LSLI)
Gaussian quadrature: numerical integration (QUAD)
Madelung constant calculation (MADC)
Radial wavefunctions (RADL)
Maxwell–Boltzmann distribution (GASD)
Electron-in-a-box (BOXS)
Monte Carlo procedures (MONC)
Angular wavefunctions (PLOT)
Eigenvalues and eigenvectors (EIGN)
Hückel molecular-orbital calculations (HUCK)
Point-group derivation (EULS and EULH)
Point-group recognition (SYMS and SYMH)

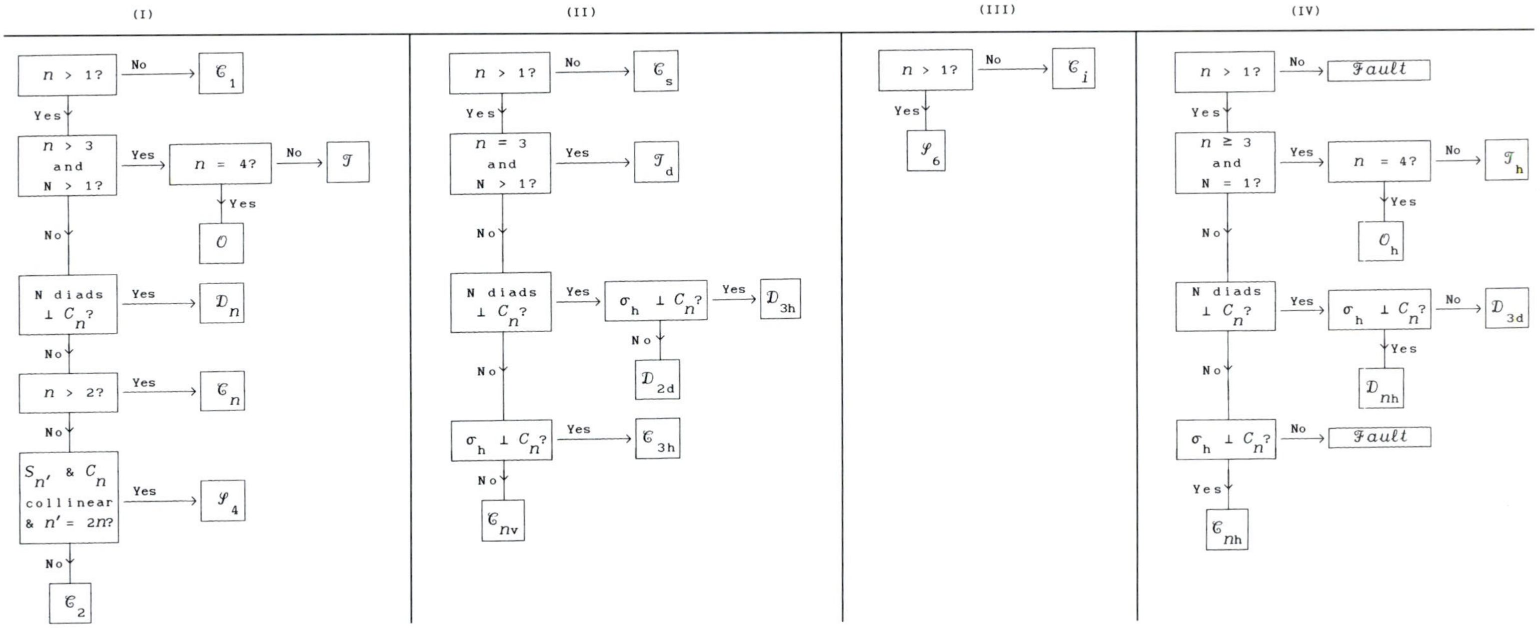

Figure A1.2 Block diagram of the scheme for point-group recognition embodied in the programs SYMS and SYMH. The point groups $\mathscr{C}_{\infty v}$ and $\mathscr{D}_{\infty h}$ fit the scheme under types (II) and (IV), respectively.

Table A1.1. Schönflies and Hermann–Mauguin symmetry notations

Schönflies	Hermann–Mauguin	Schönflies	Hermann–Mauguin
$\mathcal{C}_1$	1	$\mathcal{D}_2$	222
$\mathcal{C}_2$	2	$\mathcal{D}_3$	32
$\mathcal{C}_3$	3	$\mathcal{D}_4$	422
$\mathcal{C}_4$	4	$\mathcal{D}_6$	622
$\mathcal{C}_6$	6	$\mathcal{D}_{2h}$	mmm
$\mathcal{C}_i$	$\bar{1}$	$\mathcal{D}_{3h}$	$\bar{6}m2$
$\mathcal{C}_s(\mathcal{S}_1)$	$m\ (\bar{2})$	$\mathcal{D}_{4h}$	$\frac{4}{m}mm$
$\mathcal{S}_4$	$\bar{4}$	$\mathcal{D}_{6h}$	$\frac{6}{m}mm$
$\mathcal{S}_6$	$\bar{3}$	$\mathcal{D}_{2d}$	$\bar{4}2m$
$\mathcal{C}_{2h}$	$2/m$	$\mathcal{D}_{3d}$	$\bar{3}m$
$\mathcal{C}_{3h}\ (\mathcal{S}_3)$	$\bar{6}$	$\mathcal{T}$	23
$\mathcal{C}_{4h}$	$4/m$	$\mathcal{T}_d$	$m3$
$\mathcal{C}_{6h}$	$6/m$	$\mathcal{T}_d$	$\bar{4}3m$
$\mathcal{C}_{2v}$	$mm2$	$\mathcal{O}$	432
$\mathcal{C}_{3v}$	$3m$	$\mathcal{O}_h$	$m3m$
$\mathcal{C}_{4v}$	$4mm$	$\mathcal{C}_{\infty v}$	∞m
$\mathcal{C}_{6v}$	$6mm$	$\mathcal{D}_{\infty h}$	∞/m

Roots of polynomials (POLY)
Curve fitting and interpolation (INTP)
Computer-simulation kinetics (MONK)

Figure A1.2 is a block diagram of the scheme embodied in the point-group recognition programs SYMS and SYMH. The program accommodates the 32 crystallographic point groups, and the two groups $\mathcal{C}_{\infty v}$ and $\mathcal{D}_{\infty h}$ fit the scheme under types (II) and (IV), respectively. In this scheme n refers to the degree of proper rotation, as in $\mathcal{C}_n$, and N refers to the number of these $\mathcal{C}_n$ axes. Table A1.1 compares the Schönflies and Hermann–Mauguin point-group symmetry notations.

APPENDIX 2

Stereoviewing

The representation of molecular and crystal structures by stereoscopic pairs of drawings has become commonplace in recent years. Computer programs are available that prepare stereoviews from structural data. Two diagrams of a given object are necessary and they must correspond to the views seen by the eyes in normal vision. Correct stereoscopic viewing requires that each eye sees only the appropriate drawing, and there are several ways in which it may be accomplished[1].

1. A stereoviewer may be purchased for a modest sum. The stereoscopic drawing may be viewed directly, and the three-dimensional image appears centrally between the images of the two given drawings. Two suppliers of stereoviewers are:
   (a) Casella & Co Ltd, Regent House, Britannia Walk, London N1 7ND;
   (b) Taylor-Merchant, 212 West 35th Street, New York, NY10001.
2. The unaided eyes can be trained to defocus, so that each eye sees only the appropriate diagram. The eyes must be relaxed and look straight ahead. This process may be aided by placing a white card edgeways between the two drawings. It may be helpful to close the eyes for a moment, open them wide without attempting to focus on the diagram, and then allow them to relax. Again, the stereoview appears centrally with respect to the left and right pair of images.

[1] Instructions for constructing a stereoviewer are given by M F C Ladd, *Symmetry in Molecules and Crystals* (Ellis Horwood, 1992).

Average classical thermal energies

A3.1 Average kinetic energy

Consider a system of classical particles, each of mass m but with different speeds v. The kinetic energy ϵ_K of any particle is $\frac{1}{2}mv^2$. We shall assume that the energies of the particles follow a Boltzman distribution; then, from (5.43), we have for the average kinetic energy

$$\bar{\epsilon}_K = \overline{mv^2/2} = \frac{\int_{-\infty}^{\infty}\int_{-\infty}^{\infty}\int_{-\infty}^{\infty}(mv^2/2)\exp[-mv^2/(2k_BT)]\,dv_x dv_y dv_z}{\int_{-\infty}^{\infty}\int_{-\infty}^{\infty}\int_{-\infty}^{\infty}\exp[-mv^2/(2k_BT)]\,dv_x dv_y dv_z} \tag{A3.1}$$

Following Appendix 5, and since the integrands are even functions of v, we can write

$$\bar{\epsilon}_K = \frac{2\int_0^{\infty}(mv^2/2)\exp[-mv^2/(2k_BT)]\,v^2 dv\int_0^{\pi}\sin(\theta)\,d\theta\int_0^{2\pi}d\phi}{2\int_0^{\infty}\exp[-mv^2/(2k_BT)]\,v^2 dv\int_0^{\pi}\sin(\theta)\,d\theta\int_0^{2\pi}d\phi} \tag{A3.2}$$

which simplifies to

$$\bar{\epsilon}_K = \frac{m/2\int_0^{\infty}v^4\exp[-mv^2/(2k_BT)]\,dv}{\int_0^{\infty}v^2\exp[-mv^2/(2k_BT)]\,dv} \tag{A3.3}$$

Following Appendix 6, it is straightforward to show that the numerator and denominator in (A3.3) evaluate to $(m/4)x^{-5/2}\Gamma(5/2)$ and $(1/2)x^{-3/2}\Gamma(3/2)$, respectively, where $x = m/(2k_BT)$. Hence,

$$\bar{\epsilon}_K = \tfrac{3}{2}k_BT \tag{A3.4}$$

A3.2 Average vibrational energy

The energy of a particle of mass m and speed v at any instant, moving under one-dimensional simple harmonic oscillation of frequency ν is, from (2.1) and (2.58), $\frac{1}{2}mv^2 + 2\pi^2 m\nu^2 x^2$. Following the same treatment as in Section A3.1, we can write the average vibrational energy as

$$\bar{\epsilon}_v = \frac{m/2\int_{-\infty}^{\infty}\int_{-\infty}^{\infty}(v^2+4\pi^2\nu^2x^2)\exp\left[-m(v^2+4\pi^2\nu^2x^2)/(2k_BT)\right]\,\mathrm{d}v\,\mathrm{d}x}{\int_{-\infty}^{\infty}\int_{-\infty}^{\infty}\exp\left[-m(v^2+4\pi^2\nu^2x^2)/(2k_BT)\right]\,\mathrm{d}v\,\mathrm{d}x} \tag{A3.5}$$

It is readily confirmed that (A3.5) is equivalent to

$$\bar{\epsilon}_v = \frac{m\int_0^{\infty}v^2\exp\left[-mv^2/(2k_BT)\right]\,\mathrm{d}v}{2\int_0^{\infty}\exp\left[-mv^2/(2k_BT)\right]\,\mathrm{d}v} + \frac{4\pi^2\nu^2m\int_0^{\infty}x^2\exp\left[-4\pi^2\nu^2mx^2/(2k_BT)\right]\,\mathrm{d}x}{2\int_0^{\infty}\exp\left[-4\pi^2\nu^2mx^2/(2k_BT)\right]\,\mathrm{d}x} \tag{A3.6}$$

Each term on the right-hand side of (A3.6) solves to $k_BT/2$, whence

$$\bar{\epsilon}_v = k_BT \tag{A3.7}$$

Reduced mass

We consider a two-particle problem in which the energy of the system is determined by the distance between the two particles.

Let two particles of masses m_1 and m_2 be vibrating along a line joining their centres. The speeds of the particles at any instant are v_1 and v_2, respectively, and the distance between them is r; C is the centre of mass of the system.

The kinetic energy E_K of the system is given by

$$E_K=\tfrac{1}{2}m_1v_1^2+\tfrac{1}{2}m_2v_2^2 \qquad (A4.1)$$

and the centre of mass is defined by

$$m_1r_1+m_2r_2=0 \qquad (A4.2)$$

From Figure A4.1, noting the positive direction of r

$$r=r_2-r_1 \qquad (A4.3)$$

Therefore,

$$m_1(r_2-r)=-m_2r_2 \qquad (A4.4)$$

or

$$r_2=m_1r/(m_1+m_2) \qquad (A4.5)$$

Hence from (A4.3) and (A4.5)

$$r_1=-m_2r/(m_1+m_2) \qquad (A4.6)$$

Since

$$v_i=\frac{\mathrm{d}r_i}{\mathrm{d}t}=\dot{r}_i \qquad (A4.7)$$

$$E_K=\tfrac{1}{2}m_1\dot{r}_1^2+\tfrac{1}{2}m_2\dot{r}_2^2 \qquad (A4.8)$$

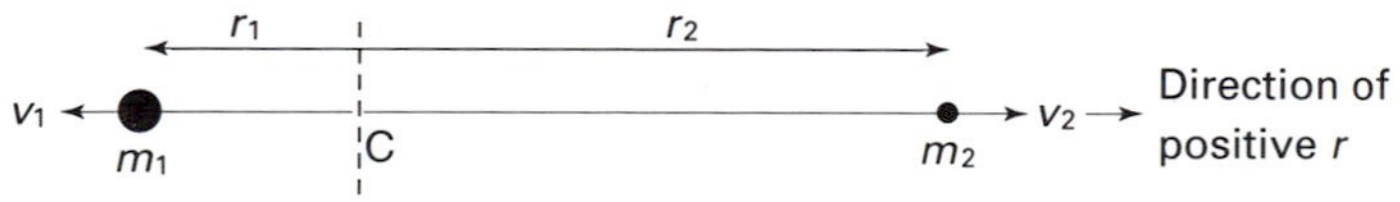

Figure A4.1 Two-particle system vibrating about a centre of mass C.

or

$$E_K = \tfrac{1}{2} m_1 m_2^2 \dot{r}^2/(m_1+m_2)^2 + \tfrac{1}{2} m_1^2 m_2 \dot{r}^2/(m_1+m_2)^2.$$

$$= \tfrac{1}{2} m_1 m_2 \dot{r}^2/(m_1+m_2) \qquad \text{(A4.9)}$$

Equation (A4.9) may be written

$$E_K = \tfrac{1}{2} \mu \dot{r}^2 \qquad \text{(A4.10)}$$

where μ, given by,

$$\mu = m_1 m_2/(m_1+m_2) \qquad \text{(A4.11)}$$

is the reduced mass of the system. The motion of the system of two particles is, thus, equivalent to the motion of one particle of mass μ about the other particle considered as a fixed centre of force. In reality neither particle is fixed: the only truly fixed point is the centre of mass, and (A4.10) gives the kinetic energy in the assumed Galilean frame of reference in which the centre of mass is stationary.

If one particle is much more massive than the other, $m_1 \gg m_2$, then the more massive particle is almost coincident with the centre of mass, and the reduced mass is almost the same as that of the lighter particle. From (A4.11)

$$1/\mu = 1/m_1 + 1/m_2 \simeq 1/m_2 \qquad \text{(A4.12)}$$

so that

$$\mu \simeq m_2 \qquad \text{(A4.13)}$$

In the hydrogen bromide molecule, the atoms are separated by a distance of 0.141 nm. The reduced relative mass is given by

$$1/\mu = 1/79.904 + 1/1.0079$$

$$\mu = 0.9952$$

so that $\mu \simeq M_r(\mathrm{H})$ to approximately 1.3%.

Spherical polar coordinates

The polar coordinates, r, θ and ϕ are defined by

$$x = r\sin(\theta)\cos(\phi)$$
$$y = r\sin(\theta)\sin(\phi) \quad \text{(A5.1)}$$
$$z = r\cos(\theta)$$

where

$$r^2 = x^2 + y^2 + z^2 \quad \text{(A5.2)}$$

In normalization problems, we need to express a volume element $d\tau$, equal to $dx\,dy\,dz$, in Cartesian coordinates, in polar coordinates. Consider the volume element $d\tau$ shown in Figure A5.1; it corresponds to the quantity $dx\,dy\,dz$. From the diagram, it is a straightforward matter to determine the magnitudes of the sides of the volume element, which may be taken to be parallelepipedal. Hence,

$$d\tau = r^2\,dr\sin(\theta)\,d\theta\,d\phi \quad \text{(A5.3)}$$

The integration limits of the variables, which correspond to x, y and z each between $-\infty$ and $+\infty$, are

$$0 \leqslant r \leqslant \infty$$
$$0 \leqslant \theta \leqslant \pi \quad \text{(A5.4)}$$
$$0 \leqslant \phi \leqslant 2\pi$$

where these limits define a sphere of infinite radius.

A5.1 Laplacian operator ∇^2

From the rules of partial differentiation, for any function $f(r)$, where $r = g(x,y,z)$, we can transform variables in the following manner:

$$\frac{\partial f}{\partial x} = \frac{\partial f/\partial r}{\partial x/\partial r} \quad \text{(A5.5)}$$

and

$$\frac{\partial^2 f}{\partial x^2} = \left[\frac{1}{\partial x/\partial r}\right]\frac{\partial}{\partial r}\left[\frac{\partial f/\partial r}{\partial x/\partial r}\right] \quad \text{(A5.6)}$$

$$= \frac{(\partial^2 f/\partial r^2)(\partial x/\partial r) - (\partial^2 x/\partial r^2)(\partial f/\partial r)}{(\partial x/\partial r)^3}$$

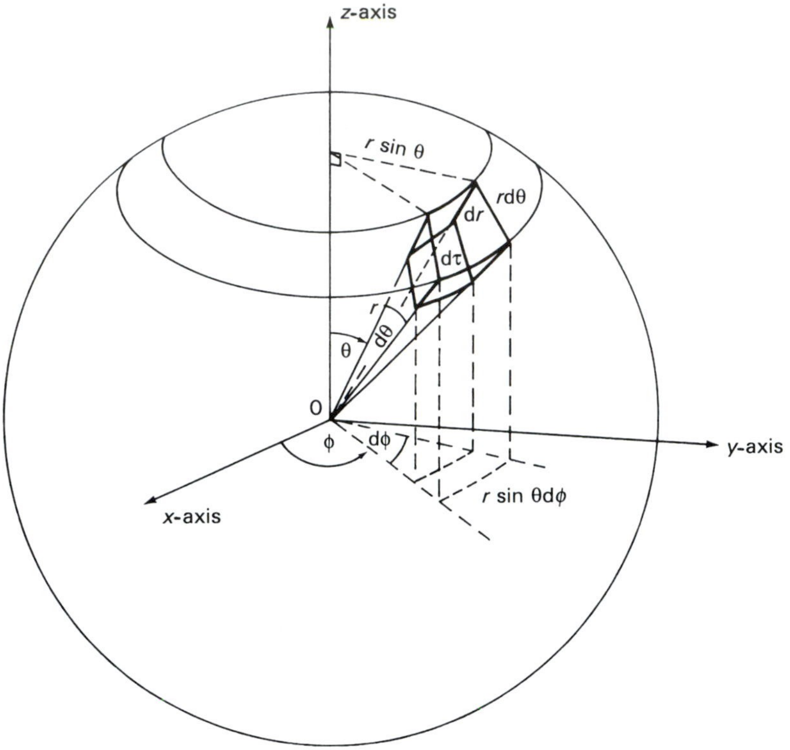

Figure A5.1 Volume element dτ in polar coordinates.

If $g(x,y,z)$ is given by (A5.2), $\partial x/\partial r = r/x$, and $\partial^2 x/\partial r^2 = (x^2 - r^2)/x^3$. Thus

$$\frac{\partial^2}{\partial x^2} = \left[\frac{r}{x}\left(\frac{\partial^2}{\partial r^2}\right) + \frac{(r^2 - x^2)}{x^3}\left(\frac{\partial}{\partial r}\right)\right] / (r/x)^3 \tag{A5.7}$$

Equations for $\partial^2/\partial y^2$ and $\partial^2/\partial z^2$ are symmetrical; hence,

$$\frac{\partial^2}{\partial x^2} + \frac{\partial^2}{\partial y^2} + \frac{\partial^2}{\partial z^2} = r^{-3}\left[r^3 \frac{\partial^2}{\partial r^2} + 2r^2 \frac{\partial}{\partial r}\right] \tag{A5.8}$$

or

$$\nabla_r^2 = \frac{\partial^2}{\partial r^2} + \frac{2}{r}\frac{\partial}{\partial r} = \frac{1}{r^2}\frac{\partial}{\partial r}\left(r^2 \frac{\partial}{\partial r}\right) \tag{A5.9}$$

In a similar manner, using (A5.1) and noting that $y/x = \tan(\phi)$, we can show that

$$\frac{\partial x}{\partial \phi} = -r\,\frac{\sin(\theta)}{\sin(\phi)}$$

$$\frac{\partial^2 x}{\partial \phi^2} = r\,\frac{\sin(\theta)\cos(\phi)}{\sin^2(\phi)}$$

$$\frac{\partial y}{\partial \phi} = r\,\frac{\sin(\theta)}{\cos(\phi)} \tag{A5.10}$$

$$\frac{\partial^2 y}{\partial \phi^2}=r\frac{\sin(\theta)\sin(\phi)}{\cos^2(\phi)}$$

$$\frac{\partial z}{\partial \phi}=\frac{\partial^2 z}{\partial \phi^2}=0$$

whence, using (A5.6)

$$\nabla_\phi^2=\frac{1}{r^2\sin^2(\theta)}\frac{\partial^2}{\partial \phi^2} \tag{A5.11}$$

It is left as an exercise to the reader to show that

$$\nabla_\theta^2=\frac{1}{r^2}\frac{\partial^2}{\partial \theta^2}+\frac{\cos(\theta)}{r^2\sin(\theta)}\frac{\partial}{\partial \theta} \tag{A5.12}$$

It is useful to put

$$\tan(\theta)=\left(\frac{x^2+y^2}{z^2}\right)^{1/2}$$

hence

$$\nabla^2=\nabla_r^2+\nabla_\theta^2+\nabla_\phi^2=r^{-2}\left[\frac{\partial}{\partial r}\left(r^2\frac{\partial}{\partial r}\right)+\frac{1}{\sin^2(\theta)}\frac{\partial^2}{\partial \phi^2}+\frac{1}{\sin(\theta)}\frac{\partial}{\partial \theta}\left(\sin(\theta)\frac{\partial}{\partial \theta}\right)\right] \tag{A5.13}$$

Gamma function

The gamma function[1] is useful in handling integrals of the type

$$\int_0^\infty x^n \exp(-ax^2)\,dx \tag{A6.1}$$

where a is a constant; they occur in several areas of physical chemistry. The gamma function $\Gamma(n)$ may be represented by the equation

$$\Gamma(n)=\int_0^\infty t^{n-1}\exp(-t)\,dt \tag{A6.2}$$

The following three results are important.

(a) For $n>0$ and integral,

$$\Gamma(n)=(n-1)! \tag{A6.3}$$

(b) For $n>0$,

$$\Gamma(n+1)=n\Gamma(n) \tag{A6.4}$$

and if n is also integral,

$$\Gamma(n+1)=n! \tag{A6.5}$$

(c)

$$\Gamma(1/2)=\sqrt{\pi} \tag{A6.6}$$

As an example, consider the solution of the integral

$$I=\int_0^\infty x^4\exp(-x^2/2)\,dx \tag{A6.7}$$

Let $x^2/2=t$, so that $x=(2t)^{1/2}$ and $dx=(2t)^{-1/2}dt$. Then,

$$I=2\sqrt{2}\int_0^\infty t^{3/2}\exp(-t)\,dt \tag{A6.8}$$

Hence,

$$I=2\sqrt{2}\,\Gamma(\tfrac{5}{2}),\text{ or }3\,(\pi/2)^{1/2} \tag{A6.9}$$

(See also (A12.7) in Appendix 12.)

[1] See, for example, H Margenau and G M Murphy, *The Mathematics of Physics and Chemistry* (van Nostrand, 1943).

Slater's rules

In quantum mechanical calculations it is often sufficiently accurate, with principal quantum numbers up to 4, to replace Z by Z_{eff}, given by (2.85), where the values for the screening constant σ are found by the following empirical rules. First, the atomic orbitals that are occupied are divided into the groups 1s|2s, 2p|3s, 3p|3d|4s, 4p|4d|4f|. Then σ is formed by summing the following contributions:

(a) From any orbital group of energy higher than that of the group containing the electron under consideration, *zero*;
(b) From each *other* electron in the group containing the electron under consideration, *0.35 per electron*, or *0.30 per electron* if the group being considered is 1s;
(c) From the electron group of next lowest energy to that containing the electron under consideration, *0.85 per electron*, if the electron under consideration is s or p, and *1.00 per electron* for all lower energy groups. If the electron under consideration is d or f, *1.00 per electron* for all lower energy electrons.

The different radial functions for electrons lead to differing contributions to the effective atomic number. For example, Figure 2.8 shows that there is a greater probability of a 2s electron being found close to the nucleus than there is of a 2p; it has a greater *penetration* through the inner shells, an effect that increases with increasing n. Some examples are given in the following table to show the working of these rules.

Atom	Z	Electron considered	σ	Z_{eff}
He	2	1s	1×0.30	1.70
Be	4	2s	$1\times0.35+2\times0.85$	1.95
C	6	1s	1×0.30	5.70
C	6	2s, 2p	$3\times0.35+2\times0.85$	3.25
Na	11	3s	$8\times0.85+2\times1.00$	2.20
Na^+	11	2s, 2p	$7\times0.35+2\times0.85$	6.85
Ni	28	3d	$7\times0.35+8\times1.00+8\times1.00+2\times1.00$	7.55

Linear least squares and the propagation of errors

A straight-line relationship may be fitted to a number of observations in excess of two by the method of least squares. Let the equation be of the form

$$y=ax+b \tag{A8.1}$$

where a and b are constants which have to be determined. For any observation i,

$$ax_i+b-y_i=e_i \tag{A8.2}$$

where e_i is an error that will be assumed both to be random and to reside in the value of the dependent variable y, the error in the independent variable x being relatively negligible. According to the principle of least squares, the best values of a and b are those for which the sum of the squares of the errors e_i is a minimum. Thus

$$\text{Min}\left(\sum_i e_i^2\right) = \text{Min}\left(\sum_i (ax_i+b-y_i)^2\right) \tag{A8.3}$$

The required minimum value may be found by differentiating the right-hand side of (A8.3) partially with respect to both a and b, setting each of the derivatives equal to zero. Hence

$$\frac{\partial\left(\sum_i e_i^2\right)}{\partial a}=2\sum_i(ax_i^2+bx_i-x_iy_i)=0 \tag{A8.4}$$

$$\frac{\partial\left(\sum_i e_i^2\right)}{\partial b}=2\sum_i(ax_i+b-y_i)=0 \tag{A8.5}$$

Thus,

$$a[x^2]+b[x]-[xy]=0 \tag{A8.6}$$

$$a[x]+bN-[y]=0 \tag{A8.7}$$

Equations (A8.6) and (A8.7) are known as the normal equations and $[x]$, for example, means $\Sigma_i x_i$ over the number N of observations. If each observation has a weight w then the normal equations become

$$a[wx^2]+b[wx]-[wxy]=0 \tag{A8.8}$$

$$a[wx]+b[w]-[wy]=0 \tag{A8.9}$$

Solving for a and b,

$$a=([w][wxy]-[wx][wy])/\Delta \tag{A8.10}$$

$$b=([wx^2][wy]-[wx][wxy])/\Delta \tag{A8.11}$$

where Δ is given by

$$\Delta=[w][wx^2]-[wx][wx] \tag{A8.12}$$

If all of the weights are unity, $[w]=N$.

The standard deviations in a and b may be estimated by the following procedure. From (A8.2),

$$[e^2]=\sum_i w_i(ax_i+b-y_i)^2 \tag{A8.13}$$

Then, we write, without proof here,[1]

$$\sigma^2(a)=\{[e^2]/(N-2)\}[w]/\Delta \tag{A8.14}$$

$$\sigma^2(b)=\{[e^2]/(N-2)\}[wx^2]/\Delta \tag{A8.15}$$

σ being an estimated standard deviation and σ^2 the corresponding variance.

It is recommended that the least-squares line be compared, when feasible, with a plot of the experimental x, y values. In the light of this inspection, certain observations may be reasoned to be unreliable. It must be remembered that a least-squares procedure will always give the best fit to the observations, including the bad ones.

A *correlation coefficient* r may be defined as

$$r=\frac{N[wxy]-[wx][wy]}{(N[wx^2]-[wx]^2)^{1/2}(N[wy^2]-[wy]^2)^{1/2}} \tag{A8.16}$$

Its value may be used as a parameter of the quality of fit; a perfect fit corresponds to $r=1$, but a linear fit that produces $r>0.95$ may be regarded as very satisfactory.

A8.1 Propagation of errors

The number of significant figures in a result is not necessarily similar to the number of significant figures in the data. Let $y=p^n$, where $p=2.0\pm0.1$. For $n=0.1$, y lies between 1.066 and 1.077, whereas for $n=4$, y lies between 13.0 and 19.4.

Consider a function $y=f(p)$ (Figure A8.1). In the small interval δp, the change δy in y is given with good accuracy by

$$\delta y=\left(\frac{\mathrm{d}y}{\mathrm{d}p}\right)\delta p \tag{A8.17}$$

Consider next any function $y=f(p_1,p_2)$ where p_1 and p_2 are independent variables. For two small independent changes δp_1 and δp_2, the changes in y are given by analogy with (A8.17) by:

$$(\delta y)_{p_1}=\left(\frac{\partial y}{\partial p_1}\right)\delta p_1 \tag{A8.18}$$

$$(\delta y)_{p_2}=\left(\frac{\partial y}{\partial p_2}\right)\delta p_2 \tag{A8.19}$$

[1] See, for example, F T Whittaker and G Robinson, *The Calculus of Observations* (Blackie, 1949).

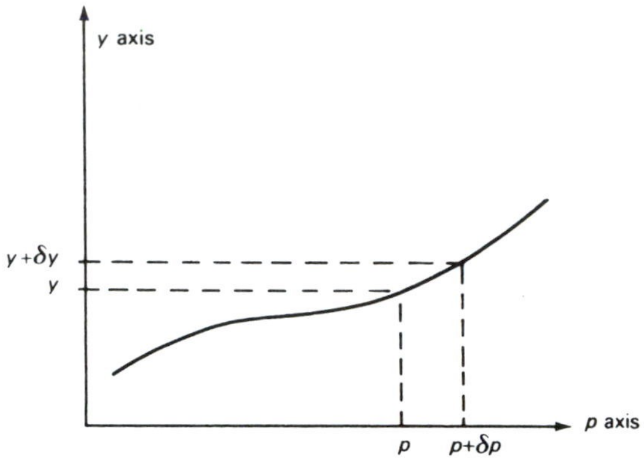

Figure A8.1 A function $y=f(p)$.

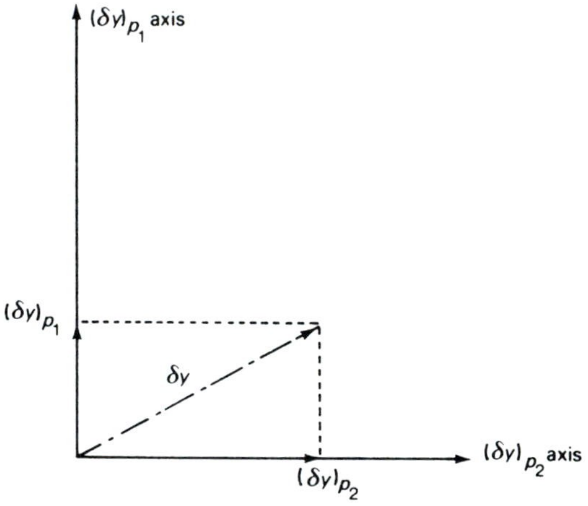

Figure A8.2 Representation of the uncorrelated errors $(\delta y)_{p_1}$ and $(\delta y)_{p_2}$.

Since we have assumed that these two variations in y are uncorrelated, they can be represented along two rectangular axes (Figure A8.2). Hence

$$(\delta y)^2=(\delta y)^2_{p_1}+(\delta y)^2_{p}=\left(\frac{\partial y}{\partial p_1}\right)^2(\delta_{p_1})^2+\left(\frac{\partial y}{\partial p_2}\right)^2(\delta_{p_2})^2 \tag{A8.20}$$

Generalizing for a function $y=f(p_j)$ $(j=1, 2, 3. \dots, n)$:

$$(\delta y)^2=\sum_{j=1}^{n}\left(\frac{\partial y}{\partial p_j}\right)^2(\delta p_j)^2 \tag{A8.21}$$

where δy is the standard deviation in y, $\sigma(y)$.

Determinants and cofactors

A9.1 Expansion of a determinant

Consider the determinant that arises in applying the HMO method to butadiene (Section 2.12.1):

$$D=\begin{vmatrix} y & 1 & 0 & 0 \\ 1 & y & 1 & 0 \\ 0 & 1 & y & 1 \\ 0 & 0 & 1 & y \end{vmatrix} \tag{A9.1}$$

In general, the value of a determinant is obtained by forming the sum of the product of each element in a given column (or row) with its *cofactor*. The cofactor $\mathscr{C}_{ij}$ of an element c_{ij} is the determinant, of one order lower, obtained by striking out the *i*th row and the *j*th column, and multiplied by $(-1)^{i+j}$. Thus, from (A9.1), using the first row, we obtain

$$D=y\begin{vmatrix} y & 1 & 0 \\ 1 & y & 1 \\ 0 & 1 & y \end{vmatrix} - \begin{vmatrix} 1 & 1 & 0 \\ 0 & y & 1 \\ 0 & 1 & y \end{vmatrix} \tag{A9.2}$$

Treating (A9.2) in a similar manner leads to

$$D=y^2\begin{vmatrix} y & 1 \\ 1 & y \end{vmatrix} - y\begin{vmatrix} 1 & 1 \\ 0 & y \end{vmatrix} - \begin{vmatrix} y & 1 \\ 1 & y \end{vmatrix} + \begin{vmatrix} 0 & 1 \\ 0 & y \end{vmatrix} \tag{A9.3}$$

$$D=y^2(y^2-1)-y^2-y^2+1=y^4-3y^2+1 \tag{A9.4}$$

A9.2 Butadiene determinant and wavefunctions

In expanding (A9.1) we employed the cofactors of the elements of the determinant. In order to complete the wavefunctions of butadiene, we need the coefficients c_1 to c_n. Generally, we have

$$c_n/c_1=\mathscr{C}_{n,1}/\mathscr{C}_{1,1} \tag{A9.5}$$

Taking c_2/c_1 as an example (we can equally well use c_n/c_1 as coefficients in place of c_n), we have from (A9.1)

$$c_2/c_1 = -\begin{vmatrix} 1 & 1 & 0 \\ 0 & y & 1 \\ 0 & 1 & y \end{vmatrix} \div \begin{vmatrix} y & 1 & 0 \\ 1 & y & 1 \\ 0 & 1 & y \end{vmatrix} = \frac{1-y^2}{y(y^2-2)} \qquad \text{(A9.6)}$$

The orthogonal butadiene molecular-orbital wavefunctions can be normalized through (2.42), by the term

$$N = \left(\sum_n (c_n/c_1)^2\right)^{1/2} \qquad \text{(A9.7)}$$

and, from further properties of determinants, the nth coefficient is given by

$$c_n = (c_n/c_1)/N \qquad \text{(A9.8)}$$

Proceeding in this way for the bonding molecular orbital of lowest energy $\alpha+1.618\beta$, $y=-1.618$. This value is inserted into (A9.6), which is solved for the ratio c_n/c_1, whence the c_n values are obtained through (A9.7) and (A9.8). The following results are obtained for the molecular orbital of energy $\alpha+1.618\beta$:

n	c_n/c_1	$(c_n/c_1)^2$	c_n
1	1.0000	1.0000	0.3717
2	1.6180	2.6180	0.6015
3	1.6180	2.6180	0.6015
4	1.0000	1.0000	0.3717
		$\surd(\Sigma)=2.6900$	

This procedure may be repeated for the remaining three molecular orbitals of butadiene.

Solution of a second-order differential equation

Consider the equation

$$d^2y/dx^2+k^2y=0 \qquad (A10.1)$$

where k^2 is a constant. Let d^2/dx^2 be represented by D^2. Then

$$D^2y+k^2y=0 \qquad (A10.2)$$

Consider next the equation

$$[(D-p_1)(D-p_2)]y=0 \qquad (A10.3)$$

Expanding (A10.3) gives

$$D^2y-D(p_1+p_2)y+p_1p_2y=0 \qquad (A10.4)$$

Comparing (A10.2) and (A10.4), we see that they will be equivalent provided that $p_2=-p_1$. Hence

$$p_1^2=-k^2 \qquad (A10.5)$$

and the two roots of (A10.5) are

$$p_1=\pm ik \qquad (A10.6)$$

Taking the terms in (A10.3) in turn

$$(D-p_1)y=0 \qquad (A10.7)$$

or

$$dy/y=p_1\,dx \qquad (A10.8)$$

On integrating (A10.8) we obtain

$$\ln(y)=p_1x+A \qquad (A10.9)$$

where A is a constant; using (A10.6)

$$y=A\exp(ikx) \qquad (A10.10)$$

In a similar manner, we have for the second root of (A10.5)

$$y=B\exp(-ikx) \qquad (A10.11)$$

and the complete solution is then written as

$$y=A\exp(ikx)+B\exp(-ikx) \qquad (A10.12)$$

It is a simple matter to show that the double differentiation of (A10.12) leads to (A10.1).

Separation of variables

Certain partial differential equations are termed *separable* and may be solved by the following type of procedure. Consider a function

$$\Psi(x,y)=\psi(x)\psi(y) \tag{A11.1}$$

It is a simple matter to show that

$$\frac{\partial^2\Psi(x,y)}{\partial x^2}=\psi(y)\frac{\partial^2\psi(x)}{\partial x^2} \tag{A11.2}$$

and

$$\frac{\partial^2\Psi(x,y)}{\partial y^2}=\psi(x)\frac{\partial^2\psi(y)}{\partial y^2} \tag{A11.3}$$

Following the form of (2.50), we may write

$$\psi(y)\frac{\partial^2\psi(x)}{\partial x^2}+\psi(x)\frac{\partial^2\psi(y)}{\partial y^2}=E\psi(x)\psi(y) \tag{A11.4}$$

Dividing by $\psi(x)\ \psi(y)$, we obtain

$$\frac{1}{\psi(x)}\frac{\partial^2\psi(x)}{\partial x^2}+\frac{1}{\psi(y)}\frac{\partial^2\psi(y)}{\partial y^2}=E \tag{A11.5}$$

Since E is a constant, both terms on the left-hand side of (A11.5) are independently constant. Hence, we may write

$$\frac{\partial^2\psi(x)}{\partial x^2}=E\psi(x) \tag{A11.6}$$

$$\frac{\partial^2\psi(y)}{\partial y^2}=E\psi(y) \tag{A11.7}$$

whence

$$E\psi(x)+E\psi(y)=E\Psi(x,y) \tag{A11.8}$$

It is straightforward to introduce $-[h^2/(8\pi m)]$ into (A11.4) and to show that the results of (2.50) to (2.53) can be so obtained.

Overlap integrals

In applying the variation principle (Section 2.11.1ff), we encountered the overlap integral $\int \psi_i \psi_j \,\mathrm{d}\tau$. Here, we will carry out the evaluation of this quantity in a fairly simple case. Consider the overlap between two hydrogen 1s atomic orbitals, 1 and 2. Taking each AO to be of the form $\psi=(\pi a_0^3)^{-1/2}\exp(-r/a_0)$, the overlap integral $\int \psi_i \psi_j \,\mathrm{d}\tau$ is given by

$$S=[1/(\pi a_0^3)]\int_0^{\infty}\int_0^{\pi}\int_0^{2\pi}\exp[-(r_1+r_2)/a_0]\,r^2\,\mathrm{d}r\sin(\theta)\,\mathrm{d}\theta\,\mathrm{d}\phi \qquad \text{(A12.1)}$$

The overlap system has cylindrical symmetry, and it is convenient to transform the integral to spheroidal coordinates μ, ν and ϕ given by

$$\mu=(r_1+r_2)/R \qquad \nu=(r_1-r_2)/R \qquad \phi=\phi \qquad \text{(A12.2)}$$

where R is the internuclear distance in units of a_0; μ ranges from 1 to ∞, ν ranges from -1 to $+1$ and ϕ ranges from 0 to 2π. The volume element $\mathrm{d}\tau$ becomes $(R^3/8)(\mu^2-\nu^2)\,\mathrm{d}\mu\,\mathrm{d}\nu\,\mathrm{d}\phi$.[1] Then (A12.1) may be rewritten as

$$S=[R^3/(8\pi)]\int_0^{2\pi}\mathrm{d}\phi\int_{-1}^{1}\int_1^{\infty}\exp(-R\mu)(\mu^2-\nu^2)\,\mathrm{d}\nu\,\mathrm{d}\mu \qquad \text{(A12.3)}$$

Integrating first over ϕ leads to

$$S=(R^3/4)\int_{-1}^{1}\mathrm{d}\nu\int_1^{\infty}\exp(-R\mu)(\mu^2-\nu^2)\,\mathrm{d}\mu \qquad \text{(A12.4)}$$

Expanding (A12.4) gives

$$S=(R^3/4)\int_1^{\infty}2\mu^2\exp(-R\mu)\,\mathrm{d}\mu-\int_{-1}^{1}\nu^2\,\mathrm{d}\nu\int_1^{\infty}\exp(-R\mu)\,\mathrm{d}\mu \qquad \text{(A12.5)}$$

which reduces to

$$S=(R^3/4)\int_1^{\infty}2\mu^2\exp(-R\mu)\,\mathrm{d}\mu-\frac{2}{3}\int_1^{\infty}\exp(-R\mu)\,\mathrm{d}\mu \qquad \text{(A12.6)}$$

The reduction formula (A12.7) then leads to the required result for S:

$$\int x^m\exp(ax)\,\mathrm{d}x=\frac{x^m\exp(ax)}{a}-\frac{m}{a}\int x^{m-1}\exp(ax)\,\mathrm{d}x \qquad \text{(A12.7)}$$

$$S=\exp(-R)(1+R+R^2/3) \qquad \text{(A12.8)}$$

[1] The formulation of $\mathrm{d}\tau$ and the calculation of S are discussed exhaustively in H Eyring, J Walter and G E Kimball, *Quantum Chemistry* (Wiley, 1963) and, with a slight change in notation, in J C Slater, *Quantum Theory of Molecules and Solids*, Volume 1 (McGraw-Hill, 1963).

Partial derivatives

A13.1 Mathematical relationships

Consider a function $z=(x,y)$; then the (total) differential df is given by

$$dz=(\partial z/\partial x)_y\,dx+(\partial z/\partial y)_x\,dy \tag{A13.1}$$

where the subscripts represent the variables held constant for the given *partial derivative*. Equation (A13.1) may be written as

$$dz=M\,dx+N\,dy \tag{A13.2}$$

where the coefficients M and N correspond to the partial derivatives of the function z. If the function z is multivariate, $z=z(x,y,u,v,\ldots)$, then (A13.1) and (A13.2) may be expanded accordingly; for example,

$$dz=M\,dx+N\,dy+P\,du+Q\,dv+\ldots \tag{A13.3}$$

Many relationships in thermodynamics involve a function such as (A13.1) or (A13.2). From (A13.1) and (A13.2), the second derivatives may be written as

$$(\partial M/\partial y)_x=\left(\frac{\partial}{\partial y}(\partial z/\partial x)_y\right)_x=\partial^2 z/(\partial y\,\partial x) \tag{A13.4}$$

$$(\partial N/\partial x)_y=\left(\frac{\partial}{\partial x}(\partial z/\partial y)_x\right)_y=\partial^2 z/(\partial x\,\partial y) \tag{A13.5}$$

Since the order of *second* differentiation is immaterial, the equality

$$(\partial M/\partial y)_x=(\partial N/\partial x)_y \tag{A13.6}$$

determines that dz is an *exact* differential (see also Section 4.3), and forms the mathematical basis for Maxwell's equations.

For any derivative, a reciprocal relation exists; thus,

$$(\partial y/\partial x)_z=1/(\partial x/\partial y)_z \tag{A13.7}$$

By rearrangement of (A13.1), we can obtain

$$dx=(\partial x/\partial z)_y\,dz-(\partial x/\partial z)_y(\partial z/\partial y)_x\,dy \tag{A13.8}$$

We can recast the original function as $x=(y,z)$; then, following (A13.1)

$$dx=(\partial x/\partial y)_z\,dy+(\partial x/\partial z)_y\,dz \tag{A13.9}$$

By comparing coefficients of dy between (A13.8) and (A13.9), and using (A13.7), we obtain the *chain equation*

$$(\partial x/\partial y)_z(\partial y/\partial z)_x(\partial z/\partial x)_y = -1 \quad \text{(A13.10)}$$

The cyclic order of x, y and z in (A13.10) should be noted. Another useful form of this equation is

$$(\partial x/\partial y)_z = -(\partial x/\partial z)_y(\partial z/\partial y)_x \quad \text{(A13.11)}$$

A13.2 Maxwell's equations

Consider the thermodynamic function

$$G = H - TS = U + pV - TS \quad \text{(A13.12)}$$

The differential is given by

$$\mathrm{d}G = \mathrm{d}U + p\,\mathrm{d}V + V\,\mathrm{d}p - T\,\mathrm{d}S - S\,\mathrm{d}T \quad \text{(A13.13)}$$

We have developed the equation

$$\mathrm{d}U = T\,\mathrm{d}S - p\,\mathrm{d}V \quad \text{(A13.14)}$$

so that with (A13.13) we have

$$\mathrm{d}G = S\,\mathrm{d}T - V\,\mathrm{d}p \quad \text{(A13.15)}$$

Since G is a state function, dG is an exact differential; hence, from (A13.6) we obtain

$$-(\partial S/\partial p)_T = (\partial V/\partial T)_p \quad \text{(A13.16)}$$

which is one of Maxwell's equations; others are developed in applications in the main body of the book.

Numerical integration

A numerical integration procedure computes the value of a definite integral from a set of values of the integrand. When these values are ordinates of a curve, the integration defines the area under the curve; we meet numerical integration often in this form.

It may happen that a curve can be simulated by a least-squares fit to an appropriate function. Then there is no problem, because we can obtain an analytical solution from the fitted function. Suppose that the curve in Figure A14.1 can be fitted by the function

$$y=2x^2+3x+2 \tag{A14.1}$$

Then

$$\int_0^{0.8} y\,dx=2x^3/3+3x^2/2+2x\Big|_0^{0.8}=2.9013\ldots \tag{A14.2}$$

A14.1 Numerical and other methods

It may not be possible to deduce a function which will represent satisfactorily a set of experimental measurements, whereupon it becomes necessary to adopt another procedure. Three methods will be described here.

Some numerical integration techniques depend on the relationship

$$\int_a^c f(x)\,dx=\int_a^b f(x)\,dx+\int_b^c f(x)\,dx \tag{A14.3}$$

One such method is embodied in Simpson's rule.

A14.1.1 Simpson's rule

In this procedure a curve such as that in Figure A14.1 is divided into an *even* number n of intervals of equal width h. Then Simpson's rule, given without proof here, states that, for any function $y=f(x)$,

$$\begin{aligned}\int_{x_0}^{x_n} y\,dx=\int_{x_0}^{x_0+nh} y\,dx\\ =(h/3)[y_0+y_n+2(y_2+y_4+y_6+\ldots+y_{n-2})\\ +4(y_1+y_3+y_5+\ldots+y_{n-1})]\end{aligned} \tag{A14.4}$$

Example

Let the curve in Figure A14.1 be divided into 8 intervals of width 0.1 units. Then, from the curve, we have the following data.

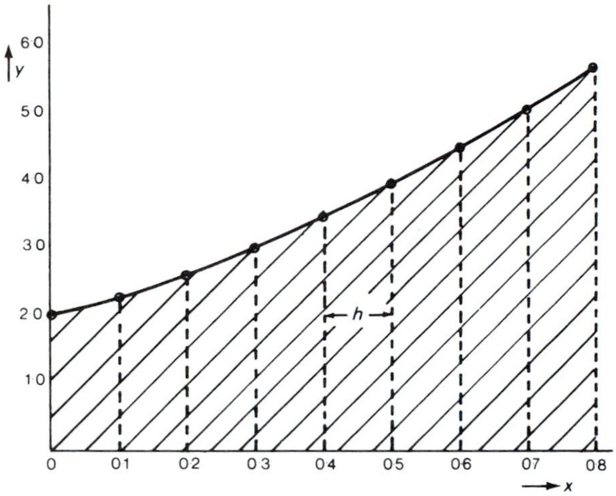

Figure A14.1 A function $y=f(x)$.

x	0.0	0.1	0.2	0.3	0.4	0.5	0.6	0.7	0.8
y	2.0	2.3	2.7	3.1	3.5	4.0	4.5	5.1	5.7

Using (A14.4),

$$\int_0^{0.8} y\,dx=(1/3)[2.0+5.7+2(2.7+3.5+4.5)]$$

$$+4(2.3+3.1+4.0+5.1)]=2.903 \qquad \text{(A14.5)}$$

which agrees with the analytical result to within 0.06%.

A14.1.2 Direct weighing (analogue) method

The curve in Figure A14.1 is drawn on good-quality graph paper. The shaded area is cut out carefully and weighed; let its mass be 0.2759 g. A certain known number of squares are cut out and weighed so as to provide a calibration relationship between mass and area. Let 500 squares have a mass of 4.7563 g and, using the graph scales, let 1 square be equal to 0.1 units of area. Then

$$\frac{\text{Mass of curve}}{\text{Area under curve}}=\frac{\text{Mass of 500 squares}}{500\times\text{scale}} \qquad \text{(A14.6)}$$

Hence,

$$\text{Area under curve}=50\times0.2759/4.7563=2.90 \qquad \text{(A14.7)}$$

which is within 0.05% of the analytical result.

Problems requiring numerical integration arise, for example, in evaluating thermodynamic functions such as ΔH or S.

A14.1.3 Gaussian quadrature

This procedure finds the area under a curve between prescribed limits, but the subdivisions need not be equally spaced. The precision is greater than with the previous two methods, and the procedure has been embodied in the program QUAD (see Appendix 1).

Fermi–Dirac statistics

Each energy state g_i, of energy E_i, in a system of electrons in a solid is determined by the four quantum numbers n_x, n_y, n_z and m_s. Electrons are indistinguishable, but any electron of given n_x, n_y, n_z and m_s occupies the corresponding energy state; thus, each state can be either empty or completely filled by one electron. Two states of the same values of n_x, n_y and n_z, occupied by a pair of electrons with $m_s = \pm\frac{1}{2}$, constitute an electron pair, or fully occupied orbital.

In Boltzmann statistics, the state of each particle in a system is determined solely by the energy that it possesses, that is to say, its energy states are nondegenerate. In quantum statistics, we are usually dealing with degenerate energy states, as the following calculation shows.

Consider an electron at 300 K. From Appendix 3, its mean kinetic energy is $\frac{3}{2}k_BT$, or 6.21×10^{-21} J. For a cubic box of side a of 1 mm, (2.56) becomes

$$E_{n_x,n_y,n_z} = \frac{h^2}{8m_e a^2}(n_x^2+n_y^2+n_z^2) \tag{A15.1}$$

Equating (A15.1) to the mean kinetic energy gives

$$n_x^2+n_y^2+n_z^2 \approx 1.0\times10^{11} \tag{A15.2}$$

Thus an electron possesses the mean classical kinetic energy, $\frac{3}{2}k_BT$, if it selects any quantum numbers satisfying (A15.2): in other words the energy levels are highly degenerate.

Consider a set of energy states or cells, g_i $(i=1, 2, \ldots, s)$ of similar energy E_s, and let the number of electrons in the set be N_s, so that N_s cells are occupied, and $g_s - N_s$ are empty. Following (6.82), we can write for the number of different ways (W_s) in which the N_s electrons can occupy g_s cells, allowing for the indistinguishability of electrons,

$$W_s = \frac{g_s!}{(g_s - N_s)!N_s!} \tag{A15.3}$$

To a given range of energies, there corresponds for each range an equation like (A15.3). Hence the total number of distinguishable arrangements (W) for an entire system of p sets is given by

$$W = \prod_{i=1}^{p} W_i = \prod_{i=1}^{p} \frac{g_i!}{(g_i - N_i)!N_i!} \tag{A15.4}$$

The most probable distribution is that which maximizes W. However the maximization is subject to the constraints

$$\sum_{i=1}^{p} N_i = N \tag{A15.5}$$

where N is the number of electrons in the entire system, and

$$\sum_{i=1}^{p} E_i N_i = E \tag{A15.6}$$

where E is the total energy of the entire system. Using Stirling's approximation with (A15.4) gives

$$\ln(W) = \sum_{i=1}^{p} [g_i \ln(g_i) - (g_i - N_i)\ln(g_i - N_i) - N_i \ln(N_i)] \tag{A15.7}$$

For a maximum,[1] $\mathrm{d}\ln(W) = 0$; thus

$$\mathrm{d}\ln(W) = \frac{\partial \ln(W)}{\partial N_1}\mathrm{d}N_1 + \frac{\partial \ln(W)}{\partial N_2}\mathrm{d}N_2 + \ldots + \frac{\partial \ln(W)}{\partial N_i}\mathrm{d}N_i + \ldots + \frac{\partial \ln(W)}{\partial N_p}\mathrm{d}N_p = 0 \tag{A15.8}$$

From (A15.5) and (A15.6), we have

$$\sum_{i=1}^{p} \mathrm{d}N_i = 0 \tag{A15.9}$$

$$\sum_{i=1}^{p} E_i \,\mathrm{d}N_i = 0 \tag{A15.10}$$

To solve (A15.8) subject to the constraints (A15.9) and (A15.10), we use Lagrange's method of undetermined multipliers.[2] Using (A15.7) in (A15.8), we obtain

$$\mathrm{d}\ln(W) = \sum_{i=1}^{p} [\ln(g_i - N_i) - \ln(N_i)]\,\mathrm{d}N_i = 0 \tag{A15.11}$$

We assign multipliers, α and β, such that

$$\sum_{i=1}^{p} [\ln(g_i - N_i) - \ln(N_i) + \alpha + \beta E_i]\,\mathrm{d}N_i = 0 \tag{A15.12}$$

Only two multipliers are needed since there are only two constraining equations. Equation (A15.12) must hold for any small change in N_i and we shall consider the simplest general variation. More than one term of N_i must be involved, in order to conform to (A15.9). If we vary only two different terms, N_i and N_j, then since $\mathrm{d}N_i = -\mathrm{d}N_j$ to satisfy (A15.9), (A15.10) cannot be satisfied simultaneously, since $E_i \neq E_j$. We conclude that the simplest general variation involves three N_i terms.

Let $\mathrm{d}N_i\,(i = 1, 2, 3)$ be nonzero. Then, the coefficients of these quantities must be zero, from (A15.12). Hence

$$\ln(g_1 - N_1) - \ln(N_1) + \alpha + \beta E_1 = 0 \tag{A15.13}$$

$$\ln(g_2 - N_2) - \ln(N_2) + \alpha + \beta E_2 = 0 \tag{A15.14}$$

Since we have chosen $\mathrm{d}N_3 \neq 0$ too,

$$\ln(g_3 - N_3) - \ln(N_3) + \alpha + \beta E_3 = 0 \tag{A15.15}$$

We can continue procedure for $\mathrm{d}N_4 \neq 0$, giving

$$\ln(g_4 - N_4) - \ln(N_4) + \alpha + \beta E_4 = 0 \tag{A15.16}$$

[1] A consideration of entropy (4.71) will convince us that we are discussing a maximum rather than a minimum.

[2] See, for example, B J McLelland, *Statistical Thermodynamics* (Wiley, 1973).

and so on. In general, for the ith state, we have

$$\ln(g_i - N_i) - \ln(N_i) + \alpha + \beta E_i = 0 \tag{A15.17}$$

This equation may be written in the form

$$\frac{g_i}{N_i} - 1 = \exp(\alpha)\exp(\beta E_i) \tag{A15.18}$$

(A15.18) is one form of the Fermi–Dirac distribution function. We need next to identify α and β: α can be determined through (A15.5), but for the moment we shall write $\exp(\alpha) = A$. At high temperatures, the number of states which are energetically accessible is very large; then $g_i/N_i \gg 1$ and

$$N_i \approx g_i A^{-1} \exp(-\beta E_i) \tag{A15.19}$$

In the high-temperature limit Boltzmann statistics apply, and we can equate β to $1/(k_B T)$:

$$N_i/g_i = \frac{1}{A\exp[E_i/(k_B T)] + 1} \tag{A15.20}$$

It is convenient to define an energy, E_F, such that

$$A = \exp[-E_F/(k_B T)] \tag{A15.21}$$

Hence,

$$\frac{N_i}{g_i} = \frac{1}{\exp[(E_i - E_F)/(k_B T)] + 1} \tag{A15.22}$$

We write $f(E_i) = N_i/g_i$, where $f(E_i)$ gives the probability that a state of energy E is occupied. Thus,

$$f(E_i) = \frac{1}{\exp[(E_i - E_F)/(k_B T)] + 1} \tag{A15.23}$$

which is a convenient form of the Fermi–Dirac distribution function.

Calculation of Madelung constants

We have discussed the Madelung constant in Section 6.8.1 and it is useful to have a simple method of calculating it for any structure, to a precision that is sufficiently high for most purposes. The method given here[1] has been applied successfully to a wide range of compounds and has been programmed for IBM-compatible personal computers. It assumes spherical charge distributions on the ions, with densities that are linear functions of the radial coordinate in reciprocal space.

The Madelung constant $\mathcal{A}$ is given by

$$\mathcal{A}=\frac{(g-Q)d}{rZ}\sum_j q_j^2-\frac{\pi r^2 d}{V}\sum_h |F_h|^2\phi(\boldsymbol{h}) \qquad \text{(A16.1)}$$

where the terms have the following meanings:

g $=26/35$;
Q a correction for termination of the $\boldsymbol{h}$ series (as below);
d a standard distance in the structure, often the nearest-neighbour distance;
r an arbitrary distance less than half (0.495) the nearest-neighbour distance;
Z the number of formula-entities in the unit cell;
q_j the charge, including sign, of the jth species in the unit cell;
V the unit-cell volume;
h the magnitude of the reciprocal lattice vector $\boldsymbol{h}$;
$\phi(\boldsymbol{h})$ $=288[\alpha\sin(\alpha)+2\cos(\alpha)-2]^2\alpha^{-10}$
α $=2\pi hr$
$\boldsymbol{F_h}$ $=\Sigma_j q_j \exp(\mathrm{i}2\pi\boldsymbol{h}\cdot\boldsymbol{x}_j)$ – the crystallographic structure factor for point atoms of form factors q_j.

The sums over j include all atoms in the unit cell and the sum over $\boldsymbol{h}$ includes all reciprocal lattice vectors hkl in a sphere of radius α; h, k and l are the components of $\boldsymbol{h}$ with respect to the reciprocal lattice axes. The series termination correction Q depends on the radius α according to the following table:

α	Q
2π	0.00030
3π	0.000090
4π	0.000012
5π	0.0000057

Termination of the series at $\alpha=2\pi$, including the correction term Q, gives results better than 0.02%, which is sufficient to match other terms in most lattice energy calculations.

[1] F Bertaut, *J. Phys. Radium* **13**, 499 (1952); D H Templeton, *J. Chem. Phys.* **23**, 1629 (1955); R E Jones and D H Templeton, *J. Chem. Phys.* **25**, 1062 (1956).

The hypsometric formula: an example of the Boltzmann distribution

This short analysis constitutes a simple approach to the Boltzmann distribution of energies. A fuller treatment employs statistical thermodynamics,[1] but the end result is similar.

Consider a rectangular column of an ideal gas of cross-sectional area A and height z, with reference to an origin O at ground level, at a uniform temperature T (Figure A17.1). The mass of a gas molecule is m and, at the height z, let the pressure of the gas be p and let there be N molecules in the volume V, ($V=Az$). We need to determine how N varies with z. Since the gas is assumed to be ideal, no intermolecular attractions need be considered. Hence, from the gas laws,

$$pV=n\mathcal{R}T \tag{A17.1}$$

The gas constant $\mathcal{R}$ is $N_A k_B$, where N_A is the Avogadro constant and k_B is the Boltzmann constant. Since $nN_A=N$

$$p=Nk_BT/V \tag{A17.2}$$

At a height $z+\mathrm{d}z$ the pressure is $p+\mathrm{d}p$. The gravitational force on a segment of width $\mathrm{d}z$ is $ADg\,\mathrm{d}z$, where D is the density of the gas and g is the gravitational acceleration. The pressure difference across the segment is $(-\mathrm{d}p/\mathrm{d}z)\,\mathrm{d}z$, the negative sign indicating that the gas pressure decreases in the positive direction of z. The hypsometric[2] force on the segment is $-A(\mathrm{d}p/\mathrm{d}z)\,\mathrm{d}z$ and at equilibrium the two forces are balanced:

$$-A(\mathrm{d}p/\mathrm{d}z)\,\mathrm{d}z=ADg\,\mathrm{d}z \tag{A17.3}$$

or

$$-\mathrm{d}p=Dg\,\mathrm{d}z \tag{A17.4}$$

Equation (A17.4) might be obtained also from a definition of pressure.

From (A17.2), at a constant T and V

$$\mathrm{d}p=(k_BT/V)\,\mathrm{d}N \tag{A17.5}$$

and using the fact that $D=mN/V$, we have

$$\frac{\mathrm{d}N}{N}=-mg\,\mathrm{d}z/(k_BT) \tag{A17.6}$$

On integration, we obtain

$$\ln N=-mgz/k_BT+\text{constant} \tag{A17.7}$$

[1] See, for example, T L Hill, *An Introduction to Statistical Thermodynamics* (Dover, 1986).

[2] A hypsometer is used to measure the heights of different parts of the earth's surface by taking the boiling point of water at the given location.

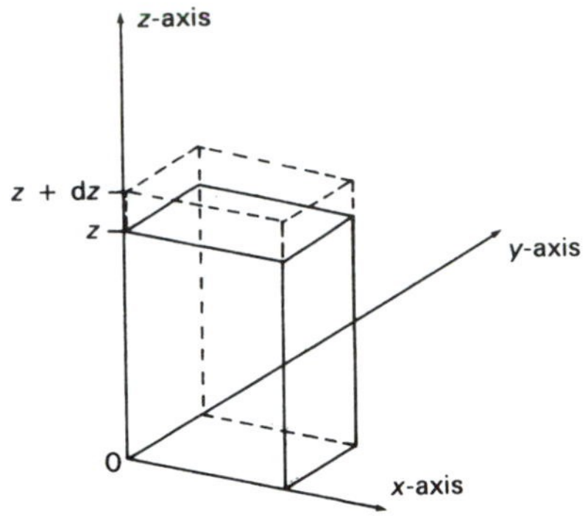

Figure A17.1 Construction for the hypsometric formula.

At $z=0$ let $N=N_0$. Thus the constant becomes $\ln N_0$ and

$$N=N_0 \exp(-mgz/k_B T) \tag{A17.8}$$

Now, mgz is the gravitational potential energy per molecule of gas at the height z. Let U_m represent this potential energy per mole of gas. Then $U_m=N_A mgz$ and

$$N=N_0 \exp(-U_m/\mathcal{R}T) \tag{A17.9}$$

As an example, consider 1 mol of air, of mean molar mass 0.0288 kg mol^{-1}, at a height of 10 km above ground level (the height of a transatlantic air liner) at 298 K. From (A17.9), N/N_0 is approximately 0.2 times its value at ground level; at 100 km it is only about 4×10^{-8} times its value at ground level.

Tables of physical data

A18.1 General thermodynamic data

All data refer to the standard state conditions of 298.15 K and 1 atm, and to the given physical state of the substance. The ΔH_f and ΔG_f values relate to the process *components in their standard state→species in its standard state*, whereas the entropy data refer to $\int_0^{298.15} [C_p(T)/T]\,dT$. Other data may be deduced from these tables: for example, $\Delta H_d^{\ominus}(NaCl)=[\Delta H_f^{\ominus}(Na^+,aq)+\Delta H_f^{\ominus}(Cl^-,aq)-\Delta H_f^{\ominus}(NaCl,s)]$.

	$\Delta H_f^{\ominus}$/kJ mol^{-1}	$\Delta G_f^{\ominus}$/kJ mol^{-1}	$S^{\ominus}$/J K^{-1} mol^{-1}
Al(s)	0	0	28.3
Al(g)	326.4	285.7	164.5
Al^{3+}(aq)	−531.0	−485.0	−321.7
$AlCl_3$(s)	−704.2	−628.8	110.7
Ba(s)	0	0	62.8
Ba(g)	180.0	146.0	170.2
Ba^{2+}(aq)	−537.6	−560.8	9.60
$BaCl_2$(s)	−858.6	−810.4	123.7
Br_2(1)	0	0	152.2
Br(g)	111.9	82.4	175.0
Br^-(aq)	−121.6	−104.0	82.4
HBr(g)	−36.4	−53.5	198.7
Ca(s)	0	0	41.4
Ca(g)	178.2	144.3	154.9
Ca^{2+}(aq)	−542.8	−553.6	−53.1
$CaCO_3$(s)	−1206.9	−1128.8	92.9
CaF_2(s)	−1219.6	−1167.3	68.9
$CaCl_2$(s)	−795.8	−748.1	104.6
C(s,graphite)	0	0	5.74
C(g)	716.7	671.3	158.1
CO(s)	−110.5	−137.2	213.7
CO_2(g)	−393.5	−394.5	213.7
CO_3^{2-}(aq)	−677.1	−527.8	−56.9
CN^-(aq)	150.6	−172.4	94.1
Cl_2(g)	0	0	223.1
Cl(g)	121.7	105.7	165.2

	ΔH_f^{-0-}/kJ mol^{-1}	ΔG_f^{-0-}/kJ mol^{-1}	S^{-0-}/J K^{-1} mol^{-1}
Cl^-(aq)	−167.2	−131.2	56.5
HCl(g)	−92.3	−95.3	186.9
Cu(s)	0	0	33.2
Cu(g)	338.3	298.6	166.4
Cu^+(aq)	71.7	50.0	40.6
Cu^{2+}(aq)	64.8	65.5	−99.6
$CuSO_4$(s)	−771.4	−661.8	109.0
$CuSO_4 \cdot 5H_2O$(s)	−2279.7	−1879.7	300.4
F_2(g)	0	0	202.8
F(g)	79.0	61.9	158.8
F^-(aq)	−332.6	−278.8	−13.8
HF(g)	−271.1	−273.2	173.8
H_2(g)	0	0	130.7
H(g)	218.0	203.3	114.7
H^+(aq)	0	0	0
H_2O(g)	−241.8	228.6	188.8
H_2O(l)	−285.8	−237.1	69.9
I_2(s)	0	0	116.1
I(g)	106.8	70.3	180.8
I^-(aq)	−55.2	−51.6	111.3
HI(g)	26.5	1.70	206.6
Fe(s)	0	0	27.3
Fe(g)	416.3	370.7	180.5
Fe^{2+}(aq)	−89.1	−78.9	−137.7
Fe^{3+}(aq)	−48.5	−4.70	−315.9
Fe_2O_3(s)	−824.2	−742.2	−87.4
Pb(s)	0	0	64.8
Pb(g)	195.0	161.9	175.4
Pb^{2+}(aq)	−1.70	−24.4	10.5
PbO(s)	−217.3	−187.9	68.7
PbO_2(s)	−277.4	−217.3	68.6
Li(s)	0	0	29.1
Li(g)	159.4	126.7	−138.8
Li^+(aq)	−278.5	−293.3	13.4
Mg(s)	0	0	32.7
Mg(g)	147.7	113.1	148.7
Mg^{2+}(aq)	−466.9	−454.8	−138.1
$MgCl_2$(s)	−641.3	−591.8	89.6
MgO(s)	−601.7	−569.4	26.9
Hg(l)	0	0	76.0
Hg(g)	61.3	31.8	175.0
Hg_2^{2+}(aq)	172.4	153.5	84.5
Hg^{2+}(aq)	171.1	164.4	−32.2
Hg_2Cl_2(s)	−265.2	−210.8	192.5

	ΔH_f^{-0-}/kJ mol^{-1}	ΔG_f^{-0-}/kJ mol^{-1}	S^{-0-}/J K^{-1} mol^{-1}
$HgCl_2$(s)	−224.3	−178.6	146.0
N_2(g)	0	0	191.6
N_2O(g)	82.1	104.2	219.9
NO(g)	90.3	86.6	210.8
NO_2(g)	33.2	51.3	240.1
N_2O_4(g)	9.16	97.9	304.3
N_2O_5(g)	11.3	115.1	355.7
NO_3^-(aq)	−205.0	−108.7	146.4
NH_3(g)	−46.1	−16.5	192.5
NH_3(aq)	−80.3	−26.5	111.3
NH_4^+(aq)	−132.5	−79.3	113.4
NH_4Cl(s)	−314.4	202.9	94.6
NH_4NO_3(s)	−365.6	−183.9	151.1
O_2(g)	0	0	205.1
OH^-(aq)	−230.0	−157.2	−10.8
P(s,white)	0	0	41.1
PH_3(g)	5.40	13.4	210.2
K(s)	0	0	64.2
K(g)	89.2	60.6	160.3
K^+(aq)	−252.4	−283.3	102.5
KF(s)	−567.3	−537.8	66.6
KCl(s)	−436.8	−409.1	82.6
Si(s)	0	0	18.8
SiO_2(s,α)	−910.9	−856.6	41.8
Ag(s)	0	0	42.6
Ag(g)	284.6	245.7	173.0
Ag^+(aq)	105.6	77.1	72.7
AgCl(s)	−127.1	−109.8	96.2
AgBr(s)	−100.4	−96.9	107.1
Na(s)	0	0	51.2
Na(g)	107.3	76.8	153.7
Na^+(aq)	−240.1	−261.9	59.0
NaCl(s)	−411.2	−384.1	72.1
NaBr(s)	−361.1	−349.0	86.8
S(s,α)	0	0	31.8
SO_2(g)	−296.8	−300.2	248.2
SO_3(g)	−395.7	−371.1	256.8
SO_4^{2-}(aq)	−909.3	−744.5	20.1
SF_6(g)	−1209.0	−1105.3	291.8
Sn(s,β)	0	0	51.6
Sn(g)	302.1	−267.3	168.5
Sn^{2+}(aq)	−8.80	−27.2	−17.0
SnO(s)	−285.8	−256.9	56.5
Zn(s)	0	0	41.6

	ΔH_f^{-0-}/kJ mol^{-1}	ΔG_f^{-0-}/kJ mol^{-1}	S^{-0-}/J K^{-1} mol^{-1}
Zn(g)	130.7	95.1	161.0
Zn^{2+}(aq)	−153.9	−147.1	−112.1
ZnO(s)	−348.3	−318.3	43.6

A18.2 Dissociation constants for weak acids and weak bases

The dissociation constants take the form $\alpha \times 10^{-\beta}$, and are dimensionless. They depend upon temperature and are quoted for 298.15 K, unless stated otherwise.

		K_1		K_2		K_3	
		α	β	α	β	α	β
Boric acid	H_3BO_3	7.3	10	1.8	13	1.6	14
Benzoic acid	$C_6H_5CO_2H$	6.5	5				
Carbonic acid	H_2CO_3	4.3	7	5.6	11		
Dichloroethanoic acid	Cl_2CHCO_2H	3.3	2				
Ethandicarboxylic acid	$(CO_2H)_2$	5.9	2	6.4	5		
Ethanoic acid	CH_3CO_2H	1.8	5				
Hydrofluoric acid	HF	3.5	4				
Methanoic acid	HCO_2H	1.8	4				
Monochloroethanoic acid	$ClCH_2CO_2H$	1.4	3				
Phenol	C_6H_5OH	1.3	10				
Phosphoric acid	H_3PO_4	7.5	3	6.2	8	2.2	13
Propan-1,2-dicarboxylic acid	$CH_2(CO_2H)_2$	1.5	3	2.0	6		
Trichloroethanoic acid	Cl_3CCO_2H	2.0	1				
Water	H_2O	1.0	14				
Ammonia	NH_3	1.8	5				
Aniline	$C_6H_5NH_2$	3.8	10				
1,4-Diaminobenzene	$C_6H_4(NH_2)_2$	1.1	8				
Diethylamine	$(C_2H_5)_2NH$	9.6	4				
Ethylamine	$C_2H_5NH_2$	5.6	4				
Hydrazine[a]	NH_2NH_2	1.7	6				
Hydroxylamine[a]	NH_2OH	1.1	8				
Piperidine	$(CH_2)_5NH$	1.6	3				
Pyridine	C_5H_5N	1.7	9				
Triethylamine	$(C_2H_5)_3N$	5.7	4				

[a] At 293.15 K

A18.3 Mean ionic activity coefficients $\gamma_\pm$ for electrolytes at 298.15 K

m/m^{-0-}	HCl	KCl	$CaCl_2$	$LaCl_3$
0.001	0.966	0.966	0.888	0.790
0.005	0.929	0.927	0.789	0.636
0.010	0.905	0.902	0.732	0.560
0.050	0.830	0.816	0.584	0.388
0.100	0.798	0.770	0.524	0.356
0.500	0.769	0.652	0.510	0.303
1.000	0.811	0.607	0.725	0.387
2.000	1.011	0.577	1.554	0.954

A18.4 Standard enthalpies of formation and combustion for carbon compounds

		ΔH_f/kJ mol^{-1}	ΔH_c/kJ mol^{-1}
Carbon	C(s,graphite)	0	−393.5
Carbon dioxide	CO_2(g)	−393.5	
Methane	CH_4(g)	−74.8	−890.0
Ethyne	C_2H_2(g)	226.7	−1300
Ethene	C_2H_4(g)	52.3	−1411
Ethane	C_2H_6(g)	−84.7	−1560
Benzene	C_6H_6(g)	82.9	−3302
Benzene	C_6H_6(l)	49.0	−3268
Ethanol	C_2H_5OH(g)	−235.1	−1409
Ethanol	C_2H_5OH(l)	−277.7	−1368
Ethanoic acid	CH_3CO_2H(l)	−484.5	−875.0
Ethandicarboxylic acid	(CO_2H)(s)	−827.2	−254.0
Benzoic acid	$C_6H_5CO_2H$(s)	−385.1	−3227

A18.5 Ionization energies for a selection of elements

The ionization energy I_q relates to the process

$$M^{(q-1)+}(g) \rightarrow M^{q+}(g) + e^-$$

at 0 K; data are listed for $q=1$ to 4. At any other temperature the process becomes enthalpic, and the quantity is increased by $\int_0^T C_p(T)\,dT$.

	Atomic number	I_1	I_2	I_3	I_4
H	1	1312			
He	2	2371	5247		
Li	3	520.1	7297	11811	
Be	4	899.1	1757	14820	20999
B	5	800.4	1462	3642	25016
C	6	1087	2352	4561	6510
N	7	1403	2856	4577	7473
O	8	1316	3391	5301	7468
F	9	1681	3375	6046	8318
Ne	10	2080	3963	6176	9376
Na	11	495.8	4565	6912	9540
Mg	12	737.6	1450	7732	10543
Al	13	577.4	1816	2744	11577
Si	14	819.2	1577	3228	4356
P	15	1061	1896	2910	4950
S	16	999.6	2258	3333	4565
Cl	17	1255	2296	3850	5163
Ar	18	1520	2665	3946	5577
K	19	418.8	3068	4439	5874
Ca	20	589.5	1145	4941	6435
Sc	21	632.6	1243	2388	7130
Ti	22	659.0	1309	2715	4181
V	23	650.2	1370	2866	4669
Cr	24	652.7	1591	2991	4845
Mn	25	717.1	1509	3251	5113
Fe	26	762.3	1561	2956	5402
Co	27	758.6	1645	3231	5113
Ni	28	736.4	1751	3489	5402
Cu	29	745.2	1958	3666	5694
Zn	30	906.3	1733	3827	5983
Se	34	941.0	2075	2902	4139
Br	35	1142	2082	3463	4845
Kr	36	1350	2371	3564	5017
Rb	37	402.9	2653	3828	5113
Sr	38	549.4	1064	4149	5498
Ag	47	730.9	2072	3483	5017

	Atomic number	I_1	I_2	I_3	I_4
Cd	48	867.3	1631	3377	5305
In	49	559.4	1820	2705	5594
Sn	50	707.5	1412	2958	3936
Sb	51	833.5	1738	2287	4238
Te	52	869.4	2079	2953	3649
I	53	1007	1834	2991	4052
Xe	54	1170	2046	3100	4498
Cs	55	375.7	2264	3377	4920
Ba	56	560.7	962.7	3570	4728
La	57	541.4	1103	1850	5017
Ce	58	666.5	1187	1939	3540
Hg	80	1007	1809	3310	6945
Tl	81	589.1	1970	2875	4874
Pb	82	715.5	1450	3095	4076
Bi	83	702.9	1862	2470	4381
Rn	84	1037	1930	2894	4247

BIBLIOGRAPHY

General

P W Atkins, *Physical Chemistry*, 5th edition, Oxford University Press, 1994.
C N Hinshelwood, *The Structure of Physical Chemistry*, Clarendon Press, 1958.
K J Laidler, *The World of Physical Chemistry*, Oxford University Press, 1990.
I M Mills (Editor), *Quantities, Units and Symbols in Physical Chemistry*, Blackwell Scientific, 1988.
E A Moelwyn-Hughes, *Physical Chemistry*, Pergamon Press, 1965.

Quantum Chemistry

P W Atkins, *Quanta: a Handbook of Concepts*, Oxford University Press, 1991.
C A Coulson, *The Shape and Structure of Molecules*, Oxford University Press, 1982.
R McWeeny, *Coulson's Valence*, Oxford University Press, 1979.
J N Murrell, S F A Kettle and J M Tedder, *The Chemical Bond*, Wiley, 1985.
L Pauling and E B Wilson, *Introduction to Quantum Mechanics*, McGraw-Hill, 1935.
G C Schatz and M A Ratner, *Quantum Mechanics in Chemistry*, Ellis Horwood, 1993.

Symmetry

G Davidson, *Group Theory for Chemists*, Macmillan Education Limited, 1991.
International Union of Crystallography, *International Tables for Crystallography*, Volume A, Reidel, 1983 [formerly *International Tables for X-ray Crystallography*, Volume I, Kynoch Press, 1965].
S F A Kettle, *Symmetry and Structure*, Wiley, 1985.
M F C Ladd, *Symmetry in Molecules and Crystals*, Ellis Horwood Limited, 1992.

Spectroscopy

R J Abraham, J Fisher and P Lofthus, *Introduction to NMR Spectroscopy*, Wiley, 1991.
D L Andrews, *Lasers in Chemistry*, Springer-Verlag, 1990.
A Ault and M R Ault, *A Handy and Systematic Catalog of NMR Spectra*, University Science Books, 1980.
C N Banwell and E M McCash, *Fundamentals of Molecular Spectroscopy*, 4th edition, McGraw-Hill, 1994.
A E Derome, *Modern NMR Techniques for Chemistry Research*, Pergamon Press, 1987.
A G Gaydon, *Dissociation Energies*, Chapman and Hall, 1968.
G Herzberg, *Infrared and Raman Spectra of Polyatomic Molecules*, Van Nostrand, 1945.
G Herzberg, *Spectra of Diatomic Molecules*, Van Nostrand, 1950.
J M Hollas, *Modern Spectroscopy*, Wiley, 1991.

T P Softley, *Atomic Spectra*, Oxford University Press, 1994.
D H Williams and I Fleming, *Spectroscopic Methods in Organic Chemistry*, 4th edition, McGraw-Hill, 1989.

X-ray Crystallographic Analysis

M F C Ladd and R A Palmer, *Structure Determination by X-ray Crystallography*, 3rd edition, Plenum Publishing Corporation, 1994.
M F C Ladd and R A Palmer (Editors), *Theory and Practice of Direct Methods in Crystallography*, Plenum Publishing Corporation, 1980.
M M Woolfson, *An Introduction to Crystallography*, Cambridge University Press, 1970.

Structural Data

L E Sutton (Editor) *Tables of Interatomic Distances and Configuration in Molecules and Ions*, Chemical Society, London, 1958, 1965.
R W G Wyckoff, *Crystal Structures*, Volumes I–VI, Wiley-Interscience, 1963–1971.

Thermodynamics and Thermodynamic Data

P W Atkins, *The Second Law*, Scientific American Books, 1984.
F H Crawford, *Heat, Thermodynamics and Statistical Physics*, Harcourt, Brace and World Inc., 1963.
E A Guggenheim, *Boltzmann's Distribution Law*, North-Holland, 1963.
T L Hill, *Introduction to Statistical Thermodynamics*, Addison-Wesley, 1960.
G N Lewis and M Randall, *Thermodynamics*, McGraw-Hill, 1961.
J N Murrell and E A Boucher, *Properties of Liquids and Solutions*, Butterworths, 1982.
P A Rock, *Chemical Thermodynamics*, University Science Books, 1983.
JANAF Thermodynamic Tables, JANAF, 1965. Also published as *J. Phys. Chem. Ref. Data*, **14**, Supplement 1, 1985.
National Bureau of Standards, Circular 500, 1952.
National Bureau of Standards, Technical Notes 270–1, 1965; 270–2, 1966; 270–3, 1968. Also published as *J. Phys. Chem. Ref. Data*, **11**, Supplement 2, 1982.

Gases

W Kauzmann, *Kinetic Theory of Gases*, Addison-Wesley, 1966.
D Tabor, *Gases, Liquids and Solids*, Cambridge University Press, 1979.

Intermolecular Forces

J N Israelachvili, *Intermolecular and Surface Forces*, Academic Press, 1992.
M Rigby, E B Smith, W A Wakeham and G C Maitland, *The Forces between Molecules*, Oxford University Press, 1986.

Liquids

M P Allen and D J Tildesley, *Computer Simulation of Liquids*, Oxford University Press, 1996.
J P Hansen and I R McDonald, *Theory of Simple Liquids*, Academic Press, 1986.
Y Marcus, *Introduction to Liquid State Chemistry*, Wiley, 1977.

Solids

A H Cottrell, *Introduction to the Modern Theory of Metals*, Institute of Metals, 1988.
P A Cox, *The Electronic Structure and Chemistry of Solids*, Oxford University Press, 1987.
A I Kitaigorodskii, *Organic Chemical Crystallography*, Consultants Bureau, 1957.
C Kittel, *Solid State Physics*, Wiley, 1976.
M F C Ladd, *Structure and Bonding in Solid State Chemistry*, Ellis Horwood, 1979.
M F C Ladd and W H Lee, in *Progress in Solid State Chemistry*, Volume I, 37ff. (1964), Volume II, 378ff. (1965), Volume III, 265ff. (1966); Pergamon Press.
H M Rosenberg, *The Solid State*, Clarendon Press, 1978.
A R West, *Basic Solid State Chemistry*, Wiley, 1988.

Phase Equilibria and Properties of Solutions

J N Murrell and E A Boucher, *Properties of Liquids and Solutions*, Wiley-Interscience, 1982.
A Reisman, *Phase Equilibria*, Academic Press, 1970.
H E Stanley, *Introduction to Phase Transitions and Critical Phenomena*, Clarendon Press, 1971.

Chemical Equilibrium

M J Blandamer, *Chemical Equilibria in Solution*, Ellis Horwood, 1992.
K G Denbigh, *The Principles of Chemical Equilibrium*, Cambridge University Press, 1971.

Electrochemistry

H S Harned and B B Owen, *The Physical Chemistry of Electrolytic Solutions*, Reinhold, 1967.
D B Hibbert, *Introduction to Electrochemistry*, Macmillan, 1993.
J Koryta, *Ion-Selective Electrodes*, Cambridge University Press, 1975.
G Prentice, *Electrochemical Engineering Principles*, Prentice-Hall, 1991.
R A Robinson and R H Stokes, *Electrolyte Solutions*, Academic Press, 1959.
C D S Tuck, *Modern Battery Construction*, Ellis Horwood, 1991.

Reaction mechanism

J H Espenson, *Chemical Kinetics and Reaction Mechanisms*, McGraw-Hill, 1981.
K J Laidler, *Chemical Kinetics*, Harper and Row, 1987.
J I Steinfeld, J S Francisco and W L Hase, *Chemical Kinetics and Dynamics*, Prentice-Hall, 1989.

Mathematics

R Courant, *Differential and Integral Calculus*, Volume II, Blackie, 1949.
D M Hirst, *Mathematics for Chemists*, Macmillan, 1983.
H Margenau and G M Murphy, *The Mathematics of Physics and Chemistry*, Van Nostrand, 1943.
L A Woodward, *Molecular Statistics for Students of Chemistry*, Clarendon Press, 1975.

ANSWERS TO NUMERICAL PROBLEMS

The reader is reminded that detailed solutions are available on the Worldwide Web, www.cup.cam.ac.uk. For those problems that do not have a numerical answer, the appropriate study areas of the text are quoted. Figure numbers (quoted in parentheses) among these answers that are prefixed with the letter S are available only on the Web.

Answers 1

1.1 116.9°.
1.2 184 K.
1.3 $+1.8e$.
1.4 0.488 nm; 2164 kg m^{-3}.
1.5 299 nm; visible to near-UV.
1.6 6.05×10^{23} mol^{-1}

Answers 2

2.1 40 m, 2 N s; 20 J.
2.2 4.114×10^{-16} J; 2.745×10^{-23} N s. (Ignoring special relativity: 4.094×10^{-16} J; 2.731×10^{-23} N s.)
2.3 5.795 μm.
2.4 Planck: 1.123×10^{-5} J m^{-3}; Rayleigh–Jeans: 3.159×10^{-2} J m^{-3}; Planck equation essential.
2.5 (a) By extrapolation:

λ/nm	546.1	365.0	312.6
V_0/V	−2.05	−0.92	−0.32

(b) 6.75×10^{-34} J Hz^{-1}; the Planck constant.

2.6 100.71 pm.
2.7 1.45×10^{6} m s^{-1}.
2.8 Section 2.3.5; $4/R_H$; 3.026×10^{-19} J.
2.9 Kinetic energy$\simeq4\times10^{-9}$ J, potential energy$\simeq2\times10^{-13}$ J; system unstable.
2.10 $n=1$: 0.818, 0.198; $n=2$: $\frac{1}{2}$, 0.00645.
2.11 1.81×10^{-7} J.
2.12 2.53×10^{-7}.
2.13 N: (He) $(2s)^2$ $(2p)^3$;
Al: (Ne) $(3s)^2$ $(3p)^1$;

Cl^-: (Ar);
K: (Ar) $(4s)^1$.
2.14 -4.357×10^{-18} J.
2.15 $[-h/(2\mu)](\nabla_1^2+\nabla_2^2)\psi-[e^2/(4\pi\epsilon_0)](Z_{eff}/r_1+Z_{eff}/r_2-1/r_{12})\psi=E\psi$, where $Z_{eff}=1.70$.
2.16 (a) $S>0$; (b) $S>0$ ($R_{2,1,0}$); $S=0$ ($R_{2,1,\pm1}$); bonding between 1s and 2s, and between 1s and $2p_z$.
2.17 (a) 0.57; (b) 0.75; (c) 2.10 (graph), 2.104 (from $d(S_{1s,2p})/d\rho=0$).
2.18 (a) Be_2:$(1s\sigma)^2$ $(1s\sigma^*)^2$ $(2s\sigma)^2$ $(2s\sigma^*)^2$, antibonding; (b) C_2:$(1s\sigma)^2$ $(1s\sigma^*)^2$ $(2s\sigma)^2$ $(2s\sigma^*)^2$ $(2p\pi)^4$; C_2 is more stable than its atoms.
2.19 NO: $(1s\sigma)^2(1s\sigma^*)^2(2s\sigma)^2(2s\sigma^*)^2(2p\pi)^4(2p\sigma)^2(2p\pi^*)^1$, $\kappa=2.5$; CN: $(1s\sigma)^2(1s\sigma^*)^2(2s\sigma)^2(2s\sigma^*)^2(2p\pi)^4(2p\sigma)^1$, $\kappa=2.5$. NO^+ stabilized over NO by loss of the unpaired electron in antibonding $2p\pi^*$; CN^- stabilized over CN by the addition of an electron to bonding $2p\sigma$.
2.20 $-0.32e$, $+0.16e$.
2.21 Section 2.11.11; use $c_p/c_s=1/\sqrt{2}$.
2.22 Sections 2.11.13 and 2.11.14; use $c_p/c_s=\sqrt{x}$.
2.23 (a) Use Solution 2.22 with $x=3$; (b) Use $(sp^3)\times\cos(109.47/2)^\circ=(sp^3)/\sqrt{3}$; $\int\psi(sp^3)(sp^3)'\,d\tau=0$.
2.24 $y^4-4y=0$, where $y=(\alpha-E)/\beta$; $E_\pi=\alpha, \alpha, \alpha\pm2\beta$; $D_\pi=0$; FMOs are α, α.
2.25 459.6 nm; visible orange.
2.26 Ne_2: $(1\sigma_g)^2$ $(1\sigma_u)^2$ $(2\sigma_g)^2$ $(2\sigma_u)^2$ $(3\sigma_g)^2$ $(1\pi_u)^4$ $(1\pi_g)^4$ $(3\sigma_u)^2$; antibonding; Ne_2 unstable; $\kappa=0$. LiH: $(1s\sigma)^2$ $(2s\sigma)^2$; σ-bonding, $\kappa=1$.
2.27 3.4%.
2.28 $\Psi_1=0.4352\psi_1+0.5573\psi_2+0.4352\psi_3+0.5573\psi_4$; $\Psi_2=0.7071\psi_1-0.7071\psi_3$; $\Psi_3=0.7071\psi_2-0.7071\psi_4$; $\Psi_4=0.5573\psi_1-0.4352\psi_2+0.5573\psi_3-0.4352\psi_4$; $p_{12}=p_{23}=p_{34}=p_{41}=0.485$; $P_{12}=1.485$, $p_{24}=0.621$; $P_{24}=1.621$. $\mathcal{F}_1=\mathcal{F}_3=0.762$; $\mathcal{F}_2=\mathcal{F}_4=0.141$; $q_1=q_3=-0.379$; $q_2=q_4=+0.379$.
2.29

	Weak field		Strong field			Weak field		Strong field	
n in d^n	Config	N	Config	N	n in d^n	Config	N	Config	N
1	t^1	1	t^1	1	6	t^4e^2	4	t^6	0
2	t^2	2	t^2	2	7	t^5e^2	3	t^6e^1	1
3	t^3	3	t^3	3	8	t^6e^2	2	t^6e^2	2
4	t^3e^1	4	t^4	2	9	t^6e^3	1	t^6e^3	1
5	t^3e^2	5	t^5	1					

2.30 Section 2.14; let C_4 and S_4 axes of each polyhedron coincide. Axes of d_{z^2} and $d_{x^2-y^2}$ midway between ligand sites; $e_g<t_{2g}$; d^{10}, diamagnetic (Figure S2.1).
2.31 5.
2.32 (a) bent, F–O–F<H–O–H in H_2O; (b) distorted tetrahedral, F–N–F<H–N–H in NH_3; (c) regular tetrahedral; (d) trigonal pyramidal, F–S–F<120° (trigonal planar).

Answers 3

3.1 (a) $\mathcal{D}_{6h}$: C_6, C_3, C_2, i, S_3, S_6, σ_h, σ_v, σ_d (C_6, C_3, S_3 and S_6 are coincident); (b) $\mathcal{D}_{2h}$: C_2, i, σ_h, σ_v; (c) $\mathcal{C}_{2v}$: C_2, σ_v.
3.2 $\mathbf{C}_2$ and $\boldsymbol{\sigma}$ commute; **E** commutes with all other elements.

3.3

$\mathscr{C}_{2h}$	E	C_2	i	σ_h
E	E	C_2	i	σ_h
C_2	C_2	E	σ_h	i
i	i	σ_h	E	C_2
σ_h	σ_h	i	C_2	E

Order 4; Abelian

$\mathscr{D}_3$	E	C_3	C_3^2	C_2	C_2'	C_2''
E	E	C_3	C_3^2	C_2	C_2'	C_2''
C_3	C_3	C_3^2	E	C_2'	C_2''	C_2
C_3^2	C_3^2	E	C_3	C_2''	C_2	C_2'
C_2	C_2	C_2''	C_2'	E	C_3^2	C_3
C_2'	C_2'	C_2	C_2''	C_3	E	C_3^2
C_2''	C_2''	C_2'	C_2	C_3^2	C_3	E

Order 6; non-Abelian

3.4 $2A_2+E$.
3.5 E, $2C_6$, $2C_3$, C_2, $3C_2'$, $3C_2''$.
3.6 (a) $A_1+B_1+B_2$; (b) $A_1'+2E_1'$; (c) A_1+2E.
3.7 $\Gamma_\sigma=A_{1g}+B_{1g}+E_u$; hybrids are sp^{2}d or d^{2}p^2; F species are A_{1g}, B_{1g}, E_u; $(a_{1g})^2$ $(b_{1g})^2$ $(e_u)^2$; (Figure S3.1).
3.8 (a) $\Psi_1=(1/\sqrt{6})(\psi_1+\psi_2+\psi_3+\psi_4+\psi_5+\psi_6)$
$\Psi_2=(1/\sqrt{12})(2\psi_1+\psi_2-\psi_3-2\psi_4-\psi_5+\psi_6)$
$\Psi_3=(1/2)(\psi_2+\psi_3-\psi_5-\psi_6)$
$\Psi_4=(1/\sqrt{12})(2\psi_1-\psi_2-\psi_3+2\psi_4-\psi_5-\psi_6)$
$\Psi_5=(1/2)(\psi_2-\psi_3+\psi_5-\psi_6)$
$\Psi_6=(1/\sqrt{6})(\psi_1-\psi_2+\psi_3-\psi_4+\psi_5-\psi_6)$

(b)				
Ψ_1	A_{2u}	$\alpha+2\beta$	Nondegenerate	Occupied bonding orbital;
Ψ_2, Ψ_3	E_{1g}	$\alpha+\beta$	Degeneracy 2	Occupied bonding orbitals;
Ψ_4, Ψ_5	E_{2u}	$\alpha-\beta$	Degeneracy 2	Nonoccupied orbitals;
Ψ_6	B_{2g}	$\alpha-2\beta$	Nondegenerate 2	Nonoccupied orbital.

(c) (Figure S3.2) $D_\pi=2\beta$.
3.9 Three fourfold axes normal to the cube faces; four threefold axes through opposite corners; six twofold axes through the centres of opposite edges; three *m*-planes normal to the fourfold axes; six *m*-planes normal to the twofold axes; centre of symmetry; *m*3*m* ($\mathbb{O}_h$).
3.10 Yes; orthorhombic C ($\equiv A$), with $\boldsymbol{a}'=\boldsymbol{b}$, $\boldsymbol{b}'=\boldsymbol{c}$, $\boldsymbol{c}'=\boldsymbol{a}$.
3.11

$$c\,2_1=\begin{bmatrix}1 & 0 & 0\\ 0 & -1 & 0\\ 0 & 0 & 1\end{bmatrix}\begin{bmatrix}0\\ 1/2\\ 1/2\end{bmatrix};\quad \bar{1}\text{ at } 0,\tfrac{1}{4},\tfrac{1}{4};\text{ no.}$$

3.12

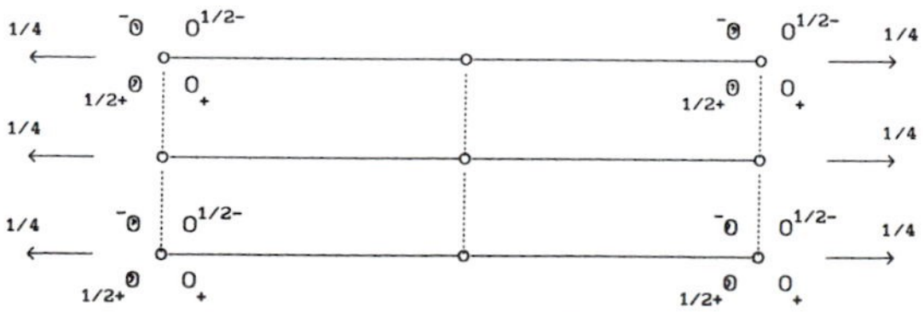

Origin on $\bar{1}$

4	g	1	$x, y, z;\ \bar{x}, \bar{y}, \bar{z};\ x, \bar{y}, \frac{1}{2}+z;\ \bar{x}, y, \frac{1}{2}-z$
2	f	2	$\frac{1}{2}, y, \frac{1}{4};\ \frac{1}{2}, \bar{y}, \frac{3}{4}$
2	e	2	$0, y, \frac{1}{4};\ 0, \bar{y}, \frac{3}{4}$
2	d	$\bar{1}$	$\frac{1}{2}, 0, 0;\ \frac{1}{2}, 0, \frac{1}{2}$
2	c	$\bar{1}$	$0, \frac{1}{2}, 0;\ 0, \frac{1}{2}, \frac{1}{2}$
2	b	$\bar{1}$	$\frac{1}{2}, \frac{1}{2}, 0;\ \frac{1}{2}, \frac{1}{2}, \frac{1}{2}$
2	a	$\bar{1}$	$0, 0, 0;\ 0, 0, \frac{1}{2}$

hkl; $l=2n$

3.13 Cl^-: $0, 0, 0;\ 0, \frac{1}{2}, \frac{1}{2};\ \frac{1}{2}, 0, \frac{1}{2};\ \frac{1}{2}, \frac{1}{2}, 0$; Na^+: $\frac{1}{2}, \frac{1}{2}, \frac{1}{2};\ \frac{1}{2}, 0, 0;\ 0, \frac{1}{2}, 0;\ 0, 0, \frac{1}{2}$; ($Cl^-$ and Na^+ may be interchanged).
3.14 9.93×10^{-21} J; far-IR.
3.15 1.81×10^{-47} kg m^{-2}; asymmetric top.
3.16 0.157 nm (C−S), 0.115 nm (C−O).
3.17 496.6 kJ mol^{-1}.
3.18 7; 30; 66.
3.19 0→1, 3958.6 cm^{-1}; 0→2, 7736.8 cm^{-1}.
3.20 5.322 eV.

3.21 3; 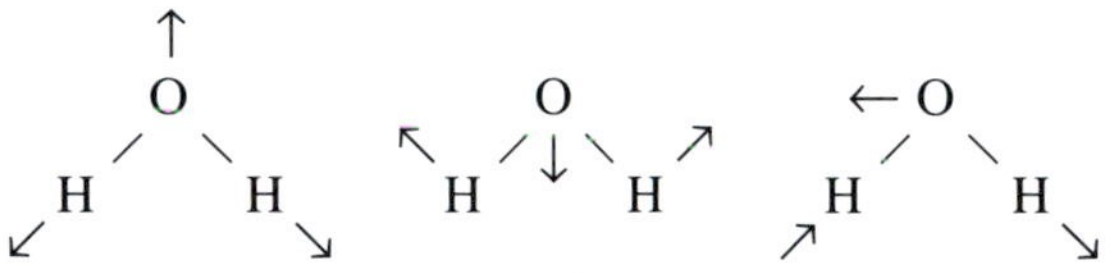 $2A_1+B_2$: 3 lines in IR and 3 (coincident) lines in Raman.

3.22 >CH=CH< dipole much less than >C=O dipole.
3.23 No double bonds; probable species $CH_3COC_2H_5$.
3.24 100 MHz.
3.25 $-CH_3$: doublet at $\delta\simeq1.0$, equal peak heights; >CH−: septet at $\delta\simeq4.3$ with peak height ratio 1 : 6 : 15 : 20 : 15 : 6 : 1.
3.26 Probable species $CH_3COC_2H_5$.
3.27 Probable species $CH_3COC_2H_5$ (molecular ion at *m*/*z* 72).
3.28 (a) Cube: (100), ($\bar{1}$00), (010), (0$\bar{1}$0), (001), (00$\bar{1}$); Tetrahedron: (111), ($\bar{1}\bar{1}$1), (1$\bar{1}\bar{1}$), ($\bar{1}$1$\bar{1}$), or ($\bar{1}$11), (1$\bar{1}$1), (11$\bar{1}$), ($\bar{1}\bar{1}\bar{1}$). (b) $r_{123}=1.395$ nm; $r_{210}=0.2750$ nm; $\theta=80.93°$.
3.29 *hkl*: $h+k=2n$; 0*kl*: $l=2n$.
3.30 0.213, 0.133, 0.0804 (all in RU), 85.93°.
3.31 Four peaks of weight 1 (symmetry-related atoms) and four peaks of weight 2 (non-symmetry-related atoms), with characteristic heavy central peak. Peaks of weight 2 lie exactly midway along the lines joining peaks of weight 1 (Figure S3.4).
3.32 (a) 2; $\pm(2x, \frac{1}{2}, 2z)$; (b) $\pm(0.422, \frac{1}{4}, 0.144)$; (c) 1.14 nm; (d) It contains peaks from symmetry-related atoms in the unit cell.
3.33 (a) Origin peak, null vectors; peak at 12.5/60, Hf–Hf; peak at 8.5/60, Hf–Si_a; y_{Hf}, 0.10(4); y_{Sia}, 0.24(6); (b) +, −, −, +, +, (+), −, (−); (c) y_{Hf}, 0.107; y_{Sia}, 0.250; y_{Sib}, 0.023; Hf–Si_b vector 'swamped' by origin peak; (d) examine the figure below; unique coordinates: Hf, 0, 0.107, $\frac{1}{4}$; Si_a, 0, 0.250, $\frac{1}{4}$; Si_b, 0, 0.023, $\frac{1}{4}$; Hf–Si_a, 0.208 nm; Hf–Si_b, 0.220 nm (Σr, 0.24 nm).

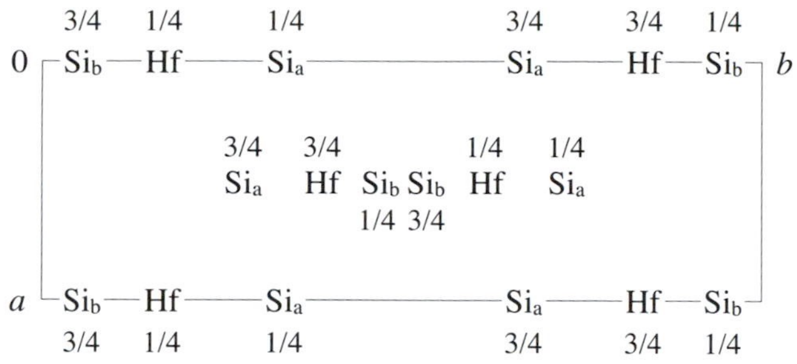

Answers 4

4.1 -34.5 kJ mol^{-1} (ΔU); 22.4 kJ mol^{-1}, endothermic.
4.2 -1413 kJ mol^{-1}.
4.3 28.2 kJ mol^{-1}.
4.4 3.10 kJ mol^{-1}.
4.5 8.36 kJ mol^{-1}; 7.11 kJ mol^{-1}.
4.6 $\Sigma_{\text{cycle}}\Delta H=0$, because H is a state property; -134 kJ mol^{-1}.
4.7 -3228.4 kJ mol^{-1}; -3227.1 kJ mol^{-1}.
4.8 Sections 4.4.1, 4.4.2, 4.2.1, 4.8, 4.8.1 and Appendix 13.
4.9 -359.9 kJ mol^{-1}.
4.10 -1426 kJ mol^{-1}.
4.11 87 J K^{-1} mol^{-1}.
4.12 5.65 kJ mol^{-1}.
4.13 Butane, -156 kJ mol^{-1}; 1-butene, -27 kJ mol^{-1}; -129 kJ mol^{-1}.
4.14 43.5 J K^{-1} mol^{-1}.
4.15 (a) Constant volume; (b, i) adiabatic; (b, ii) constant pressure and work of expansion only.
4.16 Section 4.2.1 and Appendix 13.
4.17 0.11 K.
4.18 12.6 J K^{-1} mol^{-1}.
4.19 Section 4.8; -16.49 kJ mol^{-1}; 3.56 kJ mol^{-1}; spontaneous at 25 °C.
4.20 Sections 4.6 and 4.8
4.21 5.71 kJ mol^{-1}.
4.22 2.4 J K^{-1} mol^{-1}; 0; -702 J.

Answers 5

5.1 Sections 5.2.1–5.2.5, and 5.3.1.
5.2 (a) 257 atm^{-1}; (b) 2.4×10^{10} s^{-1}.
5.3 1.15 μm.
5.4 Section 5.3.4.
5.5 Section 5.3.3.
5.6 (a) 349 m s^{-1}; (b) 698 m s^{-1}; (c) 7.75×10^{8} s^{-1}; (d) 4.78×10^{32} s^{-1}; (e) 0.90 μm.
5.7 Sections 5.3.5 and 5.3.6; 1025 K.
5.8 49.8 atm, 95.4 cm^3 mol^{-1}, 154.4 K; 0.32 nm.
5.9 (a) 1,2-difluorobenzene 2.49 D; 1,3-difluorobenzene 1.70 D; 1,4-difluorobenzene 0 (exact, by symmetry); (b) $\pm$0.411e.
5.10 0.33 ± 0.018 nm.

5.11 Sections 4.8.1 and 5.3.6; 0.042 atm^{-1}.
5.12 9.6–10.8; 0.37 nm; distribution uniform at large r.
5.13 Ethane, 0.274; ethene, 0.275; Section 5.3.7.
5.14 Ethane -0.0822 $dm^3\ mol^{-1}$, 0.00407 $dm^6\ mol^{-2}$; ethene -0.0713 $dm^3\ mol^{-1}$, 0.00326 $dm^6\ mol^{-2}$.

Answers 6

6.1 0.009 J K^{-1} mol^{-1} (electronic); 0.025 J K^{-1} mol^{-1} (lattice); 6.02 K.
6.2 0.74 (Figure S6.1).
6.3 512 nm; 1.2×10^5.
6.4 Section 5.3.3.
6.5 23.34 J K^{-1} mol^{-1}.
6.6 Sections 3.5.1 and 3.5.3, and Figure 1.2 or 6.16a.
6.7 (a) Section 6.4.5 (Figure S6.2); (b) cube of side $2\pi/a$.
6.8 ΔU_C(MgCl,s), -726 kJ mol^{-1}; ΔU_C($MgCl_2$,s), -2057 kJ mol^{-1} and compensates for $I_2 \gg I_1$.
6.9 -315.5 kJ mol^{-1}.
6.10 609 kJ mol^{-1}.
6.11 -343 kJ mol^{-1}.
6.12 4550 kg m^{-3}.
6.13 0.1945 nm, 0.1985 nm; 80.96°, 90.01°.
6.14 Section 6.12.2; Figure 6.36; 0.225.
6.15 9.1×10^{-6}.
6.16 9.70×10^{-7} $mm^2\ s^{-1}$.
6.17 191 kJ mol^{-1}; 3.42×10^7 $m^2\ s^{-1}$.
6.18 30.0 J K^{-1} mol^{-1}.
6.19 $\simeq$1.615.
6.20 Section 6.4.5 (Figure S6.5).

Answers 7

7.1	P	C	F
(a)	2	1	1
(b)	2	2	2
(c)	2	1	1
(d)	3	1	0

7.2 Sections 4.7 and 7.4.
7.3 Section 7.5.
7.4 (a) 189±7 kJ mol^{-1}; (b) 199 kJ mol^{-1}.
7.5 269.42.
7.6 60.3 kJ mol^{-1}.
7.7 33.51 Torr; 153.2 Torr.
7.8 44.12 g mol^{-1}.
7.9 45.84.
7.10 Section 7.10 (i=2.53; $\alpha_{apparent}$=0.76).
7.11 60.00.
7.12 5.09, 5.03, 4.90; activities needed for precision.
7.13 90.9±6.0 kJ mol^{-1}; 5.73±1.0 $dm^3\ mol^{-1}$.

7.14 0.153 K.
7.15 0.069 K; 0.2446 atm; $\sigma(M_r)$ 14 (from ΔT), 5 (from π).
7.16 0.091; 0.944.
7.17 (a) 1.333 g; (b) 1.440 g.
7.18 59.9%.
7.19 (a) 4; (b) 337±3.
7.20 Sections 7.6 and 7.12.

Answers 8

8.1 Section 8.2.
8.2 (a) -41.7 kJ mol^{-1}, 2.02×10^{7}; (b) -31.9 kJ mol^{-1}; (c) 7.63×10^{5}; yes, because the isomerization is exothermic.
8.3 0.142 (298 K), 0.309 (308 K); -59.3 kJ mol^{-1}.
8.4 177.9, 176.1, 175.2, 163.0, 162.7 (all in kJ mol^{-1}).
8.5 (a) increase in [products], K_p unchanged; increase in [reactants], K_p decreased; increase in [reactants], K_p unchanged; increase in [reactants], K_p unchanged (b) rates of forwards and backwards reactions: decreased; increased; increased; increased.
8.6 9.12, 3.26, 2.04.
8.7 (a) 1.08×10^{-5} mol dm^{-3} ($\gamma_\pm$0.970); (b) 1.42×10^{-7} mol dm^{-3} ($\gamma_\pm$0.774); (c) 1.68×10^{-5} mol dm^{-3} ($\gamma_\pm$0.625).
8.8 (a) 12.00; (b) 2.00; (c) 2.40; (d) 7.02.
8.9 7.46 (273 K), 7.00 (298 K), 6.51 (333 K); 55.0 kJ mol^{-1}.
8.10 (a) 3.54; (b) 8.77; (c) 5.13; (d) 7.61.
8.11 4.05; +0.13, +0.15 (Van Slyke).
8.12 0.01580 mol dm^{-3}; 1.80.
8.13 Section 8.7.4.1; 9.0236×10^{-4} mol dm^{-3}, 3.04; 1.7650×10^{-5} mol dm^{-3}, 4.75; 1.8762×10^{-9} mol dm^{-3}, 9.27.
8.14 1110 K.
8.15 73.8 atm.

Answers 9

9.1 0.7057 g.
9.2 Sections 9.2 and 9.3.7.4.
9.3 (a) 396.4 m^{-1}; (b) 0.01093 Ω^{-1} m^2 mol^{-1}.
9.4 0.04440 Ω^{-1} m^{-1}.
9.5 0.04256, 0.03908, 0.01265, 0.00917 (all in Ω^{-1} m^2 mol^{-1}).
9.6 1.002×10^{-10}.
9.7 (a) 0.178; (b) 0.597; (c) 0.503; (d) 0.557; (e) 0.512; $[H_3O]^+$ high mobility in HCl(aq) means t_+ high.
9.8 (a), (b) Sections 9.3.3 and 9.3.7.4;

	Zn/Pt,H_2	Zn/Ag	Pt,H_2/Ag
(c)	0.763 V	1.562 V	0.799 V;
(d)	-73.6 kJ mol^{-1}	-150.7 kJ mol^{-1}	-77.1 kJ mol^{-1};
(e)	Zn→Pt,H_2	Zn→Ag	Pt,H_2→Ag.

9.9 0.417 V; 6.40×10^{-4} V K^{-1}.
9.10 (a) -116.8 kJ mol^{-1}; (b) 0.459 V; unstable because $\Delta G_{\text{reaction}}$ negative.
9.11 3.99.

9.12 (a) Section 9.3.2.1; (b) 5.39 kJ mol^{-1}.
9.13 0.26838 V; 0.964, 0.949, 0.935, 0.922, 0.908; Section 8.6.4.1.
9.14 Sections 9.3.3ff; 14.24 (20 °C), 14.02 (25 °C), 13.80 (30 °C); 74.88 kJ mol^{-1}; −17.3 J K^{-1} mol^{-1}.
9.15 The cell

$$\mathrm{Hg,Zn|Zn(ClO_4)_2}\ (m_1=0.10)|\mathrm{Zn(ClO_4)_2}\ (m_2)|\mathrm{Zn,Hg}$$

was set up and the EMF with transport E_t measured at a series of concentrations m_2. The values of $\gamma_\pm$ were obtained from osmotic pressure measurements at the same concentrations. The EMF without transport E cannot be measured because there is no electrode reversible to the perchlorate ion. However, it can be calculated for a hypothetical cell without transport from the Nernst equation. At each concentration calculate E, t_- and $\sqrt{m_2}$. Determine the value for $t_{0,-}$ by linear extrapolation of t_- against $\sqrt{m_2}$.

E_t/V	0.00000	0.00879	0.01541	0.02540	0.03334
m_2/kg mol^{-1}	0.10	0.15	0.20	0.30	0.40
$\gamma_\pm$	0.573	0.560	0.556	0.565	0.588
E_t/V	0.04004	0.04607	0.05189	0.05722	0.06239
m_2/kg mol^{-1}	0.50	0.60	0.70	0.80	0.90
$\gamma_\pm$	0.620	0.661	0.710	0.769	0.838
E_t/V	0.06731	0.07687	0.08629	0.09546	0.10465
m_2/kg mol^{-1}	1.00	1.20	1.40	1.60	1.80
$\gamma_\pm$	0.916	1.111	1.367	1.695	2.126
E_t/V	0.11394	0.13718	0.16041	0.18364	0.20676
m_2/kg mol^{-1}	2.00	2.50	3.00	3.50	4.00
$\gamma_\pm$	2.694	5.11	9.88	19.58	38.9

Answers 10

10.1 One; 3.24×10^{-2} min^{-1}; 214 min.
10.2 (a) 70.6 year; (b) 6.46×10^{-6} g.
10.3 1.05×10^{-2} min^{-1}; 66.3 min.
10.4 p_{init}≃constant; 17.5 mmHg min^{-1}.
10.5 Two; 2.24×10^{-6} mmHg s^{-1}; 236.5 kJ mol^{-1}.
10.6 One; 2.07×10^{-2} min^{-1}.
10.7 1.57×10^{-2} min^{-1}; 8.97×10^{-3} min^{-1}.
10.8 Section 10.9.
10.9 Section 10.4.2.
10.10 Sections 10.6 and 10.7; (Figure S10.1).
10.11 64.5 kJ mol^{-1}; 5.2×10^{7} mol^{-1} dm^{3} s^{-1}.
10.12 37.
10.13 8 mol.
10.14 One; 0.0451 min^{-1}; 47.1 cm^{3}.
10.15 (a) Uncharged ($q_A q_B$=8.4×10^{-4}); (b) 1.27 hour^{-1}.
10.16 (a) Langmuir; (b) 0.201 cm^{3}; (c) 0.0168 mmHg^{-1}; (d) 0.186 cm^{3}.
10.17 48.3 Torr; 227 min.
10.18 Freundlich: $m=0.149c^{1/2.52}$; (units of k are g mol$^{-0.3968}$ dm$^{1.1904}$).
10.19 0.010 mol dm^{3}.
10.20 Section 10.4.3.
(a) No; see (10.38). (b) Yes; $t_{1/2}=[\mathrm{A}]_0/(2k_0)$
10.21 (a) First order; 7.2×10^{-4} s^{-1}; (b) does not hold at low pressures (Figure S10.2).

INDEX